PARTICLE PHYSICS

ONE HUNDRED YEARS OF DISCOVERIES

An Annotated Chronological Bibliography

PARTICLE PHYSICS

ONE HUNDRED YEARS OF DISCOVERIES

An Annotated Chronological Bibliography

V.V. EZHELA, B.B. FILIMONOV, S.B. LUGOVSKY,
B.V. POLISHCHUK, S.I. STRIGANOV,
Y.G. STROGANOV

COMPAS Group, Institute for High Energy Physics,
Protvino, Moscow Region, RU-142284, Russia

B. ARMSTRONG, R.M. BARNETT, D.E. GROOM,
P.S. GEE, T.G. TRIPPE, C.G. WOHL

Particle Data Group, E. O. Lawrence Berkeley National Laboratory
1 Cyclotron Rd., Berkeley, CA 94720, USA

J.D. JACKSON

Department of Physics,
University of California, Berkeley, CA 94720, USA

American Institute of Physics **Woodbury, New York**

The artwork on the cover represents the electron discovery device (cathode ray tube) used about 100 years ago (Thomson, 1897), an event observed in 1995 confirming the top quark, and a sketch of the structure within the atom showing quarks within neutrons and protons in the nucleus and showing two electrons. This structure is part of the Standard Model of Fundamental Particles; the sketch is adapted from a chart by the Contemporary Physics Education Project, 10 Bear Paw, Portola Valley, CA 94028.

Abstracts of original papers from the following journals have been reproduced with kind permission by Elsevier Science–NL: *Physics Letters B; Nuclear Physics "B"; Nuclear Instruments & Methods in Physics Research.*

Abstracts of original papers from *Zeitschrift fuer Physik* have been reproduced with kind permission by Springer-Verlag.

The copyright of Kyoto University of the abstract reprinted in this volume from the *Progress of Theoretical Physics* is acknowledged.

Abstracts from the following journals have been reproduced with kind permission by The American Physical Society: *Physical Review* and *Physical Review Letters.*

In recognition of the importance of preserving what has been written, it is a policy of the American Institute of Physics to have books published in the United States printed on acid-free paper.

AIP Press
American Institute of Physics
500 Sunnyside Boulevard
Woodbury, NY 11797-2999

Library of Congress Cataloging-in-Publication Data
Particle physics : one hundred years of discoveries : an annotated chronological bibliography / V. V. Ezhela . . . [et al.].
p. cm.
Includes bibliographical references and index.
ISBN 1-56396-642-5
1. Particles (Nuclear physics)--History--Bibliography.
I. Ezhela, V. V.
Z7144.N8P35 1996
[QC793.16] 96-9558
016.5397'2--dc20 CIP

10 9 8 7 6 5 4 3 2 1

Contents

Preface

We present an annotated chronological bibliography of papers showing the development of particle physics including the discovery of the electron in 1897 and the discovery of the top quark in 1995. For each paper there is a short description of the primary achievement and an excerpt that summarizes the paper. Early papers relating to the birth of particle physics were compiled mainly from historical works, recollections, anthologies, and chronologies [1–11]. In gathering data, we have tried to pay attention to each step in a typical discovery chain: prediction or indirect evidence $\rightarrow$ first direct evidence $\rightarrow$ confirmation $\rightarrow$ establishment. Of course, not all of these stages always occur; for example, the first evidence may have no theoretical or experimental precursor, and/or may completely establish the effect in question. Theoretical results in grand unified theories, supersymmetry, quark/lepton substructure, technicolor, string theory, *etc.*, have not been included because there has been no experimental confirmation.

References

1. E. Rutherford, *Forty Years of Physics. I. The history of radioactivity. II. The development of the theory of atomic structure. - Background to Modern Science.* Cambridge, 1940, p.47-74.
2. *Great Experiments in Physics*, Edited by M. H. Shamos, Henry Holt and Company, New York, 1960.
3. J. Six, X. Artru, *An Essay of Chronology of Particle Physics Until 1965*, Journal de Physique T.43, p.C8-465, 1982.
4. V.F. Weisskopf, *Physics in the Twentieth Century: Selected Essays.* The MIT Press, Cambridge, Massachusetts, 1972.
5. G.L. Trigg, *Landmark Experiments in Twentieth Century Physics.* Crane, Russak & Company, Inc. New York, 1975.
6. S. Weinberg, *The Discovery of Subatomic Particles.* Scientific American Library, An Imprint of W.H. Freeman & Company, New York - San Francisco, 1983.
7. B. M. Pontecorvo, *Pages in the Development of Neutrino Physics*, Uspekhi Fizicheskih Nauk **141**, 675, 1983.
8. A. Pais, *Inward Bound. On Matter and Forces in the Physical World.* Clarendon Press. Oxford University Press, Oxford, 1986.
9. R.N. Cahn, G. Goldhaber, *The Experimental Foundations of Particle Physics.* Cambridge University Press, Cambridge, 1989.
10. S.S. Schweber, *QED and the Men Who Made It: Dyson, Feynman, Schwinger, and Tomonaga*, Princeton University Press, Princeton, 1994.
11. *The Physical Review — The first hundred years: A selection of seminal papers and commentaries*, Edited by H. Henry Stroke, AIP Press, New York, 1995.

Acknowledgments

During the preparation of this book, the authors have enjoyed the opportunity to discuss aspects of the *Chronology* with the people who have made the story and the history of particle physics.

We thank Yu. M. Antipov, R. N. Cahn, S. S. Gershtein, G. Goldhaber, C. Haber, L. M. Lederman, A. A. Logunov, Yu. D. Prokoshkin, D. Silverman, V. L. Solovianov, and N. E. Tyurin for their interest, valuable inputs, criticism, and corrections.

We also thank L. Addis, S. I. Alekhin, H. Galic, G. Harper, E. A. Ludmirsky, N. Mokhov, A. S. Nikolaev, V. A. Petrov, A. P. Samokhin, and L. S. Shirshov for their help in different stages of this work.

We would like to express our appreciation to N. Arkani-Hamed and M. Graesser (of the University of California, Berkeley) for their excellent, detailed proofreading of the manuscript.

It is impossible to be absolutely correct in preparing chronologies like this. We apologize if we missed any important works or for incorrect presentation of the material. This was the result of the authors' lack of knowledge, and the responsibility is only the authors'. Fortunately the Discovery database will continue to be maintained in the future. Any corrections and suggestions are welcome and will be accepted with gratitude by the PDG and COMPAS groups.

The COMPAS Group at Protvino Institute for High Energy Physics is supported in part by the Russian Foundation for Basic Research under contracts RFBR-96-07-89230 and BAFIZ-96.

The Particle Data Group at Lawrence Berkeley Laboratory is supported by the Director, Office of Energy Research, Office of High Energy and Nuclear Physics, Division of High Energy Physics of the U.S. Department of Energy under Contract No. DE-AC03-76SF00098, and by the U.S. National Science Foundation under Agreement No. PHY-9320551.

I. Introduction

Particle physics began when there was widespread doubt in the existence of atoms, and the quantum of Planck and Einstein was yet unimagined. For a time, particle physics was part and parcel of the quest for an understanding of atoms and nuclei, especially their quantum nature. Only after fifty years did it emerge as a clearly defined sub-field of physics. The path from the beginnings of particle physics in the 1890's to the precise confirmation of the Standard Model in the 1990's is a revealing illustration of how science advances. Paths at the frontier of science are rarely straight. Along the way, brilliant insights, new experimental tools, and hard-won new data are accompanied by confusion, wrong turns, conservative dogma, or wrong experiments. This report chronicles the history of particle physics from its beginnings in gas discharges and in radioactivity to the present day with citations of virtually all the significant papers, both experimental and theoretical, right and wrong. In these pages one can see the story unfolding through its many contributors and can comprehend the immense progress in our understanding of the basic constituents and forces in Nature. These triumphant 100 years of particle physics are in many ways comparable to the preceding century of electromagnetism from Cavendish, Coulomb, and Franklin to Faraday, Maxwell, and Hertz.

The first of the fundamental particles to be discovered was the electron (1897). J.J. Thomson carried out experiments with cathode rays and established the electron's charge and mass. Beginning in 1896, the α, β, and γ rays were discovered and investigated by Becquerel, the Curies, Rutherford, Thomson, and their collaborators. While the β rays soon turned out to be electrons and the α rays helium nuclei, the γ rays were not proven to be electromagnetic radiation until 1914. Quantum mechanics had its beginnings when Planck quantized material oscillations in 1900. Soon after, Einstein formally proposed quanta of light with particle-like characteristics.

During the 1910's, the nuclear model (Rutherford) and the first quantum theory (Bohr) of the atom were developed. In the 1920's, physicists began a search for an understanding of the nucleus. Chadwick and Bieler recognized that a new kind of strong force was needed to explain what was known about the nucleus. In 1925, Goudsmit and Uhlenbeck attributed an intrinsic half-integer angular momentum (spin) to the electron, and two years later Dennison found that the proton has the same spin. During this decade, it was widely believed that nuclei were made of protons and electrons, although with the all-important development of quantum mechanics there were problems (nuclear spin and statistics) with this model. The resolution came with the discovery of the neutron by Chadwick in 1931.

By the early 1930's, physicists agreed that there were three fundamental constituents of matter: the electron, proton, and neutron. But many questions remained: What was the nature of the strong force holding the nucleus together? Why was the electron spectrum in β decay continuous? Did the antiparticles predicted by Dirac's theory exist? Each mystery was eventually resolved by the discovery of new particles. Already in 1932, Anderson discovered the positron, but not everyone immediately believed it was the antielectron. To explain the continuous β decay spectrum, Pauli proposed a new fundamental particle, the neutrino. The direct observation of neutrinos would occur only years later, but meanwhile Fermi developed the theory of β decay (1933-34), explicitly using neutrinos and proposing the weak interactions. Yukawa, during the same period, proposed a theory of the nuclear strong interaction and postulated a new particle, the pion, whose exchange between nucleons was responsible for the force.

Yukawa had predicted a mass of about 100 MeV, so the discovery in 1937 of such a particle in cosmic rays was immediately interpreted as the discovery of Yukawa's pion. It took ten years for physicists to realize that this particle did not interact strongly as it passed through matter. It was instead the muon. This discovery of a heavier clone of the electron was a stunning development. It was an early forerunner of a new generation of fundamental particles, particles with new "flavors."

The charged pion itself was found in cosmic rays in 1947, and a year later at an accelerator, marking the birth of particle physics as a field in its own right. Fermi and Yang suggested a revolutionary idea, that the pion might be a composite particle, consisting of a nucleon and an antinucleon. Though this proposal was wrong, fifteen years later the quark model explained mesons and baryons as composites.

The pion was followed two years later by the K^+ (1949), the π^0 (1950), and the Λ^0 and K^0 (1951). The development of accelerators in the early 1950's opened the door to the so-called "population explosion," a bewildering proliferation of particles. Among the particles discovered in the 1950's were the baryons Δ, $\Sigma^\pm$, Ξ^-, Σ^0, and Ξ^0, plus the antiproton and the antineutron.

The K mesons were recognized in 1955-56 by Gell-Mann and by Nishijima as having a new characteristic called "strangeness." The early 1960's saw the discovery of many more mesons such as the ρ, ω, η, K^*, ϕ, f, a_2, and η' along with many more baryons. Gell-Mann sought an understanding of this morass of particles by developing (in 1961) the "eightfold way," a precursor to the quark model, incorporating strangeness and isospin in an SU(3) group theoretic framework.

The quark model itself was proposed by Gell-Mann and by Zweig in 1964. With it came a great simplification and understanding of all the known mesons and baryons: They were all composites of only three types of quarks, the u, d, and s quarks. The s quark carried strangeness and was a constituent of "strange" particles such as the K mesons. A critical prediction of the eightfold way had been the existence of a new particle, the Ω^-, which was understood later in the quark model as a baryon made of three strange quarks (an sss state). The Ω^- was indeed discovered in 1964.

Along with all the discoveries of mesons and baryons was the critical detection of the neutrino by Cowan and Reines in 1956, demonstrating the feasibility of detecting neutrino-induced interactions.

Nishijima suggested in 1957 that there were two kinds of "lepton charge" (now called flavors), one for e and ν_e and one for μ and ν_μ. In 1963, using a beam of neutrinos from pion and K decays, Lederman, Schwartz, and Steinberger demonstrated the existence of the muon neutrino, distinct from the electron neutrino.

The next year, several physicists suggested that if the μ and ν_μ belonged to a second generation of leptons, then perhaps there was a second generation of quarks. If the d and u quarks were the first generation quarks, then the s quark belonged to the second generation, but a new heavier quark with charge $\frac{2}{3}$ would also be needed. Bjorken and Glashow called this fourth quark the "charm" quark. However, since the charm quark had not been found, the concept of flavors (generations) was not widely accepted. Six years later Glashow, Iliopoulos, and Maiani (GIM) found a much more compelling argument for the charm quark: it allowed a theory of weak interactions with flavor-conserving, but not flavor-changing, neutral currents.

Back in the 1950's, there were other developments in understanding the weak interactions. Lee and Yang proposed that parity might not be conserved in weak processes, and almost immediately experimental searches uncovered violations of parity in weak decay processes. In 1957 Landau and separately Lee, Oehme, and Yang suggested that the combined symmetry of CP (charge conjugation and parity) might still remain valid; Pais and Treiman noted that decays of neutral K's could test this idea. Also proposed (Schwinger, Bludman, and Glashow) was the existence of a new particle, the $W^\pm$ boson, that mediated the weak interactions.

In this same era, Schwinger first suggested the possibility of unifying the weak and electromagnetic interactions. Not until 1967, however, did Weinberg and Salam separately develop an actual theory for this unification, one that also required the existence of a Z^0 boson and a Higgs boson. However, for the time being, their theory attracted little attention. It focused on leptons, since the GIM model (of 1970) was needed before quarks could be included. In fact, at that time quarks were not fully accepted since they had not been observed.

To find experimental evidence for fundamental constituents such as quarks within hadrons, Bjorken and Feynman suggested looking at data in the deep-inelastic scattering of electrons by nucleons. Friedman, Kendall, Taylor, and collaborators performed this experiment in the late 1960's at the SLAC laboratory. The analysis strengthened the case for quarks as fundamental particles by providing strong evidence for point-like constituents within hadrons. In 1965, Greenberg and separately Han and Nambu suggested that quarks have a characteristic now called "color," and that hadrons were color-neutral.

In the early 1970's, Nambu and, in more detail, Fritzsch and Gell-Mann proposed a theory of the strong interaction (Quantum Chromodynamics) in which a new massless particle, the gluon, carries the strong force. Neither quarks nor gluons are color-neutral, and they cannot exist as free particles. At the end of the decade, direct evidence for gluons was found at the DESY laboratory in Hamburg. Politzer and separately Gross and Wilczek showed that in this theory the strong interactions had unusual properties (including "asymptotic freedom") that were essential to understanding data in scattering and other processes.

Improving evidence for quarks, fuller understanding of the unified theory of Glashow, Salam, and Weinberg, and the incorporation of the strong interactions, led to recognition in 1974 that there was a "Standard Model" explaining the fundamental particles and interactions.

However, there was a missing piece: The charm quark had to be found. In November 1974, experiments led by Richter and by Ting simultaneously announced unequivocal evidence for a new (very narrow) particle with a mass of 3.1 GeV. This stunning news launched a period of intense activity and led to general agreement that this J/ψ particle was a $c\bar{c}$ state.

Two years later, the D^0 meson (a $u\bar{c}$ state) was discovered by Goldhaber and Pierre, proving beyond doubt the existence of charm. The striking characteristics of the weak decays of the D mesons ($c \rightarrow s$) established the concept of a second generation (c, s) of quarks in parallel with the first (u, d).

This discovery immediately led theorists to speculate about an additional generation of quarks and leptons. While charm was being firmly established and the properties of the D mesons and the ψ family explored, Perl and collaborators found a new charged lepton (called τ) in the midst the of the D meson data. The mass of the τ lepton was coincidentally very similar to that of D mesons, a chance echo of the similarity in the masses of the hadronic pion and the leptonic muon in the 1930's.

The start of a new, third generation of leptons intensified searches for third generation quarks. At Fermilab in 1977 Lederman and collaborators found the fifth quark, the bottom or b quark. This discovery, together with an experiment by Prescott and Taylor quantitatively confirming parity violation in the scattering of polarized electrons on deuterium, were convincing evidence that the Standard Model (and in fact its simplest version) provided an excellent description of the fundamental particles and interactions.

Although the proposal of the unified electroweak theory won the 1979 Nobel Prize, not all pieces of the puzzle had been found: the $W^{\pm}$ and Z^0 bosons, the top quark, and the Higgs boson required new machines and higher energies. The work of Rubbia and van der Meer led to the 1983 discovery at the CERN laboratory in Geneva of both the $W^{\pm}$ and Z^0 bosons. Their masses and other properties have since been measured with great precision, and the Standard Model has been confirmed repeatedly.

The search for the top quark was not so straightforward. For years, new searches at ever more powerful accelerators and colliders found nothing. The lower limit on the top quark mass grew far above the 5-GeV mass of the bottom quark. Finally in 1995, after 18 years of searching, the CDF and DØ experiments at Fermilab discovered the top quark at the truly unexpectedly high mass of about 170 GeV.

After one hundred years of particle physics, mysteries still remain: Why is the top quark so much heavier than the other quarks? Why are quarks and gluons confined within hadrons? Does the Higgs boson really exist? Are there still more generations? Will the Standard Model stand up to increasingly precise data? Will theoretically attractive extensions of the Standard Model, such as supersymmetry and string theory, be indicated by new data? Are quarks and leptons truly fundamental or are they themselves also composite? The story is not over.

II. Chronological List of Discoveries

1895 Observation that cathode rays are the flow of negatively charged particles. 21
Discovery of the X rays. 21
1896 Evidence for spontaneous radioactivity effect. 22
Confirmation of the effect of spontaneous radioactivity. 22
Confirmation of the effect of spontaneous radioactivity. Observation that uranium radiation is not identical to Röntgen X rays. 23
Uranium radioactivity established. 23
1897 Discovery of the electron, the first elementary particle. 23
1899 Evidence for the α and β components of uranium radiation. 24
1900 Evidence for the γ radioactivity. 24
Confirmation of the γ radioactivity. 25
Gamma radioactivity established. 25
Discovery of a new formula for the energy spectrum of the black body radiation, Planck's radiation law. 25
Quantum hypothesis and explanation of the black body radiation spectrum. Beginnings of the quantum era in physics. 25
1901 Quantum hypothesis and final version of Planck's formula for the black body radiation spectrum. 26
First experimental evidence for pressure of the light on the solid bodies. 26
1903 Observation that α rays are the flow of doubly positive charged particles. 26
1905 Introduction of the relativity principle as an overall law of nature valid for all forces including the gravitational one. Formulation of the equation of the relativistic mechanics and transformation laws for electromagnetic field and current. Establishing of the Lorentz group as a symmetry group of nature. First proposal to modify Newtonian theory of gravity on the basis of the relativity principle. Prediction of the gravitational waves propagating with the speed of light. 27
Explanation of the photoelectric effect with use of the quantum hypothesis of Planck. Light is a flow of corpuscular objects with definite energies — Planck's quanta of energy. 28
Invention of the theory of special relativity. Beginnings of the relativistic era in physics. 28
Invention of the theory of special relativity, $E = mc^2$. Beginnings of the relativistic era in physics. 29
1906 Corpuscular-wave dualism for photons. Explanation of the photoelectric effect using the quantum hypothesis of Planck. 30
1909 First explicit identification of the photon as a genuine elementary particle possessing both energy and momentum. 31
1911 First conclusive measurement of the charge of the electron. 31
Evidence for the atomic nucleus. Rutherford model for atomic structure. 32
1912 Conclusive evidence for the cosmic rays. 32
Invention of the cloud chamber to visualize tracks of ionizing particles. 32
1913 Confirmation of cosmic rays. 33
First precise measurement of the charge of the electron and the Avogadro constant. 33
Invention of the quantum theory of atomic spectra based on the Rutherford model of atomic structure — Bohr's atom. 34
Bohr's quantum theory of atomic spectra. Evidence that radioactivity is a nuclear property. 35
Confirmation of the existence of atomic nuclei. First indication of the existence of the proton. 36
1914 The β spectrum is continuous (first observation). Indirect evidence on the existence of neutral penetrating particles. 36
1916 First conclusive measurement of energy quantization in the photoelectric effect. 36
1918 Bohr's invention of correspondence principle. 37

1919 Discovery of the proton. Evidence for the proton as a constituent of the nucleus. 38
1920 Rutherford neutron hypothesis. 38
1921 Evidence for the strong interactions. 39
1923 Direct experimental confirmation that the photon is an elementary particle, the Compton effect. 40
Suggestion of the corpuscular-wave dualism for electrons — de Broglie waves of matter particles. 40
Suggestion of the corpuscular-wave dualism for electrons. Prediction of diffraction phenomena for electrons. 41
Experimental confirmation of the ionization process predicted by Compton for a corpuscular photon. 41
1924 Discovery of new statistical counting rules for light quanta and a new derivation of Planck's radiation law. Known as Bose-Einstein quantum statistics for particles with integer spins. 41
Extension of the Bose method to the monoatomic gases. Prediction of Bose-Einstein condensation effect. 42
1925 Extension of the Bose method to the monoatomic ideal gases. Prediction of Bose-Einstein condensation effect and rediscovery of the wave properties of matter particles. 42
Introduction of an additional two–valued degree of freedom for the atomic electron. 42
Discovery of the exclusion principle — the Pauli principle. 42
Invention of the electron spin hypothesis and the notion of an elementary spin. 43
Foundation of quantum mechanics, Heisenberg approach. 43
Invention of matrix formalism for the Heisenberg quantum mechanics. Systems with one degree of freedom. 44
1926 Development of matrix formalism for the Heisenberg quantum mechanics. Systems with arbitrary many degrees of freedom. 44
Firm establishment of cosmic rays. 45
Invention of statistics for ensembles of particles obeying Pauli principle. 45
Invention of statistics for ensembles of particles obeying Pauli principle — Fermi–Dirac quantum statistics. 46
Creation of wave mechanics. Invention of the Schrödinger wave equation. 46
Wave mechanics: First applications. 46
Equivalency of quantum mechanics of Heisenberg, Born, and Jordan and wave mechanics of Schrödinger. 47
Wave mechanics: Perturbation theory and applications. 48
Wave mechanics: Further development and generalization. 50
Statistical interpretation of quantum mechanics. Quantum theory of scattering. Born approximation. 50
Statistical interpretation of quantum mechanics: Further development. 50
Rediscovery of statistics for an ensemble of fermions — the Fermi-Dirac quantum statistics. 51
Statistical interpretation of quantum mechanics: Further development. 51
1927 Experimental evidence that the electron moves as a group of de Broglie waves. 51
First steps in quantum field theory. Invention of the second quantization method. 52
Foundations of quantum electrodynamics — QED. 52
Confirmation that the β spectrum is continuous. 52
Further confirmation that the β spectrum is continuous. 53
Heisenberg discovery of the uncertainty principle. 53
Invention of spatial parity as a quantum mechanical conserved quantity. 54
Discovery of the diffraction of electrons by crystals. Confirmation of the wave properties of moving electrons. 54
1928 Confirmation of diffraction of electrons by crystals. 55
Discovery of the relativistic wave equation for the electron. Prediction of the magnetic moment of the electron. 55
Explanation of α decay as a consequence of quantum-mechanical tunneling through a potential barrier. 55
Relativistic wave equation for the electron and theory of the Zeeman effect. 56
1929 Birth of cosmic rays particle physics. Observation of energetic cosmic electrons and a shower produced by cosmic ray particle. 56
Observation that the cosmic rays at sea level consist mainly of ionizing particles. 56
First step in metrology of the general physical constants. 57
1930 Introduction of the negative energy electron sea with holes treated as positive electrons. Attempt to identify these holes with protons. 58
Firm establishment that the β spectrum is continuous. 58
Proposal for the existence of the neutral fermion emitted in nuclear beta decay – the neutrino. 59
Difficulties with identifying positive holes with protons in the Dirac theory of electrons and protons. Too small a lifetime of the ordinary atom. 59

Year	Discovery	Page
	First evidence of ultraviolet divergences occurring in the theory of quantum electrodynamics — the self-energy of the electron.	59
1931	Prediction of the anti-electron (e^+), anti-proton ($\overline{p}$), and an indication of the possible existence of magnetic monopoles.	60
	Lawrence proposal for cyclotrons.	60
	Tests of the first cyclotron.	60
	Invention of the Van de Graaff electrostatic accelerator.	61
1932	Experimental proof that the photon has spin = 1.	61
	First experimental evidence for the positron.	62
	First evidence for the neutron.	62
	Discovery of the neutron.	63
	First evidence of nuclear reactions with accelerated protons. Cockcroft–Walton accelerator.	63
	Suggestion that the neutron is a constituent of the atomic nucleus.	64
	Evidence for the deuteron.	64
	Description of the space of states for quantum systems with an arbitrary (infinite) number of particles — Fock space.	64
	Suggestion that atomic nuclei are composed of protons and neutrons. Theory of nuclear exchange forces. Invention of nucleon isotopic spin.	65
	Confirmation of fast penetrating charged particles in cosmic rays.	65
	Discovery of the deuteron.	65
	Further development of cyclotrons.	66
	Introduction of the time inversion transformation in quantum mechanics.	67
1933	Invention of electrostatic accelerators. Further development.	67
	Discovery of the positron, the first antiparticle, predicted by Dirac.	68
	e^+ and shower confirmation. First indication for e^+e^- pair production.	68
	Further development of the theory of nuclear forces.	69
	First measurement of the proton magnetic moment.	69
	Further measurement of the proton magnetic moment.	70
1934	Evidence for deuteron photodisintegration. First precise measurements of the neutron mass.	70
	Explanation of the continuous electron energy spectrum in β decay. Proposal for the neutrino.	70
	Field theory for beta decay. First estimation of the neutrino mass.	70
	Prediction of the negative anomalous magnetic moment of the neutron.	71
	First evidence for Vavilov-Čerenkov radiation.	71
	First attempt to understand Vavilov-Čerenkov radiation. Evidences that it is not a luminescence.	72
	Evidence that the spin of the deuteron is 1.	72
	First measurements of magnetic moments of the deuteron and neutron.	72
1935	Yukawa field theory of nuclear forces. Prediction of heavy quanta, the pion particles, as mediators of strong interactions.	73
1936	Breit-Wigner form of the amplitude for resonance reactions.	73
	Extension of Fermi interaction and Gamow-Teller selection rules.	74
	Hypothesis of charge independence of nuclear forces.	74
	Proposal of the isotopic spin (for the nucleon).	74
1937	Theoretical explanation of Čerenkov radiation phenomenon.	75
	Discovery of supermultiplet structure in nuclear spectroscopy.	75
	First evidence for the muon.	75
	Symmetrical theory of electrons and positrons. Majorana neutrino theory.	76
	Confirmation of the muon existence.	76
	Treatment of infrared divergence.	77
	Confirmation of the Frank-Tamm theory of the Vavilov-Čerenkov effect.	77
	Muon existence confirmation.	78
	First proposal for the S matrix formalism.	78
	Confirmation of the existence of the muon.	79
	Establishment of the neutron spin 1/2.	79
1938	First evidence for the capture of atomic electrons by the weak interaction; K-capture.	79

	Invention of the baryonic quantum number conservation law.	80
1939	First cyclotron for medical applications.	80
	High-precision measurements of proton and deuteron magnetic moments.	80
	First evidence of muon decay and first estimation of its lifetime.	81
	Further development of the metrology of fundamental physics parameters.	81
1940	First observation of muon decay.	82
	First direct measurement of the neutron magnetic moment.	83
	Theorem on the connection between spin and statistics.	83
	Kerst proposal for betatron accelerator.	84
1941	First measurement of the muon lifetime, preliminary result.	84
	First betatron.	85
	First measurement of the muon lifetime.	85
1942	Evidence for the muon exponential decay rate.	85
	Confirmation of the muon exponential decay rate.	86
1943	First Fermi nuclear reactor. Beginning of atomic energy era.	86
	Creation of the covariant quantum electrodynamic theory. Tomonaga method.	86
	Invention of the S-matrix formalism.	87
1944	Invention of the principle of phase stability for accelerators.	88
	First evidence for the K^+.	88
	Landau distribution for fast particle energy loss by ionization.	88
	Limitation of the maximal energy attainable in a betatron.	89
1945	Invention of the principle of phase stability for accelerators.	89
1946	Gamow indication on the possibility to explain the observed chemical elements abundance-curve by assumption of unequilibrium process of elements formation during a limited interval of time. Birth of the Big Bang model.	89
	Further development of the synchrotron idea.	90
	Proposal for the radiochemical method of detecting the neutrino.	91
1947	Evidence that the muon is not a strong interaction mediator.	91
	Creation of the covariant quantum electrodynamic theory. Tomonaga method.	92
	First cyclotron based on the phase stability principle – the Berkeley 184 inch synchrocyclotron.	92
	First indication of the existence of the π^-.	92
	Confirmation of the π^-.	93
	Creation of the covariant quantum electrodynamic theory. Tomonaga method.	93
	First measurements of the fine structure of the hydrogen atom, the Lamb shift.	94
	First idea about universality of the Fermi weak interactions.	94
	First theoretical calculation of the Lamb shift in non-relativistic QED.	95
	Invention of scintillation counters.	95
	First measurement of $g-2$ for the electron, preliminary result.	95
	First indication of the existence of π^- decay into μ^-.	96
	Confirmation of the π^-. First evidence for pion decay $\pi^\pm \to \mu^\pm$ neutrals.	96
	First evidence for V events. Confirmation of the existence of a charged unstable particle with a mass between those of the muon and proton.	97
1948	Nonexistence of $\mu^- \to e^-\gamma$ decay.	97
	First explicit theoretical paper on the investigation of the electromagnetic structure of a nucleus by electron-nucleus scattering.	98
	First measurement of $g-2$ for the electron.	98
	First theoretical calculation of $g-2$ for the electron.	99
	Nonexistence of $\mu^- \to e^-\gamma$ decay.	99
	Proposal to modify classical electrodynamics to a form suitable for quantization.	99
	First evidence for neutron beta decay.	100
	Creation of the covariant theory of quantum electrodynamics. Feynman method.	100
	Creation of the covariant quantum electrodynamic theory. Schwinger method.	101
	Confirmation of the nonexistence of $\mu^- \to e^-\gamma$ decay.	102
	Creation of the covariant quantum electrodynamic theory. Tomonaga method.	102

Year	Discovery	Page
	Invention of the path integral formalism for quantum mechanics.	103
1949	First evidence for three-prong kaon decay.	103
	First pion production reaction by accelerator.	103
	Covariant quantum electrodynamics: Equivalence between the Tomonaga-Schwinger method and the Feynman method and generalization.	104
	Creation of the covariant quantum electrodynamic theory. Schwinger method.	104
	Evidence for the continuous energy distribution of the e^- in the decay $\mu^- \to e^-X$.	105
	Proposal of the universality of the Fermi interaction.	106
	Confirmation of the continuous energy distribution of the e^- in the decay $\mu^- \to e^-X$. Muon spin = 1/2.	106
	Covariant quantum electrodynamics: Equivalence between the Tomonaga-Schwinger method and the Feynman method and generalization.	107
	Creation of the covariant quantum electrodynamic theory. Feynman method.	107
	Development of the covariant quantum electrodynamic theory. Feynman method.	108
	Invention of semi-conductor detectors.	108
	First composite model of pions.	109
	Creation of the spark chamber method for particle tracking.	109
1950	First evidence for the existence of the π^0.	109
	First observation of neutron beta decay.	110
	Ward identity in quantum electrodynamics.	110
	First evidence for V events.	110
	Confirmation of neutron beta decay.	111
	Invention of liquid scintillation counters.	111
	First evidence for the production of the π^0 and for $\pi^0 \to \gamma\gamma$ decay.	112
	Confirmation of the existence of the π^0, first estimation of the π^0 lifetime.	112
	Systematic treatment of the application of variational principles to the quantum theory of scattering. Invention of the Lippmann-Schwinger form of the Schrödinger equation.	113
	Mathematical proof of the validity of the Feynman rules for calculations of amplitudes in QED.	113
	First evidence for multiple hadron production in nucleon-nucleon interactions. First evidence for the forward jet of secondary particles in hadronic interactions.	114
1951	Confirmation of the π^0 and $\pi^0 \to \gamma\gamma$ decay. Direct determination of the π^- parity.	114
	Confirmation of the odd parity of the π^-.	115
	First evidence for $K^+ \to \mu^+$ neutral(s) decay.	115
	Confirmation of the existence of heavy unstable particles.	115
	Determination of the spin of the π^+.	116
	First evidence for the zero spin of the π^+.	117
	Evidence for the possible existence of the Λ hyperon. First evidence for the K^0 meson.	117
	Evidence for two types of neutral V particles, baryonic and mesonic.	118
	Confirmation of the K^+.	118
	Evidence for the possible existence of the Λ hyperon. First evidence for the K^0 meson.	118
	First results from electron-nuclei scattering experiment.	119
	Bethe-Salpeter relativistic equation for two-body bound-state problems.	119
1952	Confirmation of the existence of heavy charged unstable particles.	120
	Prediction of rising total hadronic cross sections.	120
	First indication for the $\Delta(1232\,P_{33})$ resonance.	121
	Further evidence for the $\Delta(1232\,P_{33})^{++}$ resonance.	121
	Hypothesis on associative production of V^0 particles.	121
	Invention of the strong focusing principle for accelerators.	122
	First evidence for the Ξ^- hyperon.	122
1953	First evidence for the charged Σ hyperon.	123
	Confirmation of the existence of Σ^+ and Σ^- hyperons.	123
	Direct experimental evidence for the Σ^+.	123
	First evidence for $K \to \pi\pi$ decay.	123

	First measurement of the K^0 mass.	124
	First V-event production at the Cosmotron.	124
	First evidence of charged particle tracks in a bubble chamber.	125
	Confirmation of the existence of the Λ hyperon. First evidence for associated production of heavy unstable particles.	125
	First evidence for the $\overline{\nu}_e$.	125
	Extension of isotopic multiplet structure for new unstable particles. Explanation of pairwise production of V particles.	??
	Invention of the concept of lepton quantum number.	126
	Confirmation of the existence of the Ξ^-.	126
	Further confirmation of the Ξ^-.	127
	Confirmation of the existence of the $\Delta(1232\,P_{33})$ resonance.	127
	First evidence for a hypernucleus.	128
	Invention of the Dalitz plot method to analyse spatial quantum numbers of mesons by their decays into three known particles.	128
	Invention of the renormalization group.	129
1954	**Confirmation of the decays $K^\pm \to \mu^\pm$ neutral.**	129
	Confirmation of the existence of the Σ^-. Evidence for the associated production of strange particles.	130
	Confirmation of Ξ^- cascade decay. First indication for $\overline{\Lambda}$.	130
	First indication on zero spin and odd parity for the K meson.	131
	Invention and exploration of the renormalization group concept.	131
	Forward dispersion relations for massive particles.	132
	Introduction of local gauge isotopic spin invariance in quantum field theory: Yang-Mills theory.	132
	Evidence for $K^+ \to \pi^+\pi^0$ decay.	132
	Theoretical evidence for CPT invariance in local quantum field theory.	133
1955	**First experimental evidence that neutrino is not identical with antineutrino.**	133
	Prediction of the long lived kaon K_L.	134
	Forward dispersion relations for massive particles.	134
	Confirmation of associated production of strange particles. First indication of the existence of the Σ^0.	134
	Dispersion relations for massive particles.	135
	Forward dispersion relations for massive particles.	135
	Confirmation of the existence of the K^+ and K^-.	135
	Confirmation of the existence of the K^+.	136
	First precise measurement of the K^+ mass.	136
	First measurement of the K^+ lifetime. First evidence for the equality of lifetimes of the Θ and τ mesons.	137
	K^+ lifetime measurements from decays in flight.	137
	Further measurements of charged-kaon masses. Masses of the strange K^+, Θ, and τ mesons are equal.	137
	Experimental evidence for the antiproton.	138
	Evidence that strong interactions do not modify the vector coupling constant of beta decay. Analogy between electromagnetic and weak interactions.	138
	Confirmation of the associated production of strange particles.	139
	First evidence for the odd parity of the π^0.	139
	Proposal for the $K_L \to K_S$ regeneration experiment.	139
	Beginnings of the axiomatic field theory of the S-matrix.	140
	Invention of flash tube chambers.	140
	Measurements of charged kaon decay branching fractions.	140
	First GeV linear accelerator at Stanford.	141
	Nishijima classification of strange particles with prediction of Σ^0 and Ξ^0 hyperons.	141
1956	**Confirmation of the equality of the masses of K^+ and τ^+ mesons.**	142
	Confirmation of the equality of the lifetimes of K^+ and Θ ($K^+ \to \pi^+\pi^0$) mesons.	142
	Confirmation of the existence of the antiproton.	142
	Firm establishment of the K^+ lifetime value. Establishment of equality of the lifetimes of $K^+(\to \mu^+$ neutral), $\Theta(\to \pi^+\pi^0)$, and $\tau^+(\to \pi^+\pi^+\pi^-)$ mesons.	143

	Wightman axiomatic field theory.	143
	First indication of annihilation of the antiproton in matter.	143
	First evidence of annihilation of the antiproton in emulsion.	143
	Static model for the pion–nucleon interaction.	144
	First realistic proposal for probing high energies by colliding beams of particles.	144
	First measurement of the proton electromagnetic radius.	145
	Confirmation of antiproton-nucleon annihilation.	145
	Confirmation of the detection of the $\overline{\nu}_e$.	146
	First detection of the free neutrino.	146
	First evidence for the K_L.	147
	Confirmation of the existence of the K_L.	147
	Proposals to test spatial parity conservation in weak interactions.	148
	First evidence for the antineutron.	148
	Invention of G-parity for nonstrange mesons.	149
	Development of the renormalization group method in quantum field theory.	149
	Generalization of the renormalization group equations in QED for arbitrary covariant gauge.	149
	Gell-Mann classification of strange particles with prediction of Σ^0 and Ξ^0.	149
	Formulation of the Bogolyubov axiomatic approach to the local quantum field theory. Derivation of dispersion relations in field theory for pion nucleon scattering amplitude, general case.	150
	Derivation of dispersion relations in field theory for pion nucleon forward scattering amplitude.	150
	Invention of composite model for hadrons based on three basic elements.	151
1957	**Indication of the possibility of charge conjugation violation in weak interactions.**	151
	Confirmation of the existence of the Σ^0 hyperon. First measurement of the Σ^0 mass.	152
	Postulation of γ_5 invariance for the weak interaction Lagrangian. Two–component theory of neutrino.	152
	Confirmation of antiproton-nucleon annihilation.	152
	First evidence for parity nonconservation in weak decays.	153
	Confirmation of parity violation in weak decays. Evidence of charge conjugation parity violation in weak interactions. Measurement of the μ^- magnetic moment.	153
	Two–component theory of neutrino.	154
	Confirmation of parity nonconservation in weak decays.	154
	Introduction of the CP conservation law in weak interactions and CP parity.	155
	Suggestion of the two-component theory for the neutrino.	155
	Confirmation of the existence of the K_L.	156
	Charge conjugation invariance violation in weak interactions.	156
	Further confirmation of spacial parity nonconservation in beta decays.	156
	Rigorous derivation of the dispersion relations for pion photoproduction amplitude.	157
	Further proposals for CP invariance.	157
	Introduction of lepton-family-number conservation.	157
	Evidence for limited transverse momenta in hadronic showers.	158
	Further development of the idea of the intermediate vector boson in weak interactions.	158
	First review of particle properties data.	158
	Description of the electromagnetic structure of the nucleon by form factors.	159
1958	**First collection of particle physics data in a compact and readily accessible form.**	159
	First measurement of the $K_S - K_L$ mass difference.	159
	First evidence for the $\overline{\Lambda}$.	159
	CVC and symmetry between electromagnetism and weak interaction. $\Delta S = \Delta Q$ for nonleptonic decays of the strange particles.	160
	First evidence for π_{e2} decay.	160
	Confirmation of π_{e2} decay.	161
	Proposed method for extraction of π-π interactions.	161
	First evidence of Λ beta decay.	162
	Confirmation of Λ beta decay.	162
	Proposal for the possibility of neutrino-antineutrino oscillations.	162
	First evidence for the negative ν_e helicity (ν_e is left handed).	163

Prediction of the existence and some properties of the η and η' particles on the basis of Sakata model. Invention of the $\Delta S = \Delta Q$ selection rule. 163
Theorem on asymptotic equality of hadron-hadron and antihadron-hadron interaction cross sections. 163
Confirmation of the existence of the antineutron. 164
Goldberger-Treiman relations. 164
Universal $V - A$ weak interactions. 164
Failure of universal Fermi interactions in the beta decay of hyperons. 165
Dispersion relation in two variables: the Mandelstam representation. 165
Evidence for limited transverse momenta in hadronic jets. 166
1959 Prediction of the optical activity of atomic media due to possible weak neutral currents. Prediction of the anapole moments of nuclei, due to weak interactions. 166
Principle of the spark chamber. 167
Confirmation of the existence of hadronic jets in high energy collisions: The average P_T of hadrons in jets is limited and almost independent of jet energy. 167
First evidence for the Ξ^0. 168
First observation of an enhancement in the production of like-charge pairs of pions with similar momenta. 168
First direct determination of the parity of the π^0. 169
Confirmation of the detection of the $\overline{\nu}_e$. 169
Introduction of $SU(3)$ symmetry for hadrons. Prediction of the existence of the η meson. 170
Invention of Landau singularities for perturbative amplitudes. 170
Proposed experiments to establish distinguishability of ν_e and ν_μ. Indication on the feasibility of neutrino beams with accelerators. 170
Method for extraction of pion-pion interactions. 171
Introduction of Regge poles. 171
1960 Non linear sigma model. 171
First theoretical estimations of dipole polarizabilities of nucleons. 171
Confirmation of the feasibility of neutrino beams with accelerators. 172
First evidence for the $\overline{\Xi}^0$. 172
First evidence for the $\Sigma(1385\,P_{13})$. 172
First evidence for the $\overline{\Sigma}^-$. 173
Interpretation of the enhancement in the production of like charge pairs of pions with similar momenta as an influence of Bose Einstein correlations. 173
Introduction of Regge poles. 174
First measurement of electrical polarizability of the proton. 174
1961 Prediction of the spin=1, isospin=1 resonance in the two-pion system. 175
Prediction of unavoidable massless bosons if global symmetry of the Lagrangian is spontaneously broken. 175
Invention of the gauge principle as basis to construct quantum theories of interacting fundamental fields. 175
Confirmation of $\overline{\Lambda}$ production. 175
Nambu-Jona-Lasinio nonlinear model of hadrons. 176
First conclusive measurements of the π^0 lifetime. 176
Froissart upper bound on the total cross sections of hadronic collisions. 177
Nambu-Jona-Lasinio nonlinear model of hadrons. 177
First evidence for $K_L \to K_S$ regeneration. 177
Prediction of the "radiation self-polarization" effect for electrons moving in magnetic field. 178
First evidence for the $K^*(892)$ resonance. 178
Generalization of Regge asymptotics for relativistic scattering amplitudes. 179
First evidence for the ρ meson resonance. 179
First evidence for the $\Lambda(1405\,S_{01})$ resonance. 180
Introduction of the $SU(3)$ octet structure of the known mesons and baryons. 181
First evidence for the ω meson resonance. 181
Introduction of the $SU(3)$ octet structure of the known mesons and baryons. 182
Confirmation of the existence of the ω meson. 182
Invention of equivalence of elementary hadrons and hadronic resonances on the basis of Regge trajectories. Invention of the Chew-Frautschi plot to classify hadrons. Invention of the vacuum pomeron trajectory. 183

First evidence for the η meson. Confirmation of the ω meson. 183
First introduction of the neutral intermediate boson. 184
AGS (30 GeV) at BNL — the first strong focusing proton synchrotron. 184
1962 First observation of the π^+ beta decay. First direct experimental evidence for the validity of the CVC hypothesis. 184
Determination of the Σ parity. 185
First evidence for $\Xi^- \, \overline{\Xi}^+$ pair production. 185
Confirmation of the existence of the $\overline{\Xi}^+$. 185
First evidence for the ν_μ. Evidence for more than one kind of neutrinos. 186
Introduction of the $SU(3)$ singlet-octet structure of the known mesons and octet-decuplet structure for the baryons. Prediction of the Ω^- hyperon. 186
First evidence for the $\Xi(1530\, P_{13})$ resonance. 187
Confirmation of the existence and evidence for spin zero of the η meson. 187
Confirmation of the existence of the $\Xi(1530\, P_{13})$ resonance. First evidence for the ϕ resonance. 188
Determination of the spin of the $K^*(892)$ resonance to be 1. 188
Application of Regge poles to resonances and particles. 189
Conditions for renormalizability of general gauge theories of massive vector mesons. 189
Invention of the multiperipheral model to analyze a few and many body hadronic reactions. Demonstration that multiperipheral model is capable to predict qualitatively the general features of elastic scattering, inelastic particles spectra, and topological cross sections. 190
Perturbative and general proofs of the Goldstone theorem. 190
1963 Confirmation of the ϕ meson. 190
Firm establishment of the ϕ meson. 191
$SU(3)$ and hadronic weak currents. Introduction of the Cabibbo angle; predictions for the leptonic decay rates of hyperons. 191
First evidence for a double hypernucleus. 192
Derivation of the relativistic generalization of the Lippmann-Schwinger equation for the two-body problem in quantum field theory. 192
Farther investigation of the "radiation self-polarization" effect for electrons moving in magnetic field. 192
1964 Invention of the streamer chamber. 193
Introduction of quarks as fundamental building blocks for hadrons. 193
Introduction of aces (quarks) as fundamental building blocks for hadrons. 194
Proposal for the existence of a charmed fundamental fermion. 194
Further example of a field theory with spontaneous symmetry breakdown, no massless goldstone boson, and massive vector bosons. 195
First evidence for a hyperon with strangeness -3, the Ω^-. 195
Confirmation of the Ω^- hyperon. 196
First evidence for the η' meson. 196
Confirmation of the η' meson. 197
First evidence for CP violation. 197
Introduction of the $SU(6)$ classification of hadrons. 197
Example of a field theory with spontaneous symmetry breakdown, no massless goldstone boson, and massive vector bosons. 198
Higgs mechanism of mass generation for vector gauge fields. 198
Example of a field theory with spontaneous symmetry breakdown, no massless goldstone boson, and massive vector bosons. 199
Invention of the superweak theory for CP-violation in weak interactions. 199
Introduction of the color quantum number, and colored quarks and gluons. 199
Confirmation of the existence of the Ω^- hyperon. 200
$SU(6)$ classification of hadrons. 200
Lagrangian for the electroweak synthesis, first estimations of the W mass. Salam-Ward version. 201
1965 Introduction of the color quantum number, and colored quarks and gluons. 201
Introduction of an additional quantum number (the color) to resolve conflict with Fermi statistics. Explanation of the relations between magnetic moments of baryons. 202

	First evidence of the antideuteron.	202
	Confirmation of $K_L \to K_S$ regeneration phenomenon.	202
	First evidence of the spatial-parity non-conservation in weak nuclear interactions.	202
	Evidence for large real part of the nuclear scattering amplitude.	203
1966	**Confirmation of the spatial parity nonconservation in weak nuclear interactions.**	204
	Higgs mechanism of mass generation for vector gauge fields.	204
	First cosmological upper bound on the stable neutrino masses sum.	205
	Invention of the dispersion sum rules for hadronic binary amplitudes.	205
	Invention of the idea of a vector gluon theory for strong interactions.	205
	Three triplet model for hadrons. Beginnings of the quantum chromodynamics — QCD.	206
1967	**First attempt to explain baryonic asymmetry of the observable universe.**	206
	Proposal for electron cooling of the proton and antiproton bunches in storage rings.	207
	First evidence for CP violation in $K_L \to \pi^\pm\ \mu^\mp\ \nu_\mu$ decays.	207
	Evidence for CP violation in semileptonic decays of the K_L.	207
	Lagrangian for the electroweak synthesis, first estimations of the W and Z masses.	208
	Extension of the Higgs mechanism of mass generation for non-Abelian gauge field theories. Higgs-Kibble mechanism.	208
	Faddeev-Popov method for construction of Feynman rules for Yang-Mills type of gauge theories.	209
	Generalization of the dispersion sum rules to non-decreasing hadronic amplitudes.	209
	First derivation of asymptotic bounds on the behavior of the one particle inclusive differential cross sections from general principles.	209
1968	**Lagrangian for the electroweak synthesis. Salam-Ward version.**	209
	Invention of multiwire proporional chambers.	210
	Observation of the $\phi \to e^-\ e^+$ decay.	210
1969	**Invention of Bjorken scaling behavior.**	211
	First evidence for Bjorken scaling behavior.	211
	Confirmation of Bjorken scaling behavior.	211
	Proposal for scaling behavior of the inclusive spectra of produced hadrons. Birth of the partonic picture of hadron collisions. Precise formulation of exclusive and inclusive experiments dichotomy.	212
	First conclusive evidence for scale invariance in hadronic inclusive experiments.	212
	Experimental evidence for the increasing diffraction slope parameter.	212
	Confirmation of the antimatter production in hadron nucleus collisions.	213
	Explanation of Bjorken scaling with use of the parton model.	213
1970	**Introduction of lepton-quark symmetry, proposal of a fourth (charmed) quark.**	213
	Confirmation of scale invariance phenomena in hadronic inclusive experiments.	214
	First evidence for $^3\overline{\mathrm{He}}$ production.	214
	First observation of the high mass muon pairs in hadron collisions — prototype of the experiments which lead to the discovery of the J (Ting) and Υ (Herb) as well as "Drell-Yan" analyses of quark structure functions.	214
1971	**First experimental indication of the rising total hadronic cross sections.**	215
	First evidence for the $\overline{\Omega}^+$.	215
	Rigorous proofs of renormalizability of the massless Yang-Mills quantum fields theory.	215
	Rigorous proof of renormalizability of massive Yang-Mills quantum fields theory with spontaneously broken gauge invariance.	216
1972	**Invention of Gribov-Lipatov evolution equations for perturbative parton distribution functions in scalar and vector theories. Scaling violation prediction.**	216
	Firm establishment of Bjorken scaling behavior.	217
	Universal regularization and renormalization method for gauge fields theories.	217
1973	**Evidence for cumulative effect.**	218
	Observation that CP violation can be accommodated in the standard electroweak model only if there are at least six quark flavours.	218
	Confirmation of rising total hadronic cross sections.	218
	Further confirmation of rising total hadronic cross sections.	219
	First experimental indication of the existence of weak neutral currents in pure leptonic interactions.	219

Year	Discovery	Page
	First experimental evidence for weak neutral currents.	219
	Discovery of the "asymptotic freedom" property of interacting Yang-Mills field theories.	220
	First observation of high transverse momentum hadrons at the CERN Intersecting Storage Rings.	220
	Final formulation of QCD and the Standard Model Lagrangian.	221
	Invention of the QCD Lagrangian of Yang–Mills type.	221
	Final formulation of QCD theory.	221
	Quark counting rules for asymptotic energy power low behavior of the binary hadronic amplitudes at large fixed angles.	222
1974	Confirmation of the existence of weak neutral currents.	223
	First evidence for the $J/\psi(1S)$.	223
	Confirmation of the existence of the $J/\psi(1S)$.	224
	First evidence for the $\psi(2S)$.	225
1975	First evidence for the charmed baryon $\Sigma_c(2455)^{++}$. First indication of the production of the Λ_c^+ charmed baryon.	225
	Evidence of azimuthal asymmetry in inclusive hadron production in polarized $e^+ e^-$ collisions. Confirmation of the quark-parton picture of hadron production.	226
	First indication of the τ lepton.	226
	First evidence for quark jets in $e^+ e^-$ annihilation.	226
	First evidence for the spin 4 $f_4(2050)$ resonance. Confirmation of linearity of Regge trajectories for the spin ≤ 4 resonances.	227
	Evidence for a Spin 4 Boson Resonance at 2050 MeV.	227
	Invention of the BPST-instanton – the pseudoparticle solution of the Yang-Mills equation.	227
1976	Evidence for the production of the τ lepton.	228
	First evidence for the D^0 charmed meson.	228
	Evidence for large polarization of produced hyperons in p Be collisions.	229
	First evidence for the production of D^+ and D^- charmed mesons.	229
	First evidence for the charmed antibaryon $\overline{\Lambda}_c^-$.	229
	Confirmation of muon-antineutrino scattering off electrons. Confirmation of the weak neutral current.	230
	Evidence for electron cooling.	230
	Experimental confirmation of the "radiation self-polarization" effect for electrons moving in magnetic field.	230
1977	Confirmation of the existence of the weak neutral current.	231
	First evidence for $D_S^{\pm}$, D_S^{*+}, and D_S^{*-} strange charmed mesons.	231
	Invention of Altarelli-Parisi evolution equations for quark and gluon densities in colliding hadrons.	232
	Firm establishment of the τ lepton properties.	232
	Proposal of a Peccei-Quinn spontaneously broken symmetry to explain CP conservation of strong interactions.	232
	First evidence of the $\Upsilon(1S)$ meson interpreted as a bound state of the new quarks $b\,\bar{b}$. Further indication on the existence of the third quark–lepton family.	233
	Evidence for the $\Upsilon(2S)$ resonance.	233
1978	Evidence of strong energy dependence of spin-spin correlation parameters in large angle elastic $p\,p$ scattering. "Argon spin-effect".	234
	Confirmation of the $\Upsilon(1S)$ resonance.	234
	First evidence of the weak neutral current in atomic transitions.	235
	Confirmation of weak neutral currents.	235
	Evidence for stochastic cooling.	235
	First evidence for elastic muon-neutrino scattering off electrons. Confirmation of weak neutral currents.	236
	Confirmation of the $\Upsilon(2S)$ resonance.	236
	Invention of the method of the acceleration of polarized particles to high energies - Sibirian snakes.	237
	Evidence for the nonnegligible spin effects in the strong interactions at high energies.	237
1979	Invention of the quantum chromodynamic sum rules.	238
	First evidence for the $\Upsilon(3S)$ state. Confirmation of the $\Upsilon(1S)$ and $\Upsilon(2S)$ states.	238
	Confirmation of the $\Sigma_c(2455)^{++}$ and the Λ_c^+.	238
	First evidence for the gluon jet in $e^+e^- \to 3$jet annihilations.	239
	Confirmation of weak neutral currents.	239

	Confirmation of parity nonconservation effects in atomic transitions.	239
	Confirmation of the production of gluon jets.	240
1980	**Confirmation of gluon jets.**	241
	Invention of the resonance depolarization method of the beams energy calibration in an electron-positron storage ring.	241
	First experimental determination of the gluon spin.	241
	Confirmation of the gluon spin = 1.	242
	Confirmation of the $\Upsilon(3S)$.	242
	First evidence for the $\Sigma_c(2455)^+$.	243
	First evidence for the $\Upsilon(4S)$.	243
1981	**First evidence for the B meson.**	244
	First indication of the existence of the bottom baryon Λ_b.	244
	Confirmation of the production of the B meson.	245
	First global comparison of data on weak neutral currents with minimal electroweak theory.	245
1983	**First evidence for the charged intermediate bosons W^+ and W^-.**	246
	First evidence for the $\Xi_c(2460)^+$.	246
	First evidence for the production of the charged intermediate bosons W^+ and W^-.	247
	Evidence for difference between structure functions of bound and free nucleons — EMC effect.	247
	First evidence for the neutral intermediate boson Z^0.	248
	Confirmation of Z^0 boson production. Observation of $Z^0 \to e^+e^-$ decay.	248
	Confirmation for difference between structure functions of bound and free nucleons — EMC effect.	249
1984	**Confirmation of W^+ and W^- production. First observation of $W^\pm \to \mu^\pm\nu$ decays.**	249
	Implementation of the resonance depolarization method of the beams energy calibration to the high precision measurements of heavy e^+ e^- resonances.	250
	Confirmation of Z^0 boson production. Observation of $Z^0 \to \mu^+\mu^-$ decay.	250
1985	**First evidence for the Ω_c.**	250
1986	**Firm establishment of the properties of W^+, W^- and Z^0 bosons.**	251
	First evidence for the $\Sigma_c(2455)^0$.	252
	High precision measurement of the electron g−2 factor.	252
1987	**First indication of B_S-$\overline{B}_S$ mixing.**	253
	Firm establishment of properties of the W^+, W^-, and Z^0 bosons.	253
	First evidence for the B^0-$\overline{B}^0$ mixing.	254
	First observation of the neutrino burst from supernova SN1987A.	255
	First observation of the neutrino burst from supernova SN1987A. Birth of neutrino astronomy.	255
	High precision measurements of electron and positron g−2 factors. High precision test of QED and CPT symmetry.	255
1988	**Evidence for complex spin structure of the proton, "proton spin crisis".**	256
	Confirmation of B^0-$\overline{B}^0$ mixing.	256
1989	**First evidence for the $\Xi_c(2460)^0$ and $\overline{\Xi}_c(2460)^0$.**	257
	Confirmation of B^0-$\overline{B}^0$ mixing.	257
	First evidence that the number of light neutrinos = 3.	258
	Confirmation of the $\Xi_c(2460)^+$, $\Xi_c(2460)^0$, $\overline{\Xi}_c(2460)^0$, and $\overline{\Xi}_c(2460)^-$ states.	258
1990	**Confirmation of B_S-$\overline{B}_S$ mixing.**	258
	Confirmation of the number of light neutrinos = 3.	260
	Confirmation of B_S-$\overline{B}_S$ mixing.	262
	Confirmation of the beauty baryon Λ_b.	262
1992	**First direct observation of the β decay into a bound electron state.**	263
	Confirmation of the number of light neutrinos = 3.	263
	Confirmation of B_S-$\overline{B}_S$ mixing.	264
	Precise determination of the Z^0 parameters. Confirmation of the number of light neutrinos = 3.	264
1993	**First direct and precise measurement of the B_S meson mass.**	265

III. Bibliography of Discovery Papers

PERRIN 1895 **Nobel prize**

■ Observation that cathode rays are the flow of negatively charged particles. Nobel prize to J. B. Perrin awarded in 1926 "for his work on the discontinuous structure of matter, and especially for his discovery of sedimentation equilibrium" ■

Nouvelles Propriétés des Rayons Cathodique / New Properties of the Cathode Rays

J.B. Perrin
Compt. Ren. **121** (1895) 1130;

Reprinted in
(translation into English) *A Source Book in Physics*, Ed. Gregory D. Walcott, McGraw-Hill Book Company, Inc. (1935) 580.

Excerpt On a imaginé deux hypothèses pour expliquer les propriétés des rayons cathodiques.
Les uns, avec Goldstein, Hertz ou Lenard, pensent que ce phénomène est dû, comme la lumière, à des vibrations de l'èther, ou même que c'est une lumière, à courte longueur d'onde. On conçoit bien alors que ces rayons aient une trajectoire rectiligne, excitent la phosphorescence, et impressionnent les plaques photographiques.
D'autres, avec Crookes ou J. J. Thompson, pensent que ces rayons sont formès par de la matière chargèe nègativement et cheminant avec une grande vitesse. Et l'on conçoit alors très bien leurs propriétés mécaniques, ainsi que la façon dont ils s'incurvent dans un champ magnétique.
Cette dernière hypothèse m'a suggéré quelques expèriences que je vais rèsumer sans m'inquiéter, pour le moment, de rechercher si elle rend compte de tous les faits jusqu'á présent connus, et si elle peut seule en rendre compte.
Ses partisans admettent que les rayons cathodiques sont chargés négativement; à ma connaissance, on n'a pas constaté cette électrisation; j'ai d'abord tenté de vérifier si elle existe, ou non.

Excerpt Two hypotheses have been presented to explain the properties of the cathode rays.
Some, with Goldstein, Hertz, or Lenard, think that this phenomenon, like light, results from vibrations of the ether, or even that it is light of short wave length. We then easily see that these rays might have a straight trajectory, excite phosphorescence, and act upon photographic plates.
Others, with Crookes or J. J. Thomson, think that these rays are formed of matter negatively charged and moving with great velocity. We then can easily understand their mechanical properties and also the way in which they bend in a magnetic field.
This latter hypothesis suggested to me some experiments which I shall present without troubling myself for the moment to consider if the hypothesis accounts for all the facts which are at present known, and if it alone can account for them.
Its partisans assume that the cathode rays are negatively charged; so far as I know, no one has demonstrated this electrification; I have therefore tried to determine if it exists or not. *(Extracted from the introductory part of the paper.)*

Accelerator cathode ray tube

Detectors Nonelectronic detectors

Particles studied charged^-

RÖNTGEN 1895 **Nobel prize**

■ Discovery of the X rays. Nobel prize to W. C. Röntgen awarded in 1901 "for the discovery of the remarkable rays subsequently named after him" ■

Über eine neue Art von Strahlen. Vorlaufige Mitteilung / On a New Kind of Rays. Preliminary Communication

W.C. Röntgen
Sitzber. Physik. Med. Ges. **137** (1895) 1;

Reprinted in
(translation into English) *A Source Book in Physics*, Ed. Gregory D. Walcott, McGraw-Hill Book Company, Inc. (1935) 600.
(translation into English) W. R. Nitske, *The Life of Wilhelm Conrad Röntgen*, University of Arizona Press, Tucson (1971).

Excerpt If the discharge of a fairly large induction coil be made to pass through a Hittorf vacuum tube, or through a Lenard tube, a Crookes tube, or other similar apparatus, which has been sufficiently exhausted, the tube being covered

with thin, black card board which fits it with tolerable closeness, and if the whole apparatus be placed in a completely darkened room, there is observed at each discharge a bright illumination of a paper screen covered with barium platino cyanide, placed in the vicinity of the induction coil, the fluorescence thus produced being entirely independent of the fact whether the coated or the plain surface is turned towards the discharge-'tube. This fluorescence is visible even when the paper screen is at a distance of two metres from the apparatus.
It is easy to prove that the cause of the fluorescence proceeds from the discharge apparatus, and not from any other point in the conducting circuit. *(Extracted from the introductory part of the paper.)*

Accelerator cathode ray tube

Detectors Nonelectronic detectors

BECQUEREL 1896B — Nobel prize

■ Evidence for spontaneous radioactivity effect. Nobel prize to H. Becquerel awarded in 1903 "for his discovery of spontaneous radioactivity". Co-winners M. Curie and P. Curie "for their joint researches on the radiation phenomena discovered by Becquerel" ■

Sur les Radiations Invisibles Emises par les Corps Phosphorescents / On the Invisible Radiation Emitted by Phosphorescent Substances

H. Becquerel
Compt. Ren. **122** (1896) 501;

Reprinted in
(translation into English) A. Romer, *The discovery of radioactivity and transmutation*, Dover, New York 1964.

Abstract At the last session I sketched briefly the experiments I had been led to perform in order to demonstrate the invisible radiations emitted by certain phosphorescent substances, radiations which penetrate various substances which are opaque to light.
I have been able to extend these observations, and, although I propose to continue and develop the study of these phenomena, their present interest leads me to set forth as early as today the first results I have obtained.
The experiments I shall report were made with the radiations emitted by crystalline lamellas of the double sulfate of potassium and uranium

$$[K(UO)SO_4 + H_2O],$$

a substance whose phosphorescence is very lively and whose persistence of luminosity is less than 1/100 of a second.

Accelerator Radioactive source

Detectors Nuclear emulsion

Related references

More (earlier) information
H. Becquerel, Compt. Ren. **122** (1896) 420;

Reactions
U *decay*

BECQUEREL 1896C — Nobel prize

■ Confirmation of the effect of spontaneous radioactivity. Nobel prize to H. Becquerel awarded in 1903 "for his discovery of spontaneous radioactivity". Co-winners M. Curie and P. Curie "for their joint researches on the radiation phenomena discovered by Becquerel" ■

Sur Quelques Proprietes Nouvelles des Radiations Invisibles Emises par Divers Corps Phosphorescents / On Several New Properties of Invisible Radiations Emitted by Different Phosphorescent Substances

H. Becquerel
Compt. Ren. **122** (1896) 559;

Reprinted in
(translation into English) A. Romer, *The discovery of radioactivity and transmutation*, Dover, New York 1964.

Abstract At one of the last sessions of the Academy, I announced that the invisible radiations emitted by the salts of uranium possessed the property of discharging electrified bodies. I have continued the study of this phenomenon by the use of Hurmuzescu's electroscope, and I have been able to establish, in another way than I have done by photography, that the radiations in question penetrate various opaque substances, in particular, aluminum and copper. Platinum exhibits an absorption considerably greater than that of the two preceding metals. *(Extracted from introductory part of the paper.)*

Accelerator Radioactive source

Detectors Nuclear emulsion

Related references

More (earlier) information
H. Becquerel, Compt. Ren. **122** (1896) 501;

Reactions
U *decay*

BECQUEREL 1896D **Nobel prize**

■ Confirmation of the effect of spontaneous radioactivity. Observation that uranium radiation is not identical to Röntgen X rays. Nobel prize to H. Becquerel awarded in 1903 "for his discovery of spontaneous radioactivity". Co-winners M. Curie and P. Curie "for their joint researches on the radiation phenomena discovered by Becquerel" ■

Sur les Proprietes Differentes des Radiations Invisibles Emises par les Sels d'Uranium, et du Rayonnement de la Paroi Anticathodique d'un Tube de Crookes

H. Becquerel
Compt. Ren. **122** (1896) 762;

Reprinted in
(translation into English) A. Romer, *The discovery of radioactivity and transmutation*, Dover, New York 1964.

Accelerator Radioactive source

Detectors Nuclear emulsion

Related references
More (earlier) information
H. Becquerel, Compt. Ren. **122** (1896) 501;
H. Becquerel, Compt. Ren. **122** (1896) 559,;
H. Becquerel, Compt. Ren. **122** (1896) 689;

Reactions
U *decay*

BECQUEREL 1896F **Nobel prize**

■ Uranium radioactivity established. Nobel prize to H. Becquerel awarded in 1903 "for his discovery of spontaneous radioactivity". Co-winners M. Curie and P. Curie "for their joint researches on the radiation phenomena discovered by Becquerel" ■

Emission de Radiations Nouvelles par l'Uranium Metallique / Emission of New Radiation by Metallic Uranium

H. Becquerel
Compt. Ren. **122** (1896) 1086;

Reprinted in
(translation into English) A. Romer, *The discovery of radioactivity and transmutation*, Dover, New York (1964) 19.

Excerpt All the salts of uranium I have studied, whether phosphorescent or not with respect to light, crystallized, fused, or in solution, have given comparable results. Thus I have been led to think that the effect was due to the presence in these salts of the element uranium, and that the metal would give more intense effects than its compounds would.
A few weeks ago an experiment with a commercial powder of uranium which had long been in my laboratory confirmed this prediction; the photographic effect is notably stronger than the impression produced by one of the salts of uranium, and, in particular, by uranium–potassium sulfate. *(Extracted from the introductory part of the paper.)*

Accelerator Radioactive source

Detectors Nuclear emulsion

Related references
See also
H. Becquerel, Compt. Ren. **122** (1896) 501;
H. Becquerel, Compt. Ren. **122** (1896) 559;
H. Becquerel, Compt. Ren. **122** (1896) 689;
H. Becquerel, Compt. Ren. **122** (1896) 762;

Reactions
U *decay*

THOMSON 1897 **Nobel prize**

■ Discovery of the electron, the first elementary particle. Nobel prize to J. J. Thomson awarded in 1906 "for his theoretical and experimental investigations on the conduction of electricity by gases" ■

Cathode Rays

J.J. Thomson
Phil. Mag. **44** (1897) 293; Nature **55** (1897) 453;

Reprinted in
Philosophical Magazine and Journal of Science, London, Edinburgh, and Dublin **ser. 5, v. 44, n. 269** 293.
Proceedings of the Royal Institution, **15** (1897) 419.
Great Experiments in Physics ed. by M. H. Shamos, Henry Holt and Company, New York (1960).

Excerpt The experiments discussed in this paper were undertaken in the hope of gaining some information as to the nature of the Cathode Rays. The most diverse opinions are held as to these rays; according to the almost unanimous opinion of German physicists they are due to some process in the æther to which—inasmuch as in a uniform magnetic field their course is circular and not rectilinear—no phenomenon hitherto observed is analogous: another view of these rays is that, so far from being wholly ætherial, they are in fact wholly material, and that they mark the paths of particles of matter charged with negative electricity. It would seem at first sight that it ought not to be difficult to discriminate between views so different, yet experience shows that this is not the case, as amongst the physicists who have most deeply studied the subject can be found supporters of either theory.

The electrified-particle theory has for purposes of research a great advantage over the ætherial theory, since it is definite and its consequences can be predicted; with the ætherial theory it is impossible to predict what will happen under any given circumstances, as on this theory we are dealing with hitherto unobserved phenomena in the æther, of whose laws we are ignorant. *(Extracted from the introductory part of the paper.)*

Accelerator Cathode Tube

Detectors Nonelectronic detectors

Particles studied e^-

RUTHERFORD 1899 **Nobel prize**

■ Evidence for the α and β components of uranium radiation. Nobel prize in chemistry to E. Rutherford awarded in 1908 "for his investigations into the disintegration of the elements and the chemistry of radioactive substances" ■

Uranium Radiation and the Electrical Conduction Produced by it

E. Rutherford
Phil. Mag. **47** (1899) 109;

Reprinted in
Philosophical Magazine and Journal of Science, London, Edinburgh, and Dublin, **ser.5, v.47, n.284** 109.

Excerpt It is the object of the present paper to investigate in more detail the nature of uranium radiation and the electrical conduction produced. As most of the results obtained have been interpreted on the ionization-theory of gases which was introduced to explain the electrical conduction produced by Röntgen radiation, a brief account is given of the theory and the results to which it leads.
In the course of the investigation, the following subjects have been considered:

1. Comparison of methods of investigation.
2. Refraction and polarization of uranium radiation.
3. Theory of ionization of gases.
4. Complexity of uranium radiation.
5. Comparison of the radiation from uranium and its compounds.
6. Opacity of substances for the radiation.
7. Thorium radiation.
8. Absorption of radiation by gases.
9. Variation of absorption with pressure.
10. Effect of pressure of the gas on the rate of discharge.
11. The conductivity produced in gases by complete absorption of the radiation.
12. Variation of the rate of discharge with distance between the plates.
13. Rate of recombination of the ions.
14. Velocity of the ions.
15. Fall of potential between two plates.
16. Relation between the current through the gas and electromotive force applied.
17. Production of charged gases by separation of the ions.
18. Discharging power of fine gauzes.
19. General remarks.

(Extracted from the introductory part of the paper.)

Accelerator Radioactive source

Detectors Ionization, Nuclear emulsion

Related references

See also
H. Becquerel, Compt. Ren. **122** (1896) 420;
H. Becquerel, Compt. Ren. **122** (1896) 501;
H. Becquerel, Compt. Ren. **122** (1896) 559;
H. Becquerel, Compt. Ren. **122** (1896) 689;
H. Becquerel, Compt. Ren. **122** (1896) 762;
H. Becquerel, Compt. Ren. **122** (1896) 1086;
H. Becquerel, Compt. Ren. **124** (1897) 438;
H. Becquerel, Compt. Ren. **124** (1897) 800;

Reactions
U $\rightarrow$ He X
U $\rightarrow$ e^- X

Particles studied U

VILLARD 1900

■ Evidence for the γ radioactivity ■

Sur la Réflexion et la Réfraction des Rayons Cathodiques et des Rayons Déviables du Radium

P. Villard
Compt. Ren. **130** (1900) 1010;

Accelerator Radioactive source

Detectors Nuclear emulsion

Related references

See also
M. Curie, Compt. Ren. **CXXX** (1900) 73;

Reactions
Ra $\rightarrow$ Ra γ

Particles studied γ

BECQUEREL 1900

■ Confirmation of the γ radioactivity ■

Sur la Transparence de l'Aluminium pour la Rayonnement du Radium

H. Becquerel
Compt. Ren. **130** (1900) 1154;

Accelerator Radioactive source

Detectors Nuclear emulsion

Related references

See also
P. Villard, Compt. Ren. **130** (1900) 1010;

Reactions
Ra $\rightarrow$ Ra γ

Particles studied γ

VILLARD 1900B

■ Gamma radioactivity established ■

Sur le Rayonnement du Radium / On the Rays of Radium

P. Villard
Compt. Ren. **130** (1900) 1178;

Accelerator Radioactive source

Detectors Nuclear emulsion

Related references

More (earlier) information
P. Villard, Compt. Ren. **130** (1900) 1010;
See also
H. Becquerel, Compt. Ren. **130** (1900) 1154;

Reactions
Ra $\rightarrow$ Ra γ

Particles studied γ

PLANCK 1900 — Nobel prize

■ Discovery of a new formula for the energy spectrum of the black body radiation, Planck's radiation law. Nobel prize to M. Planck awarded in 1918 "in recognition of the services he rendered to the advancement of Physics by his discovery of energy quanta" ■

Über eine Verbesserung der Wienschen Spektralgleichung / On an Improvement of the Wien's Law of Radiation

M. Planck
Verhandl. Dtsch. phys. Ges. **2** (1900) 202;

Reprinted in
(translation into Russian) M.Planck, *Izbrannye Trudy*, Nauka, M. (1975) 249.

Related references

See also
M. Thiesen, Verhandl. Dtsch. phys. Ges. **2** (1900) 67;
Analyse information from
O. Lummer and E. Jahnke, Annalen der Physik. Leipzig **3** (1900) 288;
O. Lummer and E. Pringsheim, Verhandl. Dtsch. phys. Ges. **2** (1900) 174;

PLANCK 1900B — Nobel prize

■ Quantum hypothesis and explanation of the black body radiation spectrum. Beginnings of the quantum era in physics. Nobel prize to M. Planck awarded in 1918 "in recognition of the services he rendered to the advancement of Physics by his discovery of energy quanta" ■

Zur Theorie das Gesetzes der Energieverteilung im Normalspektrum / On the Theory of Energy Distribution Law of the Normal Spectrum Radiation

M. Planck
Verhandl. Dtsch. phys. Ges. **2** (1900) 237;

Reprinted in
(translation into Russian) M.Planck, *Izbrannye Trudy*, Nauka, M. (1975) 251.

Related references

More (earlier) information
M. Planck, Verhandl. Dtsch. phys. Ges. **2** (1900) 202;
See also
M. Thiesen, Verhandl. Dtsch. phys. Ges. **2** (1900) 67;
Analyse information from
H. Rubens and F. Kurlbaum, Sitzungsber. Akad. Wiss. Berlin (1900) 929;
O. Lummer and E. Jahnke, Annalen der Physik. Leipzig **3** (1900) 288;
O. Lummer and E. Pringsheim, Verhandl. Dtsch. phys. Ges. **2** (1900) 174;

PLANCK 1901 — Nobel prize

■ Quantum hypothesis and final version of Planck's formula for the black body radiation spectrum. Nobel prize to M. Planck awarded in 1918 "in recognition of the services he rendered to the advancement of Physics by his discovery of energy quanta" ■

Über das Gesetz der Energieverteilung im Normalspektrum / On the Energy Distribution Law in the Normal Spectrum Radiation

M. Planck
Annalen der Physik. Leipzig **4** (1901) 553;

Reprinted in
(translation into Russian) M. Planck, *Izbrannye Trudy*, Nauka, M. (1975) 258.

Related references

See also
M. Thiesen, Verhandl. Dtsch. phys. Ges. **2** (1900) 66;
Analyse information from
O. Lummer and E. Pringsheim, Verhandl. Dtsch. phys. Ges. **2** (1900) 163;
H. Rubens and F. Kurlbaum, Sitzungsber. Akad. Wiss. Berlin (1900) 929;
H. Rubens, Wied. Ann. **69** (1899) 582;
F. Kurlbaum, Wied. Ann. **65** (1898) 759;
O. Lummer and E. Pringsheim, Verhandl. Dtsch. phys. Ges. **2** (1900) 176;
See also
M. Planck, Verhandl. Dtsch. phys. Ges. **2** (1900) 202;
M. Planck, Verhandl. Dtsch. phys. Ges. **2** (1900) 237;

LEBEDEV 1901

■ First experimental evidence for pressure of the light on the solid bodies ■

The Experimental Study of the Pressure of the Light

P.N. Lebedev
Annalen der Physik. Leipzig **6** (1901) 433;

Reprinted in
P. N. Lebedev, *Izbrannye Sochineniya*, red. A. K. Timiryazev, (1949) 151.

Excerpt

1. The incident beam of the light produces a pressure on the absorptive surfaces as well as on the reflective ones; These ponderomotorical forces are not connected with the convective and radiometrical forces which are due to the heating.
2. The pressure of the light is proportional to the energy of the incident beam and is independent on the colour of the light.
3. The direct measurements give the value for the pressure of the light which equal to (within accuracy of the experiment) value predicted by Maxwell and Bartoli.

Related references

See also
J. C. Maxwell, *Treatise on electricity and magnetism*;
O. Heaviside, *Electromagnetic Theory*, London **1** (1893) 334;
E. Cohn, *Das electromagnetische Feld*, Leipzig (1900) 543;
D. Holdhammer, Annalen der Physik. Leipzig **4** (1901) 834;
A. Bartoli, Nuovo Cim. **15** (1883) 195;
L. Boltzmann, Wied. Ann. **23** (1884) 33, 291, 616, **22** (1884) 292;
B. Galitzine, Wied. Ann. **47** (1892) 479;
C. E. Guillaume, Archives des Sciences phys. et nat. de Genève **31** (1894) 121;
P. Drude, Lehrbuch der Optik, Leipzig (1900) 447;
L. Euler, Historire de l'Academie de Berlin **2** (1746) 121;
De Mairan, Traitè physique et historique de l'Aurore Boréale (Sec. Ed.) Paris (1754) 371;
A. Fresnel, Ann. de Chimie et de Phys. (2)**29**, (1825) 57, 107;
W. Crooks, Phil. Trans. of the R. S. of London **164** (1874) 501, **168** (1878) 266, **170** (1879) 113;
F. Zöllner, Pogg. Ann **160** (1877) 160;
E. Nichols, Wied. Ann. **60** (1897) 405;
Angström, Wied. Ann **67** (1899) 647;
A. Schuster, Phil. Mag. **2** (1876) 313;
Bertin et Garbe, Ann. de Chim. et de Phys. (5) **11** (1877) 67;
P. Lenard and M. Wolf, Wied. Ann. **37** (1889) 455;
F. Kurlbaum, Wied. Ann. **67** (1899) 848;
F. Martens, Wied. Ann. **62** (1897) 206, **64** (1898) 625;
H. Hagen and Rubens, Annalen der Physik. Leipzig **1** (1900) 373;

RUTHERFORD 1903

■ Observation that α rays are the flow of doubly positive charged particles ■

The Magnetic and Electric Deviation of the easily absorbed Rays from Radium

E. Rutherford
Phil. Mag. **(6)5** (1903) 177;

Reprinted in
Classical Scientific Papers. Physics. Facsimile reproductions of famous scientific papers. Mills and Boon Limited, London (1964) 31.

Abstract Radium gives out three distinct types of radiation:

(1) The α rays, which are very easily absorbed by thin layers of matter, and which give rise to the greater portion of the ionization of the gas observed under the usual experimental conditions.

(2) The β rays, which consist of negatively charged particles projected with high velocity and which are similar in all respects to cathode rays produced in a vacuum-tube.

(3) The γ rays, which are non-deviable by a magnetic field, and which are of a very penetrating character.

These rays differ very widely in their power of penetrating matter. The following approximate numbers, which show the thickness of aluminium traversed before the intensity is reduced to one-half, illustrate this difference.

Radiation	Thickness of Aluminium
α rays	.0005 cm
β rays	.05 cm
γ rays	8 cm

In this paper an account will be given of some experiments which show that the α rays are deviable by a strong magnetic electric field. The deviation is in the opposite sense to that of the cathode rays, so that the radiations must consist of positively charged bodies projected with great velocity. In a previous paper (*Phil. Mag.* (Jan 1903), 113) I have given an account of the indirect experimental evidence in support of the view that the α rays consist of projected charged particles. Preliminary experiments undertaken to settle this question during the past two years gave negative results. The magnetic deviation, even in a strong magnetic field, is so small that very special methods are necessary to detect and measure it. The smallness of the magnetic deviation of the α rays, compared with that of the cathode rays in a vacuum-tube, may be judged from the fact that the α rays projected at right angles to a magnetic field of strength 10000 c.g.s. units, describe the arc of a circle of radius about 39 cm, while under the same conditions the cathode rays would describe a circle of radius about .01 cm.
In the early experiments radium of activity 1000 was used, but this did not give out strong enough rays to push the experiment to the necessary limit. The general method employed was to pass the rays through narrow slits and to observe whether the rate of discharge, due to the issuing rays, was altered by the application of a magnetic field. When, however, the rays were sent through sufficiently narrow slits to detect a small deviation of the rays, the rate of discharge of the issuing rays became too small to measure, even with a sensitive electrometer.
I have recently obtained a sample of radium of activity 19000, and using an electroscope instead of an electrometer, I have been able to extend the experiments, and to show that the α rays are all deviated by a strong magnetic field.

Accelerator Radioactive source

Detectors Ionization

Related references

See also
E. Rutherford and Grier, Phil. Mag. (Sept 1902);
F. Soddy, Proc. Chem. Soc. (1902);

POINCARE 1905

■ Introduction of the relativity principle as an overall law of nature valid for all forces including the gravitational one. Formulation of the equation of the relativistic mechanics and transformation laws for electromagnetic field and current. Establishing of the Lorentz group as a symmetry group of nature. First proposal to modify Newtonian theory of gravity on the basis of the relativity principle. Prediction of the gravitational waves propagating with the speed of light ■

Sur la Dynamique de l'Électron / On the Dynamics of the Electron

H. Poincare

Compt. Ren. **140** (1905) 1504; Rend. del Circ. Mat. di Palermo 21, 129 (1906);

Reprinted in
(translation into English) A. A. Logunov, *On the articles by Henri Poincare.* JINR Publishing Department, Dubna (1995).

Excerpt It would seem at first sight that the aberration of light and the optical and electrical effects related thereto should afford a means of determining the absolute motion of the Earth, or rather its motion relative to the ether instead of relative to the other celestial bodies. An attempt at this was made, indeed, by Fresnel, but he soon perceived that the Earth's notion does not affect the laws of refraction and reflection. Similar experiments, such as that using a waterfilled telescope, or any in which only the first-order terms relative to the aberration were considered, likewise yielded only negative results. The explanation of this was soon found; but Michelson, who devised an experiment where in the terms involving the square of the aberration should be detectable, was equally unsuccessful.
This impossibility of experimentally demonstrating the absolute motion of the earth appears to be a general law of Nature; it is reasonable to assume the existence of this law, which we shall call **the relativity postulate**, and to assume that it is universally valid. Whether this postulate, which so far is in agreement with experiment, be later confirmed or disproved by more accurate tests, it is, in any case, of interest to see what consequences follow from it. *(Extracted from the introductory part of the paper.)*

Related references

See also
H. A. Lorentz, Proceedings of the Section of Sciences, Koninklijke Akademie van Wetenschappen te Amsterdam **6** (1904) 809-831;

EINSTEIN 1905 **Nobel prize**

■ Explanation of the photoelectric effect with use of the quantum hypothesis of Planck. Light is a flow of corpuscular objects with definite energies — Planck's quanta of energy. Nobel prize to A. Einstein awarded in 1921 "for services to Theoretical Physics, and especially of the law of the photoelectric effect" ■

Über einen die Erzeugung und Verwandlung des Lichtes betreffenden heuristischen Gesichtspunkt / On a Heuristic Point of View about the Creation and Conversion of Light

A. Einstein
Annalen der Physik. Leipzig **17** (1905) 132;

Reprinted in
(translation into English) A. B. Arons and M. B. Peppard, Am. J. Phys. 33 (1965) 367.
(translation into English) *The Old Quantum Theory*, ed. by D. ter Haar, Pergamon Press (1967) 91.
Albert Einstein: *Die Hypothese der Lichtquanten.* Dokumenten der Naturwissenschaften — Abteilung Physik, **V.7** ed. A. Hermann, Stuttgart, 26.
The collected papers of Albert Einstein, v.2 The Swiss years: writings 1900-1909, edited by J. Stachel, Princeton Univ. Press, (1989) 150.

Excerpt A profound formal distinction exists between the theoretical concepts which physicists have formed regarding gases and other ponderable bodies and the Maxwellian theory of electromagnetic process in so-called empty space. While we consider the state of a body to be completely determined by the positions and velocities of a very large, yet finite, number of atoms and electrons, we make use of continuous spatial functions to describe the electromagnetic state of a given volume, and a finite number of parameters cannot be regarded as sufficient for the complete determination of such a state. According to Maxwellian theory, energy is to be considered a continuous spatial function in the case of all purely electromagnetic phenomena including light, while the energy of a ponderable object should, according to the present conceptions of physicists, be represented as a sum carried over the atoms and electrons. The energy of a ponderable body cannot be subdivided into arbitrarily many or arbitrarily small parts, while the energy of a beam of light from a point source (according to the Maxwellian theory of light or, more generally, according to any wave theory) is continuously spread over an ever increasing volume.
The wave theory of light, which operates with continuous spatial functions, has worked well in the representation of purely optical phenomena and will probably never be replaced by another theory. It should be kept in mind, however, that the optical observations refer to time averages rather than instantaneous values. In spite of the complete experimental confirmation of the theory as applied to diffraction, reflection, refraction, dispersion, etc., it is still conceivable that the theory of light which operates with continuous spatial functions may lead to contradictions with experience when it is applied to the phenomena of emission and transformation of light.
It seems to me that the observations associated with blackbody radiation, fluorescence, the production of cathode rays by ultraviolet light, and other related phenomena connected with the emission or transformation of light are more readily understood if one assumes that the energy of light is discontinuously distributed in space. In accordance with the assumption to be considered here, the energy of a light ray spreading out from a point source is not continuously distributed over an increasing space but consists of a finite number of energy quanta which are localized at points in space, which move without dividing, and which can only be produced and absorbed as complete units.
In the following I wish to present the line of thought and the facts which have led me to this point of view, hoping that this approach may be useful to some investigators in their research.

Related references

See also
M. Planck, Annalen der Physik. Leipzig **1** (1900) 99;
M. Planck, Annalen der Physik. Leipzig **4** (1901) 561;

Analyse information from
P. Lenard, Annalen der Physik. Leipzig **8** (1902) 150;
P. Lenard, Annalen der Physik. Leipzig **12** (1903) 469;

Reactions
γ atom $\rightarrow$ e^- X

EINSTEIN 1905B **Nobel prize**

■ Invention of the theory of special relativity. Beginnings of the relativistic era in physics. Nobel prize to A. Einstein awarded in 1921 "for services to Theoretical Physics, and especially of the law of the photoelectric effect" ■

Zur Elektrodynamik bewegter Körper / On the Electrodynamics of Moving Bodies

A. Einstein
Annalen der Physik. Leipzig **17** (1905) 891;

Reprinted in
The collected papers of Albert Einstein, v.2 The Swiss years: writings 1900-1909, ed. by J. Stachel, Princeton University Press, (1989) 276.
(translation into English) H. Lorentz, A. Einshtein, H. Minkowsky, and H. Weyl *The Principle of Relativity,* Methuen, London, (1923) 35.

Abstract It is known that Maxwell's electrodynamics — as

usually understood at the present time — when applied to moving bodies, leads to asymmetries which do not appear to be inherent in the phenomena. Take, for example, the reciprocal electrodynamic action of a magnet and a conductor. The observable phenomenon here depends only on the relative motion of the conductor and the magnet, whereas the customary view draws a sharp distinction between the two cases in which either the one or the other of these bodies is in motion. For if the magnet is in motion and the conductor at rest, there arises in the neighborhood of the magnet an electric field with a certain definite energy, producing a current at the places where parts of the conductor are situated. But if the magnet is stationary and the conductor in motion, no electric field arises in the neighborhood of the magnet. In the conductor, however, we find an electromotive force, to which in itself there is no corresponding energy, but which gives rise — assuming equality of relative, motion in the two cases discussed — to electric currents of the same path and intensity as those produced by the electric forces in the former case.
Examples of this sort, together with the unsuccessful attempts to discover any motion of the earth relatively to the "light medium," suggest that the phenomena of electrodynamics as well as of mechanics possess no properties corresponding to the idea of absolute rest. They suggest rather that, as has already been shown to the first order of small quantities, the same laws of electrodynamics and optics will be valid for all frames of reference for which the equations of mechanics hold good. We will raise this conjecture (the purport of which will hereafter be called the "Principle of Relativity") to the status of a postulate, and also introduce another postulate, which is only apparently irreconcilable with the former, namely, that light is always propagated in empty space with a definite velocity c which is independent of the state of motion of the emitting body. These two postulates suffice for the attainment of a simple and consistent theory of the electrodynamics of moving bodies based on Maxwell's theory for stationary bodies. The introduction of a "luminiferous ether" will prove to be superfluous inasmuch as the view here to be developed will not require an "absolutely stationary space" provided with special properties, nor assign a velocity-vector to a point of the empty space in which electromagnetic processes take place. The theory to be developed is based-like all electrodynamics-on the kinematics of the rigid body, since the assertions of any such theory have to do with the relationships between rigid bodies (systems of co-ordinates), clocks, and electromagnetic processes. Insufficient consideration of this circumstance lies at the root of the difficulties which the electrodynamics of moving bodies at present encounters.

EINSTEIN 1905C	**Nobel prize**

■ Invention of the theory of special relativity, $E = mc^2$. Beginnings of the relativistic era in physics. Nobel prize to A. Einstein awarded in 1921 "for services to Theoretical Physics, and especially of the law of the photoelectric effect" ■

Ist die Trägheit eines Körpers von seinem Energieinhalt abhängig? / Does the Inertia of a Body Depend upon its Energy Content?

A. Einstein

Annalen der Physik. Leipzig **18** (1905) 639;

Reprinted in

(translation into English) H. Lorentz, A. Einshtein, H. Minkowsky, and H. Weyl *The Principle of Relativity,* Methuen, London (1923) 67.
The collected papers of Albert Einstein, v.2 The Swiss years: writings 1900-1909, ed. J. Stachel, Princeton Univ. Press (1989) 312.

Abstract The results of the previous investigation lead to a very interesting conclusion, which is here to be deduced.
I based that investigation on the Maxwell-Hertz equations for empty space, together with the Maxwellian expression for the electromagnetic energy of space, and in addition the principle that:

The laws by which the states of physical systems alter are independent of the alternative, to which of two systems of co-ordinates, in uniform motion of parallel translation relatively to each other, these alterations of state are referred (principle of relativity).

With these principles[1] *The principle of the constancy of the velocity of light is of course contained in Maxwell's equations.* as my basis I deduced *inter alia* the following result:
Let a system of plane waves of light, referred to the system of co-ordinates (x, y, z), possess the energy l; let the direction of the ray (the wave-normal) make an angle ϕ with the axis of x of the system. If we introduce a new system of co-ordinates (ξ, η, ζ) moving in uniform parallel translation with respect to the system (x, y, z), and having its origin of coordinates in motion along the axis of x with the velocity v, then this quantity of light-measured in the system (ξ, η, ζ) — possesses the energy

$$l^* = l \frac{1 - \frac{v}{c}\cos\phi}{\sqrt{1 - v^2/c^2}}$$

where c denotes the velocity of light. We shall make use of this result in what follows.
Let there be a stationary body in the system (x, y, z), and let its energy-referred to the system (x, y, z) — be E_0. Let the

1*

energy of the body relative to the system (ξ, η, ζ), moving as above with the velocity v, be H_0.

Let this body send out, in a direction making an angle ϕ with the axis of x, plane waves of light, of energy $\frac{1}{2}L$ measured relatively to (x, y, z), and simultaneously an equal quantity of light in the opposite direction. Meanwhile the body remains at rest with respect to the system (x, y, z). The principle of energy must apply to this process, and in fact (by the principle of relativity) with respect to both systems of co-ordinates. If we call the energy of the body after the emission of light E_1 or H_1 respectively, measured relatively to the system (x, y, z) or (ξ, η, ζ) respectively, then by employing the relation given above we obtain

$$E_0 = E_1 + \frac{1}{2}L + \frac{1}{2}L,$$

$$H_0 = H_1 + \frac{1}{2}L\frac{1-\frac{v}{c}\cos\phi}{\sqrt{1-v^2/c^2}} + \frac{1}{2}L\frac{1+\frac{v}{c}\cos\phi}{\sqrt{1-v^2/c^2}}$$
$$= H_1 + \frac{L}{\sqrt{1-v^2/c^2}}.$$

By subtraction we obtain from these equations

$$H_0 - E_0 - (H_1 - E_1) = L\left\{\frac{1}{\sqrt{1-v^2/c^2}} - 1\right\}.$$

The two differences of the form H−E occurring in this expression have simple physical significations. H and E are energy values of the same body referred to two systems of co-ordinates which are in motion relatively to each other, the body being at rest in one of the two systems (system (x, y, z)). Thus it is clear that the difference H−E can differ from the kinetic energy K of the body, with respect to the other system (ξ, η, ζ), only by an additive constant C, which depends on the choice of the arbitrary additive constants of the energies H and E. Thus we may place

$$H_0 - E_0 = K_0 + C,$$

$$H_1 - E_1 = K_1 + C,$$

since C does not change during the emission of light. So we have

$$K_0 - K_1 = L\left\{\frac{1}{\sqrt{1-v^2/c^2}} - 1\right\}.$$

The kinetic energy of the body with respect to (ξ, η, ζ) diminishes as a result of the emission of light, and the amount of diminution is independent of the properties of the body. Moreover, the difference $K_0 - K_1$, like the kinetic energy of the electron, depends on the velocity.
Neglecting magnitudes of fourth and higher orders we may place

$$K_0 - K_1 = \frac{1}{2}\frac{L}{c^2}v^2.$$

From this equation it directly follows that:
If a body gives off the energy L in the form of radiation, its mass diminishes by L/c^2. The fact that the energy withdrawn from the body becomes energy of radiation evidently makes no difference, so that we are led to the more general conclusion that:
The mass of a body is a measure of its energy-content; if the energy changes by L, the mass changes in the same sense by $L/9\times 10^{20}$, the energy being measured in ergs, and the mass in grammes.
It is not impossible that with bodies whose energy-content is variable to a high degree (e.g. with radium salts) the theory may be successfully put to the test.
If the theory corresponds to the facts, radiation conveys inertia between the emitting and absorbing bodies.

Related references

More (earlier) information
A. Einstein, Annalen der Physik. Leipzig **17** (1905) 891;

EINSTEIN 1906 **Nobel prize**

■ Corpuscular-wave dualism for photons. Explanation of the photoelectric effect using the quantum hypothesis of Planck. Nobel prize to A. Einstein awarded in 1921 "for services to Theoretical Physics, and especially of the law of the photoelectric effect" ■

Zur Theorie der Lichterzeugung und Lichtabsorption / On the Theory of Light Production and Light Absorption

A. Einstein
Annalen der Physik. Leipzig **20** (1906) 199;

Reprinted in
The collected papers of Albert Einstein, v.2 The Swiss years: writings 1900-1909, ed. J. Stachel, Princeton Univ. Press (1989) 350.

Abstract In einer letztes Jahr erschienen Arbeit habe ich gezeigt, daß die Maxwellsche Theorie der Elektrizität in Verbindung mit der Elektronentheorie zu Ergebnissen führt, die mit den Erfahrungen über die Strahlung des schwarzen Körpers im Widerspruch sind. Auf einem dort dargelegten Wege wurde ich zu der Ansicht geführt, daß Licht von der Frequenz ν lediglich in Quanten von der Energie $(R/N)\beta\nu$ absorbiert und emittiert werden könne, wobei R die absolute Konstante der auf das Grammolekül angewendeten Gasgleichung, N die Anzahl der wirklichen Moleküle in einem Grammolekül, β den Exponentialkoeffizienten der Wienschen (bez. der Planckschen), Strahlungsformel und $|nu$ die Frequenz des betreffenden Lichtes bedeutet. Diese Beziehung wurde entwickelt für einen Bereich, der dem Bereich der Gültigkeit der Wienschen Strahlungsformel entspricht.

Damals schien es mir, als ob die Plancksche Theorie der Strahlung in gewisser Beziehung ein Gegenstück bildete zu meiner Arbeit.Neue Überlegungen, welche im §1 dieser Arbeit mitgeteilt sind, zeigten mir aber, daß die theoretische Grundlage, auf welcher die Strahlungstheorie von Hrn. Planck ruht, sich von der Grundlage, die sich aus der Maxwellschen Theorie und Elektronentheorie ergeben würde, unterscheidet, und zwar gerade dadurch, daß die Plancksche Theorie implizite von der eben erwähnten Lichtquantenhypothese Gebrauch macht.
Im §2 der vorliegenden Arbeit wird mit Hilfe der Lichtquantenhypothese eine Beziehung zwischen Voltaeffekt und lichtelektrischer Zerstreuung hergeleitet.

Related references

More (earlier) information
A. Einstein, Annalen der Physik. Leipzig **17** (1905) 132;
See also
A. Einstein, Annalen der Physik. Leipzig **11** (1903) 170;
M. Planck, Annalen der Physik. Leipzig **1** (1900) 99;
M. Planck, Annalen der Physik. Leipzig **4** (1901) 561;

Reactions
γ atom $\rightarrow$ e^- X

STARK 1909

■ First explicit identification of the photon as a genuine elementary particle possessing both energy and momentum ■

Zur experimentellen Entscheidung zwischen Äetherwellen und Lichtquantenhypothese. I. Röntgenstrahlung / On the Experimental Choice between the Ether Wave Hypothesis and the Light Quanta Hypothesis

J. Stark
Phys.Zeitschr. **10** (1909) 902;

MILLIKAN 1911 Nobel prize

■ First conclusive measurement of the charge of the electron. Nobel prize to R. A. Millikan awarded in 1923 "for his work on the elementary charge of electricity and on the photo-electric effect" ■

The Isolation of an Ion, a Precision Measurement of its Charge, and the Correction of Stokes's Law

R.A. Millikan
Phys. Rev. **XXXII** (1911) 349;

Excerpt In a preceding paper a method of measuring the elementary electrical charge was presented which differed essentially from methods which had been used by earlier observers only in that all of the measurements from which the charge was deduced were made upon one individual charged carrier. This modification eliminated the chief sources of uncertainty which inhered in preceding determinations by similar methods such as those made by Sir Joseph Thomson, H. A. Wilson, Ehrenhaft, and Broglie, all of whom had deduced the elementary charge from the average behavior in electrical and gravitational fields of swarms of charged particles.
The method used in the former work consisted essentially in catching ions by C. T. R. Wilson's method on droplets of water or alcohol, then isolating by a suitable arrangement a single one of these droplets, and measuring its speed first in a vertical electrical and gravitational field combined, then in a gravitational field alone.
The sources of error or uncertainty which are still inherent in the method arose from:

(1) the lack of complete stagnancy in the air through which the drop moved;
(2) the lack of perfect uniformity in the electrical field used;
(3) the gradual evaporation of the drops, rendering it impossible to hold a given drop under observation for more than a minute, or to time the drop as it fell under gravity alone through a period of more than five or six seconds;
(4) the assumption of the exact validity of Stokes's law for the drops used.

The present modification of the method is not only entirely free from all these limitations, but it constitutes an entirely new way of studying ionization and one which seems to be capable of yielding important results in a considerable number of directions.
With its aid it has already been found possible:

1. To catch upon a minute droplet of oil and to hold under observation for an indefinite length of time one single atmospheric ion or any desired number of such ions between 1 and 150.
2. To present direct and tangible demonstration, through the study of the behavior in electrical and gravitational fields of this oil drop, carrying its captured ions, of the correctness of the view advanced many years ago and supported by evidence from many sources that all electrical charges, however produced, are exact multiples of one definite, elementary, electrical charge, or in other words, that an electrical charge instead of being spread uniformly over the charged surface has a definite granular structure, consisting, in fact, of an exact number of specks, or atoms of electricity, all precise alike, peppered over the surface of the charged body.
3. To make an exact determination of the value of the elementary electrical charge which is free from all ques-

tionable theoretical assumptions and is limited in accuracy only by that attainable in the measurement of the coefficient of viscosity of air.

4. To observe directly the order of magnitude of the kinetic energy of agitation of a molecule, and thus to bring forward new direct and most convincing evidence of the correctness of the kinetic theory of matter.
5. To demonstrate that the great majority, if not all, of the ions of ionized air, of both positive and negative sign, carry the elementary electrical charge.
6. To show that Stokes's law for the motion of a small sphere through a resisting medium, breaks down as the diameter of the sphere becomes comparable with the mean free path of the molecules of the medium, and to determine the exact way in which it breaks down.

Accelerator Non accelerator experiment

Detectors Nonelectronic detectors

Related references

More (earlier) information
R. A. Millikan, Science **32** (1910) 436;
R. A. Millikan, Phil. Mag. **19** (1910) 209;

See also
J. J. Thomson, Phil. Mag. **46** (1898) 528;
J. J. Thomson, Phil. Mag. **48** (1899) 547;
J. J. Thomson, Phil. Mag. **5** (1903) 346;
H. A. Wilson, Phil. Mag. **5** (1903) 429;
F. Ehrenhaft, Phys. Zeitshr., (Mai 1909), (July 1910);
L. de Broglie, Le Radium, (July 1909);

Particles studied e^-

RUTHERFORD 1911

■ Evidence for the atomic nucleus. Rutherford model for atomic structure ■

The Scattering of α- and β- Particles by Matter and the Structure of the Atom

E. Rutherford
Phil. Mag. **21** (1911) 669;

Related references

See also
J. J. Thomson, Cambridge Liter. and Phil. Soc. **25** (1910) 5;
H. Schmidt, Annalen der Physik. Leipzig **4** (1901) 23;
H. Nagaoka, Phil. Mag. **7** (1904) 445;

Analyse information from
H. Geiger and E. Marsden, Proc. Roy. Soc. **A82** (1909) 495;
H. Geiger, Proc. Roy. Soc. **A83** (1910) 492;

Reactions
He Pb(atom) → Pb(atom) He
He Au(atom) → Au(atom) He
He Pt(atom) → Pt(atom) He
He Sn(atom) → Sn(atom) He
He Ag(atom) → Ag(atom) He
He Cu(atom) → Cu(atom) He
He Fe(atom) → Fe(atom) He
He Al(atom) → Al(atom) He
e^- Al(atom) → Al(atom) e^-
e^- Cu(atom) → Cu(atom) e^-
e^- Ag(atom) → Ag(atom) e^-
e^- Pt(atom) → Pt(atom) e^-

Particles studied nucleus

HESS 1912 Nobel prize

■ Conclusive evidence for the cosmic rays. Nobel prize to V. F. Hess awarded in 1936 "for his discovery of cosmic radiation". Co-winner C. D. Anderson "for his discovery of the positron" ■

Über Beobachtungen der durchdringenden Strahlung bei sieben Freiballonfahrten / Observation of Penetrating Radiation in Seven Balloon Flights

V.F. Hess
Phys.Zeitschr. **13** (1912) 1084;

Accelerator COSMIC

Detectors Ionization

Comments on experiment Measures the dependence of ionization upon altitude.

WILSON 1912 Nobel prize

■ Invention of the cloud chamber to visualize tracks of ionizing particles. Nobel prize to C. T. R. Wilson awarded in 1927 "for his method of making the paths of electrically charged particles visible by condensation of vapour". Co-winner A. H. Compton "for his discovery of the effect named after him" ■

On an Expansion Apparatus for Making Visible the Tracks of Ionizing Particles in Gases and Some Results Obtained by Its Use

C.T.R. Wilson
Proc. Roy. Soc. **A87** (1912) 277;

Reprinted in
Classical Scientific Papers. Physics. Facsimile reproductions of famous scientific papers. Mills and Boon Limited, London (1964) 355.

Abstract In a recent communication I described a method of making visible the tracks of ionizing particles through a

moist gas by condensing water upon the ions immediately after their liberation. At that time I had only succeeded in obtaining photographs of the clouds condensed on the ions produced along the tracks of α-particles and of the corpuscles set free by the passage of X rays through the gas. The interpretation of the photographs was complicated to a certain extent by distortion arising from the position which the camera occupied.
The expansion apparatus and the method of illuminating the clouds have both been improved in detail, and it has now been found possible to photograph the tracks of even the fastest β-particles, the individual ions being rendered visible. In the photographs of the X ray clouds the drops in many of the tracks are also individually visible; the clouds found in the α ray tracks are generally too dense to be resolved into drops. The photographs are now free from distortion. The cloud chamber has been greatly increased in size; it is now wide enough to give ample room for the longest α ray, and high enough to admit of a horizontal beam of X rays being sent through it without any risk of complications due to the proximity of the roof and floor.

Accelerator Radioactive source

Detectors Cloud chamber

Related references

More (earlier) information
C. T. R. Wilson, Proc. Roy. Soc. **A85** (1911) 285;

Reactions
^{4}He atom $\rightarrow$ *ion* X
e^- atom $\rightarrow$ *ion* X
γ atom $\rightarrow$ *ion* X

HESS 1913	**Nobel prize**

■ Confirmation of cosmic rays. Nobel prize to V. F. Hess awarded in 1936 "for his discovery of cosmic radiation". Co-winner C. D. Anderson "for his discovery of the positron" ■

Über den Ursprung der durchdringenden Strahlung / On the Origin of Penetrating Radiation

V.F. Hess
Phys.Zeitschr. **14** (1913) 610;

Accelerator COSMIC

Detectors Ionization

Comments on experiment Measures the dependence of ionization upon versus altitude.

Related references

More (earlier) information
V. F. Hess, Phys.Zeitschr. **13** (1912) 1084;

MILLIKAN 1913	**Nobel prize**

■ First precise measurement of the charge of the electron and the Avogadro constant. Nobel prize to R. A. Millikan awarded in 1923 "for his work on the elementary charge of electricity and on the photoelectric effect" ■

On the Elementary Electrical Charge and the Avogadro Constant

R.A. Millikan
Phys. Rev. **2** (1913) 109;

Reprinted in
Great Experiments in Physics (Henry Holt and Company, New York, 1960) edited by M. H. Shamos.
The Physical Review — the First Hundred Years, AIP Press (1995) 23.

Excerpt The experiments herewith reported were undertaken with the view of introducing certain improvements into the oil-drop method of determination e and N and thus obtaining a higher accuracy than had before been possible in the evaluation of these most fundamental constants...
The results of this work may be summarized in the following table in which the numbers in the error column represent in the case of the first six numbers estimated limits of uncertainty rather than the so-called "probable errors" which would be much smaller...

- *Elementary electrical charge*
  $e = 4.774 \pm 0.009 \times 10^{-10}$
- *Number of molecules per gram molecule*
  $N = 6.062 \pm 0.012 \times 10^{23}$
- *Number of gas molecules per c.c. at 0°*
  $n = 2.705 \pm 0.005 \times 10^{19}$
- *Kinetic energy of a molecule at 0°*
  $E_0 = 5.621 \pm 0.010 \times 10^{-4}$
- *Constant of molecular energy*
  $e = 2.058 \pm 0.004 \times 10^{-16}$
- *Constant of the entropy equation*
  $k = 1.372 \pm 0.002 \times 10^{-16}$

(Extracted from the introductory part of the paper.)

Accelerator Non accelerator experiment

Detectors Nonelectronic detectors

Related references

More (earlier) information
R. A. Millikan, Phys. Rev. **XXXII** (1911) 349;

Particles studied e^-

BOHR 1913	Nobel prize

■ Invention of the quantum theory of atomic spectra based on the Rutherford model of atomic structure — Bohr's atom. Nobel prize to N. Bohr awarded in 1922 "for his investigation of the structure of atoms, and of the radiation emanating from them" ■

On the Constitution of Atoms and Molecules. I.

N. Bohr

Phil. Mag. **26** (1913) 1;

Excerpt In order to explain the results of experiments on scattering of α rays by matter Prof. Rutherford has given a theory of the structure of atoms. According to this theory, the atoms consist of a positively charged nucleus surrounded by a system of electrons kept together by attractive forces from the nucleus: the total negative charge of the electrons is equal to the positive charge of the nucleus. Further, the nucleus is assumed to be the seat of the essential part of the mass of the atom, and to have linear dimensions exceedingly small compared with the linear dimensions of the whole atom. The number of electrons in an atom is deduced to be approximately equal to half the atomic weight. Great interest is to be attributed to this atom-model; for, as Rutherford has shown, the assumption of the existence of nuclei, as those in question, seems to be necessary in order to account for the results of the experiments on large angle scattering of the α rays.

In an attempt, to explain some of the properties of matter on the basis of this atom-model we meet, however, with difficulties of a serious nature arising from the apparent instability of the system of electrons: difficulties purposely avoided in atom-models previously considered, for instance, in the one proposed by Sir J. J. Thomson. According to the theory of the latter the atom consists of a sphere of uniform positive electrification, inside which the electrons move in circular orbits.

The principal difference between the atom-models proposed by Thomson and Rutherford consists in the circumstance that the forces acting on the electrons in the atom-model of Thomson allow of certain configurations and motions of the electrons for which the system is in a stable equilibrium; such configurations, however, apparently do not exist for the second atom-model. The nature of the difference in question will perhaps be most clearly seen by noticing that among the quantities characterizing the first atom a quantity appears — the radius of the positive sphere — of dimensions of a length and of the same order of magnitude as the linear extension of the atom, while such a length does not appear among the quantities characterizing the second atom. viz. the charges and masses of the electrons and the positive nucleus; nor can it be determined solely by help of the latter quantities.

The way of considering a problem of this kind has, however, undergone essential alterations in recent years owing to the development of the theory of the energy radiation, and the direct affirmation of the new assumptions introduced in this theory, found by experiments on very different phenomena such as specific heats, photoelectric effect, Röntgen rays, etc. The result of the discussion of these questions seems to be a general acknowledgment of the inadequacy of the classical electrodynamics in describing the behaviour of systems of atomic size. Whatever the alteration in the laws of motion of the electron may be, it seems necessary to introduce in the laws in question a quantity foreign to the classical electrodynamics, *i. e.* Planck's constant, or as it often is called the elementary quantum of action. By the introduction of this quantity the question of the stable configuration of the electrons in the atoms is essentially changed, as this constant is of such dimensions and magnitude that it, together with the mass and charge of the particles, can determine a length of the order of magnitude required.

This paper is an attempt to show that the application of the above ideas to Rutherford's atom-model affords a basis for a theory of the constitution of Atoms. It will further be shown that from this theory we are led to a theory of the constitution of molecules.

In the present first part of the paper the mechanism of the binding of electrons by a positive nucleus is discussed in relation to Planck's theory. It will be shown that it is possible from the point of view taken to account in a simple way for the law of the line spectrum of hydrogen. Further, reasons are given for a principal hypothesis on which the considerations contained in the following parts are based.

I wish here to express my thanks to Prof. Rutherford for his kind and encouraging interest in this work.

Related references

See also
E. Rutherford, Phil. Mag. **21** (1911) 669;
E. Rutherford, Phil. Mag. **24** (1912) 453;
M. Planck, Annalen der Physik. Leipzig **31** (1910) 758;
M. Planck, Annalen der Physik. Leipzig **37** (1912) 642;
J. J. Thomson, Phil. Mag. **7** (1904) 237;
A. Einstein, Annalen der Physik. Leipzig **17** (1905) 132;
A. Einstein, Annalen der Physik. Leipzig **22** (1907) 180;
N. Bohr, Phil. Mag. **25** (1913) 24;
W. Ritz, Phys.Zeitschr. **9** (1908) 521;
J. W. Nicholson, Month. Not. Roy. Astr. Soc. **72** (1912) 49, 139, 677, 693, 729;
A. Einstein, Annalen der Physik. Leipzig **20** (1906) 199;

Analyse information from
F. Paschen, Annalen der Physik. Leipzig **27** (1908) 565;
E. C. Pickering, Astron. Journ. **4** (1896) 369, **5** (1897) 92;
A. Fowler, Month. Not. Astr. Soc. Dec., (1912) 73;
J. J. Thomson, Phil. Mag. **23** (1912) 456;

Reactions

$atom^* \rightarrow$ atom γ

γ atom $\rightarrow atom^*$

Particles studied atom, $atom^*$

BOHR 1913B **Nobel prize**

■ Bohr's quantum theory of atomic spectra. Evidence that radioactivity is a nuclear property. Nobel prize to N. Bohr awarded in 1922 "for his investigation of the structure of atoms, and of the radiation emanating from them" ■

On the Constitution of Atoms and Molecules. II.

N. Bohr

Phil. Mag. **26** (1913) 476;

Abstract §1. General Assumptions.
Following the theory of Rutherford, we shall assume that the atoms of the elements consist of a positively charged nucleus surrounded by a cluster of electrons. The nucleus is the seat of the essential part of the mass of the atom, and has linear dimensions exceedingly small compared with the distances apart of the electrons in the surrounding cluster.
As in the previous paper, we shall assume that the cluster of electrons is formed by the successive binding by the nucleus of electrons initially nearly at rest, energy at the same time being, radiated away. This will go on until, when the total negative on the bound electrons is numerically equal to the positive charge on the nucleus, the system will be neutral and no longer able to exert sensible forces on electrons at distances from the nucleus (treat in comparison with the dimensions of the orbits of the bound electrons. We may regard the formation of helium from α rays as an observed example of a process of this kind, an α particle of this view being identical with the nucleus of a helium atom.
On account of the small dimensions of the nucleus, its internal structure will not be of sensible influence on the constitution of the cluster of electrons, and consequently will have no effect on the ordinary physical and chemical properties of the atom. The latter properties on this theory will depend entirely on the total charge and mass of the nucleus; the internal structure of the nucleus will be of influence only on the phenomena of radioactivity.
From the result of experiments on large angle scattering of α rays, Rutherford found an electric charge on the nucleus corresponding per atom to a number of electrons approximately equal to half the atomic weight. This result seems to be in agreement with the number of electrons per atom calculated from experiments on scattering of Röntgen radiation. The total experimental evidence supports the hypothesis that the actual number of electrons in a neutral atom, with a few exceptions, is equal to the number which indicates the position of the corresponding element in the series of elements arranged in order of increasing atomic weight. For example on this view, the atom of oxygen which is the eighth element of the series has eight electrons and a nucleus carrying eight unit charges.
We shall assume that the electrons are arranged at equal angular intervals in coaxial rings rotating around the nucleus. In order to determine the frequency and dimensions of the rings we shall use the main hypothesis of the first paper, viz.: that in the permanent state of an atom the angular momentum of every electron round the centre of its orbit is equal to the universal value $\frac{h}{2\pi}$ where h is Planck's constant. We shall take as a condition of stability, that the total energy of the system in the configuration in question is less than in any neighbouring configuration satisfying the same condition of the angular momentum of the electrons.
If the charge on the nucleus and the number of electrons in the different rings is known, the condition in regard to the angular momentum of the electrons will, as shown in §2, completely determine the configuration of the systems i.e., the frequency of revolution and the linear dimensions of the rings. Corresponding to different distributions of the electrons in the rings, however, there will, in general, be more than one configuration which will satisfy the condition of the angular momentum together with the condition of stability.
In §3 and §4 it will be shown that, on the general view of the formation of the atoms, we are led to indications of the arrangement of the electrons in the rings which are consistent with those suggested by the chemical properties of the corresponding element.
In §5 it will be shown that it is possible from the theory to calculate the minimum velocity of cathode rays necessary to produce the characteristic Röntgen radiation from the element, and that this is in approximate agreement with the experimental values.
In §6 the phenomena of radioactivity will be briefly considered in relation to the theory.

Related references

More (earlier) information
N. Bohr, Phil. Mag. **26** (1913) 1;
More (later) information
N. Bohr, Phil. Mag. **26** (1913) 857;
Analyse information from
H. Geiger and E. Marsden, Phil. Mag. **25** (1913) 604;
C. G. Barkla, Phil. Mag. **21** (1911) 648;
A. van den Broek, Phys.Zeitschr. **14** (1913) 32;
R. A. Millikan, Brit. Assoc. Rep. (1912) 410;
P. Gmelin, Annalen der Physik. Leipzig **28** (1909) 1086;
A. H. Bucherer, Annalen der Physik. Leipzig **37** (1912) 597;
J. J. Thomson, Phil. Mag. **23** (1912) 456;
J. J. Thomson, Phil. Mag. **24** (1912) 218;
J. Frank, Verhandl. Dtsch. phys. Ges. **12** (1910) 613;
J. Frank, G. Hertz, Verhandl. Dtsch. phys. Ges. **15** (1913) 34;
J. W. Nicholson, Month. Not. Roy. Astr. Soc. **72** (1912) 52, **73** (1913) 382;
R. Whiddington, Proc. Roy. Soc. **A85** (1911) 323;
E. Rutherford, Phil. Mag. **24** (1912) 453;
E. Rutherford, Phil. Mag. **24** (1912) 893;

Reactions

$atom^* \rightarrow$ atom γ

Particles studied atom, $atom^*$, nucleus

RUTHERFORD 1913

■ Confirmation of the existence of atomic nuclei. First indication of the existence of the proton ■

The Structure of the Atom

E. Rutherford

Nature **92** (1913) 423; Phil. Mag. **27** (1914) 488;

Abstract The present paper and the accompanying paper by Mr. Darwin deal with certain points in connection with the "nucleus" theory of the atom which were purposely omitted in my first communication on that subject (Phil. Mag. May 1911). A brief account is given of the later investigations which have been made to test the theory and of the deductions which can be drawn from them. At the same time a brief statement is given of recent observations on the passage of α particles through hydrogen, which throw important light on the dimension of the nucleus. *(Extracted from the introductory part of the paper.)*

Related references

More (earlier) information
E. Rutherford, Phil. Mag. **21** (1911) 669;

See also
F. Soddy, Jahr. d. Rad. **10** (1913) 188;
N. Bohr, Phil. Mag. **26** (1913) 857;
N. Bohr, Phil. Mag. **26** (1913) 476;

Analyse information from
H. Geiger and E. Marsden, Proc. Roy. Soc. **A82** (1909) 495;
H. Geiger and E. Marsden, Phil. Mag. **25** (1913) 604;
H. Geiger, Proc. Roy. Soc. **A83** (1910) 492;
E. Rutherford and J. Nuttall, Phil. Mag. **26** (1913) 702;
C. T. R. Wilson, Proc. Roy. Soc. **A87** (1912) 277;
Crowther, Proc. Roy. Soc. **A84** (1910) 226;
E. Rutherford and H. R. Robinson, Phil. Mag. **25** (1913) 301;
C. G. Barkla, Phil. Mag. **21** (1911) 648;
A. van den Broek, Phys.Zeitschr. **14** (1913) 32;
H. G. J. Moseley, Phil. Mag. **26** (1913) 1024;
C. G. Darwin, Phil. Mag. **25** (1913) 201;

Reactions

He Pb(atom) $\rightarrow$ Pb(atom) He
He Au(atom) $\rightarrow$ Au(atom) He
He Pt(atom) $\rightarrow$ Pt(atom) He
He Sn(atom) $\rightarrow$ Sn(atom) He
He Ag(atom) $\rightarrow$ Ag(atom) He
He Cu(atom) $\rightarrow$ Cu(atom) He
He Fe(atom) $\rightarrow$ Fe(atom) He
He Al(atom) $\rightarrow$ Al(atom) He
He H(atom) $\rightarrow$ He H(atom)
e^- Al(atom) $\rightarrow$ Al(atom) e^-
e^- Cu(atom) $\rightarrow$ Cu(atom) e^-
e^- Ag(atom) $\rightarrow$ Ag(atom) e^-
e^- Pt(atom) $\rightarrow$ Pt(atom) e^-

Particles studied nucleus, p

CHADWICK 1914

■ The β spectrum is continuous (first observation). Indirect evidence on the existence of neutral penetrating particles ■

Intensitätsverteilung im magnetischen Spektrum der β-Strahlen von Radium $B+C$ / The Intensity Distribution in Magnetic Spectrum of β-Rays of Radium $B+C$

J. Chadwick

Verhandl. Dtsch. phys. Ges. **16** (1914) 383;

Accelerator Radioactive source

Detectors Spectrometer, Ionization

Related references

See also
O. V. Baeyer and O. Hahn, Phys.Zeitschr. **11** (1910) 488;
O. V. Baeyer, O. Hahn, and L. Meitner, Phys.Zeitschr. **12** (1911) 273;
O. V. Baeyer, O. Hahn, and L. Meitner, Phys.Zeitschr. **12** (1911) 378;
J. Danysz, Compt. Ren. **153** (1911) 339;
J. Danysz, Compt. Ren. **153** (1911) 1066;
E. Rutherford and H. R. Robinson, Phil. Mag. **26** (1913) 717;
J. Danysz, Le Radium **9** (1911) 1;
H. Geiger, Verhandl. Dtsch. phys. Ges. **15** (1913) 534;

Reactions

^{214}Pb $\rightarrow$ e^- X
^{214}Bi $\rightarrow$ e^- X

MILLIKAN 1916 — Nobel prize

■ First conclusive measurement of energy quantization in the photoelectric effect. Nobel prize to R. A. Millikan awarded in 1923 "for his work on the elementary charge of electricity and on the photoelectric effect" ■

Einstein's Photoelectric Equation and Contact Electromotive Force

R.A. Millikan

Phys. Rev. **7** (1916) 18;

Excerpt The test of Einstein's photoelectric equation which I have considered and save in the case of the last, subjected to accurate experimental verification are:

1. The existence of a definite and exactly determinable maximum energy of emission of corpuscules under the influence of a given wavelength.
2. The existence of a linear relationship between photo-potential and the frequency of the incident light.
3. The exact appearance of Planck's h in the slope of the potential-frequency line. The photoelectric method is one of the most accurate available methods for fixing

this constant.

4. The agreement of the long wave-length limit with the intercept of the P.D., ν line, when the latter has been displaced by the amount of the contact E.M.F.
5. Contact E.M.F.'s are accurately given by $\frac{h}{e}(\nu_0 - \nu_0') - (V_0 - V_0')$.

Accelerator Non accelerator experiment

Detectors Nonelectronic detectors

Related references

More (earlier) information
R. A. Millikan, Phys. Rev. **IV** (1914) 73;
R. A. Millikan, Phys. Rev. **VI** (1915) 55;
R. A. Millikan, Winchester, Phys. Rev. **XXIV** (1906) 16;
R. A. Millikan, Winchester, Phil. Mag. **14** (1907) 188;
R. A. Millikan, Winchester, Phil. Mag. **14** (1907) 201;

See also
Page, Am. Jr. Sci. **36** (1913) 501;
Lienhop, Annalen der Physik. Leipzig **21** (1906) 284;
W. Schottky, Phys.Zeitschr. **15** (1914) 624;
W. Schottky, Annalen der Physik. Leipzig **44** (1914) 1011;
Kadesch, Phys. Rev. **3** (1914) 367;

Analyse information from
K. Ramsauer, Annalen der Physik. Leipzig **45** (1914) 961;
K. Ramsauer, Annalen der Physik. Leipzig **45** (1914) 1120;
O. V. Richardson, A. H. Compton, Phil. Mag. **24** (1912) 572;
O. V. Richardson, A. H. Compton, Phil. Mag. **24** (1912) 592;

Reactions

γ atom $\to e^-$ X
γ Hg(atom) $\to e^-$ X

BOHR 1918 — Nobel prize

■ Bohr's invention of correspondence principle. Nobel prize to N. Bohr awarded in 1922 "for his investigation of the structure of atoms, and of the radiation emanating from them" ■

On the Quantum Theory of Line-Spectra

N. Bohr

D. KGL. Danske Vidensk. Selsk. Skrifter, naturvidensk. og mathem. Afd. 8. Raekke, **IV.1, 1-3** 1 (1918);

Reprinted in

Sources of Quantum Mechanics. Ed. by B. L. van der Waerden, North-Holland, Amsterdam, (1967) 95.
Niels Bohr *Collected Works, V.3, The Correspondence Principle. (1918–1923)*, ed. by J. Rud Nielsen, general editor L. Rosenfeld, North Holland, (1976) 67.

Excerpt In an attempt to develop certain outlines of a theory of line-spectra based on a suitable application of the fundamental ideas introduced by Planck in his theory of temperature-radiation to the theory of the nucleus atom of Sir Ernest Rutherford, the writer has shown that it is possible in this way to obtain a simple interpretation or some of the main laws governing the line-spectra of the elements, and especially to obtain a deduction of the well known Balmer formula for the hydrogen spectrum. The theory in the form given allowed of a detailed discussion only, in the case of periodic systems, and obviously was not able to account in detail for the characteristic difference between the hydrogen spectrum and the spectra of other elements, or for the characteristic effects on the hydrogen spectrum of external electric and magnetic fields. Recently, however, a way out of this difficulty has been opened by Sommerfeld who, by introducing a suitable generalization of the theory to a simple type of non-periodic motions and by taking the small variation of the mass of the electron with its velocity into account, obtained an explanation of the fine-structure of the hydrogen lines which was found to be in brilliant conformity with the measurements. Already in his first paper on this subject, Sommerfeld pointed out that his theory evidently offered a clue to the interpretation of the more intricate structure of the spectra of other elements. Briefly afterwards Epstein and Schwarzschild, independent of each other, by adapting Sommerfeld's ideas to the treatment of a more extended class of non-periodic systems obtained a detailed explanation of the characteristic effect of an electric field on the hydrogen spectrum discovered by Stark. Subsequently Sommerfeld himself and Debye have on the same lines indicated an interpretation of the effect of a magnetic field on the hydrogen spectrum which, although no complete explanation of the observations was obtained, undoubtedly represents an important step towards a detailed understanding of this phenomenon.
In spite of the great progress involved in these investigations many difficulties of fundamental nature remained unsolved, not only as regards the limited applicability of the methods used in calculating the frequencies of the spectrum of a given system, but especially as regards the question of the polarization and intensity of the emitted spectral lines. These difficulties are intimately connected with the radical departure from the ordinary ideas of mechanics and electrodynamics involved in the main principles of the quantum theory, and with the fact that it has not been possible hitherto to replace these ideas by others forming an equally consistent and developed structure. Also in this respect, however, great progress has recently been obtained by the work of Einstein and Ehrenfest. On this state of the theory it might therefore be of interest to make an attempt to discuss the different applications from a uniform point of view, and especially to consider the underlying assumptions in their relations to ordinary mechanics and electrodynamics. Such an attempt has been made in the present paper, and it will be shown that it seems possible to throw some light on the outstanding difficulties by trying to trace the analogy between the quantum theory and the ordinary theory of radiation as closely as possible.
The paper is divided into four parts.

Part I contains a brief discussion of the general principles of the theory and deals with the application of the general theory to periodic systems of one degree of freedom and to the class of non-periodic systems referred to above.

Part II contains a detailed discussion of the theory of the hydrogen spectrum in order to illustrate the general considerations.

Part III contains a discussion of the questions arising in connection with the explanation of the spectra or other elements.

Part IV contains a general discussion of the theory of the constitution of atoms and molecules based on the application of the quantum theory to the nucleus atom.

Related references

More (earlier) information
N. Bohr, Phil. Mag. **26** (1913) 1;
N. Bohr, Phil. Mag. **26** (1913) 476;
N. Bohr, Phil. Mag. **26** (1913) 857;
N. Bohr, Phil. Mag. **27** (1914) 506;
N. Bohr, Phil. Mag. **29** (1915) 332;
N. Bohr, Phil. Mag. **30** (1915) 394;

See also
E. C. Kemble, Phys. Rev. **8** (1916) 701;
K. Schaposchnikov, Phys.Zeitschr. **15** (1914) 454;
W. Wilson, Phil. Mag. **29** (1915) 795;
W. Wilson, Phil. Mag. **31** (1916) 156;
M. Planck, Annalen der Physik. Leipzig **50** (1916) 385;
M. Planck, Verh. d. D. Phys. Ges. **17** (1915) 438, 407;

Analyse information from
A. Sommerfeld, Ber. Akad. München (1915) 425, 459, (1916) 131, (1917) 83;
P. S. Epstein, Phys.Zeitschr. **17** (1916) 148;
P. S. Epstein, Annalen der Physik. Leipzig **50** (1916) 489;
P. S. Epstein, Annalen der Physik. Leipzig **51** (1916) 168;
A. Sommerfeld, Annalen der Physik. Leipzig **51** (1916) 1;
A. Sommerfeld, Phys.Zeitschr. **17** (1916) 491;
K. Schwarzschild, Ber. Akad. Berlin (1916) 548;
P. Debye, Wolfskehl-Vor. Göttingen (1913);
P. Debye, Nachr. K. Ges. d. Wiss. Göttingen. (1916);
P. Debye, Phys.Zeitschr. **17** (1916) 507;
A. Einstein, Verh. d. D. phys. Ges. **18** (1916);
P. Ehrenfest, Proc. Acad. Amsterdam **16** (1914) 591;
P. Ehrenfest, Phys.Zeitschr. **15** (1914) 657;
P. Ehrenfest, Phys.Zeitschr. **15** (1914) 660;
P. Ehrenfest, Annalen der Physik. Leipzig **51** (1916) 327;
P. Ehrenfest, Phil. Mag. **33** (1917) 500;
A. Einstein, Phys.Zeitschr. **18** (1917) 121;
J. M. Burgers, Versl. Akad. Amsterdam **25** (1917) 849, 918, 1055;
J. M. Burgers, Annalen der Physik. Leipzig **52** (1917) 195;
J. M. Burgers, Phil. Mag. **33** (1917) 514;

RUTHERFORD 1919

■ Discovery of the proton. Evidence for the proton as a constituent of the nucleus ■

Collision of α Particles with Light Atoms IV. An Anomalous Effect in Nitrogen

E. Rutherford
Phil. Mag. **37** (1919) 581;

Accelerator Radioactive source

Detectors Counters

Related references

More (earlier) information
E. Rutherford, Phil. Mag. **37** (1919) 537;
E. Rutherford, Phil. Mag. **37** (1919) 562;
E. Rutherford, Phil. Mag. **37** (1919) 571;

Reactions
He Nit(atom) $\rightarrow$ p X

Particles studied p, nucleus

RUTHERFORD 1920

■ Rutherford neutron hypothesis ■

Nuclear Constitution of Atoms

E. Rutherford
Proc. Roy. Soc. **A97** (1920) 324;

Reprinted in
Classical Scientific Papers. Physics. Facsimile reproductions of famous scientific papers. Mills and Boon Limited, London (1964) 218.

Excerpt The conception of the nuclear constitution of atoms arose initially from attempts to account for the scattering of α-particles through large angles in traversing thin sheets of matter. (Geiger and Marsden, Proc. Roy. Soc. A82, 495 (1909)) Taking into account the large mass and velocity of the α-particles, these large deflexions were very remarkable, and indicated that very intense electric or magnetic fields exist within the atom. To account for these results, it was found necessary to assume (Rutherford, Phil. Mag. **21** (1911) 669; **27** (1914) 488) that the atom consists of a charged massive nucleus of dimensions very small compared with the ordinarily accepted magnitude of the diameter of the atom. This positively charged nucleus contains most of the mass of the atom, and is surrounded at a distance by a distribution of negative electrons equal in number to the resultant positive charge on the nucleus. Under these conditions, a very intense electric field exists close to the nucleus, and the large deflexion of the α-particle in an encounter with a single atom happens when the particle passes close to the nucleus. Assuming that the electric forces between the α-particle and the nucleus varied according, to an inverse square law in the region close to the nucleus, the writer worked out the relations connecting the number of α-particles scattered through any angle with the charge on the nucleus and the energy of the α-particle. Under the central field of force, the α-particle describes a hyperbolic orbit round the nucleus, and the magnitude of the deflection depends on the closeness of approach to the nucleus. From the data of scattering of α-particles then available, it was

deduced that the resultant charge on the nucleus was about $\frac{1}{2}Ae$, where A is the atomic weight and e the fundamental unit of charge. Geiger and Marsden (Geiger and Marsden, Phil. Mag. 25 (1913) 604) made an elaborate series of experiments to test the correctness of the theory, and confirmed the main conclusions. They found the nucleus charge was about $\frac{1}{2}Ae$, but, from the nature of the experiments, it was difficult to fix the actual value within about 20 per cent. C. G. Darwin (Darwin, Phil. Mag. 27 (1914) 499) worked out completely the deflexion of the α-particle and of the nucleus, taking into account the mass of the latter, and showed that the scattering experiments of Geiger and Marsden could not be reconciled with any law of central force, except the inverse square. The nuclear constitution of the atom was thus very strongly supported by the experiments on scattering of α-rays.
Since the atom is electrically neutral, the number of external electrons surrounding the nucleus must be equal to the number of units of resultant charge on the nucleus. It should be noted that, from the consideration of the scattering of X rays by light elements, Barkla (Phil. Mag. 21 (1911) 648) had shown, in 1911, that the number of electrons was equal to about half the atomic weight. This was deduced from the theory of scattering of Sir J. J. Thomson, in which it was assumed that each of the external electrons in an atom acted as an independent scattering unit.
Two entirely different methods had thus given similar results with regard to the number of external electrons in the atom, but the scattering of α rays had shown in addition that the positive charge must be concentrated on a massive nucleus of small dimensions. It was suggested by Van den Broek (Phys. Zeit. 14 (1913) 32) that the scattering of α-particles by the atoms was not inconsistent with the possibility that the charge on the nucleus was equal to the atomic number of the atom, *i.e.*, to the number of the atom when arranged in order of increasing atomic weight. The importance of the atomic number in fixing the properties of an atom was shown by the remarkable work of Moseley (Phil. Mag. 26 (1913) 1024) on the X ray spectra of the elements. He showed that the frequency of vibration of corresponding lines in the X ray spectra of the elements depended on the square of a number which varied by unity in successive elements. This relation received an interpretation by supposing that the nuclear charge varied by unity in passing, from atom to atom, and was given numerically by the atomic number. I can only emphasize in passing the great importance of Moseley's work, not only in fixing the number of possible elements, and the position of undetermined elements, but in showing that the properties of an atom were defined by a number which varied by unity in successive atoms. This gives a new method of regarding the periodic classification of the elements, for the atomic number, or its equivalent the nuclear charge, is of more fundamental importance than its atomic weight. In Moseley's work the frequency of vibration of the atom was not exactly proportional to N, where N is the atomic number, but to $(N-a)^2$,where a was a constant which had different values, depending on whether the K or L series of characteristic radiations were measured. It was supposed that this constant depended on the number and position of the electrons close to the nucleus.

Related references

See also

H. Geiger and E. Marsden, Proc. Roy. Soc. **A82** (1909) 495;
H. Geiger and E. Marsden, Phil. Mag. **25** (1913) 604;
E. Rutherford, Phil. Mag. **21** (1911) 669;
E. Rutherford, Phil. Mag. **27** (1914) 488;
E. Rutherford, Phil. Mag. **37** (1919) 538;
E. Rutherford, Phil. Mag. **37** (1919) 571;
C. G. Darwin, Phil. Mag. **27** (1914) 499;
C. G. Barkla, Phil. Mag. **21** (1911) 648;
A. van den Broek, Phys.Zeitschr. **14** (1913) 32;
H. G. J. Moseley, Phil. Mag. **26** (1913) 1024;
H. G. J. Moseley, Phil. Mag. **27** (1914) 703;
E. Marsden, Phil. Mag. **27** (1914) 824;
R. J. Strutt, Proc. Roy. Soc. **A80** (1908) 572;

Particles studied *n*

CHADWICK 1921

■ Evidence for the strong interactions ■

Collisions of α Particles with Hydrogen Nuclei

J. Chadwick, E.S. Bieler
Phil. Mag. **42** (1921) 923;

Excerpt In this paper the relation which hold in the collisions between α particles and H nuclei have been investigated.

(1) The angular distribution of the H particles projected by α particles of mean range 6.6 cm has been determined up to an angle of 66°.The distribution of α rays of mean ranges 8.2, 4.3, and 2.9 cm has been obtained over a smaller range of angle. It is shown that the number of H particles projected within these angles by α rays of high velocity is greatly in excess of that given by forces varying as the inverse square of the distance between the centres of the two nuclei.

(2) The variation in the number of H particles projected within a given angle with the velocity of the α rays has been observed over a wide range. It is shown that for α rays of high velocity the variation is in the opposite direction to that given by the inverse square law; for α rays of range less than 2 cm, velocity less than 1.26×10^9 per sec., however, the collision relation is about the same as that given by the inverse square law.

(3) The experimental collision relation is compared with those calculated by Darwin for various models of α particle, and the conclusion is reached that the α particle behaves in these collisions as an elastic oblate spheroid of semi-axis about 8×10^{-13} and 4×10^{-13} cm, moving

in the direction of its minor axis. Outside this surface the force varies approximately as the inverse square of the distance from the centre of the spheroid.

Accelerator Radioactive source

Detectors Counters

Related references

More (earlier) information
J. Chadwick, Phil. Mag. **40** (1920) 734;

See also
C. G. Darwin, Phil. Mag. **27** (1914) 499;
C. G. Darwin, Phil. Mag. **41** (1921) 486;

Analyse information from
E. Rutherford, Phil. Mag. **37** (1919) 537;

Reactions

He p → He p <5.6 MeV(T_{lab})

COMPTON 1923 Nobel prize

■ Direct experimental confirmation that the photon is an elementary particle, the Compton effect. Nobel prize to A. H. Compton awarded in 1927 "for his discovery of the effect named after him". Co-winner C. T. R. Wilson "for his method of making the paths of electrically charged particles visible by condensation of vapour" ■

The Spectrum of Scattered X Rays

A.H. Compton
Phys. Rev. **22** (1923) 409;

Reprinted in

Great Experiments in Physics (Henry Holt and Company, New York, 1960) edited by M. H. Shamos.
The Physical Review — the First Hundred Years, AIP Press (1995) CD-ROM.

Abstract The spectrum of molybdenum K_α rays scattered by graphite at 45°, 90° and 135° has been compared with the spectrum of the primary beam. A primary spectrum line when scattered is broken up into two lines, an "unmodified" line whose wave-length remains unchanged, and a "modified" line whose wave-length is greater than that of the primary spectrum line. Within a probable error of about 0.001 Å, the difference in the wave-lengths $(\lambda - \lambda_0)$ increases with the angle θ between the primary and the scattered rays according to the quantum relation $(\lambda - \lambda_0) = \lambda(1 - \cos\theta)$, where $\lambda = h/mc = 0.0242$ Å. This wave-length change is confirmed also by absorption measurements. The modified ray does not seem to be as homogeneous as the unmodified ray; it is less intense at small angles and more intense at large angles than is the unmodified ray.
An X ray tube of small diameter and with a water-cooled target is described, which is suitable for giving intense X rays.

Accelerator X-ray tube

Detectors Spectrometer

Related references

See also
A. H. Compton, Phys. Rev. **21** (1923) 483;
A. H. Compton, Phys. Rev. **21** (1923) 207;
P. Debye, Phys.Zeitschr. **24** (1923) 161;

Reactions

γ C → C γ

DE BROGLIE 1923 Nobel prize

■ Suggestion of the corpuscular-wave dualism for electrons — de Broglie waves of matter particles. Nobel prize to L. de Broglie awarded in 1929 "for his discovery of the wave nature of electrons" ■

Ondes et Quanta / Waves and Quanta

L. de Broglie
Compt. Ren. **177** (1923) 507;

Abstract This mathematical paper first arrives at the following result: The atom of light, equivalent on account of its total energy to a radiation of frequency ν, is the seat of a periodic internal phenomenon, which, seen by a fixed observer, has at each point of space the same phase as a wave of frequency ν propagated in the same direction with a speed sensibly equal (although very slightly superior) to the so-termed constant speed of light. The author then deals with the case of an electron describing a closed curve with a uniform speed sensibly lower than the velocity of light. The condition of stability derived is found to be that established by Bohr and Sommerfeld. When the speed varies along the trajectory, the Bohr-Einstein formula is obtained under certain conditions. *(Science Abstract, 1924, 256. H.H.Ho.)*

Related references

See also
M. Brillouin, Compt. Ren. **168** (1919) 1318;

Particles studied e^-

DE BROGLIE 1923B **Nobel prize**

■ Suggestion of the corpuscular-wave dualism for electrons. Prediction of diffraction phenomena for electrons. Nobel prize to L. de Broglie awarded in 1929 "for his discovery of the wave nature of electrons" ■

Quanta de Lumiere, Diffraction et Interferences / Light Quanta, Diffraction, and Interference

L. de Broglie
Compt. Ren. **177** (1923) 548;

Abstract The results of a previous paper (see preceding abstract) are first quoted for the case of a particle moving with velocity βc ($\beta < 1$), where c is the velocity of light, which should be regarded as a wave propagated in the same direction with a speed $c/\beta = c^2/\nu$, the frequency of this wave being equal to the total energy of the particle divided by Planck's constant h. The speed βc can then be considered as that of a group of waves of speeds c/β and frequencies $m_0c^2/h\sqrt{1-\beta^2}$. This result is then applied in a discussion of interference fringes and diffraction. *(Science Abstracts, 1924, 257. H. H. Ho.)*

Related references

See also
L. de Broglie, Compt. Ren. **177** (1923) 507;

Reactions
e^- crystal $\to$ e^- crystal

Particles studied e^-

WILSON 1923 **Nobel prize**

■ Experimental confirmation of the ionization process predicted by Compton for a corpuscular photon. Nobel prize to C. T. R. Wilson awarded in 1927 "for his method of making the paths of electrically charged particles visible by condensation of vapour". Co-winner A. H. Compton "for his discovery of the effect named after him" ■

Investigations on X Rays and β Rays by the Cloud Method. Part 1. — X Rays.

C.T.R. Wilson
Proc. Roy. Soc. **A104** (1923) 1;

Abstract The experiments so far have been confined almost entirely to air. The track of the electron ejected from an atom which emits a quantum of radiation, and that of the electron ejected from the atom which absorbs the radiation, can be identified. The primary action of X radiation ($\lambda < 0.5$Å) gives β ray tracks with initial kinetic energy comparable to a quantum of the incident radiation; and tracks of very short range; the latter electrons are ejected nearly in the direction of the primary X rays; they are probably related to the phenomena which have led to the postulation of a "J" radiation. The β rays frequently occur in pairs, or groups, of which five classes have been distinguished. The pairs probably consist of one K-electron, ejected by the primary X rays, and a second by the combined action of primary radiation, and of the K radiation from the atom from which the first electron was ejected. Range measured along track approximately is proportional to 4th power of velocity for ranges of 0.5 cm to 1.5 cm. Primary ionization is about 90 per cm for a velocity of 1010 cm per sec.; it varies approximately inversely as the square of the velocity. Total ionization per cm is about 3 or 4 times the primary. Primary ionization agrees with J. J. Thomson's theory, if the minimum energy of ejection is about 7 volts, approximately the resonance potential (not ionization) of N. The number of tracks with nuclear deflections over 90° agrees with Rutherford's theory, and gives for the charge of the nucleus 6.5e, almost identical with the actual nuclear charge of $N = 7e$. Many features of the β ray tracks, including the curvature which sometimes appears, may be due to radiations excited in atoms by the passage of the β rays. The radiations continually overtaking the β particle and affecting the subsequent collisions. For other results of the observations the original paper should be consulted. *(Science Abstracts, 1923, 2213, H. N. A.)*

Accelerator X-ray tube

Detectors Cloud chamber

Related references

More (earlier) information
C. T. R. Wilson, Proc. Roy. Soc. **A85** (1911) 285;
C. T. R. Wilson, Proc. Roy. Soc. **A87** (1912) 277;

More (later) information
C. T. R. Wilson, Proc. Roy. Soc. **A104** (1923) 192;

Reactions
γ atom $\to$ *ion* e^-

BOSE 1924

■ Discovery of new statistical counting rules for light quanta and a new derivation of Planck's radiation law. Known as Bose-Einstein quantum statistics for particles with integer spins ■

Plancks Gesetz und Lichtquantenhypothese / Plancks Law and Light Quantum Hypothesis

S.N. Bose
Annalen der Physik. Leipzig **26** (1924) 178;

Abstract Der Phasenraum eines Lichtquants in Bezug auf ein gegebenes Volumen wird in "Zellen" von der Größe h^3 aufgeteilt. Die Zahl der möglichen Verteilungen der Lichtquanten einer makroskopisch definierten Strahlung unter diesen Zellen liefert die Entropie und damit alle thermodynamischen Eigenschaften der Strahlung.

EINSTEIN 1924	**Nobel prize**

■ Extension of the Bose method to the monoatomic gases. Prediction of Bose-Einstein condensation effect. Nobel prize to A. Einstein awarded in 1921 "for services to Theoretical Physics, and especially of the law of the photoelectric effect" ■

Quantentheorie des einatomigen idealen Gases / Quantum Theory of ideal Monoatomic Gases

A. Einstein
Sitz. Ber. Preuss. Akad. Wiss. (Berlin) 22, 261 (1924);

Reprinted in
The collected papers of Albert Einstein, ed. by J. Stachel, Princeton University Press (1989).

Related references
See also
S. N. Bose, Z. Phys. **26** (1924) 178;

EINSTEIN 1925

■ Extension of the Bose method to the monoatomic ideal gases. Prediction of Bose-Einstein condensation effect and rediscovery of the wave properties of matter particles ■

Quantentheorie des einatomigen idealen Gases / Quantum Theory of ideal Monoatomic Gases

A. Einstein
Sitz. Ber. Preuss. Akad. Wiss. (Berlin) 23, 3 (1925);

Reprinted in
The collected papers of Albert Einstein, ed. by J. Stachel, Princeton University Press, (1989).

Related references
More (earlier) information
A. Einstein, Sitzber. Pr. Akad. Wiss. **22** (1924) 261;
More (later) information
A. Einstein, Sitzber. Pr. Akad. Wiss. **23** (1925) 18;
See also
S. N. Bose, Z. Phys. **26** (1924) 178;

PAULI 1925	**Nobel prize**

■ Introduction of an additional two–valued degree of freedom for the atomic electron. Nobel prize to W. Pauli awarded in 1945 "for the discovery of the Exclusion Principle, also called the Pauli Principle" ■

Über den Einfluß der Geschwindigkeitsabhängigkeit der Electronenmasse auf den Zeemaneffekt / Zeeman-Effect and the Dependence of Electron-Mass on the Velocity

W. Pauli
Z. Phys. **31** (1925) 373;

Abstract On the assumption that the electrons of the K layer contribute essentially to the impulse-moment and to the magnetic-moment of the atom, we ought to expect an influence of electron-mass variations with velocity upon the Zeeman-effect in chemically homologous elements, particularly in the alkali metals; the Zeeman-effect would then depend notably on the atomic number. From the absence of such differences, and also on other grounds, it is concluded that the above assumption cannot be supported. *(Science Abstracts. 1925. 1410. A. D.)*

Reactions
atom $\gamma^* \rightarrow$ atom γ

Particles studied e^-

PAULI 1925B	**Nobel prize**

■ Discovery of the exclusion principle — the Pauli principle. Nobel prize to W. Pauli awarded in 1945 "for the discovery of the Exclusion Principle, also called the Pauli Principle" ■

Über den Zusammenhang des Abschlusses der Elektronengruppen im Atom mit der Komplexstruktur der Spektren / Relation between the Closing In of Electron-Groups in the Atom and the Structure of Complexes in the Spectrum

W. Pauli
Z. Phys. **31** (1925) 765;

Abstract Conclusion based on some preceding work and on the Millikan-Lande representation of alkali-doublets by relativistic formulae that in these doublets and their anomalous Zeeman-effect we have a duplex character, not explainable classically, of the properties of the light-electrons, without the closed-in noble-gas configuration of the atomic residue being concerned, either by way of a body-impulse or as the seat of magneto-mechanical anomaly of the atom. Taking this as a working hypothesis, the endeavor is made to extend it to atoms other than the alkaline. It is found that, contrary to the general belief, it is possible in the case of a strong magnetic field, where we may ignore the coupling forces between atomic residue and light-electron, to give both the partial systems properties, both as regards the number of their stationary states and the values of their quantum numbers and their magnetic energy, no other than those of the atomic residue or the light-electrons in alkali metals. This leads to a general classification of the electrons in an atom by the chief quantum number n and two subsidiary quantum numbers k_1 and k_2 to which the presence of an external field adds another quantum number m_1. This classification, along with Stoner's work, leads to a general formulisation for the phenomena. *(Science Abstracts. 1925. 1875. A. D.)*

Related references

More (earlier) information
W. Pauli, Z. Phys. **31** (1925) 373;

Analyse information from
E. C. Stoner, Phil. Mag. **48** (1924) 719;

Reactions

$atom^* \rightarrow$ atom γ

Particles studied e^-

UHLENBECK 1925

■ Invention of the electron spin hypothesis and the notion of an elementary spin ■

Ersetzung der Hypothese vom unmechanischen Zwang durch eine Forderung bezüglich des inneren Verhaltens jedes einzelnen Elektrons / Replacement of the Hypothesis of Nonmechanical Connection by an Internal Degree of Freedom of the Electron

G.E. Uhlenbeck, S. Goudsmit
Naturw. **13** (1925) 953;

Particles studied e^-

HEISENBERG 1925 — Nobel prize

■ Foundation of quantum mechanics, Heisenberg approach. Nobel prize to W. Heisenberg awarded in 1932 "for the creation of quantum mechanics" ■

Über quantentheoretische Umdeutung kinematischer und mechanischer Beziehungen / Quantum-Theoretical Re-Interpretation of Kinematic and Mechanical Relations

W. Heisenberg
Z. Phys. **33** (1925) 879;

Reprinted in

W. Heisenberg, *Gesammelte Werke / Collected Works*, Ed. W. Blum, H. P. Durr, and H. Richenberg, Springer–Verlag, **Ser.A, Part I** (1985) 382.
(translation into English) *Sources of Quantum Mechanics*, Ed. B.L. van der Waerden, North Holland, Amsterdam (1967) 261.

Abstract The present paper seeks to establish a basis for theoretical quantum mechanics founded exclusively upon relationships between quantities which in principle are observable.

Abstract In this mathematical paper a basis is sought for a quantum system of mechanics, which is founded entirely upon relationships between the principal magnitudes available for observation, since former systems are based on non-observable quantities such as the electronic period. Difficulties arising from the latter conception are briefly discussed. The most important primary hypotheses concerning the new standpoint are to be found in the frequency condition and the work resulting from the Kramers dispersion theory. For the present purpose, however, treatment is limited to problems involving only one degree of freedom. *(Science Abstracts, 1925, 45. H. H. Ho.)*

Related references

See also
S. Goudsmit and R. de L. Kronig, Naturw. **13** (1925) 90;
H. Hönl, Z. Phys. **31** (1925) 340;
R. de L. Kronig, Z. Phys. **31** (1925) 885;
A. Sommerfeld and H. Hönl, Sitzungsber. d. Preuss. Akad. d. Wiss. (1925) 141;
H. N.Russell, Nature **115** (1925) 835;
B. A. Kratzer, Sitzungsber. d. Bayr. Akad. (1922) 107;
H. A. Kramers, Nature **113** (1924) 673;
M. Born, Z. Phys. **26** (1924) 379;
H. A. Kramers and W. Heisenberg, Z. Phys. **31** (1925) 681;
M. Born and P. Jordan, Z. Phys. **33** (1925) 479;
W. Kuhn, Z. Phys. **33** (1925) 408;
W. Thomas, Naturw. **13** (1925) 627;

BORN 1925

■ Invention of matrix formalism for the Heisenberg quantum mechanics. Systems with one degree of freedom ■

Zur Quantenmechanik / On Quantum Mechanics

M. Born, P. Jordan
Z. Phys. **34** (1925) 858;

Reprinted in
(translation into English) *Sources of Quantum Mechanics*, Ed. B. L. Van der Waerden, North Holland, Amsterdam (1967) 277.

Abstract Die kürzlich von Heisenberg gegebenen Ansätze werden (zunächst für Systeme von einem Freiheitsgrad) zu einer systematischen Theorie der Quantenmechanik entwickelt. Das mathematische Hilfsmittel ist die Matrizenrechnung. Nach dem diese kurz dargestellt ist, werden die mechanischen Bewegungsgleichungen aus einem Variationsprinzip abgeleitet und der Beweis geführt, daß auf Grund der Heisenbergschen Quantenbedingung der Energiesatz und die Bohrsche Frequenzbedingung aus den mechanischen Gleichungen folgen. Am Beispiel des anharmonischen Oszillators wird die Frage der Eindeutigkeit der Lösung und die Bedeutung der Phasen in den Partialschwingungen erörtert. Den Schluß bildet ein Versuch, die Gesetze des electromagnetischen Feldes der neuen Theorie einzufügen.

Abstract The recently published theoretical approach of Heisenberg is here developed into a systematic theory of quantum mechanics (in the first place for systems having one degree of freedom) with the aid of mathematical matrix methods. After a brief survey of the latter, the mechanical equations of motion are derived from a variational principle and it is shown that using Heisenberg's quantum condition, the principle of energy conservation and Bohr's frequency condition follow from the mechanical equations. Using the anharmonic oscillator as example, the question of uniqueness of the solution and of the significance of the phases of the partial vibrations is raised. The paper concludes with an attempt to incorporate electromagnetic field laws into the new theory.

Related references

See also
W. Heisenberg, Z. Phys. **33** (1925) 879;
W. Thomas, Naturw. **13** (1925) 627;
W. Kuhn, Z. Phys. **33** (1925) 408;
P. Debye, Annalen der Physik. Leipzig **30** (1909) 755;

BORN 1926

■ Development of matrix formalism for the Heisenberg quantum mechanics. Systems with arbitrary many degrees of freedom ■

Zur Quantenmechanik II / On Quantum Mechanics II

M. Born, W. Heisenberg, P. Jordan
Z. Phys. **35** (1926) 557;

Reprinted in
W. Heisenberg, *Gesammelte Werke / Collected Works*, Ed. by W. Blum, H. P. Durr, and H. Richenberg, Springer–Verlag, Ser. A, Part I, (1985) 387.
(translation into English) *Sources of Quantum Mechanics*, Ed. by B. L. van der Waerden, North Holland, Amsterdam (1967) 321.

Abstract Die aus Heisenbergs Ansätzen in Teil I dieser Arbeit entwickelte Quantenmechanik wird auf Systeme von beliebig vielen Freiheitsgraden ausgedehnt. Die Störungstheorie wird für nicht entartete und eine große Klasse entarteter Systeme durchgeführt und ihr Zusammenhang mit der Eigenwerttheorie Hermitescher Formen nachgewiesen. Die gewonnenen Resultate werden zur Ableitung der Sätze über Impuls und Drehimpuls und zur Ableitung von Auswahlregeln und Intensitätsformeln benutzt. Schließlich werden die Ansätze der Theorie auf die Statistik der Eigenschwingungen eines Hohlraumes angewendet.

Abstract The quantum mechanics developed in Part I of this paper from Heisenberg's approach is here extended to systems having arbitrarily many degrees of freedom. Perturbation theory is carried through for nondegenerate and for a large class of degenerate systems, and its connection with the eigenvalue theory of Hermitian forms is demonstrated. The results so obtained are employed in the derivation of momentum and angular momentum conservation laws, and of selection rules and intensity formulae. Finally, the theory is applied to the statistics of eigenvibrations of a black body cavity.

Related references

More (earlier) information
M. Born and P. Jordan, Z. Phys. **34** (1925) 858;

See also
W. Heisenberg, Z. Phys. **33** (1925) 879;
P. A. M. Dirac, Proc. Roy. Soc. **A109** (1925) 642;
N. Bohr, Z. Phys. **34** (1925) 142;
G. E. Uhlenbeck and S. Goudsmit, Naturw. **13** (1925) 953;
A. Einstein, Phys.Zeitschr. **10** (1909) 185;
A. Einstein, Phys.Zeitschr. **10** (1909) 817;
H. A. Kramers, Nature **113** (1924) 673;
H. A. Kramers, Nature **114** (1924) 310;
R. Ladenburg, Z. Phys. **4** (1921) 451;
R. Ladenburg and F. Reiche, Naturw. **11** (1923) 584;
M. Born, Z. Phys. **26** (1924) 379;
H. A. Kramers and W. Heisenberg, Z. Phys. **31** (1925) 681;

W. Kuhn, Z. Phys. **33** (1925) 408;
W. Thomas, Naturw. **13** (1925) 627;
F. Reiche and W. Thomas, Z. Phys. **34** (1925) 510;
M. Born and W. Heisenberg, Annalen der Physik. Leipzig **74** (1924) 1;
S. Goudsmit and R. de L. Kronig, Naturw. **13** (1925) 90;
P. Debye, Annalen der Physik. Leipzig **33** (1910) 1427;
P. Ehrenfest, Phys.Zeitschr. **7** (1906) 528;
P. Ehrenfest, Z. Phys. **34** (1925) 362;
S. N. Bose, Z. Phys. **26** (1924) 178;
H. Hönl, Z. Phys. **32** (1925) 340;
P. Jordan, Z. Phys. **33** (1925) 649;
M. Wolfke, Phys.Zeitschr. **22** (1921) 375;
W. Bothe, Z. Phys. **20** (1923) 145;
W. Bothe, Z. Phys. **23** (1924) 214;

HESS 1926 — Nobel prize

■ Firm establishment of cosmic rays. Nobel prize to V. F. Hess awarded in 1936 "for his discovery of cosmic radiation". Co-winner C. D. Anderson "for his discovery of the positron" ■

Über den Ursprung der Höhenstrahlung / Origin of Penetrating Radiation

V.F. Hess
Phys.Zeitschr. **27** (1926) 159;

Abstract A critical review of recent work on the penetrating of "ultra-gamma" radiation. Particular reference is made to the work of G. Hoffman, R. A. Millikan (Nat. Acad. Sci., Proc. 12. pp.48-55, Jan., 1926, and Ann. d. Physik, 79. 6. pp.572-582, April 6, 1926), and F. Behounek. *(Science Abstracts, 1926, 1378. A.B.W.)*

Accelerator COSMIC

Detectors Ionization

Comments on experiment Strong dependence of ionization on altitude.

Related references

More (earlier) information
V. F. Hess, Phys.Zeitschr. **13** (1912) 1084;
V. F. Hess, Phys.Zeitschr. **14** (1913) 610;
G. Hoffmann, Phys.Zeitschr. **26** (1925) 669;

FERMI 1926

■ Invention of statistics for ensembles of particles obeying Pauli principle ■

Sulla Quantizzatione del Gas Perfetto Monoatomico / Quantization of the Monatomic Perfect Gas

E. Fermi
Rend. Lincei 3, 145 (1926);

Reprinted in
Enrico Fermi. /it Collected Papers (Note e Memorie) v.I. Italy 1921-1938. The University of Chicago Press — Accademia Nazionale dei Lincei, Roma (1962).
(translation into Russian) Enrico Fermi. *Nauchnye Trudy.* I. 1921-1938 Italiya. (pod red. B.Pontecorvo), Nauka, Moskva, 1971.

Abstract The author's object is to demonstrate a method of quantizing a perfect gas which does not depend on hypotheses concerning the statistic behaviour of its molecules. Numerous attempts have recently been made to establish the equation of state of a perfect gas. The various formulae differ from each other and from the classical equation only for very low temperatures and very high densities. But that is where the deviations of the laws of real from those of perfect gases are most marked, and as, under experimental conditions, the deviations from the equation of state $pV = kT$, due to the degeneration of the gas, are, though not negligible, always somewhat less than those due to the fact that the gas is real and not perfect, the former have hitherto been masked by the latter. With a more complete knowledge of the forces acting between the molecules of a real gas it may be possible to separate the two deviations, and thus decide experimentally between the various theories of the degeneration of perfect gases. In order to quantize the molecular motion of a perfect gas, Sommerfeld's rules, subjecting the molecules to a system of forces such that their motion becomes periodic and thus quantizable, must be applicable. But to calculate systems containing elements not distinguishable from each other, Sommerfeld's rules need to be supplemented. For the choice of a hypothesis the behaviour of the heaviest hydrogen atoms, all containing more than one electron, must be examined. In the deep portions of a heavy atom the forces acting between the electrons are very small compared with those exercised by the nucleus. Sommerfeld's rules here lead to the conclusion that in the normal state of the atom many of the electrons would be situated in an orbit of a total quantum 1. Observation shows, however, that the ring K is saturated when it contains two, ring L when it contains eight electrons, etc. This fact has been interpreted by Stoner and Pauli The atom cannot thus contain two electrons whose orbits are characterized by the same quantum numbers, i.e. an electronic orbit is "occupied" when it contains one electron. Applying a similar hypothesis to the quantization of a perfect gas containing at most one molecule whose motion is characterized by certain quantum numbers, the author proposes to show that it leads to a perfectly consistent theory of such quantization, accounting for the reduction of the specific heat at low temperatures and leading to the exact value for the constant of entropy. Full mathematical details are reserved for a future paper. *(Science Abstracts, 1926, 1609. E. F.)*

Related references

See also

A. Einstein, Sitzber. Pr. Akad. Wiss. **22** (1924) 261;
A. Einstein, Sitzber. Pr. Akad. Wiss. **23** (1925) 1;
A. Einstein, Sitzber. Pr. Akad. Wiss. **23** (1925) 18;
A. Einstein, Sitzber. Pr. Akad. Wiss. **23** (1925) 49;
E. Fermi, Nuovo Cim. **1** (1924) 145;
W. Pauli, Z. Phys. **31** (1925) 765;
E. C. Stoner, Phil. Mag. **48** (1924) 719;

FERMI 1926B

■ Invention of statistics for ensembles of particles obeying Pauli principle — Fermi–Dirac quantum statistics ■

Zur Quantelung des Idealen Einatomigen Gases / On Quantization of Perfect Monatomic Gases

E. Fermi
Z. Phys. **36** (1926) 902;

Reprinted in
Enrico Fermi. *Collected Papers* (Note e Memorie) v. I. Italy 1921-1938 (1962). The University of Chicago Press — Accademia Nazionale dei Lincei, Roma.
(translation into Russian) Enrico Fermi. *Nauchnye Trudy.* I. 1921-1938 Italiya. (pod red. B.Pontecorvo), Nauka, Moskva, 1971.

Abstract Wenn der Nernstsche Wärmesatz auch für das ideale Gas seine Gültigkeit behalten soll, muß man annehmen, daß die Gesetze idealer Gase bei niedrigen Temperaturen von den klassischen abweichen. Die Ursache dieser Entartung ist in einer Quantelung der Molekularbewegungen zu suchen. Bei allen Theorien der Entartung werden immer mehr oder weniger willkürliche Annahmen über das statistische Verhalten der Moleküle, oder über ihre Quantelung gemacht. In der vorliegenden Arbeit wird nur die von Pauli zuerst ausgesprochene und auf zahlreiche spektroskopische Tatsachen begründete Annahme benutzt, daß in einem System niezwei gleichwertige Elemente vorkommen können, deren Quantenzahlen vollständig übereinstimmen. Mit dieser Hypothese werden die Zustandsgleichung und die innere Energie des idealen Gases abgeleitet; der Entropiewert für große Temperaturen stimmt mit dem Stern-Tetrodeschen überein.

Related references

More (earlier) information
E. Fermi, Lincei Rend. **3** (1926) 145;

See also
A. Einstein, Sitzber. Pr. Akad. Wiss. **22** (1924) 261;
A. Einstein, Sitzber. Pr. Akad. Wiss. **23** (1925) 49;
A. Einstein, Sitzber. Pr. Akad. Wiss. **23** (1925) 318;
E. Fermi, Nuovo Cim. **1** (1924) 145;
W. Pauli, Z. Phys. **31** (1925) 765;
B. F. Hund, Z. Phys. **33** (1925) 345;

SCHRÖDINGER 1926 — Nobel prize

■ Creation of wave mechanics. Invention of the Schrödinger wave equation. Nobel prize to E. Schrödinger awarded in 1933. Co-winner P. A. M. Dirac "for the discovery of new productive forms of atomic theory" ■

Quantizierung als Eigenwertproblem (Erste Mitteilung) / Quantization as a Problem of Proper Values. Part I.

E. Schrödinger
Annalen der Physik. Leipzig **79** (1926) 361;

Reprinted in
(translation into English) *Collected papers on wave mechanics by E. Schrödinger*, Glasgow (1928) 1.
Abhandlungen zur Wellenmechanik, Schrödinger (1927d) 1.
Gesammelte Abhandlungen / Collected Works, E.Schrödinger (1984c) 82.

Excerpt In this paper I wish to consider, first, the simple case of the hydrogen atom (non-relativistic and unperturbed), and show that the customary quantum conditions can be replaced by another postulate, in which the notion of "whole numbers", merely as such, is not introduced. Rather when integralness does appear, it arises in the same natural way as it does in the case of the *node-numbers* of a vibrating string. The new conception is capable of generalization, and strikes, I believe, very deeply at the true nature of the quantum rules. *(Extracted from the introductory part of the paper.)*

Related references

More (earlier) information
E. Schrödinger, Phys.Zeitschr. **27** (1926) 95;

See also
L. de Broglie, Ann. de Physique **3** (1925) 22;
A. Sommerfeld, *Atombau und Spectrallinien*, **4th ed.** 775;

SCHRÖDINGER 1926B — Nobel prize

■ Wave mechanics: First applications. Nobel prize to E. Schrödinger awarded in 1933. Co-winner P. A. M. Dirac "for the discovery of new productive forms of atomic theory" ■

Quantizierung als Eigenwertproblem (Zweite Mitteilung) / Quantization as a Problem of Proper Values. Part II.

E. Schrödinger
Annalen der Physik. Leipzig **79** (1926) 489;

Reprinted in
(translation into English) *Collected papers on wave mechanics*

by E. Schrödinger, Glazgow (1928) 13.
Abhandlungen zur Wellenmechanik, Schrödinger, (1927d) 17.
Gesammelte Abhandlungen / Collected Works, E.Schrödinger (1984c) 98.

Excerpt

1. The Hamiltonian Analogy between Mechanics and Optics.

2.1 The Planck oscillator. The Question of Degeneracy.

2.2 Rotator with Fixed Axis.

2.3 Rigid Rotator with Free Axis.

2.4 Non-rigid Rotator (Diatomic Molecule).

Related references

See also
A. Einstein, Verhandl. Dtsch. phys. Ges. **19** (1917) 77;
A. Einstein, Verhandl. Dtsch. phys. Ges. **19** (1917) 82;
A. Einstein, Sitzber. Pr. Akad. Wiss. **23** (1925) 9;
A. Sommerfeld and I. Runge, Annalen der Physik. Leipzig **35** (1911) 290;
P. Debye, Annalen der Physik. Leipzig **30** (1909) 755;
M. V. Laue, Annalen der Physik. Leipzig **44** (1914) 1197;
N. Bohr, Naturw. **1** (1926) 1;
Courant-Gilbert, *Methods of Mathematical Physics* **I** Berlin, Springer (1924) 36;
A. Sommerfeld, *Atombau und Spectrallinien*, **4th ed.** 775;
Analyse information from
L. de Broglie, Ann. de Physique **3** (1925) 22;
W. Heisenberg, Z. Phys. **33** (1925) 879;
M. Born and P. Jordan, Z. Phys. **34** (1925) 858;
M. Born, W. Heisenberg, and P. Jordan, Z. Phys. **35** (1926) 557;
P. A. M. Dirac, Proc. Roy. Soc. **A109** (1925) 642;

SCHRÖDINGER 1926C **Nobel prize**

■ Equivalency of quantum mechanics of Heisenberg, Born, and Jordan and wave mechanics of Schrödinger. Nobel prize to E. Schrödinger awarded in 1933. Co-winner P. A. M. Dirac "for the discovery of new productive forms of atomic theory" ■

Über das Verhältnis der Heisenberg Born Jordanischen Quantenmechanik zu der meinen / On the Relation Between the Quantum Mechanics of Heisenberg, Born, and Jordan, and that of Schrödinger

E. Schrödinger
Annalen der Physik. Leipzig **79** (1926) 734;

Reprinted in
(translation into English) *Collected Papers on Wave Mechanics by E. Schrödinger*, Glazgow (1928) 45.
Abhandlungen zur Wellenmechanik, Schrödinger, (1927d) 62.
Gesammelte Abhandlungen / Collected Works, E.Schrödinger (1984c) 98.

Excerpt Considering the extraordinary differences between the starting-points and the concepts of Heisenberg's quantum mechanics and of the theory which has been designated "undulatory" or "physical" mechanics, and has lately been described here, it is very strange that these two new theories agree *with one another* with regard to the known facts, where they differ from the old quantum theory. I refer, in particular, to the peculiar "half-integralness" which arises in connection with the oscillator and the rotator. That is really very remarkable, because starting-points, presentations, methods, and in fact the whole mathematical apparatus, seem fundamentally different. Above all, however, the departure from classical mechanics in the two theories seems to occur in diametrically opposed directions. In Heisenberg's work the classical continuous variables are replaced by systems of discrete numerical quantities (matrices), which depend on a pair of integral indices, and are defined by *algebraic* equations. The authors themselves describe the theory as a "true theory of a discontinuum". On the other hand, wave mechanics shows just the reverse tendency; it is a step from classical point-mechanics towards a *continuum-theory*. In place of a process described in terms of a finite number of dependent variables occurring in a finite number of total differential equations, we have a continuous *field-like* process in configuration space, which is governed by a single *partial* differential equation, derived from a principle of action. This principle and this differential equation replace the equations of motion *and* the quantum conditions of the older "classical quantum theory". In what follows the very intimate *inner connection* between Heisenberg's quantum mechanics and my wave mechanics will be disclosed. From the formal mathematical standpoint, one might well speak of the *identity* of the two theories. The train of thought in the proof is as follows.
Heisenberg's theory connects the solution of a problem in quantum mechanics with the solution of a system of an infinite number of algebraic equations, in which the unknowns—infinite matrices—are allied to the classical position- and momentum-co-ordinates of the mechanical system, and functions of these, and obey peculiar *calculating rules.* (The relation is this: to *one* position-, *one* momentum-co-ordinate, or to *one* function of these corresponds always *one* infinite matrix.)
I will first show (§§2 and 3) how to each function of the position- and momentum-co-ordinates there may be related a matrix in such a manner, that these matrices, *in every case, satisfy* the formal calculating rules of Born and Heisenberg (among which I also reckon the so-called "quantum condition" or "interchange rule"; see below). This relation of matrices to functions is *general*; it takes no account of the *special* mechanical system considered, but is the same for all mechanical systems. (In other words: the particular Hamilton function does not enter into the connecting law.) However, the relation is still *indefinite* of an *arbitrary* complete orthogonal system of functions having for domain *entire configuration space* (*N.B.—not "pq-space", but "q-space"*). The provisional *in-*

definiteness of the relation lies in the fact that we can assign the *auxiliary* rôle to an *arbitrary* orthogonal system.
After matrices are thus constructed in a very general way, so as to satisfy the general rules, I will show the following in §4. The *special* system of algebraic equations, which, a *special* case, connects the *matrices* of the position and impulse coordinates with the *matrix* of the Hamilton function, and which the authors call "equations of motion", will be completely solved by assigning the auxiliary rôle to a *definite* orthogonal system, namely, to the system of *proper functions* of that partial differential equation which forms the basis of my wave mechanics. The solution of the natural *boundary-value problem* of this differential equation is completely equivalent to the solution of Heisenberg's algebraic problem. *All* Heisenberg's matrix elements, which may interest us from the surmise that they define "transition probabilities" or "line intensities", can be actually evaluated *by differentiation and quadrature*, as soon as the *boundary-value problem* is solved. Moreover, in wave mechanics, these matrix elements, or quantities that are closely related to them, have the perfectly clear significance of amplitudes of the partial oscillations of the atom's electric moment. The intensity and polarization of the emitted light is thus intelligible *on the basis of the Maxwell-Lorentz theory.* A short preliminary sketch of this relationship is given in §5.

Related references

More (earlier) information
E. Schrödinger, Annalen der Physik. Leipzig **79** (1926) 361;
E. Schrödinger, Annalen der Physik. Leipzig **79** (1926) 489;

See also
P. A. M. Dirac, Proc. Roy. Soc. **A109** (1925) 642;
K. Lanczos, Z. Phys. **35** (1926) 812;
Courant-Gilbert, *Methods of Mathematical Physics,* **I**. Berlin, Springer (1924) 36;
P. A. M. Dirac, Proc. Roy. Soc. **A110** (1926) 561;

Analyse information from
W. Heisenberg, Z. Phys. **33** (1925) 879;
M. Born and P. Jordan, Z. Phys. **34** (1925) 858;
M. Born, W. Heisenberg, and P. Jordan, Z. Phys. **35** (1926) 557;
L. de Broglie, Ann. de Physique **3** (1925) 22;
A. Einstein, Sitzber. Pr. Akad. Wiss. **23** (1925) 9;

SCHRÖDINGER 1926D **Nobel prize**

■ Wave mechanics: Perturbation theory and applications. Nobel prize to E. Schrödinger awarded in 1933. Co-winner P. A. M. Dirac "for the discovery of new productive forms of atomic theory" ■

Quantizierung als Eigenwertproblem (Dritte Mitteilung: Störungstheorie, mit Anwendung auf den Starkeffekt der Balmerlinien) / Quantization as a Problem of Proper Values. Part III: Perturbation Theory, with Applications to the Stark Effect of Balmer lines

E. Schrödinger
Annalen der Physik. Leipzig **80** (1926) 437;

Reprinted in
Abhandlungen zur Wellenmechanik, Schrödinger, (1927d) 85.
Gesammelte Abhandlungen / Collected Works E.Schrödinger (1984c) 166.
(translation into English) *Collected Papers on Wave Mechanics by E. Schrödinger*, Glazgow (1928) 62.

Excerpt As has already been mentioned at the end of the preceding paper, (*last two paragraphs of Part II.*) the available range of application of the proper value theory can by comparatively elementary methods be considerably increased beyond the "directly soluble problems"; for proper values and functions can readily be approximately determined for *such* boundary value problems as are sufficiently closely related to a directly soluble problem. In analogy with ordinary mechanics, let us call the method in question the *perturbation* method. It is based upon the important *property of continuity* possessed by proper values and functions, (Courant-Hilbert, chap. VI, §§2, 4, p. 337) principally, for our purpose, upon their continuous dependence on the *coefficients* of the differential equation, and less upon the extent of the domain and on the boundary conditions, since in our case the domain ("entire q-space") and the boundary conditions ("remaining finite") are generally the same for the unperturbed and perturbed problems.
The method is essentially the same as that used by Lord Rayleigh in investigating (Courant-Hilbert, chap. V. §5, 2, p.241) the vibrations of a string with *small inhomogeneities* in his *Theory of Sound* (2nd edit., vol. I., pp. 115-118, London, 1894). This was a particularly simple case, as the differential equation of the unperturbed problem had *constant* coefficients, and only the perturbing terms were arbitrary functions along the string. A complete generalization is possible not merely with regard to these points, but also for the specially important case of *several* independent variables, i.e. for *partial* differential equations, in which *multiple proper values* appear in the unperturbed problem, and where the addition

of a perturbing term causes the *splitting up* of such values and is of the greatest interest in well - known spectroscopic questions (Zeeman effect, Stark effect, Multiplicities). In the development of the perturbation theory in the following Section I., which really yields nothing new to the mathematician, I put less value on generalizing to the *widest possible extent* than on bringing forward the very simple rudiments in the clearest possible manner. From the latter, any desired generalization arises almost automatically when needed. In Section II., as an example, the Stark effect is discussed and, indeed, by *two* methods, of which the *first* is analogous to Epstein's method, by which he first solved (P. S. Epstein, Ann. d. Phys. **50** (1916) 489) the problem on the basis of classical mechanics, supplemented by quantum conditions, while the *second*, which is much more general, is analogous to the method of secular perturbations. (N. Bohr, Kopenhagener Akademie **(8)IV., 1, 2** (1918) 69 *et seq.*) The *first* method will be utilized to show that in wave mechanics also the perturbed problem can be "separated" in parabolic co-ordinates, and the perturbation theory will first be applied to the ordinary differential equations into which the original vibration equation is split up. The theory thus merely takes over the task which on the old theory devolved on Sommerfeld's elegant complex integration for the calculation of the quantum integrals. (A. Sommerfeld, *Atombau,* **4th ed.** 772). In the *second* method, it is found that in the case of the Stark effect an exact separation coordinate system exists, quite by accident, for the perturbed problem also, and the perturbation theory is applied directly to the *partial* differential equation. This latter proceeding proves to be more troublesome in wave mechanics, although it is theoretically superior, being more capable of generalization.

Also the problem of the intensity of the components in the Stark effect will be shortly discussed in Section II. Tables will be calculated, which, as a whole, agree even better with experiment than the well known ones calculated by Kramers with the help of the correspondence principle. (H. A. Kramers, *Kopenhagener Akademie* **(8), III., 3** (1919) 287).

The application (not yet completed) to the Zeeman effect will naturally be of much greater interest. It seems to be indissolubly linked with a correct formulation in the language of wave mechanics of the relativistic problem, because in the four-dimensional formulation the vector-potential automatically ranks equally with the scalar. It was already mentioned in Part I. that the relativistic hydrogen atom may indeed be treated without further discussion, but that it leads to "half-integral" azimuthal quanta, and thus contradicts experience. Therefore "something must still be missing". Since then I have learnt *what* is lacking from the most important publications of G. E. Uhlenbeck and S. Goudsmit, (*Physica, 1925; Die Naturwissenschaften, 1926; Nature*, 20th Feb. 1926; cf. also L. H. Thomas, *Nature*, 10th April, 1926.) and then from oral and written communications from Paris (P. Langevin) and Copenhagen (W. Pauli), viz., in the language of the theory of electronic orbits, the *angular momentum* of the electron round its axis, which gives it a *magnetic moment.* The utterances of these investigators, together with two highly significant papers by Slater (Proc. Amer. Nat. Acad. **11** (1925) 732) and by Sommerfeld and Unsöld (Ztschr. f. Phys. **36** (1926) 259) dealing with the Balmer spectrum, leave no doubt that, by the introduction of the paradoxical yet happy conception of the spinning electron, the orbital theory will be able to master the disquieting difficulties which have latterly begun to accumulate (anomalous Zeeman effect; Paschen-Back effect of the Balmer lines; irregular and regular Röntgen doublets; analogy of the latter with the alkali doublets, etc.). We shall be obliged to attempt to take over the idea of Uhlenbeck and Goudsmit into wave mechanics. I believe that the latter is a very fertile soil for this idea, since in it the electron is not considered as a point charge, but as continuously flowing through space, and so the unpleasing conception of a "rotating point-charge" is avoided. In the present paper, however, the taking over of the idea is not yet attempted.

To the third section, as "mathematical appendix", have been relegated numerous uninteresting calculations—mainly quadratures of products of proper functions, required in the second section.

Related references

More (earlier) information
E. Schrödinger, Annalen der Physik. Leipzig **79** (1926) 489;
E. Schrödinger, Naturw. **14** (1926) 664;

See also
Courant-Gilbert, *Methods of Mathematical Physics,* **I**. Berlin, Springer (1924) 26;
A. Sommerfeld, *Atombau und Spectrallinien* **4th ed.** 772;
L. H. Thomas, Nature, 10th Apr., (1926);
A. Sommerfeld and A. Unsoeld, Z. Phys. **36** (1926) 259;
N. Bohr, Naturw. **1** (1926) 1;
W. Pauli, Z. Phys. **36** (1926) 336;
J. C. Slater, Proc. Amer. Nat. Acad. **11** (1925) 732;

Analyse information from
P. S. Epstein, Annalen der Physik. Leipzig **50** (1916) 489;
N. Bohr, Kopenhagener Akademie (8) **IV.** (1918) 69;
H. A. Kramers, Kopenhagener Akademie (8) **III** (1919) 287;
G. E. Uhlenbeck and S. Goudsmit, Naturw. **13** (1925) 953;
W. Heisenberg, Z. Phys. **33** (1925) 879;
M. Born and P. Jordan, Z. Phys. **34** (1925) 867;
M. Born and P. Jordan, Z. Phys. **34** (1925) 886;
J. Stark, Annalen der Physik. Leipzig **43** (1914) 1001;
J. Stark, Annalen der Physik. Leipzig **48** (1915) 193;
R. Ladenburg, Z. Phys. **38** (1926) 249;
R. Ladenburg and F. Reiche, Naturw. **11** (1923) 584;

SCHRÖDINGER 1926E **Nobel prize**

■ Wave mechanics: Further development and generalization. Nobel prize to E. Schrödinger awarded in 1933. Co-winner P. A. M. Dirac "for the discovery of new productive forms of atomic theory" ■

Quantizierung als Eigenwertproblem (Vierte Mitteilung) / Quantization as a Problem of Proper Values. Part IV.

E. Schrödinger
Annalen der Physik. Leipzig **81** (1926) 109;

Reprinted in
Abhandlungen zur Wellenmechanik, Schrödinger (1927d) 139.
Gesammelte Abhandlungen / Collected Works E.Schrödinger (1984c) 220.
(translation into English) *Collected papers on wave mechanics by E. Schrödinger*, Glasgow (1928) 102.

Abstract

§1. Elimination of the energy-parameter from the vibration equation. The real wave equation. Non-conservative systems.

§2. Extension of the perturbation theory to perturbations which explicitly contain the time. Theory of dispersion.

§3. Supplementing §2. Excited atoms, degenerate systems, continuous spectrum.

§4. Discussion of the resonance case.

§5. Generalization for an arbitrary perturbation.

§6. Relativistic-magnetic generalization of the fundamental equations.

§7. On the physical significance of the field scalar.

Related references

More (earlier) information
E. Schrödinger, Annalen der Physik. Leipzig **79** (1926) 361;
E. Schrödinger, Annalen der Physik. Leipzig **79** (1926) 489;
E. Schrödinger, Annalen der Physik. Leipzig **79** (1926) 734;
E. Schrödinger, Annalen der Physik. Leipzig **80** (1926) 437;

See also
H. Weyl, Math. Ann. **69** (1910) 220, Goet. Nachr. (1910);
E. Fermi, Z. Phys. **29** (1924) 315;

Analyse information from
H. A. Kramers, Nature (May. 1924) 10, Nature (Aug. 1924) 30;
H. A. Kramers and W. Heisenberg, Z. Phys. **31** (1925) 681;
M. Born, W. Heisenberg, and P. Jordan, Z. Phys. **35** (1926) 572;
K. F. Herzfeld and K. L. Wolf, Annalen der Physik. Leipzig **76** (1925) 567;
H. Kollmann and H. Mark, Naturw. **14** (1926) 648;

BORN 1926B **Nobel prize**

■ Statistical interpretation of quantum mechanics. Quantum theory of scattering. Born approximation. Nobel prize to Max Born awarded in 1954 "for his fundamental research in quantum mechanics, especially for his statistical interpretation of the wave-function". Co-winner W. Bothe "for the coincidence method and his discoveries made therewith" ■

Zur Quantenmechanik der Stoßvorgänge / Quantum Mechanics of Collision

M. Born
Z. Phys. **37** (1926) 863;

Reprinted in
Dokumente der Naturwissenschaft, **I** (1962) 48.
(translation into English) *On the Quantum Mechanics of Collisions. Quantum Theory of Measurement*, ed. J. A. Wheeler, W. H. Zurek Princeton Univ. Press (1983) 52.

Abstract Durch eine Untersuchung der Stoßvorgänge wird die Auffassung entwickelt, daß die Quantenmechanik in der Schrödingerschen Form nicht nur die stationären Zustände, sondern auch die Quantensprünge zu beschreiben gestattet.

Abstract Through the investigation of collisions it is argued that quantum mechanics in the Schrödinger form allows one to describe not only stationary states but also quantum jumps.

BORN 1926C **Nobel prize**

■ Statistical interpretation of quantum mechanics: Further development. Nobel prize to Max Born awarded in 1954 "for his fundamental research in quantum mechanics, especially for his statistical interpretation of the wave-function". Co-winner W. Bothe "for the coincidence method and his discoveries made therewith" ■

Quantenmechanik der Stoßvorgänge / Quantum Mechanics of Collision

M. Born
Z. Phys. **38** (1926) 803;

Abstract Die Schrödingersche Form der Quantenmechanik erlaubt in natürlicher Weise die Häufigkeit eines Zustandes zu definieren mit Hilfe der Intensität der zugeordneten Eigenschwingung. Diese Auffassung führt zu einer Theorie der Stoßvorgänge, bei der die Übergangswahrscheinlichkeiten durch das asymptotische Verhalten aperiodischer Lösungen bestimmt werden.

Abstract The Schrödinger form of quantum mechanics permits the definition in a natural way of the frequency of a state with the aid of the intensity of the characteristic oscillation. This conception leads to a theory of impact processes by which the transition probabilities are determined by the asymptotic behavior of aperiodic solutions. *(Science Abstracts, 1927, 43.)*

Related references

More (earlier) information
M. Born, Z. Phys. **37** (1926) 863;

See also
W. Heisenberg, Z. Phys. **33** (1925) 879;
M. Born and P. Jordan, Z. Phys. **34** (1925) 858;
M. Born, W. Heisenberg, and P. Jordan, Z. Phys. **35** (1926) 557;
P. A. M. Dirac, Proc. Roy. Soc. **A109** (1925) 642;
P. A. M. Dirac, Proc. Roy. Soc. **A110** (1926) 561;
E. Schrödinger, Annalen der Physik. Leipzig **79** (1926) 361;
E. Schrödinger, Annalen der Physik. Leipzig **79** (1926) 489;
E. Schrödinger, Annalen der Physik. Leipzig **79** (1926) 734;

DIRAC 1926

■ Rediscovery of statistics for an ensemble of fermions — the Fermi-Dirac quantum statistics ■

On the Theory of Quantum Mechanics

P.A.M. Dirac
Proc. Roy. Soc. **A112** (1926) 661;

Abstract The present theory is shown to account for the absorption and stimulated emission of radiation, and also shows that the elements of the matrices representing the total polarization determine the transition probabilities. One cannot take spontaneous emission into account without a more elaborate theory involving the positions of the various atoms and the interference of their individual emissions, as the effects will depend upon whether the atoms are distributed at random, or arranged in a crystal lattice, or all confined in a volume small compared with a wave-length. The last alternative mentioned, which is of no practical interest, appears to be the simplest theoretically. It should be observed that we get the simple Einstein results only because we have averaged over all initial phases of the atoms. *(Science Abstracts, 1927, 294. G. W. de T.)*

BORN 1926D — Nobel prize

■ Statistical interpretation of quantum mechanics: Further development. Nobel prize to Max Born awarded in 1954 "for his fundamental research in quantum mechanics, especially for his statistical interpretation of the wave-function". Co-winner W. Bothe "for the coincidence method and his discoveries made therewith" ■

Das Adiabatenprinzip in der Quantenmechanik / The Adiabatic Principle in Quantum Mechanics

M. Born
Z. Phys. **40** (1926) 167;

Abstract Auf Grund der statistischen Auffassung der Quantenmechanik, die kürzlich an Hand der Stoßvorgänge entwickelt wurde, läßt sich der dem Ehrenfestschen Adiabatenprinzip analoge Satz formulieren und beweisen.

Related references

More (earlier) information
M. Born, Z. Phys. **38** (1926) 803;

See also
E. Schrödinger, Annalen der Physik. Leipzig **81** (1926) 109;
E. Schrödinger, Annalen der Physik. Leipzig **80** (1926) 437;
O. Klein, Z. Phys. **37** (1926) 904;
P. A. M. Dirac, Proc. Roy. Soc. **A112** (1926) 661;
M. Born and P. Jordan, Z. Phys. **33** (1925) 479;
M. Born, W. Heisenberg, and P. Jordan, Z. Phys. **35** (1926) 557;
P. Jordan, Z. Phys. **33** (1925) 506;

DAVISSON 1927 — Nobel prize

■ Experimental evidence that the electron moves as a group of de Broglie waves. Nobel Prize to C. J. Davisson awarded in 1937 and to co-winner G. P. Thomson "for their experimental discovery of the diffraction of electrons by crystals" ■

The Scattering of Electrons by a Single Crystal of Nickel

C.J. Davisson, L.H. Germer
Nature **119** (1927) 558;

Excerpt In a series of experiments now in progress, we are directing a narrow beam of electrons normally against a target cut from a single crystal of nickel, and are measuring the intensity of scattering (number of electrons per unit solid angle with speeds near that of the bombarding electrons) in various directions in front of the target. The experimental arrangement is such that the intensity of scattering can be measured in any latitude from the equator (plane of the target) to within 20° of the pole (incident beam) and in any azimuth. ...

...If the incident electron beam were replaced by a beam of monochromatic X rays of adjustable wave length, very similar phenomena would, of course, be observed. ...
...These results are highly suggestive, of course, of the ideas underlying the theory of wave mechanics, and we naturally inquire if the wave length of the X ray beam which we thus associate with a beam of electrons is in fact the h/mv of L. de Broglie. The comparison may be made, as it happens, without assuming a particular correspondence between X ray and electron beams, and without use of the contraction factor. *(Extracted from the introductory part of the paper.)*

Accelerator Electron-Gun

Detectors Counters

Related references

More (later) information
C. J. Davisson and L. H. Germer, Phys. Rev. **30** (1927) 705;

Reactions
e^- crystal $\rightarrow$ e^- crystal

DIRAC 1927

■ First steps in quantum field theory. Invention of the second quantization method ■

The Quantum Theory of the Emission and Absorption of Radiation

P.A.M. Dirac
Proc. Roy. Soc. **A114** (1927) 243;

Reprinted in
(translation into Russian) *Einsteinovskij Sbornik 1984–1985*, s. 215, M., Nauka, 1988.

Excerpt The problem is treated of an assembly of similar systems satisfying the Einstein-Bose statistical mechanics, which interact with another different system, a Hamiltonian function being obtained to describe the motion. The theory is applied to the interaction of an assembly of light-quanta with an ordinary atom, and it is shown that it gives Einstein's laws for the emission and absorption of radiation.
The interaction of an atom with electromagnetic waves is then considered, and it is shown that if one takes the energies and phases of the waves to be q-numbers satisfying the proper quantum conditions instead of c-numbers, the Hamiltonian function takes the same form as in the light-quantum treatment. The theory leads to the correct expressions for Einstein's A's and B's.

Related references

More (earlier) information
P. A. M. Dirac, Proc. Roy. Soc. **A111** (1926) 281;
P. A. M. Dirac, Proc. Roy. Soc. **A112** (1926) 661;
P. A. M. Dirac, Proc. Roy. Soc. **A113** (1926) 621;

See also
P. Jordan, Z. Phys. **40** (1926) 809;
F. London, Z. Phys. **40** (1926) 193;
M. Born, Z. Phys. **40** (1926) 167;
M. Born, Z. Phys. **38** (1926) 803;
M. Born, Nachr. Gessel. d. Wiss. Gottingen, (1926) 146;
O. Klein and Rosseland, Z. Phys. **4** (1921) 46;
A. Einstein and P. Ehrenfest, Z. Phys. **19** (1923) 301;
W. Pauli, Z. Phys. **18** (1923) 272;

DIRAC 1927B

■ Foundations of quantum electrodynamics — QED ■

The Quantum Theory of Dispersion

P.A.M. Dirac
Proc. Roy. Soc. **A114** (1927) 710;

Related references

More (earlier) information
P. A. M. Dirac, Proc. Roy. Soc. **A114** (1927) 243;

See also
E. Schrödinger, Annalen der Physik. Leipzig **81** (1926) 109;
W. Gordon, Z. Phys. **40** (1926) 117;
O. Klein, Z. Phys. **41** (1927) 407;
H. A. Kramers and W. Heisenberg, Z. Phys. **31** (1925) 681;
M. Born, W. Heisenberg, and P. Jordan, Z. Phys. **35** (1926) 557;

ELLIS 1927

■ Confirmation that the β spectrum is continuous ■

The Continuous Spectrum of β Rays

C.D. Ellis, W.A. Wooster
Nature **119** (1927) 563;

Excerpt The continuous spectrum of the β-rays arising from radio-active bodies is a matter of great importance in the study of their disintegration. Two opposite views have been held about the origin of this continuous spectrum. It has been suggested that, as in the α-ray case, the nucleus, at each disintegration, emits an electron having a fixed characteristic energy, and that this process is identical for different atoms of the same body. The continuous spectrum given by these disintegration electrons is then explained as being due to secondary effects, into the nature of which we need not enter here. The alternative theory supposes that the process of emission of the electron is not the same for different atoms, and that the continuous spectrum is a fundamental characteristic of the type of atom disintegrating. Discussing of these views has hitherto been concerned with the problem of whether or not certain specified secondary effects could produce the observed heterogenity, and although no satisfactory explanation has yet been given by the assumption of secondary effects, it was most important to clear up the problem by a direct method.

There is a ready means of distinguishing between the two views, since in one case a given quantity of energy would be emitted at each disintegration equal to or greater than the maximum energy observed in the electrons escaping from the atom, whereas in the second case the average energy per disintegration would be expected to equal the average energy of the particles emitted. If we were to measure the total energy given out by a known amount of material, as, for example, by enclosing it in a thick-walled calorimeter, then in the first case the heating effect should lead to an average energy per disintegration equal to or greater than the fastest electron emitted, no matter in what way this energy was afterwards split up by secondary effects. Since on the second hypothesis no secondary effects are presumed to be present, the heating effect should correspond simply to the average kinetic energy of the particles forming the continuous spectrum.
To avoid complications due to α-rays or to γ-rays from parent or successive atoms, we measured the heating effect in a thick-walled calorimeter of a known quantity of radium E. This measurement proved difficult because of the small rate of evolution of heat, but by taking special precautions it has been possible to show that the average energy emitted at each disintegration of radium E is 340,000−30,000 volts. This result is a striking confirmation of the hypothesis that the continuous spectrum is emitted as such from the nucleus, since the average energy of the particles as determined by ionization measurements over the whole spectrum gives a value about 390,000 volts, whereas if the energy emitted per disintegration were equal to that of the fastest β-rays, the corresponding value of the heating would be three times as large — in fact, 1,050,000 volts.
Many interesting points are raised by the question of how a nucleus, otherwise quantized, can emit electrons with velocities varying over a wide range, but consideration of these will be deferred until the publication of the full results.

Accelerator Radioactive source

Detectors Ionization, Calorimeter

Related references

More (later) information
C. D. Ellis and W. A. Wooster, Proc. Roy. Soc. **A117** (1927) 109;

Reactions
$^{210}\mathrm{Bi} \rightarrow {}^{210}\mathrm{Po}\ e^- \ \mathrm{X}$

ELLIS 1927B

■ Further confirmation that the β spectrum is continuous ■

Average Energy of Disintegration of Radium E

C.D. Ellis, W.A. Wooster
Proc. Roy. Soc. **A117** (1927) 109;

Abstract The experiments described in this paper show that the average energy of disintegration of Ra E is about 350000 volts, and this energy is liberated in such a form that the major portion, 344000 volts, is stopped by 1.2 mm of lead and the remainder has an absorption coefficient in lead of 5.9 cm^{-1}. The interpretation of this result is as follows. The main energy is due to the disintegration electrons, and the small extra radiation is probably continuous γ-radiation of a relatively hard type, emitted by a few of the disintegrated electrons which suffer close collisions with the planetary electrons in their escape from the atom. These results have been confirmed by independent methods. The authors generalize the results obtained for RaE to all γ ray bodies, thereby settling a long controversy. We must conclude that in a γ ray disintegration the nucleus can break up with emission of an amount of energy that varies within wide limits. A simple hypothesis is put forward to explain this curious result. *(Science Abstracts, 1928, 664. A. B. W.)*

Accelerator Radioactive source

Detectors Ionization, Calorimeter

Related references

More (earlier) information
C. D. Ellis and W. A. Wooster, Nature **119** (1927) 563;
Proc. Camb. Phil. Soc. **22** (1925) 400, 849;

Reactions
$^{210}\mathrm{Bi} \rightarrow {}^{210}\mathrm{Po}\ e^- \ \mathrm{X}$

HEISENBERG 1927

■ Heisenberg discovery of the uncertainty principle ■

Über den anschaulichen Inhalt der quantentheoretischen Kinematik und Mechanik / The Actual Content of Quantum Theoretical Kinematics and Mechanics

W. Heisenberg
Z. Phys. **43** (1927) 172;

Reprinted in
W. Heisenberg, *Gesammelte Werke / Collected Works*, Ed. by W. Blum, H. P. Durr, and H. Richenberg, Springer–Verlag, Ser. A, Part I, (1985) 478.
(translation into English) *The Physical Content of Quantum Kinematics and Mechanics* see in: *Quantum Theory of Measurement*, ed. by J. A. Wheeler, W. H. Zurek, Princeton University Press, N.J. (1983) 62.
Dokumente der Naturwissenschaft **4** (1963) 9.

Abstract First we define the terms *velocity, energy,* etc. (for example, for an electron) which remain valid in quantum mechanics. It is shown that canonically conjugate quantities can be determined simultaneously only with a characteristic indeterminacy (§1). This indeterminacy is the real basis for the occurrence of statistical relations in quantum mechanics. Its mathematical formulation is given by the Dirac-Jordan theory (§2). Starting from the basic principles thus obtained, we show how microscopic processes can be understood by way of quantum mechanics (§3). To illustrate the theory, a few special *gedanken experiments* are discussed (§4).

Related references

More (earlier) information
M. Born, W. Heisenberg, and P. Jordan, Z. Phys. **35** (1926) 557;

See also
N. Bohr, Z. Phys. **13** (1923) 117;
P. A. M. Dirac, Proc. Roy. Soc. **A112** (1926) 661;
P. A. M. Dirac, Proc. Roy. Soc. **A113** (1926) 621;
P. A. M. Dirac, Proc. Roy. Soc. **A114** (1927) 243;
A. Einstein, Sitzber. Pr. Akad. Wiss. **23** (1925) 3;
W. Heisenberg, Z. Phys. **40** (1926) 501;
W. Pauli, Z. Phys. **41** (1927) 81;
P. Jordan, Z. Phys. **37** (1926) 376;
P. Jordan, Z. Phys. **40** (1926) 661;
P. Jordan, Z. Phys. **40** (1926) 809;
P. Jordan, Naturw. **15** (1927) 105;
M. Born, Z. Phys. **38** (1926) 803;
M. Born, Z. Phys. **40** (1926) 167;
M. Born, Naturw. **15** (1927) 238;
E. Schrödinger, Naturw. **14** (1926) 664;
P. Ehrenfest and G. Breit, Z. Phys. **9** (1922) 207;
P. Ehrenfest and R. C. Tolman, Phys. Rev. **24** (1924) 287;

WIGNER 1927B — Nobel prize

■ Invention of spatial parity as a quantum mechanical conserved quantity. Nobel prize to E. P. Wigner awarded in 1963 "for his contributions to the theory of the atomic nucleus and the elementary particles, particularly through the discovery and application of fundamental symmetry principles". Co-winners M. Goeppert-Mayer and J. H. D. Jensen "for their discoveries of nuclear shell structure" ■

Über die Erhaltungssätze in der Quantenmechanik / On the Conservation Laws of Quantum Mechanics

E.P. Wigner
Nachr. Ges. Wiss. Goett. P.375 (1927);

Related references

More (earlier) information
E. P. Wigner, Z. Phys. **43** (1927) 624;

Reactions

$atom^* \rightarrow \text{atom}\ \gamma$

DAVISSON 1927B — Nobel prize

■ Discovery of the diffraction of electrons by crystals. Confirmation of the wave properties of moving electrons. Nobel Prize to C. J. Davisson awarded in 1937 and to co-winner G. P. Thomson "for their experimental discovery of the diffraction of electrons by crystals" ■

Diffraction of Electrons by a Crystal of Nickel

C.J. Davisson, L.H. Germer
Phys. Rev. **30** (1927) 705;

Reprinted in
The Physical Review — the First Hundred Years, AIP Press (1995) 48.

Abstract *The intensity of scattering* of a homogeneous beam of electrons of adjustable speed incident upon a single crystal of nickel has been measured as a function of direction. The crystal is cut parallel to a set of its {111}–planes and bombardment is at normal incidence. The distribution in latitude and azimuth has been determined for such scattered electrons as have lost little or none of their incident energy.
Electron beams resulting from diffraction by a nickel crystal. —Electrons of the above class are scattered in all directions at all speeds of bombardment, but at and near critical speeds sets of three or of six sharply defined beams of electrons issue from the crystal in its principal azimuths. Thirty such sets of beams have been observed for bombarding potentials below 370 volts. Six of these sets are due to scattering by adsorbed gas; they are not found when the crystal is thoroughly degassed. Of the twenty-four sets due to scattering by the gas-free crystal, twenty are associated with twenty sets of Laue beams that would issue from the crystal within the range of observation if the incident beam were a beam of heterogeneous x rays, three that occur near grazing are accounted for as diffraction beams due to scattering from a single {111}–layer of nickel atoms, and one set of low intensity has not been accounted for. Missing beams number eight. These are beams whose occurrence is required by the correlations mentioned above, but which have not been found. The intensities expected for these beams are all low.
The spacing factor concerned in electron diffraction by a nickel crystal. —The electron beams associated with Laue beams do not coincide with these beams in position, but occur as if the crystal were contracted normally to its surface. The spacing factor describing this contraction varies from 0.7 for electrons of lowest speed to 0.9 for electrons whose speed corresponds to a potential difference of 370 volts.
Equivalent wave-lengths of the electron beams may be calculated from the diffraction data in the usual way. These turn out to be in acceptable agreement with the values of h/mv of

the undulatory mechanics.
Diffraction beams due to adsorbed gas are observed except when the crystal has been thoroughly cleaned by heating. Six sets of beams of this class have been found; three of these appear only when the crystal is heavily coated with gas; the other three only when the amount of adsorbed gas is slight. The structure of the gas film giving rise to the latter beams has been deduced.

Accelerator Electron-Gun

Detectors Counters

Related references

More (earlier) information
C. J. Davisson and L. H. Germer, Nature **119** (1927) 558;
C. J. Davisson and Kunsman, Science **64** (1921) 522;
C. J. Davisson and Kunsman, Phys. Rev. **22** (1923) 242;
W. Elsasser, Naturw. **13** (1921) 711;

Reactions
e^- crystal $\to$ e^- crystal

THOMSON 1928 — Nobel prize

■ Confirmation of diffraction of electrons by crystals. Nobel prize to G. P. Thomson awarded in 1937 and to co-winner C. L. Davisson "for their experimental discovery of the diffraction of electrons by crystals" ■

Experiments on Diffraction of Cathode Rays

G.P. Thomson
Proc. Roy. Soc. **A117** (1928) 600;

Excerpt

1. Experiments are described giving the pattern formed by cathode rays scattered by thin films of aluminium, gold, celluloid, and an unknown substance.
2. These patterns are closely similar to those obtained with X rays in the "powder method."
3. The sizes of patterns agree to 5 per cent with those predicted on the de Broglie theory of wave mechanics, regarding the phenomenon as one of diffraction of the phase waves associated with the electrons.

Accelerator cathode ray tube

Detectors Nuclear emulsion

Related references

More (earlier) information
G. P. Thomson, Nature **119** (1927) 890;

Analyse information from
C. J. Davisson and L. H. Germer, Nature **119** (1927) 558;
C. J. Davisson and Kunsman, Phys. Rev. **22** (1923) 242;

Reactions
e^- crystal $\to$ e^- crystal

DIRAC 1928 — Nobel prize

■ Discovery of the relativistic wave equation for the electron. Prediction of the magnetic moment of the electron. Nobel prize to P. A. M. Dirac awarded in 1933. Co-winner E. Schrödinger "for the discovery of new productive forms of atomic theory" ■

The Quantum Theory of the Electron

P.A.M. Dirac
Proc. Roy. Soc. **A117** (1928) 610;

Abstract By employing for a point electron in an arbitrary electromagnetic field the simplest Hamiltonian which satisfies the requirements of both relativity and the general transformation theory of quantum mechanics, an explanation is obtained of all the "duplexity" phenomena without further assumption. The spinning electron model devised to account for the observed duplexity may be regarded as true as a first approximation. The present theory, in fact, indicates that the electron will behave as though possessing the magnetic moment assumed in the model, in addition to an electric moment, which is, however, imaginary and of doubtful physical significance (if any). The theory is applied to the case of motion in a central field, and the energy levels are shown to agree to a first approximation with those given by Darwin and by Pauli in which the spin of the electron is allowed for. *(Science Abstracts. 1928. 1168. W.S.S.)*

Particles studied e^-

GAMOW 1928

■ Explanation of α decay as a consequence of quantum-mechanical tunneling through a potential barrier ■

Zur Quantentheorie des Atomkernes / On Quantum Theory of Atomic Nuclei

G. Gamow
Z. Phys. **51** (1928) 204;

Abstract Es wird der Versuch gemacht, die Prozesse der α-Ausschtrahlung auf Grund der Wellenmechanik näher zu untersuchen und den experimentell festgestellten Zusammenhang zwischen Zerfallskonstante und Energie der α-Partikel theoretisch zu erhalten.

Related references

See also
J. Frenkel, Z. Phys. **37** (1926) 243;
E. Rutherford, Phil. Mag. **4** (1927) 580;
J. R. Oppenheimer, Phys. Rev. **31** (1928) 66;

Nodheim, Z. Phys. **46** (1927) 833;
H. Geiger and J. Nuttall, Phil. Mag. **23** (1912) 439;
Swinne, Phys.Zeitschr. **13** (1912) 14;
Jacobsen, Phil. Mag. **47** (1924) 23;

Reactions
nucleus $\rightarrow$ nucleus He

Particles studied nucleus

DIRAC 1928B Nobel prize

■ Relativistic wave equation for the electron and theory of the Zeeman effect. Nobel prize to P. A. M. Dirac awarded in 1933. Co-winner E. Schrödinger "for the discovery of new productive forms of atomic theory" ■

The Quantum Theory of the Electron. Pt II

P.A.M. Dirac
Proc. Roy. Soc. **A118** (1928) 351;

Abstract The theory enunciated in Part I is further developed, commencing with a proof of the conservation theorem, which states that the change in probability of the electron being in a given volume during a given time is equal to the probability of its having crossed the boundary. The selection rule for j, the quantum number determining the magnitude of the resultant angular momentum for an electron in a central field of force, is worked out and found to be exactly equivalent to the two selection rules for j and k, of the usual theory, which agree with experiment. The relative intensities of the Zeeman components of the lines in a combination doublet and the alteration of the energy levels by a magnetic field are investigated and the results shown to be in agreement with those of previous theories based on the spinning electron model. *(Science Abstracts. 1928. 1507. W. S. S.)*

Related references
More (earlier) information
P. A. M. Dirac, Proc. Roy. Soc. **A117** (1928) 610;

Particles studied e^-

SKOBELZYN 1929

■ Birth of cosmic rays particle physics. Observation of energetic cosmic electrons and a shower produced by cosmic ray particle ■

Über eine neue Art sehr schneller β–Strahlen / A New Type of Very Fast Beta Rays

D.V. Skobelzyn
Z. Phys. **54** (1929) 686;

Reprinted in
Selected Papers of Soviet Physicists, Usp. Fiz. Nauk **93** (1967) 331.

Abstract From about 600 pictures obtained with a Wilson chamber in the uniform magnetic field, 32 pictures were found with tracks originated outside of the Wilson chamber and not affected noticeably by the magnetic field. One has to assign to these tracks energies greater than 15000 eV. Approximately calculated ionization effect of these tracks was about 1, the angular distribution shows a sharp excess of tracks directed to large angles with respect to the horizontal plane. One should assign these β rays to the secondary electrons created by Hess ultra-γ rays. It should be stressed that simultaneous appearance of several such tracks occurred from common centers. Possible effects important to theme thods of measuring of "high altitude rays" and anomalies of "transition zones" are discussed.

Accelerator COSMIC

Detectors Cloud chamber

Related references
More (earlier) information
D. V. Skobelzyn, Z. Phys. **28** (1924) 278;
D. V. Skobelzyn, Z. Phys. **43** (1927) 354;
See also
L. Myssowsky and L. Tuwim, Z. Phys. **36** (1926) 615;
W. Kolhoerster, Z. Phys. **34** (1925) 147;
O. Klein and Y. Nishina, Z. Phys. **52** (1928) 853;
R. A. Millikan and G. H. Cameron, Phys. Rev. **28** (1926) 865;
W. Kolhoerster and W. Bothe, Naturw. **16** (1928) 1045;
R. A. Millikan and G. H. Cameron, Phys. Rev. **31** (1928) 921;
E. Rutherford, Proc. Roy. Soc. **A122** (1929) 15;
Analyse information from
E. Steinke, Z. Phys. **42** (1927) 593;
E. Steinke, Z. Phys. **48** (1928) 656;
E. Steinke, Z. Phys. **48** (1928) 671;

Reactions
e^- >0.015 MeV(P_{lab})

Particles studied shower

BOTHE 1929 Nobel prize

■ Observation that the cosmic rays at sea level consist mainly of ionizing particles. Nobel prize to W. Bothe awarded in 1954 "for his coincidence method and his discoveries made therewith". Co-winner M. Born "for his fundamental research in quantum mechanics, especially for his statistical interpretation of the wave-function" ■

Das Weßen der Höhenstrahlung / Nature of High-Altitude Radiation

W. Bothe, W. Kolhorster
Z. Phys. **56** (1929) 751;

Abstract When two Geiger-Müller counting tubes are placed

near to one another it is found that, amongst the deflections produced by high altitude radiation, there is a noticeable fraction which take place simultaneously in both tubes. From the way in which these simultaneous deflection depend on the relative positions of the two tubes, and from their large number, it is possible to conclude that each pair is due to the passage of a single corpuscle through both counting tubes. The absorption coefficient of this corpuscular radiation has been determined by placing layers of absorbing material between the two tubes, and measuring the diminution of the number of coincidences. The result obtained is that the corpuscular radiation is just as strongly absorbed as the high-altitude radiation itself. The conclusion is that the high-altitude radiation, as far as it has manifested itself in the phenomena so far observed, is of corpuscular nature. Its probable properties are discussed from this point of view. *(Science Abstracts, 1929, 675. H.N.A.)*

Accelerator COSMIC

Detectors Counters, Ionization

Related references

See also
H. Geiger and Mueller, Phys.Zeitschr. **29** (1928) 839;
R. A. Millikan and G. H. Cameron, Phys. Rev. **31** (1928) 921;
O. Klein and Y. Nishina, Z. Phys. **52** (1928) 853;
D. V. Skobelzyn, Z. Phys. **54** (1929) 686;

BIRGE 1929

■ First step in metrology of the general physical constants ■

Probable Values of the General Physical Constants

R.T. Birge

Rev. of Mod. Phys. **1** (1929) 1;

Excerpt Some of the most important results of physical science are embodied, directly or indirectly, in the numerical magnitudes of various universal constants, and the accurate determination of such constants has engaged the time and labor of many of the world's most eminent scientists. Some of these constants can be evaluated by various methods. Each has been investigated by various persons, at various times, and each investigation normally produces a numerical result more or less different from that of any other investigation. Under such conditions there arises a general and continuous need for a searching examination of the most probable value of each important constant. The need is general since every physical scientist uses such constants. The need is continuous since the most probable value of to-day is not that of to-morrow, because of the never-ending progress of scientific research. These remarks appear to the writer so self-evident that the mere statement of them may be deemed superfluous. However, in spite of these facts, an investigation of the values of general constants in current use in the literature reveals a surprising lack of consistency, both in regard to the actually adopted values and to the origin of such values. This is probably due to the fact that it is almost impossible to find a critical study of the best values, sufficiently up-to-date to be really reliable, and sufficiently detailed to explain the inconsistencies found among older tables.
The situation is much better in the case of selected groups of constants. Thus the best value of the atomic weight of each element is determined annually by certain atomic weight committees, and the need of such a list of atomic weights is obvious to every chemist. There is certainly a similar need in the case of the even more important constants such as the velocity of light, the charge of the electron, the Planck constant h, etc. In attempting to respond to this need, the writer has become only too well aware of the intrinsic difficulties involved, but at the same time he has become increasingly convinced of the existence of the need itself. The present investigation was undertaken only at the express request of others, and the results given here should be considered more as a presentation of the situation than as a final solution of the problem. To obtain a satisfactory and thoroughly reliable judgment in such matters, there is required the unbiased cooperation of many persons situated in scientific laboratories throughout the world. *(Extracted from the introductory part of the paper.)*

Related references

More (earlier) information
R. T. Birge, Phys. Rev. **33** (1929) 265;
See also
E. J. G. de Bray, Nature **120** (1927) 602;
A. A. Michelson, Astr. Jour. **65** (1927) 1;
R. A. Millikan, Phil. Mag. **34** (1917) 1;
R. A. Millikan, Phys. Rev. **22** (1923) 1;
A. H. Compton, H. N. Beets, and O. K. DeFoe, Phys. Rev. **25** (1925) 625;
O. K. De Foe and A. H. Compton, Phys. Rev. **25** (1925) 621;
A. P. R. Wadlund, Phys. Rev. **32** (1928) 841;
H. D. Babcock, Astr. Jour. **58** (1923) 149;
H. D. Babcock, Phys. Rev. **33** (1929) 268A;
H. D. Babcock, Astr. Jour. **69** (1929) 43;
W. V. Houston, Phys. Rev. **30** (1927) 608;
F. Pashen, Annalen der Physik. Leipzig **50** (1916) 901;
R. T. Birge, Phys. Rev. **17** (1921) 589;
R. T. Birge, Nature **111** (1923) 287;
R. T. Birge, Phys. Rev. **14** (1919) 361;
E. O. Lawrence, Phys. Rev. **28** (1926) 947;
R. T. Birge, Science **64** (1926) 180;
F. C. Blake and W. Duane, Phys. Rev. **10** (1917) 624;
W. Duane, H. H. Palmer, and Chi-Sun Yeh, Proc. Nat. Acad. Sci. (USA) **7** (1921) 237;
E. Wagner, Phys.Zeitschr. **21** (1920) 621;
P. Lukirsky and S. Prilezhaev, Z. Phys. **49** (1928) 236;
R. A. Millikan, Phys. Rev. **7** (1916) 255;
W. W. Coblentz, Proc. Nat. Acad. Sci. (USA) **3** (1917) 504;
K. Hoffman, Z. Phys. **14** (1923) 301;
A. Kussman, Z. Phys. **25** (1924) 58;
W. Gerlach, Annalen der Physik. Leipzig **50** (1916) 259;
G. Michel and A. Kussman, Z. Phys. **18** (1923) 263;
L. Strum, Z. Phys. **51** (1928) 287;
F. E. Hoare, Phil. Mag. **6** (1928) 828;
H. M. Sharp, Phys. Rev. **26** (1925) 691;
B. Davis and D. P. Mitchel, Phys. Rev. **32** (1928) 331;
G. N. Lewis and E. Q. Adams, Phys. Rev. **3** (1914) 92;

A. S. Eddington, Proc. Roy. Soc. **A122** (1929) 358;
R. T. Birge, Nature **123** (1929) 318;

DIRAC 1930

■ Introduction of the negative energy electron sea with holes treated as positive electrons. Attempt to identify these holes with protons ■

A Theory of Electrons and Protons

P.A.M. Dirac
Proc. Roy. Soc. **A126** (1930) 360;

Abstract The difficulty of the negative energy electron is discussed and a Solution is proposed. Such electrons move in an external field as though carrying a positive charge. The most stable states for an electron — the states of lowest energy —are those with negative energy and very high velocity. All the electrons in the world will tend to fall into these states with emission of radiation. Assume there are so many electrons in the world that all the states of negative energy are occupied except perhaps a few of small velocity. Any electrons with positive energy will now have very little chance of jumping into negative-energy states and their behaviour will be as usually observed. There will be an infinite number of electrons in negative-energy states, but if their distribution is exactly uniform they will be completely unobservable. Only the small departures from exact uniformity, brought about by some of the negative-energy states being unoccupied, can be observed. The properties of the vacant states or "holes" are discussed. These holes will be things of positive energy and in this respect like ordinary particles. The motion of one of these holes in an external electromagnetic field will be the same as that of the negative-energy electron that would fill it, and will thus correspond to its possessing a charge $+e$. This leads to the assumption that the holes in the distribution of negative-energy electrons are the protons. When an electron of positive energy drops into a hole and fills it up an electron and proton disappear together with emission of radiation. This theory is applied to the problem of the scattering of radiation by an electron. *(Science Abstracts. 1930. 2416. J.J.S.)*

Particles studied e^+, p

MEITNER 1930

■ Firm establishment that the β spectrum is continuous ■

Über eine absolute Bestimmung der Energie der primären β-Strahlen von Radium E / Absolute Determination of Energy of Primary β Rays from Radium E

L. Meitner, W. Orthmann
Z. Phys. **60** (1930) 143;

Abstract Es wird durch eine Wärmemessung die Energie der primären β–Strahlen von Radium E-Präparaten festgestellt, deren Radiumäquivalent, also die pro Sekunde zerfallende Anzahl von Atomen, durch eine besondere Untersuchung ermittelt wird, um die von einem einzelnen Kernelektron mitgeführte Energie zu erhalten. Außerdem wird gezeigt, daß Radium E innerhalb eines weit umfassenden Wellenlängenbereiches kein kontinuierliches γ-Strahlenspektrum besitzt.

Abstract The method employed was a calorimetric one, and the mean value of the energy of a primary β-particle was determined. The number of radium E atoms disintegrating per sec was arrived at by preparing standard preparations, calibrated in radium equivalents. The most probable value as obtained is 337,000 volts ± 6%, which agrees well with the value 344,000 volts ± 10% obtained by Ellis and Wooster. The actual velocities of the β-particles vary from about one million down to 200,000 volts, and a continuous magnetic spectrum is obtained. This distribution of velocity and energy exists when the β-particles leave the atomic nuclei. A special experiment showed that RaE does not emit a continuous γ-ray spectrum, which might compensate for the inhomogeneous character of the disintegration energy given out in the form of β-radiation. From the experiments it can be deduced that the number of atoms disintegrating in a second in a gramme of radium is 3.68×10^{10}, with a maximum error of 5%. *(Science Abstracts. 1930, 2580. H. N. A.)*

Related references

See also
C. D. Ellis, Proc. Roy. Soc. **A101** (1922) 1;
L. Meitner, Z. Phys. **11** (1922) 35;
W. Gurney and E. U. Condon, Nature **122** (1928) 438;
O. Erbacher and K. Philipp, Z. Phys. **51** (1928) 309;
B. Heiman and W. Marckwald, Phys.Zeitschr. **14** (1913) 303;
H. J. Braddick and H. M. Cave, Proc. Roy. Soc. **A121** (1928) 367;
D. V. Skobelzyn, Z. Phys. **58** (1929) 595;
G. Gamow, Z. Phys. **51** (1928) 204;
Analyse information from
C. D. Ellis and W. A. Wooster, Proc. Roy. Soc. **A117** (1927) 109;

Reactions
$^{210}\text{Bi} \rightarrow {}^{210}\text{Po}\ e^-\ \text{X}$

PAULI 1930

■ Proposal for the existence of the neutral fermion emitted in nuclear beta decay – the neutrino ■

Open Letter to Radioactive Persons

W. Pauli
LETTER (1930);

Reprinted in
Lecture of W. Pauli *Zur älteren und neueren Geschichte des Neutrinos* published in Vierteljahresschr. Naturforsch. Ges. Zürich, **102** (1957) 387.
(translation into English) Physics Today **31** (1978) 27.
(translation into Russian) W. Pauli *Fizicheskie ocherki*, sbornik statey, "Nauka," M. 1975 (red. Y. A. Smorodinsky).

Reactions
nucleus $\rightarrow$ nucleus e^- neutral

Particles studied neutral, n, ν

TAMM 1930

■ Difficulties with identifying positive holes with protons in the Dirac theory of electrons and protons. Too small a lifetime of the ordinary atom ■

On Free Electron Interaction with Radiation in the Dirac Theory of the Electron and in Quantum Electrodynamics

I.E. Tamm
Z. Phys. **62** (1930) 545;

Reprinted in
(translation into Russian) I. E. Tamm, *Sobranie nauchnykh trudov*, Nauka Moskva, 1975, t. II, s. 24.

Abstract Dirac's wave equations are used to determine the wave functions of the free election (a simple method for manipulating these wave functions is introduced) and the radiation field is also quantized. In this way a formula is derived for the scattered radiation from free electrons which is identical with the Klein-Nishina result, but which does not rest on an appeal to the correspondence principle to evaluate the radiated intensity from the distribution of the electron current. It appears that quantum jumps to intermediate states of negative energy play a determining part in the scattering process. In addition, the probability of a spontaneous transition of an electron from a positive to a negative energy level is evaluated and shown to equal to the classically estimated probability of collision between two electrons of relative velocity c. Such a transition would correspond in Dirac's recent theory of the proton to the transformation of the matter of the electron and proton into radiation. Applying the result obtained to an atom, an impossibly short life period for the atom is obtained (of the order 10^{-3} sec), and this appears to be a fundamental difficulty for the new Dirac theory of the proton if the assumptions of the present analysis are justified. *(Science Abstracts. 1930. 3645. W. S. S.)*

Related references

More (earlier) information
I. E. Tamm, Z. Phys. **60** (1930) 345;

See also
E. Schrödinger, Annalen der Physik. Leipzig **81** (1926) 109;
O. Klein, Z. Phys. **41** (1927) 407;
O. Klein, Z. Phys. **53** (1929) 157;
O. Klein and Y. Nishina, Z. Phys. **52** (1928) 853;
P. A. M. Dirac, Proc. Roy. Soc. **A126** (1930) 360;
P. A. M. Dirac, Proc. Roy. Soc. **A114** (1927) 243;
P. A. M. Dirac, Proc. Roy. Soc. **A114** (1927) 710;
P. A. M. Dirac, Proc. Roy. Soc. **A117** (1928) 610;
W. Heisenberg and W. Pauli, Z. Phys. **56** (1929) 1;
E. Fermi, Rend. Accad. nat. Lincei **31** (1922) 184;

Reactions
$\gamma\, e^- \rightarrow e^-\, \gamma$

Particles studied e^+, p

OPPENHEIMER 1930

■ First evidence of ultraviolet divergences occurring in the theory of quantum electrodynamics — the self-energy of the electron ■

Note on the Theory of the Interaction of Field and Matter

J.R. Oppenheimer
Phys. Rev. **35** (1930) 461;

Abstract The paper develops a method for the systematic integration of the relativistic wave equations for the coupling of electrons and protons with each other and with the electromagnetic field. It is shown that, when the velocity of light is made infinite, these equations reduce to the Schrödinger equation in configuration space for the many body problem. It is further shown that it is impossible on the present theory to eliminate the interaction of a charge with its own field, and that the theory leads to false predictions when it is applied to compute the energy levels and the frequency of the absorption and emission lines of an atom.

Related references

See also
W. Heisenberg and W. Pauli, Z. Phys. **56** (1929) 1;

DIRAC 1931

■ Prediction of the anti-electron (e^+), anti-proton ($\overline{p}$), and an indication of the possible existence of magnetic monopoles ■

Quantized Singularities in the Electromagnetic Field

P.A.M. Dirac
Proc. Roy. Soc. **A133** (1931) 60;

Abstract This paper puts forward a new idea concerning negative energies of electrons, and is concerned essentially with the reason for the existence of an elementary electric charge, which, by experiment, is known to exist. The theory when worked out gives a connection between the elementary electric charge and the elementary magnetic pole; it shows a symmetry between electricity and magnetism quite foreign to current views, but does not, however, give a theoretical value for e, the elementary electric charge. The paper also shows that quantum mechanics really does not preclude the existence of isolated magnetic poles, and discusses why these poles are not observed. *(Science Abstract, 1931, 340. H. L. B.)*

Reactions
$\gamma\,\gamma \rightarrow e^-\,e^+$ >1.2 MeV(E_{cm})

Particles studied $e^+, \overline{p}$, monopole

LAWRENCE 1931 — Nobel prize

■ Lawrence proposal for cyclotrons. Nobel prize to E. O. Lawrence awarded in 1939 "for the invention and development of the cyclotron and for results obtained with it, especially with regard to artificial radioactive elements" ■

A Method for Producing High Speed Hydrogen Ions without the Use of High Voltages

E.O. Lawrence, M.S. Livingston
Phys. Rev. **37** (1931) 1707;

Excerpt A method for producing high speed hydrogen ions without the use of high voltages was described at the September meeting of the National Academy of Sciences. (Science 72, 376 (1930)) The hydrogen ions are set in resonance with a high frequency oscillating voltage between two hollow semicircular plates in a vacuum, and are made to spiral around in semicircular paths inside these plates by a magnetic field. Each time the ions pass from the interior of one plate to that of the other they gain energy corresponding to the voltage across the plates. This method has now been tried out with the following results: Using a magnet with pole faces 10 cm in diameter and giving a field of 12700 gauss, 80000 volt hydrogen molecule ions have been produced using 2000 volt high frequency oscillations on the plates. A voltage amplification (the ratio of the equivalent voltage of the ions produced to the high frequency voltage applied to the plates) of 82 has been obtained. These preliminary experiments indicate clearly that there are no difficulties in the way of producing one million volt ions in this manner. A larger magnet is under construction for this purpose.

Related references

More (earlier) information
E. O. Lawrence and W. E. Edlefsen, Science **72** (1930) 376;

More (later) information
E. O. Lawrence and M. S. Livingston, Phys. Rev. **38** (1931) 834;
E. O. Lawrence and M. S. Livingston, Phys. Rev. **40** (1932) 19;

LAWRENCE 1931B — Nobel prize

■ Tests of the first cyclotron. Nobel prize to E. O. Lawrence awarded in 1939 "for the invention and development of the cyclotron and for results obtained with it, especially with regard to artifical radioactive elements" ■

The Production of High Speed Protons Without the use of High Voltages

E.O. Lawrence, M.S. Livingston
Phys. Rev. **38** (1931) 834;

Excerpt A method for the production of high speed protons without the use of high voltages was described before the meeting of the National Academy of Sciences last September (Lawrence and Edlefson, Science 72, 376-377, 1930). Later before the American Physical Society (Lawrence and Livingston, Phys. Rev 37, 1707, 1931) results of a preliminary study of the practicability of this method were presented. In this preliminary experimental work 80000-volt hydrogen molecule ions were successfully produced in a vacuum tube in which the maximum applied potential was less than 2000 volts, and the conclusion of the experiments was that there are no serious difficulties in the way of producing 1000000-volt protons in this indirect manner.
This important conclusion has now been confirmed. A magnet having pole faces nine inches in diameter and producing a field of 15000 gauss has recently been constructed and with its aid protons and hydrogen molecule ions having energies in excess of one half million volt-electrons have been produced. The magnitudes of the high speed hydrogen ion currents turned out to be surprisingly large, being in excess of one-tenth of one microampere. The proton currents were about one-tenth this value.
The voltage amplification obtained in the present experiment was approximately one hundred. That is to say, about five thousand volts were applied to the tube for the production of five hundred thousand volt ions. This amplification was lim-

ited by the slit system used to select out the high speed ions, and can be greatly increased by better design of this part of the tube.

There can be little doubt that one million volt ions will be produced with intensities as great as here recorded when the present experimental tube is enlarged to make full use of the magnet. This alteration is now being carried out.

These experiments make it evident that with quite ordinary laboratory facilities proton beams having great enough energies for nuclear studies can be readily produced with intensities far exceeding the intensities of beams of alpha-particles from radioactive sources.

Possibly the most interesting consequence of these experiments is that it appears now that the production of 10000000-volt protons can be readily accomplished when a suitably larger magnet and high frequency oscillator are available. The importance of the production of protons of such speeds can hardly be overestimated and it is our hope that the necessary equipment for doing this will be made available to us.

Related references

More (later) information
E. O. Lawrence and M. S. Livingston, Phys. Rev. **40** (1932) 19;
See also
H. Rose, Z. Phys. **64** (1930) 1;
W. Bothe and H. Becker, Z. Phys. **66** (1930) 289;
G. Beck, Naturw. **18** (1930) 896;
C. Y. Chao, Phys. Rev. **36** (1930) 1519;
W. Gurney and E. U. Condon, Phys. Rev. **33** (1929) 127;
G. Gamow, Z. Phys. **51** (1928) 204;
G. Gamow, Z. Phys. **52** (1928) 514;
J. Chadwick, J. E. R. Constable, and E. C. Pollard, Proc. Roy. Soc. **A130** (1930) 463;
C. C. Lauritsen and R. D. Bennett, Phys. Rev. **32** (1928) 850;
M. A. Tuve et al., Phys. Rev. **35** (1930) 66;
M. A. Tuve et al., Phys. Rev. **39** (1932) 384;
A. Brash and J. Lande, Z. Phys. **70** (1931) 10;
D. H. Sloan and E. O. Lawrence, Phys. Rev. **38** (1931) 2021;
E. O. Lawrence and W. E. Edlefsen, Science **72** (1930) 376;
E. O. Lawrence and M. S. Livingston, Phys. Rev. **37** (1931) 1707;

VAN DE GRAAFF 1931

■ Invention of the Van de Graaff electrostatic accelerator ■

A 1500000 Volt Electrostatic Generator

R.J. Van de Graaff
Phys. Rev. **38** (1931) 1919;

Reprinted in
The Physical Review — the First Hundred Years, AIP Press (1995) 255.

Excerpt The application of extremely high potentials to discharge tubes affords a powerful means for the investigation of the atomic nucleus and other fundamental problems. The electrostatic generator here described was developed to supply suitable potentials for such investigations. In recent preliminary trials, spark-gap measurements showed a potential of approximately 1,500,000 volts, the only apparent limit being brush discharge from the whole surface of the 24-inch spherical electrodes. The generator has the basic advantage of supplying a direct steady potential, thus eliminating certain difficulties inherent in the application of non-steady high potentials. The machine is simple, inexpensive, and portable. An ordinary lamp socket furnishes the only power needed. The apparatus is composed of two identical units, generating opposite potentials. The high potential electrode of each unit consists of a 24-inch hollow copper sphere mounted upon a 7 foot upright Pyrex rod. Each sphere is charged by a silk belt running between a pulley in its interior and a grounded motor driven pulley at the base of the rod. The ascending surface of the belt is charged near the lower pulley by a brush discharge, maintained by a 10,000 volt transformer kenotron set, and is subsequently discharged by points inside the sphere.

RAMAN 1932

■ Experimental proof that the photon has spin = 1 ■

Experimental Proof of the Spin of the Photon

C.V. Raman, S. Bhagavantam
Nature **129** (1932) 22;

Excerpt In a paper under this title which has recently appeared, we have described and discussed observations which have led us to the conclusion that the light quantum possesses an intrinsic spin equal to one Bohr unit of angular momentum. In the four weeks which have elapsed since that paper was put into print, the experimental technique has been much improved in the direction of attaining greater precision. It appears desirable forthwith to report our newer results, which confirm the conclusion stated above. *(Extracted from the introductory part of the paper.)*

Accelerator Non accelerator experiment

Detectors Spectrometer

Related references

More (earlier) information
C. V. Raman and S. Bhagavantam, Ind. Jour. Phys. **6** (1931) 353;
C. V. Raman and S. Bhagavantam, Nature **128** (1931) 576;
C. V. Raman and S. Bhagavantam, Nature **128** (1931) 727;

Particles studied γ

ANDERSON 1932	Nobel prize

■ First experimental evidence for the positron. Nobel prize to C. D. Anderson awarded in 1936 "for his discovery of the positron". Co-winner V. F. Hess "for his discovery of cosmic rays" ■

The Apparent Existence of Easily Deflectable Positives

C.D. Anderson
Science **76** (1932) 238;

Accelerator COSMIC

Detectors Cloud chamber

Particles studied e^+

CHADWICK 1932	Nobel prize

■ First evidence for the neutron. Nobel prize to J. Chadwick awarded in 1935 "for his discovery of the neutron" ■

Possible Existence of a Neutron

J. Chadwick
Nature **129** (1932) 312;

Reprinted in
R. N. Cahn and G. Goldhaber, *The Experimental Foundations of Particle Physics*, Cambridge Univ. Press (1991) 9.

Excerpt It has been shown by Bothe and others that beryllium when bombarded by α-particles of polonium emits a radiation of great penetrating power, which has an absorption coefficient in lead of about $0.3(\text{cm})^{-1}$. Recently Mme. Curie-Joliot and M. Joliot found, when measuring the ionization produced by this beryllium radiation in a vessel with a thin window, that this ionization increased when matter containing hydrogen was placed in front of the window. The effect appeared to be due to the ejection of protons with velocities up to maximum of nearly 3×10^9 cm per sec. They suggested that the transference of energy to the proton was by a process similar to the Compton effect, and estimated that the beryllium radiation had a quantum energy of 50×10^6 electron volts.
I have made some experiments using the valve counter to examine the properties of this radiation excited in beryllium. The valve counter consists of a small ionization chamber connected to an amplifier, and the sudden production of ions by the entry of a particle, such as a proton or α-particle, is recorded by the deflexion of an oscillograph. These experiments have shown that the radiation ejects particles from hydrogen, helium, lithium, beryllium, carbon, air, and argon. The particles ejected from hydrogen behave, as regards range and ionizing power, like protons with speeds up to about 3.2×10^9 cm per sec. The particles from the other elements have a large ionizing power, and appear to be in each case recoil atoms of the elements.
If we ascribe the ejection of the proton to a Compton recoil from a quantum of 52×10^6 electron volts, then the nitrogen recoil atom arising by a similar process should have an energy not greater than about 400000 volts, should produce not greater than about 10000 ions, and have a range in air at N.T.P. of about 1.3 mm. Actually, some of the recoil atoms in nitrogen produce at least 30000 ions. In collaboration with Dr. Feather, I have observed the recoil atoms in an expansion chamber, and their range, estimated visually, was sometimes as much as 3 mm at N.T.P.
These results, and others I have obtained in the course of the work, are very difficult to explain on the assumption that the radiation from beryllium is a quantum radiation, if energy and momentum are to be conserved in the collisions. The difficulties disappear, however, if it be assumed that the radiation consists of particles of mass 1 and charge 0, or neutron. From the energy relations of this process the velocity of the neutron emitted in the forward direction may well be about 3×10^9 cm per sec. The collisions of this neutron with the atoms through which it passes give rise to the recoil atoms, and the observed energies of the recoil atoms are in fair agreement with this view. Moreover, I have observed that the protons ejected from hydrogen by the radiation emitted in the opposite direction to that of the exciting α-particle appear to have a much smaller range than those ejected by the forward radiation. This again receives a simple explanation on the neutron hypothesis.
If it be supposed that the radiation consists of quanta, then the capture of the α-particle by the Be^9 nucleus will form a C^{12} nucleus. The mass defect of C^{12} is known with sufficient accuracy to show that the energy of the quantum emitted in this process cannot be greater than about 14×10^6 volts. It is difficult to make such a quantum responsible for the effects observed.
It is to be expected that many of the effects of a neutron in passing through matter should resemble those of a quantum of high energy, and it is not easy to reach the final decision between the two hypotheses. Up to the present, all the evidence is in favour of the neutron, while the quantum hypothesis can only be upheld if the conservation of energy and momentum be relinquished at some point.

Accelerator Radioactive source

Detectors Cloud chamber, Counters

Reactions
He Be $\rightarrow$ nucleus n
$n\ p \rightarrow p\ n$

Particles studied n

CHADWICK 1932B **Nobel prize**

■ Discovery of the neutron. Nobel prize to J. Chadwick awarded in 1935 "for his discovery of the neutron" ■

The Existence of a Neutron

J. Chadwick
Proc. Roy. Soc. **A136** (1932) 692;

Reprinted in
Great Experiments in Physics ed. M. H. Shamos, Henry Holt and Company, New York (1960).
Classical Scientific Papers. Physics. Facsimile reproductions of famous scientific papers. Mills and Boon Limited, London (1964) 246.

Excerpt The properties of the penetrating radiation emitted from beryllium (and boron) when bombarded by the α-particles of polonium have been examined. It is concluded that the radiation consists, not of quanta as hitherto supposed, but of neutrons, particles of mass 1, and charge 0. Evidence is given to show that the mass of the neutron is probably between 1.005 and 1.008. This suggests that the neutron consists of a proton and an electron in close combination, the binding energy being about 1 to 2 $\times$ 10^6 electron volts. From experiments on the passage of the neutrons through matter the frequency of their collisions with atomic nuclei and with electrons is discussed.

Accelerator Radioactive source

Detectors Cloud chamber, Counters

Related references

More (earlier) information
J. Chadwick, Nature **129** (1932) 312;
J. Chadwick, J. E. R. Constable, and E. C. Pollard, Proc. Roy. Soc. **A130** (1930) 463;
See also
W. Bothe and H. Becker, Z. Phys. **66** (1930) 289;
E. Rutherford, Proc. Roy. Soc. **A97** (1920) 374;
G. Gamow and Houtermans, Z. Phys. **52** (1928) 453;
I. Curie and F. Joliot, Compt. Ren. **194** (1932) 273;
I. Curie and F. Joliot, Compt. Ren. **194** (1932) 876;
I. Curie and F. Joliot, Compt. Ren. **194** (1932) 708;
D. L. Webster, Proc. Roy. Soc. **A136** (1932) 428;
I. Curie, Compt. Ren. **193** (1931) 1412;
J. K. Roberts, Proc. Roy. Soc. **A102** (1922) 72;
J. L. Glasson, Phil. Mag. **42** (1921) 596;
F. Rasetti, Naturw. **20** (1932) 252;
P. Auger, Compt. Ren. **194** (1932) 877;
Massey, Nature **129** (1932) 691;
Massey, Nature **129** (1932) 469;
Jenkins and McKellar, Phys. Rev. **39** (1932) 549;

Reactions
He Be $\to$ nucleus n
$n\,p \to p\,n$

Particles studied n

COCKCROFT 1932 **Nobel prize**

■ First evidence of nuclear reactions with accelerated protons. Cockcroft–Walton accelerator. Nobel prize to J. D. Cockcroft and E. T. S. Walton awarded in 1951 "for their pioneer work on the transmutation of atomic nuclei by artificially accelerated atomic particles" ■

Disintegration of Lithium by Swift Protons

J.D. Cockcroft, E.T.S. Walton
Nature **129** (1932) 649;

Excerpt In a previous letter to this journal we have described a method of producing a steady stream of swift protons of energies up to 600 kilovolts by the application of high potentials, and have described experiments to measure the range of travel of these protons outside the tube. We have employed the same method to examine the effect of the bombardment of a layer of lithium by a stream of these ions, the lithium being placed inside the tube at 45° to the beam. A mica window of stopping power of 2 cm of air was sealed on to the side of tube, and the existence of radiation from the lithium was investigated by the scintillation method outside the tube. The thickness of the mica window was much more than sufficient to prevent any scattered protons from escaping into the air even at the highest voltage used. On applying an accelerating potential of the order of 125 kilovolts, a number of bright scintillations were at once observed, the numbers increasing rapidly with voltage up to the highest voltage used, namely 400 kilovolts. At this point many hundreds of scintillations per minute were observed using a proton current of a few microampers. No scintillations were observed when the proton was cut off or when the lithium was shielded from it by a metal screen. The range of the particles was measured by introducing mica screens in the path of the rays, and found to be about eight centimetres in air and not to vary appreciably with voltage. To throw light on the nature of these particles, experiments were made with a Shimizu expansion chamber, when a number of tracks resembling those of α-particles were observed and of range agreeing closely with that determined by the scintillations. It is estimated that at 250 kilovolts, one particle is produced for approximately 10^9 protons. The brightness of the scintillations and the density of the tracks observed in the expansion chamber suggest that the particles are normal α-particles. If this point of view turns out to be correct, it seems not unlikely that the lithium isotope of mass 7 occasionally captures a proton and the resulting nucleus of mass 8 breaks into two α-particles, each of mass four and each with an energy about eight million electron volts. The evolution of energy on this view is about sixteen million electron volts per disintegration, agreeing approximately with that to be expected from the decrease of atomic mass involved in such a disintegration. Experiments are in progress to determine the effect on other elements when bombarded by a stream of swift protons and other particles.

Accelerator Cockcroft-Walton accelerator

Detectors Counters

Related references

More (earlier) information
J. D. Cockcroft and E. T. S. Walton, Nature **129** (1932) 242;

Reactions

$p\ ^7Li \rightarrow 2He$ 0.125,0.250 MeV(T_{lab})

IWANENKO 1932

■ Suggestion that the neutron is a constituent of the atomic nucleus ■

The Neutron Hypothesis

D.D. Iwanenko
Nature **129** (1932) 798;

Excerpt Dr. J. Chadwick's explanation of the mysterious beryllium radiation is very attractive to theoretical physicists. Is it not possible to admit that neutrons play also an important role in the building of nuclei, the nuclei electrons being *all* packed in α-particles or neutrons? The lack of a theory of nuclei makes, of course, this assumption rather uncertain, but perhaps it sounds not so improbable if we remember that the nuclei electrons profoundly change their properties when entering into the nuclei, and lose, so to say, their individuality, for example, their spin and magnetic moment.
The chief point of interest is how far the neutrons can be considered as elementary particles (something like protons and electrons). It is easy to calculate the number of α-particles, protons, and neutrons for a given nucleus, and form in this way an idea about the momentum of nucleus (assuming for the neutron a moment $\frac{1}{2}$). It is curious that berillium nuclei do not possess free protons but only α-particles and neutrons.

Particles studied nucleus

UREY 1932

■ Evidence for the deuteron ■

A Hydrogen Isotope of Mass 2

H.C. Urey, F.G. Brickwedde, G.M. Murphy
Phys. Rev. **39** (1932) 164;

Reprinted in

The Physical Review — the First Hundred Years, AIP Press (1995) CD-ROM.

Excerpt The proton-electron plot of known atomic nuclei shows some rather marked regularities among atoms of lower atomic number. Up to O^{16} a simple step-wise figure appears into which the nuclear species H^2, H^3 and H^5 could be fitted very nicely. Birge and Menzel have shown that the discrepancy between the chemical atomic weight of hydrogen and Aston's value by the mass spectrograph could be accounted for by the assumption of a hydrogen isotope of mass 2 present to the extent of 1 part in 4500 parts of hydrogen of mass 1. *(Extracted from the introductory part of the paper.)*

Accelerator Radioactive source

Detectors Spectrometer

Comments on experiment From the analysis of the atomic spectra in a discharge tube.

Related references

See also
J. Urey, Am. Chem. Soc. **53** (1931) 2872;
R. T. Birge and D. H. Menzel, Phys. Rev. **37** (1931) 1669;
Finkelnburg, Z. Phys. **52** (1928) 57;
Connelly, Proc. Phys. Soc. **42** (1929) 28;
W. H. Keesom and H. van Dijk, Proc. Acad. Sci. Amsterdam **34** (1931) 52;

Particles studied deuteron

FOCK 1932

■ Description of the space of states for quantum systems with an arbitrary (infinite) number of particles — Fock space ■

Konfigurationsraum und zweite Quantelung / Configuration Space and Second Quantization

V.A. Fock
Z. Phys. **75** (1932) 622;

Reprinted in

(translation into Russian) V. A. Fock, *Raboty po kvantovoj teorii polya*, Izdatel'stvo Leningradskogo Universiteta, 1957.

Abstract Es wird der Zusammenhang zwischen der Methode der gequantelten Wellenfunktionen und der Koordinatenraummethode untersucht. Die Operatoren der zweiten Quantelung werden in einer Folge von Konfigurationersräumen für 1, 2, ... usw. Teilchen dargestelt. Die gewonnene Darstellung ermöglicht eine einfache Ableitung der Hartreeschen Gleichungen mit Austausch.

Related references

See also
P. A. M. Dirac, Proc. Roy. Soc. **A114** (1927) 243;
P. Jordan and O. Klein, Z. Phys. **45** (1927) 751;
P. Jordan, Z. Phys. **45** (1927) 766;

P. Jordan, Z. Phys. **44** (1927) 473;
P. Jordan and E. P. Wigner, Z. Phys. **47** (1928) 631;

HEISENBERG 1932

■ Suggestion that atomic nuclei are composed of protons and neutrons. Theory of nuclear exchange forces. Invention of nucleon isotopic spin ■

Über den Bau der Atomkerne. I / On the Structure of Atomic Nuclei. I.

W. Heisenberg
Z. Phys. **77** (1932) 1;

Reprinted in
(translation into English) D. M. Brink. *Nuclear Forces*. Pergamon Press (1965) 214.

Abstract We discuss the implications of our assumption that atomic nuclei are made up of protons and neutrons and do not contain electrons.

1. The Hamiltonian function of the nucleus.
2. The relation of charge and mass and the special stability of the He-nucleus.

3-5. The stability of nuclei and radioactive decay series.

6. Discussion of the physical assumptions.

The experiments of Curie and Joliot and their interpretation by Chadwick have shown that in the structure of nuclei a new, fundamental component, the neutron, plays an important part. This suggests that atomic nuclei are composed of protons and neutrons, but do not contain any electrons. If this is correct, it means a very considerable simplification of nuclear theory. The fundamental difficulties of the theory of β-decay and the statistics of the nitrogen-nucleus can then be reduced to the question: In what way can a neutron decay into proton and electron and what statistics does it satisfy? The structure of nuclei, however, can be described, according to the laws of quantum mechanics, in terms of the interactions between protons and neutrons.

Related references

More (later) information
W. Heisenberg, Z. Phys. **78** (1932) 156;
W. Heisenberg, Z. Phys. **80** (1933) 587;

See also
D. D. Iwanenko, Nature **129** (1932) 798;
N. Bohr, *Faraday Lecture*, J. Chem. Soc., (1932) 349;

Analyse information from
I. Curie and F. Joliot, Compt. Ren. **194** (1932) 273;
I. Curie and F. Joliot, Compt. Ren. **194** (1932) 876;
J. Chadwick, Nature **129** (1932) 312;
H. C. Urey, F. G. Brickwedde, and G. M. Murphy, Phys. Rev. **39** (1932) 164;
H. C. Urey, F. G. Brickwedde, and G. M. Murphy, Phys. Rev. **40** (1932) 1;
H. C. Urey, F. G. Brickwedde, and G. M. Murphy, Phys. Rev. **40** (1932) 464;

Particles studied nucleus

BLACKETT 1932 — Nobel prize

■ Confirmation of fast penetrating charged particles in cosmic rays. Nobel prize to P. M. S. Blackett awarded in 1948 "for his development of the Wilson cloud chamber method and his discoveries therewith in the field of nuclear physics and cosmic radiation" ■

Photography of Penetrating Corpuscular Radiation

P.M.S. Blackett, G.P.S. Occhialini
Nature **130** (1932) 363;

Excerpt Since Skobelzyn (Skobelzyn, Z. Phys. **54** (1929) 686) discovered the tracks of particles of high energy on photographs taken with a Wilson cloud chamber, this method has been used by him and others in a number of investigations (Skobelzyn, Comptes rendus **194** (1932) 118) of the nature of penetrating radiation. Such work is laborious, since these tracks occur in only a small fraction of the total number of expansions made. We have found it possible to obtain good photographs of these high energy particles by arranging that the simultaneous discharge of two Geiger-Müller counters due to the passage of one of these particles shall operate the expansion itself. *(Extracted from the introductory part of the paper.)*

Accelerator COSMIC

Detectors Cloud chamber

Related references

Analyse information from
D. V. Skobelzyn, Z. Phys. **54** (1929) 686;
D. V. Skobelzyn, Compt. Ren. **194** (1932) 118;
R. A. Millikan and C. D. Anderson, Phys. Rev. **40** (1932) 325;
T. H. Johnson, W. Fleisher, and J. C. Street, Phys. Rev. **40** (1932) 1048;

UREY 1932B

■ Discovery of the deuteron ■

A Hydrogen Isotope of Mass 2 and its Concentration

H.C. Urey, F.G. Brickwedde, G.M. Murphy
Phys. Rev. **40** (1932) 1;

Reprinted in
The Physical Review — the First Hundred Years, AIP Press (1995) 75.

Excerpt In a recent paper Birge and Menzel pointed out that

if hydrogen had an isotope with mass number two present to the extent of one part in 4500, it would explain the discrepancy which exists between the atomic weights of hydrogen as determined chemically and with the mass spectrograph, when reduced to the same standard. Systematic arrangements of atomic nuclei require the existence of isotopes of hydrogen H^2 and H^3 and helium He^5 to give them a completed appearance when they are extrapolated to the limit of nuclei with small proton and electron numbers. An isotope of hydrogen with mass number two has been found present to the extent of one part in about 4000 in ordinary hydrogen; no evidence for H^3 was obtained. The vapor pressures of pure crystals containing only a single species of the isotopic molecules H^1H^1, H^1H^2, H^1H^3 were calculated after postulating:

1. that the rotational and vibrational energies of the molecules are the same in the solid and gaseous states;
2. that in the Debye theory of the solid state, the Θ's are inversely proportional to the square roots of the molecular masses;
3. that the free energy of the gas is given by the free energy equation of an ideal monatomic gas; and
4. that there is a zero point lattice energy equal to $(9/8)R\Theta$ per mole.

The calculated vapor pressures of the three isotopic molecules in equilibrium with their solids at the triple point for ordinary hydrogen are in the ratio $p_{11}:p_{12}:p_{13} = 1:0.37:0.29$. The isotope was concentrated in three samples of gas by evaporating large quantities of liquid hydrogen and collecting the gas which evaporated from the last two or three cc. Sample I was collected from the end portion of six liters evaporated at atmospheric pressure and samples II and III from four liters, each, evaporated at a pressure only a few millimeters above the triple point.
These samples and ordinary hydrogen were investigated for the visible, atomic Balmer series spectra of H^2 and H^3 from a hydrogen discharge tube run in the condition favorable for the enhancement of the atomic spectrum and for the repression of the molecular spectrum, using the second order of a 21 foot grating with a dispersion of 1.31Åper mm. When with ordinary hydrogen, the times of exposure required to just record the strong H^1 lines were increased 4000 times, very faint lines appeared at the calculated positions for the H^2 lines accompanying $H^1\beta$, $H^1\gamma$ and $H^1\delta$ on the short wavelength side and separated from them by between 1 and 2Å. These lines do not agree in wave–length with any known molecular lines and they do not appear on the plates taken with the discharge tube operating under conditions favorable for the production of a strong molecular spectrum and the repression of the atomic spectrum. With ordinary hydrogen they were so weak that it was difficult to be sure that they were not irregular ghosts of the strongly overexposed atomic lines. Samples II and III evaporated near the triple point show these lines and another near $H^1\alpha$ greatly enhanced relative to the H^1 lines over those with ordinary hydrogen showing that these new lines are not ghosts, and that a considerable increase in the concentration of the isotope had been effected. With sample I, evaporated at the boiling point, no appreciable increase in concentration was detected. The new lines agree in wave-length with those calculated for an H^2 isotope.
The H^2 lines are broad as is to be expected for close unresolved doublets, but they are not as broad and diffuse as the H^1 lines, probably due to the smaller Doppler broadening. The $H^2\alpha$ line is resolved into a close doublet with a separation that agrees within the accuracy of the measurements with the observed separation for $H^1\alpha$.
Relative abundances were estimated by comparing the times required to just record photographically the corresponding H^1 and H^2 lines. The relative abundance of H^2 and H^1 in natural hydrogen is estimated to be about 1:4000 and in the concentrated samples about five times as great.

Related references

See also

R. T. Birge and D. H. Menzel, Phys. Rev. **37** (1931) 1669;
W. H. Keesom and H. van Dijk, Proc. Acad. Sci. Amsterdam **34** (1931) 42;
F. W. Aston, Proc. Roy. Soc. **A115** (1927) 487;
G. Beck, Z. Phys. **47** (1928) 407;
H. A. Barton, Phys. Rev. **35** (1930) 408;
F. A. Lindemann and A. W. Aston, Phil. Mag. **37** (1919) 523;
F. A. Lindemann, Phil. Mag. **38** (1919) 173;

Particles studied deuteron

LAWRENCE 1932 — Nobel prize

■ Further development of cyclotrons. Nobel prize to E. O. Lawrence awarded in 1939 "for the invention and development of the cyclotron and for results obtained with it, especially with regard to artifical radioactive elements" ■

The Production of High Speed Light Ions without the Use of High Voltages

E.O. Lawrence, M.S. Livingston

Phys. Rev. **40** (1932) 19;

Reprinted in

The Physical Review — the First Hundred Years, AIP Press (1995) 256.

Abstract The study of the nucleus would be greatly facilitated by the development of sources of high speed ions, particularly protons and helium ions, having kinetic energies in excess of 1000000 volt-electrons; for it appears that such swiftly moving particles are best suited to the task of nuclear excitation. The straightforward method of accelerating ions through the requisite differences of potential presents great experimental difficulties associated with the high electric fields necessarily involved. The present paper reports the development of a method that avoids these difficulties by means of the multiple acceleration of ions to high speeds without the use of high voltages. The method is as follows: Semi-circular hollow plates, not unlike duants of an electrometer, are mounted with their diametral edges adjacent, in a vacuum and in a uniform magnetic field that is normal to the plane of the plates. High frequency oscillations are applied to the plate electrodes producing an oscillating electric field over the diametral region between them. As a result during one half cycle the electric field accelerates ions, formed in the diametral region, into the interior of one of the electrodes, where they are bent around on circular paths by the magnetic field and eventually emerge again into the region between the electrodes. The magnetic field is adjusted so that the time required for traversal of a semi-circular path within the electrodes equals a half period of the oscillations. In consequence, when the ions return to the region between the electrodes, the electric field will have reversed direction, and the ions thus receive second increments of velocity on passing into the other electrode. Because the path radii within the electrodes are proportional to the velocities of the ions, the time required for a traversal of a semi-circular path is independent of their velocities. Hence if the ions take exactly one half cycle on their first semi-circles, they do likewise on all succeeding ones and therefore spiral around in resonance with the oscillating field until they reach the periphery of the apparatus. Their final kinetic energies are as many times greater than that corresponding to the voltage applied to the electrodes as the number of times they have crossed from one electrode to the other. This method is primarily designed for the acceleration of light ions and in the present experiments particular attention has been given to the production of high speed protons because of their presumably unique utility for experimental investigations of the atomic nucleus. Using a magnet with pole faces 11 inches in diameter, a current of 10^{-9} ampere of 1220000 volt-protons has been produced in a tube to which the maximum applied voltage was only 4000 volts. There are two features of the developed experimental method which have contributed largely to its success. First there is the focussing action of the electric and magnetic fields which prevents serious loss of ions as they are accelerated. In consequence of this, the magnitudes of the high speed ion currents obtainable in this indirect manner are comparable with those conceivably obtainable by direct high voltage methods. Moreover, the focussing action results in the generation of very narrow beams of ions—less than 1 mm cross-sectional diameter—which are ideal for experimental studies of collision processes. Of hardly less importance is the second feature of the method which is the simple and highly effective means for the correction of the magnetic field along the paths of the ions. This makes it possible, indeed easy, to operate the tube effectively with a very high amplification factor (i.e., ratio of final equivalent voltage of accelerated ions to applied voltage). In consequence, this method in its present stage of development constitutes a highly reliable and experimentally convenient source of high speed ions requiring relatively modest laboratory equipment. Moreover, the present experiments indicate that this indirect method of multiple acceleration now makes practicable the production in the laboratory of protons having kinetic energies in excess of 10000000 volt-electrons. With this in mind, a magnet having pole faces 114 cm in diameter is being installed in our laboratory.

Related references

See also
H. Rose, Z. Phys. **64** (1930) 1;
W. Bothe and H. Becker, Z. Phys. **66** (1930) 289;
G. Beck, Naturw. **18** (1930) 896;
C. Y. Chao, Phys. Rev. **36** (1930) 1519;
W. Gurney and E. U. Condon, Phys. Rev. **33** (1929) 127;
G. Gamow, Z. Phys. **51** (1928) 204;
G. Gamow, Z. Phys. **52** (1928) 514;
J. Chadwick, J. E. R. Constable, and E. C. Pollard, Proc. Roy. Soc. **A130** (1930) 463;
C. C. Lauritsen and R. D. Bennett, Phys. Rev. **32** (1928) 850;
M. A. Tuve et al., Phys. Rev. **35** (1930) 66;
M. A. Tuve et al., Phys. Rev. **39** (1932) 384;
A. Brash and J. Lande, Z. Phys. **70** (1931) 10;
D. H. Sloan and E. O. Lawrence, Phys. Rev. **38** (1931) 2021;
E. O. Lawrence and W. E. Edlefsen, Science **72** (1930) 376;
E. O. Lawrence and M. S. Livingston, Phys. Rev. **37** (1931) 1707;

WIGNER 1932 — Nobel prize

■ Introduction of the time inversion transformation in quantum mechanics. Nobel prize to E. P. Wigner awarded in 1963 "for his contributions to the theory of the atomic nucleus and the elementary particles, particularly through the discovery and application of fundamental symmetry principles". Co-winners M. Goeppert-Mayer and J. H. D. Jensen "for their discoveries of nuclear shell structure" ■

Über die Operation der Zeitumkehr in der Quantenmechanik / On the Time Inversion Operation in Quantum Mechanics

E.P. Wigner
Nachr. Akad. Ges. Wiss. Göttingen 31, p.546 (1932);

VAN DE GRAAFF 1933

■ Invention of electrostatic accelerators. Further development ■

The Electrostatic Production of High Voltage for Nuclear Investigations

R.J. Van de Graaff, K.T. Compton, L.C. Van Atta
Phys. Rev. **43** (1933) 149;

Abstract The developments in nuclear physics emphasize the need of a new technique adapted to deliver enormous energies in concentrated form in order to penetrate or disrupt atomic nuclei. This may be achieved by a generator of current at very high voltage. Economy, freedom from the inherent defects of an impulsive, alternating or rippling source and the logic of simplicity point to an electrostatic generator as a suitable tool for this technique. Any such generator needs a conducting terminal, its insulating support and a means for conveying electricity to the terminal. These needs are naturally met by a hollow metal sphere supported on an insulator and charged by a belt conveying electricity from earth potential and depositing it within the interior of the sphere. Four models of such a generator are described, three being successive developments of generators operating in air, and designed respectively for 80000, 1500000 and 10000000 volts, and the fourth being an essentially similar generator operating in a highly evacuated tank. Methods are described for depositing electric charge on the belts either by external or by self-excitation. The upper limit to the attainable voltage is set by the breakdown strength of the insulating medium surrounding the sphere, and by its size. The upper limit to the current is set by the rate at which belt area enters the sphere, carrying a surface density of charge whose upper limit is that which causes a breakdown field in the surrounding medium, e.g., 30000 volts per cm if the medium is air at atmospheric pressure. The voltage and the current each vary as the breakdown strength of the surrounding medium and the power output as its square. Also the voltage, current and power vary respectively as the 1st, 2nd and 3rd powers of the linear dimensions.

Related references

More (earlier) information
R. J. Van de Graaff, Phys. Rev. **38** (1931) 1931;

See also
J. D. Cockcroft and E. T. S. Walton, Nature **129** (1932) 649;
J. D. Cockcroft and E. T. S. Walton, Proc. Roy. Soc. **A137** (1932) 229;
H. A. Barton, Mueller, and L. C. Van Atta, Phys. Rev. **42** (1932) 901A;
L. C. Van Atta, R. J. Van de Graaff, and H. A. Barton, Phys. Rev. **43** (1933) 158;

ANDERSON 1933 — Nobel prize

■ Discovery of the positron, the first antiparticle, predicted by Dirac. Nobel prize to C. D. Anderson awarded in 1936 "for his discovery of the positron". Co-winner V. F. Hess "for his discovery of cosmic rays" ■

The Positive Electron

C.D. Anderson
Phys. Rev. **43** (1933) 491;

Reprinted in
R. N. Cahn and G. Goldhaber, *The Experimental Foundations of Particle Physics*, Cambridge Univ. Press (1991) 10.
The Physical Review — the First Hundred Years, AIP Press (1995) 610.

Abstract Out of the group of 1300 photographs of cosmic-ray tracks in a vertical Wilson chamber 15 tracks were of positive particles which could not have a mass as great as that of the proton. From an examination of the energy-loss and ionization produced it is concluded that the charge is less than twice, and is probably exactly equal to, that of the proton. If these particle carry unit positive charge the curvatures and ionizations produced require the mass to be less than twenty times the electron mass. These particles will be called positrons. Because they occur in groups associated with other tracks it is concluded that they must be secondary particles ejected from atomic nuclei. *Editor..*

Accelerator COSMIC

Detectors Cloud chamber

Related references

More (earlier) information
C. D. Anderson, Science **76** (1932) 238;

See also
C. D. Anderson, Phys. Rev. **43** (1933) 381A;

Particles studied e^+

BLACKETT 1933 — Nobel prize

■ e^+ and shower confirmation. First indication for e^+e^- pair production. Nobel prize to P. M. S. Blackett awarded in 1948 "for his development of the Wilson cloud chamber method and his discoveries therewith in the field of nuclear physics and cosmic radiation" ■

Some Photographs of the Tracks of Penetrating Radiation

P.M.S. Blackett, G.P.S. Occhialini
Proc. Roy. Soc. **A139** (1933) 699;

Excerpt

1. A short description is given of a method of making particles of high energy take their own cloud photographs.
2. The most striking features of some 500 photographs taken by this method are described, and the nature of the showers of particles producing the complex tracks is discussed.
3. A consideration of the range, ionization, curvature and direction of the particles leads to a confirmation of the view put forward by Anderson that particles must exist with a positive charge but with a mass comparable with that of an electron rather with that of a proton.

4. The frequency of occurrence of the showers is discussed, and also their possible relation to the bursts of ionization observed by Hoffman, Steinke and others.
5. The origin of the positive and negative electrons in the showers is discussed, and the conclusion is reached that they are best considered as being created during a collision process.

Accelerator COSMIC

Detectors Cloud chamber, Counters

Related references

More (earlier) information
P. M. S. Blackett and G. P. S. Occhialini, Nature **130** (1932) 363;
See also
D. V. Skobelzyn, Z. Phys. **54** (1929) 686;
D. V. Skobelzyn, Compt. Ren. **194** (1932) 118;
D. V. Skobelzyn, Compt. Ren. **195** (1932) 315;
C. D. Anderson, Phys. Rev. **41** (1932) 405;
C. D. Anderson, Science **76** (1932) 238;
P. Auger and D. V. Skobelzyn, Compt. Ren. **189** (1929) 55;
B. Rossi, Phys.Zeitschr. **33** (1932) 304;
B. Rossi, Acad. Licei **15** (1932) 734;
Mott-Smith and Locher, Phys. Rev. **38** (1931) 1399;
Mott-Smith and Locher, Phys. Rev. **39** (1932) 883;
T. H. Johnson, W. Fleischer, and J. C. Street, Phys. Rev. **40** (1932) 1048;
T. H. Johnson and J. C. Street, Phys. Rev. **40** (1932) 635;
L. O. Grounfahl, Phys. Rev. **40** (1932) 635;
Williams and Terroux, Proc. Roy. Soc. **A126** (1930) 289;
H. A. Bethe, Z. Phys. **76** (1932) 293;
P. M. S. Blackett, Proc. Roy. Soc. **A135** (1932) 132;
R. A. Millikan and C. D. Anderson, Phys. Rev. **40** (1932) 325;
E. Steinke and Schindler, Z. Phys. **75** (1932) 115;
E. Steinke and Schindler, Naturw. **20** (1932) 491;
Messerschmidt, Z. Phys. **78** (1932) 668;
D. D. Iwanenko, Physik Zeits. Sow. **1** (1932) 820;
Mandel, Physik Zeits. Sow. **2** (1932) 286;
F. Perrin, Compt. Ren. **195** (1932) 236;
W. Heisenberg, Z. Phys. **77** (1932) 1;
W. Heisenberg, Annalen der Physik. Leipzig **13** (1932) 430;
W. Heisenberg, Naturw. **21** (1932) 365;
N. Bohr, *Report of Congress in Rome* (1931);
C. T. R. Wilson, Proc. Camb. Phil. Soc. **22** (1925) 534;
P. A. M. Dirac, Proc. Roy. Soc. **A126** (1930) 360;
P. A. M. Dirac, Proc. Roy. Soc. **A133** (1931) 60;
P. A. M. Dirac, Proc. Camb. Phil. Soc. **26** (1930) 361;
C. T. R. Wilson, Proc. Phys. Soc. **31** (1925) 32D;
H. Weil, *Gruppentheorie und Quantenmechanik* 2nd ed., (1931) 234;
E. Rutherford, J. Chadwick, and Ellis, *Radiation from Radioactive Substances* 443;
C. Y. Chao, Phys. Rev. **36** (1931) 1519;
L. Meitner and Hupfield, Naturw. **19** (1931) 775;
Gray and Tarrant, Proc. Roy. Soc. **A136** (1932) 662;
J. Chadwick, Proc. Roy. Soc. **A136** (1932) 692;

Reactions

unspec nucleus $\to$ shower X
unspec nucleus $\to$ e^- e^+ X

Particles studied e^+

MAJORANA 1933

■ Further development of the theory of nuclear forces ■

Nuclear Theory

E. Majorana
Z. Phys. **82** (1933) 137;

Abstract Following Heisenberg it is assumed that the stability of the nucleus can be ascribed to attractive forces overcoming the mutual repulsive forces of nuclear protons. They may be divided up into attractions between protons and neutrons and between neutrons and neutrons. It is assumed that normally the former of these is the more important. In heavy nuclei the repulsive forces tend to become large compared with the attractive forces. The author develops the mathematical theory of such a nucleus, a statistical method being used for this purpose. *(Science Abstracts, 1933, 2923. J. E. R. C.)*

Particles studied nucleus

FRISCH 1933 — Nobel prize

■ First measurement of the proton magnetic moment. Nobel prize to O. Stern awarded in 1943 “for his contributions to the development of the molecular ray method and his discovery of the magnetic moment of the proton” ■

Über die magnetische Ablenkung von Wasserstoffmolekülen und das magnetische Moment des Protons. I / Magnetic Deviation of Hydrogen Molecules and the Magnetic Moment of the Proton. I.

R. Frisch, O. Stern
Z. Phys. **85** (1933) 4;

Abstract Experimental details are given of the method for the determination of the magnetic moment of hydrogen positive rays by magnetic deviation. Measurements with parahydrogen give a magnetic moment of 1 nuclear magneton; ortho-hydrogen a value of 2 to 3. *(Science Abstracts 1933, 5075. W. R. A.)*

Accelerator Non accelerator experiment

Detectors Counters

Particles studied p

ESTERMAN 1933 — Nobel prize

■ Further measurement of the proton magnetic moment. Nobel prize to O. Stern awarded in 1943 "for his contributions to the development of the molecular ray method and his discovery of the magnetic momentof the proton" ■

Über die magnetische Ablenkung von Wasserstoffmolekülen und das magnetische Moment des Protons. II / Magnetic Deviation of Hydrogen Molecules and the Magnetic Moment of the Proton. II.

I. Esterman, O. Stern
Z. Phys. **85** (1933) 17;

Abstract Accurate investigation of the deviation of hydrogen molecular rays gives the value of 2.5 nuclear magnetons (±10%) for the magnetic moment of the proton, and the value 0.8-0.9 for the rotation moment of the hydrogen molecule. Deviation curves for ortho and para hydrogen are given. *(Science Abstracts 1933, 5076, W.R.A.)*

Accelerator Non accelerator experiment

Detectors Counters

Particles studied p

CHADWICK 1934

■ Evidence for deuteron photodisintegration. First precise measurements of the neutron mass ■

A Nuclear Photoeffect: Disintegration of the Diplon by γ Rays

J. Chadwick, M. Goldhaber
Nature **134** (1934) 237;

Excerpt By analogy with the excitation and ionization of atoms by light, one might expect that any complex nucleus should be excited or "ionized", that is, disintegrated, by γ-rays of suitable energy. Disintegration would be much easier to detect than excitation. The necessary condition to make disintegration possible is that the energy of the γ-ray must be greater than the binding energy of the emitted particle. The γ-rays of thorium C$''$ of $h\nu = 2.62 \times 10^6$ electron volts are the most energetic which are available in sufficient intensity, and therefore one might expect to produce disintegration with emission of a heavy particle, such as a neutron, proton, etc., only of those nuclei which have a small or negative mass defect; for example, D^2, Be^9, and the radioactive nuclei which emit α-particles. The emission of a positive or negative electron from a nucleus under the influence of γ-rays would be difficult to detect unless the resulting nucleus were radioactive.
Heavy hydrogen was chosen as the element first to be examined, because the diplon has a small mass defect and also because it is the simplest of all nuclear systems and its properties are as important in nuclear theory as the hydrogen atom is in atomic theory. *(Extracted from the introductory part of the paper.)*

Accelerator Radioactive source

Detectors Cloud chamber, Counters

Reactions
γ deuteron $\rightarrow$ $p\ n$ — 1.8,2.62 MeV(P_{lab})

Particles studied n

PAULI 1934

■ Explanation of the continuous electron energy spectrum in β decay. Proposal for the neutrino ■

Structure et proprietes des Noyaux Atomique / Structure and Properties of the Atomic Nucleus

W. Pauli
PARIS-34, P.324 (1934);

Reactions
nucleus $\rightarrow$ nucleus e^- $\overline{\nu}_e$

Particles studied ν

FERMI 1934

■ Field theory for beta decay. First estimation of the neutrino mass ■

Versuch einer Theorie der β–Strahlen.I / Towards the Theory of β-Rays

E. Fermi
Z. Phys. **88** (1934) 161; Nuovo Cim. **11** (1934) 1;

Reprinted in
Enrico Fermi. *Collected Papers* (Note e Memorie) v.I. Italy 1921 – 1938 (1962). The University of Chicago Press – Accademia Nazionale dei Lincei, Roma.
(translation into Russian) Enrico Fermi. *Nauchnye Trudy.* I. 1921-1938 Italiya. (pod red. B.Pontecorvo), Nauka, Moskva, 1971, str. 525.

Abstract Eine quantitative Theorie des β-Zerfalls wird vorgeschlagen, in welcher man die Existenz des Neutrinos annimmt, und die Emission der Elektronen und Neutrinos aus einem

Kern beim β-Zerfall mit einer ähnlichen Methode behandelt, wie die Emission eines Licht-quants aus einem angeregten Atom in der Strahlungstheorie. Formeln für die Lebensdauer und für die Form des emittierten kontinuierlichen β-Strahlenspektrums werden abgeleitet und mit der Erfahrung verglichen.

Abstract A quantitative theory of the emission of β-rays is explained. This admits the existence of the "neutrino," a new particle proposed by Pauli having no electric charge and mass of the order of magnitude of that of the electron or less. The emission of the electrons and of the neutrinos from a nucleus on the occasion of the β-disintegration is treated by a method similar to that followed in the theory of radiation to describe the emission of a quantum of light from an excited atom. Formulae are deduced for the mean life and for the form of the continuous spectrum of the β-rays and these are compared with experimental data. *(Science Abstracts. 1934. 1547. J.J.S.)*

Related references

More (earlier) information
E. Fermi, La Ricerca Scientifica, **IV-II** (1933) 491;
E. Fermi, Nuovo Cim. **11** (1934) 1;
E. Fermi, Rev. of Mod. Phys. **4** (1932) 87;

See also
W. Heisenberg, Z. Phys. **77** (1932) 1;
W. Heisenberg, Annalen der Physik. Leipzig **10** (1931) 888;
P. Jordan and O. Klein, Z. Phys. **45** (1927) 751;
F. Perrin, Compt. Ren. **197** (1933) 1625;
R. H. Hulme, Proc. Roy. Soc. **A133** (1931) 381;

Analyse information from
B. W. Sargent, Proc. Roy. Soc. **A139** (1933) 659;
B. W. Sargent, Proc. Camb. Phil. Soc. **28** (1932) 538;

Reactions

nucleus $\rightarrow$ nucleus $e^- \ \overline{\nu}_e$

$n \rightarrow p \ e^- \ \overline{\nu}_e$

Particles studied $\overline{\nu}_e$

ALTSHULER 1934

■ Prediction of the negative anomalous magnetic moment of the neutron ■

Magnetic Moment of the Neutron

C.A. Altshuler, I.E. Tamm
Doklady Akad. Nauk SSSR **8** (1934) 455;

Reprinted in

translated into russian Neutron. K pyatidesyatiletiyu otkrytiya. Moskva (1983) 254.

Abstract Using methods similar to those used by Lande, an estimate is made of the magnetic moment of the neutron by a study of the hyperfine structure of some atomic spectra. The elements studied are Cd, Sn, Ba, Hg, Pb, Sr, Ga, and Sb. It is deduced that the magnetic moment is – 1/2 nuclear magneton. *(Science Abstracts, 1934, 2265, J. E. R. C.)*

Related references

See also
A. Landé, Phys. Rev. **44** (1933) 1028;
B. Venkatasachar and T. Subbaraya, Z. Phys. **85** (1933) 264;
H. Schüler and H. Westmeyer, Naturw. **21** (1933) 674;
R. F. Bacher, Phys. Rev. **43** (1933) 1001;

Particles studied n

CERENKOV 1934 Nobel prize

■ First evidence for Vavilov-Čerenkov radiation. Nobel prize to P. A. Čerenkov awarded in 1958. Co-winners I. M. Frank and I. E. Tamm "for the discovery and the interpretation of the Čerenkov effect" ■

Visible Emission of Clean Liquids by Action of γ Radiation

P.A. Cerenkov
Doklady Akad. Nauk SSSR **2** (1934) 451;

Reprinted in

Selected Papers of Soviet Physicists, Usp. Fiz. Nauk **93** (1967) 385.

Abstract In due course of investigations of luminescence invoked in uranil salt solutions by γ-rays we have found that all used purified liquids (20 liquids), when irradiated by γ-rays, produce a weak visible light. This phenomenon is not connected with contaminations, as experiments with liquids of different purity showed.

Accelerator Radioactive source

Detectors Photon spectrometer

Related references

More (later) information
I. E. Tamm and I. M. Frank, Doklady Akad. Nauk SSSR **14** (1937) 107;

See also
P. A. Čerenkov, Bull. l'Acad. Sci. l'URSS **7** (1933) 919;
P. A. Čerenkov, Newcommer, Journ. Amer. Chem. Soc, **42** (1920) 1997;
S. I. Vavilov, Doklady Akad. Nauk SSSR **2** (1934) 457;

VAVILOV 1934

■ First attempt to understand Vavilov-Čerenkov radiation. Evidences that it is not a luminescence ■

On Possible Causes of Dark Blue γ Radiation in Liquids

S.I. Vavilov
Doklady Akad. Nauk SSSR **2** (1934) 457;

Reprinted in
Selected Papers of Soviet Physicists, Usp. Fiz. Nauk **93** (1967) 383.

Accelerator Radioactive source

Detectors Photon spectrometer

Related references

More (earlier) information
P. A. Čerenkov, Doklady Akad. Nauk SSSR **2** (1934) 451;

See also
S. I. Vavilov, Physik Zeits. Sow. **5** (1934) 369;
S. I. Vavilov and I. M. Frank, Z. Phys. **69** (1931) 100;
O. Klein and Y. Nishina, Z. Phys. **52** (1929) 853;

MURPHY 1934

■ Evidence that the spin of the deuteron is 1 ■

The Nuclear Spin of the Deuterium

G.M. Murphy, H. Johnston
Phys. Rev. **45** (1934) 761(A108);

Excerpt The relative intensities of 29 lines have been measured in the α-bands of the molecular spectrum of deuterium. The bands lie between 5939 and 6291Å and were photographed in the second order of a 21 foot grating by using a sample of gas containing more than 90 percent deuterium. The latter was kindly furnished by Professor H. C. Urey and the analysis of the bands by Professor G. H. Dieke. A tungsten filament lamp and a set of 8 neutral wire screens were used for putting density marks on each plate. They were put on the same plate with the molecular spectrum using the same exposure time for each. The usual type of photographic density curves were plotted and from them the intensities of the molecular lines determined. The values of g_s and g_a, the statistical weight due to nuclear spin for the symmetric and antisymmetric levels were obtained for each branch by a least squares solution of the equation, $\ln I/i = \ln Cg - BJ(J+l)/kT$ where I is the measured intensity of the line, i is the transition probability, g is the statistical weight for the nucleus, C is a constant and the other quantities have their usual significance. The results for Q and R branches of the (0,0), (1,1), (2,2) and (3,3) bands using 2 different plates gave 1.95 ± 0.06 and 2.02 ± 0.04 as the ratio of g_s/g_a. Since the symmetric levels are more intense, the nucleus obeys Bose-Einstein statistics and the nuclear spin is 1.

Accelerator Non accelerator experiment

Detectors Photon spectrometer

Reactions
$deut(atom)^* \rightarrow$ deut(atom) γ

Particles studied deuteron

ESTERMAN 1934

■ First measurements of magnetic moments of the deuteron and neutron ■

Magnetic Moment of the Deuton

I. Esterman, O. Stern
Phys. Rev. **45** (1934) 761(A109);

Excerpt Molecular rays of heavy hydrogen molecules were deflected in the Stern-Gerlach experiment. From these experiments we derived the magnetic moment of the deuton in the same way as the magnetic moment of the proton in our previous work with ordinary hydrogen. Our preliminary results suggest a value between 0.5 and 1.0 nuclear magnetons for the magnetic moment of the deuton. The result of our previous work is 2.5 nuclear magnetons for the proton. If we assume the deuton to be composed of one proton and one neutron, and if we simply assume addition of the magnetic moments, we would get a value between -1.5 and -2 nuclear magnetons for the magnetic moment of the neutron.

Accelerator Non accelerator experiment

Detectors Counters

Reactions
deut(atom) $\gamma^* \rightarrow$ deut(atom) γ^*

Particles studied deuteron, n

YUKAWA 1935 **Nobel prize**

■ Yukawa field theory of nuclear forces. Prediction of heavy quanta, the pion particles, as mediators of strong interactions. Nobel prize to H. Yukawa awarded in 1949 "for his prediction of the existence of mesons on the basis of theoretical work on nuclear forces" ■

On the Interaction of Elementary Particles

H. Yukawa
Proc.Phys.Math.Soc.Jap. **17** (1935) 48;

Reprinted in
R. T. Beyer. *Foundation of Nuclear Physics*, Dover, New York (1949).
D. M. Brink. *Nuclear Forces.* Pergamon Press (1965) 214.

Excerpt At the present stage of the quantum theory little is known about the nature of interactions of elementary particles. Heisenberg considered the interaction of "Platzwechsel" between the neutron and the proton to be of importance to the nuclear structure.
Recently Fermi treated the problem of β-disintegration on the hypothesis of "neutrino." According to this theory, the neutron and the proton can interact by emitting and absorbing a pair of neutrino and electron. Unfortunately the interaction energy calculated on such assumption is much too small to account for the binding energy of neutrons and protons in the nucleus. To remove this defect, it seems natural to modify the theory of Heisenberg and Fermi in the following way. The transition of a heavy particle from neutron state to proton state is not always accompanied by the emission of light particles, i.e. neutrino and an electron, but energy liberated by the transition is taken up sometimes by another particle, which in turn will be transformed from proton state into neutron state. If the probability of occurrence of the latter process is much large that of the former, the interaction between the neutron and the proton will be much large than in the case of Fermi, whereas the probability of emission of light particles is not affected essentially.
Now such interaction between the elementary particles can be described by means of a field of force, just as the interaction between the charged particles is described by the electromagnetic field. The above considerations shows that the interaction of heavy particles with this field is much larger than that of light particles with it.
In the quantum field theory this field should be accompanied by a new sort of quantum, just as the electromagnetic field is accompanied by the photon.
In this paper the possible nature of this field and the quantum accompanying it will be discussed briefly and also their bearing on the nuclear structure will be considered.
Besides such an exchange force and ordinary electric and magnetic forces there may be other forces between the elementary particles, but we disregard the latter for the moment.
Fuller account will be made in the next paper.

Related references

See also
W. Heisenberg, Z. Phys. **77** (1932) 1;
W. Heisenberg, Z. Phys. **78** (1932) 156;
W. Heisenberg, Z. Phys. **80** (1933) 587;
E. Fermi, Z. Phys. **88** (1934) 161;
I. E. Tamm, Nature **133** (1934) 981;
D. D. Iwanenko, Nature **133** (1934) 981;
Analyse information from
T. W. Bonner, Phys. Rev. **45** (1934) 606;

Particles studied $\pi^{\pm}$

BREIT 1936

■ Breit-Wigner form of the amplitude for resonance reactions ■

Capture of Slow Neutrons

G. Breit, E.P. Wigner
Phys. Rev. **49** (1936) 519;

Reprinted in
The Physical Review — the First Hundred Years, AIP Press (1995) 265.

Abstract Current theories of the large cross sections of slow neutrons are contradicted by frequent absence of strong scattering in good absorbers as well as the existence of resonance bands. These facts can be accounted for by supposing that in addition to the usual effect there exist transitions to virtual excitation states of the nucleus in which not only the captured neutron but, in addition to this, one of the particles of the original nucleus is in an excited state. Radiation damping due to the emission of γ-rays broadens the resonance and reduces scattering in comparison with absorption by a large factor. Interaction with the nucleus is most probable through the s part of the incident wave. The higher the resonance region, the smaller will be the absorption. For a resonance region at 50 volts the cross section at resonance may be as high as 10^{-19} cm^2 and 0.5×10^{-20} cm^2 at thermal energy. The estimated probability of having a nuclear level in the low energy region is sufficiently high to make the explanation reasonable. Temperature effects and absorption of filtered radiation point to the existence of bands which fit in with the present theory.

Related references

See also
H. A. Bethe, Phys. Rev. **47** (1935) 747;
E. Amaldi et al., Proc. Roy. Soc. **A149** (1935) 522;
F. Perrin and W. Elsasser, Compt. Ren. **200** (1935) 450;
G. Beck and Horsley, Phys. Rev. **47** (1935) 510;
P. B. Moon and R. R. Tillman, Nature **135** (1935) 904;
Bjerge and Westcott, Proc. Roy. Soc. **A150** (1935) 709;

Artsimovich, Kourchatov, Myssovskii, and Palibin, Compt. Ren. **200** (1935) 2159;
Ridenour and Yost, Phys. Rev. **48** (1935) 383;
L. Szilard, Nature **136** (1935) 950;
J. H. Van Vleck, Phys. Rev. **48** (1935) 367;
O. K. Rice, Phys. Rev. **33** (1929) 748;
O. K. Rice, Phys. Rev. **35** (1930) 1551;
O. K. Rice, Phys. Rev. **38** (1931) 1943;
M. Polanyi and E. P. Wigner, Z. Phys. **33** (1925) 429;
P. A. M. Dirac, Z. Phys. **44** (1927) 594;
V. F. Weisskopf and E. P. Wigner, Z. Phys. **63** (1930) 54;
G. Breit, Rev. of Mod. Phys. **5** (1933) 91;
G. Breit, Rev. of Mod. Phys. **5** (1933) 117;
G. Breit and I. S. Lowen, Phys. Rev. **46** (1934) 590;
G. Breit and F. L. Yost, Phys. Rev. **48** (1935) 203;
J. R. Dunning et al., Phys. Rev. **48** (1935) 265;
A. C. G. Mitchel and E. J. Murphy, Phys. Rev. **48** (1935) 653;
F. Rasetti, E. Segrè et al., Phys. Rev. **49** (1936) 103;
P. B. Moon and R. R. Tillman, Proc. Roy. Soc. **A153** (1936) 476;
O. R. Frisch, G. Hevesy, and H. A. C. McKay, Nature **137** (1936) 149;
L. R. Hafstad and M. A. Tuve, Phys. Rev. **48** (1935) 306;
P. Savel, Compt. Ren. **198** (1934) 1404;
J. Chadwick and M. Goldhaber, Proc. Roy. Soc. **A151** (1935) 479;

GAMOW 1936

■ Extension of Fermi interaction and Gamow-Teller selection rules ■

Selection Rules for the β-Disintegration

G. Gamow, E. Teller
Phys. Rev. **49** (1936) 895;

Reprinted in
The Physical Review — the First Hundred Years, AIP Press (1995) CD-ROM.

Abstract

§1. The selection rules for β-transformations are stated on the basis of the neutrino theory outlined by Fermi. If it is assumed that the spins of the heavy particles have a direct effect on the disintegration these rules are modified.

§2. It is shown that whereas the original selection rules of Fermi lead to difficulties if one tries to assign spins to the members of the thorium family the modified selection rules are in agreement with the available experimental evidence.

Related references

See also
E. Fermi, Z. Phys. **88** (1934) 161;
B. W. Sargent, Proc. Roy. Soc. **A139** (1933) 659;
E. J. Konopinski and G. E. Uhlenbeck, Phys. Rev. **48** (1935) 7;
W. Heisenberg, Z. Phys. **77** (1932) 1;
E. Majorana, Z. Phys. **82** (1933) 137;
G. Gamow, Proc. Roy. Soc. **A146** (1934) 217;
G. Gamow, Phys.Zeitschr. **35** (1934) 533;
C. D. Ellis and N. F. Mott, Proc. Roy. Soc. **A139** (1933) 369;
N. F. Mott and Taylor, Proc. Roy. Soc. **A138** (1932) 665;
Fisk and Taylor, Proc. Roy. Soc. (1934) 146;
G. Gamow and Rosenblum, Compt. Ren. **197** (1933) 1620;

BREIT 1936B

■ Hypothesis of charge independence of nuclear forces ■

Theory of Scattering of Protons by Protons

G. Breit, E.U. Condon, R.D. Present
Phys. Rev. **50** (1936) 825;

Reprinted in
The Physical Review — the First Hundred Years, AIP Press (1995) CD-ROM.

Excerpt The experiments of Tuve, Heydenburg and Hafstad and those of White are discussed by means of the standard theory of scattering in central fields. The theoretical formulas are presented in a form convenient for numerical computation and are supplemented by tables. *(Extracted from the introductory part of the paper.)*

Related references

Analyse information from
W. G. White, Phys. Rev. **47** (1935) 573;
W. H. Wells, Phys. Rev. **47** (1935) 591;
J. R. Dunning, G. B. Pergam, and G. A. Fink, Phys. Rev. **47** (1935) 970;
M. White, Phys. Rev. **49** (1936) 309;
M. A. Tuve, N. P. Heydenburg, and L. R. Hafstad, Phys. Rev. **49** (1936) 402;
M. A. Tuve, N. P. Heydenburg, and L. R. Hafstad, Phys. Rev. **50** (1936) 806;
Bjerge and Westcott, Proc. Roy. Soc. **A150** (1935) 709;
E. Fermi, La Ricerca Scientifica **I** (1936) 1;

Reactions

$n\,p \to p\,n$	<0.75 MeV(T_{lab})
$p\,p \to 2p$	<0.75 MeV(T_{lab})

CASSEN 1936

■ Proposal of the isotopic spin (for the nucleon) ■

On Nuclear Forces

B. Cassen, E.U. Condon
Phys. Rev. **50** (1936) 846;

Reprinted in
The Physical Review — the First Hundred Years, AIP Press (1995) CD-ROM.

Abstract The various types of exchange forces that are being used in current discussions of nuclear structure may all be simply expressed in terms of a formalism which attributes five coordinates to each "heavy" particle and applies the Pauli exclusion principle to all the particles in the system. The simplest assumption for the interaction law is that which implies equality of proton-proton and neutron-neutron forces and also equality with the proton-neutron forces of corresponding symmetry. This is in accord with the empirical knowledge of these interactions at present.

Related references

See also
W. Heisenberg, Z. Phys. **77** (1932) 1;
E. Fermi, Z. Phys. **88** (1934) 161;

TAMM 1937 — Nobel prize

■ Theoretical explanation of Čerenkov radiation phenomenon. Nobel prize to P. A. Čerenkov, I. M. Frank, and I. E. Tamm awarded in 1958 "for the discovery and the interpretation of the Čerenkov effect" ■

Coherent Radiation of Fast Electrons in a Medium

I.E. Tamm, I.M. Frank
Doklady Akad. Nauk SSSR **14** (1937) 107;

Reprinted in
Selected Papers of Soviet Physicists, Usp. Fiz. Nauk **93** (1967) 388.

Abstract The observations of Čerenkov on the visible radiation emitted during the passage of fast electrons through matter are explained qualitatively and quantitatively on the assumption that an electron moving with uniform velocity radiates light if its velocity is greater than that of light in the medium traversed. In this case, that is when $\beta n > 1$ where n is the refractive index of the medium, the total energy W radiated by an electron through the surface of a cylinder of length 1, whose axis is in the direction of motion of the electron, is given by W= $(e^2 l/c^2) \int w dw(1 - 1/\beta^2 n^2)$ where w is a characteristic frequency of a molecule of the medium. In the visible region about 10 photons are emitted by each fast electron. *(Science Abstract, 1937, 2082. F.C.C.)*

Related references

See also
P. A. Čerenkov, Doklady Akad. Nauk SSSR **2** (1934) 451;
P. A. Čerenkov, Doklady Akad. Nauk SSSR **3** (1936) 414;
S. I. Vavilov, Doklady Akad. Nauk SSSR **2** (1934) 457;
A. Sommerfeld, Gött. Nachrichten **99** (1904) 363, 201;

WIGNER 1937 — Nobel prize

■ Discovery of supermultiplet structure in nuclear spectroscopy. Nobel prize to E. P. Wigner awarded in 1963 "for his contributions to the theory of the atomic nucleus and the elementary particles, particularly through the discovery and application of fundamental symmetry principles". Co-winners M. Goeppert-Mayer and J. H. D. Jensen "for their discoveries of nuclear shell structure" ■

On the Consequences of the Symmetry of the Nuclear Hamiltonian on the Spectroscopy of Nuclei

E.P. Wigner
Phys. Rev. **51** (1937) 106;

Reprinted in
The Physical Review — the First Hundred Years, AIP Press (1995) CD-ROM.

Abstract The structure of the multiplets of nuclear terms is investigated, using as first approximation a Hamiltonian which does not involve the ordinary spin and corresponds to equal forces between all nuclear constituents, protons and neutrons. The multiplets turn out to have a rather complicated structure, instead of the S of atomic spectroscopy, one has three quantum numbers S, T, Y. The second approximation can either introduce spin forces (method 2), or else can discriminate between protons and neutrons (method 3). The last approximation discriminates between protons and neutrons in method 2 and takes the spin forces into account in method 3. The method 2 is worked out schematically and is shown to explain qualitatively the table of stable nuclei to about Mo.

Related references

See also
M. A. Tuve, N. P. Heydenburg, and L. R. Hafstad, Phys. Rev. **50** (1936) 806;
G. Breit, E. U. Condon, and R. D. Present, Phys. Rev. **50** (1936) 825;
W. Heisenberg, Z. Phys. **77** (1932) 1;

Particles studied nucleus

NEDDERMEYER 1937

■ First evidence for the muon ■

Note on the Nature of Cosmic-Ray Particles

S.H. Neddermeyer, C.D. Anderson
Phys. Rev. **51** (1937) 884;

Reprinted in
R. N. Cahn and G. Goldhaber, The Experimental Foundations of Particle Physics, Cambridge Univ. Press (1991) 32.

The Physical Review — the First Hundred Years, AIP Press (1995) CD-ROM.

Excerpt Measurements (Anderson and Neddermeyer, Phys. Rev. **50** (1936) 263) of the energy loss of the particles occurring in the cosmic–ray showers have shown that this loss is proportional to the incident energy and within the range of the measurements, up to about 400 MeV is in approximate agreement with values calculated theoretically for electrons by Bethe and Heitler. These measurements were taken using a thin plate of lead (0.35 cm), and the observed individual losses were found to vary from an amount below experimental detection up to the whole initial energy of the particle, with a mean fractional loss of about 0.5. If these measurements are correct it is evident that in a much thicker layer of heavy material multiple losses should become much more important, and the probability of observing a particle loss less than a large fraction of its initial energy should be very small. For the purpose of testing this inference and also for checking our previous measurements (Anderson and Neddermeyer, *Report of London Conference*, **V.1** (1934) 179) which had shown the presence of some particles less massive than protons but more penetrating than electrons obeying the Bethe-Heitler theory, we have taken about 6000 counter-tripped photographs with a 1 cm plate of platinum placed across the center of the cloud chamber. *(Extracted from the introductory part of the paper.)*

Accelerator COSMIC

Detectors Cloud chamber

Related references

More (earlier) information
C. D. Anderson and S. H. Neddermeyer, *Report of London Conference*, **1** (1934) 179, Phys. Rev. **50** (1936) 263;

See also
J. Crussard and L. Leprince-Ringuet, Compt. Ren. **204** (1937) 240;

Particles studied μ^+, μ^-

MAJORANA 1937

■ Symmetrical theory of electrons and positrons. Majorana neutrino theory ■

Symmetrical Theory of Electrons and Positrons

E. Majorana
Nuovo Cim. **14** (1937) 171;

Abstract A new method of quantization is proposed which allows Dirac's theory of the positron to be built up in such a form that there is complete symmetry between the positive and negative charge throughout the formalism of the theory, while in Dirac's original form this symmetry applied only to the results of the theory, which had to be obtained by using ambiguous mathematical operations such as subtraction of infinities. It is also claimed that the new method of quantization is capable of describing a neutral particle without states of negative kinetic energy and without introducing a "mirror image" like the positron. *(Science Abstracts. 1937, 4685. R. P.)*

STREET 1937

■ Confirmation of the muon existence ■

Penetrating Corpuscular Component of the Cosmic Radiation

J.C. Street, E.G. Stevenson
Phys. Rev. **51** (1937) 1005;

Excerpt The confirmation of the validity of the Heitler theory of radiation and pair formation both by energy loss measurements (Anderson and Neddermeyer, Phys. Rev. **50** (1936) 263) and by the nature of shower formation (Carlson and Oppenheimer, Phys. Rev. **51** (1937) 220) leaves no explanation for the penetrating corpuscular rays except to assume that they are protons. However, the evidence against protons is strong. (C. G. and D. D. Montgomery, Phys. Rev. **51** (1937) 220) To investigate the penetrating rays a vertical column was set up as follows: a counter, 10 cm Pb, a second counter, a cloud chamber in a magnetic field, a third and fourth counter, 3 cm Pb, and finally a cloud chamber containing 3 separated lead plates 1 cm thick. The counter telescope selected particles directed toward the visible region of the lower chamber where their absorption and shower production was observed. The distribution of the particles with respect to range and Hρ values was as follows:

$H\rho \times 10^{-6}$	% of total tracks	Fraction with range > 3 cm Pb	Fraction with range > 6 cm Pb
>5	49	1	1
2.5 to 5	19	.9	.7
1.5 to 2.5	23	.9	.7
0.7 to 1.5	9	.7	.3

Only a single shower has been observed in 500 traversals of 1 cm of lead. The ionization density of protons with Hρ < 2.5×10^6 distinguishes them, and two were observed. From the data in the table it is evident that the penetrating particles cannot be described as electrons obeying the Heitler theory nor can an appreciable fraction be protons.

Accelerator COSMIC

Detectors Cloud chamber

Related references

See also
C. D. Anderson and S. H. Neddermeyer, Phys. Rev. **50** (1936) 263;
J. F. Carlson and J. R. Oppenheimer, Phys. Rev. **51** (1937) 220;

C. G. Montgomery and D. D. Montgomery, Phys. Rev. **50** (1936) 975;

Particles studied μ^+, μ^-

BLOCH 1937

■ Treatment of infrared divergence ■

Note on the Radiation Field of the Electron

F. Bloch, A. Nordsieck
Phys. Rev. **52** (1937) 54;

Reprinted in
The Physical Review — the First Hundred Years, AIP Press (1995) CD-ROM.

Abstract Previous methods of treating radiative corrections in nonstationary processes such as the scattering of an electron in an atomic field or the emission of a β-ray, by an expansion in powers of $e^2/\hbar c$, are defective in that they predict infinite low frequency corrections to the transition probabilities. This difficulty can be avoided by a method developed here which is based on the alternative assumption that $e^2\omega/mc^3$, $\hbar\omega/mc^2$ and $\hbar\omega/c\delta p$ (ω=angular frequency of radiation, δp = change in momentum of electron) are small compared to unity. In contrast to the expansion in powers of $e^2/\hbar c$, this permits the transition to the classical limit $\hbar = 0$. External perturbations on the electron are treated in the Born approximation. It is shown that for frequencies such that the above three parameters are negligible the quantum mechanical calculation yields just the directly reinterpreted results of the classical formulae, namely that the total probability of a given change in the motion of the electron is unaffected by the interaction with radiation, and that the mean number of emitted quanta is infinite in such a way that the mean radiated energy is equal to the energy radiated classically in the corresponding trajectory.

Related references

See also
N. F. Mott, Proc. Camb. Phil. Soc. **27** (1931) 255;
A. Sommerfeld, Annalen der Physik. Leipzig **11** (1931) 257;
H. Bethe and W. Heitler, Proc. Roy. Soc. **A146** (1934) 83;
J. K. Knipp and G. E. Uhlenbeck, Physica **3** (1936) 425;
F. Bloch, Phys. Rev. **50** (1936) 272;
W. Pauli, Handbuch der Phys. **24/1** (1933) 266;

CERENKOV 1937 — Nobel prize

■ Confirmation of the Frank-Tamm theory of the Vavilov-Čerenkov effect. Nobel prize to P. A. Čerenkov awarded in 1958. Co-winners I. M. Frank and I. E. Tamm "for the discovery and the interpretation of the Čerenkov effect" ■

Visible Radiation Produced by Electron Moving in a Medium with Velocities Exceeding that of Light

P.A. Cerenkov
Phys. Rev. **52** (1937) 378;

Excerpt In a note published in 1934 as well as in the subsequent publications the present author reported his discovery of feeble visible radiation emitted by pure liquids under the action of fast electrons (β-particles of radioactive elements or Compton electrons liberated in liquids in the process of scattering of γ-rays). This radiation was a novel phenomenon, which could not be identified with any of the kinds of luminescence then known, as the theory of luminescence failed to account for a number of unusual properties (insensitiveness to the action of quenching agents, anomalous polarization, marked spacial asymmetry, etc.) exhibited by the radiation in question. In 1934 the earliest results obtained in the experiments with γ-rays led S. I. Vavilov to interpret the radiation observed as a result of the retardation of the Compton electrons liberated in liquids by γ-rays. A comprehensive quantitative theory subsequently advanced by I. M. Frank and I. E. Tamm afforded an exhaustive interpretation of all the peculiarities of the new phenomenon, including its most remarkable characteristic — the asymmetry.
According to their theory, an electron moving in a medium of refractive index n with a velocity exceeding that of light in the same medium ($\beta > 1/n$) is liable to emit light which must be propagated in a direction forming an angle θ with the path of the electron, this angle being determined by the equation:

$$\cos\theta = 1/\beta n \tag{1}$$

where β is the ratio of the electron velocity to that of light in vacuum.
A successful experimental verification of formula (1) was only performed with water for which, at the moment of publication of the above theory, data were already available which had been obtained by visual observations by the method of quenching.
We recently performed additional experiments in which the intensity of radiation was recorded photographically, the records being taken simultaneously for all the angles θ lying in a plane passing through the primary electron beam.
All the results obtained are in good agreement with I. M. Frank and I. E. Tamm's theory of the coherent radiation of electrons

moving in a medium. *(Extracted from the introductory part of the paper.)*

Accelerator Radioactive source

Detectors Photon spectrometer

Related references

See also
P. A. Čerenkov, Doklady Akad. Nauk SSSR **2** (1934) 451;
P. A. Čerenkov, Doklady Akad. Nauk SSSR **12** (1936) 413;
P. A. Čerenkov, Doklady Akad. Nauk SSSR **14** (1937) 102;
P. A. Čerenkov, Doklady Akad. Nauk SSSR **14** (1937) 105;
P. A. Čerenkov, Bull. of Ac. of Sci. USSR Phys. Ser. **7** (1933) 919;
S. I. Vavilov, Doklady Akad. Nauk SSSR **2** (1934) 457;
I. M. Frank and I. E. Tamm, Doklady Akad. Nauk SSSR **14** (1937) 109;
E. Brumberg and S. I. Vavilov, Doklady Akad. Nauk SSSR **3** (1934) 405;

STREET 1937B

■ Muon existence confirmation ■

New Evidence for the Existence of a Particle of Mass Intermediate Between the Proton and Electron

J.C. Street, E.G. Stevenson
Phys. Rev. **52** (1937) 1003;

Reprinted in

R. N. Cahn and G. Goldhaber, *The Experimental Foundations of Particle Physics*, Cambridge Univ. Press (1991) 35.
The Physical Review — the First Hundred Years, AIP Press (1995) CD-ROM.

Excerpt Anderson and Neddermeyer have shown that, for energies up to 300 and 400 MeV, the cosmic-ray shower particles have energy losses in lead plates corresponding to those predicted by theory for electrons. Recent studies of range and energy loss indicate that the singly occurring cosmic-ray corpuscles, even in the energy range below 400 MeV, are more penetrating than shower particles of corresponding magnetic deflection. Thus the natural assumption have been expressed: the shower particles are electrons, the theory describing their energy losses is satisfactory, and the singly occurring particles are not electrons. The experiments cited above have shown from consideration of the specific ionization that the penetrating rays are not protons. The suggestion has been made that they are particles of electronic charge, and of mass intermediate between those of the proton and electron. If this is true, it should be possible to distinguish clearly such a particle from an electron or proton by observing its track density and magnetic deflection near the end of its range, although it is to be expected that the fraction of the total range in which the distinction can be made is very small. *(Extracted from the introductory part of the paper.)*

Accelerator COSMIC

Detectors Cloud chamber

Related references

See also
C. D. Anderson and S. H. Neddermeyer, Phys. Rev. **50** (1936) 263;
J. C. Street and E. G. Stevenson, Phys. Rev. **51** (1937) 1005;
S. H. Neddermeyer and C. D. Anderson, Phys. Rev. **51** (1937) 885;

Particles studied $\mu^{\pm}, \mu^{+}, \mu^{-}$

WHEELER 1937

■ First proposal for the S matrix formalism ■

On the Mathematical Description of Light Nuclei by the Method of Resonating Group Structure

J.A. Wheeler
Phys. Rev. **52** (1937) 1107;

Abstract The wave function for the composite nucleus is written as a properly antisymmetrized combination of partial wave functions, corresponding to various possible ways of distributing the neutrons and protons into various groups, such as alpha-particles, *di*-neutrons, etc. The dependence of the total wave function on the intergroup separations is determined by the variation principle. The analysis is carried out in detail for the case that the configurations considered contain only two groups. Integral equations are derived for the functions of separation. The associated Fredholm determinant completely determines the stable energy values of the system (Eq. (33)), Eq. (48) connects the asymptotic behavior of an arbitrary particular solution with that of solutions possessing a standard asymptotic form. With its help, the Fredholm determinant also determines all scattering and disintegration cross sections (Eqs. (50) · (54) and (57)), without the necessity of actually obtaining the intergroup wave functions. The expressions (43) and (60) obtained for the cross sections, taking account of spin effects, have general validity. Details of the application of the method of resonating group structure to actual problems are discussed.

Related references

See also
E. Schrödinger, Annalen der Physik. Leipzig **79** (1926) 361;
R. Courant and D. Hilbert, *Modern der Mathematischen Physik I* (Berlin, 1931) 159;
J. A. Wheeler, Phys. Rev. **51** (1937) 683;

NISHINA 1937

■ Confirmation of the existence of the muon ■

On the Nature of Cosmic-Ray Particles

Y. Nishina, M. Takeuchi, T. Ichimiya
Phys. Rev. **52** (1937) 1198;

Excerpt Various authors have taken the view that cosmic-ray particles consist of two or more kinds of corpuscules. According to Compton and Bethe, and Auger the soft component near sea level is thus composed of electrons and the penetrating one of protons. Assuming the theory of showers by Bhabha and Heitler to be correct, we ought to be able to distinguish cosmic-ray electrons from protons, if they exist at all, by observing whether or not the particles suffer a large loss of energy and often produce showers on colliding with a lead plate of a suitable thickness.
We carried out such experiments with a lead bar 1.5 cm thick mounted in the middle of a Wilson chamber 40 cm in diameter, which is placed in a magnetic field of about 17,000 oersteds. The operation of the chamber is actuated by the coincidence of two Geiger-Müller tube counters mounted above the chamber, the distance between the counters being about 50 cm. *(Extracted from the introductory part of the paper.)*

Accelerator COSMIC

Detectors Cloud chamber, Counters

Related references

See also
A. H. Compton and H. A. Bethe, Nature **134** (1934) 734;
P. Auger, Jour. de Phys. **6** (1935) 226;
C. D. Anderson and S. H. Neddermeyer, Phys. Rev. **50** (1936) 268;
C. D. Anderson and S. H. Neddermeyer, Phys. Rev. **51** (1937) 884;
J. Clay, Physica **3** (1936) 338;
L. Leprince-Ringuet, Jour. de Phys. **7** (1936) 70;
J. Crussard and L. Leprince-Ringuet, Compt. Ren. **204** (1937) 240;
J. Crussard and L. Leprince-Ringuet, Jour. de Phys. **8** (1937) 215;
H. J. Bhabha and W. Heitler, Proc. Roy. Soc. **A159** (1937) 432;
J. F. Carlson and J. R. Oppenheimer, Phys. Rev. **51** (1937) 220;
J. C. Street et al., Bull.Am.Phys.Soc. **12** (1937) 13;

Particles studied μ^+, μ^-

SCHWINGER 1937

■ Establishment of the neutron spin 1/2 ■

On the Spin of the Neutron

J. Schwinger
Phys. Rev. **52** (1937) 1250;

Reprinted in
The Physical Review — the First Hundred Years, AIP Press (1995) CD-ROM.

Excerpt The intrinsic angular momenta of the proton and the deuteron imply a neutron spin (in units of $\hbar$) of either $\frac{1}{2}$ or $\frac{3}{2}$. The usual assumption of $\frac{1}{2}$ for the neutron spin is based entirely upon arguments of simplicity, since either of these two possible values is consistent with data on nuclear spins. In view of the importance of the neutron spin in nuclear theory, it would be desirable to determine this quantity by direct experiment. It has recently been shown that experiments on the scattering of neutrons by ortho- and parahydrogen would enable one to obtain information about the spin dependence and the range of the neutron-proton interaction. It is the purpose of this note to point out that such experiments also permit the determination of the neutron spin... The experiments of both Dunning and Stern and their collaborators show that the scattering cross section of ortho-H_2 at liquid-air neutron temperatures (T= 100°K) is much larger than the corresponding para-H_2 cross section. It has already been pointed out that this result is in agreement with the theoretical expectations for a virtual singlet state. Assuming a neutron spin of $\frac{3}{2}\hbar$ the theoretical value of the ratio $\sigma_{ortho}/\sigma_{para}$, at an energy of $3kT/2 = 0.012$ eV, is 3.11 for a virtual quintet state and 1.09 for a real quintet state. In either case, the two cross sections are quite comparable in magnitude, in contradiction with experiment. On the basis of these experiments the conclusion must be drawn that the intrinsic angular momentum of the neutron is, in reality, $\frac{1}{2}\hbar$. *(Extracted from the introductory part of the paper.)*

Related references

See also
J. Schwinger and E. Teller, Phys. Rev. **52** (1937) 286;
E. Fermi, La Ricerca Scientifica **VII-II** (1936) 13;
Analyse information from
J. Halpern, I. Esterman, O. C. Simpson, and O. Stern, Phys. Rev. **52** (1937) 142;

Particles studied n

ALVAREZ 1938

■ First evidence for the capture of atomic electrons by the weak interaction; K-capture ■

The Capture of Orbital Electrons by Nuclei

L.W. Alvarez
Phys. Rev. **54** (1938) 486;

Reprinted in
The Physical Review — the First Hundred Years, AIP Press (1995) CD-ROM.

Abstract The simple theory of electron capture is outlined and three methods for its detection are suggested. The first experimental evidence for the process (in activated titanium) is described. A rigorous experimental proof of the hypothesis is given for the case of ^{67}Ga. A summary of several isotopes whose properties are best explained on this hypothesis is appended. The properties of ^{67}Ga are described in considerable detail, and include the first evidence for internal conversion in artificially radioactive atoms.

Accelerator Radioactive source

Detectors Counters

Related references

More (earlier) information
L. W. Alvarez, Phys. Rev. **53** (1938) 606;
L. W. Alvarez, Phys. Rev. **52** (1937) 134;
L. W. Alvarez, Phys. Rev. **53** (1938) 213;
L. W. Alvarez, Phys. Rev. **53** (1938) 326;

See also
H. Yukawa and S. Sakata, Proc.Phys.Math.Soc.Jap. **17** (1935) 467;
H. Yukawa and S. Sakata, Proc.Phys.Math.Soc.Jap. **18** (1936) 128;
Mercier, Nature **139** (1937) 797;
Hoyle, Nature **140** (1937) 235;
G. E. Uhlenbeck and Kuiper, Physica **4** (1937) 601;
Roberts and N. P. Heydenburg, Phys. Rev. **53** (1938) 374;
Laslett, Phys. Rev. **52** (1937) 529;
W. E. Lamb, Phys. Rev. **50** (1936) 388;
Jacobsen, Nature **139** (1937) 879;
Walke, Phys. Rev. **51** (1937) 1011;
R. C. Williams and E. Pickup, Nature **141** (1938) 199;

Reactions
^{67}Ga(atom) $\rightarrow$ ^{67}Zn(atom) ν_e γ

STÜCKELBERG 1938

■ Invention of the baryonic quantum number conservation law ■

Die Wechselwirkunskräfte in der Elektrodynamik und in der Feldtheorie der Kernkräfte (Teil II und III) / Interaction in Electrodynamics and in the Field Theory of Nuclear Forces (Part II and III)

E.C.G. Stückelberg
Helv. Phys. Acta **11** (1938) 299;

Excerpt "Apart from ... the conservation law of electricity there exists evidently a further conservation law: No transmutations of heavy particles (neutron and proton) into light particles (electron and neutrino) have yet been observed in any transformation of matter. We shall therefore demand a conservation law of heavy charge." *(Quotation is taken from* A. Pais, *Inward Bound of the Matter and Forces in the Particle World, Clarendon Press · Oxford. Oxford University Press · New York, 1988)*

Related references

More (earlier) information
H. Weyl, Z. Phys. **56** (1929) 330;

More (later) information
E. P. Wigner, Proc. Am. Phil. Soc. **93**, 521 (1949);

LAWRENCE 1939

■ First cyclotron for medical applications ■

Initial Performance of the 60-Inch Cyclotron of the William H. Crocker Radiation Laboratory, University of California

E.O. Lawrence et al.
Phys. Rev. **56** (1939) 124;

Abstract During the past few weeks, we have been engaged in the adjustments of the 60-inch cyclotron of the William H. Crocker Radiation Laboratory, and at this time we wish to report its initial performance.
As is always fruitful in first turning on a cyclotron, we looked for resonance in the first instance by placing a Geiger counter nearby. With hydrogen, proton resonance was found close to the expected value of magnet current, and we proceeded to build up the intensity of the resonance effect by adjusting the magnetic field with shims. Following this adjustment, the beam to the target was observed and further shimming yielded 25 microamperes at eight-million volt protons. Probe measurements indicated about 100 microamperes circulating within the dees.
Next the hydrogen was replaced with deuterium at a pressure about one tenth that normally used in our 37-inch cyclotron, and the procedure of adjustment was repeated. Again resonance of deuterons was first observed with a Geiger counter and shims were used to build up the effect. *(Extracted from the introductory part of the paper.)*

KELLOGG 1939 Nobel prize

■ High-precision measurements of proton and deuteron magnetic moments. Nobel prize to I. I. Rabi awarded in 1944 "for his resonance method for recording the magnetic properties of atomic nuclei" ■

The Magnetic Moment of the Proton and the Deuteron. The Radiofrequency Spectrum of ^2H in Various Magnetic Fields

J.M.B. Kellogg, I.I. Rabi, N.F. Ramsey, J.R. Zacharias
Phys. Rev. **56** (1939) 728;

Reprinted in
The Physical Review — the First Hundred Years, AIP Press (1995) CD-ROM.

Abstract The molecular-beam magnetic-resonance method for measuring nuclear magnetic moments has been applied to the proton and the deuteron. In this method the nuclear moment is obtained by observing the Larmor frequency of precession

($\nu = \mu H/hI$) in a uniform magnetic field. For this purpose HD and D_2 molecules are most suitable because they are largely in the state of zero rotational momentum. Very sharp resonance minima are observed which makes it possible to show that the observed values of ν/H are independent of H, and to make a very accurate determination of the ratio μ_P/μ_D. With molecules of orthohydrogen in the first rotational state a radiofrequency spectrum of six resonance minima was obtained. This spectrum when analyzed yields a set of nine energy levels from which are obtained

1. the proton moment from its Larmor precession frequency;
2. the proton moment from the magnitude of the dipole interaction between the two proton magnetic moments (the directly measured quantity is μ_P/r^3); and
3. the value of the spin orbit interaction constant of the proton moment with the rotation of the molecule or the magnetic field H' produced by the rotation of the molecule at the position of the nucleus.

The numerical results are $\mu_P = 2.785 \pm 0.02$ nuclear magnetons, $\mu_D = 0.855 \pm 0.006$ nuclear magneton; $(\mu_P/\mu_D) = 3.257 \pm 0.001$; $H' = 27.2 \pm 0.3$ gauss; $\mu_P/r^3 = 34.1 \pm 0.3$ gauss which gives $\mu_P = 2.785 \pm 0.03$ nuclear magnetons. To within experimental error there is no disagreement of the results of these direct measurements with those from atomic beam measurements of the h.f.s. $\Delta\nu$ of the ground states of H and D.

Accelerator Non accelerator experiment

Detectors Nonelectronic detectors

Related references

More (earlier) information
I. I. Rabi, S. Millman, P. Kusch, and J. R. Zacharias, Phys. Rev. **55** (1939) 526;
J. M. B. Kellogg, I. I. Rabi, and J. R. Zacharias, Phys. Rev. **50** (1936) 472;

See also
I. Esterman and O. Stern, Z. Phys. **85** (1933) 17;
B. G. Lasarew and Shubnikow, Physik Zeits. Sowj. **11** (1937) 445;
I. Esterman, O. C. Simpson, and O. Stern, Phys. Rev. **52** (1937) 535;
J. H. Manley, Phys. Rev. **49** (1936) 921;
S. Millman, I. I. Rabi, and J. R. Zacharias, Phys. Rev. **53** (1938) 384;
I. I. Rabi, Phys. Rev. **51** (1937) 652;

Particles studied p, deuteron

ROSSI 1939

■ First evidence of muon decay and first estimation of its lifetime ■

The Disintegration of Mesotrons

B. Rossi, H. Van Norman Hilbery, J.B. Hoag
Phys. Rev. **56** (1939) 837;

Excerpt In order to test the hypothesis of the spontaneous decay of mesotrons we have compared the absorption of the mesotron component of cosmic radiation in air and in carbon.
The mesotrons were detected by the coincidences of three Geiger-Müller tubes arranged in a vertical plane. The counters were shielded with 10 cm of lead on each side to prevent coincidences from the air showers. Also, 12.7 cm of lead was placed between the counters in order to cut off the soft component. *(Extracted from the introductory part of the paper.)*

Accelerator COSMIC

Detectors Counters

Reactions
$\mu^{\pm} \to$ charged (neutrals)

Particles studied μ^-, μ^+

DUNNINGTON 1939

■ Further development of the metrology of fundamental physics parameters ■

A Re-evaluation of the Atomic Constants

F.G. Dunnington
Phys. Rev. **55** (1939) 683;

Excerpt The experimental work on all significant determinations of the atomic constants e, m and h has been reexamined and the results recalculated with two changes:

a) all assumptions as to values of combinations of atomic constants have been eliminated so that the results given represent what the experiments actually yield and
b) a consistent set of auxiliary constants has been used throughout.

A Birge-Bond diagram will be given to present the results graphically and illustrate clearly the discrepancy. A least squares solution of these results has been made *without* using the Rydberg formula. The results are:

$$e = (4.8025 \pm 0.0007) \times 10^{-10} \text{e.s.u.},$$
$$m = (9.1073 \pm 0.0024) \times 10^{-28} \text{g},$$
$$h = (6.6133 \pm 0.0034) \times 10^{-27} \text{erg} \cdot \text{sec.},$$
$$e/m_0 = (1.7590 \pm 0.0004) \times 10^{7} \text{e.m.u.},$$
$$h/e = (1.3771 \pm 0.0007) \times 10^{-17} \text{e.s.u.}$$

As to where the discrepancy originates, this solution together with an analysis of each type of measurement as to the fundamental laws involved indicates that either:

the existing body of experimental results is substantially correct and the Rydberg formula is in error, or the Rydberg formula is correct and all the measurements of h/e and the radiation constants are in error, the error being in some cases presumably experimental, and in other cases in the theory involved.

Related references

See also
W. N. Bond, Proc. Roy. Soc. **49** (1937) 205;
P. J. Rigden, Phil. Mag. **25** (1938) 961;
G. B. Banerjea and B. Pattanaik, Nature **141** (1938) 1016;
E. L. Harrington, Phys. Rev. **8** (1916) 738;
R. A. Millikan, Phil. Mag. **34** (1917) 1;
E. Backlin and H. Flemberg, Nature **137** (1936) 655;
E. O. Lawrence, Phys. Rev. **28** (1926) 947;
L. C. Van Atta, Phys. Rev. **38** (1931) 876;
L. C. Van Atta, Phys. Rev. **39** (1932) 1012;
A. E. Shaw, Phys. Rev. **44** (1933) 1009;
J. E. Roberts and R. Whiddington, Phil. Mag. **12** (1931) 962;
R. Whiddington and E. G. Woodroofe, Phil. Mag. **20** (1935) 1109;
C. Muller, Z. Phys. **82** (1933) 1;
F. E. Hoare, Phil. Mag. **6** (1928) 828;
F. E. Hoare, Phil. Mag. **13** (1932) 380;
W. V. Houston and Y. M. Hsieh, Phys. Rev. **45** (1934) 263;
F. H. Spedding, C. D. Shane, and Grace, Phys. Rev. **47** (1935) 38;
R. T. Birge, Rev. of Mod. Phys. **1** (1929) 14;
R. T. Birge, Rev. of Mod. Phys. **1** (1929) 45;
R. T. Birge, Rev. of Mod. Phys. **1** (1929) 54;
R. T. Birge, Rev. of Mod. Phys. **1** (1929) 60;
R. T. Birge, Phys. Rev. **40** (1932) 213;
R. T. Birge, Phys. Rev. **40** (1932) 228;
R. T. Birge, Phys. Rev. **52** (1937) 241;
R. T. Birge, Phys. Rev. **54** (1938) 972;
R. T. Birge, Nature **134** (1934) 771;
R. T. Birge, Nature **137** (1936) 187;
J. W. M. DuMond and V. Bollman, Phys. Rev. **50** (1936) 524;
J. W. M. DuMond and V. Bollman, Phys. Rev. **51** (1937) 400;
J. W. M. DuMond and V. Bollman, Phys. Rev. **54** (1938) 1005;
R. Ladenburg, Annalen der Physik. Leipzig **28** (1937) 458;
J. W. M. DuMond, Phys. Rev. **52** (1937) 1251;
R. A. Millikan, Annalen der Physik. Leipzig **32** (1938) 34;
J. A. Bearden, Phys. Rev. **48** (1935) 385;
J. A. Bearden, Phys. Rev. **48** (1935) 698;
P. Haglund, Z. Phys. **94** (1935) 369;
F. Tyren, Z. Phys. **109** (1938) 722;
F. G. Dunnington, Phys. Rev. **52** (1937) 498;
R. C. Williams, Phys. Rev. **54** (1938) 568;
R. C. Gibbs and R. C. Williams, Phys. Rev. **48** (1935) 971;
A. E. Shaw, Phys. Rev. **54** (1938) 193;
C. D. Shane and F. H. Spedding, Phys. Rev. **47** (1935) 33;
L. Livingston and H. A. Bethe, Rev. of Mod. Phys. **9** (1937) 373;
W. V. Houston, Phys. Rev. **30** (1927) 608;
L. E. Kinsler and W. V. Houston, Phys. Rev. **45** (1934) 104;
L. E. Kinsler and W. V. Houston, Phys. Rev. **46** (1934) 533;
C. T. Perry and E. L. Chaffee, Phys. Rev. **36** (1930) 904;
F. Kirchner, Annalen der Physik. Leipzig **8** (1931) 975;
F. Kirchner, Annalen der Physik. Leipzig **12** (1932) 503;
P. A. Ross and P. Kirkpatrick, Phys. Rev. **45** (1934) 223;
P. A. Ross and P. Kirkpatrick, Phys. Rev. **45** (1934) 454;
P. A. Ross and P. Kirkpatrick, Phys. Rev. **46** (1934) 668;
G. Schaitberger, Annalen der Physik. Leipzig **24** (1935) 84;
H. Feder, Annalen der Physik. Leipzig **1** (1929) 497;
E. Brunner, Phys. Rev. **53** (1938) 457;
J. Gnan, Annalen der Physik. Leipzig **20** (1934) 361;
R. V. McIbon and E. Rupp, Annalen der Physik. Leipzig **13** (1932) 725;
F. Bloch, Phys. Rev. **46** (1934) 674;
H. R. Robinson, Andrews, and Irons, Proc. Roy. Soc. **A143** (1933) 48;
H. R. Robinson and C. J. B. Clews, Proc. Roy. Soc. **A149** (1935) 587;
H. R. Robinson, Proc. Roy. Soc. **46** (1934) 693;
H. R. Robinson, Phil. Mag. **22** (1936) 1129;
W. A. Wooster, Proc. Roy. Soc. **A114** (1927) 729;
G. G. Kretschniar, Phys. Rev. **43** (1933) 417;
L. W. Alvarez, Phys. Rev. **47** (1935) 636;
G. Kellstrom, Phil. Mag. **23** (1937) 313;
W. V. Houston, Phys. Rev. **51** (1937) 446;
W. V. Houston, Phys. Rev. **52** (1937) 751;

WILLIAMS 1940

■ First observation of muon decay ■

Evidence for Transformation of Mesotrons into Electrons

E.J. Williams, G.E. Roberts
Nature **145** (1940) 102;

Excerpt One of the outstanding questions regarding the mesotron is that of its ultimate fate. Certain properties of this particle are remarkably like those of the hypothetical particle assumed by Yukawa in his theory of nuclear forces and β-disintegration, and this has led to the view that the two may be identical. Within a rather large experimental error they have the same mass, and both are unstable in the free state, having an average life of the order of 10^{-6} seconds. The disappearance of the particle of Yukawa's theory at the end of its life takes place through its transformation into an electron and a neutrino, and it is regarding this that hitherto there has been no evidence of a parallel between it and the mesotron of cosmic rays. In fact, existing experimental evidence has rather gone to show that mesotrons suffer at the end of their life, some other fate than befalls the Yukawa particle. With the object of obtaining information on this crucial point we constructed a large cloud-chamber (24 in. diameter, 20 in. deep) which, with its large sensitive period and volume, might catch a cosmic ray mesotron coming to the end of its range in the gas of the chamber. A recent photograph taken with this shows a mesotron track terminating in the gas as desired. From its end there emerges a fast electron track, the kinetic energy of which is very much greater than the kinetic energy of the mesotron, but is comparable with its mass energy. This indicates that the mesotron transforms into an electron, in which case the remarkable parallel between the mesotron and the Yukawa particle is taken one stage further. In terms of Yukawa's theory, the phenomenon observed may be described as a disintegration of the mesotron with the emission of an electron, thus constituting the most elementary form of β-disintegration. *(Extracted from the introductory part of the paper.)*

Accelerator COSMIC

Detectors Cloud chamber

Reactions
$\mu^- \to e^-$ neutral (neutrals)

Particles studied μ^-, μ^+

ALVAREZ 1940

■ First direct measurement of the neutron magnetic moment ■

A Quantitative Determination of the Neutron Moment in Absolute Nuclear Magnetons

L.W. Alvarez, F. Bloch
Phys. Rev. **57** (1940) 111;

Reprinted in
The Physical Review — the First Hundred Years, AIP Press (1995) CD-ROM.

Abstract The magnetic resonance method of determining nuclear magnetic moments in molecular beams, recently described by Rabi and his collaborators, has been extended to allow the determination of the neutron moment. In place of deflection by inhomogeneous magnetic fields, magnetic scattering is used to produce and analyze the polarized beam of neutrons. Partial depolarization of the neutron beam is observed when the Larmor precessional frequency of the neutrons in a strong field is in resonance with a weak oscillating magnetic field. A knowledge of the frequency and field when the resonance is observed, plus the assumption that the neutron spin is 1/2 yields the moment directly. The theory of the experiment is developed in some detail, and a description of the apparatus is given. A new method of evaluating magnetic moments in all experiments using the resonance method is described. It is shown that the magnetic moment of any nucleus may be determined directly in absolute nuclear magnetons merely by a measurement of the *ratio* of two magnetic fields. These two fields are

(a) that at which resonance occurs in a Rabi type experiment for a certain frequency, and

(b) that at which protons are accelerated in a cyclotron operated on the nth harmonic of that frequency.

The magnetic moment is then (for $J = \frac{1}{2}$), $\mu = H_b/nH_a$. n is an integer and H_b/H_a may be determined by null methods with arbitrary precision. The final result of a long series of experiments during which 200 million neutrons were counted is that the magnetic moment of the neutron, $\mu_n = 1.93 \pm 0.02$ absolute nuclear magnetons. A brief discussion of the significance of this result is presented.

Accelerator LBL 88 inch cyclotron

Detectors Counters

Related references

See also
O. R. Frisch and O. Stern, Z. Phys. **85** (1933) 4;
I. Esterman and O. Stern, Z. Phys. **85** (1933) 17;
O. Stern, Z. Phys. **89** (1934) 665;
I. I. Rabi, J. M. B. Kellogg, and J. R. Zacharias, Phys. Rev. **45** (1934) 761;
I. I. Rabi, Phys. Rev. **51** (1937) 652;
P. N. Powers, H. Carroll, and J. R. Dunning, Phys. Rev. **51** (1937) 1112;
P. N. Powers, H. Carroll, H. Beyer, and J. R. Dunning, Phys. Rev. **52** (1937) 38;
I. I. Rabi, J. R. Zacharias, S. Millman, and P. Kusch, Phys. Rev. **53** (1938) 318;
O. R. Frisch, H. von Halban, and J. Koch, Phys. Rev. **53** (1938) 719;
P. N. Powers, Phys. Rev. **54** (1938) 827;
I. I. Rabi, S. Millman, P. Kusch, and J. R. Zacharias, Phys. Rev. **55** (1939) 526;
J. M. B. Kellogg, I. I. Rabi, N. F. Ramsey, and J. R. Zacharias, Phys. Rev. **55** (1939) 318;

Reactions
$n\,\gamma^* \rightarrow n\,\gamma^*$

Particles studied n

PAULI 1940B

■ Theorem on the connection between spin and statistics ■

The Connection Between Spin and Statistics

W. Pauli
Phys. Rev. **58** (1940) 716;

Reprinted in
Collected Scientific Papers by Wolfgang Pauli, ed. R. Kronig and V. F. Weisskopf, Interscience Publishers, John Willey and Sons, Inc. **v.2** (1964) 911.
The Physical Review — the First Hundred Years, AIP Press (1995) CD-ROM.

Abstract In the following paper we conclude for the relativistically invariant wave equation for free particles: From postulate (I), according to which the energy must be positive, the necessity of *Fermi-Dirac* statistics for particles with arbitrary half–integral spin; from postulate (II), according to which observables on different space-time points with a space-like distance are commutable, the necessity of *Einstein-Bose* statistics for particles for arbitrary integral spin.

Related references

More (earlier) information
W. Pauli and F. J. Belinfante, Physica **7** (1940) 177;

See also
B. L. van der Waerden, *Die gruppentheoretishe Methode in der Quantentheorie*, Berlin (1932);
M. Fierz and W. Pauli, Proc. Roy. Soc. **A173** (1939) 211;
M. Fierz, Helv. Phys. Acta **12** (1939) 3;
L. de Broglie, Compt. Ren. **208** (1939) 1697;
L. de Broglie, Compt. Ren. **209** (1939) 265;

KERST 1940

■ Kerst proposal for betatron accelerator ■

Acceleration of Electrons by Magnetic Induction

D.W. Kerst
Phys. Rev. **58** (1940) 841;

Reprinted in
The Physical Review — the First Hundred Years, AIP Press (1995) CD-ROM.

Excerpt For some time it has been realized that it might be possible to make use of the electromotive force induced by a changing magnetic flux to accelerate charged particles traveling in an orbit around the changing flux. Although previous attempts to accelerate electrons by this means have been unsuccessful, (Wideroe, Archiv f. Elektrotechnik **21** (1938) 400, E. T. S. Walton, Proc. Camb. Phil. Soc. **25** (1929) 469) careful examination showed that it should be possible to get good magnetic focusing by the proper arrangement of a magnetic field to guide the electrons around the changing flux and that if the rate of change of flux within the orbit were sufficiently high it would be possible to capture electrons in usable orbits and that vacuum requirements should not be difficult to satisfy.
It seamed feasible to attempt the experiment with a600-cycle per second magnetic field, since a sufficiently high rate of change of of flux would be obtained and since it seemed that it would not be necessary to have a vacuum better than 10^{-6} millimeter of mercury in the acceleration chamber, in spite of the fact that at this frequency the length of the electron path would be of the order of 10^7 centimeters.
To hold the electrons in the acceleration chamber for such a long path it is necessary to fulfill the condition that $\phi = 2\pi R_0^2 H$, where ϕ is the flux enclosed by the orbit and H is the magnetic field at the orbit which causes the electrons to travel in a circle of radius R_0. When this condition is satisfied, the electron orbit neither shrinks nor expands, and the electrons can be accelerated by increasing ϕ and H together.
A laminated electromagnet with pole faces 8 inches in diameter, which satisfied all the necessary conditions, was constructed. The stable orbit was shrunk from R_0 toward the position of a tungsten target by causing saturation of the portion of the magnetic circuit which supplied the flux through the center of the orbit. X-rays produced by the impact of the electrons upon the target showed that the accelerator operated, and a lead collimator in front of a Geiger-Müller counter showed that the only portion of the acceleration chamber from which x-rays came was the target.
By taking the sweep voltage for an oscillograph from a coil surrounding the core of the magnet and putting the pulses from the Geiger-Müller counter circuit on the vertical deflection plates, the phase of the magnetic field at which the electrons struck the target could be determined. It was possible to hold the electrons in the acceleration chamber for one-fourth of a cycle during which the magnetic field changed from a low value to its maximum. Conservative estimates of the magnetic field at the target when the electrons strike it indicated that the energy of the electrons was about 2.2 MeV. This estimate was substantiated by a comparison of the absorption of the X-rays in lead with published data on the absorption of X-rays produced by 2-million-volt electrons. (D. L. Northrup and L. C. Van Atta, Am. J. Röntgenology and Radium Therapy **41** (1939) 633). After filtering the X-rays from the accelerator through about 1.8 cm of lead, their absorption coefficient is 0.57 cm^{-1}. A correction had to be made for scattering of X-rays from the magnetic yoke. Since the absorption coefficient for X-rays produced by 2.0 MeV electrons is 0.62 cm^{-1}, the electrons in the new accelerator must have reached about 2.35 MeV energy before striking the target. The absorption measurements were taken with Lauritsen electroscopes, and calibration of the electroscopes showed that the intensity of the radiation was greater than the intensity of the gamma-rays from 10 millicuries of radium.
Of several suggestions which have been made for naming the apparatus, induction accelerator seems to be the shortest descriptive one.
It has been a great help to be able to discuss the theoretical aspects of the accelerator with Professor R. Serber and Professor H. M. Mott-Smith.

RASETTI 1941

■ First measurement of the muon lifetime, preliminary result ■

Mean Life of Slow Mesotrons

F. Rasetti
Phys. Rev. **59** (1941) 613;

Abstract The author recently described an experiment which enabled him to detect the disintegration electrons emitted by mesotrons at the end of their range.
This is a preliminary report of new results obtained with an improved arrangement which was designed to measure a decay curve of mesotrons at rest. *(Extracted from introductory part of the paper.)*

Accelerator COSMIC

Detectors Counters

Related references

See also
S. Tomonaga and G. Araki, Phys. Rev. **58** (1940) 90;

Reactions
$\mu^- \rightarrow e^- X$

$\mu^+ \rightarrow e^+ X$

Particles studied μ^-, μ^+

KERST 1941

■ First betatron ■

The Acceleration of Electrons by Magnetic Induction

D.W. Kerst
Phys. Rev. **60** (1941) 47;

Abstract Apparatus with which electrons have been accelerated to an energy of 2.3 MeV by means of the electric field accompanying a changing magnetic field is described. Stable circular orbits are formed in a magnetic field, and the changing flux within the orbits accelerates the electrons. As the magnetic field reaches its peak value, saturation of the iron supplying flux through the orbit causes the electrons to spiral inward toward a tungsten target. The X-rays produced have an intensity approximately equal to that of the gamma-rays from one gram of radium; and, because of the tendency of the X-rays to proceed in the direction of the electrons, a pronounced beam is formed.

Related references

See also
G. Breit and M. A. Tuve, *Carnegie Institute Year Book* **27** (1927) 209;
R. Wideroe, Archiv f. Elektrotechnik **21** (1928) 400;
E. T. S. Walton, Proc. Camb. Phil. Soc. **25** (1929) 469;
W. W. Jassinsky, Archiv f. Elektrotechnik **30** (1936) 500;
D. W. Kerst, Phys. Rev. **58** (1940) 841;
D. W. Kerst, Phys. Rev. **59** (1941) 110;
D. L. Northrup and L. C. Van Atta, Am. J. Röntgenology and Radium Therapy **41** (1939) 633;

RASETTI 1941B

■ First measurement of the muon lifetime ■

Disintegration of Slow Mesotrons

F. Rasetti
Phys. Rev. **60** (1941) 198;

Abstract The disintegration of mesotrons at the end of their range was investigated by means of an improved arrangement of the type already described by the author. The absorption of a mesotron by a block of aluminium or iron is recorded by a system of coincidence and anticoincidence counters. Another system of counters and circuits registers the delayed emission of a particle, which is interpreted as the disintegration electron associated with the absorbed mesotron. The present apparatus enables one to determine the time distribution of the emitted particles and hence the mean life of the decay process, independently of the effects produced by the scattering of mesotrons. The mean life is found to be 1.5 ± 0.3 microseconds, in substantial agreement with the value deduced from the atmospheric absorption effect. The absolute number of disintegration electrons per absorbed mesotron has also been determined (for an Al absorber) and found to be about one-half. This result suggests that, in agreement with theoretical predictions, positive mesotrons undergo spontaneous decay, while the negative ones react with nuclear particles.

Accelerator COSMIC

Detectors Counters

Related references

More (earlier) information
F. Rasetti, Phys. Rev. **59** (1941) 613;
F. Rasetti, Phys. Rev. **59** (1941) 706;
Analyse information from
B. Rossi and D. B. Hall, Phys. Rev. **59** (1941) 223;
W. H. Nielsen, C. M. Ryerson, L. W. Nordheim, and K. Z. Morgan, Phys. Rev. **59** (1941) 547;
E. J. Williams and G. E. Roberts, Nature **145** (1940) 102;

Reactions
$\mu^- \rightarrow e^- X$
$\mu^+ \rightarrow e^+ X$

Particles studied μ^-, μ^+

ROSSI 1942

■ Evidence for the muon exponential decay rate ■

Experimental Determination of the Disintegration Curve of Mesotrons

B. Rossi, N. Nereson
Phys. Rev. **62** (1942) 417;

Abstract The disintegration curve of mesotrons has been experimentally determined by investigating the delayed emission of disintegration electrons which takes place after the absorption of mesotrons by matter. Within the experimental errors, the disintegration curve is exponential and corresponds to a mean lifetime of 2.3 ± 0.2 microseconds.

Accelerator COSMIC

Detectors Counters

Related references

Analyse information from
C. G. Montgomery, N. F. Ramsey, D. H. Cowie, and D. D. Montgomery, Phys. Rev. **56** (1939) 635;
F. Rasetti, Phys. Rev. **60** (1941) 198;
M. D. Souza Santos, Phys. Rev. **62** (1942) 178;
J. A. Wheeler and R. Ladenburg, Phys. Rev. **60** (1941) 754;
B. Rossi and D. B. Hall, Phys. Rev. **59** (1941) 223;
K. Greisen, J. C. Stearns, D. K. Froman, and P. G. Koontz, Phys. Rev. **61** (1942) 675;
W. H. Nielsen, C. M. Ryerson, L. W. Nordheim, and K. Z. Morgan,

Phys. Rev. **59** (1941) 547;

Reactions
$\mu^- \rightarrow e^- X$
$\mu^+ \rightarrow e^+ X$

Particles studied μ^+, μ^-

NERESON 1942

■ Confirmation of the muon exponential decay rate ■

Further Measurements on Disintegration Curve of Mesotrons

N. Nereson, B. Rossi
Phys. Rev. **64** (1942) 199;

Abstract Further measurements have been made on the disintegration curve of mesotrons. An extensive set of data gives a mean lifetime of 2.15 ± 0.07 microseconds as calculated from the differential disintegration curve.

Accelerator COSMIC

Detectors Counters

Related references

More (earlier) information
B. Rossi and N. Nereson, Phys. Rev. **62** (1942) 417;

See also
F. Rasetti, Phys. Rev. **60** (1941) 198;
B. Rossi and K. Greisen, Rev. of Mod. Phys. **13** (1941) 240;

Reactions
$\mu^- \rightarrow e^- X$
$\mu^+ \rightarrow e^+ X$

Particles studied μ^+, μ^-

FERMI 1943

■ First Fermi nuclear reactor. Beginning of atomic energy era ■

Experimental Production of a Divergent Chain Reaction

E. Fermi
Internal report for the Metallurgical Laboratory of University of Chicago (1943);

Reprinted in
American Journal of Physics **20** (1952) 536.
Enrico Fermi. *Collected Papers (Note e Memorie)* v. I. Italy 1921 - 1938 (1962). The University of Chicago Press - Accademia Nazionale dei Lincei, Roma.
(translation into Russian) Enrico Fermi. *Nauchnye Trudy.* I. 1921-1938 Italiya. (red. B.Pontecorvo), Nauka, Moskva, 1971.

Excerpt This report gives a description of the construction and operation of a chain reacting pile. The pile was constructed in the West Stands Laboratory during the months of October and November 1942 and was operated for the first time on December 2, 1942.
It will appear from its description that an experiment of this kind requires the collaboration of a large number of physicists. The two groups of Zinn and Anderson took charge of the preparation of the materials and of the actual construction of the pile; the group of Wilson prepared the measuring equipment and the automatic controls. The details of this work are given by the members of the two groups in the appendices.
A large share of the credit for the experiment goes also to all the services of the Metallurgical Laboratory and in particular to the groups responsible for the development of the production and the testing of the materials. The exceptionally high purity requirements of graphite and uranium which were needed in very large amounts made the procurement of suitable materials the greatest single difficulty in all the development. *(Extracted from the introductory part of the paper.)*

TOMONAGA 1943 — Nobel prize

■ Creation of the covariant quantum electrodynamic theory. Tomonaga method. Nobel prize to S. Tomonaga awarded in 1965. Co-winners J. S. Schwinger and R. P. Feynman "for their fundamental work in quantum electrodynamics, with deep-ploughing consequences for the physics of elementary particles" ■

On a Relativistically Invariant Formulation of the Quantum Theory of Wave Fields

S. Tomonaga
Riken Iho 22, 525 (1943);Progr. of Theor. Phys. **1** (1946) 27;

Reprinted in
(translation into English) Prog. Theor. Phys. **1** (1946) 27.

Excerpt We have thus shown that the quantum theory of wave fields can be really brought into a form which reveals directly the invariance of the theory against Lorentz transformations. The reason why the ordinary formalism of the quantum field-theory is so unsatisfactory lies in the fact that one has built up this theory in the way which is too much analogous to the ordinary nonrelativistic mechanics. In this ordinary formalism of the quantum theory of fields the theory is divided into two distinct sections: the section giving the kinematical relations between various quantities at the same

instant of time, and the section determining the causal relations between quantities at different instants of time. Thus the commutation relations (1) belong to the first section and the Schrödinger equation (2) to the second.
As stated before, this way of separating the theory into two sections is very unrelativistic, since here the concept "same instant of time" plays a distinct role.
Also in our formalism the theory is divided into two sections. But now the separation is introduced in another place: In our formalism the theory consists of two sections, one of which gives the laws of behavior of the fields when they are left alone, and the other of which gives the laws determining the deviation from this behavior due to interactions. This way of separating the theory can be carried out relativistically.
Although in this way the theory can be brought into more satisfactory form, no new contents are added thereby. So, the well known divergence difficulties of the theory are inherited also by our theory. *(Extracted from the introductory part of the paper.)*

Related references

See also
H. Yukawa, Kagaku **12** (1924) 251, 282, 322;
P. A. M. Dirac, Physik Zeits. Sow. **3** (1933) 64;
W. Pauli, Solvey Berichte (1939);
F. Bloch, Physik Zeits. Sow. **5** (1934) 301;
E. C. G. Stückelberg, Helv. Phys. Acta **11** (1928) 225;
W. Heisenberg, Z. Phys. **110** (1938) 251;
P. A. M. Dirac, Proc. Roy. Soc. **A136** (1932) 453;
W. Heisenberg and W. Pauli, Z. Phys. **56** (1929) 1;

HEISENBERG 1943

■ Invention of the S-matrix formalism ■

Die "beobachtbaren Größen" in der Theorie der Elementarteilchen. / The "Observable Quantities" in the Theory of Elementary Particles

W. Heisenberg
Z. Phys. **120** (1943) 513;

Reprinted in
(translation into English) UCRL-TRANS 808 (1962).

Abstract Die bekannten Divergenzschwierigkeiten in der Theorie der Elementarteilchen zeigen, daß die zukünftige Theorie in ihren Grundlagen eine universelle Konstante von der Dimension einer Länge enthalten wird, die in die bisherige Form der Theorie offenbar nicht widerspruchsfrei eingebaut werden kann. Im Hinblick auf diese spätere Abänderung der Theorie versucht die vorliegende Arbeit, aus dem Begriffsgebäude der Quantentheorie der Wellenfelder diejenige Begriffe herauszuschälen, die von der zukünftigen Änderung wahrscheinlich nicht betroffen werden und die daher einen Bestandteil auch der zukünftigen Theorie bilden werden. Die Arbeit gliedert sich in folgende Abschnitte:

I. Die beobachtbaren Größen und ihre mathematische Darstellung.
   a) Formulierung der Grundannahmen.
   b) Die mathematische Darstellung der beobachtbaren Größen.
   c) Die Singularitäten im Impulsraum.
   d) Die charakteristische Matrix.
   e) Die Emission mehrerer Teilchen.

II. Die Eigenschaften der Matrix S.
   a) Die allgemeinen quantenmechanischen Eigenschaften von S.
   b) Die singulären Teile der Matrix S.
   c) Berücksichtigung von Spin und Statistik der verschiedenen Elementarteilchen.
   d) Das relativistische Verhalten der Matrix S.

III. Beziehungen zwischen den beobachtbaren Größen.
   a) Das durch die Matrix S gegebene Eigenwertproblem.
   b) Beziehungen zwischen Streu- und Emissionskoeffizienten.
   c) Beziehungen zwischen den beobachtbaren Größen, die nicht aus den allgemeinen Eigenschaften der Matrix S folgen.

Abstract The known divergence difficulties in the theory of elementary particles show that the further theory will contain in its fundamentals a universal constant of the dimension of a length, which evidently cannot be built into the present form of the theory without contradiction. With a view to this later modification of the theory, the present paper tries to isolate from the conceptual structure of the quantum theory of wave fields those concepts which will probably not be affected by the future modification and which therefore will be partially of the future theory. The paper is divided into the following sections:

II. The observable quantities and their mathematical representation.
   a) Formulation of the basic assumptions.
   b) The mathematical representation of the observable quantities.
   c) The singularities in the momentum space.
   d) The characteristic matrix.
   e) The emission of several particles.

II. The properties of matrix S.
   a) The general quantum-mechanical properties of S.

b) The singular parts of matrix S.

c) Consideration of spin and statistics of various elementary particles.

d) The relativistic behavior of matrix S.

III. Relations between the observable quantities.

a) The inherent value problem given by matrix S.

b) Relations between the observable quantities which do not follow from the general properties of matrix S.

Related references

More (later) information
W. Heisenberg, Z. Phys. **120** (1943) 673;
W. Heisenberg, Z. Phys. **123** (1944) 93;
W. Heisenberg, Annalen der Physik. Leipzig **32** (1938) 20;

See also
V. F. Weisskopf, Z. Phys. **89** (1934) 27;
A. March, Z. Phys. **104** (1936) 93;
A. March, Z. Phys. **104** (1936) 161;
A. March, Z. Phys. **105** (1937) 620;
G. Watagin, Z. Phys. **88** (1934) 82;
G. Watagin, Z. Phys. **92** (1934) 547;

VEKSLER 1944

■ Invention of the principle of phase stability for accelerators ■

A New Method of the Acceleration of Relativistic Particles

V.I. Veksler
Doklady Akad. Nauk SSSR **43** (1944) 346;

Reprinted in
Selected Papers of Soviet Physicists, Usp. Fiz. Nauk **93** (1967) 521.

Excerpt It is generally believed that the Lawrence method of resonant acceleration of heavy particles is not applicable to electron acceleration due to relativistic variation of electron mass with its velocity. This difficulty is resolved in the accelerator proposed by Videroe and realized by Kerst. With Kerst apparatus 20 MeV energy electron beam is already obtained and evidently 100 MeV electron beams will be achieved soon. However, further increasing of electron energy by the Kerst method will meet tremendous technical difficulties. It is worth to point out on one new possibility to obtain relativistic particles based on one simple generalization of the resonant method. *(Extracted from the introductory part of the paper.)*

Related references

More (later) information
V. I. Veksler, Doklady Akad. Nauk SSSR **44** (1944) 393;
V. I. Veksler, Phys. Rev. **69** (1946) 244;
V. I. Veksler, J. Phys. USSR **9** (1945) 153;

See also
E. O. Lawrence, Phys. Rev. **50** (1936) 1134;
R. Wideroe, Archiv f. Elektrotechnik **21** (1928) 400;
D. W. Kerst, Phys. Rev. **60** (1941) 47;

LEPRINCE-RINGUET 1944

■ First evidence for the K^+ ■

Existence probable d'une particule de masse 990 m_e dans le rayonnement cosmique / On the Probable Existence of Cosmic Ray Particle with Mass 990 m_e

L. Leprince-Ringuet, M. Lheritier
Compt. Ren. **219** (1944) 618;

Reprinted in
R. N. Cahn and G. Goldhaber, *The Experimental Foundations of Particle Physics*, Cambridge Univ. Press (1991) 66.

Abstract An estimate of mass, 990 $m_e \pm 12\%$ is obtained by the method of elastic collision for a penetrating particle photographed at 1000 m altitude. Other particles in the same series of photographs show collisions consistent with the normal meson mass. *(Science Abstracts, 1947, 2927. J. G. W.)*

Accelerator COSMIC

Detectors Cloud chamber

Related references

See also
L. Leprince-Ringuet, S. Gorodetzky, E. Nageotti, and R. Foy, Compt. Ren. **211** (1940) 382;
L. Leprince-Ringuet, S. Gorodetzky, E. Nageotti, and R. Foy, Phys. Rev. **59** (1941) 460;
L. Leprince-Ringuet, S. Gorodetzky, E. Nageotti, and R. Foy, Journal de Physique **2** (1941) 63;
S. Gorodetzky, *These*, Paris (1942);
R. Richard-Foy, Cahiers de Physique **2** (1942) 65, Compt. Ren. **213** (1941) 1;

Particles studied K^+

LANDAU 1944

■ Landau distribution for fast particle energy loss by ionization ■

On the Energy Loss of Fast Particles by Ionization

L.D. Landau
Jour. Phys. USSR 8, 201 (1944);

Reprinted in
(translation into English) *Collected Papers of L. D. Landau* ed. D. Ter Haar, Gordon and Breach, Science Publishers, London (1965) 417.

Abstract The energy distribution function has been determined for fast particles which have traversed a layer of matter of a given thickness and lost energy in the latter as a result of ionization collisions.

Related references

See also
L. Livingston and H. A. Bethe, Rev. of Mod. Phys. **9** (1937) 245;

IWANENKO 1944

■ Limitation of the maximal energy attainable in a betatron ■

On the Maximal Energy Attainable in a Betatron

D.D. Iwanenko, I.Y. Pomeranchuk
Doklady Akad. Nauk SSSR **44** (1944) 364; Phys. Rev. **65** (1944) 343;

Reprinted in
The Physical Review — the First Hundred Years, AIP Press (1995) CD-ROM.

Excerpt By means of a recently constructed induction accelerator-betatron, Kerst succeeded in obtaining electrons up to 20 MeV. The principle of operation of the betatron is the acceleration of electrons by a tangential electric field produced by a changing magnetic flux, which is connected with the magnetic field keeping electrons on the orbit by a simple relation. In contrast to a cyclotron, whose applicability is essentially limited to the non-relativistic region on the ground of defocusing of orbits due to the change of mass at high energies, there is no such limitation for the betatron.
We may point out, however, that quite another circumstance would lead as well to the existence of a limitation for maximal energy attainable in a betatron. This is the radiation of electrons in the magnetic field. *(Extracted from the introductory part of the paper.)*

Related references

See also
D. W. Kerst, Phys. Rev. **61** (1942) 93;
I. Y. Pomeranchuk, J. Phys. **2** (1940) 65;
D. W. Kerst and R. Serber, Phys. Rev. **60** (1941) 53;

MCMILLAN 1945

■ Invention of the principle of phase stability for accelerators ■

The Synchrotron — A Proposed High Energy Particle Accelerator

E.M. McMillan
Phys. Rev. **68** (1945) 143;

Reprinted in
The Physical Review — the First Hundred Years, AIP Press (1995) CD-ROM.

Excerpt One of the most successful methods for accelerating charged particles to very high energies involves the repeated application of an oscillating electric field, as in the cyclotron. If a very large number of individual accelerations is required, there may be difficulty in keeping the particles in step with the electric field. In the case of the cyclotron this difficulty appears when the relativistic mass change causes an appreciable variation in the angular velocity of the particles.
The device proposed here makes use of a "phase stability" possessed by certain orbits in a cyclotron. Consider, for example, a particle whose energy is such that its angular velocity is just right to match the frequency of the electric field. This will be called the equilibrium energy. Suppose further that the particle crosses the accelerating gaps just as the electric field passes through zero, changing in such a sense that an earlier arrival of the particle would result in an acceleration. This orbit is obviously stationary. To show that it is stable, suppose that a displacement in phase is made such that the particle arrives at the gaps too early. It is then accelerated; the increase in energy causes a decrease in angular velocity, which makes the time of arrival tend to become later. A similar argument shows that a change of energy from the equilibrium value tends, to correct itself. These displaced orbits will continue to oscillate, with both phase and energy varying about their equilibrium values.
In order to accelerate the particles it is now necessary to change the value of the equilibrium energy, which can be done by varying either the magnetic field or the frequency. While the equilibrium energy is changing, the phase of the motion will shift ahead just enough to provide the necessary accelerating force; the similarity of this behavior to that of a synchronous motor suggested the name of the device. *(Extracted from the introductory part of the paper.)*

GAMOW 1946

■ Gamow indication on the possibility to explain the observed chemical elements abundance-curve by assumption of unequilibrium process of elements formation during a limited interval of time. Birth of the Big Bang model ■

Expanding Universe and the Origin of Elements

G. Gamow
Phys. Rev. **70** (1946) 572;

Excerpt It is generally agreed at present that the relative abundance of various chemical elements were determined by physical conditions existing in the universe during the early stages of its expansion, when the temperature and density were sufficiently high to secure appreciable reaction-rates for the light as well as for the heavy nuclei.

In all the so-far published attempts in this direction the observed abundance-curve is supposed to represent some equilibrium state determined by nuclear binding energies at some very high temperature and density. This point of view encounters, however, serious difficulties in the comparison with empirical facts. Indeed, since binding energy is, in a first approximation, a linear function of atomic weight, any such equilibrium theory would necessarily lead to a rapid exponential decrease of abundance through the entire natural sequence of elements. It is known, however, that whereas such a rapid decrease actually takes place for the first half of chemical elements, the abundance of heavier nuclei remains nearly constant. Attempts have been made to explain this discrepancy by the assumption that heavy elements were formed at higher temperatures, and that their abundances were already "frozen" when the adjustment of lighter elements was taking place. Such an explanation, however, can be easily ruled out if one remembers that at the temperatures and densities in question (about $10^{10\circ}$K, and $10^6 g/cm^3$) nuclear transformations are mostly caused by the process of absorption and re-evaporation of free neutrons so that their rates are essentially the same for the light and for the heavy elements. Thus it appears that the only way of explaining the observed abundance-curve lies in the assumption of some kind of unequilibrium process taking place during a limited interval of time.

The above conclusion finds a strong support in the study of the expansion process itself. *(Extracted from the introductory part of the paper.)*

<u>*Related references*</u>

See also

von Weizsäcker, Phys.Zeitschr. **39** (1938) 633;
Chandrasekhar and Heinrich, Astr. Jour. **95** (1942) 288;
G. Watagin, Phys. Rev. **66** (1944) 149;
Goldschmidt, *Verteilung der Elemente* (Oslo, 1938);
R. Tolman, *Relativity, Thermodynamics and Cosmology* (Oxford Press, New York, 1934);
Hubble, *The Realm of the Nebulae* (Yale University Press, New Haven, 1936);
G. Gamow and E. Teller, Phys. Rev. **55** (1939) 654;

MCMILLAN 1946

■ Further development of the synchrotron idea ■

Resonance Acceleration of Charged Particles

E.M. McMillan
Phys. Rev. **70** (1946) 800;

<u>*Excerpt*</u> The most fruitful means now known for accelerating particles to very high energies are:

(1) to give repeated small pushes by an alternating electric field, as in the cyclotron or linear accelerator, and

(2) to give a steady push produced by a varying magnetic flux, as in the betatron.

The first of these, which may be called "resonance acceleration" gives promise of reaching higher ultimate energies, and will be the subject of this paper. The problem of keeping the particles in step with the accelerating electric field for a very large number of cycles seems at first sight to be a formidable one, not only because of intrinsic difficulties such as the well-known relativistic limit of the cyclotron, but also because of practical limits on the accuracy of construction of machines. However, an appreciation of the property of "phase stability" shows the way to a simple solution of this problem. If the particle is moving in a circular orbit, the angular velocity must just match the applied electric frequency for resonance to be maintained; if the motion is in a straight line, there is a corresponding relation between the linear velocity, and the product of the electrode repeat length by the frequency. It can now be shown that, in general, if the angular or linear velocity varies with the energy, it is not necessary to make the exact match mentioned above, since particles started sufficiently near the right velocity will fall into a stable motion in which the velocity, energy, and phase oscillate about equilibrium values. (The phase describes the time relation between the arrival of the particle at an acceleration gap and the field across the gap.) The equilibrium energy is that value which achieves the exact match between velocity and frequency. If this energy is now made to vary slowly by changing the frequency or magnetic field in the circular case, the stability will remain and the actual energy will oscillate about the varying equilibrium value. In the linear case, the variation is produced by the changing repeat length as the particle travels down the tube. Thus acceleration is accomplished with complete stability of the motion at all times, provided that the rate of variation is not too great, and that adequate focusing of the beam is provided. The focusing problem in the linear case has been discussed by Dr. Alvarez; in the circular case it is the same as in the cyclotron and betatron. This stability takes care of relativistic difficulties as well as tolerances in the machine. The circular orbit machines can be divided into three types:

(1) magnetic field variation (synchrotron), which is most suitable for accelerating electrons;

(2) frequency variation (synchro-cyclotron), most suitable for ions; and

(3) variation of both quantities.

The third type has the advantage that the size of the orbit can be held constant by maintaining a certain relation between the field and frequency, but it involves all the practical difficulties inherent in both of the other types. Methods of obtaining the variations and details of design will be illustrated by descriptions of the 300 MeV synchrotron and the 184″ synchro-cyclotron now under construction at Berkeley.

There will also be a discussion of future possibilities, leading to the eventual attainment of the billion-volt range.

PONTECORVO 1946

■ Proposal for the radiochemical method of detecting the neutrino ■

Inverse β-Decay

B. Pontecorvo
Chalk River Laboratory Report PD-205 (1946);

Reprinted in
Neutrino Astrophysics, John N. Bahcall, Cambridge Univ. Press (1989).

Excerpt The goal of present article is to show that the experimental observation of the inverse β-process induced by neutrino is possible with the recent experimental equipment and to suggest a method which can do experimental observation feasible. *(Extracted from the introductory part of the paper.)*

CONVERSI 1947

■ Evidence that the muon is not a strong interaction mediator ■

On the Disintegration of Negative Mesons

M. Conversi, E. Pancini, O. Piccioni
Phys. Rev. **71** (1947) 209;

Reprinted in
R. N. Cahn and G. Goldhaber, *The Experimental Foundations of Particle Physics*, Cambridge Univ. Press (1991) 37.
The Physical Review — the First Hundred Years, AIP Press (1995) 817.

Excerpt In a previous Letter to the Editor, we gave a first account of an investigation of the difference in behavior between positive and negative mesons stopped in dense materials. Tomonaga and Araki showed that, because of the Coulomb field of the nucleus, the capture probability for negative mesons at rest would be much greater than their decay probability, while for positive mesons the opposite should be the case. If this is true, then practically all the decay processes which one observes should be owing to positive mesons. Several workers have measured the ratio η between the number of the disintegration electrons and the number of mesons stopped in dense materials. Using aluminum, brass, and iron, these workers found values of η close to 0.5 which, if one assumes that the primary radiation consists of approximately equal numbers of positive and negative mesons, support the above theoretical prediction. Auger, Maze, and Chaminade, on the contrary, found η to be close to 1.0, using aluminum as absorber.
Last year we succeeded in obtaining evidence of different behavior of positive and negative mesons stopped in 3 cm of iron as an absorber by using magnetized iron plates to concentrate mesons of the same sign while keeping away mesons of the opposite sign (at least for mesons of such energy that would be stopped in 3 cm of iron). We obtained results in agreement with the prediction of Tomonaga and Araki. After some improvements intended to increase the counting rate and improve our discrimination against the "mesons of the opposite sign," we continued the measurements using, successively, iron and carbon as absorbers. The recording equipment was one which two of us had previously used in a measurement of the meson's mean life. *(Extracted from the introductory part of the paper.)*

Accelerator COSMIC

Detectors Counters

Related references

More (earlier) information
M. Conversi, E. Pancini, and O. Piccioni, Phys. Rev. **68** (1945) 232;
See also
S. Tomonaga and G. Araki, Phys. Rev. **58** (1940) 90;
Analyse information from
F. Rasetti, Phys. Rev. **60** (1941) 198;
B. Rossi and N. Nereson, Phys. Rev. **62** (1942) 417;
M. Conversi and O. Piccioni, Nuovo Cim. **2** (1944) 40;
M. Conversi and O. Piccioni, Nuovo Cim. **2** (1944) 71;
M. Conversi and O. Piccioni, Phys. Rev. **70** (1946) 859;
M. Conversi and O. Piccioni, Phys. Rev. **70** (1946) 874;
P. Auger, R. Maze, and Chaminade, Compt. Ren. **213** (1941) 381;
Y. Nishina, M. Takeuchi, and T. Ichimiya, Phys. Rev. **55** (1939) 585;
H. Maier-Leibnitz, Z. Phys. **112** (1939) 569;
T. H. Johnson and P. Shutt, Phys. Rev. **61** (1942) 380;
H. Jones, Rev. of Mod. Phys. **11** (1939) 235;
D. J. Hughes, Phys. Rev. **57** (1940) 592;
G. Bernardini et al., Phys. Rev. **68** (1945) 109;

Reactions
$\mu^- \rightarrow e^- X$

Particles studied μ^-

KOBA 1947 Nobel prize

■ Creation of the covariant quantum electrodynamic theory. Tomonaga method. Nobel prize to S. Tomonaga awarded in 1965. Co-winners J. S. Schwinger and R. P. Feynman "for their fundamental work in quantum electrodynamics, with deep-ploughing consequences for the physics of elementary particles" ■

On a Relativistically Invariant Formulation of the Quantum Theory of Wave Fields. II. Case of Interacting Electromagnetic and Electron Fields

Z. Koba, T. Tati, S. Tomonaga
Progr. of Theor. Phys. **2** (1947) 101;

Abstract The aim of the present paper is to apply our method of formulation to the quantum electrodynamics which deals with the electromagnetic field interacting with the electron field.
Although we have in the case of quantum electrodynamics a perfectly relativistic formalism, that is Dirac's many-time theory, (P. A. M. Dirac, V. Fock and B. Podolsky, Physik Zeits. Sow. **2** (1932) 468) it will be nevertheless of some importance to formulate the theory also in our formalism, since the former, in which the states of electrons are described in the configuration space, applies only to the case where the number of the electrons does not change, and is thus incapable of dealing with such phenomena as the emission of β-rays or the decay of mesons emitting electrons, without introducing a rather unfavourable complication into the theory at the sacrifice of its beautifulness. It is, therefore, desirable to treat the electrons as a quantized field and not as particles, since the former treatment fits the cases better in which the number of the electrons really changes.
It is already nearly known in I how to apply our formalism to the quantum electrodynamics except the well-known complication due to the existence of the so-called auxiliary condition. One of the main tasks in the present paper is thus to find the auxiliary condition in our formalism and to eliminate it in a relativistically invariant way on the lines which has been reported by Hayakawa, Miyamoto and one of the authors in another place.

Related references

More (earlier) information
S. Tomonaga, Riken Iho **22** (1943) 525;
S. Tomonaga, Progr. of Theor. Phys. **1** (1946) 27;
See also
P. A. M. Dirac, V. A. Fock, and B. Podolsky, Physik Zeits. Sow. **2** (1932) 468;

BROBECK 1947

■ First cyclotron based on the phase stability principle – the Berkeley 184 inch synchrocyclotron ■

Initial Performance of the, 184-Inch Cyclotron of the University of California

W.M. Brobeck et al.
Phys. Rev. **71** (1947) 449;

Excerpt The successful application of the *principle of phase stability* to a cyclotron has been reported from this Laboratory. The satisfactory results of these experiments, performed on the 37-inch cyclotron, led to the decision to complete the 184-inch cyclotron then under construction as a frequency modulated machine. It is the purpose of this note to describe briefly the equipment and experiments by which deuteron and alpha-particle beams of approximately 200 and 400 MeV, respectively, have been produced. *(Extracted from the introductory part of the paper.)*

Related references

See also
E. M. McMillan, Phys. Rev. **68** (1945) 143;
V. I. Veksler, J. Phys. USSR **9** (1945) 153;
J. R. Richardson, K. R. MacKenzie, E. J. Lofgren, and B. T. Wright, Phys. Rev. **69** (1946) 669;
F. H. Schmidt, Rev. Sci. Inst. **17** (1946) 301;

Reactions

deuteron	180 MeV(T_{lab})
^{4}He	400 MeV(T_{lab})

PERKINS 1947

■ First indication of the existence of the π^- ■

Nuclear Disintegration by Meson Capture

D.H. Perkins
Nature **159** (1947) 126;

Reprinted in
R. N. Cahn and G. Goldhaber, *The Experimental Foundations of Particle Physics*, Cambridge Univ. Press (1991) 39.

Abstract Recently, multiple nuclear disintegration "stars", produced by cosmic radiation, have been investigated by the photographic emulsion technique. Plates coated with 50 μ Ilford B.1 emulsions, were exposed in aircraft for several hours at 30,000 ft. One of these disintegrations was of particular interest, for whereas all stars previously observed had been initiated by radiation not producing ionizing tracks in the emulsion, the one in question appears to be due to nuclear capture of a charged particle, presumably a slow meson. *(Extracted from the introductory part of the paper.)*

Accelerator COSMIC

Detectors Nuclear emulsion

Related references

See also
Williams, Proc. Roy. Soc. **A169** (1938) 531;
C. F. Powell, G. P. S. Occhialini, Liversey and Chilton, J. Sci. Instr. **23** (1946) 102;

Reactions

π^- nucleus $\rightarrow$ nucleus ^{3}H $2p$

Particles studied π^-

OCCHIALINI 1947	Nobel prize

■ Confirmation of the π^-. Nobel prize to C. F. Powell awarded in 1950 "for his development of the photographic method of studying nuclear processes and his discoveries regarding mesons made with his method" ■

Nuclear Disintegrations Produced by Slow Charged Particles of Small Mass

G.P.S. Occhialini, C.F. Powell
Nature **159** (1947) 186;

Abstract In studying photographic plates exposed to the cosmic rays, we have found a number of multiple disintegrations each of which appears to have been produced by the entry of a slow charged particle into a nucleus. *(Extracted from the introductory part of the paper.)*

Accelerator COSMIC

Detectors Nuclear emulsion

Related references

See also
A. Shapiro, Rev. of Mod. Phys. **13** (1941) 58;
Analyse information from
A. P. Zhdanov, N. A. Prefilov, and M. Y. Deisenrod, Phys. Rev. **65** (1944) 202;
S. Tomonaga and G. Araki, Phys. Rev. **58** (1940) 90;
F. Rasetti, Phys. Rev. **59** (1941) 706;
F. Rasetti, Phys. Rev. **60** (1941) 109;
G. C. Wick, M. Conversi, and E. Pancini, Phys. Rev. **60** (1941) 535;
G. Bernardini et al., Phys. Rev. **68** (1945) 109;
M. Conversi, E. Pancini, and O. Piccioni, Phys. Rev. **68** (1945) 232;

Reactions

π^- nucleus $\rightarrow$ 2nucleus

Particles studied π^-

KOBA 1947B	Nobel prize

■ Creation of the covariant quantum electrodynamic theory. Tomonaga method. Nobel prize to S. Tomonaga awarded in 1965. Co-winners J. S. Schwinger and R. P. Feynman "for their fundamental work in quantum electrodynamics, with deep-ploughing consequences for the physics of elementary particles" ■

On a Relativistically Invariant Formulation of the Quantum Theory of Wave Fields. III. Case of Interacting Electromagnetic and Electron Fields

Z. Koba, T. Tati, S. Tomonaga
Progr. of Theor. Phys. **2** (1947) 198;

Excerpt We have thus shown that it is in fact possible to describe the behavior of the electromagnetic field interacting with electrons, in a perfectly relativistic form according to our general scheme developed in I. But if we wish to apply similar methods to more general cases, for instance, to the cases of the meson field interacting with the electromagnetic field or the nucleon field, some generalization is necessary. In fact, in the case of the quantum electrodynamics the situation is exceptionally simple owing to the following two facts:

(i) the interaction, energy density $H_{I,II}(P)$ is a scalar function of the fields and

(ii) the integrability condition $[H_{I,II}(P), H_{I,II}(P\prime)] = 0$ is satisfied from the beginning.

These simplifying facts do no longer hold in the more general cases mentioned above. However, if some generalization of the formalism is made, it is possible to develop the similar theory also in these more complicated cases as will be shown in the later paper. The development of such a theory seems to us of interest, not only because we obtain thus formally more satisfactory theory, but also because in this way we can hope that some new aspects of the difficulties underlying the current quantum theory of the fields would reveal itself. *(Extracted from concluding remarks of the paper.)*

Related references

More (earlier) information
S. Tomonaga, Riken Iho **22** (1943) 525;
S. Tomonaga, Progr. of Theor. Phys. **1** (1946) 27;
Z. Koba, T. Tati, and S. Tomonaga, Progr. of Theor. Phys. **2** (1947) 101;
See also
W. Heisenberg, Z. Phys. **110** (1938) 251;
T. Miyazima, Riken Iho **23** (1944) 27;
P. A. M. Dirac, V. A. Fock, and B. Podolsky, Physik Zeits. Sow. **2** (1932) 468;

LAMB 1947 Nobel prize

■ First measurements of the fine structure of the hydrogen atom, the Lamb shift. Nobel prize to W. E. Lamb awarded in 1955 "for his discoveries concerning the fine structure of the hydrogen spectrum. Co-winner P. Kusch "for his precision determination of the magnetic moment of the electron" ■

Fine Structure of the Hydrogen Atom by a Microwave Method

W.E. Lamb, R.C. Retherford
Phys. Rev. **72** (1947) 241;

Reprinted in
The Physical Review — the First Hundred Years, AIP Press (1995) 148, 195.

Excerpt The spectrum of the simplest atom, hydrogen, has a fine structure[1] which according to the Dirac wave equation for an electron moving in a Coulomb field is due to the combined effects of relativistic variation of mass with velocity and spin-orbit coupling. It has been considered one of the great triumphs of Dirac's theory that it gave the "right" fine structure of the energy levels. However, the experimental attempts to obtain a really detailed confirmation through a study of the Balmer lines have been frustrated by the large Doppler effect of the lines in comparison to the small splitting of the lower or $n = 2$ states. The various spectroscopic workers have alternated between finding confirmations[2] of the theory and discrepancies[3] of as much as eight percent. More accurate information would clearly provide a delicate test of the form of the correct relativistic wave equation, as well as information on the possibility of line shifts due to coupling of the atom with the radiation field and clues to the nature of any non-Coulombic interaction between the elementary particles: electron and proton... *(Extracted from the introductory part of the paper.)*

Summary Hydrogen atoms are bombarded by an electron stream. Metastable $2^2S_{1/2}$ atoms are detected by electron ejection from a metal target. The current is reduced if radiofrequency radiation is applied, for which $h\nu$ corresponds to the energy difference between one of the Zeeman components of $2^2S_{1/2}$ and any component of the P levels. The results indicate that the $2^2S_{1/2}$, state is higher than the $2^2P_{1/2}$ by ~ 0.033 cm^{-1}, thereby showing that Dirac's theory is not exact. The method will be extended to measurements of hyperfine structures, etc.

[1] (H. E. White, *Introduction to Atomic Spectra* (McGraw-Hill Book Company, New York, 1934), Chap. 8).

[2] (J. W. Drinkwater et al., Proc. Roy. Soc. **174** (1940) 164).

[3] (W. V. Houston, Phys. Rev. **51** (1937) 446).

(Science Abstracts, 1947, 3605. L. P.)

Accelerator Non accelerator experiment

Detectors Nonelectronic detectors

Related references

More (later) information
W. E. Lamb and R. C. Retherford, Phys. Rev. **79** (1950) 549;

See also
J. W. Drinkwater, O. V. Richardson, and W. E. Williams, Proc. Roy. Soc. **A174** (1940) 164;
W. V. Houston, Phys. Rev. **51** (1937) 446;
R. C. Williams, Phys. Rev. **54** (1938) 558;
H. A. Bethe, Handbuch der Phys. **XXVI/1** (1933) 495f;
G. Breit and E. Teller, Astr. Jour. **91** (1940) 215;

Reactions
H(atom) $\to$ H(atom) γ

Particles studied H(atom)

PONTECORVO 1947

■ First idea about universality of the Fermi weak interactions ■

Nuclear Capture of Mesons and the Meson Decay

B. Pontecorvo
Phys. Rev. **72** (1947) 246;

Excerpt The experiment of Conversi, Panchini, and Piccioni indicates that the probability of capture of a meson by nuclei is much smaller than would be expected on the basis of Yukawa theory. Gamow has suggested that the nuclear forces are due exclusively to the exchange of neutral mesons, the processes involving charged mesons and the β-processes having probabilities which are smaller by a factor of about 10^{12}.
We notice that the probability ($\sim 10^6$) of capture of a bound negative meson is of the order of the probability of ordinary K-capture processes, when allowance is made for the difference in the disintegration energy and the difference in the volumes of the K-shell and of meson orbit. We assume that this is significant and wish to discuss the possibility of a fundamental analogy between β-processes and processes of emission or absorption of charged mesons. *(Extracted from the introductory part of the paper.)*

Related references

See also
T. Sigurgeirson and A. Yamakawa, Phys. Rev. **71** (1947) 319;
E. Fermi, E. Teller, and V. Weisskopf, Phys. Rev. **71** (1947) 314;
J. A. Wheeler, Phys. Rev. **71** (1947) 320;
W. Nordheim, Phys. Rev. **59** (1941) 544;
R. E. Marshak, Phys. Rev. **57** (1940) 1101;

G. Cocconi, A. Loveredo, and V. Tongiorgi, Phys. Rev. **70** (1946) 852;
G. Gamow and E. Teller, Phys. Rev. **51** (1937) 289;
Analyse information from
M. Conversi, E. Pancini, and O. Piccioni, Phys. Rev. **71** (1947) 209;

BETHE 1947

■ First theoretical calculation of the Lamb shift in non-relativistic QED ■

The Electromagnetic Shift of Energy Levels

H.A. Bethe
Phys. Rev. **72** (1947) 339;

Reprinted in
The Physical Review — the First Hundred Years, AIP Press (1995) 949.

Abstract The reported difference of energy of the *2s* and *2p* levels of H, in contradiction to the Dirac theory, is explained as an interaction energy of the electron with the radiation field. The calculated and observed values are in good agreement. The correction used is the difference of the interaction energies of the bound electron with the field, and that of a free electron with the same kinetic energy, since the latter is interpreted as part of the electronic mass. Calculation using non-relativistic radiation theory shows that this difference is still logarithmically divergent and ∞. A finite value is obtained by cutting off the spectrum at an upper energy limit of mc^2, since a relativistic calculation using hole theory is likely to give a finite result not differing essentially from this value. *(Science Abstracts, 1947, 3608. G. J. K.)*

Related references

See also
E. A. Uehling, Phys. Rev. **48** (1935) 55;
Analyse information from
W. E. Lamb and R. C. Retherford, Phys. Rev. **72** (1947) 241;
W. V. Houston, Phys. Rev. **51** (1937) 446;
R. C. Williams, Phys. Rev. **54** (1938) 558;
E. C. Kemble and R. D. Present, Phys. Rev. **44** (1933) 1031;
S. Pasternack, Phys. Rev. **54** (1938) 113;

MARSHALL 1947

■ Invention of scintillation counters ■

The Photo-Multiplier Radiation Detector

F. Marshall, J.W. Coltman
Phys. Rev. **72** (1947) 528;

Excerpt The photo-multiplier X ray detector, consisting of a multiplier photo-tube to measure the light emitted by a fluorescent screen, has been extended in its use to the measurement of alpha-, beta-, and gamma-rays, high energy electrons and protons, and neutrons. Improved light-gathering methods and high frequency circuits have resulted in registration of individual quanta of all these radiations as pulses rising above the noise level of the photo-multiplier's input-stage dark current of 10^5 electrons per second, permitting use of a discriminator-counter circuit. This removes the sensitivity limitation inherent when the instrument is used only for current measurement. At high intensities the detector can be switched from pulse counting to current measurement, with a wide range of overlap for calibration. Employing commercial photo-tubes and screens, this detector is simple, rugged, compact, and spectacularly fast. It competes favorably in sensitivity with all other detectors, including the Geiger counter, and it can cover a tremendous range of intensities. It is most suitable for use with narrow beams of radiation. The thin-window technique is unnecessary for beta-rays, since the phosphor is exposed. Dr. Kuan-Han Sun is responsible for extending the detector to neutron measurement, using boron impregnated phosphors to convert neutrons to alpha-particles.

KUSCH 1947 — Nobel prize

■ First measurement of $g-2$ for the electron, preliminary result. Nobel prize to P. Kusch awarded in 1955 "for his precision determination of the magnetic moment of the electron". Co-winner W. E. Lamb "for his discoveries concerning the fine structure of the hydrogen spectrum" ■

Precision Measurement of the Ratio of the Atomic "g Values" in the $^2P_{3/2}$ and $^2P_{1/2}$ States of Gallium

P. Kusch, H.M. Foley
Phys. Rev. **72** (1947) 1256;

Reprinted in
The Physical Review — the First Hundred Years, AIP Press (1995) 152.

Excerpt The measurement of the frequencies associated with the Zeeman splittings of the energy levels in two different atomic states in a constant magnetic field permits a determination of the ratio of the g_j values of the atomic states. This determination involves only an accurate measurement of the frequencies, and does not require a knowledge of the value of the constant magnetic field.
Using the atomic beam magnetic resonance technique we have measured six lines in the Zeeman spectrum of the $^2P_{1/2}$ state, and five lines of the $^2P_{3/2}$ state of gallium at a field strength of 380 gauss. The spectrum is, of course, complicated by the level splittings produced by the nuclear magnetic moments and electric quadrupole moments. At the field strength employed in this experiment the nuclear energy level pattern is of an intermediate Paschen-Back character.
The procedure employed in the observations was to make a se-

ries of alternate measurements of the frequencies of the lines in the $^2P_{1/2}$, and $^2P_{3/2}$ states. In this way the effect of a drift in magnetic field was minimized ... *(Extracted from the introductory part of the paper.)*

Accelerator Non accelerator experiment

Detectors Nonelectronic detectors

Related references

See also
N. A. Razetti, Phys. Rev. **57** (1940) 753;

Particles studied e^-

LATTES 1947	Nobel prize

■ First indication of the existence of π^- decay into μ^-. Nobel prize to C. F. Powell awarded in 1950 "for his development of the photographic method of studying nuclear processes and his discoveries regarding mesons made with his method" ■

Processes Involving Charged Mesons

C.M.G. Lattes, H. Muirhead, G.P.S. Occhialini, C.F. Powell
Nature **159** (1947) 694;

Excerpt In recent investigations with the photographic method, it has been shown that slow charged particles of small mass, present as a component of the cosmic radiation at high altitudes, can enter nuclei and produce disintegrations with the emission of heavy particles. It is convenient to apply the term "meson" to any particle with a mass intermediate between that of a proton and an electron. In continuing our experiments we have found evidence of mesons which, at the end of their range, produce secondary mesons. We have also observed transmutations in which slow mesons are ejected from disintegrating nuclei. Several features of these processes remain to be elucidated, but we present the following account of the experiments because the results appear to bear closely on the important problem of developing a satisfactory meson theory of nuclear forces. *(Extracted from the introductory part of the paper.)*

Accelerator COSMIC

Detectors Nuclear emulsion

Reactions
π^- nucleus $\to$ 2nucleus
π^- nucleus $\to$ 2nucleus meson
nucleus $\to$ nucleus meson
nucleus $\to$ 2nucleus meson
$\pi^- \to \mu^-$ neutral

Particles studied π^-

LATTES 1947B	Nobel prize

■ Confirmation of the π^-. First evidence for pion decay $\pi^\pm \to \mu^\pm$ neutrals. Nobel prize to C. F. Powell awarded in 1950 "for his development of the photographic method of studying nuclear processes and his discoveries regarding mesons made with his method" ■

Observation on the Tracks of Slow Mesons in Photographic Emulsions

C.M.G. Lattes, G.P.S. Occhialini, C.F. Powell
Nature **160** (1947) 453;

Reprinted in
R. N. Cahn and G. Goldhaber, *The Experimental Foundations of Particle Physics*, Cambridge Univ. Press (1991) 41.

Excerpt In recent experiments, it has been shown that charged mesons, brought to rest in photographic emulsions, sometimes lead to the production of secondary mesons. We have now extended these observations by examining plates exposed in the Bolivian Andes at a height of 5,500 m, and have found, in all, forty examples of the process leading to the production of secondary mesons. In eleven of these, the secondary particle is brought to rest in the emulsion so that its range can be determined. In Part I of this article, the measurements made on these tracks are described, and it is shown that they provide evidence for the existence of mesons of different mass. In Part 2, we present further evidence on the production of mesons, which allows us to show that many of the observed mesons are locally generated in the "explosive" disintegration of nuclei, and to discuss the relationship of the different types of mesons observed in photographic plates to the penetrating component of the cosmic radiation investigated in experiments with Wilson chambers and counters.
Our preliminary measurements appear to indicate, therefore, that the emission of the secondary meson cannot be regarded as due to a spontaneous decay of the primary particle, in which the momentum balance is provided by a photon, or by a particle of small rest-mass. On the other hand, the results are consistent with the view that a neutral particle of approximately the same rest-mass as the μ-meson is emitted. A final conclusion may become possible when further examples of the μ-decay, giving favorable conditions for grain-counts, have been discovered. *(Extracted from the introductory part of the paper.)*

Accelerator COSMIC

Detectors Nuclear emulsion

Related references

More (earlier) information
G. P. S. Occhialini and C. F. Powell, Nature **159** (1947) 93;
G. P. S. Occhialini and C. F. Powell, Nature **159** (1947) 186;

C. M. G. Lattes et al., Nature **159** (1947) 694;

More (later) information
C. M. G. Lattes, G. P. S. Occhialini, and C. F. Powell, Nature **160** (1947) 486;

Reactions
$\pi^{\pm} \rightarrow \mu^{\pm}$ neutral

Particles studied π^-, π^+

ROCHESTER 1947

■ First evidence for V events. Confirmation of the existence of a charged unstable particle with a mass between those of the muon and proton ■

Evidence for the Existence of New Unstable Elementary Particles

G.D. Rochester, C.C. Butler
Nature **160** (1947) 855;

Reprinted in
R. N. Cahn and G. Goldhaber, *The Experimental Foundations of Particle Physics*, Cambridge Univ. Press (1991) 69.

Excerpt Among some fifty counter-controlled cloud-chamber photographs of penetrating showers which we have obtained during the past year as part of an investigation of the nature of penetrating particles occurring in cosmic ray showers under lead, there are two photographs containing forked tracks of a very striking character. These photographs have been selected from five thousand photographs taken in an effective time of operation of 1,500 hours.
...We conclude, therefore, that the two forked tracks do not represent collision processes, but do represent spontaneous transformations. They represent a type of process with which we are already familiar in the decay of the meson into an electron and an assumed neutrino, and the presumed decay of the heavy meson recently discovered by Lattes, Occhialini and Powell (Nature **160** (1947) 453). *(Extracted from the introductory part of the paper.)*

Accelerator COSMIC

Detectors Cloud chamber

Comments on experiment 2 events with forked tracks from 5000 photographs.

Related references

See also
C. M. G. Lattes, G. P. S. Occhialini, and C. F. Powell, Nature **160** (1947) 453;
C. M. G. Lattes, G. P. S. Occhialini, and C. F. Powell, Nature **160** (1947) 486;
L. Leprince-Ringuet and L'heritier, J. Phys. Radium (ser.8) **7** (1947) 66, 69;
L. Janossy, G. D. Rochester, and D. Broadbent, Nature **155** (1945) 142;
H. A. Bethe, Phys. Rev. **70** (1946) 821;
J. Daudin, Anneles de Physique, 11e serie, 19(Avril-Juin), 1944;

Reactions
vee $\rightarrow$ 2charged
neutral $\rightarrow$ 2charged
charged $\rightarrow$ charged (neutrals)

HINCKS 1948

■ Nonexistence of $\mu^- \rightarrow e^- \gamma$ decay ■

Search for Gamma-Radiation in the 2.2-Microsecond Meson Decay Process

E.P. Hincks, B. Pontecorvo
Phys. Rev. **73** (1948) 257;

Excerpt The meson decay process which is identified by a mean life of 2.2 microseconds has been usually thought of as consisting of the emission of an electron and a single neutrino, as suggested by the well-known Yukawa explanation of the ordinary beta-process in nuclei. However, the Yukawa theory is at variance with the results of the experiment of Conversi, Pancini, and Piccioni, and since there remains no strong justification for the electron-neutrino hypothesis, a direct experiment to test an alternative hypothesis — *that the decay process consists of the emission of an electron and a photon, each of about 50 MeV* has been performed. *(Extracted from the introductory part of the paper.)*

Accelerator COSMIC

Detectors Counters

Related references

Analyse information from
N. Nereson and B. Rossi, Phys. Rev. **64** (1942) 199;
M. Conversi and O. Piccioni, Phys. Rev. **70** (1946) 859;
M. Conversi, E. Pancini, and O. Piccioni, Phys. Rev. **71** (1947) 209;
J. A. Wheeler, Phys. Rev. **71** (1947) 320;
B. Pontecorvo, Phys. Rev. **72** (1947) 246;
G. Bernardini, B. N. Cacciapuoti, and B. Querzoli, Nuovo Cim. **3** (1946) 349;
C. D. Anderson, R. V. Adams, P. E. Lloyd, and R. R. Rau, Phys. Rev. **72** (1947) 724;
C. M. G. Lattes, G. P. S. Occhialini, and C. F. Powell, Nature **160** (1947) 453;
C. M. G. Lattes, G. P. S. Occhialini, and C. F. Powell, Nature **160** (1947) 486;

Reactions
$\mu^- \rightarrow e^- \gamma$

Particles studied μ^-, e^-

ROSE 1948

■ First explicit theoretical paper on the investigation of the electromagnetic structure of a nucleus by electron-nucleus scattering ■

The Charge Distribution in Nuclei and the Scattering of High Energy Electrons

M.E. Rose
Phys. Rev. **73** (1948) 279;

Abstract It is pointed out that the finite size of the nucleus will give rise to large deviations from Mott scattering when the change in wave-length of the electrons is of order of the nuclear dimensions. This deviation from Mott scattering at large scattering angles therefore provides a possibility for determination of the shape of the charge distribution and size of nuclei. In the case of a spherically symmetric charge distribution the nuclear charge density is immediately obtained from the observed angular distribution by a Fourier transform. The effects of competing processes, inelastic collisions with nuclear excitation or disintegration, atomic excitation or ionization and bremstrahlung are considered. It is shown that the first two competing effects may be disregarded if the electron energy is in the neighbourhood of 50 MeV, the angle of scattering large (but not near π) and if the scattered electron has an energy equal to or nearly equal to the primary energy. With the latter condition fulfilled the bremstrahlung is reduced by the same factor as the elastic scattering and the two processes are indistinguishable.

Related references

See also
H. A. Bethe, Handbuch der Phys. **XXVI/1** (1933) 495f;

Reactions
e^- nucleus $\rightarrow$ nucleus e^- 50 MeV(P_{lab})

Particles studied nucleus

FOLEY 1948 Nobel prize

■ First measurement of $g-2$ for the electron. Nobel prize to P. Kusch awarded in 1955 "for his precision determination of the magnetic moment of the electron". Co-winner W. E. Lamb "for his discoveries concerning the fine structure of the hydrogen spectrum" ■

On the Intrinsic Moment of the Electron

H.M. Foley, P. Kusch
Phys. Rev. **73** (1948) 412;

Excerpt In a previous letter[1] we have reported the observation that the ratio of the g_J values of the $^2P_{3/2}$ and $^2P_{1/2}$ states of gallium has the value 2.00344; the value 2 for this ratio follows from Russell-Saunders Coupling and the conventional spin and orbital gyromagnetic ratios. If each of these states is exactly described by Russell-Saunders coupling, this observation can only be explained by setting $(\delta_S - 2\delta_L) = 0.00229 \pm 0.00008$, where the electron spin g value is $g_S = 2 + \delta_S$, and the orbital momentum g value is $g_L = 1 + \delta_L$. Since each of these atomic states may be separately subject to configuration interaction perturbations, the interpretation of this result was not entirely clear.
A determination has now been made of the ratio of the g_J values of Na in the $^2S_{1/2}$ state and of Ga in the $^2P_{1/2}$ state. The experimental procedure was similar to that previously described.[1] The known hyperfine interaction constants of gallium[2] and sodium[3] were employed in the analysis of the data. We find for this ratio the value 3.00732 ± 0.00018 instead of the value 3. This result can be explained by making $(\delta_S - 2\delta_L) = 0.00244 \pm 0.00006$.
The agreement between the values of $(\delta_S - 2\delta_L)$ obtained by the two experiments makes it unlikely that one can account for the effect by perturbation of the states. The effect of configuration interaction on the g_J value of sodium is presumably negligible.[4] To explain our observed effect without modification of the conventional values of g_S or g_L introduces the rather unlikely requirement that both states of gallium be perturbed, and by amounts just great enough to give the agreement noted above.
From any experiment in which the ratio of the g_J values of atomic states is determined, it is possible to determine only the quantity $(\delta_S - 2\delta_L)$. If, on the basis of the correspondence principle we set δ_L equal to zero, we may state the result of our first experiment as

$$g_S = 2.00229 \pm 0.00008$$

and that of our recent experiment as

$$g_S = 2.00244 \pm 0.00006.$$

It is not possible, at the present time, to state whether the apparent discrepancy between these values is real. It is conceivable that some small perturbation of the states would give rise to a discrepancy of the indicated magnitude.
These results are not in agreement with the rec ent suggestion by Breit[6] as to the magnitude of the intrinsic moment of the electron.

[1] P. Kusch and H. M. Foley, Phys. Rev. **72** (1947) 1256.
[2] G. E. Becker and P. Kusch, to be published.
[3] S. Millman and P. Kusch. Phys. Rev. **58** (1940) 438.
[4] M. Phillips. Phys. Rev. **60** (1941) 100.
[5] Dr. J. Schwinger has very kindly informed us in advance of publication of his conclusion from theoretical studies that δ_L is zero whereas δ_S may not vanish.

[6]G. Breit, Phys. Rev. **72** (1947) 984.

Related references

More (earlier) information
P. Kusch and H. M. Foley, Phys. Rev. **72** (1947) 1256;

See also
S. Millman and P. Kusch, Phys. Rev. **60** (1941) 100;
G. Breit, Phys. Rev. **72** (1947) 984;

Particles studied e^-

SCHWINGER 1948B

■ First theoretical calculation of $g - 2$ for the electron ■

On Quantum-Electrodynamics and the Magnetic Moment of the Electron

J. Schwinger
Phys. Rev. **73** (1948) 416;

Reprinted in
The Physical Review — the First Hundred Years, AIP Press (1995) 167.

Excerpt Attempts to evaluate radiative corrections to electron phenomena have heretofore been beset by divergence difficulties, attributable to self-energy and vacuum polarization effects. Electrodynamics unquestionably requires revision at ultra-relativistic energies, but is presumably accurate at moderate relativistic energies. It would be desirable, therefore, to isolate those aspects of the current theory that essentially involve high energies, and are subject to modification by a more satisfactory theory, from aspects that involve only moderate energies and are thus relatively trustworthy. This goal has been achieved by transforming the Hamiltonian of current hole theory electrodynamics to exhibit explicitly the logarithmically divergent self-energy of a free electron, which arises from the virtual emission and absorption of light quanta. The electromagnetic self-energy of a free electron can by ascribed to an electromagnetic mass, which must be added to the mechanical mass of the electron. Indeed, the only meaningful statements of the theory involve this combination of masses, which is the experimental mass of a free electron. It might appear, from this point of view, that the divergence of the electromagnetic mass is unobjectionable, since the individual contributions to the experimental mass are unobservable. However, the transformation of the Hamiltonian is based on the assumption of a weak interaction between matter and radiation, which requires that the electromagnetic mass be a small correction ($\sim (e^2/\hbar c)m_0$) to the mechanical mass m_o.

Summary The simplest example of a radiative correction is that for the energy of an electron in an external magnetic field. The detailed application of the theory shows that the radiative correction to the magnetic interaction energy, corresponds to an additional magnetic moment associated with the electron spin, of magnitude $\delta\mu/\mu = (\frac{1}{2}\pi)e^2/\hbar c = 0.001162$. It is indeed gratifying that recently acquired experimental data confirm this prediction. *(Extracted from the introductory part of the paper.)*

Related references

See also
J. E. Nafe, E. B. Nelson, and I. I. Rabi, Phys. Rev. **71** (1947) 914;
D. E. Nagle, R. S. Julian, and J. R. Zacharias, Phys. Rev. **72** (1947) 971;
G. Breit, Phys. Rev. **71** (1947) 984;
P. Kusch and H. M. Foley, Phys. Rev. **72** (1947) 1256;
W. E. Lamb and R. C. Retherford, Phys. Rev. **72** (1947) 241;
J. E. Mack and N. Austern, Phys. Rev. **72** (1947) 972;
H. A. Bethe, Phys. Rev. **72** (1947) 339;

Particles studied e^-, e^+

SARD 1948

■ Nonexistence of $\mu^- \to e^- \gamma$ decay ■

Test of the Hypothesis that the Sea-Level Cosmic-Ray Meson Disintegrates into a Photon and an Electron

R.D. Sard, E.J. Althauss
Phys. Rev. **73** (1948) 1251;

Excerpt The hypothesis has been tested by searching for delayed coincidences between the 50 MeV photon, detected by its materialization in a Pb sheet and the arrival of the meson that gets stopped. *(Extracted from the introductory part of the paper.)*

Accelerator COSMIC

Detectors Counters

Related references

Analyse information from
B. Rossi, M. Sands, and R. D. Sard, Phys. Rev. **72** (1947) 120;

Reactions
$\mu^- \to e^- \gamma$

Particles studied μ^-

FEYNMAN 1948

■ Proposal to modify classical electrodynamics to a form suitable for quantization ■

A Relativistic Cut-Off for Classical Electrodynamics

R.P. Feynman
Phys. Rev. **74** (1948) 939;

Abstract Ordinarily it is assumed that interaction between charges occurs along light cones, that is, only where the four-dimensional interval $s^2 = t^2 - r^2$ is exactly zero. We discuss the modifications produced if, as in the theory of F. Bopp, substantial interaction is assumed to occur over a narrow range of s^2 around zero. This has no practical effect on the interaction of charges which are distant from one another by several electron radii. The action of a charge on itself is finite and behaves as electromagnetic mass for accelerations which are not excessive. There also results a classical representation of the phenomena of pair production in sufficiently strong fields.

Related references

See also
M. Born and L. Infeld, Proc. Roy. Soc. **A144** (1934) 425;
P. A. M. Dirac, Proc. Roy. Soc. **A167** (1938) 148;
C. J. Eliezer, Rev. of Mod. Phys. **19** (1947) 147;
J. A. Wheeler and R. P. Feynman, Rev. of Mod. Phys. **17** (1945) 157;
F. Bopp, Annalen der Physik. Leipzig **42** (1942) 573;
B. Podolsky and P. Schwed, Rev. of Mod. Phys. **20** (1948) 40;
N. Rosen, Phys. Rev. **72** (1947) 298;

SNELL 1948

■ First evidence for neutron beta decay ■

On the Decay of the Neutron

A.H. Snell, L.C. Miller
Phys. Rev. **74** (1948) 1217;

Excerpt A collimated beam of neutrons, three inches in diameter, emerges from the nuclear reactor and passes axially through a thin-walled, aluminum, evacuated cylindrical tank. A transverse magnetic field behind the thin entrance window cleans the beam of secondary electrons. Inside the vacuum, axially arranged, an open-sided cylindrical electrode is held at +4000 volts with respect to ground. Opposite the open side a smoothed graphite plate is held at −4400 volts. The field between these electrodes accelerates and focuses protons which may result from decay of neutrons, so that they pass through a $2\frac{7}{8} \times 1\frac{5}{8}$ inch aperture in the center of the graphite plate, and strike the first dynode of a secondary electron multiplier. The first dynode is specially enlarged so as to cover the aperture. Readings are taken

(1) with and without a thin B^{20} shutter in the neutron beam;
(2) with and without a thin foil over the multiplier aperture;
(3) with and without the accelerating voltage.

In a total counting rate of about 300 per min., about 100 are sensitive to operations (1), (2), and (3). In the absence of the accelerating field or with the foil (2) in, operation (1) does not change the counting rate. Assuming all of the 100 c.p.m. to be due to decay protons, preliminary estimates of the collecting and counting efficiency (10 percent) and of the number of neutrons in the sample (4×10^4) give for the neutron a half-life of about 30 minutes. It is at present much safer however to say that the neutron half-life must exceed 15 minutes. Coincidences are presently being sought between the disintegration betas and the collected protons.

Accelerator Nuclear reactor

Detectors Counters

Reactions
$n \to p\, e^-$ neutral

Particles studied n

FEYNMAN 1948B — Nobel prize

■ Creation of the covariant theory of quantum electrodynamics. Feynman method. Nobel prize to R. P. Feynman awarded in 1965. Co-winners J. S. Schwinger and S. Tomonaga "for their fundamental work in quantum electrodynamics, with deep-ploughing consequences for the physics of elementary particles" ■

Relativistic Cut-Off for Quantum Electrodynamics

R.P. Feynman
Phys. Rev. **74** (1948) 1430;

Abstract A relativistic cut-off of high frequency quanta, similar to that suggested by Bopp, is shown to produce a finite invariant self-energy for a free electron. The electromagnetic line shift for a bound electron comes out as given by Bethe and Weisskopf's wave packet prescription. The scattering of an electron in a potential, without radiation, is discussed. The cross section remains finite. The problem of polarization of the vacuum is not solved. Otherwise, the results will in general agree essentially with those calculated by the prescription of Schwinger. An alternative cut-off procedure analogous to one proposed by Wataghin, which eliminates high frequency intermediate states, is shown to do the same things but to offer to solve vacuum polarization problems as well.

Related references

More (earlier) information
R. P. Feynman, Phys. Rev. **74** (1948) 939;
See also
H. A. Bethe, Phys. Rev. **72** (1947) 339;
H. A. Bethe, Phys. Rev. **73** (1948) 1271A;
H. W. Lewis, Phys. Rev. **73** (1948) 173;
V. Weisskopf, Phys. Rev. **56** (1939) 72;
F. Bloch and A. Nordsieck, Phys. Rev. **52** (1937) 54;
E. A. Uehling, Phys. Rev. **48** (1935) 55;
G. Wataghin, Z. Phys. **88** (1934) 92;
R. Serber, Phys. Rev. **48** (1935) 49;
W. Pauli, Handbuch der Phys. **24/1** (1933) 233;
S. M. Dancoff, Phys. Rev. **55** (1939) 959;
J. Schwinger, Phys. Rev. **73** (1948) 415A;
J. Schwinger and V. F. Weisskopf, Phys. Rev. **73** (1948) 1272A;
Analyse information from
F. Bopp, Annalen der Physik. Leipzig **42** (1942) 573;

SCHWINGER 1948 Nobel prize

■ Creation of the covariant quantum electrodynamic theory. Schwinger method. Nobel prize to J. S. Schwinger awarded in 1965. Co-winners S. Tomonaga and R. P. Feynman "for their fundamental work in quantum electrodynamics, with deep-ploughing consequences for the physics of elementary particles" ■

Quantum Electrodynamics. I. A Covariant Formulation

J. Schwinger
Phys. Rev. **74** (1948) 1439;

Reprinted in
The Physical Review — the First Hundred Years, AIP Press (1995) CD-ROM.

Abstract Attempts to avoid the divergence difficulties of quantum electrodynamics by mutilation of the theory have been uniformly unsuccessful. The lack of convergence does indicate that a revision of electrodynamic concepts at ultrarelativistic energies is indeed necessary, but no appreciable alteration of the theory for moderate relativistic energies can be tolerated. The elementary phenomena in which divergencies occur, in consequence of virtual transitions involving particles with unlimited energy, are the polarization of the vacuum and the self-energy of the electron, effects which essentially express the interaction of the electromagnetic and matter fields with their own vacuum fluctuations. The basic result of these fluctuation interactions is to alter the constants characterizing the properties of the individual fields, and their mutual coupling, albeit by infinite factors. The question is naturally posed whether all divergencies can be isolated in such unobservable renormalization factors; more specifically, we inquire whether quantum electrodynamics can account unambiguously for the recently observed deviations from the Dirac electron theory, without the introduction of fundamentally new concepts. This paper, the first in a series devoted to the above question, is occupied with the formulation of a completely covariant electrodynamics. Manifest covariance with respect to Lorentz and gauge transformations is essential in a divergent theory since the use of a particular reference system or gauge in the course of calculation can result in a loss of covariance in view of the ambiguities that may be the concomitant of infinities. It is remarked, in the first section, that the customary canonical commutation relations, which fail to exhibit the desired covariance since they refer to field variables at equal times and different points of space, can be put in covariant form by replacing the four-dimensional surface t=const. by a space-like surface. The latter is such that light signals cannot be propagated between any two points on the surface. In this manner, a formulation of quantum electrodynamics is constructed in the Heisenberg representation, which is obviously covariant in all its aspects. It is not entirely suitable, however, as a practical means of treating electrodynamic questions, since commutators of field quantities at points separated by a time-like interval can be constructed only by solving the equations of motion. This situation is to be contrasted with that of the Schrödinger representation, in which all operators refer to the same time, thus providing a distinct separation between kinematical and dynamical aspects. A formulation that retains the evident covariance of the Heisenberg representation, and yet offers something akin to the advantage of the Schrödinger representation can be based on the distinction between the properties of non-interacting fields, and the effects of coupling between fields. In the second section, we construct a canonical transformation that changes the field equations in the Heisenberg representation into those of non-interacting fields, and therefore describes the coupling between fields in terms of a varying state vector. It is then a simple matter to evaluate commutators of field quantities at arbitrary space-time points. One thus obtains an obviously covariant and practical form of quantum electrodynamics, expressed in a mixed Heisenberg-Schrödinger representation, which is called the interaction representation. The third section is devoted to a discussion of the covariant elimination of the longitudinal field, in which the customary distinction between longitudinal and transverse fields is replaced by a suitable covariant definition. The fourth section is concerned with the description of collision processes in terms of an invariant collision operator, which is the unitary operator that determines the over-all change in state of a system as the result of interaction. It is shown that the collision operator is simply related to the Hermitian reaction operator, for which a variational principle is constructed.

Related references

See also
P. A. M. Dirac, Proc. Camb. Phil. Soc. **30** (1934) 150;
W. Heisenberg, Z. Phys. **90** (1934) 209;
W. Heitler and H. W. Peng, Proc. Camb. Phil. Soc. **38** (1942) 296;
R. Serber, Phys. Rev. **49** (1936) 545;
H. A. Bethe and J. R. Oppenheimer, Phys. Rev. **70** (1946) 451;
V. F. Weisskopf, Phys. Rev. **56** (1939) 72;
W. E. Lamb and R. C. Retherford, Phys. Rev. **72** (1947) 241;
J. E. Mack and N. Austern, Phys. Rev. **72** (1947) 972;
J. E. Nafe, E. B. Nelson, and I. I. Rabi, Phys. Rev. **71** (1947) 914;
D. E. Nagle, R. S. Julian, and J. R. Zacharias, Phys. Rev. **72** (1947) 971;
P. Kusch and H. M. Foley, Phys. Rev. **72** (1947) 1256;
H. M. Foley and P. Kusch, Phys. Rev. **73** (1948) 412;
H. A. Bethe, Phys. Rev. **72** (1947) 339;
J. Schwinger, Phys. Rev. **72** (1947) 742;
J. Schwinger, Phys. Rev. **73** (1948) 416;
S. Tomonaga, Progr. of Theor. Phys. **1** (1946) 27;
P. A. M. Dirac, Phys. Rev. **73** (1948) 1092;

PICCIONI 1948

■ Confirmation of the nonexistence of $\mu^- \to e^- \gamma$ decay ■

Search for Photons from Meson-Capture

O. Piccioni

Phys. Rev. **74** (1948) 1754;

Abstract High energy photons (about 50 MeV) associated with mesons stopped in iron have been searched for. The experimental evidence shows that no such photons arise from the capture of negative mesons or from the free decay of positive ones.

Accelerator COSMIC

Detectors Counters, Photon spectrometer

Related references

More (earlier) information
O. Piccioni, Phys. Rev. **73** (1948) 411;

See also
J. A. Wheeler, Phys. Rev. **71** (1947) 320;
F. Rasetti, Phys. Rev. **60** (1941) 198;
Rossi and Greisen, Rev. of Mod. Phys. **13** (1941) 240;

Analyse information from
S. Tomonaga and G. Araki, Phys. Rev. **58** (1940) 90;
M. Conversi, E. Pancini, and O. Piccioni, Phys. Rev. **68** (1945) 232;
E. G. Valley, Phys. Rev. **72** (1947) 772;
C. M. G. Lattes and E. Gardner, Bull.Am.Phys.Soc. **23** (1948) 42;
R. D. Sard and E. J. Althaus, Bull.Am.Phys.Soc. **23** (1948) 20;
E. P. Hincks and B. Pontecorvo, Phys. Rev. **73** (1948) 257;

Reactions

$\mu^- \to e^- \gamma$

Particles studied μ^-, e^-

TATI 1948 Nobel prize

■ Creation of the covariant quantum electrodynamic theory. Tomonaga method. Nobel prize to S. Tomonaga awarded in 1965. Co-winners J. S. Schwinger and R. P. Feynman "for their fundamental work in quantum electrodynamics, with deep-ploughing consequences for the physics of elementary particles" ■

A Self-Consistent Subtraction Method in the Quantum Field Theory. I.

T. Tati, S. Tomonaga

Progr. of Theor. Phys. **3** (1948) 391;

Abstract As is well known the present formalism of the quantum field theory contains a fundamental defect which reveals itself most directly in the infinite self-energy of elementary particles. The infinities of the same nature appear also when one deals with collision problems involving field reactions, e.g. radiative correction to the cross section for the elastic scattering of an electron in an external field of force. As was discussed by Pauli and Fierz in the nonrelativistic case and by Dancoff in the relativistic, the correction due to the radiation reaction in the latter problem turned out infinite. Such a circumstance implies that the satisfactory solution of this problem would be only reached when the fundamental difficulty of the current quantum field theory has found its ultimate solution.
In such a situation, one used to resort to some procedure to get rid of the difficulty, such as cutting off of high frequency effects or subtracting infinite terms by a suitable prescription. But it goes without saying that such procedures are only makeshifts far from true solution having neither theoretical basis nor any connection with experimental facts.
A remarkable progress was brought about; however, when Lamb and Retherford confirmed the level-shift of the hydrogen atom by their ingenious experiment and Schwinger, Weisskopf, Oppenheimer and Bethe gave its theoretical interpretation in terms of the radiative reaction. Especially Bethe proposed a method how to manage the infinity in this problem and enabled us to treat the field-reaction problem for the first time in close connection with reliable experimental data ...
It becomes thus desirable to obtain a relativistic generalization of the canonical transformation of Pauli and Fierz and find a general prescription to separate the infinite terms which can be amalgamated into the mass of the electron. In this and the following papers it will be shown that such a generalization is in fact possible, and the field reaction problems can be treated in a consistent way without touching the inherent infinities of the current quantum field theory.

Related references

More (earlier) information
S. Tomonaga, Progr. of Theor. Phys. **1** (1946) 27;
Z. Koba, T. Tati, and S. Tomonaga, Progr. of Theor. Phys. **2** (1947) 101;
Z. Koba, T. Tati, and S. Tomonaga, Progr. of Theor. Phys. **2** (1947) 198;

See also
W. Pauli and M. Fierz, Nuovo Cim. **15** (1938) 167;
S. M. Dancoff, Phys. Rev. **55** (1939) 959;
W. E. Lamb and R. C. Retherford, Phys. Rev. **72** (1947) 241;
H. A. Bethe, Phys. Rev. **72** (1947) 339;
F. Bloch and A. Nordsieck, Phys. Rev. **52** (1937) 54;
Z. Koba and S. Tomonaga, Progr. of Theor. Phys. **2** (1947) 218;
Z. Koba and S. Tomonaga, Progr. of Theor. Phys. **3** (1948) 208;
Z. Koba and G. Takeda, Progr. of Theor. Phys. **3** (1948) 98;
Z. Koba and G. Takeda, Progr. of Theor. Phys. **3** (1948) 203;
Z. Koba and G. Takeda, Progr. of Theor. Phys. **3** (1948) 387;
H. W. Lewis, Phys. Rev. **73** (1948) 173;
S. T. Epstein, Phys. Rev. **75** (1949) 177;
J. Schwinger, Phys. Rev. **73** (1948) 416;
V. F. Weisskopf, Phys. Rev. **56** (1939) 72;

FEYNMAN 1948C

■ Invention of the path integral formalism for quantum mechanics ■

Space-Time Approach to Non-Relativistic Quantum Mechanics

R.P. Feynman
Rev. of Mod. Phys. **20** (1948) 367;

Abstract Non-relativistic quantum mechanics is formulated here in a different way. It is, however, mathematically equivalent to the familiar formulation. In quantum mechanics the probability of an event which can happen in several different ways is the absolute square of a sum of complex contributions, one from each alternative way. The probability that a particle will be found to have a path $x(t)$ lying somewhere within a region of space time is the square of a sum of contributions, one from each path in the region. The contribution from a single path is postulated to be an exponential whose (imaginary) phase is the classical action (in units of $\hbar$) for the path in question. The total contribution from all paths reaching x, t from the past is the wave function $\psi(x, t)$. This is shown to satisfy Schrödinger's equation. The relation to matrix and operator algebra is discussed. Applications are indicated, in particular to eliminate the coordinates of the field oscillators from the equations of quantum electrodynamics.

Related references

See also
P. A. M. Dirac, Physik Zeits. Sow. **3** (1933) 64;
P. A. M. Dirac, *The Principles of Quantum Mechanics*, the Clarendon Press, Oxford (1935);
P. A. M. Dirac, Rev. of Mod. Phys. **17** (1945) 195;
W. Heisenberg, *The Physical Principles of the Quantum Theory* University of Chicago Press, Chicago (1930);
E. Schrödinger, Annalen der Physik. Leipzig **79** (1926) 489;

BROWN 1949B — Nobel prize

■ First evidence for three-prong kaon decay. Nobel prize to C. F. Powell awarded in 1950 "for his development of the photographic method of studying nuclear processes and his discoveries regarding mesons made with his method" ■

Observations with Electron Sensitive Plates Exposed to Cosmic Radiation. II. Further Evidence for the Existence of Unstable Charged Particles of Mass ~ 1000 m_e and Observations on their Mode of Decay

R.H. Brown et al.
Nature **163** (1949) 82;

Reprinted in
R. N. Cahn and G. Goldhaber, *The Experimental Foundations of Particle Physics*, Cambridge Univ. Press (1991) 72.

Abstract An event is described which offers further evidence for the existence of mesons of mass > π-particles. A particle, of which the mass, as determined by grain counts, is $\sim 1000 m_e$, comes to rest in the photographic emulsion and leads to the emission of three charged particles, one of which is apparently a π-meson. Evidence is given which suggests that the event corresponds to the spontaneous disintegration of the heavy particle into 3 charged mesons, whose masses are in the 200 - 300 m_e, region. *(Science Abstract, 1949, 1974, U. C.)*

Accelerator COSMIC

Detectors Nuclear emulsion

Related references

More (earlier) information
R. H. Brown et al., Nature **163** (1949) 47;
See also
L. Livingston and H. A. Bethe, Rev. of Mod. Phys. **9** (1937) 263;
Analyse information from
L. Leprince-Ringuet, Compt. Ren. **226** (1948) 1897;
G. D. Rochester and C. C. Butler, Nature **160** (1947) 855;
Alichanian, Alichanov, and J. Weissenberg, Zh. Eksp. Teor. Fiz. **18** (1948) 301;
U. Camerini, H. Muirhead, C. F. Powell, and D. M. Ritson, Nature **162** (1948) 433;
Goldschmidt-Clermont, C. F. Powell, and D. M. Ritson, Proc. Roy. Soc. **61** (1948) 138;
C. M. G. Lattes, G. P. S. Occhialini, and C. F. Powell, Proc. Roy. Soc. **61** (1948) 173;
C. Dilworth, G. P. S. Occhialini, and Payne, Nature **162** (1948) 102;
G. P. S. Occhialini and C. F. Powell, Nature **162** (1948) 168;
O. Halpern and H. Hall, Phys. Rev. **73** (1948) 477;

Reactions

$K^+ \rightarrow$ 2charged$^+$ charged$^-$
$K^+ \rightarrow 2\pi^+ \pi^-$

BURFENING 1949

■ First pion production reaction by accelerator ■

Positive Mesons Produced by the 184-Inch Berkeley Cyclotron

J. Burfening, E. Gardner, C.M.G. Lattes
Phys. Rev. **75** (1949) 382;

Abstract Positive mesons produced by 380 MeV α-particles in the 184-inch Berkeley cyclotron have been detected by means of photographic plates. The experimental arrangement is similar to that used for detecting negative mesons except that the plates are placed in a position to receive positive instead of negative particles from the target. Heavy positive mesons are observed to decay into secondary mesons in the manner described by Lattes, Occhialini, and Powell. Relative numbers of positive and negative mesons coming from a target are found by placing plates symmetrically on opposite sides of the target. Preliminary results indicate that for a $\frac{1}{16}$-inch carbon target there are about one-fourth as many heavy positive mesons as heavy negative ones for meson energies of 2-3 MeV in the laboratory system.

Accelerator LBL-CYC-184IN

Detectors Nuclear emulsion

Related references

See also
E. Gardner and C. M. G. Lattes, Science **107** (1948) 270;
W. M. Brobeck, E. O. Lawrence et al., Phys. Rev. **71** (1947) 449;

Reactions
^{4}He nucleus $\rightarrow$ π^+ X 380 MeV(T_{lab})

DYSON 1949

■ Covariant quantum electrodynamics: Equivalence between the Tomonaga-Schwinger method and the Feynman method and generalization ■

The Radiation Theories of Tomonaga, Schwinger, and Feynman

F.J. Dyson
Phys. Rev. **75** (1949) 486;

Reprinted in
The Physical Review — the First Hundred Years, AIP Press (1995) 952.

Abstract A unified development of the subject of quantum electrodynamics is outlined, embodying the main features both of the Tomonaga-Schwinger and of the Feynman radiation theory. The theory is carried to a point further than that reached by these authors, in the discussion of higher order radiative reactions and vacuum polarization phenomena. However, the theory of these higher order processes is a program rather than a definitive theory, since no general proof of the convergence of these effects is attempted.
The chief results obtained are

(a) a demonstration of the equivalence of the Feynman and Schwinger theories, and

(b) a considerable simplification of the procedure involved in applying the Schwinger theory to particular problems, the simplification being the greater the more complicated the problem.

Related references

See also
S. Tomonaga, Progr. of Theor. Phys. **1** (1946) 27;
S. Tomonaga, Phys. Rev. **74** (1948) 224;
Z. Koba, T. Tati, and S. Tomonaga, Progr. of Theor. Phys. **2** (1947) 101;
Z. Koba, T. Tati, and S. Tomonaga, Progr. of Theor. Phys. **2** (1947) 198;
S. Kanesawa and S. Tomonaga, Progr. of Theor. Phys. **3** (1948) 1;
J. Schwinger, Phys. Rev. **73** (1948) 416;
J. Schwinger, Phys. Rev. **74** (1948) 1439;
R. P. Feynman, Rev. of Mod. Phys. **20** (1948) 367;
R. P. Feynman, Phys. Rev. **74** (1948) 939;
R. P. Feynman, Phys. Rev. **74** (1948) 1430;
J. A. Wheeler and R. P. Feynman, Rev. of Mod. Phys. **17** (1945) 157;
Z. Koba and G. Takeda, Progr. of Theor. Phys. **3** (1948) 205;
W. Heisenberg, Z. Phys. **120** (1943) 513;
W. Heisenberg, Z. Phys. **120** (1943) 673;
W. Heisenberg, Zeit. Naturforschung **1** (1946) 608;
W. Pauli, Rev. of Mod. Phys. **13** (1941) 203;
R. Serber, Phys. Rev. **48** (1935) 49;
E. A. Uehling, Phys. Rev. **48** (1935) 55;

SCHWINGER 1949 Nobel prize

■ Creation of the covariant quantum electrodynamic theory. Schwinger method. Nobel prize to J. S. Schwinger awarded in 1965. Co-winners S. Tomonaga and R. P. Feynman "for their fundamental work in quantum electrodynamics, with deep-ploughing consequences for the physics of elementary particles" ■

Quantum Electrodynamics. II. Vacuum Polarization and Self–Energy

J. Schwinger
Phys. Rev. **75** (1949) 651;

Reprinted in
The Physical Review — the First Hundred Years, AIP Press (1995) CD-ROM.

Abstract The covariant formulation of quantum electrodynamics, developed in a previous paper, is here applied to two elementary problems—the polarization of the vacuum and the self-energies of the electron and photon. In the first section the vacuum of the non-interacting electromagnetic and matter fields is covariantly defined as that state for which the eigenvalue of an arbitrary time-like component of the energy-momentum four-vector is an absolute minimum. It is remarked that this definition must be compatible with the requirement that the vacuum expectation values of a physical quantity in various coordinate systems should be, not only covariantly related, but identical, since the vacuum has a significance that is independent of the coordinate system. In order to construct a suitable characterization of the vacuum state vector, a covariant decomposition of the field operators into positive and negative frequency components is introduced, and the properties of these associated fields developed. It is shown that the state vector for the electromagnetic vacuum is annihilated by the positive frequency part of the transverse four-vector potential, while that for the matter vacuum is annihilated by the positive frequency part of the Dirac spinor and of its charge conjugate. These defining properties of the vacuum state vector are employed in the calculation of the vacuum expectation values of quadratic field quantities, specifically the energy-momentum tensors of the independent electromagnetic and matter fields, and the current four-vector. It is inferred that the electromagnetic energy-momentum tensor and the current vector must vanish in the vacuum, while the matter field energy-momentum tensor vanishes in the vacuum only by the addition of a suitable multiple of the unit tensor. The second section treats the induction of a current in the vacuum by an external electromagnetic field. It is supposed that the latter does not produce actual electron-positron pairs; that is, we consider only the phenomenon of virtual pair creation. This restriction is introduced by requiring that the establishment and subsequent removal of the external field produce no net change in state for the matter field. It is demonstrated, in a general manner, that the induced current at a given space-time point involves the external current in the vicinity of that point, and not the electromagnetic potentials. This gauge invariant result shows that a light wave, propagating at remote distances from its source, induces no current in the vacuum and is therefore undisturbed in its passage through space. The absence of a light quantum self-energy effect is thus indicated. The current induced at a point consists, more precisely, of two parts: a logarithmically divergent multiple of the external current at that point, which produces an unobservable renormalization of charge, and a more involved finite contribution, which is the physically significant induced current. The latter agrees with the results of previous investigations. The modification of the matter field properties arising from interaction with the vacuum fluctuations of the electromagnetic field is considered in the third section. The analysis is carried out with two alternative formulations, one employing the complete electromagnetic potential together with a supplementary condition, the other using the transverse potential, with the variables of the supplementary condition eliminated. It is noted that no real processes are produced by the first order coupling between the fields. Accordingly, alternative equations of motion for the state vector are constructed, from which the first order interaction term has been eliminated and replaced by the second order coupling which it generates. The latter includes the self action of individual particles and light quanta, the interaction of different particles, and a coupling between particles and light quanta which produces such effects as Compton scattering and two quantum pair annihilation. It is concluded from a comparison of the alternative procedures that, for the treatment of virtual light quantum processes, the separate consideration of longitudinal and transverse fields is an inadvisable complication. The light quantum self-energy term is shown to vanish, while that for a particle has the anticipated form for a change in proper mass, although the latter is logarithmically divergent, in agreement with previous calculations. To confirm the identification of the self-energy effect with a change in proper mass, it is shown that the result of removing this term from the state vector equation of motion is to alter the matter field equations of motion in the expected manner. It is verified, finally, that the energy and momentum modifications produced by self-interaction effects are entirely accounted for by the addition of the electromagnetic proper mass to the mechanical proper mass—an unobservable mass renormalization. An appendix is devoted to the construction of several invariant functions associated with the electromagnetic and matter fields.

Related references

More (earlier) information
J. Schwinger, Phys. Rev. **74** (1948) 1439;

See also
R. Serber, Phys. Rev. **48** (1935) 49;
P. A. M. Dirac, 7e Conseil Solvay **203** (1934);
W. Heisenberg, Z. Phys. **90** (1934) 209;
E. A. Uehling, Phys. Rev. **48** (1935) 55;
W. Pauli, M. E. Rose, Phys. Rev. **49** (1936) 462;
V. F. Weisskopf, Z. Phys. **89** (1934) 27;
V. F. Weisskopf, Z. Phys. **90** (1934) 817;
V. F. Weisskopf, Phys. Rev. **56** (1939) 72;

HINCKS 1949

■ Evidence for the continuous energy distribution of the e^- in the decay $\mu^- \to e^- X$ ■

The Penetration of μ-Meson Decay Electrons and Their Bremsstrahlung Radiation

E.P. Hincks, B. Pontecorvo
Phys. Rev. **75** (1949) 698;

Excerpt Measurements of the penetration of the charged particles from the 2.2-μsec meson decay arrangement of counters and delayed coincidence circuits previously described have been extended with absorbers of lead and aluminum, in addition to carbon. The results confirm our previous conclusion that at least a substantial number of particles have an energy, > 25 MeV. Although detailed analysis of the absorption curves does not seem justified, a soft and a hard component are evident. In lead, for example, the "soft" component is essentially absorbed by about 20 g/cm^2, and the "hard" one is easily detected after 38 g/cm^2, where its intensity ($\sim$ 1 count/day) is about ten times the casual rate.
While our results will be published in full later, we report here an investigation of the "hard component". *(Extracted from the introductory part of the paper.)*

Accelerator COSMIC

Detectors Counters, Photon spectrometer

Related references

More (earlier) information
E. P. Hincks and B. Pontecorvo, Phys. Rev. **74** (1948) 697;

See also
J. Steinberger, Phys. Rev. **74** (1948) 500;

Reactions

$\mu^- \rightarrow e^-$ 2neutral

LEE 1949

■ Proposal of the universality of the Fermi interaction ■

Interaction of Mesons with Nucleons and Light Particles

T.D. Lee, M.N. Rosenbluth, C.N. Yang
Phys. Rev. **75** (1949) 905;

Excerpt We have been making a phenomenological study of the various experiments which have been done in recent years on the interaction between the various types of particles. In the course of this investigation two interesting points have come to light.
First, we found that if the decay of the μ-mesons and the capture of the μ^--mesons by nuclei are described by the reactions[1]

$$\mu \rightarrow e\ \nu\ \nu \quad (e = \text{electron},\ \nu = \text{neutrino})$$

$$\mu^-\ P \rightarrow N\ \nu \quad (P = \text{proton},\ N = \text{neutron}),$$

and that the Fermi type interactions are assumed to be responsible for these processes, the coupling constants would have the values

$$g_{\mu e} \sim 3 \times 10^{-48} \text{erg cm}^3$$

and

$$g_{\mu P} \sim 2 \times 10^{-49} \text{erg cm}^3$$

respectively. These values are so determined as to fit the experimental lifetime[2] of the μ-mesons and the capture probability of the μ^--mesons by nuclei.[3] It is remarkable that the three independent experiments: the β-decay of the nucleons and the μ-mesons and the interaction of the nucleons with the μ-mesons lead to coupling constants of the same order of magnitude.
One can perhaps attempt to explain the equality of these interactions in a manner analogous to that used for the Coulomb interactions, i.e. by assuming these interactions to be transmitted through an intermediate field with respect to which all particles have the same "charge." The "quanta" of such a field would have a very short lifetime and would have escaped detection.
Second, if we assume the π-mesons to have integral spin and assume direct couplings for the processes

$$\pi \rightarrow \mu + \text{anti } \nu$$

$$N \rightarrow P + \pi^-$$

with coupling constants determined from the lifetime of the π-mesons[4] and the strength of nuclear forces,[5] the interaction between the μ-mesons and the nucleons can be *quantitatively* explained as a second-order interaction through the virtual creation and annihilation of π-mesons.
After the completion of our work Mr. A. Ore has kindly informed us that similar considerations have been carried out by J. A. Wheeler and J. Tiomno.

[1] The masses of the π- and μ-mesons are taken to be

$$m_\pi = 286 m_e, \quad m_\mu = 212 m_e.$$

[2] B. Rossi, Rev. Mod. Phys. **20** (1948) 537.

[3] B. Rossi, Rev. Mod. Phys. **20** (1948) 537. In the calculation for the capture process the Fermi model for the nucleus is assumed and only single particle excitations are considered. See M. Rosenbluth, Phys. Rev. **75** (1949) 532.

[4] J. R. Richardson, Phys. Rev. **74** (1948) 1720.

[5] H. Bethe, Phys. Rev. **57** (1940) 390.

Related references

See also
B. Rossi, Rev. of Mod. Phys. **20** (1948) 537;
M. N. Rosenbluth, Phys. Rev. **75** (1949) 532;
J. R. Richardson, Phys. Rev. **74** (1948) 1720;
H. A. Bethe, Phys. Rev. **57** (1940) 390;

LEIGHTON 1949

■ Confirmation of the continuous energy distribution of the e^- in the decay $\mu^- \rightarrow e^-$X. Muon spin = 1/2 ■

The Energy Spectrum of the Decay Particles and the Mass and Spin of the Mesotron

R.B. Leighton, C.D. Anderson, A.J. Seriff
Phys. Rev. **75** (1949) 1432; Phys. Rev. **76** (1949) 159;

Abstract Energy values determined from measurements of 75 cloud-chamber tracks of decay particles of cosmic-ray mesotrons at sea level, in a magnetic field of 7250 gauss, are here reported. The observed spectrum extends from 9 MeV to 55 MeV with an apparently continuous distribution of intermediate energy values and a mean energy of 34 MeV. The shape of the spectrum and the value of its upper limit are strong evidence that the mesotron has half-integral spin. The value of the observed upper limit of the energy spectrum corresponds to a mass value of the mesotron equal to 217 ± 4 electron masses.

Accelerator COSMIC

Detectors Cloud chamber

Related references

More (earlier) information
R. B. Leighton, C. D. Anderson, and A. J. Seriff, Phys. Rev. **75** (1949) 1466;

See also
J. Tiomno and J. A. Wheeler, Rev. of Mod. Phys. **21** (1949) 144;
S. H. Neddermeyer and C. D. Anderson, Rev. of Mod. Phys. **11** (1939) 201;

Analyse information from
R. D. Sard and E. J. Althaus, Phys. Rev. **75** (1949) 1251;
R. W. Thompson, Phys. Rev. **74** (1948) 490;
R. W. Thompson, Phys. Rev. **75** (1949) 1279;
C. D. Anderson, R. V. Adams, P. E. Lloyd, and R. R. Rau, Phys. Rev. **72** (1947) 724;
E. C. Fowler, R. L. Cool, and J. C. Street, Phys. Rev. **74** (1948) 101;
J. L. Zar, J. Hershkowitz, and E. Berezin, Phys. Rev. **74** (1948) 111;
J. Steinberger, Phys. Rev. **74** (1948) 500;
E. P. Hincks and B. Pontecorvo, Phys. Rev. **73** (1948) 257;
E. P. Hincks and B. Pontecorvo, Phys. Rev. **74** (1948) 697;
E. P. Hincks and B. Pontecorvo, Phys. Rev. **75** (1949) 698;
M. H. Shamos and A. Russek, Phys. Rev. **74** (1948) 1545;
K. C. Wang and S. B. Jones, Phys. Rev. **74** (1948) 1547;
R. H. Brown et al., Nature **163** (1949) 47;
C. T. R. Wilson and J. G. Wilson, Proc. Roy. Soc. **A148** (1935) 523;
J. C. Fletcher and H. K. Forster, Phys. Rev. **75** (1949) 204;
A. S. Bishop, Phys. Rev. **75** (1949) 1468A;
O. Piccioni, Phys. Rev. **74** (1948) 1754;

Reactions
$\mu^- \rightarrow e^-$ 2neutral

Particles studied μ^-, μ^+

DYSON 1949B

■ Covariant quantum electrodynamics: Equivalence between the Tomonaga-Schwinger method and the Feynman method and generalization ■

The S-Matrix in Quantum Electrodynamics

F.J. Dyson
Phys. Rev. **75** (1949) 1736;

Reprinted in
The Physical Review — the First Hundred Years, AIP Press (1995) CD-ROM.

Abstract The covariant quantum electrodynamics of Tomonaga, Schwinger, and Feynman is used as the basis for a general treatment of scattering problems involving electrons, positrons, and photons. Scattering processes, including the creation and annihilation of particles, are completely described by the S-matrix of Heisenberg. It is shown that the elements of this matrix can be calculated, by a consistent use of perturbation theory, to any desired order in the fine-structure constant. Detailed rules are given for carrying out such calculations, and it is shown that divergencies arising from higher order radiative corrections can be removed from the S-matrix by a consistent use of the ideas of mass and charge renormalization.
Not considered in this paper are the problems of extending the treatment to include bound-state phenomena, and of proving the convergence of the theory as the order of perturbation itself tends to infinity.

Related references

More (earlier) information
F. J. Dyson, Phys. Rev. **75** (1949) 486;

See also
E. C. G. Stückelberg and D. Rivier, Phys. Rev. **74** (1948) 218;
W. H. Furry, Phys. Rev. **51** (1937) 125;
H. W. Lewis, Phys. Rev. **73** (1948) 173;

Analyse information from
S. Tomonaga, Progr. of Theor. Phys. **1** (1946) 27;
Z. Koba, T. Tati, and S. Tomonaga, Progr. of Theor. Phys. **2** (1947) 101;
Z. Koba, T. Tati, and S. Tomonaga, Progr. of Theor. Phys. **2** (1947) 198;
S. Kanesawa and S. Tomonaga, Progr. of Theor. Phys. **3** (1948) 1;
S. Kanesawa and S. Tomonaga, Progr. of Theor. Phys. **3** (1948) 101;
S. Tomonaga, Phys. Rev. **74** (1948) 224;
Ito, Z. Koba, and S. Tomonaga, Progr. of Theor. Phys. **3** (1948) 276;
Z. Koba and S. Tomonaga, Progr. of Theor. Phys. **3** (1948) 290;
J. Schwinger, Phys. Rev. **73** (1948) 416;
J. Schwinger, Phys. Rev. **74** (1948) 1439;
J. Schwinger, Phys. Rev. **75** (1949) 651;
R. P. Feynman, Phys. Rev. **74** (1948) 1430;

FEYNMAN 1949

■ Creation of the covariant quantum electrodynamic theory. Feynman method ■

The Theory of Positrons

R.P. Feynman
Phys. Rev. **76** (1949) 749;

Reprinted in
The Physical Review — the First Hundred Years, AIP Press (1995) CD-ROM.

Abstract The problem of the behavior of positrons and electrons in given external potentials, neglecting their mutual interaction, is analyzed by replacing the theory of holes by a reinterpretation of the solutions of the Dirac equation. It is possible to write down a complete solution of the problem in terms of boundary conditions on the wave function, and this solution contains automatically all the possibilities of virtual (and real) pair formation and annihilation together with the ordinary scattering processes, including the correct relative signs of the various terms.
In this solution, the "negative energy states" appear in a form which may be pictured (as by Stückelberg) in space-time as waves traveling away from the external potential backwards in time. Experimentally, such a wave corresponds to a positron approaching the potential and annihilating the electron. A particle moving forward in time (electron) in a potential may be scattered forward in time (ordinary scattering) or backward (pair annihilation). When moving backward (positron) it may be scattered backward in time (positron scattering) or forward (pair production). For such a particle the amplitude for transition from an initial to a final state is analyzed to any order in the potential by considering it to undergo a sequence of such scatterings.
The amplitude for a process involving many such particles is the product of the transition amplitudes for each particle.

The exclusion principle requires that antisymmetric combinations of amplitudes be chosen for those complete processes which differ only by exchange of particles. It seems that a consistent interpretation is only possible if the exclusion principle is adopted. The exclusion principle need not be taken into account in intermediate states. Vacuum problems do not arise for charges which do not interact with one another, but these are analyzed nevertheless in anticipation of application to quantum electrodynamics.
The results are also expressed in momentum-energy variables. Equivalence to the second quantization theory of holes is proved in an appendix.

Related references

See also
R. P. Feynman, Rev. of Mod. Phys. **20** (1948) 367;
R. P. Feynman, Phys. Rev. **74** (1948) 939;
F. J. Dyson, Phys. Rev. **75** (1949) 486;
W. Pauli, Phys. Rev. **58** (1940) 716;
W. Pauli, Rev. of Mod. Phys. **13** (1941) 203;
E. C. C. Stückelberg, Helv. Phys. Acta **15** (1942) 23;
W. H. Furry, Phys. Rev. **51** (1937) 125;
G. Wentzel, *Einführung in die Quantentheorie der Wellenfelder*, Franz Deuticke, Leipzig **Chapter V** (1943);

FEYNMAN 1949B — Nobel prize

■ Development of the covariant quantum electrodynamic theory. Feynman method. Nobel prize to R. P. Feynman awarded in 1965. Co-winners J. S. Schwinger and S. Tomonaga "for their fundamental work in quantum electrodynamics, with deep-ploughing consequences for the physics of elementary particles" ■

Space-Time Approach to Quantum Electrodynamics

R.P. Feynman
Phys. Rev. **76** (1949) 769;

Reprinted in
The Physical Review — the First Hundred Years, AIP Press (1995) 969.

Abstract In this paper two things are done.

(1) It is shown that a considerable simplification can be attained in writing down matrix elements for complex processes in electrodynamics. Further, a physical point of view is available which permits them to be written down directly for any specific problem. Being simply a restatement of conventional electrodynamics, however, the matrix elements diverge for complex processes.

(2) Electrodynamics is modified by altering the interaction of electrons at short distances. All matrix elements are now finite, with the exception of those relating to problems of vacuum polarization.

The latter are evaluated in a manner suggested by Pauli and Bethe, which gives finite results for these matrices also. The only effects sensitive to the modification are changes in mass and charge of the electrons. Such changes could not be directly observed. Phenomena directly observable are insensitive to the details of the modification used (except at extreme energies). For such phenomena, a limit can be taken as the range of the modification goes to zero. The results then agree with those of Schwinger. A complete, unambiguous, and presumably consistent method is therefore available for the calculation of all processes involving electrons and photons.
The simplification in writing the expressions results from an emphasis on the over-all space-time view resulting from a study of the solution of the equations of electrodynamics. The relation of this to the more conventional Hamiltonian point of view is discussed. It would be very difficult to make the modification which is proposed if one insisted on having the equations in Hamiltonian form.
The methods apply as well to charges obeying the Klein-Gordon equation, and to the various meson theories of nuclear forces. Illustrative examples are given. Although a modification like that used in electrodynamics can make all matrices finite for all of the meson theories, for some of the theories it is no longer true that all directly observable phenomena are insensitive to the details of the modification used.
The actual evaluation of integrals appearing in the matrix elements may be facilitated, in the simpler cases, by methods described in the appendix.

Related references

More (earlier) information
R. P. Feynman, Phys. Rev. **76** (1949) 749;
R. P. Feynman, Phys. Rev. **74** (1948) 939;
R. P. Feynman, Phys. Rev. **74** (1948) 1430;
R. P. Feynman, Rev. of Mod. Phys. **20** (1948) 367;
See also
J. Schwinger, Phys. Rev. **74** (1948) 1439;
J. Schwinger, Phys. Rev. **75** (1949) 651;
F. J. Dyson, Phys. Rev. **75** (1949) 486;
J. A. Wheeler and R. P. Feynman, Rev. of Mod. Phys. **17** (1945) 157;
J. B. French and V. F. Weisskopf, Phys. Rev. **75** (1949) 1240;
N. H. Kroll and W. E. Lamb, Phys. Rev. **75** (1949) 388;
H. W. Lewis, Phys. Rev. **73** (1948) 173;
E. A. Uehling, Phys. Rev. **48** (1935) 55;
W. Pauli, Rev. of Mod. Phys. **13** (1941) 203;
M. Slotnick and W. Heitler, Phys. Rev. **75** (1949) 1645;
H. A. Bethe, Bull.Am.Phys.Soc. **24** (1949) 3;
F. Bloch and A. Nordsieck, Phys. Rev. **52** (1937) 54;

MCKAY 1949

■ Invention of semi-conductor detectors ■

A Germanium Counter

K.G. McKay
Phys. Rev. **76** (1949) 1537;

Excerpt When insulating crystals, such as diamonds, are used as crystal counters of nuclear particles, a complicating factor is the trapping of mobile charge carriers in the crystal. This results in a broad pulse height distribution and the development of internal space charge fields. These effects can be greatly minimized for bombarding particles of high specific ionization by using a very thin crystal together with a high electric field. A method of obtaining the equivalent of this is to use a barrier layer in a semiconductor. The properties of the barrier layer in germanium under a point contact are such that one can calculate under what circumstances it could be used as a counter. *(Extracted from the introductory part of the paper.)*

Detectors Semiconductor

Related references

See also
J. Bardeen and W. H. Brattain, Phys. Rev. **74** (1948) 230;
J. A. Becker and J. N. Shive, Elec. Eng. **68** (1949) 215;
W. Shockley, G. L. Pearson, and M. Sparks, Phys. Rev. **76** (1949) 180;

FERMI 1949

■ First composite model of pions ■

Are Mesons Elementary Particles?

E. Fermi, C.N. Yang
Phys. Rev. **76** (1949) 1739;

Reprinted in
Enrico Fermi. *Collected Papers (Note e Memorie)* v. I. Italy 1921 – 1938 (1962). The University of Chicago Press – Accademia Nazionale dei Lincei, Roma.
(translation into Russian) Enrico Fermi. *Nauchnye Trudy.* I. 1921-1938 Italiya. (pod red. B.Pontecorvo), Nauka, Moskva, 1971.

Abstract The hypothesis that π-mesons may be composite particles formed by the association of a nucleon with an anti-nucleon is discussed. From an extremely crude discussion of the model it appears that such a meson would have in most respects properties similar to those of the meson of the Yukawa theory.

Related references

See also
G. Wentzel, Rev. of Mod. Phys. **19** (1947) 1;
K. M. Case, Phys. Rev. **76** (1949) 14;

KEUFFEL 1949

■ Creation of the spark chamber method for particle tracking ■

Parallel-Plate Counters

J.W. Keuffel
Rev. Sci. Inst. **20** (1949) 202;

Abstract The counter characteristics of a discharge tube using plane-parallel electrodes have been investigated, particularly with regard to the short time lags inherent in the streamer type of spark which occur with such a geometry at near-atmospheric pressure. Construction details for parallel-plate counters with good counter characteristics are given. Spurious counts were minimized by an argon-xylene filling mixture and the use of a univibrator quench circuit. The uncertainty in the reaction time of the counters is $\pm 5 \times 10^{-9}$ sec.

Detectors Spark chamber

Related references

See also
C. W. Sherwin, Rev. Sci. Inst. **19** (1948) 111;
B. Rossi and N. Nereson, Phys. Rev. **62** (1942) 417;
H. J. White, Phys. Rev. **49** (1936) 507;
R. R. Wilson, Phys. Rev. **50** (1936) 1082;
M. Newman, Phys. Rev. **52** (1937) 652;
L. B. Loeb, *Fundamental Processes of Electrical Discharge in Gases*, John Wiley and Sons, Inc., New York (1939) 426;
L. Madansky and R. W. Pidd, Phys. Rev. **73** (1948) 1215;
H. Paetow, Z. Phys. **111** (1939) 770;
B. Howland, C. A. Schroeder, and J. D. Shipman, Rev. Sci. Inst. **18** (1947) 551;

BJORKLUND 1950

■ First evidence for the existence of the π^0 ■

High Energy Photons from Proton-Nucleon Collisions

R. Bjorklund, W.E. Crandall, B.J. Moyer, H.F. York
Phys. Rev. **77** (1950) 213;

Abstract High energy photons (up to 200 MeV) are emitted by various cyclotron targets when these are bombarded by protons with energies greater than 180 MeV. Spectral distributions, angular distributions, and yields are given for various proton energies and targets. Various possible origins are discussed in the light of the experimental observations.

Accelerator LBL-CYC-184IN

Detectors Counters

Related references

See also

H. W. Lewis, J. R. Oppenheimer, and S. A. Wouthuysen, Phys. Rev. **73** (1948) 127;
C. Y. Chao, Phys. Rev. **75** (1949) 581;
W. B. Fretter, Phys. Rev. **76** (1949) 511;
S. Hayakawa, Phys. Rev. **75** (1949) 1759;
J. Ashkin and R. E. Marshak, Phys. Rev. **76** (1949) 58;
L. I. Schiff, Phys. Rev. **76** (1949) 89;
MacDaniel, von Dardel, and W. D. Walker, Phys. Rev. **72** (1947) 985;

Reactions

p Be $\to \gamma$ X	340,350 MeV(T_{lab})
p C $\to \gamma$ X	180,230,290,340 MeV(T_{lab})
p Cu $\to \gamma$ X	340 MeV(T_{lab})
p Ta $\to \gamma$ X	340 MeV(T_{lab})
p nucleon $\to$ meson^0 X	
meson^0 $\to 2\gamma$	

Particles studied meson^0, π^0

ROBSON 1950

■ First observation of neutron beta decay ■

Radioactive Decay of the Neutron

J.M. Robson
Phys. Rev. **77** (1950) 747(A);

Excerpt The positive particle from the radioactive decay of the neutron has been identified as a proton from a measurement of charge to mass. A collimated beam of neutrons emerging from the Chalk River pile passes between two electrodes in an evacuated tank. One electrode is held at a positive potential, up to 20 KeV, while the other electrode is grounded and forms the entrance aperture to a thin lens magnetic spectrometer, the axis of which is perpendicular to the beam of neutrons. The positive decay particles can be focused on the first electrode of an electron multiplier. The background counting rate is 60 c.p.m. A peak of 80 c.p.m. is observed above background when the magnetic field is adjusted for protons of energy expected from the electrostatic field. When a thin boron shutter is placed in the neutron beam, the proton peak disappears. Preliminary estimates of the collecting and focusing efficiency and the neutron flux indicate a minimum half-life of 9 minutes and a maximum of 18 minutes for the neutron.

Accelerator Nuclear reactor

Detectors Spectrometer

Reactions
$n \to p$ X

Particles studied n

WARD 1950

■ Ward identity in quantum electrodynamics ■

An Identity in Quantum Electrodynamics

J.C. Ward
Phys. Rev. **78** (1950) 182;

Reprinted in
The Physical Review — the First Hundred Years, AIP Press (1995) CD-ROM.

Abstract Theoretical verification of Dyson's conjecture leading to the finiteness of renormalized charge, proof that the infinite constants Z_1 and Z_2 are equal. *(Science Abstracts, 1950, 6241. J. R. Maddox)*

Related references
See also
F. J. Dyson, Phys. Rev. **75** (1949) 1736;

SERIFF 1950

■ Further evidence for V events ■

Cloud-Chamber Observations of the New Unstable Cosmic-Ray Particles

A.J. Seriff et al.
Phys. Rev. **78** (1950) 290;

Abstract Thirty-four "forked tracks" similar to two previously reported by Rochester and Butler have been observed in 11,000 cloud-chamber photographs of cosmic-ray penetrating showers, confirming their conclusion that new unstable neutral and charged particles exist. The lifetime of the neutral particles is found to be about 3×10^{-10} sec, and some information as to the nature of the decay products is given.

Accelerator COSMIC

Detectors Cloud chamber

Related references
More (earlier) information
R. B. Leighton, C. D. Anderson, and A. J. Seriff, Phys. Rev. **75** (1949) 1432;
Analyse information from
G. D. Rochester and C. C. Butler, Nature **160** (1947) 855;

Reactions
charged $\to$ charged (neutrals)
neutral $\to$ charged$^+$ charged$^-$

SNELL 1950

■ Confirmation of neutron beta decay ■

Radioactive Decay of the Neutron

A.H. Snell, F. Pleasonton, R.V. McCord
Phys. Rev. **78** (1950) 310;

Reprinted in
(translation into Russian) Usp. Fiz. Nauk **42** (1950) 311.
The Physical Review — the First Hundred Years, AIP Press (1995) CD-ROM.

Abstract Positive results have been obtained with an apparatus designed to record electron-proton coincidences. The observations are consistent with neutrons in free flight transforming into protons, and electrons of max. energy less than 0.9 MeV. The half life has not yet been evaluated accurately but probably lies between 10 and 30 minutes. *(Science Abstracts, 1950, 6745. C. O. Baines.)*

Accelerator Nuclear reactor

Detectors Counters

Comments on experiment Triple coincidence with 0.25 μsec proton signal delay.

Related references

See also
J. Chadwick and M. Goldhaber, Proc. Roy. Soc. **A151** (1935) 479;

Reactions
$n \to p\, e^-$ neutral

Particles studied n

ROBSON 1950B

■ Further confirmation of neutron beta decay ■

Radioactive Decay of the Neutron

J.M. Robson
Phys. Rev. **78** (1950) 311;

Reprinted in
(translation into Russian) Usp. Fiz. Nauk **42** (1950) 311.
The Physical Review — the First Hundred Years, AIP Press (1995) CD-ROM.

Excerpt By using a thin magnetic lens spectrometer in conjunction with an electrostatic field, the positive particle from the radioactive decay of the neutron has been identified as a proton.
The number of protons collected has been estimated to give a value for the half life of the neutron lying between 9 and 25 min. *(Extracted from the introductory part of the paper.)*

Accelerator Nuclear reactor

Detectors Spectrometer

Related references

More (earlier) information
J. M. Robson, Rev. Sci. Inst. **19** (1948) 865;

See also
V. K. Zworykin and J. A. Rajchman, Proc. I.R.E. **27** (1939) 558;

Reactions
$n \to p\, \mathrm{X}$

Particles studied n

REYNOLDS 1950

■ Invention of liquid scintillation counters ■

Liquid Scintillation Counters

G.T. Reynolds, F.B. Harrison, G. Salvini
Phys. Rev. **78** (1950) 488;

Excerpt Experiments have been initiated to investigate the scintillation properties of various liquids by studying the light flashes due to ionizing particles. Measurements have been made to date on the following solutions and liquids: (a) naphthalene in benzene; (b) naphthalene plus anthracene in benzene; (c) terphenyl in benzene; (d) dibenzyl in ether; (e) anthracene plus naphthalene in xylene; (f) terphenyl in xylene, (g) benzene, (h) ether, (i) xylene, and (j) melted dibenzyl. *(Extracted from the introductory part of the paper.)*

Related references

See also
R. Hofstadter, Liebson, and H.Elliot, Phys. Rev. **78** (1950) 81;
Ageno, Chizzotto, and Querzoli, Atti Accad. naz Lincei **6** (1949) 81;

KALLMANN 1950

■ Another invention of liquid scintillation counters ■

Scintillation Counting with Solutions

H. Kallmann
Phys. Rev. **78** (1950) 621;

Excerpt It is a drawback of scintillation counting with crystals that large-sized crystals are found not to be very transparent. They exhibit a considerable light scattering which gives rise to additional absorption. This drawback could be overcome if instead of crystals one were able to use liquids, which can be applied in large thicknesses without considerable absorption. In the course of the investigation of the fluorescent efficiency of liquid solutions when excited by gamma-radiation and by neutrons, special solutions were used which exhibited a fluorescent efficiency high enough to make them applicable for counting work. *(Extracted from the introductory part of the paper.)*

STEINBERGER 1950

■ First evidence for the production of the π^0 and for $\pi^0 \to \gamma\gamma$ decay ■

Evidence for the Production of Neutral Mesons by Photons

J. Steinberger, W.K.H. Panofsky, J. Steller
Phys. Rev. **78** (1950) 802;

Reprinted in
(translation into Russian) Usp. Fiz. Nauk **42** (1950) 313.
R. N. Cahn and G. Goldhaber, *The Experimental Foundations of Particle Physics*, Cambridge Univ. Press (1991) 45.
The Physical Review — the First Hundred Years, AIP Press (1995) 820.

Abstract In the bombardment of nuclei by 330 MeV X-rays, multiple gamma-rays are emitted. From their angular correlation it is deduced that they are emitted in pairs in the disintegration of neutral particles moving with relativistic velocities and therefore of intermediate mass. The neutral mesons are produced with cross sections similar to those for the charged mesons and with an angular distribution peaked more in in the forward direction. The production cross section in hydrogen and the production cross section per nucleon in C and Be are comparable.

Accelerator LBL-CYC-184IN

Detectors Counters

Related references

See also
R. J. Finkelstein, Phys. Rev. **72** (1947) 414;
H. Fukuda and Y. Miamoto, Progr. of Theor. Phys. **4** (1949) 347;
Ozaki, Oneda, and Sasaki, Progr. of Theor. Phys. **4** (1949) 524;
C. N. Yang, Phys. Rev. **77** (1950) 242;
W. B. Fretter, Phys. Rev. **75** (1949) 41;
J. Steinberger and A. S. Bishop, Phys. Rev. **78** (1950) 493;
E. M. McMillan, Peterson, and White, Science **110** (1949) 579;
W. B. Fretter, Phys. Rev. **76** (1949) 511;
J. R. Oppenheimer, Phys. Rev. **71** (1947) 462;
J. Steinberger, Phys. Rev. **76** (1949) 1180;

Analyse information from
R. Bjorklund et al., Phys. Rev. **77** (1950) 213;

Reactions

$\gamma\, p \to p$ mult$[\gamma]$	175-330 MeV(P_{lab})
γ Be $\to$ Be mult$[\gamma]$	175-330 MeV(P_{lab})
γ C $\to$ C mult$[\gamma]$	175-330 MeV(P_{lab})
meson$^0 \to 2\gamma$	
$\pi^0 \to 2\gamma$	

Particles studied π^0

CARLSON 1950

■ Confirmation of the existence of the π^0, first estimation of the π^0 lifetime ■

Nuclear Transmutations Produced by Cosmic-Ray Particles of Great Energy. — Part V. The Neutral Meson

A.G. Carlson, J.E. Hooper, D.T. King
Phil. Mag. **41** (1950) 701;

Excerpt The spectrum of the γ-radiation in the atmosphere at 70,000 ft has been determined by observation on the scattering of pairs of fast electrons recorded in photographic emulsions exposed in high-flying balloons. The detailed form of the spectrum is consistent with the assumption that the γ–rays originate by the decay of neutral mesons. It is found that the mass of the neutral-mesons is $295 \pm 20m_e$, and that they are created in nuclear explosions with an "energy spectrum" similar to that of the charged π–particles. A method is described for determining the lifetime of the neutral mesons and their frequency of occurrence compared with charged π–particles. The lifetime, τ_{π^0}, is less than 5×10^{-14} sec. It may be possible to determine the lifetime, by observations of greater statistical weight, if it is longer than 2×10^{-14} sec. The ratio of the number of neutral mesons to charged mesons, produced in nuclear explosions of great energy, is equal to 0.45 ± 0.10.

Accelerator COSMIC

Detectors Nuclear emulsion

Related references

More (earlier) information
U. Camerini et al., Phil. Mag. **40** (1949) 1073;

See also
R. Bjorklund et al., Phys. Rev. **77** (1950) 213;
M. F. Kaplon, B. Peters, and H. L. Brandt, Helv. Phys. Acta **23** (1950) 24;
O. Piccioni, Phys. Rev. **77** (1950) 1;
P. H. Fowler, Phil. Mag. **41** (1950) 169;

Reactions

unspec nucleus $\to \gamma$ X
unspec nucleus $\to \pi^0$ X
$\pi^0 \to 2\gamma$

Particles studied π^0

LIPPMANN 1950

■ Systematic treatment of the application of variational principles to the quantum theory of scattering. Invention of the Lippmann-Schwinger form of the Schrödinger equation ■

Variational Principles for Scattering Processes. I

B.A. Lippmann, J. Schwinger
Phys. Rev. **79** (1950) 469;

Reprinted in
The Physical Review — the First Hundred Years, AIP Press (1995) 183.

Abstract A systematic treatment is presented of the application of variational principles to the quantum theory of scattering.
Starting from the time-dependent theory, a pair of variational principles is provided for the approximate calculation of the unitary (collision) operator that describes the connection between the initial and final states of the system. An equivalent formulation of the theory is obtained by expressing the collision operator in terms of an Hermitian (reaction) operator: variational principles for the reaction operator follow. The time-independent theory, including variational principles for the operators now used to describe transitions, emerges from the time dependent theory by restricting the discussion to stationary states. Specialization to the case of scattering by a central force field establishes the connection with the conventional phase shift analyses and results in a variational principle for the phase shift.
As an illustration, the results of Fermi and Breit on the scattering of slow neutrons by bound protons are deduced by variational methods.

Related references

More (later) information
B. A. Lippmann, Phys. Rev. **79** (1950) 481;

See also
R. E. Marshak, Phys. Rev. **71** (1947) 688;
H. Levin and J. Schwinger, Phys. Rev. **74** (1948) 958;
H. Levin and J. Schwinger, Phys. Rev. **75** (1949) 1423;
J. Schwinger, *Lectures on Nuclear Physics*, Harvard University (1947);
J. Schwinger, Phys. Rev. **72** (1947) 742;
J. M. Blatt, Phys. Rev. **74** (1948) 92;
W. Kohn, Phys. Rev. **74** (1948) 1763;
J. M. Blatt and J. D. Jackson, Phys. Rev. **76** (1949) 18;
I. E. Tamm, Zh. Eksp. Teor. Fiz. **18** (1948) 337;
I. E. Tamm, Zh. Eksp. Teor. Fiz. **19** (1949) 74;
E. Fermi, La Ricerca Scient. **VII-II** (1936) 13;
G. Breit, Phys. Rev. **71** (1947) 215;
G. Breit and P. L. Zilcel, Phys. Rev. **71** (1947) 232;
G. Breit, P. L. Zilcel, and Darling, Phys. Rev. **72** (1947) 576;

FEYNMAN 1950 — Nobel prize

■ Mathematical proof of the validity of the Feynman rules for calculations of amplitudes in QED. Nobel prize to R. P. Feynman awarded in 1965. Co-winners J. S. Schwinger and S. Tomonaga "for their fundamental work in quantum electrodynamics, with deep-ploughing consequences for the physics of elementary particles" ■

Mathematical Formulation of the Quantum Theory of Electromagnetic Interaction

R.P. Feynman
Phys. Rev. **80** (1950) 440;

Reprinted in
The Physical Review — the First Hundred Years, AIP Press (1995) 990.

Abstract The validity of the rules given in previous papers for the solution of problems in quantum electrodynamics is established. Starting with Fermi's formulation of the field as a set of harmonic oscillators, the effect of the oscillators is integrated out in the Lagrangian form of quantum mechanics. There results an expression for the effect of all virtual photons valid to all orders in $e^2/\hbar c$. It is shown that evaluation of this expression as a power series in $e^2/\hbar c$ gives just the terms expected by the aforementioned rules.
In addition, a relation is established between the amplitude for a given process in an arbitrary unquantized potential and in a quantum electrodynamical field. This relation permits a simple general statement of the laws of quantum electrodynamics.
A description, in Lagrangian quantum-mechanical form, of particles satisfying the Klein-Gordon equation is given in an Appendix. It involves the use of an extra parameter analogous to proper time to describe the trajectory of the particle in four dimensions.
A second Appendix discusses, in the special case of photons, the problem of finding what real processes are implied by the formula for virtual processes.
Problems of the divergencies of electrodynamics are not discussed.

Related references

See also
R. P. Feynman, Phys. Rev. **76** (1949) 749;
R. P. Feynman, Phys. Rev. **76** (1949) 769;
R. P. Feynman, Rev. of Mod. Phys. **20** (1948) 367;
S. Tomonaga, Phys. Rev. **74** (1948) 224;
S. Kanesawa and S. Tomonaga, Progr. of Theor. Phys. **3** (1948) 101;
J. Schwinger, Phys. Rev. **76** (1949) 790;
F. J. Dyson, Phys. Rev. **75** (1949) 486;
F. J. Dyson, Phys. Rev. **75** (1949) 1736;
W. Pauli and F. Villars, Rev. of Mod. Phys. **21** (1949) 434;
E. Fermi, Rev. of Mod. Phys. **4** (1932) 87;
W. Heitler, *The Quantum Theory of Radiation*, Oxford Univ. Press, London (1944);

Y. Nambu, Progr. of Theor. Phys. **5** (1950) 82;
V. Fock, Physik Zeits. Sow. **12** (1937) 404;

LORD 1950

■ First evidence for multiple hadron production in nucleon-nucleon interactions. First evidence for the forward jet of secondary particles in hadronic interactions ■

Evidence for the Multiple Production of Mesons in a Single Nucleon-Nucleon Collision

J.J. Lord, J. Fainberg, M. Schein
Phys. Rev. **80** (1950) 970;

Abstract Evidence has been obtained for the multiple production of mesons in a single nucleon-nucleon collision in a photographic emulsion exposed to the cosmic radiation at an altitude of 95,000ft. The nuclear encounter in which the mesons were created was produced by a primary proton of 3×10^{13} eV energy. Directly in line with the incident proton 7 particles of minimum ionization were emitted in a central core with an angular divergence of 0.003 radian. In addition 8 other minimum ionization particles were emitted in a wider diffuse cone of 0.13 radian angular divergence. Only one track had the appearance of a fragment which, however, could have been a proton of 10 MeV energy. Most of the particles in the central core had energies in excess of 250 BeV, while those in the diffuse cone were of much lower energies as determined by small angle scattering measurements. A pair of very small angular divergence was produced in the central core 4800 microns from the point of origin of the nuclear interaction. Assuming that the pair were produced by a gamma-ray from a decay of a neutral meson, an upper limit of 2×10^{-15} second was deduced for its meanlife. Both the angular and the energy distribution of the emitted particles is in good agreement with the assumption that in the center-of-mass system the mesons are emitted in two distinct cones of angular width of about 30° forward and backward with the reference to the direction of the primary proton. The average multiplicity of 15 agrees with the recent calculation by Fermi, and according to his prediction about one-half of the particles could be made up of nucleon-antinucleon pairs.

Accelerator COSMIC

Detectors Nuclear emulsion

Related references

See also
R. H. Brown et al., Phil. Mag. **40** (1949) 862;
R. Bjorklund, Crandall, Moyer, and York, Phys. Rev. **77** (1950) 213;

Analyse information from
J. J. Lord and M. Schein, Phys. Rev. **77** (1950) 19;
J. Hornbostell and E. O. Salant, Phys. Rev. **76** (1949) 859;
L. Leprince-Ringuet et al., Compt. Ren. **229** (1949) 163;
M. F. Kaplon, B. Peters, and H. L. Brandt, Phys. Rev. **76** (1949) 1735;
P. H. Fowler, Phil. Mag. **41** (1950) 169;
W. Heitler and L. Janossy, Proc. Roy. Soc. **62** (1949) 669;
H. W. Lewis, J. R. Oppenheimer, and S. A. Wouthuysen, Phys. Rev. **73** (1948) 127;
W. Heisenberg, Nature **164** (1949) 65;

Reactions
p nucleus $\rightarrow$ jet X   30 TeV(P_{lab})

Particles studied jet

PANOFSKY 1951

■ Confirmation of the π^0 and $\pi^0 \rightarrow \gamma\gamma$ decay. Direct determination of the π^- parity ■

The Gamma-Ray Spectrum Resulting from Capture of Negative π^- Mesons in Hydrogen and Deuterium

W.K.H. Panofsky, R.L. Aamodt, J. Hadley
Phys. Rev. **81** (1951) 565;

Reprinted in
The Physical Review — the First Hundred Years, AIP Press (1995) 824.

Abstract π^- mesons produced in an internal wolfram target bombarded by 330 MeV protons in the 184-inch cyclotron are absorbed in a high pressure hydrogen target. The resulting gamma-ray spectrum is analyzed outside the shielding of the cyclotron by means of a 30-channel electron-positron pair spectrometer. The principal results are as follows.

(1) The gamma-rays result from two competing reactions: $\pi^-\ p \rightarrow n\ \gamma$ and $\pi^-\ p \rightarrow n\ \pi^0$; $\pi^0 \rightarrow 2\gamma$.

(2) The ratio between the π^0 yield to the single gamma-ray yield is $= 0.944 \pm 0.20$.

(3) The mass difference between the π^- meson and the π^0 meson is given by 10.6 ± 2.0 electron masses.

(4) The π^- mass is 275.2 ± 2.5 electron masses.

The large mass difference between π^- and π^0 precludes the conclusion that the unexpectedly small π^0 to γ ratio is due to the small amount of momentum space available for π^0 emission. It rather indicates that π^0 emission is slowed down by the nature of the coupling of the π^0 field to the nucleons. The experiment has been repeated by substituting D_2 for H_2 in the vessel. The result is that the reaction π^- deuteron $\rightarrow 2n$ and π^- deuteron $\rightarrow 2n\ \gamma$ compete in the ratio 2 : 1. The reaction π^- deuteron $\rightarrow 2n\ \pi^0$ is absent.

Accelerator LBL-CYC-184IN

Detectors Spectrometer

Related references

More (earlier) information
W. K. H. Panofsky, R. L. Aamodt, and H. F. York, Phys. Rev.

78 (1950) 825;
J. Steinberger, W. K. H. Panofsky, and J. Steller, Phys. Rev. **78** (1950) 802;

See also
M. Conversi, E. Pancini, and O. Piccioni, Phys. Rev. **68** (1945) 232;
Manon, H. Muirhead, and Rachat, Phil. Mag. **41** (1950) 583;
F. Adelman and S. B. Jones, Science **111** (1950) 226;
W. Cheston and L. Goldfarb, Phys. Rev. **78** (1950) 320A;
R. Bjorklund et al., Phys. Rev. **77** (1950) 213;
W. B. Fretter, Phys. Rev. **73** (1948) 41;
W. B. Fretter, Phys. Rev. **76** (1949) 216;
C. Y. Chao, Phys. Rev. **75** (1949) 581;
B. P. Gregory, B. Rossi, and Tinlot, Phys. Rev. **77** (1950) 299;
M. F. Kaplon, B. Peters, and H. L. Brandt, Phys. Rev. **76** (1949) 1735;
Carlson, Cooper, and King, Phil. Mag. **41** (1950) 701;
W. K. H. Panofsky, *Post Deadline Paper*, New York Meeting A.P.S. (1950);

Reactions
$\pi^- \; p \rightarrow n \; \gamma$
$\pi^- \; p \rightarrow n \; 2\gamma$
π^- deuteron $\rightarrow 2n$
π^- deuteron $\rightarrow 2n \; \gamma$
π^- deuteron $\rightarrow 2n \; \pi^0$
$\pi^0 \rightarrow 2\gamma$

Particles studied π^-, π^0

BRUECKNER 1951

■ Confirmation of the odd parity of the π^- ■

The Capture of π-Mesons in Deuterium

K.A. Brueckner, R. Serber, K.M. Watson
Phys. Rev. **81** (1951) 575;

Abstract Panofsky, Aamodt, and Hadley find that the capture of π-mesons in deuterium leads to two neutron emission in 70 percent of the cases, and to two neutrons and a γ-ray in the other 30 percent. The former process is forbidden for capture of scalar mesons from the K shell. A detailed balancing argument, based on the observed cross section for $2p \rightarrow$ deuteronπ^+, permits calculation of the non-radiative capture rate from orbits of higher angular momentum. It is shown that this is too small to account for the observation, and the possibility that the meson is a scalar is thus ruled out. This conclusion could only be avoided by postulating the existence of an interaction between meson and nucleons sufficiently large to alter radically the magnitude of the p-wave functions near the nucleons. For pseudoscalar mesons, non-radiative capture from the K shell is permitted, and the ratio of radiative to non-radiative transitions can be related to the ratio of meson photoproduction cross section to the cross section for $2p \rightarrow$ deuteronπ^+. The result is compatible with the type of meson-nucleon coupling which might be expected for pseudoscalar mesons, but uncertainties in the measured quantities are too large to permit a sharper conclusion.

Related references

Analyse information from
W. K. H. Panofsky, R. L. Aamodt, and J. Hadley, Phys. Rev. **81** (1951) 565;
B. Ferreti, *Report on the Int. Conf. on Low Temperature and Fundamental Particles*, Cambridge (1946) 75;
S. Tamor, Phys. Rev. **78** (1950) 221;

Particles studied π^-

O'CEALLAIGH 1951

■ First evidence for $K^+ \rightarrow \mu^+$ neutral(s) decay ■

Masses and Modes of Decay of Heavy Mesons — Part I.

C. O'Ceallaigh
Phil. Mag. **42** (1951) 1032;

Excerpt Observations by the photographic method are described which establish the existence of heavy charged mesons, of mass about 1200 m_e, which decay with the emission of a single charged particle. In one case, it is shown that the secondary charged particle is a μ-meson of energy 5.9 MeV. In another, it is a particle of charge $|e|$ and of energy about 200 MeV. If the two particles are of the same type with similar modes of decay, then the charged secondary particle must be accompanied by at least two neutral particles. The question whether the k-particles are identical with the τ-mesons which decay into three charged particles, is discussed.

Accelerator COSMIC

Detectors Nuclear emulsion

Reactions
$K^+ \rightarrow \mu^+$ neutral (neutrals)
$\mu^+ \rightarrow e^+$ 2neutral

Particles studied K^+

FRETTER 1951

■ Confirmation of the existence of heavy unstable particles ■

Some Observations on the New Unstable Cosmic-Ray Particles

W.B. Fretter
Phys. Rev. **82** (1951) 294;

Excerpt A large cloud chamber containing several lead plates has been used to observe the unstable particles already reported. The probability of observation has been increased by making the distance between the lead plates two inches instead of 0.75 inch as in our previous experiments. Twenty cases have been seen at sea level where a neutral particle changed into two charged particles in the gas and three cases of sudden changes of direction in the gas have been observed. In one event produced by a neutral, one of the secondary particles was heavily ionizing (estimated at about 4x minimum), passed through a 0.25-inch lead plate without noticeable change in ionization or scattering, and thus must have had a mass larger than that of a π-meson. These events all occur in penetrating showers and the findings of the previous observers are generally confirmed.

Accelerator COSMIC

Detectors Cloud chamber

Data comments 23 events.

Related references

Analyse information from
G. D. Rochester and C. C. Butler, Nature **160** (1947) 855;
A. J. Seriff et al., Phys. Rev. **78** (1950) 290;

Reactions
neutral $\rightarrow$ 2charged
charged $\rightarrow$ charged (neutrals)

BRIDGE 1951

■ Further confirmation of the existence of heavy unstable particles ■

A Cloud-Chamber Study of The New Unstable Particles

H.S. Bridge, M. Annis
Phys. Rev. **82** (1951) 445;

Abstract In a set of pictures taken with a large multiple-plate cloud chamber at 700 g cm^{-2} atmospheric depth, we have observed 10 examples of tracks that deviate suddenly in the gas of the chamber. These are similar to the "V" tracks reported by Rochester and Butler, by Seriff, Leighton, Hsiao, Cowan, and Anderson, and recently by Fretter. Of these 10 events, 4 can be interpreted as neutral-particle decays, 4 as charged-particle decays, while 2 are uncertain as to type.
Of the 4 neutral-particle decays, 2 show one of the decay products ionizing heavily and stopping in a plate. In 2 cases it is possible to say from ionization, range, and scattering that one of the decay products is probably a meson. *(Extracted from the introductory part of the paper.)*

Accelerator COSMIC

Detectors Cloud chamber

Related references

Analyse information from
G. D. Rochester and C. C. Butler, Nature **160** (1947) 855;
A. J. Seriff et al., Phys. Rev. **78** (1950) 290;
W. B. Fretter, Phys. Rev. **82** (1951) 294;
U. Camerini, P. H. Fowler, W. O. Lock, and H. Muirhead, Phil. Mag. **41** (1950) 413;

Reactions
neutral $\rightarrow$ 2charged
charged $\rightarrow$ charged (neutrals)

DURBIN 1951C

■ Determination of the spin of the π^+ ■

The Spin of the Pion via the Reaction π^+ deuteron $\rightarrow$ $p\,p$

R. Durbin, H. Loar, J. Steinberger
Phys. Rev. **83** (1951) 646;

Reprinted in
R. N. Cahn and G. Goldhaber, *The Experimental Foundations of Particle Physics*, Cambridge Univ. Press (1991) 49.
The Physical Review — the First Hundred Years, AIP Press (1995) 834.

Abstract It is possible to determine the spin of the pion by comparing the forward and backward rates of the reaction π^+ deuteron $\rightleftharpoons p\,p$. The backward rate has been measured in Berkeley. We have measured the forward rate. Comparison of the two results shows the spin to be zero. In the light of other recent experimental results the meson is then pseudoscalar.

Accelerator Nevis labs cyclotron

Detectors Counters

Related references

See also
C. N. Yang, Phys. Rev. **77** (1950) 242;
K. A. Brueckner, R. Serber, and K. M. Watson, Phys. Rev. **81** (1951) 575;
Analyse information from
J. Steinberger, W. K. H. Panofsky, and J. Steller, Phys. Rev. **78** (1950) 802;
W. K. H. Panofsky, R. L. Aamodt, and J. Hadley, Phys. Rev. **81** (1951) 565;
F. S. Crawford, K. Crowe, and M. L. Stevenson, Phys. Rev. **82** (1951) 97;
A. S. Bishop, J. Steinberger, and L. J. Cook, Phys. Rev. **80** (1950) 291;
Cartwright, Richman, M. M. Whitehead, and Wilcox, Phys. Rev. **81** (1951) 652;
V. Z. Peterson, Phys. Rev. **79** (1950) 407;
H. Primakoff and F. Villars, Phys. Rev. **83** (1951) 686;

Reactions
π^+ deuteron $\rightarrow$ $2p$ 75 MeV(T_{lab})

Particles studied π^+

CLARK 1951

■ Another determination of the spin of the π^+ ■

Cross Section for the Reaction π^+ deuteron $\rightarrow p\, p$ and the Spin of the π^+ Meson

D. Clark, A. Roberts, R. Wilson
Phys. Rev. **83** (1951) 649;

Excerpt The application of detailed balancing to the determination of the spin of the π^+ meson from the reaction π^+ deuteron $\rightarrow p\, p$ and its inverse has been suggested by Marshak and Cheston (Rochester, High Energy Conference, December, 1950) and independently by Johnson (private communication). The detailed balancing argument requires the comparison of either differential or total cross sections of both the meson-producing and meson-absorbing reactions at the same energy in the center-of-mass system. The reaction $p\, p \rightarrow \pi^+$ deuteron has been studied for 340 MeV protons by Richman and others. (Cartwright et al., Phys. Rev. **81** (1951) 652; Crawford, Crowe, and Stevenson, Phys. Rev. **82** (1951) 97) The best data are for the differential cross section at 0°, and other data are available at 18°, 30°, and 60°. From these data, limits on the angular distribution can be obtained and the total cross section computed. We have now measured the total cross section for the meson-absorbing reaction.
...We conclude that the spin of the π^+ meson is zero. *(Extracted from the introductory part of the paper.)*

Accelerator Rochester cyclotron

Detectors Combination

Related references

See also
R. E. Marshak, *Rochester High Energy Conference*, December (1950);
Cartwright, Richman, M. M. Whitehead, and Wilcox, Phys. Rev. **81** (1951) 652;
F. S. Crawford, K. Crowe, and M. L. Stevenson, Phys. Rev. **82** (1951) 97;
D. Clark, Phys. Rev. **82** (1951) 97;
W. K. H. Panofsky, R. L. Aamodt, and J. Hadley, Phys. Rev. **82** (1951) 97;
S. Tamor and R. E. Marshak, Phys. Rev. **80** (1950) 766;

Reactions
π^+ deuteron $\rightarrow 2p$

Particles studied π^+

ARMENTEROS 1951

■ Evidence for the possible existence of the Λ hyperon. First evidence for the K^0 meson ■

Decay of V Particles

R. Armenteros et al.
Nature **167** (1951) 501;

Excerpt ...we can summarize the main results of the present investigations as follows.

(a) The occurrence of V-tracks in the chamber has again been confirmed.

(b) Probably not more than 10 per cent of the V-tracks can be explained by processes other than the spontaneous decay of new unstable particles.

(c) Momentum measurements on fifty-four secondary particles from neutral V-decays have been made and, together with ionization estimates, make it possible to identify with certainty both protons and mesons.

(d) Assuming that only two particles are produced by the neutral decays, two schemes are suggested to explain the photographs: $V^0 \rightarrow p^+\, \pi^-$, with the V^0 mass in the range $(2,000 - 2,500) \times m_e$, $V^0 \rightarrow \pi^+\, \pi^-$ with the V^0 mass about $1,000\ m_e$. The relative frequency of these two processes will be discussed in a later paper, but the evidence is already strong for the existence of two types of V^0-particle.

(e) Six charged decays have been observed, but no definite conclusion have been reached about the mass of the charged V-particles.

(Extracted from the introductory part of the paper.)

Accelerator COSMIC

Detectors Cloud chamber

Reactions
unspec nucleus $\rightarrow$ vee X
vee $\rightarrow p\, \pi^-$
vee $\rightarrow \pi^+\, \pi^-$
$\Lambda \rightarrow p\, \pi^-$
$K^0 \rightarrow \pi^+\, \pi^-$

Particles studied Λ, K^0

LEIGHTON 1951

■ Evidence for two types of neutral V particles, baryonic and mesonic ■

On the Decay of Neutral V-Particles

R.B. Leighton, S.D. Wanlass, W.L. Alford
Phys. Rev. **83** (1951) 843;

Abstract Nine measured examples of V^0 decay show all negative particles mesons, one appearing to undergo $\pi^- \mu$ decay, and most positive particles protons with probably two examples of lighter positives. The Q-values deduced for the $V^0 \to p\ \pi^-$ reaction are regarded as inconsistent with a unique two-body decay, and the interpretation of the results on the scheme $V^0 \to$nucleon $\pi\ \pi$ is examined. *(Science Abstracts, 1951, 8810. J. G. Wilson.)*

Accelerator COSMIC

Detectors Cloud chamber

Related references

Analyse information from
G. D. Rochester and C. C. Butler, Nature **160** (1947) 855;
A. J. Seriff et al., Phys. Rev. **78** (1950) 290;
R. Armenteros et al., Nature **167** (1951) 501;
R. W. Thompson, H. O. Cohn, and R. S. Flum, Phys. Rev. **83** (1951) 175;

Reactions
neutral $\to p\ \pi^-$
neutral $\to \pi^+\ \pi^-$

Particles studied neutral, vee

FOWLER 1951

■ Confirmation of the K^+ ■

Masses and Modes of Decay of Heavy Mesons — Part II. τ-Particles.

P.H. Fowler, M.G.K. Menon, C.F. Powell, O. Rochat
Phil. Mag. **42** (1951) 1040;

Excerpt Further evidence which establishes the existence of heavy charged mesons which decay into three charged particles is presented. It is shown that the three secondary particles are probably π-particles, and that the mass of the parent mesons, called τ-particles, is $966 \pm 8 m_e$. The frequency of occurrence of κ- and τ-particles at the Jungfraujoch, in comparison with the number of π-particles, is discussed, and it is shown that in the nuclear collisions involving protons of energy greater than 10 BeV, a large fraction of the energy lost in the creation of mesons must appear as κ- and τ-particles. The evidence suggests that the life-time of the heavy mesons is about 10^{-9} sec. The most favorable conditions for the detection of the particles by the photographic method are discussed.

Accelerator COSMIC

Detectors Nuclear emulsion

Related references

Analyse information from
R. H. Brown et al., Nature **163** (1949) 82;
Harding, Phil. Mag. **41** (1950) 405;
Alvarez, *Proc. Harwell Nuclear Physics Conference* (1950) 1;
C. M. G. Lattes, P. H. Fowler, and Cuer, Proc. Roy. Soc. **59** (1947) 883;
C. O'Ceallaigh, Phil. Mag. **42** (1951) 1032;
Bradner, F. M. Smith, W. H. Barkas, and A. S. Bishop, Phys. Rev. **77** (1950) 462;

Reactions
$K^+ \to 2\pi^+\ \pi^-$

Particles studied K^+

ARMENTEROS 1951B

■ Evidence for the possible existence of the Λ hyperon. First evidence for the K^0 meson ■

The Properties of Neutral V-Particles

R. Armenteros, K.H. Barker, C.C. Butler, A. Cachon
Phil. Mag. **42** (1951) 1113;

Abstract In a cloud-chamber investigation at 2867 m altitude 70 V-tracks due to decay of neutral V-particles have been observed. An analysis has been made on the basis of the two-body decay process: $V_1^0 \to p\pi^-$ and $V_2^0 \to \pi^+\pi^-$ and it is concluded that the presented limited data are consistent with these two processes. The ratio of the numbers of the two types of V-particles is found to be $N_{V_1^0}/N_{V_2^0} = 1.6 \pm 0.5$ and accurate mass estimate have been obtained for 12 examples of the first type of decay and for eight examples of the second type. The results are consistent with the unique mass values $(2203 \pm 12)\ m_e$ and $(796 \pm 27)\ m_e$ respectively. The possibility that there are neutral secondary particles among the decay products cannot be excluded. However, some evidence is presented that seems to exclude an explanation of the observation in terms of two modes of decay of a heavy particle each giving a nucleon and two mesons. Finally, it is concluded that there are probably two types of neutral V-particles.

Accelerator COSMIC

Detectors Cloud chamber

Related references

More (earlier) information
R. Armenteros et al., Nature **167** (1951) 501;

Reactions

unspec nucleus $\rightarrow$ vee X
vee $\rightarrow p\,\pi^-$
vee $\rightarrow \pi^+\,\pi^-$
$\Lambda \rightarrow p\,\pi^-$
$K^0 \rightarrow \pi^+\,\pi^-$

Particles studied Λ, K^0

LYMAN 1951

■ First results from electron-nuclei scattering experiment ■

Scattering of 15.7 MeV Electrons by Nuclei

E.M. Lyman, A.O. Hanson, M.B. Scott
Phys. Rev. **84** (1951) 626;

Abstract Electrons removed from the 20 MeV betatron are focused to a 0.08 inch spot about 10 feet from the betatron by a magnetic lens. The electrons impinge on thin foils at the center of a highly evacuated scattering chamber having a diameter of 20 inches. Elastically scattered electrons, selected by a $\frac{3}{8}$ inch×2 inch aperture, are focused by means of a 75° magnetic analyzer with 3 percent energy resolution and are detected by coincidence Geiger counters. Corrections are applied for multiple scattering and for energy losses which remove the electrons from the range of energies accepted by the detector arrangement. The scattering cross section for gold at 150° is found to be about 2.6 times that given by Mott's formula in the Born approximation and about one-half of that expected for the scattering by a point nucleus. This result is in good agreement with the calculations for electrons of this energy if the nuclear charge is assumed to be distributed uniformly throughout the nuclear volume.
The results for the scattering from C, Al, Cu, and Ag are also in agreement with the assumption of a uniformly distributed nuclear charge within the uncertainties involved in the theory and the experimental results.

Accelerator Ill-Betatron

Detectors Counters

Related references

See also
Randels, Chao, and Crane, Phys. Rev. **68** (1945) 64;
R. J. Van de Graaff, Buechner, and H. Feshbach, Phys. Rev. **69** (1946) 452;
Buechner, R. J. Van de Graaff et al., Phys. Rev. **72** (1947) 678;
F. C. Champion and R. R. Roy, Proc. Roy. Soc. **61** (1948) 532;
W. Siegrist, Helv. Phys. Acta **16** (1943) 471;
Alichanian, Alichanow, and Weissenberg, J. Phys. (USSR) **9** (1945) 280;
W. Bothe, Zeit. Naturforschung **4a** (1949) 88;
W. A. McKinley and H. Feshbach, Phys. Rev. **74** (1948) 1759;
M. E. Rose, Phys. Rev. **73** (1948) 279;
E. M. Lyman, A. O. Hanson, and M. B. Scott, Phys. Rev. **79** (1950) 228;
E. M. Lyman, A. O. Hanson, and M. B. Scott, Phys. Rev. **81** (1951) 309;
L. R. B. Elton, Phys. Rev. **79** (1950) 412;
L. R. B. Elton, Proc. Roy. Soc. **63** (1950) 1115;

Reactions

e^- nucleus $\rightarrow$ nucleus e^-	15.7 MeV(E_{lab})
e^- C $\rightarrow$ C e^-	15.7 MeV(E_{lab})
e^- Al $\rightarrow$ Al e^-	15.7 MeV(E_{lab})
e^- Cu $\rightarrow$ Cu e^-	15.7 MeV(E_{lab})
e^- Ag $\rightarrow$ Ag e^-	15.7 MeV(E_{lab})
e^- Au $\rightarrow$ Au e^-	15.7 MeV(E_{lab})

Particles studied nucleus

SALPETER 1951

■ Bethe-Salpeter relativistic equation for two-body bound-state problems ■

A Relativistic Equation for Bound-State Problems

E.E. Salpeter, H.A. Bethe
Phys. Rev. **84** (1951) 1232;

Reprinted in
The Physical Review — the First Hundred Years, AIP Press (1995) CD-ROM.

Abstract The relativistic S-matrix formalism of Feynman is applied to the bound-state problem for two interacting Fermi-Dirac particles. The bound state is described by a wave function depending on separate times for each of the two particles. Two alternative integral equations for this wave function are derived with kernels in the form of an expansion in powers of g^2, the dimensionless coupling constant for the interaction. Each term in these expansions gives Lorentz-invariant equations. The validity and physical significance of these equations is discussed. In the extreme nonrelativistic approximation and to lowest order in g^2 they reduce to the appropriate Schrödinger equation.
One of these integral equations is applied to the deuteron ground state using scalar mesons of mass μ with scalar coupling. For neutral mesons the Lorentz-invariant interaction is transformed into the sum of the instantaneous Yukawa interaction and a retarded correction term. The value obtained for g^2 differs only by a fraction proportional to $(\mu/M)^2$ from that obtained by using a phenomenological Yukawa potential. For a purely charged meson theory a correction term is obtained by a direct solution of the relativistic integral equation using only the first term in the expansion of the kernel. This correction is due to the fact that a nucleon can emit, or absorb, positive and negative mesons only alternately. The constant g^2 is increased by a fraction of $1.1(\mu/M)$ or 15 percent.

Related references

See also
R. P. Feynman, Phys. Rev. **76** (1949) 749;

R. P. Feynman, Phys. Rev. **76** (1949) 769;
P. A. M. Dirac, V. A. Fock, and B. Podolsky, Phys. Zh. USSR **2** (1932) 468;
F. Bloch, Phys. Z. USSR **5** (1934) 301;
F. J. Dyson, Phys. Rev. **75** (1949) 486;
F. J. Dyson, Phys. Rev. **75** (1949) 1736;
E. Fermi and C. N. Yang, Phys. Rev. **76** (1949) 1739;
S. M. Dancoff, Phys. Rev. **78** (1950) 382;
E. E. Salpeter, Phys. Rev. **82** (1951) 60;
E. E. Salpeter, Phys. Rev. **84** (1951) 1226;
F. M. Smith et al., Phys. Rev. **78** (1950) 86;

ANNIS 1952

■ Confirmation of the existence of heavy charged unstable particles ■

S-Particles

M. Annis et al.
Nuovo Cim. **9** (1952) 624;

Abstract Discusses two events observed in a multiplate cloud chamber operated at 3260 m and triggered by a penetrating shower detector. In each case a charged particle comes to rest in a lead plate and gives rise to a single secondary of near minimum ionization. It is concluded from the observed range and scattering that the secondary particle cannot be an electron. The mass of the primary is estimated as 1420^{+910}_{-450} electron masses. In addition a further three events of this kind have been observed where it is not certain that the secondary particle is not a decay electron but the primary appears to be heavier than a π or μ-meson. If it is assumed that these three events also represent the decay of S-particles the mass of the primary deduced from all five events is 1780^{+690}_{-450} electron masses. *(Science Abstracts, 1952, 8483. H. Elliot)*

Accelerator COSMIC

Detectors Cloud chamber

Related references

More (earlier) information
H. S. Bridge and M. Annis, Phys. Rev. **82** (1951) 445;
B. Rossi, H. S. Bridge, and M. Annis, Rend. Lincei **11** (1951) 73;

Analyse information from
C. O'Ceallaigh, Phil. Mag. **42** (1951) 1032;

Reactions
charged $\rightarrow$ charged (neutrals)

Particles studied heavy

HEISENBERG 1952

■ Prediction of rising total hadronic cross sections ■

Mesonenerzeugung als Stoßwellenproblem / Production of Mesons as a Shock Wave Problem

W. Heisenberg
Z. Phys. **133** (1952) 65;

Reprinted in
W. Heisenberg, *Gesammelte Werke / Collected Works*, Ed. by W. Blum, H. P. Durr, and H. Richenberg, Springer–Verlag, Ser. A, Part I, (1985) 382.

Abstract Die Erzeugung vieler Mesonen beim Zusammenstoß zweier Nukleonen wird als ein Stoßwellenvorgang beschrieben, der von einer nichtlinearen Wellengleichung dargestellt wird. Die quantentheoretischen Züge des Vorgangs können dabei näherungsweise nach dem Korrespondenzprinzip berücksichtigt werden, da es sich um einen "Vorgang hoher Quantenzahl" handelt. Aus der Diskussion der Lösungen der nichtlinearen Wellengleichung ergeben sich Aussagen über die Energie und Winkelverteilung der verschiedenen Mesonensorten.

Related references

More (earlier) information
W. Heisenberg, Z. Phys. **113** (1939) 61;
W. Heisenberg, Z. Phys. **126** (1946) 519;

See also
M. Born, Proc. Roy. Soc. **A143** (1933) 410;
M. Born and L. Infeld, Proc. Roy. Soc. **A150** (1935) 141;
E. Fermi, Progr. of Theor. Phys. **5** (1950) 570;
M. Born and L. Infeld, Proc. Roy. Soc. **A147** (1934) 522;
M. Born and L. Infeld, Proc. Roy. Soc. **A144** (1934) 425;
E. Fermi, Phys. Rev. **81** (1951) 683;
M. Teucher, Naturw. **37** (1950) 260;
L. J. Schiff, Phys. Rev. **84** (1951) 1;
W. E. Thirring, Zeit. Naturforschung **7a** (1952) 63;
M. Teucher, Naturw. **39** (1952) 68;

Analyse information from
U. Camerini, P. H. Fowler, W. O. Lock, and H. Muirhead, Phil. Mag. **41** (1950) 413;
J. J. Lord, J. Fainberg, and M. Schein, Phys. Rev. **80** (1950) 970;
D. Clark, Phys. Rev. **81** (1951) 313;
V. D. Hopper, S. Biswas, and J. F. Darby, Phys. Rev. **84** (1951) 457;
E. Pickup and L. Voyvodic, Phys. Rev. **82** (1951) 265;
E. Pickup and L. Voyvodic, Phys. Rev. **84** (1951) 1190;

Reactions

p nucleon $\rightarrow$ mult[meson] X	<30 TeV(P_{lab})
$p\,p \rightarrow$ X	>1 GeV(P_{lab})

ANDERSON 1952B

■ First indication for the $\Delta(1232\,P_{33})$ resonance ■

Total Cross Sections of Negative Pions in Hydrogen

H.L. Anderson et al.
Phys. Rev. **85** (1952) 934;

Abstract The interaction of negative pions and protons has been investigated by Steinberger and co-workers for pions of 85 MeV energy by transmission measurements and by Shutt and co-workers for pions of 55 MeV by direct observation of pion tracks in a Wilson chamber. Both measurements indicate a surprisingly low value for the cross section in this range of energies. We have undertaken to extend the total cross section measurements to higher energies. *(Extracted from the introductory part of the paper.)*

Accelerator Chicago cyclotron

Detectors Counters

Related references

Analyse information from
P. J. Chedester, A. M. Isaacs, R. G. Sachs, and J. Steinberger, Phys. Rev. **82** (1951) 958;
R. P. Shutt et al., Phys. Rev. **84** (1951) 1247;

Reactions

$\pi^- \; p \to X$ 80.-230. MeV(T_{lab})

Particles studied $\Delta(1232\,P_{33})$

ANDERSON 1952E

■ Further evidence for the $\Delta(1232\,P_{33})^{++}$ resonance ■

Total Cross Sections of Positive Pions in Hydrogen

H.L. Anderson, E. Fermi, E.A. Long, D.E. Nagle
Phys. Rev. **85** (1952) 936;

Reprinted in
R. N. Cahn and G. Goldhaber, *The Experimental Foundations of Particle Physics*, Cambridge Univ. Press (1991) 121.
The Physical Review — the First Hundred Years, AIP Press (1995) 837.

Excerpt In this letter and in the two preceding ones, the three processes:

(1) scattering of positive pions,

(2) scattering of negative pions with exchange of charge, and

(3) scattering of negative pions without exchange of charge

have been investigated. It appears that over a rather wide range of energies, from about 80 to 150 M eV, the cross section for process (1) is the largest, for process (2) is intermediate, and for process (3) is the smallest. Furthermore, the cross sections of both positive and negative pions increase rather rapidly with the energy. Whether the cross sections level off at a high value or go through a maximum, as might be expected if there should be a resonance, is impossible to determine from our present experimental evidence. *(Extracted from the introductory part of the paper.)*

Accelerator Chicago cyclotron

Detectors Counters

Related references

More (earlier) information
H. L. Anderson et al., Phys. Rev. **85** (1952) 934;
E. Fermi et al., Phys. Rev. **85** (1952) 935;

Reactions

$\pi^+ \; p \to X$ 56.-136. MeV(T_{lab})

Particles studied $\Delta(1232\,P_{33})^{++}$

PAIS 1952

■ Hypothesis on associative production of V^0 particles ■

Some Remarks on the V-Particles

A. Pais
Phys. Rev. **86** (1952) 663;

Reprinted in
The Physical Review — the First Hundred Years, AIP Press (1995) CD-ROM.

Abstract It is qualitatively investigated whether the abundance of V-particle production can be reconciled with their long lifetime by using only interactions of a conventional structure. This is possible, provided a V-particle is produced together with another heavy unstable particle (Sec. II). Two distinct groups of interactions are needed: for one, the coupling is strong (II); for the other, it is very weak (III). Two kinds of V-particles are considered, Fermions of mass $\sim 2200m$ and Bosons ($\sim 800m$). The arguments are somewhat different, according to whether the latter are nonpseudoscalar (III) or pseudoscalar (V). The competition with processes involving μ-mesons is discussed (IV). Possible connections with the τ-meson are commented on in Sec. V. The preliminary nature of the present analysis is stressed (VI).

Related references

Analyse information from
G. D. Rochester and C. C. Butler, Nature **160** (1947) 855;
A. J. Seriff et al., Phys. Rev. **78** (1950) 290;
G. McCusker and D. H. Miller, Nuovo Cim. **8** (1951) 289;
V. D. Hopper and S. Biswas, Phys. Rev. **80** (1950) 1099;
R. Armenteros et al., Nature **167** (1951) 501;

H. S. Bridge and M. Annis, Phys. Rev. **82** (1951) 445;
R. W. Thompson, H. O. Cohn, and R. S. Flum, Phys. Rev. **83** (1951) 175;
R. B. Leighton, S. D. Wanlass, and W. L. Alford, Phys. Rev. **83** (1951) 843;
W. B. Fretter, Phys. Rev. **83** (1951) 1053;
R. Armenteros, K. H. Barker, C. C. Butler, and A. Cachon, Phil. Mag. **42** (1951) 1113;

COURANT 1952

■ Invention of the strong focusing principle for accelerators ■

The Strong-Focusing Synchrotron — A New High Energy Accelerator

E.D. Courant, M.S. Livingston, H.S. Snyder
Phys. Rev. **88** (1952) 1190;

Reprinted in
The Physical Review — the First Hundred Years, AIP Press (1995) CD-ROM.

Abstract Strong focusing forces result from the alternation of large positive and negative n-values in successive sectors of the magnetic guide field in a synchrotron. This sequence of alternately converging and diverging magnetic lenses of equal strength is itself converging, and leads to significant reductions in oscillation amplitude, both for radial and axial displacements. The mechanism of phase-stable synchronous acceleration still applies, with a large reduction in the amplitude of the associated radial synchronous oscillations. To illustrate, a design is proposed for a 30 BeV proton accelerator with an orbit radius of 300 ft, and with a small magnet having an aperture of 1 × 2 inches. Tolerances on nearly all design parameters are less critical than for the equivalent uniform-n machine. A generalization of this focusing principle leads to small, efficient focusing magnets for ion and electron beams. Relations for the focal length of a double-focusing magnet are presented, from which the design parameters for such linear systems can be determined.

Related references

See also
D. W. Kerst and R. Serber, Phys. Rev. **60** (1941) 53;
N. M. Blachman and E. D. Courant, Rev. Sci. Inst. **20** (1949) 596;
L. H. Thomas, Phys. Rev. **54** (1938) 580;
L. H. Thomas, Phys. Rev. **54** (1938) 588;

ARMENTEROS 1952

■ First evidence for the Ξ^- hyperon ■

The Properties of Charged V Particles

R. Armenteros et al.
Phil. Mag. **43** (1952) 597;

Abstract Twenty one examples of the decay of charged V-particles have been found among 25000 counter-controlled cloud-chamber photographs obtained on the Pic-du-Midi. These events have been analysed and compared with the data obtained by other groups. The measurements show that, if the majority of the $V^{\pm}$-tracks are produced by one type of particle having a single mode of decay, there must be at least two neutral secondary particles, in addition to the charged secondary particle. The latter is probably a μ- or a τ-meson and the combined mass of the neutral particles must be less than $1000m_e$. Since the κ-particles, of mass about $1200m_e$, which were discovered in photographic emulsions, also decay into a charged meson and at least two neutral particles, it is plausible to assume provisionally, that the $V^{\pm}$-particles and the κ-particles are the same. On this assumption most of the available data concerning $V^{\pm}$ and κ-particles can be explained if the unstable particle has a mass of about $1200m_e$ and decays into a charged μ-meson, a π^0-meson and a neutrino, or alternatively into a charged μ-meson and two neutrinos. As yet no direct evidence has been found for the existence of a π^0-meson in the decay scheme. One negatively charged V-particle probably produced a neutral V-track and a charged meson.

Accelerator COSMIC

Detectors Cloud chamber

Related references

More (earlier) information
R. Armenteros et al., Nature **167** (1951) 501;
R. Armenteros, K. H. Barker, C. C. Butler, and A. Cachon, Phil. Mag. **42** (1951) 1113;
See also
H. S. Bridge and M. Annis, Phys. Rev. **82** (1951) 445;
P. Astbury et al., *Report given at Bristol Conference in December* (1951);
C. C. Butler, W. G. V. Rosser, and K. H. Barker, Proc. Roy. Soc. **63** (1950) 145;
C. C. Butler, *Progress in Cosmic Ray Physics*, (North Holland Publishing Co.) (1951) 110;
P. H. Fowler, M. G. K. Menon, C. F. Powell, and O. Rochat, Phil. Mag. **42** (1951) 1040;
W. B. Fretter, Phys. Rev. **83** (1951) 1053;
O. Kofoed-Hansen, Phil. Mag. **42** (1951) 1411;
R. B. Leighton, S. D. Wanlass, and W. L. Alford, Phys. Rev. **83** (1951) 843;
C. O'Ceallaigh, Phil. Mag. **42** (1951) 1032;
G. D. Rochester and C. C. Butler, Nature **160** (1947) 855;
A. J. Seriff et al., Phys. Rev. **78** (1950) 290;
R. W. Thompson, H. O. Cohn, and R. S. Flum, Phys. Rev. **83** (1951) 175;
J. Trembley, *Report given at Bristol Conference in December* (1951);
M. G. K. Menon, *Report given at Bristol Conference in December* (1951);

Particles studied Ξ^-

BONETTI 1953

■ First evidence for the charged Σ hyperon ■

Observation of the Decay at Rest of a Heavy Particle

A. Bonetti, R. Levi-Setti, M. Panetti, G. Tomasini
Nuovo Cim. **10** (1953) 345;

Excerpt During an investigation on meson decays in nuclear emulsions exposed at high altitude, carried on as a collaboration between the Genoa and Milan group, a particular event has been observed by G. Tomasini, suggesting the existence of an unstable charged particle heavier than the proton. *(Extracted from the introductory part of the paper.)*

Accelerator COSMIC

Detectors Nuclear emulsion

Reactions
charged $\rightarrow$ charged (neutrals)

Particles studied Σ

BONETTI 1953B

■ Confirmation of the existence of Σ^+ and Σ^- hyperons ■

On the Existence of Unstable Charged Particles of Hyperonic Mass

A. Bonetti, R. Levi-Setti, M. Panetti, G. Tomasini
Nuovo Cim. **10** (1953) 1736;

Excerpt Further evidence is presented in favour of the existence of charged unstable particles of hyperprotonic mass. In two cases the decay product is a fast light particle, in one case it is a slow proton. The decay schemes: $\Omega_\pi^\pm \rightarrow n\ \pi^\pm + Q$; $\Omega_p^+ \rightarrow p\ \pi^0 + Q$ are proposed for the two types of events respectively. They are possible alternative modes of decay of the same type of H-particle. The evidence available from photographic emulsion and cloud chamber experiments is discussed.

Accelerator COSMIC

Detectors Nuclear emulsion

Related references

More (earlier) information
A. Bonetti et al., Nuovo Cim. **10** (1953) 345;

Analyse information from
M. Ceccarelli and M. Merlin, Nuovo Cim. **10** (1953) 1207;
H. S. Bridge and M. Annis, Phys. Rev. **82** (1951) 445;
C. M. York, R. B. Leighton, and E. K. Bjornerud, Phys. Rev. **90** (1953) 167;

Reactions
$\Sigma^+ \rightarrow n\ \pi^+$
$\Sigma^- \rightarrow n\ \pi^-$
$\Sigma^+ \rightarrow p\ \pi^0$

Particles studied Σ^+, Σ^-

YORK 1953

■ Direct experimental evidence for the Σ^+ ■

Direct Experimental Evidence for the Existence of a Heavy Positive V Particle

C.M. York, R.B. Leighton, E.K. Bjornerud
Phys. Rev. **90** (1953) 167;

Abstract The indications that of 53 V-particles observed in a cloud chamber some are longer lived (10^{-9} sec) than others (10^{-10} sec) is supported by an analysis of two V events with a secondary of mass 1300-2300m_e, and consistent with the decay of a heavy V_1^+ into $p\ \pi^0 + (Q \sim 40$ MeV). *(Science Abstracts, 1953, 5104. W. J. Swiatecki)*

Accelerator COSMIC

Detectors Cloud chamber

Related references

Analyse information from
R. B. Leighton and S. D. Wanlass, Phys. Rev. **86** (1952) 426;
R. Armenteros et al., Phil. Mag. **43** (1952) 597;
R. Armenteros, K. H. Barker, C. C. Butler, and A. Cachon, Phil. Mag. **42** (1951) 1113;
R. B. Leighton, S. D. Wanlass, and C. D. Anderson, Phys. Rev. **89** (1953) 148;

Reactions
charged$^+$ $\rightarrow p\ \pi^0$

Particles studied Σ^+

THOMPSON 1953

■ First evidence for $K \rightarrow \pi\pi$ decay ■

The Disintegration of V^0 Particles

R.W. Thompson et al.
Phys. Rev. **90** (1953) 329;

Reprinted in
The Physical Review — the First Hundred Years, AIP Press (1995) CD-ROM.

Abstract Many V^0 particles appear to decay into a proton and negative pion with a Q value in the neighborhood of 35 MeV. However, it has been apparent for some time that there are cases in which the positive fragment is considerably less

massive than a proton. Various conjectures as to the decay scheme in such cases have been made: the one ($V^0 \to \pi\ \pi$) most frequently discussed having been introduced in the nature of a simplifying assumption. The California Institute of Technology group reports a few cases which may be so interpreted.
The object of this note is to report preliminary results obtained with a new magnet chamber at sea level. The chamber has a field of 7000 gauss over an illuminated volume 22 in. high, 11 in. wide, and 5 in. deep and is actuated by penetrating showers. Measurements of no field tracks indicate that the maximum detectable momentum for long tracks is well in excess of 10^{11} eV/c. *(Extracted from the introductory part of the paper.)*

Accelerator COSMIC

Detectors Cloud chamber

Related references

Analyse information from
R. Armenteros et al., Nature **167** (1951) 501;
R. W. Thompson, H. O. Cohn, and R. S. Flum, Phys. Rev. **83** (1951) 175;
R. B. Leighton, S. D. Wanlass, and W. L. Alford, Phys. Rev. **83** (1951) 843;
W. B. Fretter, May, and Nakada, Phys. Rev. **89** (1953) 168;
R. Armenteros, K. H. Barker, C. C. Butler, and A. Cachon, Phil. Mag. **42** (1951) 1113;
R. B. Leighton, S. D. Wanlass, and C. D. Anderson, Phys. Rev. **89** (1953) 148;

Reactions
$K^0 \to \pi^+ \pi^-$
$K^+ \to \pi^+ \pi^0$

Particles studied K^0, K^+

THOMPSON 1953B

■ First measurement of the K^0 mass ■

An Unusual Example of V^0 Decay

R.W. Thompson et al.
Phys. Rev. **90** (1953) 1122;

Reprinted in
R. N. Cahn and G. Goldhaber, *The Experimental Foundations of Particle Physics*, Cambridge Univ. Press (1991) 73.

Abstract Discussion of a V^0 event in a cloud chamber showing a pos., 43 cm, 0.67 ± 0.02 BeV/c, min. ionization track at $79.2° \pm 0.4°$ to a neg. 0.094 ± 0.008 BeV/c track of 2 - 3 times min. ionization. *(Science Abstracts. 1953, 7172. W. J. Swiatecki)*

Accelerator COSMIC

Detectors Cloud chamber

Related references

More (earlier) information
R. W. Thompson et al., Phys. Rev. **90** (1953) 329;

See also
R. B. Leighton, S. D. Wanlass, and C. D. Anderson, Phys. Rev. **89** (1953) 148;
D. H. Perkins, in the *Proc. of the Third Ann. Rochester Conf. on High Energy Physics*, Interscience, New York (1953);

Reactions
neutral $\to$ charged$^+$ charged$^-$
$K^0 \to \pi^+ \pi^-$

Particles studied K^0

FOWLER 1953C

■ First V-event production at the Cosmotron ■

Observation of V^0 Particles Produced at the Cosmotron

W.B. Fowler, R.P. Shutt, A.M. Thorndike, W.L. Whittemore
Phys. Rev. **90** (1953) 1126;

Excerpt Two definite examples of V^0 particles similar to those found in cosmic rays by many workers have been observed in a cloud chamber exposed to a neutron beam from the Cosmotron. These two cases, in addition to several other less definite ones, were found in a total of about 4000 photographs scanned up to date. Further work is in progress. *(Extracted from the introductory part of the paper.)*

Accelerator BNL-COSMOTRON

Detectors Cloud chamber

Related references

See also
R. B. Leighton, S. D. Wanlass, and C. D. Anderson, Phys. Rev. **89** (1953) 148;
W. B. Fretter, May, and Nakada, Phys. Rev. **89** (1953) 168;
R. W. Thompson et al., Phys. Rev. **90** (1953) 329;
A. Pais, Phys. Rev. **86** (1952) 663;
W. B. Fowler, *Proc. of the Third Ann. Rochester Conf. on High Energy Physics*, Interscience Publishing Company, New York (1953);

Reactions
n nucleus $\to$ vee X <2.2 GeV(T_{lab})

Particles studied vee

GLASER 1953 **Nobel prize**

■ First evidence of charged particle tracks in a bubble chamber. Nobel prize to D. A. Glaser awarded in 1960 "for the invention of bubble chambers" ■

Bubble Chamber Tracks of Penetrating Cosmic Ray Particles

D.A. Glaser
Phys. Rev. **91** (1953) 762;

Reprinted in
The Physical Review — the First Hundred Years, AIP Press (1995) CD-ROM.

Excerpt Tracks of penetrating cosmic ray particles passing through an ether-filled bubble chamber under 10 cm of lead have been recorded by flash photography triggered by a two-fold vertical coincidence telescope. The bubble chamber consisted of a heavy-walled cylindrical Pyrex bulb 3 cm long and 1 cm inside diameter, which communicates with a pressure-regulating device by means of a Pyrex capillary tube 45 cm long. A thermostated temperature bath of mineral oil surrounded the bulb, maintaining the temperature constant within 0.5°C in the range 138°C to 143°C. The pressure-regulating device consisted of a brass cylinder of length 2 cm and inside diameter 3 cm. One end of the cylinder was scaled with a flexible diaphragm of $\frac{1}{8}$-in. Neoprene faced with Teflon to confine the ether and permit variation of its pressure by controlling the pressure of compressed gas on the outside of the diaphragm.
To prepare for taking a picture of a track, the ether was compressed by admitting compressed nitrogen to the pressure regulator at a pressure of 300 pounds per square inch so that no vapor bubbles remained in the system. Then the gas was allowed to escape, so that the ether suddenly became highly superheated at atmospheric pressure. On the average the liquid remained quietly in this unstable condition for several seconds until a violent eruptive boiling occurred. If a coincidence of the vertical counter telescope occurred during this waiting time, a picture was taken by means of a xenon discharge flash lamp. About 5 seconds were required to recompress the ether in preparation for the next event ... *(Extracted from introductory part of the paper.)*

Accelerator COSMIC

Detectors Heavy-liquid bubble chamber

Reactions
charged atom $\rightarrow$ *ion* e^- charged

FOWLER 1953B

■ Confirmation of the existence of the Λ hyperon. First evidence for associated production of heavy unstable particles ■

Production of V_1^0 Particles by Negative Pions in Hydrogen

W.B. Fowler, R.P. Shutt, A.M. Thorndike, W.L. Whittemore
Phys. Rev. **91** (1953) 1287;

Reprinted in
The Physical Review — the First Hundred Years, AIP Press (1995) CD-ROM.

Abstract Two diffusion cloud chamber events are discussed, produced by 1.5 BeV π^--particles from the Brookhaven cosmotron. The lifetimes of the V_1^0 are 4×10^{-11} and 3×10^{-11} sec respectively. Momentum measurements obtained for the decay products in one of the events (450 ± 80 MeV/c for the proton and 210 ± 70 MeV/c for π^-) are in agreement with the accepted Q value of 37 MeV. *(Science Abstracts, 1954, 781. G. Field)*

Accelerator BNL-COSMOTRON

Detectors Cloud chamber

Related references

See also
R. B. Leighton, S. D. Wanlass, and C. D. Anderson, Phys. Rev. **89** (1953) 148;
A. Pais, Phys. Rev. **86** (1952) 663;

Reactions

$\pi^- \; p \rightarrow \Lambda$ neutral	1.63 GeV(P_{lab})
$\pi^- \; p \rightarrow \Lambda \; K^0$	1.63 GeV(P_{lab})
$\Lambda \rightarrow p \; \pi^-$	
$K^0 \rightarrow \pi^+ \; \pi^-$	

Particles studied Λ

REINES 1953B **Nobel prize**

■ First evidence for the $\overline{\nu}_e$. Nobel prize to F. Reines awarded in 1995 "for the detection of the neutrino". Co-winner M. Perl "for the discovery of the tau lepton" ■

Detection of the Free Neutrino

F. Reines, C.L. Cowan
Phys. Rev. **92** (1953) 830;

Reprinted in
The Physical Review — the First Hundred Years, AIP Press

(1995) CD-ROM.

Excerpt An experiment has been performed to detect the free neutrino. It appears probable that this aim has been accomplished although further confirmatory work is in progress. *(Extracted from the introductory part of the paper.)*

Accelerator Nuclear reactor

Detectors Counters

Related references

More (earlier) information
F. Reines and C. L. Cowan, Phys. Rev. **90** (1953) 492;
C. L. Cowan, F. Reines et al., Phys. Rev. **90** (1953) 493;

More (later) information
C. L. Cowan, F. Reines et al., Science **124** (1956) 103;

Reactions
$\overline{\nu}_e\ p \rightarrow n\ e^+$ 1.8-6.0 MeV(E_{lab})

Particles studied $\nu_e, \overline{\nu}_e$

GELL-MANN 1953 Nobel prize

■ Extension of isotopic multiplet structure for new unstable particles. Explanation of pairwise production of V particles. Nobel prize to M. Gell-Mann awarded in 1969 "for his contributions and discoveries concerning the classification of elementary particles and their reactions" ■

Isotopic Spin and New Unstable Particles

M. Gell-Mann
Phys. Rev. **92** (1953) 833;

Reprinted in
The Physical Review — the First Hundred Years, AIP Press (1995) CD-ROM.

Excerpt In connection with the work of Peaslee and of Pais, the author would like to put forward an alternative hypothesis that he has considered for some time, and which, like that of Pais, overcomes the difficulty posed by electromagnetic interactions. Let us suppose that the new unstable particles are fermions with integral isotopic spin and bosons with half–integral isotopic spin. For example, the V_1 particles may form an isotopic triplet, consisting of V_1^+, V_1^0, and V_1^-. The τ^+ and V_4^0 may form an isotopic doublet, which we may call τ^+ and τ^0. *(Extracted from the introductory part of the paper.)*

Related references

See also
D. C. Peaslee, Phys. Rev. **86** (1952) 127;
A. Pais, Phys. Rev. **86** (1952) 663;

Particles studied strange

KONOPINSKI 1953

■ Invention of the concept of lepton quantum number ■

The Universal Fermi Interaction

E.J. Konopinski, H.M. Mahmoud
Phys. Rev. **92** (1953) 1045;

Abstract On the basis of the hypothesis that the same form of interaction acts among any spin-$\frac{1}{2}$ particles, it is interesting to apply the interaction law found for β decay to the muon processes. The application is beset by two types of ambiguity. The first is due to the uncertainty in measured values of coupling constants, and particularly their signs. The second arises from the various ways in which the correspondence between the particles of μ and β decay may be taken. Arguments are presented that the unique correspondence established if two *like* neutrinos are ejected in μ decay is the correct one.
It is argued that previous interpretations of the Universal Fermi Interaction have been unjustifiably broad. Only processes in which two normal particles (*vs* antiparticles) are annihilated, and two created, should be expected. The positive muon must be treated as the normal-particle (if the neutron, proton and negatron are) in order to avoid the expectation that muon capture by a proton may yield electrons, contrary to experimental facts. The conclusion that two like neutrinos are ejected in μ decay follows immediately.

Related references

See also
H. M. Mahmoud and E. J. Konopinski, Phys. Rev. **88** (1952) 1266;
J. S. Allen and W. K. Jentschke, Phys. Rev. **89** (1953) 902;
B. M. Rustad and S. L. Ruby, Phys. Rev. **89** (1953) 880;
M. E. Rose, Phys. Rev. **90** (1953) 1123;
A. Petschek and R. E. Marshak, Phys. Rev. **85** (1952) 698;
E. J. Konopinski and L. M. Langer, Ann. Rev. Nucl. Sci. **2** (1952) 261;
O. Klein, Nature **161** (1948) 897;
L. Michel, Phys. Rev. **86** (1952) 814;

ANDERSON 1953B

■ Confirmation of the existence of the Ξ^- ■

Cascade Decay of V Particles. I.

C.D. Anderson, E.W. Cowan, R.B. Leighton, V.A.J. Van Lint
Phys. Rev. **92** (1953) 1089 (H3);

Reprinted in
Contributed paper to the APS Meeting, September, 1-5, 1953, Albuquerque.

Excerpt Three cloud-chamber exposures of a charged V particle apparently decaying into a V^0 particle in a magnetic field have been taken at Pasadena. In a similar previous case, the V^0 particle could not be classified as to type. The new data provide strong evidence for the reality of cascade decays, and identify the V^0 particle as a V_1^0 (decaying into $p\pi^-$). In all three cases the parent charged V particle is negative and emits a negative π or μ meson. The above conclusions are based on measurements of

1. relative positions of decay points in the cloud chamber,
2. "coplanarity,"
3. momenta and ionizations of secondary particles, and
4. position and number of nuclear events in which V particles might originate.

In all three cases the line of flight of the V^0 particle makes a much smaller angle with the positive decay product than with the negative, $\theta_+/\theta_- = 0.16, 0.22$, and 0.41, thus indicating the decay products of the V^0 particle to have unequal masses.

Accelerator COSMIC

Detectors Cloud chamber

Related references

More (later) information
E. W. Cowan et al., Phys. Rev. **92** (1953) 1089;

Analyse information from
R. Armenteros et al., Phil. Mag. **43** (1952) 597;

Reactions

vee $\rightarrow$ $p\,\pi^-$
charged$^-$ $\rightarrow$ vee charged$^-$

Particles studied Ξ^-

COWAN 1953

■ Further confirmation of the Ξ^- ■

Cascade Decay of V Particles. II.

E.W. Cowan, C.D. Anderson, R.B. Leighton, V.A.J. Van Lint
Phys. Rev. **92** (1953) 1089 (H4);

Reprinted in
Contributed paper to the APS Meeting, September, 1-5, 1953, Albuquerque.

Excerpt In one of the three cases of the preceding abstracts (ANDERSON 1953B), the positive particle in the V^0 decay has a momentum of 880 ± 100 MeV/c, and ionization 1.7 times minimum, determining by counting droplets. This is consistent with a proton but not with a π meson, indicating the V^0 particle is not of the type that decays into two π mesons. The Q value is 39 ± 10 MeV assuming $\pi\,'$ decay, and 36 ± 10 MeV assuming $\pi\ \pi$ decay. The latter value is much lower than the usual range of values for $\pi\ \pi$ decay. It can further be shown, on the assumption of a two-body decay, using only the momenta and included angle of the two decay particles that the observed angular asymmetry about the line of flight of the neutral particle is too great to correspond to a decay into two π mesons. The charged secondary of the V^- decay has a momentum of 64 ± 10 MeV/c, an estimated ionization of $2-4$ times minimum, and is therefore either a π or μ mesons. On the assumption that the V^- particle decays into a V_1^0 and a π or μ meson, the best Q value, obtained from 2 of the 3 cases, is 60 ± 15 MeV for a π meson and 62 ± 15 MeV for a μ meson secondary. The calculated mass-values of the charged V particle on these two assumptions are 2570 ± 30 and 2510 ± 30 m_e, respectively.

Accelerator COSMIC

Detectors Cloud chamber

Related references

More (earlier) information
C. D. Anderson, E. W. Cowan et al., Phys. Rev. **92** (1953) 1089H3;

Analyse information from
R. Armenteros et al., Phil. Mag. **43** (1952) 597;

Reactions

vee $\rightarrow$ $p\,\pi^-$
charged$^-$ $\rightarrow$ vee charged$^-$

Particles studied Ξ^-

LINDENBAUM 1953

■ Confirmation of the existence of the $\Delta(1232\,P_{33})$ resonance ■

The Interaction Cross Section of Hydrogen and Heavier Elements for 450 MeV Negative and 340 MeV Positive Pions

S.J. Lindenbaum, L.C.L. Yuan
Phys. Rev. **92** (1953) 1578;

Excerpt The hydrogen interaction cross section for pions of kinetic energies up to about 150-200 MeV have been measured by Anderson, Fermi, and others. An investigation of the behavior of these cross sections for negative and positive ions from 150 MeV to about 700 MeV is now in progress using fast scintillation counter telescopes in the external meson beams of the 2.3 BeV Brookhaven Cosmotron. This letter will report the cross sections obtained for 450 MeV negative and 340 MeV positive pions. *(Extracted from the introductory part of the paper.)*

Accelerator BNL-COSMOTRON

Detectors Counters

Related references

See also
Fernbach, R. Serber, and S. Taylor, Phys. Rev. **75** (1949) 1352;
K. A. Brueckner, Phys. Rev. **86** (1952) 109;
G. F. Chew, Phys. Rev. **89** (1953) 591;
Analyse information from
H. L. Anderson et al., Phys. Rev. **85** (1952) 934;
H. L. Anderson et al., Phys. Rev. **85** (1952) 936;
H. L. Anderson et al., Phys. Rev. **86** (1952) 793;
H. L. Anderson et al., Phys. Rev. **91** (1953) 155;
P. J. Chedester, A. M. Isaacs, R. G. Sachs, and J. Steinberger, Phys. Rev. **82** (1951) 958;
A. M. Isaacs, R. G. Sachs, and J. Steinberger, Phys. Rev. **85** (1952) 803;
V. E. Barnes, D. Clark, C. T. Perry, and Angel, Phys. Rev. **87** (1952) 669;

Reactions

$\pi^+ p \to X$	340 MeV(T_{lab})
$\pi^- p \to X$	450 MeV(T_{lab})
$\pi^- C \to X$	450 MeV(T_{lab})
$\pi^- Cu \to X$	450 MeV(T_{lab})

Particles studied $\Delta(1232\, P_{33})$

DANYSZ 1953

■ First evidence for a hypernucleus ■

Delayed Disintegration of a Heavy Nuclear Fragment

M. Danysz, J. Pniewski
Phil. Mag. **44** (1953) 348;

Excerpt A remarkable coincidence of two events recorded in a photographic emulsion has recently been observed in this laboratory. It occurred in a G5 emulsion, 600μ thick, which had been exposed to cosmic radiation at an altitude of 85 000 feet, and consists of two stars marked A and B in the photomicrograph reproduced in Plate 13. The centre of the star B coincides with the end of the track of a heavy fragment ejected from the star A. If the coincidence is not accidental, it must be considered as an example of the delayed disintegration of a heavy fragment. The probability of a fortuitous coincidence is very small, and it therefore seemed appropriate to analyse the events more closely. It is clear, of course, that any novel conclusions drawn from a single observation should be treated with proper reserve. *(Extracted from the introductory part of the paper.)*

Accelerator COSMIC

Detectors Nuclear emulsion

Particles studied hypernucleus

DALITZ 1953

■ Invention of the Dalitz plot method to analyse spatial quantum numbers of mesons by their decays into three known particles ■

On the Analysis of τ-Meson Data and the Nature of the τ-Meson

R.H. Dalitz
Phil. Mag. **44** (1953) 1068;

Abstract A convenient method of representation is proposed (§2) for data on τ-meson decay configurations, applicable when the unlike outgoing π-meson is not distinguished. The relation between the spin and-parity of the τ-meson and the distribution of decay configuration is obtained for some simple cases. The hypothesis that the τ- and χ-mesons are identical requires a non-zero spin for this particle and the available data on τ-meson decay does not exclude this possibility. However, observations in which the unlike outgoing τ-meson is not distinguished are relatively ineffective in discriminating between the various possibilities. The distortions which strong meson-meson attraction may produce in τ-decay configurations are discussed; the present data offers no evidence on this effect.

Related references

See also
D. Rochester and C. C. Butler, *Report on Progress in Physics* **16** London, Physical Society (1953) 364;
K. M. Watson, Phys. Rev. **88** (1952) 1163;
K. M. Watson and R. N. Stuart, Phys. Rev. **82** (1951) 738;
R. H. Dalitz, Proc. Roy. Soc. **66** (1953) 710;
Analyse information from
R. Armenteros et al., Nature **167** (1951) 501;
R. H. Brown et al., Nature **163** (1949) 82;
K. A. Brueckner and K. M. Watson, Phys. Rev. **87** (1952) 621;
R. R. Daniel, J. H. Davies, J. H. Mulvey, and D. H. Perkins, Phil. Mag. **43** (1952) 753;
K. H. Barker, Proc. Roy. Soc. **A221** (1954) 328;
M. Danysz, W. O. Lock, and G. Yekutieli, Nature **169** (1952) 364;
M. G. K. Menon and C. O'Ceallaigh, Proc. Roy. Soc. **A221** (1954) 292;
R. W. Thompson et al., Phys. Rev. **90** (1953) 329;

Reactions

$\text{meson}^- \to \pi^+ 2\pi^-$
$\text{meson}^0 \to \pi^+ \pi^0 \pi^-$
$\text{meson}^+ \to 2\pi^+ \pi^-$

Particles studied meson

STÜCKELBERG 1953

■ Invention of the renormalization group ■

La normalization des constantes dans la theorie des quanta / On the Normalizations of the Quantum Theory Constants

E.C.G. Stückelberg, A. Petermann
Helv. Phys. Acta **26** (1953) 499;

Excerpt This article propose a mathematical foundation to the method previously employed* to give a definite meaning to the products of invariant distributions such as $(\Delta^{(1)}_{x-y} D^{(s)}_{x-y} +^{(s)(1)})$, $(\Delta^{(1)}_{x-y}\Delta^{(2)}_{y-z}D^{(s)}_{x-z} + \ldots)$, etc. in terms of arbitrary constants $c_1, c_2 \ldots c_{r(n)}$. The n'th approximation $S^{(n)}$ of the $S[V]$ *matrix* (defined for a given space time-region V) depends on these $r(n)$ arbitrary constants in addition to the arbitrary physical parameters (masses κ, μ, and coupling constants $e, g \ldots$). In the introduction (§1), we see that a definite physical meaning can be given to the masses κ, μ. A coupling parameter, however can only be specified in terms of a chosen development of a *function* $S(xy.., \kappa, .., c_1..)$ of physical significance. However, the terms of the actual correspondence development (in terms of e^2) $S = S_2 + S_4 + \ldots$ have no physical meaning. Therefore the coefficient e^2 in S_2 has only a mathematical significance. It requires that the functions of $xy..S_2, S_4, ..S_n$ have all be specified. As this specification involves the $c_i's$, we must expect that a group of infinitesimal operations $P_i = (\frac{\partial}{\partial c_i})_{c=o}$ exists, satisfying

$$P_i = h_{ie}(\kappa, \mu, e)\frac{\partial S(\kappa, \mu, e, 00\ldots)}{\partial e},$$

admitting thus a *renormalization* of e.

* E.C.G.Stückelberg and D.Rivier, E.C.G.Stückelberg and T.A.Green.

(Extracted from the introductory part of the paper.)

Related references

See also
E. C. G. Stückelberg and D. Rivier, Helv. Phys. Acta **23** (1950) 236;
E. C. G. Stückelberg and T. A. Green, Helv. Phys. Acta **24** (1951) 153;
A. Petermann and E. C. G. Stückelberg, Phys. Rev. **82** (1951) 548;
E. C. G. Stückelberg, Phys. Rev. **81** (1951) 130;
E. C. G. Stückelberg, Helv. Phys. Acta **14** (1941) 51;
L. Schwartz, *Théorie des distributions.* **V.I, II** Paris (1950, 1951);
E. E. Salpeter and H. A. Bethe, Phys. Rev. **84** (1951) 1232;
R. P. Feynman, Phys. Rev. **76** (1949) 749;
R. P. Feynman, Phys. Rev. **76** (1949) 769;
H. Weyl, *Raum-Zeit-Materie*, **5 ed.** Berlin (1951) 128;
L. L. Foldy, Phys. Rev. **84** (1951) 168;
S. N. Gupta, Proc. Phys. Soc. **53** (1950) 681;
K. Bleuler, Helv. Phys. Acta **23** (1950) 567;
H. Fukuda and T. Kinoshita, Progr. of Theor. Phys. **5** (1950) 1024;
A. Salam, Phys. Rev. **82** (1951) 217;
Z. Koba et al., Progr. of Theor. Phys. **6** (1951) 849;

GREGORY 1954

■ Confirmation of the decays $K^{\pm} \to \mu^{\pm}$ neutral ■

Etude des Mesons K Charges, au Moyen de Deux Chambres de Wilson Superposées / Study of Charged K-mesons by means of two Superposed Wilson Chambers

B.P. Gregory et al.
Nuovo Cim. **11** (1954) 292;

Excerpt In this paper we present some results on charged heavy mesons. They were obtained with an experimental arrangement consisting essentially of two large cloud-chambers placed one above the other. The top chamber had a magnetic field while the bottom chamber was of the multiplate type. The contents of the different sections of the paper can be summarized as follows:

1. The double cloud-chamber method is compared with other more conventional methods.
2. The apparatus is briefly described.
3. The behaviour in the multiplate chamber of the charged secondaries of $V^{\pm}$-decays occurring in the top chamber is discussed. π-mesons have been identified through their nuclear interactions. The presence of μ-mesons is also strongly suggested by: (a) the long mean-free-path of the secondaries and (b) one direct mass measurement.
4. S and slow $V^{\pm}$-events observed in the multiplate chamber are discussed. It is shown that one should distinguish between two groups of decays: One with secondaries of range smaller than 20 g/cm^2 of Pb, the other with larger range secondaries. Although the first group can be due to τ-mesons, the second group cannot be so explained, at least not phenomenologically.
5. The results of mass measurements — from range and momentum — on the primaries of the second group of events are presented. The measurements agree with a unique value of (914 ± 20) m$_e$; which appears to be different from the τ-mass.
6. We show that some, and probably most, of the secondaries of this second group of events are μ-mesons. This conclusion is based on the dynamics of the decay and results from the measured mass of the primary particle and the observed ranges of the secondaries. The observation of one large angle scattering indicates, however, that π-mesons may also be present among the secondaries.
7. The data on the range of the charged secondaries are analysed. If we exclude the possible π^- secondary, our

results are in good agreement with a μ-meson of unique momentum. However, a continuous spectrum, sharply peaked in the high momentum region, is not excluded.

8. The almost complete absence of negative primaries for S-events is reported.
9. The results are compared with those obtained elsewhere. They agree with other cloud-chamber results assuming the existence of π-secondaries with a range of 66 g/cm^2 of Pb. The spectrum obtained for the μ-secondaries does not fit with that of the secondaries of k-particles found in photographic emulsions.
10. The conclusions are summarized.

Accelerator COSMIC

Detectors Cloud chamber

Related references

Analyse information from
G. D. Rochester and C. C. Butler, Nature **160** (1947) 855;
M. Annis et al., Nuovo Cim. **9** (1952) 624;
C. O'Ceallaigh, Phil. Mag. **42** (1951) 1032;
R. W. Thompson, A. V. Buskirk, C. J. Karzmark, and R. H. Rediker, Phys. Rev. **92** (1953) 209;
D. Lal, Y. Pal, and B. Peters, Phys. Rev. **92** (1953) 438;
W. B. Fretter et al., Phys. Rev. **92** (1953) 1583;
H. S. Bridge and M. Annis, Phys. Rev. **82** (1951) 445;
R. H. Brown et al., Nature **163** (1949) 82;
E. Amaldi et al., Nuovo Cim. **10** (1953) 937;

Reactions

$K^- \rightarrow \mu^- \nu$
$K^+ \rightarrow \mu^+ \nu$

Particles studied K^+, K^-

FOWLER 1954C

■ Confirmation of the existence of the Σ^-. Evidence for the associated production of strange particles ■

Production of Heavy Unstable Particles by Negative Pions

W.B. Fowler, R.P. Shutt, A.M. Thorndike, W.L. Whittemore
Phys. Rev. **93** (1954) 861;

Reprinted in

R. N. Cahn and G. Goldhaber, *The Experimental Foundations of Particle Physics*, Cambridge Univ. Press (1991) 75.
The Physical Review — the First Hundred Years, AIP Press (1995) 838.

Abstract In addition to two previously discussed cloud-chamber examples of V-particle production by 1.5 BeV π^- mesons from the Cosmotron, four further examples are discussed here. In two of the new examples a $\Lambda^0(V_1^0)$ and a $\vartheta^0(V_4^0)$ are seen to decay in a geometry indicating that they were produced together in a $\pi^- p$ collision. A third example is best interpreted as production of a $\Lambda^-(V_1^-)$ together with a $K^+(V_2^+)$ by a π^- colliding with a proton. A fourth example shows of $\sim$ 1 millibarn for V-particle production is inferred from the number of $\pi^- p$ collisions observed.

Accelerator BNL-COSMOTRON

Detectors Cloud chamber

Comments on experiment 26000 photographs, 4 events with associative strange particle production.

Related references

More (earlier) information
W. B. Fowler et al., Phys. Rev. **91** (1953) 1287;
See also
G. D. Rochester and C. C. Butler, Rep.Prog.Phys. **16** (1953) 364;
G. Ascoli, Phys. Rev. **90** (1953) 1079;
D. C. Peaslee, Phys. Rev. **86** (1952) 127;
E. Fermi, Progr. of Theor. Phys. **5** (1950) 570;
Analyse information from
A. Pais, Phys. Rev. **86** (1952) 663;
M. Gell-Mann, Phys. Rev. **92** (1953) 833;
M. Goldhaber, Phys. Rev. **92** (1953) 1279;
R. B. Leighton, S. D. Wanlass, and C. D. Anderson, Phys. Rev. **89** (1953) 148;
W. B. Fretter, May, and Nakada, Phys. Rev. **89** (1953) 168;
C. C. Butler, Repts. Progr. in Phys. **16** (1953) 364;
C. M. York, R. B. Leighton, and E. K. Bjornerud, Phys. Rev. **90** (1953) 167;
R. W. Thompson et al., Phys. Rev. **90** (1953) 1122;
R. L. Cool, L. Madansky, and O. Piccioni, Phys. Rev. **93** (1954) 637;

Reactions

$\pi^- p \rightarrow \Lambda K^0$ 1.5 GeV(T$_{\text{lab}}$)
$\pi^- p \rightarrow \Sigma^- K^+$ 1.5 GeV(T$_{\text{lab}}$)
$\Lambda \rightarrow p \pi^-$
$K^0 \rightarrow \pi^+ \pi^-$
$\Sigma^- \rightarrow n \pi^-$

Particles studied Σ^-

COWAN 1954

■ Confirmation of Ξ^- cascade decay. First indication for $\overline{\Lambda}$ ■

A V-Decay Event with a Heavy Negative Secondary, and Identification of the Secondary V-Decay Event in a Cascade

E.W. Cowan
Phys. Rev. **94** (1954) 161;

Abstract Two cosmic-ray decay events have been photographed in a cloud chamber under conditions that yield mass values from combined magnetic-field momentum measurements and ionization measurements from droplet counting. A method has been developed for assigning meaningful probable errors to the ionization measurements. The first event is interpreted as the decay of a neutral V particle into a positive π meson and a negative particle of mass $1850 \pm 250 m_e$. On the assumption of a two-body decay, the Q value for the decay is 11.7 ± 4 MeV. The second event is a cascade decay that can be summarized by the following reaction:

$$Y^- \rightarrow 67 \pm 12\text{MeV} \; + \; \pi^- \Lambda^0 \rightarrow 40 \pm 13\text{MeV} \; + \; \pi^- p.$$

The proton of the Λ^0 decay is identified by a measured mass of $2050 \pm 350 m_e$. On the assumption of a two body decay, the mass of the primary V particle is $2600 \pm 34 m_e$.

Accelerator COSMIC

Detectors Cloud chamber

Reactions
$\Lambda \rightarrow p \, \pi^-$
neutral $\rightarrow \pi^+$ hadron$^-$
$\Xi^- \rightarrow \Lambda \, \pi^-$

Particles studied Ξ^-

DALITZ 1954

■ First indication on zero spin and odd parity for the K meson ■

Decay of τ Mesons of Known Charge

R.H. Dalitz
Phys. Rev. **94** (1954) 1046;

Reprinted in
American Physical Society, Chicago Meeting, Phys. Rev. **93** (1954) 914.
The Physical Review — the First Hundred Years, AIP Press (1995) CD-ROM.

Abstract The experimental data on the 3π decay of τ mesons is summarized on a convenient two-dimensional plot, both (a) when the π-meson charges are known and (b) when they are not. Some events may be included in plot (a) only if the parent τ meson is assumed positive and arguments supporting this identification for τ mesons decaying in an emulsion are discussed. The dependence of this plot on the τ-meson spin (j) and parity (w) is discussed in general terms and those features depending particularly on w and on its relation with j are emphasized — for example, if the density of events does not vanish at the bottom of the plot, the τ meson must have odd parity and even spin. Simple estimates of the distribution, using only the lowest allowable angular momenta and a "short range" approximation, may be modified by final-state meson-meson attractions, whose effects are discussed qualitatively. The available data are insufficient for any strong conclusion to be drawn but rather suggest even spin and odd parity for the τ meson; the need for careful assessment of geometrical bias in the selection of experimental material is stressed.

Related references

More (earlier) information
R. H. Dalitz, Phil. Mag. **44** (1953) 1068;

See also
R. H. Dalitz, Proc. Roy. Soc. **66** (1953) 710;

Analyse information from
D. Lal, Y. Pal, and B. Peters, Phys. Rev. **92** (1953) 438;
R. B. Leighton and S. D. Wanlass, Phys. Rev. **86** (1952) 426;
V. A. J. van Lint and G. H. Trilling, Phys. Rev. **92** (1953) 1089;
J. Crussard, M. F. Kaplon, J. Klarmann, and J. H. Noon, Phys. Rev. **93** (1954) 253;
R. H. Brown et al., Nature **163** (1949) 82;
M. G. K. Menon and C. O'Ceallaigh, Proc. Roy. Soc. **A221** (1954) 292;
K. H. Barker, Proc. Roy. Soc. **A221** (1954) 328;
R. W. Thompson et al., Phys. Rev. **90** (1953) 329;

Reactions
$K^\pm \rightarrow \pi^+ \, \pi^- \, \pi^\pm$
$K^\pm \rightarrow 2\pi^0 \, \pi^\pm$

Particles studied K^+, K^-

GELL-MANN 1954B

■ Invention and exploration of the renormalization group concept ■

Quantum Electrodynamics at Small Distances

M. Gell-Mann, F.E. Low
Phys. Rev. **95** (1954) 1300;

Reprinted in
The Physical Review — the First Hundred Years, AIP Press (1995) CD-ROM.

Abstract The renormalized propagation functions D_{FC} and S_{FC} for photons and electrons, respectively, are investigated for momenta much greater than the mass of the electron. It is found that in this region the individual terms of the perturbation series to all orders in the coupling constant take on very simple asymptotic forms. An attempt to sum the entire series is only partially successful. It is found that the series satisfy certain functional equations by virtue of the renormalizability of the theory. If photon self-energy parts are omitted from the series, so that $D_{FC} = D_F$, then S_{FC} has the asymptotic form $A[p^2/m^2]^n[i\gamma \cdot p]^{-1}$, where $A = A(e_1{}^2)$ and $n = n(e_1{}^2)$. When all diagrams are included, less specific results are found. One conclusion is that the *shape* of the charge distribution surrounding a test charge in the vacuum does not, at small distances, depend on the coupling constant except through a scale factor. The behavior of the propagation functions for large momenta is related to the magnitude of the renormalization constants in the theory. Thus it is shown that the unrenormalized coupling constant $e_0{}^2/4\pi\hbar c$,

which appears in perturbation theory as a power series in the renormalized coupling constant $e_1{}^2/4\pi\hbar c$ with divergent coefficients, may behave in either of two ways:
(a) It may really be infinite as perturbation theory indicates;
(b) It may be a finite number independent of $e_1{}^2/4\pi\hbar c$.

Related references

See also
R. Serber, Phys. Rev. **48** (1935) 49;
E. A. Uehling, Phys. Rev. **48** (1935) 55;
P. A. M. Dirac, Proc. Camb. Phil. Soc. **30** (1934) 150;
W. Heisenberg, Z. Phys. **90** (1934) 209;
J. Schwinger, Phys. Rev. **75** (1949) 651;
F. J. Dyson, Phys. Rev. **75** (1949) 1756;
A. Salam, Phys. Rev. **84** (1951) 426;
J. C. Ward, Proc. Roy. Soc. **A64** (1951) 54;
J. C. Ward, Phys. Rev. **84** (1951) 897;
G. Källén, Helv. Phys. Acta **25** (1952) 417;
J. C. Maxwell, Phil. Mag. **19** (1860) 19;
S. F. Edwards, Phys. Rev. **90** (1953) 284;
G. Källén, Kgl. Danske Videnskab. Selskab, Mat.-fiz. Medd. **27** (1953) 12;
S. W. Gupta, Proc. Roy. Soc. **63** (1950) 681;
K. Bleuler, Helv. Phys. Acta **23** (1950) 567;

GELL-MANN 1954

■ Forward dispersion relations for massive particles ■

Use of Causality Conditions in Quantum Theory

M. Gell-Mann, M.L. Goldberger, W.E. Thirring
Phys. Rev. **95** (1954) 1612;

Abstract The limitations on scattering amplitudes imposed by causality requirements are deduced from the demand that the commutator of field operators vanish if the operators are taken at points with space-like separations. The problems of the scattering of spin-zero particles by a force center and the scattering of photons by a quantized matter field are discussed. The causality requirements lead in a natural way to the well-known dispersion relation of Kramers and Kronig. A new sum rule for the nuclear photoeffect is derived and the scattering of photons by nucleons is discussed.

Related references

See also
R. de L. Kronig, J. Opt. Soc. Am. **12** (1926) 547, Physica **12** (1946) 543;
H. A. Kramers, Atti. Congr. Intern. Fisici, Como. **2** (1927) 545;
W. Schutzer and J. Tiomno, Phys. Rev. **83** (1951) 249;
N. G. van Kampen, Phys. Rev. **91** (1953) 1267;
N. G. van Kampen, Phys. Rev. **89** (1953) 1072;

YANG 1954

■ Introduction of local gauge isotopic spin invariance in quantum field theory: Yang-Mills theory ■

Conservation of Isotopic Spin and Isotopic Gauge Invariance

C.N. Yang, R.L. Mills
Phys. Rev. **96** (1954) 191;

Reprinted in
The Physical Review — the First Hundred Years, AIP Press (1995) 1008.

Abstract It is pointed out that the usual principle of invariance under isotopic spin rotation is not consistent with the concept of localized fields. The possibility is explored of having invariance under local isotopic spin rotations. This leads to formulating a principle of isotopic gauge invariance and the existence of a **b** field which has the same relation to the isotopic spin that the electromagnetic field has to the electric charge. The **b** field satisfies nonlinear differential equations. The quanta of the **b** field are particles with spin unity, isotopic spin unity, and electric charge $\pm e$ or zero.

Related references

See also
W. Heisenberg, Z. Phys. **77** (1932) 1;
G. Breit, E. U. Condon, and R. D. Present, Phys. Rev. **50** (1936) 825;
E. P. Wigner, Phys. Rev. **51** (1937) 106;
B. Cassen and E. U. Condon, Phys. Rev. **50** (1936) 846;
W. Pauli, Rev. of Mod. Phys. **13** (1941) 203;
M. Gell-Mann, Phys. Rev. **92** (1953) 833;
F. J. Dyson, Phys. Rev. **75** (1949) 1736;
J. Schwinger, Phys. Rev. **76** (1949) 790;
F. J. Dyson, Phys. Rev. **75** (1949) 486;
W. Heisenberg and W. Pauli, Z. Phys. **56** (1929) 1;
Analyse information from
C. C. Lauritsen, Ann. Rev. Nucl. Sci. **1** (1952) 67;
D. R. Inglis, Rev. of Mod. Phys. **25** (1953) 390;
R. H. Hildebrand, Phys. Rev. **89** (1953) 1090;

HODSON 1954

■ Evidence for $K^+ \to \pi^+\pi^0$ decay ■

Cloud-Chamber Evidence for a Charged Counterpart of the Θ^0 Particle

A.L. Hodson et al.
Phys. Rev. **96** (1954) 1089;

Abstract A photograph obtained in a Wilson cloud chamber, operated in a magnetic field, shows the following unusual event: A positive particle, produced in an interaction above the cloud chamber, decays in flight into a positive particle less massive than a K meson. Four other lightly ionizing particles also originate from the decay point; these appear in the form of two small-angle pairs, each pair consisting of one positive and one negative particle. The observed momenta and ionizations are consistent with three of these four particles being either electrons or mesons; the fourth must be an electron. The event may be interpreted as the decay:

$$K^+ \to \begin{pmatrix} \pi^+ \\ \mu^+ \end{pmatrix} \pi^0 + Q \text{ MeV},$$

followed by the decay of the π^0 meson into four electrons (a possible but hitherto unobserved mode of decay):

$$\pi^0 \to e^+ \; e^- \; e^+ \; e^-.$$

This interpretation leads to a remarkable internal consistency of the data and is supported by the following experimental facts:

(i) There is good overall transverse-momentum balance in two mutually perpendicular planes.

(ii) If we assume that all the particles in the above group of four are electrons resulting from the decay of a neutral particle the mass of the latter, determined from energy-momentum balance, is $(255^{+15}_{-10})m_e$, in good agreement with the known mass of the π^0 meson.

(iii) A transformation to the rest system of the π^0 meson shows that in this frame the four electrons come off as two small-angle pairs traveling in opposite directions. This is the most probable configuration in the four-electron decay of a π^0 meson.

The Q values calculated for the K^+ decay are: $Q(\pi^+, \pi^0) = (213^{+15}_{-10})$ MeV and $Q(\mu^+, \pi^0) = (207^{+15}_{-10})$ MeV. Comparison of the first Q values with that for the θ^0 particle $Q(\pi^+, \pi^-) = 214 \pm 5$ MeV, suggests that the unstable meson observed in this event and designated phenomenologically above as a K^+ meson may be a charged counterpart of the θ^0 particle.
This interpretation may explain at least some of the cases, observed by other workers, in which γ rays appear to be associated with S particles.
Other possible interpretations of the event are considered.

Accelerator COSMIC

Detectors Cloud chamber

Related references

See also
D. R. Harris and A. L. Hodson, Phys. Rev. **95** (1954) 661A;

Reactions
$K^+ \to \pi^+ \, \pi^0$

Particles studied K^+

LUDERS 1954

■ Theoretical evidence for CPT invariance in local quantum field theory ■

On the Equivalence of Invariance under Time Reversal and under Particle-Antiparticle Conjugation for Relativistic Field Theories

G. Luders
Dan. Mat. Fys. Medd. **28(5)** (1954) 1 (1954);

Abstract For relativistic field theories, in a sense specified in section 2, the invariance under time reversal "of the second kind" (time reversal including particle-antiparticle conjugation) is proved mathematically. Consequently, the postulate of invariance under time reversal ("of the first kind") is, for field theories of this type, completely equivalent to the postulate of invariance under particle-antiparticle conjugation.

Related references

See also
G. Luders, R. Oehme, and W. E. Thirring, Zeit. Naturforschung **7** (1952) 213;
A. Pais and R. Jost, Phys. Rev. **87** (1952) 871;
G. Luders, Z. Phys. **133** (1952) 325;
L. C. Biedenharn and M. E. Rose, Phys. Rev. **83** (1951) 459;
H. A. Tolhoek and S. R. de Groot, Phys. Rev. **84** (1951) 151;
S. Watanabe, Phys. Rev. **84** (1951) 1008;
F. Coester, Phys. Rev. **84** (1951) 1259;
H. A. Kramers, Proc. Acad. Sci. Amsterdam **40** (1937) 814;
W. Pauli, Rev. of Mod. Phys. **13** (1941) 203;
J. Schwinger, Phys. Rev. **74** (1948) 1439;
E. Majorana, Nuovo Cim. **14** (1937) 171;
W. Heisenberg, Z. Phys. **90** (1934) 209;
W. Heisenberg, Z. Phys. **92** (1934) 692;
M. Fierz, Z. Phys. **104** (1936) 553;

DAVIES 1955B

■ First experimental evidence that neutrino is not identical with antineutrino ■

Attempt to Detect the Antineutrinos from a Nuclear Reactor by the ^{37}Cl$(\overline{\nu}_e,e^-)^{37}$Ar Reaction

R. Davies
Phys. Rev. **97** (1955) 766;

Abstract Tanks containing 200 and 3900 liters of carbon tetrachloride were irradiated outside of the shield of the Brookhaven reactor in an attempt to induce the reaction ^{37}Cl$(\overline{\nu}_e,e^-)^{37}$Ar with fission product antineutrinos. The experiments serve to place an upper limit on the antineutrino capture cross section for the reaction of $2 \cdot 10^{-42}$ cm^2 per atom. Cosmic ray induced ^{37}Ar was observed and production rate measured at 14100 feet altitude and sea level. Measurements with the 3900-liter container shielded from cosmic rays with 19 feet of earth permit placing an upper limit on the neutrino flux of the sun.

Accelerator Nuclear reactor

Detectors RADIOCHEMICAL

Related references

See also
H. R. Crane, Rev. of Mod. Phys. **20** (1948) 278;
J. H. Barrett, Phys. Rev. **79** (1950) 907;
Cowah, Reines, and Harrison, Phys. Rev. **96** (1954) 1294;
F. Reines and C. L. Cowan, Phys. Rev. **92** (1953) 830;
B. Pontecorvo, Chalk River Laboratory Report PD-205 (1946);
L. W. Alvarez, UCRL-328 (1949);
E. J. Konopinsky and L. M. Langer, Ann. Rev. Nucl. Sci. **2** (1953) 261;
K. Way and E. Wigner, Phys. Rev. **73** (1948) 1318;
E. E. Salpeter Ann. Rev. Nucl. Sci. **2** (1953) 41;
F. G. Houtermans and W. Thirring, Helv. Phys. Acta **27** (1954) 81;

Reactions
$\overline{\nu}_e\ ^{37}\mathrm{Cl} \rightarrow\ ^{37}\mathrm{Ar}\ e^-$

Particles studied $\nu_e, \overline{\nu}_e$

Accelerator COSMIC

Detectors RADIOCHEMICAL

Reactions
$\nu_e\ ^{37}\mathrm{Cl} \rightarrow\ ^{37}\mathrm{Ar}\ e^-$

Accelerator COSM-SUN

Detectors RADIOCHEMICAL

Reactions
ν_e

GELL-MANN 1955

■ Prediction of the long lived kaon K_L ■

Behavior of Neutral Particles under Charge Conjugation

M. Gell-Mann, A. Pais
Phys. Rev. **97** (1955) 1387;

Reprinted in
The Physical Review — the First Hundred Years, AIP Press (1995) 1013.

Abstract Some properties are discussed of the θ^0, a heavy boson that is known to decay by the process $\theta^0 \rightarrow \pi^+\pi^-$. According to certain schemes proposed for the interpretation of hyperons and K particles, the θ^0 possesses an antiparticle $\overline{\theta}^0$ distinct from itself. Some theoretical implications of this situation are discussed with special reference to charge conjugation invariance. The application of such invariance in familiar instances is surveyed in Sec. I. It is then shown in Sec. II that, within the framework of the tentative schemes under consideration, the θ^0 must be considered as a "particle mixture" exhibiting two distinct lifetimes, that each lifetime is associated with a different set of decay modes, and that no more than half of all θ^0's undergo the familiar decay into two pions. Some experimental consequences of this picture are mentioned.

Related references

See also
A. Pais and R. Jost, Phys. Rev. **87** (1952) 871;
L. Wolfenstein and D. G. Ravenhall, Phys. Rev. **88** (1952) 279;
L. Michel, Nuovo Cim. **10** (1953) 319;

Particles studied K_L

KARPLUS 1955

■ Forward dispersion relations for massive particles ■

Applications of Causality to Scattering

R. Karplus, M.A. Ruderman
Phys. Rev. **98** (1955) 771;

Abstract The optical dispersion relations are extended to the scattering of massive particles and are applied to the nuclear interaction of the pion-nucleon system. The sign of the forward scattering amplitude is unambiguously inferred from measured total cross sections and found to agree with that determined from Coulomb interference.

Related references

See also
W. Heisenberg, Z. Phys. **120** (1943) 513;
W. Heisenberg, Z. Phys. **120** (1943) 673;
R. de L. Kronig, J. Opt. Soc. Am. **12** (1926) 547, Physica **12** 1946) 543;
W. Schutzer and J. Tiomno, Phys. Rev. **83** (1951) 249;
N. G. van Kampen, Phys. Rev. **89** (1953) 1072;
N. G. van Kampen, Phys. Rev. **91** (1953) 1267;
M. Gell-Mann, M. L. Goldberger, and W. E. Thirring, Phys. Rev. **95** (1954) 1612;
H. A. Kramers, Atti. Congr. Intern. Fisici, Como. **2** (1927) 545;

WALKER 1955D

■ Confirmation of associated production of strange particles. First indication of the existence of the Σ^0 ■

Λ-Θ Production in π^- p Collisions at 1 BeV

W.D. Walker
Phys. Rev. **98** (1955) 1407;

Abstract Two examples of Λ^0-θ^0 production have been found in the collision of 1.0 Bev π^- mesons with protons. Data on these two cases are given. The two cases are qualitatively similar, yet there are quantitative differences in terms of the angles of production of the θ^0 and Λ^0 particles. One case is internally consistent with the production of a Λ^0 and θ^0

directly. The other case, in which the Λ^0 and θ^0 particles seem slightly noncoplanar with the incoming π^- meson, is not internally consistent with the direct production of a Λ^0 and θ^0. It is likely that either the θ^0 or the Λ^0 or both θ^0 and Λ^0 are the products of decay of heavier parents. It is also possible that a light particle (γ or ν) may be produced simultaneously with the θ^0 and Λ^0. If one supposes that the Λ^0 is the product of decay of a heavier parent then the Q of such a parent, were it to decay into a proton and π^- meson, is calculated to be 117 ± 30 MeV. This Q value is close to that of the charged hyperons. A short-lived neutral hyperon of this mass has been predicted by Gell-Mann.

Accelerator BNL-COSMOTRON

Detectors Cloud chamber

Related references

See also
M. Gell-Mann, Phys. Rev. **92** (1953) 833;

Analyse information from
W. B. Fowler et al., Phys. Rev. **91** (1953) 1287;
W. B. Fowler et al., Phys. Rev. **93** (1954) 861;
V. A. J. van Lint, G. H. Trilling, R. B. Leighton, and C. D. Anderson, Phys. Rev. **95** (1954) 295;
R. W. Thompson et al., Phys. Rev. **90** (1953) 1122;
R. W. Thompson, Burwell, Huggert, and C. J. Karzmark, Phys. Rev. **95** (1954) 1576;

Reactions
$\pi^- \, p \to \Lambda \, K^0$ 1.11 GeV(E_{lab})
$\pi^- \, p \to \Sigma^0 \, K^0$ 1.11 GeV(E_{lab})
$\Lambda \to p \, \pi^-$
$K^0 \to \pi^+ \, \pi^-$
$\Sigma^0 \to \Lambda \, \gamma$

Particles studied Σ^0

GOLDBERGER 1955

■ Dispersion relations for massive particles ■

Causality Conditions and Dispersion Relations. I. Boson Fields

M.L. Goldberger
Phys. Rev. **99** (1955) 979;

Abstract The dispersion relations of Kramers and Kronig as generalized for charged and neutral Bose particles with finite rest mass are derived in a new way using the formalism of quantum field theory. The alternative forms of dispersion relations obtained by making various assumptions on the high-frequency limit of total cross sections are used to obtain information about the high-frequency behavior of the total cross section for the scattering of γ rays by electrons.

Related references

See also
M. Gell-Mann, M. L. Goldberger, and W. E. Thirring, Phys. Rev. **95** (1954) 1612;
M. L. Goldberger, Phys. Rev. **97** (1955) 508;
R. Karplus and M. A. Ruderman, Phys. Rev. **98** (1955) 771;

GOLDBERGER 1955B

■ Forward dispersion relations for massive particles ■

Application of Dispersion Relations to Pion-Nucleon Scattering

M.L. Goldberger, H. Miyazawa, R. Oehme
Phys. Rev. **99** (1955) 986;

Abstract The generalized Kramers-Kronig dispersion relations for charged bosons are used to treat the problem of pion-nucleon scattering. The complications associated with the charge of the pions are discussed. The importance of a "bound state" corresponding to the neutron is emphasized and its contribution to the scattering amplitude is computed rigorously, assuming only that pions are pseudoscalar and that the interaction with nucleons is charge-independent. The connection between our exact dispersion relations and the approximate equations for pion-nucleon scattering given by Low is discussed. A rigorous effective-range relation is derived.

Related references

See also
M. L. Goldberger, Phys. Rev. **99** (1955) 979;
F. E. Low, Phys. Rev. **97** (1955) 1392;
C. F. Chew and F. E. Low, *Fifth Ann. Rochester Conf. on High Energy Nuclear Physics* (1955);
J. Orear, Phys. Rev. **96** (1954) 176;

BRIDGE 1955

■ Confirmation of the existence of the K^+ and K^- ■

Evidence for Heavy Mesons with the Decay Processes $K_{\pi 2} \to \pi^\pm \, \pi^0$ and $K_{\mu 2} \to \mu^\pm \, \nu$ from Observations with a Multiplate Cloud Chamber

H.S. Bridge, H.C. de Staebler, B. Rossi, B.V. Sreekantan
Nuovo Cim. **1** (1955) 874;

Excerpt Analysis of the data on S-events observed in the M.I.T. multiplate cloud chamber shows that these events represent the decay processes of two kinds of heavy mesons. The decay processes are of two type: $K_{\pi 2} \to \pi\pi^0$ and $K_{\mu 2} \to \mu\nu$.

Accelerator COSMIC

Detectors Cloud chamber

Related references

More (earlier) information
H. S. Bridge, C. Peyrou, B. Rossi, and R. Stafford, Phys. Rev. **90** (1953) 921;
H. S. Bridge et al., Nuovo Cim. **12** (1954) 81;

Analyse information from
H. S. Bridge, H. Courant, H. C. de Staebler, and B. Rossi, Phys. Rev. **91** (1953) 1024;
B. P. Gregory et al., Nuovo Cim. **11** (1954) 292;
M. G. K. Menon and C. O'Ceallaigh, Proc. Roy. Soc. **A221** (1954) 292;
R. D. Hill, E. O. Salant, and M. Widgoff, Bull.Am.Phys.Soc. **29** (1954) 32;
A. L. Hodson et al., Phys. Rev. **96** (1954) 1089;

Reactions
meson $\to \pi^0\ \pi^\pm$
meson $\to \mu^\pm\ \nu$

Particles studied meson, K^+, K^-

ARMENTEROS 1955

■ Confirmation of the existence of the K^+ ■

Further Discussion of the K_μ Decay Mode

R. Armenteros et al.
Nuovo Cim. **1** (1955) 915;

Excerpt After a description of the methods used in the measurements of momentum and range, the data on 46 S-events obtained with a double cloud-chamber experiment are examined. Momentum-range measurements on 22 positive S's give a mass consistent with a single value at $(928 \pm 13)\ m_e$. The analysis of the ranges of the secondaries shows, however, that not all S-events can be explained by a single decay mode. A large proportion of the secondaries correspond to a light meson of unique range (75.7 ± 1.7) g $\cdot$ cm^{-2} Cu. The comparison of this range with the measured masses of the primary particles proves that these secondaries are μ-mesons. The accompanying neutral secondary is shown at the same time to be a particle of zero mass. The absence of cascade showers associated with S-events with long range secondaries shows that the neutral secondary is a neutrino. The decay mode $K_\mu \to \mu\nu$ previously proposed is more firmly established. Two independent mass estimates are obtained for the K_μ-particle. One — from momentum-range measurements — gives directly $(906 \pm 27)\ m_e$. Another — from the mode of decay and the range of the μ-secondary — gives indirectly $(941 \pm 11)\ m_e$. A final best mass of $(935 \pm 15)\ m_e$ is obtained by the combination of the two estimates after taking into account possible additional errors. The K_μ particle is then shown to be essentially positive and to constitute $(66 \pm 18)\%$ of the S's which passed through the top chamber. No precise information can be obtained as to the mode of decay of S-particles which are known not to belong to the K_μ-decay mode.

Accelerator COSMIC

Detectors Cloud chamber

Related references

See also
M. Annis, W. Cheston, and H. Primakoff, Rev. of Mod. Phys. **25** (1953) 818;
R. Armenteros et al., Nuovo Cim. Suppl. **12** (1954) 327;
H. A. Bethe, Phys. Rev. **70** (1946) 821;
H. S. Bridge et al., Nuovo Cim. **12** (1954) 81;
R. H. Brown et al., Nature **163** (1949) 82;
M. W. Friedlander, D. Keefe, M. G. K. Menon, and L. van Rossum, Phil. Mag. **45** (1954) 1043;
B. P. Gregory et al., Nuovo Cim. **11** (1954) 292;
A. L. Hodson et al., Phys. Rev. **96** (1954) 1089;
M. G. K. Menon and C. O'Ceallaigh, Proc. Roy. Soc. **A221** (1954) 292;
C. O'Ceallaigh, Phil. Mag. **42** (1951) 1032;
V. A. Regener, Phys. Rev. **84** (1951) 161;

Reactions
$K^- \to \mu^-\ \nu$
$K^+ \to \mu^+\ \nu$

Particles studied K^+, K^-

CHUPP 1955

■ First precise measurement of the K^+ mass ■

K-Meson Mass from K-Hydrogen Scattering Event

W.W. Chupp et al.
Phys. Rev. **99** (1955) 1042;

Excerpt One of the basic problems in the classification and understanding of K-mesons is the relation between the various modes of decay ($K_{\pi 2} \equiv \theta^+$, $K_{\mu 2}$, $K_{\mu 3}$, K_{e3}, and τ). There has been much discussion recently on the K-meson mass and to what extent the masses of the various K-mesons (or perhaps just different decay modes) differ from that of the τ meson $(965.3\ m_e)$.
In this connection, we want to report here on a K_L^+-particle which underwent a scattering from hydrogen in a nuclear emulsion stack and thus enabled us to obtain a rather good mass measurement. The mass value obtained as the weighted mean of two independent methods is 973 ± 12 m$_e$. *(Extracted from the introductory part of the paper.)*

Accelerator LBL-Bevatron

Detectors Nuclear emulsion

Related references

See also
L. T. Kerth et al., Bull.Am.Phys.Soc. **30** (1955) 41;
R. M. Sternheimer, Phys. Rev. **91** (1953) 256;
J. R. Fleming and J. J. Lord, Phys. Rev. **92** (1953) 511;

Reactions
$K^+\ p \to p\ K^+$ 399.-423. MeV(P_{lab})

Particles studied K^+

ALVAREZ 1955

■ First measurement of the K^+ lifetime. First evidence for the equality of lifetimes of the Θ and τ mesons ■

The Lifetime of the τ-Meson

L.W. Alvarez, S. Goldhaber
Nuovo Cim. **2** (1955) 344;

Excerpt Now that K-mesons are available in large numbers from proton synchrotrons, experiments will soon yield precise values for the lifetime, or lifetimes, of the K-mesons. Exposures of emulsions to K-particles have been made by several groups at Berkeley, under quite different conditions, so far as distance from the target and magnetic resolution are concerned. If one knew the relative integrated currents on the targets for exposures with long and short flight paths, and if geometrical and resolution factors were properly taken into account, these experiments would yield a lifetime. Until recently, such an intercomparison of the results of the various experiments has appeared impossible. A method has now been found to tie the results of the various experiments together; this note describes the method and presents the lifetime so determined. *(Extracted from the introductory part of the paper.)*

Accelerator LBL-Bevatron

Detectors Nuclear emulsion

Data comments 10 τ-decay events without magnetic field and 60 with magnetic field.

Related references

Analyse information from
H. H. Heckman, G. Goldhaber, and F. M. Smith, Bull.Am.Phys.Soc. **30** (1955) 63;
L. T. Kerth et al., Bull.Am.Phys.Soc. **30** (1955) 41;
R. W. Birge et al., preprint UCRL-3009 (1955);
W. W. Chupp et al., preprint UCRL-3031 (1955);

Reactions
$K^+ \rightarrow 2\pi^+ \pi^-$

Particles studied K^+

ILOFF 1955

■ K^+ lifetime measurements from decays in flight ■

Mean Lifetime of Positive K Mesons

E.L. Iloff et al.
Phys. Rev. **99** (1955) 1617;

Excerpt Mean lifetimes for heavy mesons from cosmic rays have been reported by various groups using cloud chambers and Cerenkov counters. (L. Mezetti and J. W. Keufel, Phys. Rev. **95** (1954) 859; Barker et al., Phil. Mag., (7) **46** (1955) 307) We have carried out a measurement of the mean lifetime of artificially produced K^+ mesons by making use of their decay in flight in nuclear emulsion.
"Along the track" scanning of K^+ mesons in nuclear emulsions has shown a number of interactions in flight; in addition, 19 events have been found in which there is a single outgoing track of grain density less than that of the incoming K^+ meson. In each of these events the mass of the incoming particle was determined by grain counting and multiple-scattering measurements and its identity as a K^+ particle was thus established. If these events were due to interactions in flight, one would expect to find some stars with a lightly ionizing track coming out together with one or more black evaporation prongs. No such stars were observed. Also, none of the interactions in flight so far seen give off a visible L meson. It therefore seems reasonable to identify all events of this type as the decay of K^+ mesons in flight. From the number of decays in flight found and the proper slowing-down time of all the K^+ mesons followed, a mean lifetime is obtained....
From the 19 decays in flight observed we find a mean lifetime for K^+ mesons of

$$\tau_{K^+} = 1.01^{+0.33}_{-0.21} \times 10^{-8}\text{sec}.$$

(Extracted from the introductory part of the paper.)

Accelerator LBL-Bevatron

Detectors Nuclear emulsion

Related references

See also
L. Mezetti and J. W. Keuffel, Phys. Rev. **95** (1954) 859;
K. H. Barker et al., Phil. Mag. **46** (1955) 307;
L. T. Kerth, D. H. Stork, R. P. Haddock, and M. M. Whitehead, Phys. Rev. **99** (1955) 641;

Reactions
$K^+ \rightarrow$ charged X 335-360,390-450 MeV(P_{lab})

Particles studied K^+

BIRGE 1955B

■ Further measurements of charged-kaon masses. Masses of the strange K^+, Θ, and τ mesons are equal ■

Mass Values of the K Mesons

R.W. Birge, J.R. Peterson, D.H. Stork, M.N. Whitehead
Phys. Rev. **100** (1955) 430;

Abstract Additional mass values are reported for a large number of K-mesons found in two emulsion stacks exposed to 114 MeV and 170 MeV focused K-mesons from the Bevatron. For 459 K_L-mesons and 55 τ-mesons in stack 1 the mean values are, 971 ± 1.3 and 978 ± 4 electron masses respectively, and for $177 K_L$-mesons and 16 τ-mesons in stack 2, the values are 962 ± 1.9 and $965.4 \pm 3.8 m_e$. *(Science Abstracts, 1956, 1622. D. J. Prowse)*

Accelerator LBL-Bevatron

Detectors Nuclear emulsion, Spectrometer

Related references

More (earlier) information
R. W. Birge et al., Phys. Rev. **99** (1955) 329;

Analyse information from
L. T. Kerth, D. H. Stork, R. P. Haddock, and M. M. Whitehead, Phys. Rev. **99** (1955) 641;

Reactions

$K^+ \to 2\pi^+ \pi^-$	114,170 MeV(T_{lab})
$K^+ \to \pi^+ \pi^0$	114,170 MeV(T_{lab})
$K^+ \to \pi^+ 2\pi^0$	114,170 MeV(T_{lab})

Particles studied K^+

CHAMBERLAIN 1955 Nobel prize

■ Experimental evidence for the antiproton. Nobel prize to O. Chamberlain and E. Segrè awarded in 1959 "for their discovery of the antiproton" ■

Observation of Antiprotons

O. Chamberlain, E. Segrè, C. Wiegand, T. Ypsilantis
Phys. Rev. **100** (1955) 947;

Reprinted in
(translation into Russian) Usp. Fiz. Nauk **58** (1956) 685.
R. N. Cahn and G. Goldhaber, *The Experimental Foundations of Particle Physics*, Cambridge Univ. Press (1991) 92.
The Physical Review — the First Hundred Years, AIP Press (1995) 847.

Excerpt One of the striking features of Dirac's theory of the electron was the appearance of solutions to his equations which required the existence of an antiparticle, later identified as the positron.
The extension of the Dirac theory to the proton requires the existence of an antiproton, a particle which bears to the proton the same relationship as the positron to the electron. However, until experimental proof of the existence of the antiproton was obtained, it might be questioned whether a proton is a Dirac particle in the same sense as is the electron. For instance, the anomalous magnetic moment of the proton indicates that the simple Dirac equation does not give a complete description of the proton.
The experimental demonstration of the existence of antiprotons was thus one of the objects considered in the planning of the Bevatron. The minimum laboratory kinetic energy for the formation of an antiproton in a nucleon-nucleon collision is 5.6 BeV. If the target nucleon is in a nucleus and has some momentum, the threshold is lowered. Assuming a Fermi energy of 25 MeV, one may calculate that the threshold for formation of a proton-antiproton pair is approximately 4.3 BeV. Another, two-step process that has been considered by Feldman has an even lower threshold.
There have been several experimental events recorded in cosmic-ray investigations which might be due to antiprotons, although no sure conclusion can be drawn from them at present. With this background of information we have performed an experiment directed to the production and detection of the antiproton. It is based upon the determination of the mass of negative particles originating at the Bevatron target. This determination depends on the simultaneous measurement of their momentum and velocity. Since the antiprotons must be selected from a heavy background of pions it has been necessary to measure the velocity by more than one method. To date, sixty antiprotons have been detected. *(Extracted from the introductory part of the paper.)*

Accelerator LBL-Bevatron

Detectors Counters, Optical spark chamber

Related references

See also
G. Feldman, Phys. Rev. **95** (1954) 1967;
J. Marshall, Ann. Rev. Nucl. Sci. **4** (1954) 141;

Analyse information from
E. Hayward, Phys. Rev. **72** (1947) 937;
E. Amaldi et al., Nuovo Cim. **1** (1955) 492;
H. S. Bridge, H. Courant, H. C. de Staebler, and B. Rossi, Phys. Rev. **95** (1954) 1101;

Reactions

$p\,\text{Cu} \to \pi^- X$	4.2,5.1,6.2 GeV(E_{lab})
$p\,\text{Cu} \to \bar{p} X$	4.2,5.1,6.2 GeV(E_{lab})

Particles studied $\bar{p}$

GERSHTEIN 1955

■ Evidence that strong interactions do not modify the vector coupling constant of beta decay. Analogy between electromagnetic and weak interactions ■

On Corrections from Mesons to the Theory of β-Decay

S.S. Gershtein, Y.B. Zeldovich
Zh. Eksp. Teor. Fiz. **29** (1955) 698; JETP **2** (1956) 576;

Abstract Corrections from virtual mesons are calculated by Feynman techniques.
(Science Abstracts, 1956, 2358. G. E. Brown)

Related references

See also
R. P. Feynman and M. Gell-Mann, Phys. Rev. **109** (1958) 193;
M. Gell-Mann, Phys. Rev. **111** (1958) 362;

Reactions

$n \rightarrow p\, e^- \,\overline{\nu}_e$

SORRELS 1955

■ Confirmation of the associated production of strange particles ■

Associated Production of Ξ^- with Two Θ^0 Particles

J.D. Sorrels, R.B. Leighton, C.D. Anderson
Phys. Rev. **100** (1955) 1457;

Abstract A cosmic-ray event is described in which a negative cascade particle and two neutral heavy mesons appear to be produced in a single nuclear interaction above a cloud chamber. It is suggested that this event may be an example of the associated production of a Ξ^- particle with two θ^0 particles according to the scheme of Gell-Mann.

Accelerator COSMIC

Detectors Cloud chamber

Related references

See also
M. Gell-Mann, Phys. Rev. **92** (1953) 833;

Analyse information from
W. B. Fowler et al., Phys. Rev. **91** (1953) 1287;
W. B. Fowler et al., Phys. Rev. **93** (1954) 861;
W. B. Fowler et al., Phys. Rev. **98** (1955) 121;
R. W. Thompson, Burwell, Huggert, and C. J. Karzmark, Phys. Rev. **95** (1954) 1576;
D. Lal, Y. Pal, and B. Peters, Proc. Indian Acad. Sci. **38** (1953) 398;
C. Dahanayake, P. E. Francois, Y. Fujimoto, and P. Iredale, Phil. Mag. **45** (1954) 855;
W. H. Arnold, J. Ballam, G. K. Lindenberg, and V. A. J. van Lint, Phys. Rev. **98** (1955) 838;
W. B. Fretter and F. W. Friesen, Phys. Rev. **96** (1954) 853;
E. W. Cowan, Phys. Rev. **94** (1954) 161;

Reactions

unspec nucleus $\rightarrow \Xi^-\ 2K_S\ X$
$\Lambda \rightarrow p\,\pi^-$
$\Xi^- \rightarrow \Lambda\,\pi^-$
$K_S \rightarrow \pi^+\,\pi^-$

CHINOWSKY 1955

■ First evidence for the odd parity of the π^0 ■

Reaction π^- deuteron $\rightarrow n\, n\, \pi^0$: Parity of the Neutral Meson

W. Chinowsky, J. Steinberger
Phys. Rev. **100** (1955) 1476;

Abstract The branching ratio between the capture reactions π^- deuteron $\rightarrow 2n\,\pi^0$ and π^- deuteron $\rightarrow 2n\,\gamma$ been determined to be $< 0.10\%$. Comparison with the previously measured branching ratio between the corresponding processes in hydrogen provides strong evidence for pseudoscalar π^0 parity.

Accelerator Nevis labs cyclotron

Detectors Counters

Related references

More (earlier) information
D. Bodansky et al., Phys. Rev. **93** (1954) 1367;
W. Chinowsky and J. Steinberger, Phys. Rev. **95** (1954) 1561;

Analyse information from
W. K. H. Panofsky, R. L. Aamodt, and J. Hadley, Phys. Rev. **81** (1951) 565;
R. Durbin, H. Loar, and J. Steinberger, Phys. Rev. **83** (1951) 646;
D. Clark, A. Roberts, and R. Wilson, Phys. Rev. **83** (1951) 649;

Reactions

π^- deuteron $\rightarrow 2n\,\gamma$ $<$0.38 GeV(P_{lab})
π^- deuteron $\rightarrow 2n\,\pi^0$ $<$0.38 GeV(P_{lab})
$\pi^0 \rightarrow 2\gamma$

Particles studied π^0

PAIS 1955

■ Proposal for the $K_L \rightarrow K_S$ regeneration experiment ■

Note on the Decay and Absorption of the θ^0

A. Pais, O. Piccioni
Phys. Rev. **100** (1955) 1487;

Reprinted in
The Physical Review — the First Hundred Years, AIP Press (1995) CD-ROM.

Abstract A suggestion is made on how to verify experimentally a recent theoretical suggestion that the θ^0 meson is a "particle mixture".

Related references

More (earlier) information
M. Gell-Mann and A. Pais, Phys. Rev. **97** (1955) 1387;

Analyse information from
W. B. Fowler et al., Phys. Rev. **91** (1953) 1287;
W. B. Fowler et al., Phys. Rev. **93** (1954) 861;

Reactions

K_L nucleon $\rightarrow$ nucleon K_S

Particles studied K_L, K_S

LEHMANN 1955

■ Beginnings of the axiomatic field theory of the S-matrix ■

Zur Formulierung quantisierter Feldtheorien / A New Formulation of Quantized Field Theories

H. Lehmann, K. Symanzik, W. Zimmermann
Nuovo Cim. **1** (1955) 205;

Excerpt A new formulation of quantized field theories is proposed. Starting from some general requirements we derive a set of equations which determine the matrix-elements of field operators and the S-Matrix. These equations contain no renormalization constants, but only experimental masses and coupling parameters. The main advantage over the conventional formulation is thus the elimination of all divergent terms in the basic equations. This means that no renormalization problem arises. The formulation is here restricted to theories which do not involve stable bound states. For simplicity we derive the equations for spin 0 particles, however the extension to other cases (e.g. quantum electrodynamics) is obvious. The solutions of the equations are discussed in a power-series expansion. They are then identical with the renormalized expressions of the conventional formulation. However, the equations set up here are not restricted to the application of perturbation theory.

Related references

See also
E. P. Wigner, Annals of Math. **30** (1939) 149;
M. Gell-Mann, M. L. Goldberger, and W. E. Thirring, Phys. Rev. **95** (1954) 1612;
G. Kallen, Helv. Phys. Acta **25** (1952) 417;
W. Zimmermann, Nuovo Cim. Suppl. **11** (1954) 43;
W. Zimmermann, Nuovo Cim. **11** (1954) 416;
K. Symanzik, Zeit. Naturforschung **10A** (1954) 809;

CONVERSI 1955

■ Invention of flash tube chambers ■

The "Hodoscope Chamber": a New Instrument for Nuclear Research

M. Conversi, A. Gozzini
Nuovo Cim. **2** (1955) 189;

Excerpt A new instrument, which may be defined as a "hodoscope chamber", has been developed for the detection of events occurring in nuclear or electromagnetic interactions. It is based on the following principle. If a strong electric field is produced soon after the passage of an ionizing particle in a region of space filled with a gas, then a luminous discharge occurs all over this region as a consequence of the processes following the acceleration undergone by the electrons freed by the impinging particle.
The hodoscope chamber that we are going to describe is essentially a parallel plate condenser filled with a large number of thin glass tubes containing neon at a pressure of 35 cm Hg ... *(Extracted from the introductory part of the paper.)*

Related references

More (earlier) information
M. Conversi et al., *Int. Conf. on Elementary Particles*, Pisa, (12-18th June, 1955);

DAVIES 1955

■ Measurements of charged kaon decay branching fractions ■

On the Masses and Modes of Decay of Heavy Mesons Produced by Cosmic Radiation

G-Stack collaboration; J.H. Davies et al.
Nuovo Cim. **2** (1955) 1063;

Excerpt A large emulsion stack, exposed at high altitude, has been used to study the decay-modes of K-particles which produce a single charged secondary. The 5 decay modes, $K_\mu, \chi, \kappa, K_\beta$ and τ' have been recognized from the nature and energy of the charged secondary. The masses of the parent particles of the two-body decay modes, K_μ and χ, found from the mean range of the charged secondary are (976 ± 7) m_e and (969 ± 3) m_e respectively. Independent values derived from measurements of the scattering of the secondary track near the decay point are (954 ± 17) and (972 ± 3) m_e respectively. The reliability of these mass values is discussed. An examination of the relative frequency of the decay modes in these experimental conditions indicates that the K_μ- and χ-modes constitute respectively about 67% and 20% of all the K-particle decays: the K_β, κ and τ' modes being present to about 9%, 3% and 1% respectively.

Accelerator COSMIC

Detectors Nuclear emulsion

Related references

See also
S. C. Fung, N. Mohler, A. Pevzner, and D. Ritson, *Mimeografed report of the Piza Conference* (1955) 201;
R. W. Birge et al., *Mimeografed report of the Piza Conference* (1955) 151;
R. H. Dalitz, *Proc. of the Rochester Conf.* (1955);
G. Puppi, Nuovo Cim. Suppl. **11** (1954) 438;
F. Anderson, G. Lawlor, and T. E. Nevin, Nuovo Cim. **2** (1955) 608;
F. M. Smith, W. Birnbaum, and W. H. Barkas, Phys. Rev. **91** (1953) 765;
G. Baroni et al., preprint CERN BS/9 (1954);
W. H. Barkas and L. Yong, Phys. Rev. **98** (1955) 605;
W. F. Fry, G. R. White, Phys. Rev. **90** (1953) 207;
O. Heinz, Phys. Rev. **94** (1954) 1728;
L. Voyvodic and E. Pickup, Phys. Rev. **85** (1952) 91;
K. Gottstein et al., Phil. Mag. **42** (1951) 708;
Analyse information from

R. H. Brown et al., Nature **163** (1949) 82;
C. O'Ceallaigh, Phil. Mag. **42** (1951) 1032;
M. G. K. Menon and C. O'Ceallaigh, Proc. Roy. Soc. **A221** (1954) 292;
H. S. Bridge et al., Nuovo Cim. **1** (1955) 874;
M. Baldo et al., Nuovo Cim. **1** (1955) 1180;
A. L. Hodson et al., Phys. Rev. **96** (1954) 1089;
B. P. Gregory et al., Nuovo Cim. **11** (1954) 292;
R. Armenteros et al., Nuovo Cim. **1** (1955) 915;
R. H. Dalitz, Proc. Roy. Soc. **66** (1953) 710;
A. Pais, Phys. Rev. **86** (1952) 663;
J. Cruesard, M. F. Kaplon, J. Klarmann, and J. H. Noon, Phys. Rev. **93** (1954) 253;
M. W. Friedlander, D. Keefe, M. G. K. Menon, and I. Van Rossum, Phil. Mag. **45** (1954) 1043;
C. Dahanayake, P. E. Francois, and Y. Fujimoto, Phil. Mag. **45** (1954) 1219;
R. H. W. Johnston and C. O'Ceallaigh, Phil. Mag. **46** (1955) 393;

Reactions
$K^+ \to \pi^+ \pi^0$
$K^- \to \mu^- \nu$
$K^+ \to \mu^+ \nu$
$K^+ \to \pi^+ 2\pi^0$
$K^- \to \pi^0 \pi^-$
$K^+ \to e^+$ 2neutral
$K^- \to e^-$ 2neutral
$K^+ \to \mu^+$ 2neutral
$K^- \to \mu^-$ 2neutral

Particles studied K^+, K^-

CHODOROW 1955

■ First GeV linear accelerator at Stanford ■

Stanford High-Energy Linear Electron Accelerator (Mark III)

M. Chodorow et al.
Rev. Sci. Inst. **26** (1955) 134;

Excerpt This paper describes the design, construction, and early tests of the high-energy linear electron accelerator which has been constructed at Stanford University. *(Extracted from the introductory part of the paper.)*

Data comments This paper gives a summary of a linear accelerator work at Stanford which began in about 1935 with W. W. Hansen. It was the development of a klystron during the war which allowed the development of the high energy LINAC after the war. Partial operation of the 1 GeV LINAC was begun at 1951, at reduced energy.

Related references

See also
Ginzton, Hansen, and Kennedy, Rev. Sci. Inst. **19** (1948) 89;
J. C. Slater, Rev. of Mod. Phys. **20** (1948) 473;
E. L. Chu and W. W. Hansen, J. Appl. Phys. **8** (1947) 996;
G. E. Becker and D. A. Caswell, Rev. Sci. Inst. **22** (1951) 402;
D. H. Sloan and E. O. Lawrence, Phys. Rev. **38** (1931) 2021;
J. W. Beams and L. B. Soddy, Phys. Rev. **45** (1934) 287;
J. W. Beams and H. Trotter, Phys. Rev. **45** (1934) 849;
Haxby et al., Phys. Rev. **70** (1946) 797A;
Schultz et al., Phys. Rev. **72** (1947) 346A;
W. F. Fry, Harvie, L. B. Mullett, and W. Walkinshaw, Nature **160** (1947) 351;
W. Walkinshaw, Proc. Roy. Soc. **61** (1948) 246;
L. B. Mullett and B. G. Loach, Proc. Roy. Soc. **61** (1948) 271;
B. Rossi and K. Greisen, Rev. of Mod. Phys. **13** (1941) 268;
B. Rossi and K. Greisen, Rev. of Mod. Phys. **13** (1941) 240;
L. Brillouin, Phys. Rev. **67** (1945) 260;
J. R. Woodyard, Phys. Rev. **69** (1946) 50;
J. R. Blewett, Phys. Rev. **88** (1952) 1197;
J. A. McIntyre and W. K. H. Panofsky, Rev. Sci. Inst. **25** (1954) 287;
G. W. Tautfest and H. R. Fechter, Rev. Sci. Inst. **26** (1955) 229;
J. S. Levinger, Phys. Rev. **84** (1951) 43;

NISHIJIMA 1955

■ Nishijima classification of strange particles with prediction of Σ^0 and Ξ^0 hyperons ■

Charge Independence Theory of V Particles

K. Nishijima
Progr. of Theor. Phys. **13** (1955) 285;

Abstract Based on the charge independence hypothesis the properties of V particles are theoretically investigated. It is found that the curious behaviors of these unstable particles are most simply interpreted in terms of the η-charge conservation law which directly results from the C.I. hypothesis and suitable isotopic spin assignments to these particles. The topics which are discussed in this paper are

(a) the isotopic spin assignments to V particles,
(b) the concept of the η-charge,
(c) the η-charge conservation law which is to incorporate with the even-odd rule, the so-called "cascade" decay of some hyperons, and the positive excess of long-lived K particles,
(d) the interpretation of heavy nuclear fragments,
(e) the possible models of τ-mesons.

Related references

See also
K. Nishijima, Progr. of Theor. Phys. **8** (1952) 401;
K. Nishijima, Progr. of Theor. Phys. **10** (1953) 549;
K. Nishijima, Progr. of Theor. Phys. **12** (1954) 279;
R. P. Feynman, Phys. Rev. **76** (1949) 749;
R. P. Feynman, Phys. Rev. **76** (1949) 769;
E. E. Salpeter and H. A. Bethe, Phys. Rev. **84** (1951) 1232;
G. C. Wick, Phys. Rev. **80** (1950) 268;
G. C. Wick, Phys. Rev. **96** (1954) 1124;
R. E. Cutkosky, Phys. Rev. **96** (1954) 1135;
D. C. Hayashi and Y. Munakara, Progr. of Theor. Phys. **7** (1951) 481;
J. Schwinger, Proc. Natl. Acad. Sci. **37** (1951) 452, 455;
M. Gell-Mann and F. E. Low, Phys. Rev. **84** (1951) 350;

FUNG 1956

■ Confirmation of the equality of the masses of K^+ and τ^+ mesons ■

K^+ — τ^+ Mass Difference

S. Fung, A. Pevsner, D.M. Ritson, R. Mohler
Phys. Rev. **101** (1956) 493;

Excerpt Theoretical evidence indicates the existence of a heavy meson in addition to the τ meson. The availability of analyzed K-meson beams at the Berkeley Bevatron allows a comparison of τ and K-meson masses with much greater statistical accuracy than was hitherto possible. This letter reports on the result of such a comparison between 743 K-particles and 65 τ mesons. *(Extracted from the introductory part of the paper.)*

Accelerator LBL-Bevatron

Detectors Nuclear emulsion

Reactions

$K^+ \to$ charged (neutrals)	316-340 MeV(P_{lab})
$K^+ \to 2\pi^+ \pi^-$	316-340 MeV(P_{lab})

Particles studied K^+

FITCH 1956

■ Confirmation of the equality of the lifetimes of K^+ and Θ ($K^+ \to \pi^+\pi^0$) mesons ■

Mean Life of K^+ Mesons

V.L. Fitch, R. Motley
Phys. Rev. **101** (1956) 496;

Excerpt Currently the best evidence supports the view that there are at least two types of K-mesons in the mass range of 900 to 1000 m_e. The τ meson appears not to have both the spin and parity of the $K_{\pi 2}$ when the experimental data on τ decay are compared with the analysis of Dalitz. On the other hand, there is good evidence from range and momentum measurements on the parent particle and from Q-value measurements that the masses of the $K_{\mu 2}$, the $K_{\pi 2}$ and τ are the same to within one percent. This situation has led us to investigate the lifetime of the K^+-meson as a function of its decay mode. Except in the case of τ decay, previous measurements of the lifetime were made irrespective of the decay mode. These are summarized in the report of the Pisa Conference. This letter reports the results of measurements made on the decay of the $K^+_{\mu 2} \to \mu^+ \nu$ and $K^+_{\pi 2} \to \pi^+ \pi^0$. *(Extracted from the introductory part of the paper.)*

Accelerator BNL-COSMOTRON

Detectors Counters

Data comments 246 $K^+ \to \pi^+ \pi^0$, and 393 $K^+ \to \mu^+ \nu$.

Reactions

$K^+ \to \pi^+ \pi^0$	465 MeV(P_{lab})
$K^+ \to \mu^+ \nu$	465 MeV(P_{lab})

Particles studied K^+

BRABANT 1956C

■ Confirmation of the existence of the antiproton ■

Terminal Observations on Antiprotons

J.M. Brabant et al.
Phys. Rev. **101** (1956) 498;

Excerpt Recently Chamberlain, Segrè, Wiegand, and Ypsilantis have observed negatively charged particles of mass $1840 \pm 90 m_e$, emerging from a target of the Berkeley Bevatron. In their experiment, protons of 6.2 BeV energy bombarded a Cu target, and secondary particles of unit negative charge emitted near 0° were selected in a momentum orbit of 1.20 ± 0.02 BeV/c by the system of deflecting and focusing magnets described in reference 1. Their additional measurement of flight time over a 40-ft portion of the path allowed the identification of mass within the limits mentioned above, and certain requirements of response in special Čerenkov counters assisted in rejecting background events. The fact that each of these unique particles was accompanied by about 4.4×10^4 negative pions within the defined momentum channel emphasizes the importance of background rejection.
Since it is required of an antiproton that it be capable of annihilation in combination with a nucleon, it is significant to observe the passage through matter of particles purported to be antiprotons, and particularly to examine the region of their range endings for evidences of large energy release. The first aim of the experiment described here was to show that the proton-mass particles produce events different from those associated with passage of the negative pion beam. If such observations can be made on a quantitative basis they can presumably insure the identity of these particles as antiprotons in distinction from combinations of K mesons, hyperons, or unknown objects that could demonstrate the proper charge, mass, and lifetime. Annihilation is expected to occur in several modes, but the immediate products may include pions, photons, and possibly K mesons; and the identities and multiplicities of these product particles may vary. *(Extracted from the introductory part of the paper.)*

Accelerator LBL-Bevatron

Detectors Counters

Reactions
$\overline{p}$ nucleus $\rightarrow$ mult[charged] X
$\overline{p}$ nucleon $\rightarrow$ annihil

Particles studied $\overline{p}$

ALVAREZ 1956

■ Firm establishment of the K^+ lifetime value. Establishment of equality of the lifetimes of $K^+(\rightarrow \mu^+$ neutral), $\Theta(\rightarrow \pi^+\pi^0)$, and $\tau^+(\rightarrow \pi^+\pi^+\pi^-)$ mesons ■

Lifetime of K Mesons

L.W. Alvarez, F.S. Crawford, M.L. Good, M.L. Stevenson
Phys. Rev. **101** (1956) 503;

Excerpt Since the various species of K mesons produced at the Bevatron are found to have masses equal within the rather small experimental error, it becomes a critical matter to see if the lifetimes of the different species are different (as one would expect if they have separate identities) or if the lifetimes are all the same (as they would be if there is but one primary K meson which has several alternate modes of decay). This letter describes preliminary results of a counter experiment investigating this point. *(Extracted from the introductory part of the paper.)*

Accelerator LBL-Bevatron

Detectors Spectrometer, Counters

Related references

Analyse information from
R. W. Birge et al., Phys. Rev. **99** (1955) 329;
L. T. Kerth, D. H. Stork, R. P. Haddock, and M. M. Whitehead, Phys. Rev. **99** (1955) 641;
L. W. Alvarez and S. Goldhaber, Nuovo Cim. **2** (1955) 344;

Reactions
$K^+ \rightarrow 2\pi^+ \pi^-$
$K^+ \rightarrow \pi^+ \pi^0$
$K^+ \rightarrow \mu^+ \nu_\mu$

Particles studied K^+

WIGHTMAN 1956

■ Wightman axiomatic field theory ■

Quantum Field Theory in Terms of Vacuum Expectation Values

A.S. Wightman
Phys. Rev. **101** (1956) 860;

Abstract Vacuum expectation values of products of neutral scalar field operators are discussed. The properties of these distributions arising from Lorentz invariance, the absence of negative energy states and the positive definiteness of the scalar product are determined. The vacuum expectation values are shown to be boundary values of analytic functions. Local commutativity of the field is shown to be equivalent to a symmetry property of the analytic functions. The problem of determining a theory of a neutral scalar field given its vacuum expectation values is posed and solved.

HILL 1956

■ First indication of annihilation of the antiproton in matter ■

Nuclear Emulsion Observation of Annihilation of an Antiproton

R.O. Hill, S.D. Johansson, F.T. Gardner
Phys. Rev. **101** (1956) 907;

Excerpt An event, which we believe has a reasonable probability of representing the creation and subsequent annihilation of an antiproton, has been observed in emulsions exposed directly in the proton beam of the Berkeley Bevatron. *(Extracted from the introductory part of the paper.)*

Accelerator LBL-Bevatron

Detectors Nuclear emulsion

Related references

See also
A. Husain and E. Pickup, Phys. Rev. **98** (1955) 136;
M. F. Kaplon, J. Klarmann, and G.Yekutieli, Phys. Rev. **99** (1955) 1528;
J. R. Fleming and J. J. Lord, Phys. Rev. **92** (1953) 511;
D. Fox, Phys. Rev. **94** (1954) 499;
R. N. Thorn, Phys. Rev. **94** (1954) 501;
G. Feldman, Phys. Rev. **95** (1954) 1967;

Reactions
p nucleus $\rightarrow \overline{p}$ X 6.2 GeV(T_{lab})
$\overline{p}$ nucleus $\rightarrow$ mult[charged] X
$\overline{p}$ nucleon $\rightarrow$ annihil

CHAMBERLAIN 1956E Nobel prize

■ First evidence of annihilation of the antiproton in emulsion. Nobel prize to O. Chamberlain and E. Segrè awarded in 1959 "for their discovery of the antiproton" ■

Antiproton Star Observed in Emulsion

O. Chamberlain et al.
Phys. Rev. **101** (1956) 909;

Excerpt In connection with the antiproton investigation at the Bevatron, we planned and carried out a photographic-

emulsion exposure in a magnetically selected beam of negative particles. The magnetic system was identical to the first half (one deflecting magnet and one magnetic lens) of the system used in the antiproton experiment of Chamberlain, Segrè, Wiegand, and Ypsilantis. The selected particles left the copper target in the forward direction with momentum 1.09 BeV/c.
Cosmic-ray events possibly due to antiprotons had been observed previously by Hayward, Cowan, Bridge, Courant, De Staebler, and Rossi, and (in nuclear emulsion) by Amaldi, Castagnoli, Cortini, Franzinetti, and Manfredini. We were hopeful of finding events similar to the last one in our experiment as reported here. *(Extracted from the introductory part of the paper.)*

Accelerator LBL-Bevatron

Detectors Nuclear emulsion

Related references

Analyse information from
O. Chamberlain, E. Segrè, C. Wiegand, and T. Ypsilantis, Phys. Rev. **100** (1955) 947;
E. Hayward, Phys. Rev. **72** (1947) 937;
E. W. Cowan, Phys. Rev. **94** (1954) 161;
H. S. Bridge, H. Courant, H. C. de Staebler, and B. Rossi, Phys. Rev. **95** (1954) 1101;
E. Amaldi et al., Nuovo Cim. **1** (1955) 492;

Reactions

$\bar{p}$ nucleon $\to$ annihil 1.09 GeV(P_{lab})
$\bar{p}$ nucleus $\to$ mult[hadron] X 1.09 GeV(P_{lab})

CHEW 1956

■ Static model for the pion–nucleon interaction ■

Effective-Range Approach to the Low-Energy p-Wave Pion-Nucleon Interaction

G.F. Chew, F.E. Low
Phys. Rev. **101** (1956) 1570;

Abstract The theory of p-wave pion-nucleon scattering is reexamined using the formalism recently proposed by one of the authors (F.E.L.). On the basis of the cut-off Yukawa theory without nuclear recoil it is found, for not too high values of the coupling constant, that:

(a) For each p-wave phase shift a certain function of the cotangent should be approximately linear at low energies and should extrapolate to the Born approximation at zero total energy. The value of the renormalized unrationalized coupling constant determined in this way from experiment is $f^2 = 0.08$. A special feature of the predicted energy dependence of the phase shifts is that δ_{33} is positive and the other p phase shifts are negative.

(b) The so-called "crossing theorem" requires a relation between the four p phase shifts, so that in addition to the coupling constant only two further constants are needed to completely specify the low-energy behavior.

(c) The direction of the energy variation in the $(3,3)$ state is such that a resonance will occur for a sufficiently large cut-off ω_{max}. Rough estimates indicate that $\omega_{max} \sim 6$ will produce a resonance at the energy required by experiment.

It is argued that the results (a) and (b) are very probably also consequences of a relativistic theory but that (c) may not be.

Related references

See also
G. F. Chew, Phys. Rev. **94** (1954) 1748;
G. F. Chew, Phys. Rev. **95** (1954) 1699;
F. E. Low, Phys. Rev. **97** (1955) 1392;
G. C. Wick, Rev. of Mod. Phys. **27** (1955) 339;
H. Lehmann, K. Symanzik, and W. Zimmermann, Nuovo Cim. **1** (1955) 1;
B. A. Lippmann and J. Schwinger, Phys. Rev. **79** (1950) 469;
P. A. M. Dirac, *The Principles of Quantum Mechanics*, Oxford University Press, New York (1947) 198;
M. Gell-Mann and M. L. Goldberger, in *Proc. VI Annual Rochester Conf.* Univ. of Rochester Press, (1954);
N. M. Kroll and M. A. Ruderman, Phys. Rev. **93** (1954) 233;
T. D. Lee, Phys. Rev. **95** (1954) 1329;
Castilejo, Dalitz, and Dyson, Phys. Rev. **101** (1956) 453;
S. J. Lindenbaum and L. C. L. Yuan, Phys. Rev. **100** (1955) 306;
J. I. Friedman, Lee, and Christian, Phys. Rev. **100** (1955) 1494;
R. Karplus and M. A. Ruderman, Phys. Rev. **98** (1955) 771;
M. L. Goldberger, H. Miyazawa, and R. Oehme, Phys. Rev. **99** (1955) 979;
R. Oehme, Phys. Rev. **100** (1955) 1503;

KERST 1956

■ First realistic proposal for probing high energies by colliding beams of particles ■

Attainment of Very High Energy by Means of Intersecting Beams of Particles

D.W. Kerst et al.
Phys. Rev. **102** (1956) 590;

Reprinted in
The Physical Review — the First Hundred Years, AIP Press (1995) CD-ROM.

Excerpt In planning accelerators of higher and higher energy, it is well appreciated that the energy which will be available for interactions in the center-of-mass coordinate system will increase only as the square root of the energy of the accelerator. The possibility of producing interactions in stationary coordinates by directing beams against each other has often been considered, but the intensities of beams so far available have made the idea impractical. Fixed-field alternating-gradient accelerators offer the possibility of obtaining sufficiently intense beams so that it may now be reasonable to reconsider directing two beams of approximately equal energy at each other. In this circumstance, two 21.6 BeV accelerators are equivalent to one machine of 1000 BeV. *(Extracted from introductory part of the paper.)*

Related references

See also

K. R. Simon, Phys. Rev. **98** (1955) 1152A1;
L. W. Jones et al., Phys. Rev. **98** (1955) 1153A1;
K. M. Terwilliger et al., Phys. Rev. **98** (1955) 1153A2;
D. W. Kerst et al., Phys. Rev. **98** (1955) 1153A3;

MCALLISTER 1956 — Nobel prize

■ First measurement of the proton electromagnetic radius. Nobel prize to R. Hofstadter awarded in 1961 "for his pioneering studies of electron scattering in atomic nuclei and for his thereby achieved discoveries concerning the structure of the nucleons". Co-winner R. Mössbauer "for his researches concerning the resonance absorption of gamma radiation and his discovery in this connection of the effect which bears his name" ■

Elastic Scattering of 188 MeV Electrons from Proton and the Alpha Particle

R.W. McAllister, R. Hofstadter
Phys. Rev. **102** (1956) 851;

Reprinted in

R. N. Cahn and G. Goldhaber, *The Experimental Foundations of Particle Physics*, Cambridge Univ. Press (1991) 234.

Abstract The elastic scattering of 188 MeV electrons from gaseous targets of hydrogen and helium has been studied. Elastic profiles has been obtained at laboratory angles between 35° and 138°. The areas under such curves, within energy limits of ±1.5 MeV of the peak, have been measured and the results plotted against angle. In the case of hydrogen, a comparison has been made with the theoretical predictions of the Mott formula for elastic scattering and also with a modified Mott formula (due to Rosenbluth) taking into account both the anomalous magnetic moment of the proton and a finite size effect. The comparison shows that a finite size of the proton will account for the results and the present experiment fixes this size. The root-mean-square radii of charge and magnetic moment are each $(0.74 \pm 0.24) \times 10^{-13}$ cm. In obtaining these results it is assumed that the usual laws of electromagnetic interaction and the Coulomb law are valid at distances less than 10^{-13} cm and the charge and moment radii are equal. In helium, large effects of the finite size of the alpha-particle are observed and the rms radius of the alpha particle is found to be $(1.6 \pm 0.1) \times 10^{-13}$ cm.

Accelerator Stanford linear electron accelerator

Detectors Spectrometer

Related references

More (earlier) information

R. Hofstadter, H. R. Fetcher, and J. A. McIntyre, Phys. Rev. **91** (1953) 422;
R. Hofstadter, H. R. Fetcher, and J. A. McIntyre, Phys. Rev. **92** (1953) 978;
R. Hofstadter, B. Hahn, A. W. Knudsen, and J. A. McIntyre, Phys. Rev. **95** (1954) 512;
R. Hofstadter and R. W. McAllister, Phys. Rev. **98** (1955) 217;

See also

Yennie, Wilson, and D. G. Ravenhall, Phys. Rev. **92** (1953) 1325;
J. A. McIntyre and R. Hofstadter, Phys. Rev. **98** (1955) 158;
G. W. Tautfest and H. R. Fechter, Phys. Rev. **96** (1954) 35;
G. W. Tautfest and H. R. Fechter, Rev. Sci. Inst. **26** (1955) 229;
M. N. Rosenbluth, Phys. Rev. **79** (1950) 615;

Reactions

$e^- \, p \to p \, e^-$	0.188 GeV(E_{lab})
$e^- \, \mathrm{He} \to \mathrm{He} \, e^-$	0.188 GeV(E_{lab})

Particles studied p, He

CHAMBERLAIN 1956F — Nobel prize

■ Confirmation of antiproton-nucleon annihilation. Nobel prize to O. Chamberlain and E. Segrè awarded in 1959 "for their discovery of the antiproton" ■

Example of an Antiproton-Nucleon Annihilation

O. Chamberlain et al.
Phys. Rev. **102** (1956) 921;

Reprinted in

R. N. Cahn and G. Goldhaber, *The Experimental Foundations of Particle Physics*, Cambridge Univ. Press (1991) 96.
The Physical Review — the First Hundred Years, AIP Press (1995) 851.

Excerpt The existence of antiprotons has recently be demonstrated at the Berkeley Bevatron by Counter experiment. The antiprotons were found among the momentum-analyzed (1190 MeV/c) negative particles emitted by a copper target bombarded by 6.2 BeV Protons. Concurrently with the counter experiment, stacks of nuclear emulsions were exposed in the beam adjusted to accept 1090 MeV/c negative particles in an experiment designed to observe the properties of antiprotons when coming to rest. This required a 132-g/cm^2 copper absorber to slow down the antiprotons sufficiently to stop them in the emulsion stack. Only one antiproton was found in stacks in which seven were expected, assuming a geometric interaction cross section for antiprotons in copper. It has now been found that the cross section in copper is about twice geometric, which explains this low yield.
In view of this result a new irradiation was planned in which

(1) no absorbing material preceded the stack,

(2) the range of the antiprotons ended in the stack, and

(3) antiprotons and mesons were easily distinguishable by grain density at the entrance of the stack. In order to achieve these three results it was necessary to select antiprotons of lower momentum, even if these should be admixed with a larger number of π^- than at higher momenta.

In the present experiment we exposed a stack in the same beam used previously, adjusted for a momentum of 700 MeV/c instead of 1090 MeV/c. Since the previous work had indicated that the most troublesome background was due to ordinary protons, the particles were also passed through a clearing magnetic field just prior to their entrance into the emulsion stacks. *(Extracted from the introductory part of the paper.)*

Accelerator LBL-Bevatron

Detectors Nuclear emulsion

Data comments One event described.

Related references

Analyse information from
O. Chamberlain, E. Segrè, C. Wiegand, and T. Ypsilantis, Phys. Rev. **100** (1955) 947;
O. Chamberlain et al., Phys. Rev. **101** (1956) 909;

Reactions
$\overline{p}$ nucleon $\rightarrow$ annihil
$\overline{p}$ nucleus $\rightarrow$ nucleus 5π

COWAN 1956 — Nobel prize

■ Confirmation of the detection of the $\overline{\nu}_e$. Nobel prize to F. Reines awarded in 1995 "for the detection of the neutrino". Co-winner M. Perl "for the discovery of the tau lepton" ■

Detection of the Free Neutrino: A Confirmation

C.L. Cowan et al.
Science **124** (1956) 103;

Excerpt A tentative identification of the free neutrino was made as an experiment performed at Hanford in 1953. In that work the reaction:

$$\overline{\nu}\ p^+ \rightarrow \beta^+\ n^0 \qquad (1)$$

was employed wherein the intense neutrino flux from fission-fragment decay in a large reactor was incident on a detector containing many target protons in a hydrogenous liquid scintillator. The reaction products were detected as a delayed pulse pair; the first pulse being due to the slowing down and annihilation of the positron and the second to capture of the moderated neutron in cadmium dissolved in the scintillator. To identify the observed signal as neutrino-induced, the energies of the two pulses, their time-delay spectrum, the dependance of the signal rate on reactor power, and its magnitude, as compared with the predicted rate were used. The calculated effectiveness of the shielding employed, together with neutron measurements made with emulsions external to the shield, seemed to rule out reactor neutrons and gamma radiation as the cause of the signal. Although a high background was experienced due to both the reactor and to cosmic radiation, it was felt that an identification of the free neutrino had probably been made.
To carry this work to a more definitive conclusion, a second experiment was designed, and the equipment was taken to the Savannah River Plant of the US Atomic Energy Commission, where the present work was done. This work confirms the results obtained at Hanford and so verifies the neutrino hypothesis suggested by Pauli and incorporated in a quantitative theory of beta decay by Fermi.
In this experiment, a detailed check of each term of Eq.1 was made using a detector consisting of a multiple-layer (club-sandwich) arrangement of scintillation counters and target tanks. This arrangement permits the observation of prompt spatial coincidences characteristic of positron annihilation radiation and of the multiple gamma ray burst due to neutron capture in cadmium as well as the delayed coincidences described in the first paragraph. *(Extracted from the introductory part of the paper.)*

Accelerator Nuclear reactor

Detectors Counters

Related references

See also
F. Reines and C. L. Cowan, Phys. Rev. **90** (1953) 492;
F. Reines and C. L. Cowan, Phys. Rev. **92** (1953) 830;
C. L. Cowan, F. Reines et al., Phys. Rev. **90** (1953) 493;

Reactions
$\overline{\nu}_e\ p \rightarrow n\ e^+$ 1.8-6.0 MeV(E_{lab})

Particles studied $\nu_e, \overline{\nu}_e$

REINES 1956 — Nobel prize

■ First detection of the free neutrino. Nobel prize to F. Reines awarded in 1995 "for the detection of the neutrino". Co-winner M. Perl "for the discovery of the tau lepton" ■

The Neutrino

F. Reines, C.L. Cowan
Nature **178** (1956) 446;

Reprinted in
(translation into Russian) Usp. Fiz. Nauk **62** (1957) 391.

Abstract A review of the properties of the neutrino including the results of some important new experiments by the authors. Using a very large scintillation counter near a fission reactor an upper limit of 10^{-9} Bohr electron magnetons has been placed on the magnetic moment. The non-identity of the neutrino and antineutrino has again been demonstrated by measurement of the lifetime of Nd^{150}. A lower limit of 4×10^{18} years was obtained which is to be compared with 1.3×10^{15} years expected if the neutrinos are identical as proposed by Majorana and Furry. The reaction $\nu_e\ p^+ \to \beta^+\ n$ had been investigated previously using the neutrino flux from a large reactor, the target protons being supplied by a large liquid scintillator; the reaction was observed to take place but the signal/noise ratio was very small. A new experiment has been performed in which the positron annihilation pulses are detected by two scintillators in coincidence, the γ-rays following the capture of the slow neutron by cadmium in the target are detected in coincidence several microseconds later. A counting rate of 3 per hour was observed which gives a cross section of about $6 \times 10^{-44} \mathrm{cm}^2$ which is in agreement with that predicted. A discussion of the neutrino-deuteron reaction shows interesting possibilities: a possible direct measurement of the ratio of the Gamow-Teller and Fermi coupling constants in β-decay; the existence of a bi-neutron might also be detected if such a particle exists in a bound state. *(Science Abstracts, 1956, 2719. D. J. A. Prowse)*

Accelerator Nuclear reactor

Detectors Counters

Related references

More (earlier) information
F. Reines and C. L. Cowan, Phys. Rev. **90** (1953) 492;
C. L. Cowan, F. Reines et al., Phys. Rev. **90** (1953) 493;
C. L. Cowan, F. Reines et al., Science **124** (1956) 103;

Reactions

$\overline{\nu}_e\ p \to n\ e^+$ 1.8-6.0 MeV(E_{lab})

Particles studied $\nu_e, \overline{\nu}_e$

LANDE 1956

■ First evidence for the K_L ■

Observation of Long-Lived V Particles

K. Lande et al.
Phys. Rev. **103** (1956) 1901;

Reprinted in

R. N. Cahn and G. Goldhaber, *The Experimental Foundations of Particle Physics*, Cambridge Univ. Press (1991) 205.
The Physical Review — the First Hundred Years, AIP Press (1995) 854.

Excerpt The application of rigorous charge conjugation invariance to strange particle interactions has led to the prediction of rather startling properties for the θ^0-meson state. Some of these are:

(I) the existence of a second neutral particle, θ_2^0, for which two-pion decay is prohibited;

(II) the consequent existence of a second lifetime, considerably longer than that for two-pion decay of the θ_1^0 ($\sim 1 \times 10^{-10}$ sec);

(III) a complicated time dependence for the nuclear interaction properties. The only additional assumption in this "particle mixture" theory is the nonidentity of θ^0 and its antiparticle.

These theoretical considerations have stimulated us to undertake a search for long-lived neutral particles. To this end, the Columbia 36-in. magnet cloud chamber was exposed to the neutral radiation emitted from a copper target at an angle of 68° to the 3 BeV external proton beam of the Brookhaven Cosmotron. *(Extracted from the introductory part of the paper.)*

Accelerator BNL-COSMOTRON

Detectors Cloud chamber

Related references

See also
M. Gell-Mann and A. Pais, Phys. Rev. **97** (1955) 1387;
A. Pais and O. Piccioni, Phys. Rev. **100** (1955) 932;
G. Snow, Phys. Rev. **103** (1956) 1111;
S. B. Treiman and R. G. Sachs, Phys. Rev. **103** (1956) 1545;
K. M. Case, Phys. Rev. **103** (1956) 1449;
O. Piccioni et al., Rev. Sci. Inst. **26** (1955) 232;
Blumenfeld, N. E. Booth, L. M. Lederman, and W. Chinowsky, Phys. Rev. **102** (1956) 1184;
M. M. Block, Harth, and R. M. Sternheimer, Phys. Rev. **100** (1955) 324;
T. D. Lee and C. N. Yang, Phys. Rev. **102** (1956) 290;
Kadyk, G. H. Trilling, R. B. Leighton, and C. D. Anderson, Bull.Am.Phys.Soc. **1** (1956) 251;
J. Ballam, Grisaru, and S. B. Treiman, Phys. Rev. **101** (1956) 1438;
Slaughter, M. M. Block, and Harth, Bull.Am.Phys.Soc. **1** (1956) 186;

Reactions

$p\ \mathrm{Cu} \to$ neutral X 3 GeV(T_{lab})
$K^0 \to \pi^+\ \pi^0\ \pi^-$

Particles studied K_L

FRY 1956

■ Confirmation of the existence of the K_L ■

Evidence for a Long-Lived Neutral Unstable Particle

W.F. Fry, J. Schneps, M.S. Swami
Phys. Rev. **103** (1956) 1904;

Reprinted in

R. N. Cahn and G. Goldhaber, *The Experimental Foundations of Particle Physics*, Cambridge Univ. Press (1991) 208.

Abstract In the course of a scan of emulsion exposed to the Bevatron 90° 280 MeV/c momentum K-beam, four events were found which involve the apparent production of a single strange particle in a low-energy interaction. They are interpreted as due to interactions of long-lived neutral K-particles produced in the target with frequency comparable to that of K^+-meson production. "Ordinary" θ^0 particles would have had to live about 20 mean lifetimes. Energetic neutron stars are rare in the stack. *(Science Abstracts, 1957, 2684. R. H. W. Johnston)*

Accelerator LBL-Bevatron

Detectors Nuclear emulsion

Data comments ${}^4He_S$ – ${}_\Lambda He^4$ hyperfragment.

Related references

More (earlier) information

W. F. Fry, J. Schneps, and M. S. Swami, Phys. Rev. **101** (1956) 1526;

See also

M. Gell-Mann and A. Pais, Phys. Rev. **97** (1955) 1387;
G. A. Snow, Phys. Rev. **103** (1956) 1111;

Reactions

p nucleus $\to$ neutral X	6 GeV(T_{lab})
neutral nucleus $\to$ ${}^4He_S$ X	
${}^4He_S$ $\to$ ${}^3He\, p\, \pi^-$	
neutral nucleus $\to$ K^- X	

Particles studied K_L

LEE 1956 — Nobel prize

■ Proposals to test spatial parity conservation in weak interactions. Nobel prize to T. D. Lee and C. N. Yang awarded in 1957 "for their penetrating investigation of the so-called parity laws, which has led to important discoveries regarding the elementary particles" ■

Question of Parity Conservation in Weak Interactions

T.D. Lee, C.N. Yang
Phys. Rev. **104** (1956) 254;

Reprinted in
The Physical Review — the First Hundred Years, AIP Press (1995) 1016.

Abstract The question of parity conservation in β decays and in hyperon and meson decays is examined. Possible experiments are suggested which might test parity conservation in these interactions.

Related references

See also

T. D. Lee and C. N. Yang, Phys. Rev. **102** (1956) 290;

Analyse information from

M. M. Whitehead et al., Bull.Am.Phys.Soc. **1** (1956) 184;
G. Harris, J. Orear, and S. Taylor, Phys. Rev. **100** (1955) 932;
V. L. Fitch and R. Motley, Phys. Rev. **101** (1956) 496;
L. W. Alvarez et al., Phys. Rev. **101** (1956) 503;
R. H. Dalitz, Phil. Mag. **44** (1953) 1068;
E. Fabri, Nuovo Cim. **11** (1954) 479;
J. Orear, G. Harris, and S. Taylor, Phys. Rev. **102** (1956) 1676;
O. Chamberlain, E. Segrè, R. D. Tripp, and T. Ypsilantis, Phys. Rev. **93** (1954) 1430;
E. M. Purcell and N. F. Ramsey, Phys. Rev. **78** (1950) 807;

Reactions

nucleus $\to$ nucleus $e^-\, \overline{\nu}_e$	
nucleus $\to$ nucleus $e^+\, \nu_e$	
$\Lambda \to p\, \pi^-$	>100 MeV(P_{lab})
$\pi^-\, p \to \Lambda < p\, \pi^- > K^0$	>100 MeV(P_{lab})
$\pi^- \to \mu^- < e^-\, \nu_\mu\, \overline{\nu}_e > \overline{\nu}_\mu$	

CORK 1956

■ First evidence for the antineutron ■

Antineutrons Produced from Antiprotons in Charge-Exchange Collisions

B. Cork, G.R. Lambertson, O. Piccioni, W.A. Wenzel
Phys. Rev. **104** (1956) 1193;

Reprinted in
(translation into Russian) Usp. Fiz. Nauk **62** (1957) 385.
R. N. Cahn and G. Goldhaber, *The Experimental Foundations of Particle Physics*, Cambridge Univ. Press (1991) 99.
The Physical Review — the First Hundred Years, AIP Press (1995) 858.

Excerpt The principle of invariance under charge conjugation gained strong support when it was found that the Bevatron produces antiprotons. Another prediction of the same theory which could be tested experimentally was the existence of the antineutron. Additional interest arises from the fact that charge conjugation has somewhat less obvious consequences when applied to neutral particles than it has when applied to particles with electric charge.
The purpose of this experiment was to detect the annihilation of antineutrons produced by charge exchange from antiprotons. *(Extracted from the introductory part of the paper.)*

Accelerator LBL-Bevatron

Detectors Counters

Related references

Analyse information from

O. Chamberlain, E. Segrè, C. Wiegand, and T. Ypsilantis, Phys. Rev. **100** (1955) 947;
J. M. Brabant et al., Phys. Rev. **101** (1956) 498;
O. Chamberlain et al., Phys. Rev. **102** (1956) 921;

Reactions

p Be $\to$ $\overline{p}$ X	6.2 GeV(T_{lab})
$\overline{p}$ nucleus $\to$ $\overline{n}$ X	1.4 GeV(P_{lab})

$\overline{n}\, p \rightarrow$ annihil
$\overline{n}\, n \rightarrow$ annihil

Particles studied $\overline{n}$

LEE 1956B

■ Invention of G-parity for nonstrange mesons ■

Charge Conjugation, a New Quantum Number G, and Selection Rules Concerning a Nucleon-Antinucleon System

T.D. Lee, C.N. Yang
Nuovo Cim. **3** (1956) 749;

Excerpt A new quantum number is introduced which leads to selection rules concerning transitions between state with heavy particles number = 0. They are applied to the system nucleon-antinucleon; it is also shown that they can be extended to include the conservation of strangeness.

Related references

See also
L. Wolfenstein and D. G. Ravenhall, Phys. Rev. **88** (1952) 279;
A. Pais and R. Jost, Phys. Rev. **87** (1952) 871;
L. Michel, Nuovo Cim. **10** (1953) 319;
D. Amati and B. Vitale, Nuovo Cim. **2** (1955) 719;
T. Nakano and K. Nishijima, Progr. of Theor. Phys. **10** (1953) 581;

Particles studied π^+, π^0, π^-, meson

BOGOLYUBOV 1956

■ Development of the renormalization group method in quantum field theory ■

Charge Renormalization Group in Quantum Field Theory

N.N. Bogolyubov, D.V. Shirkov
Nuovo Cim. **3** (1956) 845;

Reprinted in
N. N. Bogolyubov, D. V. Shirkov, *Introduction to the Theory of Quantized Fields*, New York, (1959).

Excerpt Lie differential equations are obtained for the multiplicative charge renormalization group in quantum electrodynamics and in pseudoscalar meson theory with two coupling constants. By the employment of these equations there have been found the asymptotic expressions for the electrodynamical propagation functions in the "ultraviolet" and in the "infrared" regions. The asymptotic high momenta behaviour of meson theory propagation functions is also discussed for the weak coupling case.

Related references

See also
S. F. Edwards, Phys. Rev. **90** (1953) 284;
L. D. Landau, A. A. Abrikosov, and I. M. Halatnikov, Doklady Akad. Nauk SSSR **95** (1954) 497;
L. D. Landau, A. A. Abrikosov, and I. M. Halatnikov, Doklady Akad. Nauk SSSR **95** (1954) 773;
L. D. Landau, A. A. Abrikosov, and I. M. Halatnikov, Doklady Akad. Nauk SSSR **95** (1954) 1177;
L. D. Landau, A. A. Abrikosov, and I. M. Halatnikov, Doklady Akad. Nauk SSSR **96** (1954) 261;
M. Gell-Mann and F. E. Low, Phys. Rev. **95** (1954) 1300;
E. C. G. Stückelberg and A. Peterman, Helv. Phys. Acta **26** (1953) 499;
F. J. Dyson, Phys. Rev. **75** (1949) 1736;

LOGUNOV 1956

■ Generalization of the renormalization group equations in QED for arbitrary covariant gauge ■

On the Generalization of Renormalization Group

A.A. Logunov
Zh. Eksp. Teor. Fiz. **30** (1956) 793;

Excerpt The main aim of this note is the generalization of the group Lee equations for the Green functions to the case of arbitrary longitudinal photonic contraction. *(Extracted from the introductory part of the paper.)*

Related references

More (later) information
A. A. Logunov, Doklady Akad. Nauk SSSR **109** (1956) 740;

See also
N. N. Bogolyubov and D. V. Shirkov, Doklady Akad. Nauk SSSR **103** (1955) 203;
N. N. Bogolyubov and D. V. Shirkov, Doklady Akad. Nauk SSSR **103** (1955) 391;
N. N. Bogolyubov and D. V. Shirkov, Usp. Fiz. Nauk **57** (1955) 87;
L. D. Landau, A. A. Abrikosov, and I. M. Khalatnikov, Doklady Akad. Nauk SSSR **95** (1954) 497;
L. D. Landau, A. A. Abrikosov, and I. M. Khalatnikov, Doklady Akad. Nauk SSSR **95** (1954) 773;
L. D. Landau, A. A. Abrikosov, and I. M. Khalatnikov, Doklady Akad. Nauk SSSR **96** (1954) 261;

GELL-MANN 1956

■ Gell-Mann classification of strange particles with prediction of Σ^0 and Ξ^0 ■

The Interpretation of the New Particles as Displaced Charge Multiplets

M. Gell-Mann
Nuovo Cim. Suppl. **4** (1956) 848;

Excerpt The purpose of this communication is to present a coherent summary of the author's theoretical proposals[1] concerning the new unstable particles.
Section 2 is devoted to some background material on elemen-

tary particles; the object there is to introduce the point of view adopted in the work that follows. In Section 3 the fundamental ideas about displaced multiplets are given, and in the succeeding section these are applied to the interpretation of known particles. A scheme is thus set up, which is used in Section 5 to predict certain results of experiments involving the new particles.

[1] Much of the work to be presented is contained in the following publications:
M. Gell-Mann, Phys. Rev. **92** (1953) 833; M. Gell-Mann and A. Pais, Proceedings of the Glasgow Conference, 1954, and in an unpublished note: M. Gell-Mann: *On the Classification of Particles* (circulated in preprint form, August, 1953). In these references, however, the proposals under discussion are mentioned only briefly and are sometimes buried in a mass of other material. Here they are treated alone and in some detail. Practically the same proposals have been put forward in Japan. See T. Nakano and K. Nishijima: Prog. Theor. Phys. **10** (1953) 581 and K. Nishijima (to be published).

Related references

See also
R. P. Feynman and G. Speisman, Phys. Rev. **94** (1954) 500;
Y. Nambu, K. Nishijima, and Y. Yamaguchi, Progr. of Theor. Phys. **6** (1951) 651;
S. Oneda, Progr. of Theor. Phys. **6** (1951) 633;
A. Pais, Phys. Rev. **86** (1952) 663;
G. C. Wick, Accad. Lincei, Atti. **21** (1935) 170;
M. Gell-Mann and A. Pais, Phys. Rev. **97** (1955) 1387;
W. B. Fowler et al., Phys. Rev. **98** (1955) 121;
A. Pais and O. Piccioni, Phys. Rev. **100** (1955) 1487;

BOGOLYUBOV 1956B

■ Formulation of the Bogolyubov axiomatic approach to the local quantum field theory. Derivation of dispersion relations in field theory for pion nucleon scattering amplitude, general case ■

Problems of the Theory of Dispersion Relations

N.N. Bogolyubov, B.V. Medvedev, M.K. Polivanov
SEATTLE-56 (1956);Forts. der Phys. **6** (1958) 169;

Reprinted in
N. N. Bogolyubov, *Izbrannnye trudy*, Naukova Dumka, Kiev **3** (1971) 340.
The International Conference on Theoretical Physics, Seattle, Sept. (1956).
N. N. Bogolyubov, B. V. Medvedev, M. K. Polivanov, (extended version). *Voprosy Teorii Dispersionnykh Sootnoshenij*, Fizmatgiz, M. (1958).

Related references

More (earlier) information
N. N. Bogolyubov and D. V. Shirkov, Doklady Akad. Nauk SSSR **103** (1955) 203;

See also
R. Kronig, J. Opt. Soc. Am. **12** (1926) 547;
H. A. Kramers, Atti Congr. Intern. Fisici **2** (1927) 545;
W. Heisenberg, Z. Phys. **120** (1943) 513;
W. Heisenberg, Z. Phys. **120** (1943) 673;
W. Heisenberg, Zeit. Naturforschung **1** (1940) 608;
Hu Ning, Phys. Rev. **74** (1948) 131;
M. G. Krein, Doklady Akad. Nauk SSSR **105** (1955) 433;
C. J. Goebel, R. Karplus, and M. A. Ruderman, Phys. Rev. **100** (1955) 240;
M. Gell-Mann, M. L. Goldberger, and W. E. Thirring, Phys. Rev. **95** (1954) 1612;
M. L. Goldberger, Phys. Rev. **97** (1955) 508;
M. L. Goldberger, Phys. Rev. **99** (1955) 979;
R. Karplus and M. A. Ruderman, Phys. Rev. **98** (1955) 771;
M. L. Goldberger, H.Miyazawa, and R. Oehme, Phys. Rev. **99** (1955) 986;
R. Oehme, Phys. Rev. **100** (1955) 1503;
R. Oehme, Phys. Rev. **102** (1956) 1174;
N. N. Bogolyubov and D. V. Shirkov, Usp. Fiz. Nauk **55** (1955) 149;
R. Haag, Dan. Mat. Fys. Medd. **29** (1955) 12;
A. Klein, Progr. of Theor. Phys. **14** (1955) 580;
H. Lehmann, Nuovo Cim. **11** (1954) 342;
G. Källen, Helv. Phys. Acta **25** (1952) 417;
L. D. Landau and I. Y. Pomeranchuk, Doklady Akad. Nauk SSSR **102** (1955) 489;
N. N. Bogolyubov and D. V. Shirkov, *Introduction to the Theory of Quantized Fields*, New York, (1959);
K. Symanzik and R. Jost, *Report at the Seattle conf.* (1956);
M. L. Goldberger, F. E. Low, G. F. Chew, and Y. Namby, *Report at the Seattle conf.* (1956);
N. N. Bogolyubov, V. L. Bonch-Bruevich, and B. V. Medvedev, Doklady Akad. Nauk SSSR **75** (1950) 681;
N. N. Bogolyubov and V. S. Vladimirov, Izv. AN SSSR Ser. mat. **22** (1957) 15;
N. N. Bogolyubov, S. M. Bilenkii, and A. A. Logunov, Nucl. Phys. **5** (1957) 383;
R. Jost and H. Lehmann, Nuovo Cim. **5** (1957) 1598;
M. L. Goldberger, Y. Nambu, and R. Oehme, Ann.Phys. **21** (1957) 226;
A. A. Logunov and A. N. Tavkhelidze, Zh. Eksp. Teor. Fiz. **32** (1957) 1393;
A. A. Logunov and A. N. Tavkhelidze, Nucl. Phys. **8** (1958) 374;
A. A. Logunov and A. N. Tavkhelidze, Nuovo Cim. **10** (1958) 943;
A. A. Logunov, Nucl. Phys. **10** (1959) 71;
W. Kibble, Proc. Roy. Soc. **244** (1958) 355;
H. J. Bremmermann, R. Oehme, and J. L. Taylor, Phys. Rev. **109** (1958) 2178;
F. J. Dyson, Phys. Rev. **110** (1958) 1460;

SYMANZIK 1956

■ Derivation of dispersion relations in field theory for pion nucleon forward scattering amplitude ■

Derivation of Dispersion Relations for Forward Scattering

K. Symanzik
SEATTLE-56 (1956);Phys. Rev. **105** (1957) 743;

Abstract The dispersion relation for forward meson-nucleon scattering is derived in the simplified case of scalar neutral particles. Use is made of the local property of the nucleon field and of certain features of the mass spectrum. In addition, it is assumed that the only singularities of certain matrix elements of the nucleon field commutator are derivatives of finite order of δ functions on the light cone. Under some further assumptions of existence, the dispersion relations for the derivatives of the scattering amplitude with respect angle at zero angle can be derived.

Related references

N. N. Bogolyubov, B. V. Medvedev, and M. K. Polivanov, *Report at the Seattle conf.* (1956);

See also

M. L. Goldberger, Phys. Rev. **99** (1955) 979;
M. L. Goldberger, H.Miyazawa, and R. Oehme, Phys. Rev. **99** (1955) 986;
Anderson, Davidon, Kruse, Phys. Rev. **100** (1955) 339;
Lehmann, Symanzik, and Zimmermann, Nuovo Cim. **1** (1955) 205;
K. Nishijima, Progr. of Theor. Phys. **10** (1953) 549;
K. Nishijima, Progr. of Theor. Phys. **13** (1955) 305;
H. Lehmann, Nuovo Cim. **11** (1954) 342;
G. F. Chew, *Encyclopedia of Physics*, Springer-Verlag, Berlin (1958);
M. Gell-Mann, M. L. Goldberger, and W. E. Thirring, Phys. Rev. **95** (1954) 1612;
E. C. Titchmarsch, *Theory of Fourier Integrals*, Oxford University Press, Oxford (1937) 11;
K. Nishijima, Progr. of Theor. Phys. **12** (1954) 279;
F. E. Low, Phys. Rev. **97** (1955) 1392;
M. L. Goldberger, Phys. Rev. **99** (1955) 979;
R. Oehme, Phys. Rev. **100** (1955) 1503;
R. Oehme, Phys. Rev. **102** (1956) 1174;

SAKATA 1956

■ Invention of composite model for hadrons based on three basic elements ■

On a Composite Model for New Particles

S. Sakata

Progr. of Theor. Phys. **16** (1956) 686;

Reprinted in

Shoichi Sakata Scientific Works. Publication Committee of Scientific Papers of Prof. Shoichi Sakata, Horei Printed Co., Ltd, Tokyo, (1977).

Abstract Recently, Nishijima-Gell-Mann's rule for the systematization of new particles has achieved a great success to account for various facts obtained from the experiments with cosmic rays and with high energy accelerators. Nevertheless, it would be desirable from the theoretical standpoint to find out a more profound meaning hidden behind this rule. The purpose of this work is concerned with this point. *(Extracted from the introductory part of the paper.)*

Related references

See also

T. Nakano and K. Nishijima, Progr. of Theor. Phys. **10** (1953) 581;
K. Nishijima, Progr. of Theor. Phys. **12** (1954) 107;
K. Nishijima, Progr. of Theor. Phys. **13** (1954) 285;
M. Gell-Mann, Phys. Rev. **92** (1953) 833;
D. Iwanenko, Nature **129** (1932) 798;
W. Heisenberg, Z. Phys. **77** (1932) 1;
W. Heisenberg, Zeit. Naturforschung **9A** (1954) 291;
W. Heisenberg, Zeit. Naturforschung **10A** (1955) 425;
M. A. Markov, *O sistematike elementarnykh chastits*, Izd. AN SSSR, (1955);
E. Fermi and C. N. Yang, Phys. Rev. **76** (1949) 1739;
S. Tanaka, Progr. of Theor. Phys. **16** (1956) 625;
Z. Maki, Progr. of Theor. Phys. **16** (1956) 667;
K. Matsumoto, Progr. of Theor. Phys. **16** (1956) 583;
M. A. Markov, Usp. Fiz. Nauk **51** (1953) 317;
L. Landau, Doklady Akad. Nauk SSSR **95** (1954) 497;
L. Landau, Doklady Akad. Nauk SSSR **95** (1954) 733;
L. Landau, Doklady Akad. Nauk SSSR **95** (1954) 1177;
L. Landau, Doklady Akad. Nauk SSSR **96** (1955) 261;
L. Landau, Doklady Akad. Nauk SSSR **102** (1955) 489;
S. Kamefuchi and H. Umezawa, Progr. of Theor. Phys. **15** (1956) 298;
S. Kamefuchi and H. Umezawa, Nuovo Cim. **3** (1956) 1060;

IOFFE 1957

■ Indication of the possibility of charge conjugation violation in weak interactions ■

The Problem of Parity Non-conservation in Weak Interaction

B.L. Ioffe, L.B. Okun, A.P. Rudik

Zh. Eksp. Teor. Fiz. **32** (1957) 396; JETP **5** (1957) 328;

Excerpt One of the possible theoretical explanations of the paradox of the θ and τ-decay of K-mesons (Orear, Harris, and Taylor, Phys. Rev. **104** (1956) 1676) is the hypothesis of parity non-conservation in weak interactions. Lee and Yang (Phys. Rev. **104** (1956) 254) showed that parity non-conservation in weak interactions could not have been detected by any experiments which have yet been performed (except, of course, the K-meson decay experiments), and they proposed several experiments by which the conservation of parity could be tested. However, Yang and Lee did not require that the weak interactions be invariant under time-reversal or charge-conjugation.* If one assumes that space-parity is not conserved, and that the θ and τ particles are identical, then the existence of a long-lived K^0-particle can be explained by supposing either that charge-parity or time-parity is conserved. The analysis of correlation experiments by Lee and Yang assumes that time-parity is conserved and charge-parity is not conserved.
We have found that, in the absence of conservation of space-parity, there exists an experimental test to decide whether charge-parity or time-parity is conserved in weak interactions. Namely, if time-parity is conserved, the long-lived (odd with respect to time-reversal) K^0-particle can decay into 3 pions or into $3\pi^0$-mesons forming an S-state, while this process is forbidden if charge-parity is conserved. We shall show that if one assumes conservation of charge-parity one is led to results quite different from those of Lee and Yang.

* Pauli has discussed (W. Pauli, *Niels Bohr and the Development of Physics*, Pergamon Press, London, 1955) these problems in a general way. It is important to observe that, when parity is not conserved, charge conjugation ceases to be equivalent to time reversal. Pauli proved that, assuming the ordinary relation between spin and statistics, the Lagrangian must be invariant under the combined operation of charge-conjugation and inversion of all four coordinates.

(Extracted from the introductory part of the paper.)

Related references

See also
T. D. Lee and C. N. Yang, Phys. Rev. **104** (1956) 254;
W. Pauli, *Niels Bohr and the Development of Physics*, Pergamon Press, London (1955);
H. A. Tolhoek and S. R. de Groot, Phys. Rev. **84** (1951) 150;

Analyse information from
J. Orear, G. Harris, and S. Taylor, Phys. Rev. **102** (1956) 1676;
K. Lande et al., Phys. Rev. **103** (1956) 1901;
W. F. Fry, J. Schneps, and M. S. Swami, Phys. Rev. **103** (1956) 1904;

PLANO 1957

■ Confirmation of the existence of the Σ^0 hyperon. First measurement of the Σ^0 mass ■

Demonstration of the Existence of the Σ^0 Hyperon and a Measurement of its Mass

R. Plano, N.P. Samios, M. Schwartz, J. Steinberger
Nuovo Cim. **5** (1957) 216;

Excerpt Three events, demonstrating the existence of the Σ^0 hyperon, have been found in a propane bubble chamber. The Q-value for the decay $\Sigma^0 \to \Lambda^0\gamma + Q$ has been measured to be (73.0 ± 3.5) MeV.

Accelerator BNL-COSMOTRON

Detectors Heavy-liquid bubble chamber

Related references

See also
T. Nakano and K. Nishijima, Progr. of Theor. Phys. **10** (1953) 581;
M. Gell-Mann and A. Pais, *Proc. of the Glasgow Conference.*;
W. W. Chupp, G. Goldhaber, S. Goldhaber, and F. B. Webb, *Pisa Conference.*;
M. Ceccarelli et al., *Pisa Conference* (1955);

Analyse information from
W. B. Fowler et al., Phys. Rev. **98** (1955) 121;
W. D. Walker and W. D. Shephard, Phys. Rev. **101** (1956) 1810;
R. Budde et al., Phys. Rev. **103** (1956) 1827;

Reactions

$\pi^- p \to \Sigma^0 K^0$ 1.15 GeV(T_{lab})
$\Sigma^0 \to \Lambda\gamma$
$\Lambda \to p\pi^-$

Particles studied Σ^0

SALAM 1957

■ Postulation of γ_5 invariance for the weak interaction Lagrangian. Two–component theory of neutrino ■

On Parity Conservation and Neutrino Mass

A. Salam
Nuovo Cim. **5** (1957) 299;

Abstract Yang and Lee* have recently suggested that present experimental evidence does not exclude the possibility that parity is not conserved in β-decay. If future experiments confirm this, it may be possible to relate parity-violation in neutrino-decays to the vanishing of neutrino mass and self-mass. The argument is as follows: the free neutrino Lagrangian is invariant for the substitution $\psi_\nu \to \gamma_5\psi_\nu$ ($\overline{\psi}_\nu \to -\overline{\psi}_\nu\gamma_5$). If it is further postulated that neutrino interactions produce no self-mass, one way to secure this is to require that the total Lagrangian also remain invariant for the same substitution (so that $\overline{\psi}_\nu\psi_\nu \to -\overline{\psi}_\nu\psi_\nu$) while other fields (barring degeneracies which we consider later) remain unchanged. In so far as ψ_ν and $\gamma_5\psi_\nu$ have opposite intrinsic parity, most neutrino interactions would then violate parity conservation.

* C. N. Yang and T. D. Lee. The author is indebted for a pre-print.

Related references

Analyse information from
J. Tiomno, Nuovo Cim. **1** (1955) 226;

Particles studied $\nu, \overline{\nu}$

BARKAS 1957

■ Confirmation of antiproton-nucleon annihilation ■

Antiproton-Nucleon Annihilation Process (Antiproton Collaboration Experiment)

W.H. Barkas et al.
Phys. Rev. **105** (1957) 1037;

Abstract Thirty-five antiproton stars have been found in an emulsion stack exposed to a 700 MeV/c negative particle beam. Of these antiprotons, 21 annihilate in flight and three give large-angle scatters ($\theta > 15°, T_{\overline{p}} > 50$ MeV), while 14 annihilate at rest. From the interactions in flight we obtain the total cross section for antiproton interaction: $\sigma_{\overline{p}}/\sigma_0 = 2.9 \pm 0.7$, where $\sigma_0 = \pi R_0^2$ and $R_0 = 1.2 \times 10^{-13} A^{1/3}$ cm. This cross section was measured at an average antiproton energy of $\overline{T_{\overline{p}}} = 140$ MeV.
We also find that the antiproton-nucleon annihilation proceeds primarily through pion production with occasional emission of K particles. On the average 5.3 ± 0.4 pions are produced in the primary process; of these, 1 pion is absorbed and

0.3 inelastically scattered. From the small fraction of pions absorbed, we conclude that the annihilation occurs mainly at the surface of the nucleus at a distance larger than the conventional radius.
A total energy balance of particles emitted in the annihilation gives a ratio of charged to neutral pions consistent with charge independence. Conversely, assuming charge independence, we conclude that the energy going into electromagnetic radiation or neutrinos is small.
Comparisons with the Fermi statistical model and the Lepore-Neuman statistical model have been made. Good agreement with the experimental results on the annihilation process can be obtained through appropriate choice of the interaction volume parameters.
Several different estimates of the antiproton mass are in good agreement and suggest strongly that the antiproton mass is the same as the proton mass within an accuracy of 2.5%. A study of the elastic scattering of the antiprotons down to angles of 2° suggests a possible destructive interference between nuclear and Coulomb scattering.

Accelerator LBL-Bevatron

Detectors Nuclear emulsion

Related references

More (earlier) information
O. Chamberlain et al., Phys. Rev. **101** (1956) 909;
O. Chamberlain et al., Phys. Rev. **102** (1956) 921;
O. Chamberlain et al., Phys. Rev. **102** (1956) 1637;
O. Chamberlain, E. Segrè, C. Wiegand, and T. Ypsilantis, Phys. Rev. **100** (1955) 947;

See also
B. Hahn, D. G. Ravenhall, and R. Hofstadter, Phys. Rev. **101** (1956) 1131;
Melkanhoff, Mozkowski, Nodvik, and Saxon, Phys. Rev. **101** (1956) 507;
J. M. Brabant et al., Phys. Rev. **101** (1956) 498;

Reactions

$\overline{p}$ nucleon $\to$ annihil   700 MeV(P_{lab})

WU 1957

■ First evidence for parity nonconservation in weak decays ■

Experimental Test of Parity Conservation in Beta Decay

C.S. Wu et al.
Phys. Rev. **105** (1957) 1413;

Reprinted in

R. N. Cahn and G. Goldhaber, *The Experimental Foundations of Particle Physics*, Cambridge Univ. Press (1991) 171.
The Physical Review — the First Hundred Years, AIP Press (1995) 361.

Excerpt In a recent paper on the question of parity in weak interactions, Lee and Yang critically surveyed the experimental information concerning this question and reached the conclusion that there is no existing evidence either to support or to refute parity conservation in weak interactions. They proposed a number of experiments on beta decays and hyperon and meson decays which would provide the necessary evidence for parity conservation or nonconservation. In beta decay, one could measure the angular distribution of the electrons coming from beta decays of polarized nuclei. If an asymmetry in the distribution between θ and 180° - θ (where θ is the angle between the orientation of the parent nuclei and the momentum of the electrons) is observed, it provides unequivocal proof that parity is not conserved in beta decay. This asymmetry effect has been observed in the case of oriented Co^{60}. *(Extracted from the introductory part of the paper.)*

Accelerator Radioactive source

Detectors Counters

Related references

More (later) information
T. D. Lee, R. Oehme, and C. N. Yang, Phys. Rev. **106** (1957) 340;

See also
T. D. Lee and C. N. Yang, Phys. Rev. **104** (1956) 254;
E. Ambler et al., Phil. Mag. **44** (1953) 216;

Reactions

$^{60}Co \to {}^{60}Ni\ e^-\ \overline{\nu}_e$

GARWIN 1957

■ Confirmation of parity violation in weak decays. Evidence of charge conjugation parity violation in weak interactions. Measurement of the μ^- magnetic moment ■

Observation of the Failure of Conservation of Parity and Charge Conjugation in Meson Decays: The Magnetic Moment of the Free Muon

R.L. Garwin, L.M. Lederman, M. Weinrich
Phys. Rev. **105** (1957) 1415;

Reprinted in

R. N. Cahn and G. Goldhaber, *The Experimental Foundations of Particle Physics*, Cambridge Univ. Press (1991) 173.
The Physical Review — the First Hundred Years, AIP Press (1995) 863.

Excerpt Lee and Yang have proposed that the long held space-time principles of invariance under charge conjugation, time reversal, and space reflection (parity) are violated by the "weak" interactions responsible for decay of nuclei, mesons, and strange particles. Their hypothesis, born out of the τ - θ puzzle, was accompanied by the suggestion that confirmation should be sought (among other places) in the study of the successive reactions

$$\pi^+ \to \mu^+ \nu, \quad (1)$$

$$\mu^+ \to e^+ 2\nu. \quad (2)$$

They have pointed out that parity nonconservation implies a polarization of the spin of the muon emitted from stopped pions in (1) along the direction of motion and that furthermore, the angular distribution of electrons in (2) should serve as an analyzer for the muon polarization. They also point out that the longitudinal polarization of the muons offers a natural way of determining the magnetic moment. Confirmation of this proposal in the form of preliminary results on β decay of oriented nuclei by Wu *et al.* reached us before this experiment was begun.
By stopping, in carbon, the μ^+ beam formed by forward decay in flight of π^+ mesons inside the cyclotron, we have performed the meson experiment, which establishes the following facts:

I. A large asymmetry is found for the electrons in (2), establishing that our μ^+ beam is strongly polarized.

II. The angular distribution of the electrons is given by $1 + a\cos\theta$, where θ is measured from the velocity vector ofthe incident μ's. We find $a = -\frac{1}{3}$ with an estimated error of 10%.

III. In reactions (1) and (2), parity is not conserved.

IV. By a theorem of Lee, Oehme, and Yang, the observed asymmetry proves that invariance under charge conjugation is violated.

V. The g value (ratio of magnetic moment to spin) for the (free) μ^+ particle is found to be $+2.00 \pm 0.10$.

VI. The measured g value and the angular distribution in (2) lead to the very strong probability that the spin of the μ^+ is $\frac{1}{2}$.

VII. The energy dependence of the observed asymmetry is not strong.

VIII. Negative muons stopped in carbon show an asymmetry (also peaked backwards) of $a \sim -1/20$, i.e., about 1.5% of that for μ^+.

IX. The magnetic moment of the μ^-, bound in carbon, is found to be negative and agrees within limited accuracy with that of the μ^+.

X. Large asymmetries are found for the electrons from polarized μ^+ beams stopped in polyethylene and calcium. Nuclear emulsion yields an asymmetry of about half that observed in carbon.

(Extracted from the introductory part of the paper.)

Accelerator Nevis labs cyclotron

Detectors Counters

Related references

More (later) information
T. D. Lee, R. Oehme, and C. N. Yang, Phys. Rev. **106** (1957) 340;

See also
T. D. Lee and C. N. Yang, Phys. Rev. **104** (1956) 254;
R. H. Dalitz, Phil. Mag. **44** (1953) 1068;
F. J. Belinfante, Phys. Rev. **92** (1953) 997;
M. Weinrich and L. M. Lederman, *Proceedings of the CERN Symposium*, Geneva (1956);

Analyse information from
C. S. Wu et al., Phys. Rev. **105** (1957) 1413;
V. L. Fitch and J. Rainwater, Phys. Rev. **92** (1953) 789;

Reactions
$\mu^+ \to e^+ 2\nu$
$\mu^- \to e^- 2\nu$

Particles studied μ^+, μ^-

LEE 1957B

■ Two–component theory of neutrino ■

Parity Nonconservation and Two-Component Theory of the Neutrino

T.D. Lee, C.N. Yang
Phys. Rev. **105** (1957) 1671;

Reprinted in
The Physical Review — the First Hundred Years, AIP Press (1995) 1021.

Abstract A two-component theory of the neutrino is discussed. The theory is possible only if parity is not conserved in interactions involving the neutrino. Various experimental implications are analyzed. Some general remarks concerning nonconservation are made.

Related references

More (earlier) information
T. D. Lee and C. N. Yang, Phys. Rev. **104** (1956) 254;

See also
W. Pauli, Handbuch der Phys. **24/1** (1933) 226;
G. Luders, Kgl. Dansk. Videnskab. Selskab., Mat.-Fys. Medd., **28**, No. 5 (1954);
W. Pauli, *Niels Bohr and the Development of Physics*, Pergamon Press, London (1955);
L. Michel, Proc. Roy. Soc. **63** (1950) 514;
C. P. Sargent, M. C. Rinehart, L. M. Lederman, and K. C. Rogers, Phys. Rev. **99** (1955) 885;
E. J. Konopinski and H. M. Mahmoud, Phys. Rev. **92** (1953) 1045;

FRIEDMAN 1957

■ Confirmation of parity nonconservation in weak decays ■

Nuclear Emulsion Evidence for Parity Nonconservation in the Decay Chain $\pi^+ \to \mu^+ \to e^+$

J.I. Friedman, V.L. Telegdi
Phys. Rev. **105** (1957) 1681;

Reprinted in

R. N. Cahn and G. Goldhaber, *The Experimental Foundations of Particle Physics*, Cambridge Univ. Press (1991) 176.
The Physical Review — the First Hundred Years, AIP Press (1995) CD-ROM.

Abstract Preliminary data on 1300 decays in magnetically-shielded emulsions exposed to the Chicago synchrocyclotron are consistent with an angular distribution W(θ) = 1 - 0.12 cos θ. The formation of "muonium" (μ^+ e^-) may be an important cause of depolarization in such experiments. *(Science Abstracts, 1957, 5643. K. H. Barker.)*

Accelerator Chicago cyclotron

Detectors Nuclear emulsion

Related references

More (later) information
T. D. Lee, R. Oehme, and C. N. Yang, Phys. Rev. **106** (1957) 340;
See also
S. Berko and F. Hereford, Rev. of Mod. Phys. **28** (1956) 299;
V. L. Telegdi, J. C. Seng, and D. D. Yovanovitch, Phys. Rev. **104** (1956) 867;
T. D. Lee and C. N. Yang, Phys. Rev. **104** (1956) 254;
Analyse information from
R. L. Garwin, L. M. Lederman, and M. Weinrich, Phys. Rev. **105** (1957) 1415;

Reactions

$\pi^+ \rightarrow \mu^+ \nu_\mu$
$\mu^+ \rightarrow e^+ \overline{\nu}_\mu \nu_e$

LANDAU 1957

■ Introduction of the CP conservation law in weak interactions and CP parity ■

On the Conservation Laws in Weak Interactions

L.D. Landau

Zh. Eksp. Teor. Fiz. **32** (1957) 405; Nucl. Phys. **3** (1957) 127;

Reprinted in

L. D. Landau, *Sobranie Trudov*, Nauka, Mos. 1969, t. 2, s. 349 (red. E. M. Lifshits i I. M. Khalatnikov).

Abstract A variant of the theory is proposed in which non-conservation of parity can be introduced without assuming symmetry of space with respect to inversion.
Various possible consequences of non-conservation of parity are considered which pertain to the properties of the neutrino and in this connection some processes involving neutrinos are examined on the assumption that the neutrino mass is exactly zero.

Related references

See also
T. D. Lee and C. N. Yang, Phys. Rev. **102** (1956) 290;
T. D. Lee and C. N. Yang, Phys. Rev. **104** (1956) 254;
M. Gell-Mann and A. Pais, Phys. Rev. **97** (1955) 1387;
Analyse information from
K. Lande et al., Phys. Rev. **103** (1956) 1901;
C. P. Sargent, M. C. Rinehart, L. M. Lederman, and K. C. Rogers, Phys. Rev. **99** (1955) 885;
A. Bonetti et al., Nuovo Cim. **3** (1956) 33;

Particles studied K_L, K_S

LANDAU 1957B

■ Suggestion of the two-component theory for the neutrino ■

On one Possibility for Polarization Properties of the Neutrino

L.D. Landau

Zh. Eksp. Teor. Fiz. **32** (1957) 407; JETP **5** (1957) 337;

Excerpt If the law of conservation of parity is abandoned, then new properties of the neutrino become possible. In the case of zero mass, the Dirac equation separates into two uncoupled pairs of equations. In the usual theory it is impossible to restrict attention to one pair of equations, since the two pairs are interchanged by a space-inversion. But if we require only invariance under combined inversion (L. D. Landau, J. Exp. Theor. Phys. (USSR) **32** (1957) 405), then we can suppose that the neutrino is described by a single pair of equations. In ordinary language, this implies that the neutrino is always polarized along (or always opposite to) the direction of its motion. The antineutrino is then polarized always in the opposite sense. In this scheme the neutrino is not a truly neutral particle, in agreement with the observed absence of double β-decay and especially with the experiments on induced β-decay. We call this kind of neutrino a longitudinally polarized neutrino, or a longitudinal neutrino for short.
In the usual theory the neutrino mass is zero "accidentally." And if one takes into account the neutrino interactions, a non-zero rest-mass appear automatically, although it is of negligible magnitude. The mass of a longitudinal neutrino is automatically zero, and this fact is not disturbed by an interactions.
If we assume the neutrino to be longitudinal, the number of possible types of weak interaction operator is greatly reduced. *(Extracted from the introductory part of the paper.)*

Related references

See also
L. D. Landau, Zh. Eksp. Teor. Fiz. **32** (1957) 405;
L. Michel, Proc. Roy. Soc. **A63** (1950) 514;
L. Michel, Proc. Roy. Soc. **A63** (1950) 1371;
T. D. Lee and C. N. Yang, Phys. Rev. **104** (1956) 254;
Analyse information from
C. P. Sargent, M. C. Rinehart, L. M. Lederman, and K. C. Rogers, Phys. Rev. **99** (1955) 885;
A. Bonetti et al., Nuovo Cim. **3** (1956) 33;

Particles studied $\nu, \overline{\nu}$

LANDE 1957

■ Confirmation of the existence of the K_L ■

Report on Long-Lived K^0 Mesons

K. Lande, L.M. Lederman, W. Chinowsky
Phys. Rev. **105** (1957) 1925;

Excerpt The experiment previously reported (Lande et al., Phys. Rev. **103** (1956) 1901) which established the existence of a long-lived neutral V particle is being continued. In this report, we give evidence that

1. strengthens the previous surmise that these are indeed K-mass particles with decay modes primarily into $\pi e \nu$ and $\pi \mu \nu$,
2. establishes rather convincingly the existence of the π^+ π^- π^0 mode (Cooper et al., Nuovo cimento **4** (1956) 1433)
3. provides additional evidence for the particle mixture theory. (M. Gell-Mann and A. Pais, Phys. Rev. **97** (1955) 1387).

(Extracted from the introductory part of the paper.)

Accelerator BNL-COSMOTRON

Detectors Cloud chamber

Related references

More (earlier) information
K. Lande et al., Phys. Rev. **103** (1956) 1901;

See also
Cooper et al., Nuovo Cim. **4** (1955) 1433;
M. Gell-Mann and A. Pais, Phys. Rev. **97** (1955) 1387;
D. O. Caldwell and Y. Pal, Rev. Sci. Inst. **27** (1956) 633;
R. H. Dalitz, Phys. Rev. **99** (1955) 915;

Reactions
$K^0 \rightarrow \pi^+ \pi^0 \pi^-$

Particles studied K_L

LEE 1957

■ Charge conjugation invariance violation in weak interactions ■

Remarks on Possible Noninvariance under Time Reversal and Charge Conjugation

T.D. Lee, R. Oehme, C.N. Yang
Phys. Rev. **106** (1957) 340;

Abstract Interrelations between the nonconservation properties of parity, time reversal, and charge conjugation are discussed. The results are stated in two theorems. The experimental implications for the K-$\overline{K}$ complex are discussed in the last section.

Related references

See also
G. Luders, Kgl. Dansk. Videnskab. Selskab., Mat.-Fys. Medd., **28**, No. 5 (1954);
W. Pauli, *Niels Bohr and the Development of Physics*, Pergamon Press, London (1955);
J. Schwinger, Phys. Rev. **91** (1953) 720;
J. Schwinger, Phys. Rev. **91** (1953) 723;
V. F. Weisskopf and E. P. Wigner, Z. Phys. **63** (1930) 54;
V. F. Weisskopf and E. P. Wigner, Z. Phys. **65** (1930) 18;
M. Gell-Mann and A. Pais, Phys. Rev. **97** (1955) 1387;

Analyse information from
T. D. Lee and C. N. Yang, Phys. Rev. **104** (1956) 254;
K. Lande et al., Phys. Rev. **103** (1956) 1901;

FRAUENFELDER 1957

■ Further confirmation of spatial parity nonconservation in beta decays ■

Parity and the Polarization of Electrons from ^{60}Co

H. Frauenfelder et al.
Phys. Rev. **106** (1957) 386;

Reprinted in
R. N. Cahn and G. Goldhaber, *The Experimental Foundations of Particle Physics*, Cambridge Univ. Press (1991) 178.

Excerpt Lee and Yang (Phys. Rev. **104** (1956) 254) recently proposed that parity may not be conserved in weak interactions and suggested various experiments to verify their hypothesis. Two of the experiments have since been performed with positive result — the asymmetry of the electron emission from aligned nuclei and polarization of muons. In a second paper, Lee and Yang discuss a two-component theory of the neutrino and consider some more experimental tests. Among these, they list the measurement of momentum and polarization of electrons emitted in beta decay. If parity is not conserved, the electrons should be longitudinally polarized. For tensor and scalar interaction, the degree of polarization is simply equal to (v/c). We have found this polarization in the case of Co^{60}.

Accelerator Radioactive source

Detectors Counters

Related references

See also
T. D. Lee and C. N. Yang, Phys. Rev. **104** (1956) 254;
C. S. Wu et al., Phys. Rev. **105** (1957) 1413;
J. I. Friedman and V. L. Telegdi, Phys. Rev. **105** (1957) 1681;
R. L. Garwin, L. M. Lederman, and M. Weinrich, Phys. Rev. **105** (1957) 1415;
H. A. Tolhoek, Rev. of Mod. Phys. **28** (1956) 277;
N. Sherman, Phys. Rev. **103** (1956) 1601;

Reactions
$^{60}\mathrm{Co} \rightarrow e^- \bar{\nu}_e \mathrm{X}$

LOGUNOV 1957

■ Rigorous derivation of the dispersion relations for pion photoproduction amplitude ■

Dispersion Relations for Photoproduction of Mesons on Nucleons

A.A. Logunov, A.N. Tavkhelidze
Zh. Eksp. Teor. Fiz. **32** (1957) 1393;

Abstract Dispersion relations for π-meson photoproduction reactions on nucleons have been obtained. The spin and isotopic structure of the reaction amplitude has been established. The unobservable energy range has been separated. It is proved that the dispersion relations are inhomogeneous.

Related references

More (earlier) information
A. A. Logunov and B. M. Stepanov, Doklady Akad. Nauk SSSR **110** (1957) 368;
A. A. Logunov, B. M. Stepanov, and A. N. Tavkhelidze, Doklady Akad. Nauk SSSR **112** (1957) 45;
More (later) information
A. A. Logunov, A. N. Tavkhelidze, and L. D. Soloviev, Nucl. Phys. **4** (1957) 427;
See also
M. L. Goldberger, Phys. Rev. **99** (1955) 979;
Koba, Katani, and Nakai, Progr. of Theor. Phys. **6** (1951) 849;
G. F. Chew and F. E. Low, Phys. Rev. **101** (1956) 1579;
B. L. Ioffe, Zh. Eksp. Teor. Fiz. **31** (1956) 583;

PAIS 1957

■ Further proposals for CP invariance ■

Three-Pion Decay Modes of Neutral K Mesons

A. Pais, S.B. Treiman
Phys. Rev. **106** (1957) 1106;

Excerpt It is the purpose of this note to point out certain effects which obtain in neutral K-meson decay into pions if it is true that invariance under CP and T, rather than under C, P, T separately, is valid for all weak interactions (C = charge conjugation, P = space inversion, T = time reversal). The effects concern not so much the 2π- as the 3π-decay mode. *(Extracted from the introductory part of the paper.)*

Related references

See also
T. D. Lee and C. N. Yang, Phys. Rev. **105** (1957) 1671;
L. D. Landau, Nucl. Phys. **3** (1957) 127;
A. Salam, Nuovo Cim. **5** (1957) 299;
R. Gatto, Nuovo Cim. **3** (1956) 318;
R. Gatto, Phys. Rev. **106** (1957) 168;
H. W. Wyld and S. B. Treiman, Phys. Rev. **106** (1957) 169;
M. Gell-Mann and A. Pais, Phys. Rev. **97** (1955) 1387;
R. H. Dalitz, Phil. Mag. **44** (1953) 1068;
R. H. Dalitz, Phil. Mag. **94** (1954) 1046;
G. A. Snow, Phys. Rev. **103** (1956) 1111;
G. Wentzel, Phys. Rev. **101** (1956) 1215;
S. Oneda, Nucl. Phys. **3** (1957) 97;

NISHIJIMA 1957

■ Introduction of lepton-family-number conservation ■

Vanishing of the Neutrino Rest Mass

K. Nishijima
Phys. Rev. **108** (1957) 907;

Abstract The requirement that the mass of the neutrino be exactly zero may by satisfied by demanding invariance under many transformations besides the one proposed by Salam. One such, which involves a transformation on the muon (electron) field as well, leads to the selection rule: neutrinoless transitions involving an odd number of muons and an odd number of electrons are strictly forbidden. No such process has been observed. The theory of beta-decay is unaffected but, in muon decay, the predominant mode must be $\mu^+ \rightarrow e^+ \nu_R \nu_L$ in order to obtain the observed spectrum, so that the theory must include two kinds of two-component neutrinos. If leptons are conserved, there will be two kinds of "lepton charge", one carried by the muons and one group of neutrinos and the other by the electrons and the other type of neutrinos, and each will be conserved separately. *(Science Abstracts, 1958, 1827. P. K. Kabir.)*

Related references

See also
E. J. Konopinski and H. M. Mahmoud, Phys. Rev. **92** (1953) 1045;

Reactions
$\mu^- \rightarrow e^- \gamma$
$\mu^- \rightarrow 2e^- e^+$
$\mu^- p \rightarrow p e^-$
$K^+ \rightarrow \pi^+ \mu^+ e^-$
$K^+ \rightarrow \pi^+ \mu^- e^+$
$K^0 \rightarrow \mu^+ e^-$
$K^0 \rightarrow \mu^- e^+$

Particles studied μ^+, μ^-, e^+, e^-

MILEKHIN 1957

■ Evidence for limited transverse momenta in hadronic showers ■

Hydrodynamical Interpretation of a Characteristic of Large Showers Recorded in Photographic Emulsions

G.A. Milekhin, I.L. Rozental
Zh. Eksp. Teor. Fiz. **33** (1957) 187;

Abstract The experimental distributions in the transverse momentum components of secondary particles is compared with the predictions of the hydrodynamical theory of multiple particle production. It is found that the predictions of the one dimensional theory for a final temperature of $T_{fin} = mc^2/k$ (m is the π-meson mass) agrees satisfactorily with the experimental data. This permits one to draw some conclusions concerning the nature of π-π interaction.

Accelerator COSMIC

Detectors Nuclear emulsion

Related references

See also
I. L. Rozental and D. S. Chernavsky, Usp. Fiz. Nauk **52** (1954) 185;
L. D. Landau, Izv. Akad. Nauk SSSR, Fiz. **17** (1953) 51;
I. L. Rosental, Zh. Eksp. Teor. Fiz. **31** (1956) 278;
S. Z. Belenky, Doklady Akad. Nauk SSSR **99** (1954) 523;
Z. Koba, Progr. of Theor. Phys. **15** (1956) 461;
Analyse information from
A. de Benedetti et al., Nuovo Cim. **4** (1956) 1142;
I. M. Gramenitsky et al., Zh. Eksp. Teor. Fiz. **32** (1957) 936;

Reactions
hadron nucleus $\rightarrow$ shower X
shower $\rightarrow$ hadron X

LEE 1957C

■ Further development of the idea of the intermediate vector boson in weak interactions ■

Possible Nonlocal Effects in μ Decay

T.D. Lee, C.N. Yang
Phys. Rev. **108** (1957) 1611;

Abstract Possible nonlocal effects in μ decay are investigated phenomenologically. It is shown that the possible difference between the experimental ρ value from $\frac{3}{4}$ can be attributed to such nonlocal phenomena.

Related references

See also
T. D. Lee and C. N. Yang, Phys. Rev. **105** (1957) 1671;
A. Salam, Nuovo Cim. **5** (1957) 299;
R. L. Garwin, L. M. Lederman, and M. Weinrich, Phys. Rev. **105** (1957) 1415;
L. D. Landau, Nucl. Phys. **3** (1957) 127;
J. I. Friedman and V. L. Telegdi, Phys. Rev. **105** (1957) 1681;
H. Yukawa, Proc.Phys.Math.Soc.Jap. **17** (1935) 48;
Analyse information from
C. P. Sargent, M. C. Rinehart, L. M. Lederman, and K. C. Rogers, Phys. Rev. **99** (1955) 885;
K. Crowe, Bull.Am.Phys.Soc. **2** (1957) 206;
L. Michel, Proc. Roy. Soc. **63** (1950) 514;

GELL-MANN 1957

■ First review of particle properties data ■

Hyperons and Heavy Mesons (Systematics and Decay)

M. Gell-Mann, A.H. Rosenfeld
Ann. Rev. Nucl. Sci. **7** (1957) 407;

Excerpt We attempt, in this article, to summarize the information now available, both experimental and theoretical, on the classification and decays of hyperons and heavy mesons. Our principal emphasis is on the "weak interactions" responsible for the slow decays of these particles. The "strong interactions" involved in production and scattering phenomena form a separate topic, which we do not discuss at length. We do, however, mention the hyperfragments, the study of which bears on both kinds of interactions.

Related references

See also
R. H. Dalitz, *Reports in Progress in Physics*, To be published by the Physical Society, London, (1957);
M. Gell-Mann and E. P. Rozenbaum, Sci. American **197** (1957) 72;
L. B. Okun', Usp. Fiz. Nauk **61** (1957) 535;
J. Schwinger, Phys. Rev. **82** (1951) 914;
G. Luders, Kgl. Dansk. Videnskab. Selskab., Mat.-Fys. Medd., **28**, No. 5 (1954);
G. Luders and B. Zumino, Phys. Rev. **106** (1957) 385;
M. Gell-Mann, Nuovo Cim. Suppl. **4** (1956) 848;
M. Gell-Mann, Phys. Rev. **92** (1953) 833;
T. D. Lee and C. N. Yang, Phys. Rev. **104** (1956) 254;
H. Postma et al., Physica **23** (1957) 259;
C. S. Wu et al., Phys. Rev. **105** (1957) 1413;
L. D. Landau, Nucl. Phys. **3** (1957) 127;
F. Coester, Phys. Rev. **89** (1953) 619;
E. Fermi, Nuovo Cim. Suppl. **2** (1955) 54;
M. Gell-Mann and K. M. Watson, Ann. Rev. Nucl. Sci. **5** (1954) 219;
H. A. Bethe and F. de Hoffman, *Mesons and Fields*, II, Row, Peterson and Co., Evanston, Ill., Sect. 3.1 (1955) 434;
T. Nakano and K. Nishijima, Progr. of Theor. Phys. **10** (1953) 581;

YENNIE 1957

■ Description of the electromagnetic structure of the nucleon by form factors ■

Electromagnetic Structure of Nucleons

D.R. Yennie, M.M. Levy, D.G. Ravenhall
Rev. of Mod. Phys. **29** (1957) 144;

Abstract The theoretical implications of various experiments relating to the electromagnetic structure of nucleons are examined in the light of current field theory. It is concluded either that the nucleon core is about three times as large as would be expected from intuitive considerations of meson theory, or that there is some inconsistency in the present field theory.

Related references

See also
R. P. Feynman, Phys. Rev. **74** (1948) 939;

BARKAS 1958

■ First collection of particle physics data in a compact and readily accessible form ■

Data for Elementary-Particle Physics

W.H. Barkas, A.H. Rosenfeld
UCRL-8030 (1958);

Abstract Elementary-particle data and certain other reference information frequently are needed by research workers in high-energy physics in a compact and readily accessible form. For the use of students and staff members in the Radiation Laboratory we have attempted to meet this need. In this summary we have tried to employ units and concepts natural to this field, and to drop those that are irrelevant or obsolete, Slightly older versions of Tables I and V have already appeared in Gell-Mann's and Rosenfeld's review of elementary particles.

(M. Gell-Mann and A. H. Rosenfeld, Ann. Rev. Nuclear Sci. **7** (1957) 407).

Related references

See also
W. H. Barkas, Barnett, Cüer et al., *The Range-Energy Relation in Emulsion.* Part 1. Range Measurements, UCRL-3768 (April 1957);
W. H. Barkas, *The Range-Energy Relation in Emulsion.* Part 2. The Theoretical Range, UCRL-3769 (April 1957);
M. Gell-Mann and A. H. Rosenfeld, Ann. Rev. Nucl. Sci. **7** (1957) 407;
Cohen, K. Crowe, and DuMond, *Fundamental Constants of Physics*, Interscience, New York (1957);
Henri, A. Shapiro, and Way, to be published in Revs. Modern Phys.;
H. A. Bethe and J. Ashkin, Part II of *Experimental Nuclear Physics*, E. Segrè, Ed. Wiley, New York, (1953);
J. A. Wheeler and W. E. Lamb, Phys. Rev. **55** (1939) 858;
J. H. Davies, H. A. Bethe, and Maximon, Phys. Rev. **93** (1954) 788;

BOLDT 1958B

■ First measurement of the $K_S - K_L$ mass difference ■

Θ_1^0-Θ_2^0 Mass Difference

E. Boldt, D.O. Caldwell, Y. Pal
Phys. Rev. Lett. **1** (1958) 150;

Excerpt We have observed directly the change with time of the nature of the neutral particle produced in association with the Λ hyperon. The "particle," initially the Θ^0, is observed to interact with matter to produce another hyperon only when at some distance from its origin, and never when close to that origin. This observation is a direct confirmation of the particle-mixture theory for the Θ^0 meson of Gell-Mann and Pais, and when quantitatively interpreted in terms of that theory, yields an estimate of the mass difference between the Θ_1^0 (short-lived) and Θ_2^0 (long-lived) components of the Θ^0.

Accelerator BNL-COSMOTRON

Detectors Cloud chamber

Related references

See also
M. Gell-Mann and A. Pais, Phys. Rev. **97** (1955) 1387;
W. F. Fry and R. G. Sachs, Phys. Rev. **109** (1958) 2212;

Reactions
π^- Fe $\rightarrow \Lambda K^0$ X   1.5 GeV(E_{lab})
$K^0 \rightarrow \overline{K}^0$
$\overline{K}^0$ nucleus $\rightarrow$ hyperon X

Particles studied K_L, K_S

PROWSE 1958

■ First evidence for the $\overline{\Lambda}$ ■

Anti-Lambda Hyperon

D.J. Prowse, M. Baldo-Ceolin
Phys. Rev. Lett. **1** (1958) 179;

Excerpt An event has recently been found in an emulsion stack exposed to a 4.6 ± 0.3 BeV π^--meson beam at the Berkeley Bevatron which is interpreted as the decay of an anti-lambda hyperon, $\overline{\Lambda}^0$. The threshold for production with a free nucleon is 4.73 BeV and extends down to ~ 4.3 BeV with a bound nucleon, the production reaction being of the type

$$\pi^- \; p \rightarrow \Lambda^0 \, \overline{\Lambda}^0 \, n$$

The side of the stack was exposed to a beam of 10^6 π mesons per sq cm, the intensity falling off by a factor of 10^3 in a distance of 8 in. across the emulsion sheets. The emulsions were scanned for stopping π^+ mesons and these were followed back to their origins if it appeared that they were produced in the region of high π^--meson flux. Most tracks which were successfully traced back came from primary π^--meson stars but 3 traced back to decaying K^+ mesons and one to a "V" type event. The other branch of the "V" event was traced after 3 cm range into a large star from which 3 shower particles were emitted. The visible energy in this star is 783 MeV including the rest masses of the 3 shower particles which were all identified as π mesons; one was arrested and decayed into a μ meson, the other two were identified by the Δg^* *vs* ΔR and by the $\overline{\alpha}$ *vs* g^* methods. *(Extracted from the introductory part of the paper.)*

Accelerator LBL-Bevatron

Detectors Nuclear emulsion

Data comments One event.

Related references

More (earlier) information
M. Baldo-Ceolin and D. J. Prowse, Bull.Am.Phys.Soc. **3** (1958) 163;

See also
G. Alexander and R. H. W. Johnston, Nuovo Cim. **5** (1957) 263;
W. H. Barkas et al., Nuovo Cim. **8** (1958) 186;

Reactions

π^- nucleus $\to \overline{p}\,\pi^+$ X	4.6 GeV(T_{lab})
$\pi^-\,p \to n\,\Lambda\,\overline{\Lambda}$	4.6 GeV(T_{lab})
$\overline{\Lambda} \to \overline{p}\,\pi^+$	

Particles studied $\overline{\Lambda}$

FEYNMAN 1958

■ CVC and symmetry between electromagnetism and weak interaction. $\Delta S = \Delta Q$ for nonleptonic decays of the strange particles ■

Theory of the Fermi Interaction

R.P. Feynman, M. Gell-Mann
Phys. Rev. **109** (1958) 193;

Reprinted in
The Physical Review — the First Hundred Years, AIP Press (1995) CD-ROM.

Abstract The representation of Fermi particles by two-component Pauli spinors satisfying a second order differential equation and the suggestion that in β decay these spinors act without gradient couplings leads to an essentially unique weak four-fermion coupling. It is equivalent to equal amounts of vector and axial vector coupling with two-component neutrinos and conservation of leptons. (The relative sign is not determined theoretically.) It is taken to be "universal"; the lifetime of the μ agrees to within the experimental errors of 2%. The vector part of the coupling is, by analogy with electric charge, assumed to be not renormalized by virtual mesons. This requires, for example, that pions are also "charged" in the sense that there is a direct interaction in which, say, a π^0 goes to π^- and an electron goes to a neutrino. The weak decays of strange particles will result qualitatively if the universality is extended to include a coupling involving a Λ or Σ fermion. Parity is then not conserved even for those decays like $K \to 2\pi$ or 3π which involve no neutrinos. The theory is at variance with the measured angular correlation of electron and neutrino in He^6, and with the fact that fewer than 10^{-4} pion decay into electron and neutrino.

Related references

See also
A. Salam, Nuovo Cim. **5** (1957) 299;
L. D. Landau, Nucl. Phys. **3** (1957) 127;
T. D. Lee and C. N. Yang, Phys. Rev. **105** (1957) 1671;
R. P. Feynman, Rev. of Mod. Phys. **20** (1948) 367;
R. P. Feynman, *Proc. of the Seventh Ann. Rochester Conf. on High Energy Nuclear Physics*, Interscience Publishers, Inc., New York (1957);
Bromley, Almquvist, Gove et al., Phys. Rev. **105** (1957) 957;
B. M. Rustad and S. L. Ruby, Phys. Rev. **97** (1955) 991;
M. T. Burgy, Epstein, Krohn et al., Phys. Rev. **107** (1957) 1731;
R. E. Behrends and C. Fronsdal, Phys. Rev. **106** (1957) 345;
W. B. Herrmansfeldt, D. R. Maxson, P. Stahelin, and J. S. Allen, Phys. Rev. **107** (1957) 641;

FAZZINI 1958

■ First evidence for π_{e2} decay ■

Electron Decay of the Pion

T. Fazzini et al.
Phys. Rev. Lett. **1** (1958) 247;

Excerpt It was predicted some years ago by Ruderman and Finkelstein that if the decay of the pion into an electron and a neutrino goes through an axial vector interaction, then it should occur at a rate 1.3×10^{-4} of the normal decay into a muon and a neutrino. This conclusion has not been changed in the light of recent work on the nonconservation of parity in weak interactions. However, experiments by Lokanathan and Steinberger and by Anderson and Lattes failed to show the existence of the electron mode of decay. Interest has been revived in a search for this decay by the evidence for the validity of a universal Fermi interaction, which, with the single exception of the π e decay, is good. Theoretical attempts to remove this discrepancy have been made by a number of authors.
The experiment is made difficult by the presence of a large background of electrons from the π μ e chain of decay. However, these electrons can be distinguished in three ways. Firstly, they have a continuous spectrum with a maximum kinetic

energy of 52.3 MeV, compared with the line spectrum of electrons from π e decay of 69.3 MeV. Secondly, the π e electrons should show a simple exponential decay with the mean life identical to that of the π μ decay, while the π μ e electron time distribution would show a two-stage radioactive decay, with a fast rise (approx. π μ mean life) and a slow fall (approx. μ e mean life). Lastly, three charged particles can be seen in the π μ e chain, while only two are shown in the π e decay.
We have used, like Lokanathan and Steinberger, a range telescope to search for the high energy π e electron. *(Extracted from the introductory part of the paper.)*

Accelerator CERN-SC synchrocyclotron

Detectors Counters

Related references

See also
M. A. Ruderman and R. J. Finkelstein, Phys. Rev. **76** (1949) 1458;
R. P. Feynman and M. Gell-Mann, Phys. Rev. **109** (1958) 193;
Analyse information from
S. Lokanathan and J. Steinberger, Nuovo Cim. Suppl. **2** (1955) 151;
H. L. Anderson and C. M. G. Lattes, Nuovo Cim. **6** (1957) 1356;

Reactions
$\pi^+ \rightarrow e^+ \nu_e$
$\pi^- \rightarrow e^- \overline{\nu}_e$

Particles studied π^+, π^-

IMPEDUGLIA 1958

■ Confirmation of π_{e2} decay ■

β Decay of the Pion

J. Impeduglia et al.
Phys. Rev. Lett. **1** (1958) 249;

Excerpt The electron-neutrino decay mode of the pion has been the object of several unsuccessful searches. The latest and most sensitive of these puts an upper limit of about 10^{-5} for the relative frequency of this process.
On the other hand, there has recently been spectacular progress in the understanding of nuclear β decay and of μ decay in terms of a "universal" $V-A$ theory. It has been emphasized, especially by Feynman, that in a universal $V-A$ theory the ratio of pion β decay to μ decay should be 1/8000, and that deviations from this value are difficult and awkward to accommodate theoretically.
There is, therefore, considerable interest in the announcement of the CERN group that positive evidence for the π e decay in the theoretically expected order of magnitude has been found. In view of the disagreement between these results and those of Anderson and Lattes, there may be some interest in the progress of our experiment on this decay. *(Extracted from the introductory part of the paper.)*

Accelerator Nevis labs cyclotron

Detectors Hydrogen bubble chamber

Related references

See also
R. P. Feynman and M. Gell-Mann, Phys. Rev. **109** (1958) 193;
Analyse information from
H. L. Friedman and J. Rainwater, Phys. Rev. **84** (1951) 684;
S. Lokanathan and J. Steinberger, Nuovo Cim. Suppl. **2** (1955) 151;
H. L. Anderson and C. M. G. Lattes, Nuovo Cim. **6** (1957) 1356;
M. Goldhaber, L. Grodzins, and A. W. Sunyar, Phys. Rev. **109** (1958) 1015;
E. C. G. Sudarshan and R. E. Marshak, Phys. Rev. **109** (1958) 1860;
T. Fazzini et al., Phys. Rev. Lett. **1** (1958) 247;

Reactions
$\pi^+ \rightarrow e^+ \nu_e$

Particles studied π^+

GOEBEL 1958

■ Proposed method for extraction of π-π interactions ■

Determination of the π-π Interaction Strength from π-nucleon Scattering

C. Goebel
Phys. Rev. Lett. **1** (1958) 337;

Abstract The class of Feynman diagrams contributing to $\pi N \rightarrow N\pi\pi$ which have a single pion exchange, exhibit a pole at the (unphysical) value of the momentum transfer $\Delta^2 = -\mu^2$. It is shown how the residue at this pole determines the π π scattering cross section. Assuming other terms do not interfere too badly, this residue should be detectable at high energy scattering with small nucleon recoil. *(Science Abstracts, 1959, 546. J. C. Taylor)*

Related references

See also
G. F. Chew, preprint UCRL-2883;
W. D. Walker, Phys. Rev. **108** (1957) 872;
W. D. Walker et al., Bull.Am.Phys.Soc. **3** (1958) 104;
G. Maenchen et al., Phys. Rev. **108** (1957) 850;
S. D. Drell and F. Zachariazen, Phys. Rev. **104** (1956) 236;
R. M. Sternheimer, Phys. Rev. **101** (1956) 384;
G. F. Chew, M. L. Goldberger, F. E. Low, and Y. Nambu, Phys. Rev. **106** (1957) 1345;
F. J. Dyson, Phys. Rev. **99** (1955) 1037;
G. Takeda, Phys. Rev. **100** (1955) 440;
W. D. Walker, Hushfar, and W. D. Shephard, Phys. Rev. **104** (1956) 526;
E. L. Feinberg and I. Y. Pomeranchuk, Nuovo Cim. Suppl. **4** (1956) 652;

Reactions
π nucleon $\rightarrow$ nucleon 2π

CRAWFORD 1958

■ First evidence of Λ beta decay ■

Beta Decay of the Λ

F.S. Crawford et al.
Phys. Rev. Lett. **1** (1958) 377;

Excerpt In the course of studying a large number of decays of hyperons, produced in our hydrogen bubble chamber by 1.23 BeV/c pions, we have found an unambiguous case of a Λ undergoing beta decay. To our knowledge this is the first leptonic Λ decay seen. *(Extracted from the introductory part of the paper.)*

Accelerator LBL-Bevatron

Detectors Hydrogen bubble chamber

Related references

Analyse information from
J. Hornbostell and E. O. Salant, Phys. Rev. **102** (1956) 502;
R. P. Feynman and M. Gell-Mann, Phys. Rev. **109** (1958) 193;

Reactions
$\pi^- \, p \to \Sigma^0 \, K^0$ 1.23 GeV(P_{lab})
$\Sigma^0 \to \Lambda \, \gamma$
$\Lambda \to p \, e^- \, \overline{\nu}_e$

Particles studied Λ

NORDIN 1958

■ Confirmation of Λ beta decay ■

Leptonic Decays of Hyperons

P. Nordin et al.
Phys. Rev. Lett. **1** (1958) 380;

Excerpt A few days after learning of the event reported in the preceding Letter one of us noticed another Λ beta decay. This second event was found in the film taken in the course of an experiment in which an electrostatically separated K^- beam was passed into the Berkeley 15-inch hydrogen bubble chamber. *(Extracted from the introductory part of the paper.)*

Accelerator LBL-Bevatron

Detectors Hydrogen bubble chamber

Related references

Analyse information from
J. Hornbostell and E. O. Salant, Phys. Rev. **102** (1956) 502;
F. S. Crawford et al., Phys. Rev. Lett. **1** (1958) 377;
R. P. Feynman and M. Gell-Mann, Phys. Rev. **109** (1958) 193;

Reactions
$K^- \, p \to \Lambda \, \pi^0$
$\Lambda \to p \, e^- \, \overline{\nu}_e$

Particles studied Λ

PONTECORVO 1958

■ Proposal for the possibility of neutrino-antineutrino oscillations ■

Inverse β-Processes and Lepton Charge Nonconservation

B. Pontecorvo
Zh. Eksp. Teor. Fiz. **34** (1958) 247;

Excerpt Recently the question was discussed whether there exist other "mixed" neutral particles beside the K^0 mesons, i.e., particles that differ from the corresponding antiparticles, with the transitions between particle and antiparticle states not being strictly forbidden. It was noted that the neutrino might be such a mixed particle, and consequently there exists the possibility of real neutrino $\rightleftharpoons$ antineutrino transitions in vacuum, provided that lepton (neutrino) charge is not conserved. In the present note we make a more detailed study of this possibility, in which interest has been renewed owing to recent experiments dealing with inverse beta processes. *(Extracted from the introductory part of the paper.)*

Related references

More (earlier) information
B. Pontecorvo, Zh. Eksp. Teor. Fiz. **33** (1957) 549;
See also
M. Gell-Mann and A. Pais, Phys. Rev. **97** (1955) 1387;
A. Pais and O. Piccioni, Phys. Rev. **100** (1955) 1487;
Y. B. Zeldovich, Doklady Akad. Nauk SSSR **86** (1952) 505;
L. D. Landau, Zh. Eksp. Teor. Fiz. **32** (1957) 405;
C. L. Cowan, F. Reines et al., Science **124** (1956) 103;
M. Awshalom, Phys. Rev. **101** (1956) 1041;
E. I. Dobrokhotov, B. P. Lazarenko, and S. Y. Lukyanov, Doklady Akad. Nauk SSSR **110** (1956) 966;
L. B. Okun' and B. Pontecorvo, Zh. Eksp. Teor. Fiz. **32** (1957) 1587;
G. C. Hanna and B. Pontecorvo, Phys. Rev. **75** (1949) 983;
S. C. Curran, J. Angus, and A. L. Cockcroft, Phys. Rev. **76** (1949) 853;
L. M. Langer and B. J. D. Moffat, Phys. Rev. **88** (1952) 689;
R. C. Cornelius et al., Phys. Rev. **92** (1953) 1521;

Reactions
$\nu_e \to \overline{\nu}_e$
$\overline{\nu}_e \to \nu_e$

GOLDHABER 1958C

■ First evidence for the negative ν_e helicity (ν_e is left handed) ■

Helicity of Neutrinos

M. Goldhaber, L. Grodzins, A.W. Sunyar
Phys. Rev. **109** (1958) 1015;

Reprinted in
R. N. Cahn and G. Goldhaber, *The Experimental Foundations of Particle Physics*, Cambridge Univ. Press (1991) 180.
The Physical Review — the First Hundred Years, AIP Press (1995) CD-ROM.

Excerpt Combined analysis of circular polarization and resonant scattering of γ rays following orbital electron capture measures the helicity of the neutrino. We have carried out such a measurement with En^{152m}, which decays by orbital electron capture. If we assume the most plausible spin-parity assignment for this isomer compatible with its decay scheme, 0−, we find that the neutrino is "left-handed," i.e., $\sigma_\nu \cdot \hat{p}_\nu = -1$ (negative helicity). *(Extracted from the introductory part of the paper.)*

Accelerator Radioactive source

Detectors Counters

Related references

More (earlier) information
M. Goldhaber, L. Grodzins, and A. W. Sunyar, Phys. Rev. **106** (1957) 826;
M. Goldhaber, L. Grodzins, and A. W. Sunyar, Phys. Rev. **109** (1958) 1015;

See also
T. D. Lee and C. N. Yang, Phys. Rev. **105** (1957) 1671;
E. C. G. Sudarshan and R. E. Marshak, Phys. Rev. **109** (1958) 1860;
R. P. Feynman and M. Gell-Mann, Phys. Rev. **109** (1958) 193;

Analyse information from
L. A. Page and M. Heinberg, Phys. Rev. **106** (1957) 1220;
E. Ambler et al., Phys. Rev. **106** (1957) 1361;
H. Postma et al., Physica **23** (1957) 259;
W. B. Herrmansfeldt, D. R. Maxson, P. Stahelin, and J. S. Allen, Phys. Rev. **107** (1957) 641;
B. M. Rustad and S. L. Ruby, Phys. Rev. **97** (1955) 991;
G. Culligan et al., Nature **180** (1957) 751;
M. T. Burgy, Epstein, Krohn et al., Phys. Rev. **107** (1957) 1731;

Reactions
$e^- \, {}^{152}\text{Eu} \to {}^{152}\text{Sm}^* \, \nu_e$
${}^{152}\text{Sm}^* \to {}^{152}\text{Sm} \, \gamma$

Particles studied ν

OKUN 1958

■ Prediction of the existence and some properties of the η and η' particles on the basis of Sakata model. Invention of the $\Delta S = \Delta Q$ selection rule ■

Some Remarks on the Compound Model of Elementary Particles

L.B. Okun
Zh. Eksp. Teor. Fiz. **34** (1958) 469;

Abstract Some properties are considered of the compound model of elementary particles in which Λ-hyperons are assumed to be the primary particles.

Related references

See also
E. Fermi and C. N. Yang, Phys. Rev. **76** (1949) 1739;
M. Goldhaber, Phys. Rev. **101** (1956) 433;
G. Dierdi, Zh. Eksp. Teor. Fiz. **32** (1957) 152;
Y. B. Zeldovich, Zh. Eksp. Teor. Fiz. **33** (1957) 829;
B. Neganov, Zh. Eksp. Teor. Fiz. **33** (1957) 260;
L. B. Okun, Zh. Eksp. Teor. Fiz. **32** (1957) 400;
M. A. Markov, *O sistematike elementarnykh chastits*, Izd. AN SSSR, (1955);
S. Sakata, Progr. of Theor. Phys. **16** (1956) 686;
R. W. King, D. C. Peaslee, Phys. Rev. **106** (1957) 360;
M. Gell-Mann, Phys. Rev. **106** (1957) 1296;
M. Gell-Mann, Nuovo Cim. **5** (1957) 758;
L. B. Okun and B. Pontecorvo, Zh. Eksp. Teor. Fiz. **32** (1957) 1587;

POMERANCHUK 1958

■ Theorem on asymptotic equality of hadron-hadron and antihadron-hadron interaction cross sections ■

Equality between the Interaction Cross Sections of High Energy Nucleons and Antinucleons

I.Y. Pomeranchuk
Zh. Eksp. Teor. Fiz. **34** (1958) 725;

Abstract On basis of the dispersion relations the equality between the interaction cross sections of high energy nucleons and antinucleons is demonstrated.

Related references

See also
V. Y. Fainberg and E. S. Fradkin, Doklady Akad. Nauk SSSR **109** (1956) 507;
B. L. Ioffe, Zh. Eksp. Teor. Fiz. **31** (1956) 583;
I. Y. Pomeranchuk, Zh. Eksp. Teor. Fiz. **30** (1956) 423;
L. B. Okun' and I. Y. Pomeranchuk, Zh. Eksp. Teor. Fiz. **30** (1956) 424;

Analyse information from
O. Chamberlain et al., Phys. Rev. **102** (1956) 1637;
B. Cork et al., Phys. Rev. **107** (1957) 248;

Reactions
$p\, p \to X$ >10 GeV(P_{lab})

$\pi^- p \to X$	>10 GeV(P_{lab})
$\pi^+ p \to X$	>10 GeV(P_{lab})
$\overline{p} p \to X$	>10 GeV(P_{lab})
$n p \to X$	>10 GeV(P_{lab})
$\overline{n} p \to X$	>10 GeV(P_{lab})
$K^+ p \to X$	>10 GeV(P_{lab})
$K^- p \to X$	>10 GeV(P_{lab})
$\Lambda p \to X$	>10 GeV(P_{lab})
$\overline{\Lambda} p \to X$	>10 GeV(P_{lab})
$\Sigma p \to X$	>10 GeV(P_{lab})
$\overline{\Sigma} p \to X$	>10 GeV(P_{lab})
$\Xi p \to X$	>10 GeV(P_{lab})
$\overline{\Xi} p \to X$	>10 GeV(P_{lab})

AGNEW 1958

■ Confirmation of the existence of the antineutron ■

$\overline{p}$-p Elastic and Charge Exchange Scattering at About 120 MeV

L. Agnew et al.
Phys. Rev. **110** (1958) 994;

Excerpt We have observed 478 antiprotons entering the chamber. One such event is shown. The complete analysis of these events is in progress; however, a preliminary report on the $\overline{p}$-p elastic scattering is now available. A total of 33 $\overline{p}$-p elastic collisions have been observed over a path length of 179 meters. The scatterings occur between 30 and 215 MeV with an average energy of 120 MeV. Only $\overline{p}$-p scatterings with angles $> 7.5°$ (lab) were identifiable because otherwise the energy of the recoil proton was too small to make a visible track, so that these scatterings are indistinguishable from $\overline{p}$-C. *(Extracted from the introductory part of the paper.)*

Accelerator LBL-Bevatron

Detectors Heavy-liquid bubble chamber

Related references

See also
J. R. Fulco, Phys. Rev. **110** (1958) 784;
J. S. Ball and G. F. Chew, Phys. Rev. **109** (1958) 1385;

Reactions

$\overline{p} p \to p \overline{p}$	.12 GeV(T_{lab})
$\overline{p} p \to n \overline{n}$	.12 GeV(T_{lab})
$\overline{n} p \to 3\pi^+ 2\pi^-$	

GOLDBERGER 1958

■ Goldberger-Treiman relations ■

Decay of the π Meson

M.L. Goldberger, S.B. Treiman
Phys. Rev. **110** (1958) 1178;

Reprinted in
The Physical Review — the First Hundred Years, AIP Press (1995) CD-ROM.

Abstract A quantitative study of $\pi \to \mu\ \nu$ decay is presented using the techniques of dispersion theory. The discussion is based on a model in which the decay occurs through pion disintegration into a nucleon-antinucleon pair, the latter annihilating via a Fermi interaction to produce the leptons. The weak vertex contains effectively both axial vector and pseudoscalar couplings even if one adopts the point of view of a universal axial vector and vector Fermi interaction. The pion-nucleon vertex which enters our model is also calculated using dispersion techniques. Under the assumption that this vertex is damped for large momentum transfers, we obtain a result for the pion lifetime largely independent of the detailed properties of the vertex and one which is in very close agreement with experiment. The precise prediction of our theory depends on the energy dependence of the complex phase shift for nucleon-antinucleon scattering in the $^1S_0$ isotopic triplet state.

Related references

See also
M. A. Ruderman and R. J. Finkelstein, Phys. Rev. **76** (1949) 1458;
S. B. Treiman and H. W. Wyld, Phys. Rev. **101** (1956) 1552;
H. Lehmann, K. Symanzik, and W. Zimmermann, Nuovo Cim. **2** (1955) 425;

Analyse information from
H. L. Anderson and C. M. G. Lattes, Nuovo Cim. **6** (1957) 1356;
Cassels, Rigby, A. M. Wetherell, and Wormald, Proc. Roy. Soc. **70** (1957) 729;

SAKURAI 1958

■ Universal $V - A$ weak interactions ■

Mass Reversal and Weak Interactions

J.J. Sakurai
Nuovo Cim. **7** (1958) 649;

Excerpt An attempt is made to construct a theory of all weak processes from a single invariance principle. The requirement that the weak interaction Hamiltonian be invariant under the reversal of the sign of the mass in the Dirac equation separately for *any* fermion field leads to a unique form of the four-fermion interaction responsible for weak processes. The

Hamiltonian contains equal amounts of V and A, and is necessarily invariant under time reversal but non-invariant under parity regardless of whether or not the neutrino field is involved. (Our results are identical to those obtained by Feynman and Gell-Mann, and by Sudarshan and Marshak using somewhat different approaches.) The possible existence of an intermediate *charged* boson field responsible for all weak processes is discussed with reference to several weak processes which have not been observed. Most experiments in P decay and the π μ e sequence are consistent with the predictions of our theory with the important exception of the ^{6}He recoil experiment. Various implications of our theory which can be experimentally tested are examined with special attention to the decay interactions of hyperons and K-mesons.

Related references

See also
A. Salam, Nuovo Cim. **5** (1957) 299;
L. D. Landau, Nucl. Phys. **3** (1957) 127;
T. D. Lee and C. N. Yang, Phys. Rev. **104** (1956) 254;
T. D. Lee and C. N. Yang, Phys. Rev. **106** (1957) 1671;
S. B. Treiman and H. W. Wyld, Phys. Rev. **106** (1957) 1320;
F. Eisler et al., Phys. Rev. **107** (1957) 324;
J. A. McLennan Jr., Phys. Rev. **106** (1957) 821;
K. M. Case, Phys. Rev. **107** (1957) 307;
J. Tiomno, Nuovo Cim. **1** (1955) 226;
S. Hori and A. Wakasa, Nuovo Cim. **6** (1957) 304;
M. Fierz, Z. Phys. **104** (1937) 553;
S. Ogawa, Progr. of Theor. Phys. **15** (1956) 487;
R. P. Feynman, Phys. Rev. **76** (1949) 749;
H. Frauenfelder et al., Phys. Rev. **106** (1957) 386;
M. Deutsch et al., Phys. Rev. **107** (1957) 1733;
C. S. Wu et al., Phys. Rev. **105** (1957) 1413;
F. Boehm and A. H. Wapstra, Phys. Rev. **107** (1957) 1462;
M. T. Burgy et al., Phys. Rev. **107** (1957) 1731;
J. M. Robson, Phys. Rev. **100** (1955) 933;
D. R. Maxson, J. S. Allen, and W. K. Jentschke, Phys. Rev. **97** (1955) 109;
M. L. Good and E. J. Lauer, Phys. Rev. **105** (1957) 213;
W. B. Herrmansfeldt, D. R. Maxson, P. Stahelin, and J. S. Allen, Phys. Rev. **107** (1957) 641;
B. M. Rustad and S. L. Ruby, Phys. Rev. **97** (1955) 991;
S. Oneda, S. Hori, and A. Wakasa, Progr. of Theor. Phys. **15** (1956) 302;
J. J. Sakurai, Phys. Rev. **108** (1957) 491;
S. Lokanathan and J. Steinberger, Nuovo Cim. Suppl. **2** (1955) 151;
A. Pais and S. B. Treiman, Phys. Rev. **105** (1957) 1616;
S. Furuichi et al., Progr. of Theor. Phys. **4** (1949) 171;

EISLER 1958

■ Failure of universal Fermi interactions in the beta decay of hyperons ■

Leptonic Decay Modes of the Hyperons

F. Eisler et al.
Phys. Rev. **112** (1958) 979;

Abstract We have searched for the leptonic decay of the Λ and Σ^{-}. The sensitivity of the experiment was such that 5-6 events should have been found according to the predictions of the "universal" $V - A$ model of β decay. No examples of leptonic decay were observed.

Related references

See also
R. P. Feynman and M. Gell-Mann, Phys. Rev. **109** (1958) 193;
R. E. Marshak and E. C. G. Sudarshan, *Venice Conference on Elementary Particles*, September, 1957 (unpublished);
Bromley, Almqvist, Gove et al., Phys. Rev. **105** (1957) 957;
Analyse information from
F. S. Crawford et al., Phys. Rev. **108** (1957) 1102;
F. Eisler et al., Phys. Rev. **108** (1957) 1353;

MANDELSTAM 1958

■ Dispersion relation in two variables: the Mandelstam representation ■

Determination of the Pion-Nucleon Scattering Amplitude from Dispersion Relations and Unitarity. General Theory

S. Mandelstam
Phys. Rev. **112** (1958) 1344;

Abstract A method is proposed for using relativistic dispersion relations, together with unitarity, to determine the pion-nucleon scattering amplitude. The usual dispersion relations by themselves are not sufficient, and we have to assume a representation which exhibits the analytic properties of the scattering amplitude as a function of the energy and the momentum transfer. Unitarity conditions for the two reactions $\pi N \to \pi N$ and $N\overline{N} \to 2\pi$ will be required, and they will be approximated by neglecting states with more than two particles. The method makes use of an iteration procedure analogous to that used by Chew and Low for the corresponding problem in the static theory. One has to introduce two coupling constants; the pion-pion coupling constant can be found by fitting the sum of the threshold scattering lengths with experiment. It is hoped that this method avoids some of the formal difficulties of the Tamm-Dancoff and Bethe-Salpeter methods and, in particular, the existence of ghost states. The assumptions introduced are justified in perturbation theory.
As an incidental result, we find the precise limits of the region for which the absorptive part of the scattering amplitude is an analytic function of the momentum transfer, and hence the boundaries of the region in which the partial-wave expansion is valid.

Related references

See also
G. F. Chew, M. L. Goldberger, F. E. Low, and Y. Nambu, Phys. Rev. **106** (1957) 1345;
G. F. Chew, R. Karplus, S. Gasiorowicz, and F. Zachariazen, Phys. Rev. **110** (1958) 265;
M. Gell-Mann, *Proc. of the Sixth Ann. Rochester Conf. High Energy Physics*, Interscience Publishers, Inc., New York, Sec. III, (1956) 30;
G. F. Chew and F. E. Low, Phys. Rev. **101** (1956) 1570;

G. Salzman and F. Salzman, Phys. Rev. **108** (1957) 1619;
L. Castillejo, R. H. Dalitz, and F. J. Dyson, Phys. Rev. **101** (1956) 453;
K. W. Ford, Phys. Rev. **105** (1957) 320;

Reactions

p nucleus $\rightarrow$ jet X	1-100 TeV(P_{lab})
He nucleus $\rightarrow$ jet X	1-100 TeV(P_{lab})

EDWARDS 1958

■ Evidence for limited transverse momenta in hadronic jets ■

Analysis of Nuclear Interactions of Energies between 1000 and 100000 BeV

B. Edwards et al.
Phil. Mag. **(7)3** (1958) 237;

Abstract Twenty nuclear interactions produced by protons and ten produced by α-particles of energy above 1000 BeV have been analysed.
The proportion of pions among the secondaries is found to be 80% for the core and less than 70% for the wide angle tracks. The average transverse momentum resulting from our measurements is $P_T = 0.5$ BeV/c for pions and $P_T = 1-2$ BeV/c for heavy particles. It appears to be independent of angle of emission and primary energy. The multiplicity of the interaction, n_s shows no variation with primary energy: it varies, however, with the anisotropy of the angular distribution, and shows wide fluctuations at a fixed primary energy. The inelasticity, K, of the collisions is close to unity for secondary interactions of mean energy $\sim$ 100 BeV and about 0.2 for jets produced by protons of energy $\sim$ 100000 BeV. At a fixed energy, K does not appear to be strongly dependent on n_S. None of the current theories appears capable of giving a satisfactory explanation of the experimental results, and an alternative model is proposed.

Accelerator COSMIC

Detectors Nuclear emulsion

Related references

See also
C. Castagnoli et al., Nuovo Cim. **10** (1956) 1539;
G. Cocconi, Phys. Rev. **93** (1954) 1107;
B. Edwards, A. Eglier, M. W. Friedlander, and A. A. Kamal, Nuovo Cim. **5** (1957) 1188;
R. R. Daniel, J. H. Davies, J. H. Mulvey, and D. H. Perkins, Phil. Mag. **43** (1952) 753;
E. Fermi, Progr. of Theor. Phys. **5** (1950) 570;
W. Heisenberg, Z. Phys. **133** (1953) 65;
L. D. Landau, Doklady Akad. Nauk SSSR **87** (1953) 51;
K. Pinkau, Phil. Mag. **2** (1957) 1389;
F. Roesler and C. B. A. McCusker, Nuovo Cim. **10** (1953) 127;
H. W. Lewis, J. R. Oppenheimer, and S. A. Wouthuysen, Phys. Rev. **73** (1948) 127;
W. L. Kraushaar and L. J. Marks, Phys. Rev. **93** (1954) 326;
R. G. Glasser, D. M. Haskin, M. Schein, and J. J. Lord, Phys. Rev. **99** (1955) 1555;
Analyse information from
W. H. Barkas et al., Phys. Rev. **105** (1957) 1037;
H. J. Bhabha, Proc. Roy. Soc. **A219** (1953) 293;
F. A. Brisbout et al., Phil. Mag. **1** (1956) 605;

ZELDOVICH 1959

■ Prediction of the optical activity of atomic media due to possible weak neutral currents. Prediction of the anapole moments of nuclei, due to weak interactions ■

Parity Nonconservation in Electron Scattering and in other Effects in the First Order of the Weak Interaction Coupling Constant

Y.B. Zeldovich
Zh. Eksp. Teor. Fiz. **36** (1959) 964;

Reprinted in
Ya. B. Zeldovich, *Izbrannye trudy. Chastitsy, Yadra, Vselennaya*, Nauka, M. 1985 s. 72 (red. Yu. B. Khariton).

Excerpt We assume that besides the weak interaction that causes beta decay,

$$g(\overline{P}ON)(\overline{e}^- O\nu) + \text{Herm.conj.}, \qquad (1)$$

there exists an interaction

$$g(\overline{P}OP)(\overline{e}^- Oe^-) \qquad (2)$$

with $g \approx 10^{-49}$ and the operator $O = \gamma_\mu(1 + i\gamma_5)$ characteristic of processes in which parity is not conserved. Such an interaction has been repeatedly discussed in the past in connection with the problem of the isotope shift of electron levels (I. E. Tamm). On an analogous interaction between the neutron and the electron, see references (L. L. Foldy, Phys. Rev. **87** (1952) 693) and (J. L. Lopes, Nucl. Phys. **8** (1958) 234). New experimental possibilities arise in connection with the nonconservation of parity in the interaction (2).
Then in the scattering of electrons by protons the interaction (2) will interfere with the Coulomb scattering, and the nonconservation of parity will appear in terms of the first order in the small quantity g. Owing to this it becomes possible to test the hypothesis used here experimentally and to determine the sign of g ...
Finally, the interaction (2) leads to a rotation of the plane of polarization of visible light by any substance not containing molecules optically active in the ordinary sense of the words. The rotation of the plane of polarization also occurs because the weak interaction mixes atomic electronic states of different parity.
Assumptions have been put forward about a direct interaction $g(\overline{e}^- O\nu)(\overline{\nu}Oe^-)$, which would lead to a scattering of neutrinos by electrons, and also about a weak interaction of four nucleons, which leads to parity nonconservation in first order in g in nuclear reactions and the stationary states of nuclei.

The four-nucleon interaction has as a consequence that odd nuclei (spin $\neq 0$) will have an "anapole" moment proportional to g. *(Extracted from the introductory part of the paper.)*

Related references

See also
R. P. Feynman and M. Gell-Mann, Phys. Rev. **109** (1958) 193;
Y. B. Zeldovich, Zh. Eksp. Teor. Fiz. **33** (1957) 1531;
L. L. Foldy, Phys. Rev. **87** (1952) 693;
J. L. Lopes, Nucl. Phys. **8** (1958) 234;

Reactions
e^- nucleus $\rightarrow$ nucleus e^- >1 GeV(P_{lab})

Reactions
γ atom $\rightarrow$ atom γ

Particles studied nucleus

FUKUI 1959

■ Principle of the spark chamber ■

A New Type of Particle Detector: the "Discharge Chamber"

S. Fukui, S. Miyamoto
Nuovo Cim. **11** (1959) 113;

Abstract Recently, the devices based on gaseous discharges which can determine the trajectory of ionizing particles, such as the hodoscope chamber or the triggered spark counter have been reported. However, these have some inherent imperfections, in spite of their wide usefulness as detectors. Regarding the former, even taking into account the considerations of F. Ashton et al., one cannot expect a *very precise determination* of the trajectory of the particle, owing to the finite dimensions of the glass tubes and their dead spaces, and in the latter it is difficult to determine all trajectories when more than two particles pass through simultaneously because the discharge, which grows in advance of others, decreases the intensity of the electric field very rapidly. We describe here the possibility of constructing a new type of detector also based on the gaseous discharge. This detector secures the precision of the trajectory of the particle and works for a group of particles.

Detectors Spark chamber

Related references

See also
M. Conversi and A. Gozzini, Nuovo Cim. **2** (1955) 189;
M. Conversi et al., Nuovo Cim. Suppl. **4** (1956) 234;
G. Barsanti et al., *Report at the Geneva International Conference* (1956);
M. Gardener, S. Kisdnasami, E. Rossle, and A. W. Wolfendale, Proc. Roy. Soc. **B70** (1957) 687;
T. E. Cranshaw and J. F. De Beer, Nuovo Cim. **5** (1957) 1107;
F. Ashton, S. Kisdnasami, and A. W. Wolfendale, Nuovo Cim. **8** (1958) 615;

MINAKAWA 1959

■ Confirmation of the existence of hadronic jets in high energy collisions: The average P_T of hadrons in jets is limited and almost independent of jet energy ■

Observation of High Energy Jets with Emulsion Chambers. I.—Transverse Momentum of the π^0 Meson

Japan emulsion collaboration; O. Minakawa et al.
Nuovo Cim. Suppl. **11** (1959) 125;

Excerpt

1. Introduction.
2. Experimental procedure.
   1) Design of emulsion chamber.
   2) Balloon flight. Scanning.
3. Analysis of the event.
4. Experimental results.
   1) Table of the events.
   2) Spectrum of the transverse momentum.
5. Summary and discussions.

APPENDIX I. Three dimensional electron shower theory without Landau's approximation.

APPENDIX II. Experimental test on the accuracy of the cascade shower theory.

APPENDIX III. Design of emulsions chamber.

APPENDIX IV. Statistical treatment of eliminating ambiguity of shower axis.

Accelerator COSMIC

Detectors Nuclear emulsion

Related references

More (earlier) information
The Japanese Emulsion Group, Nuovo Cim. Suppl. **8** (1958) 761;
Analyse information from
M. F. Kaplon, B. Peters, and H. L. Brandt, Phys. Rev. **76** (1949) 1735;
M. F. Kaplon, B. Peters, and H. L. Brandt, Helv. Phys. Acta **23** (1950) 24;
L. Leprince-Ringuet et al., Compt. Ren. **229** (1949) 163;
E. Lohrmann, Nuovo Cim. **5** (1957) 1074;
F. Roesler and C. B. A. McCusker, Nuovo Cim. **10** (1953) 127;
W. Heitler and C. H. Terreaux, Proc. Roy. Soc. **A64** (1953) 929;
G. Cocconi, Phys. Rev. **93** (1954) 1107;
Y. Terashima, Progr. of Theor. Phys. **13** (1955) 1;
R. R. Daniel, J. H. Davies, J. H. Mulvey, and D. H. Perkins, Phil. Mag. **43** (1952) 753;
F. A. Brisbout et al., Phil. Mag. **1** (1956) 605;
M. Koshiba and M. F. Kaplon, Phys. Rev. **97** (1955) 193;
D. Lal, Y. Pal, and Rama, Nuovo Cim. Suppl. **12** (1954) 347;

M. F. Kaplon, B. Peters, and D. M. Ritson, Phys. Rev. **85** (1952) 900;
Soryushiron Kenkyu **12**, 1, 24;
B. Edwards et al., Phil. Mag. **(7)3** (7)3,) ;
S. Nasegawa, J. Nishimura, and Y. Nishimura, Nuovo Cim. **6** (1957) 979;
I. M. Gramenitsky et al., Zh. Eksp. Teor. Fiz. **32** (1957) 936;
P. Ciok, M. Danysz, and J. Guerula, Nuovo Cim. **6** (1957) 1409;
G. A. Milekhin and I. L. Rozental, Zh. Eksp. Teor. Fiz. **33** (1957) 187;
A. de Benedetti et al., Nuovo Cim. **4** (1956) 1142;
N. A. Dobrotin, G. T. Zacepin, S. I. Nikolskij, and G. B. Hristiansen, Nuovo Cim. **3** (1956) 634;
Z. Koba, Progr. of Theor. Phys. **15** (1956) 461;
L. D. Landau, Doklady Akad. Nauk SSSR **87** (1953) 51;
E. Fermi, Progr. of Theor. Phys. **5** (1950) 570;
J. Nishimura and K. Kamata, Progr. of Theor. Phys. **7** (1951) 185;

Reactions

unspec nucleus $\rightarrow$ jet X	>100 GeV(P_{lab})
jet $\rightarrow$ π^0 X	

Particles studied jet

ALVAREZ 1959

■ First evidence for the Ξ^0 ■

Neutral Cascade Hyperon Event

L.W. Alvarez et al.
Phys. Rev. Lett. **2** (1959) 215;

Excerpt The existence of a neutral cascade hyperon Ξ^0 has been predicted theoretically, on the basis of the strangeness theory of Gell-Mann and Nishijima, as the neutral counterpart of the negative cascade hyperon, Ξ^-, which decays by $\Xi^- \rightarrow \pi^- \Lambda$.
In an attempt to establish the existence of this particle the Lawrence Radiation Laboratory 15-in. hydrogen bubble chamber was operated in a separated beam of (1.15 ± 0.02) BeV/c K^- mesons produced by the Bevatron. *(Extracted from the introductory part of the paper.)*

Accelerator LBL-Bevatron

Detectors Hydrogen bubble chamber

Related references

See also
M. Gell-Mann, Nuovo Cim. Suppl. **4** (1956) 848;
K. Nishijima, Progr. of Theor. Phys. **13** (1955) 285;
R. Armenteros et al., Phil. Mag. **43** (1952) 597;
C. Franzinetti, G. Morpurgo, Nuovo Cim. Suppl. **6** (1957) 565;
Coombes et al., Phys. Rev. **112** (1958) 1303;
F. S. Crawford et al., Phys. Rev. Lett. **1** (1958) 377;
F. S. Crawford et al., Phys. Rev. Lett. **2** (1959) 110;
P. Nordin et al., Phys. Rev. Lett. **1** (1958) 380;
A. H. Rosenfeld et al., Phys. Rev. Lett. **2** (1959) 112;
M. Gell-Mann and A. H. Rosenfeld, Ann. Rev. Nucl. Sci. **7** (1957) 410;

Reactions

$K^- p \rightarrow$ 2vee	1.15 GeV(P_{lab})
$K^- p \rightarrow \Xi^0 K_S$	1.15 GeV(P_{lab})
$\Lambda \rightarrow p \pi^-$	
$K_S \rightarrow \pi^+ \pi^-$	
$\Xi^0 \rightarrow \Lambda$ neutral	

Particles studied Ξ^0

GOLDHABER 1959

■ First observation of an enhancement in the production of like-charge pairs of pions with similar momenta ■

Pion-Pion Correlations in Antiproton Annihilation Events

G. Goldhaber et al.
Phys. Rev. Lett. **3** (1959) 181;

Excerpt We have observed angular correlation effects between pions emitted from antiproton annihilation events. This experiment was carried out with a separated antiproton beam of momentum $p_{\overline{p}} = 1.05$ BeV/c. A total of 2500 annihilation events were observed in 20000 pictures taken with the Lawrence Radiation Laboratory 30-in. propane bubble chamber.
Pion pairs formed by the charged pions emitted in an antiproton-annihilation event can be considered in two groups: viz., like pairs (in the isotopic-spin state $I = 2$) and unlike pairs (in the isotopic-spin states $I = 0$, 1, or 2). We searched for correlation effects in these separate groups. Our results show that the distribution of the angles between pions of like charges is strikingly different from the distribution of the angles between pions of unlike charges. *(Extracted from the introductory part of the paper.)*

Accelerator LBL-Bevatron

Detectors Heavy-liquid bubble chamber

Related references

See also
L. W. Alvarez et al., Phys. Rev. Lett. **2** (1959) 215;
P. P. Strivastava and E. C. G. Sudarshan, Phys. Rev. **110** (1958) 765;
O. Chamberlain et al., Phys. Rev. **113** (1959) 1615;
W. R. Frazer and J. R. Fulco, Phys. Rev. Lett. **2** (1959) 365;

Reactions

$\overline{p}\, p \rightarrow 2\pi^+ 2\pi^-$	1.05 GeV(P_{lab})
$\overline{p}\, p \rightarrow 3\pi^+ 3\pi^-$	1.05 GeV(P_{lab})

PLANO 1959

■ First direct determination of the parity of the π^0 ■

Parity of the Neutral Pion

R. Plano et al.
Phys. Rev. Lett. **3** (1959) 525;

Excerpt Here we present a preliminary report of a study of certain angular correlations in the decay $\pi^0 \rightarrow e^+ e^- e^+ e^-$. This decay is expected theoretically, and is calculable on the basis of quantum electrodynamics. It is the internal conversion of the two photons of the normal decay. The planes of the pairs "remember" the polarization of the virtual intermediate photons; the correlation of these planes reflects then the polarization correlation of Yang. The correlation of the planes in double internal conversion has been calculated by Kroll and Wada.
The experimental difficulty is due to the rarity of the decay (1/30000) in combination with the necessity of establishing at least the relative orientation of the planes of the two pairs. In this experiment, the π^0's are produced in the capture reaction $\pi^- p \rightarrow \pi^0 n$. About 60% of the negative pions coming to rest in hydrogen undergo this reaction, emitting 4 MeV π^0 mesons. A liquid hydrogen bubble chamber, 30 cm diam., 15 cm deep, in a field of 5500 gauss, is exposed to a stopping π^--beam. 700000 pictures have been taken, and there are $\sim$ 15 stopping mesons per picture. We expect therefore that $\sim$ 200 events will be found eventually. We report here an analysis of the first 103 events, since we believe that this already permits a convincing conclusion! *(Extracted from the introductory part of the paper.)*

Accelerator Nevis labs cyclotron

Detectors Hydrogen bubble chamber

Related references

See also
K. A. Brueckner, R. Serber, and K. M. Watson, Phys. Rev. **81** (1951) 575;
C. N. Yang, Phys. Rev. **77** (1950) 242;
H. M. Kroll and W. Wada, Phys. Rev. **98** (1955) 1355;

Analyse information from
W. K. H. Panofsky, R. L. Aamodt, and J. Hadley, Phys. Rev. **81** (1951) 565;
W. Chinowsky and J. Steinberger, Phys. Rev. **95** (1954) 1561;
W. Chinowsky and J. Steinberger, Phys. Rev. **100** (1955) 1476;

Reactions

$\pi^- p \rightarrow n \pi^0$	0 GeV(P_{lab})
$\pi^0 \rightarrow 2e^- 2e^+$	4 MeV(P_{lab})

Particles studied π^0

REINES 1959 — Nobel prize

■ Confirmation of the detection of the $\overline{\nu}_e$. Nobel prize to F. Reines awarded in 1995 "for the detection of the neutrino". Co-winner M. Perl "for the discovery of the tau lepton" ■

Free Antineutrino Absorption Cross Section. I. Measurement of the Free Antineutrino Absorption Cross Section by Protons

F. Reines, C.L. Cowan
Phys. Rev. **113** (1959) 273;

Reprinted in
R. N. Cahn and G. Goldhaber, *The Experimental Foundations of Particle Physics*, Cambridge Univ. Press (1991) 183.
The Physical Review — the First Hundred Years, AIP Press (1995) 866.

Abstract The cross section for the reaction $p(\overline{\nu},\beta^+)n$ was measured using antineutrinos from a powerful fission reactor at the Savannah River Plant of the United States Atomic Energy Commission. Target protons were provided by a 1.4·1000 liter liquid scintillation detector in which the scintillation solution (triethylbenzene, terphenyl, and POPOP) was loaded with a cadmium compound (cadmium octoate) to allow the detection of the reaction by means of the delayed coincidence technique. The first pulse of the pair was caused by the slowing down and annihilation of the positron, the second by the capture of the neutron in cadmium following its moderation by the scintillator protons. A second giant scintillation detector without cadmium loading was used above the first to provide an anticoincidence signal against events induced by the cosmic rays. The antineutrino signal was related to the reactor by means of runs taken while the reactor was on and off. Reactor radiations other than antineutrinos were ruled out as the cause of the signal by a differential shielding experiment. The signal rate was 36 ± 4 events/hr and the signal-to-noise ratio was 1/5, where half the noise was correlated and cosmic-ray associated and about half was due to non-reactor-associated accidental coincidences. The cross section per fission $\overline{\nu}$ (assuming 6.1 $\overline{\nu}$ per fission) for the inverse beta decay of the proton was measured to be $(11.\pm2.6)\cdot10^{-44}$ cm^2/fission. These values are consistent with prediction based on the two-component theory of the neutrino.

Accelerator Nuclear reactor

Detectors Counters

Related references

More (earlier) information
F. Reines and C. L. Cowan, Phys. Rev. **90** (1953) 492;
C. L. Cowan, F. Reines et al., Science **124** (1956) 103;

See also

Ronzio, C. L. Cowan and F. Reines, Rev. Sci. Inst. **29** (1958) 146;

Reactions

$\overline{\nu}_e\, p \rightarrow n\, e^+$ 1.8-6.0 MeV(E_{lab})

Particles studied $\overline{\nu}_e$

IKEDA 1959

■ Introduction of $SU(3)$ symmetry for hadrons. Prediction of the existence of the η meson ■

A Possible Symmetry in Sakata's Model for Bosons-Baryons System

M. Ikeda, S. Ogawa, Y. Ohnuki

Progr. of Theor. Phys. **22** (1959) 715;

Abstract In this paper we study a possible symmetry in Sakata's model for the strongly interacting particles. In the limiting case in which the basic particles, proton, p, neutron, n and Λ-particle, Λ, have an equal mass, our theory holds the invariance under the exchange of p and Λ or n and Λ in addition to the usual charge independence and the conservation of electrical and hyperonic charge.
From our theory the following are obtained: (a) iso-singlet $\pi_0{}'$-meson state, which is a pseudo-scalar, exists, (b) the spin of Ξ-particle may be $(3/2)^+$ and (c) several resonating states in K- and π-nucleon scattering are anticipated to exist.

Related references

See also
S. Oneda, Progr. of Theor. Phys. **9** (1953) 327;
S. Oneda and H. Umezawa, Progr. of Theor. Phys. **9** (1953) 685;
K. Iwata et al., Progr. of Theor. Phys. **13** (1955) 19;
T. Nakano and K. Nishijima, Progr. of Theor. Phys. **10** (1953) 581;
S. Sakata, Progr. of Theor. Phys. **16** (1953) 686;
S. Tanaka, Progr. of Theor. Phys. **16** (1953) 625;
S. Matsumoto, Progr. of Theor. Phys. **16** (1953) 583;
Z. Maki, Progr. of Theor. Phys. **16** (1953) 667;
E. Fermi and C. N. Yang, Phys. Rev. **76** (1949) 1739;
S. Ogawa, Progr. of Theor. Phys. **21** (1959) 209;

Particles studied meson, baryon, η

LANDAU 1959

■ Invention of Landau singularities for perturbative amplitudes ■

On Analytic Properties of Vertex Parts in Quantum Field Theory

L.D. Landau

JETP **37** (1959) 62; Nucl. Phys. **13** (1959) 181; JETP **10** (1960) 45;

Abstract A general method of finding the singularities of quantum field theory values on the basis of graph technique is evolved.

Related references

See also
Y. Nambu, Nuovo Cim. **6** (1957) 1064;
R. Karplus, C. M. Sommerfeld, and E. H. Wichman, Phys. Rev. **111** (1958) 1187;

PONTECORVO 1959

■ Proposed experiments to establish distinguishability of ν_e and ν_μ. Indication on the feasibility of neutrino beams with accelerators ■

Electron and Muon Neutrinos

B. Pontecorvo

Zh. Eksp. Teor. Fiz. **37** (1959) 1751; JETP **10** (1960) 1236;

Abstract Some processes due to free neutrinos which heretofore had not been considered are discussed. Especial attention is payied to those processes which in principle may help to solve the problem concerning the existence of two neutral lepton pairs (electron pair (ν_e and $\overline{\nu}_e$) and muon pair (ν_μ and $\overline{\nu}_\mu$)).
To solve the fundamental question whether ν_μ and ν_e are identical particles, a method is proposed which in essence is analogous to the method employed for solving the problem of the distinguishability of the neutrino and antineutrino or K^0 and $\overline{K}^0$ mesons. In principle the problem can be solved if it is demonstrated experimentally that a $\overline{\nu}_\mu$ beam is capable of inducing transitions which $\overline{\nu}_e$ particle can certainly induce (e.g. the $\overline{\nu}_\mu p \rightarrow ne^+$ reaction).
The experiment suggested above although difficult, should be feasible with accelerators capable of producing more intense beams than those produced by present day accelerators.

Related references

See also
H. A. Bethe and R. Pierls, Nature **133** (1934) 532;
L. W. Alvarez, UCRL-328 (1949);
S. Sakata, Progr. of Theor. Phys. **16** (1953) 686;
A. G. Cameron, Ann. Rev. Nucl. Sci. **8** (1958) 1;
S. Oneda and J. C. Pati, Phys. Rev. Lett. **2** (1959) 125;
R. P. Feynman and M. Gell-Mann, Phys. Rev. **109** (1958) 193;
L. B. Okun', Zh. Eksp. Teor. Fiz. **34** (1958) 469;
G. Feinberg, Phys. Rev. **110** (1958) 1482;
B. Pontecorvo, REPORT-PD-205 (1946), Zh. Eksp. Teor. Fiz. **33** (1957) 549;
D. Berley, J. Lee, and M. Bardon, Phys. Rev. Lett. **2** (1959) 357;
B. Pontecorvo, Zh. Eksp. Teor. Fiz. **36** (1959) 1615;
Analyse information from
F. Reines and C. L. Cowan, Phys. Rev. **90** (1953) 492;
F. Reines and C. L. Cowan, Phys. Rev. **113** (1959) 273;
R. Davis, Phys. Rev. **86** (1952) 976;
W. A. Love et al., Phys. Rev. Lett. **2** (1959) 107;
M. Goldhaber, L. Grodzins, and A. W. Sunyar, Phys. Rev. **109** (1958) 1015;
M. P. Balandin et al., Zh. Eksp. Teor. Fiz. **29** (1955) 265;

Particles studied $\nu_e, \overline{\nu}_e, \nu_\mu, \overline{\nu}_\mu$

CHEW 1959

■ Method for extraction of pion-pion interactions ■

Unstable Particles as Targets in Scattering Experiments

G.F. Chew, F.E. Low

Phys. Rev. **113** (1959) 1640;

Abstract A general method is suggested for analyzing the scattering of particle A by particle B, leading to three or more final particles, in order to obtain the cross section for the interaction of A with a particle which is virtually contained in B. Binding complications are absent if a plausible assumption about the location and residues of poles in the S-matrix is accepted. The method is useful for unstable particles from which free targets cannot be made; the special examples of pion and neutron targets are discussed in detail.

Related references

See also
W. K. H. Panofsky and E. A. Allton, Phys. Rev. **110** (1958) 1155;

REGGE 1959

■ Introduction of Regge poles ■

Introduction to Complex Orbital Momenta

T. Regge

Nuovo Cim. **14** (1959) 951;

Excerpt In this paper the orbital momentum j, until now considered as an integer discrete parameter in the radial Schrödinger wave equations, is allowed to take complex values. The purpose of such an enlargement is not purely academic but opens new possibilities in discussing the connection between potentials and scattering amplitudes. In particular it is shown that under reasonable assumptions, fulfilled by most field theoretical potentials, the scattering amplitude at some fixed energy determines the potential uniquely, when it exists. Moreover for special classes of potentials $V(x)$, which are analytically continuable into a function $V(z), z = x + iy$, regular and suitable bounded in $x > 0$, the scattering amplitude has the remarkable property of being continuable for arbitrary negative and large cosine of the scattering angle and therefore for arbitrary large real and positive transmitted momentum. The range of validity of the dispersion relations is therefore much enlarged.

Related references

See also
E. P. Wigner and J. Neumann, Annals of Math. **59** (1954) 418;
M. N. Khuri, Phys. Rev. **107** (1957) 1148;
I. M. Gel'fand and Levitan, *Amer. Math. Soc. Trans.*, Sec.2.1.250 (1955);
R. Jost and W. Kohn, Math. Phys. Medd. **27(9)** (1953);
L. D. Faddeev, Soviet Physics Doklady **3** (1959) 747;
S. Gasiorowicz and H. P. Noyes, Nuovo Cim. **10** (1958) 78;

GÜRSEY 1960

■ Non linear sigma model ■

On the Symmetries of Strong and Weak Interactions

F. Gürsey

Nuovo Cim. **16** (1960) 230;

Excerpt A model of strong interactions is proposed which admits a group $(G_4 \times H)$ in the limit of a suitably defined doublet approximation, G_4 being a 4-dimensional extension of the isotopic spin group and H the hypercharge gauge group. Weak interactions having phenomenological chirality invariance properties or obeying the $\Delta I = \frac{1}{2}$ rule are shown to be invariant under an unitary subgroup of the group $(G_4 \times H)$.

Related references

See also
B. D'Espagnat and J. Prentki, Nucl. Phys. **11** (1959) 700;
B. D'Espagnat and J. Prentki, Phys. Rev. **114** (1959) 1366;
R. L. Cool, B. Cork, J. W. Cronin, and W. A. Wenzel, Phys. Rev. **119** (1960) 912;
F. S. Crawford et al., Phys. Rev. Lett. **2** (1959) 266;
A. Pais, Phys. Rev. **110** (1958) 1480;
A. Pais, Phys. Rev. **110** (1958) 574;
A. Pais, Phys. Rev. **112** (1958) 624;

BALDIN 1960

■ First theoretical estimations of dipole polarizabilities of nucleons ■

Polarizability of Nucleons

A.M. Baldin

Nucl. Phys. **18** (1960) 310;

Abstract Estimates of dipole polarizabilities of nucleons and the values they involve are given on the basis of data on photoproduction of π-mesons and the Compton effect on nucleons. It is indicated that no upper estimate of neutron polarizability exists at present. The preliminary experimental data now available may be interpreted as indicating that neutron has an abnormally large polarizability. The effect leading to the inapplicability of the impulse approximation for describing the reaction γ deuteron $\rightarrow p\ n\ \gamma$ are estimated. It is pointed out that the measurement of the cross section of the reaction γ deuteron $\rightarrow \gamma$ deuteron would yield an answer for the value of neutron dipole polarizability.

Related references

See also

F. Low, Phys. Rev. **96** (1954) 1428;
M. Gell-Mann and H. Z. Goldberger, Phys. Rev. **96** (1954) 1433;
A. Klein, Phys. Rev. **99** (1955) 998;
M. Gell-Mann, N. L. Goldberg, and W. E. Thirring, Phys. Rev. **95** (1954) 1612;
R. M. Thaler, Phys. Rev. **114** (1959) 827;
Y. A. Alexandrov and I. I. Bondarenko, Zh. Eksp. Teor. Fiz. **31** (1956) 726;
L. G. Hyman et al., Phys. Rev. Lett. **3** (1959) 93;
R. H. Capps, Phys. Rev. **106** (1957) 1031;
B. S. Barashenkov, I. P. Stakhanov, and Y. A. Alexandrov, Zh. Eksp. Teor. Fiz. **32** (1957) 154;
G. Breit and N. Z. Rustgi, Phys. Rev. **114** (1959) 830;
N. N. Bogolyubov and D. V. Shirkov, Doklady Akad. Nauk SSSR **113** (1957) 529;

Analyse information from
B. B. Govorkov et al., Doklady Akad. Nauk SSSR **111** (1956) 988;
G. E. Pugh et al., Phys. Rev. **105** (1957) 982;

SCHWARTZ 1960 — Nobel prize

■ Confirmation of the feasibility of neutrino beams with accelerators. Nobel prize to L. M. Lederman, M. Schwartz, and J. Steinberger awarded in 1988 "for the neutrino beam method and the demonstration of the doublet structure of the leptons through the discovery of the muon neutrino" ■

Feasibility of Using High Energy Neutrinos to Study the Weak Interactions

M. Schwartz
Phys. Rev. Lett. **4** (1960) 306;

Reprinted in
The Physical Review — the First Hundred Years, AIP Press (1995) CD-ROM.

Excerpt For many years, the question to how to investigate the behavior of the weak interactions at high energies has been one of considerable interest. It is the purpose of this note to show that experiments pointed in this direction, though not quite feasible with presently existing equipment, are within the capabilities of present technology and should be possible within the next decade.
We propose the use of high-energy neutrinos as a probe to investigate the weak interactions.
A natural source of high-energy neutrinos are high-energy pions. Such pions will produce neutrinos whose laboratory energy will range with equal probability from zero to 45% of the pion energy, and whose direction will tend very much toward the pion direction. *(Extracted from the introductory part of the paper.)*

Related references

See also
B. Pontecorvo, Zh. Eksp. Teor. Fiz. **37** (1959) 1751;

Reactions

ν_μ	>400 MeV(P_{lab})
$\overline{\nu}_\mu$	>400 MeV(P_{lab})

BUTTON 1960

■ First evidence for the $\overline{\Sigma}^0$ ■

Evidence for the Reaction $\overline{p}\, p \to \overline{\Sigma}^0\, \Lambda$

J. Button et al.
Phys. Rev. Lett. **4** (1960) 530;

Excerpt We have found an event which may be interpreted as the reaction $\overline{p}\, p \to \overline{\Sigma}^0\, \Lambda$. The event was produced in the 72-in. liquid hydrogen bubble chamber. A highly purified beam of antiprotons of 1.99 ± 0.03 BeV/c was produced by using three velocity-selecting spectrometers. The approximate composition of this beam under normal operating conditions was $1.0\overline{p} : 1.5\pi^- : 1.9\mu^- : 0.015K^-$. A description of the beam will be presented in a later publication. *(Extracted from the introductory part of the paper.)*

Accelerator LBL-Bevatron

Detectors Hydrogen bubble chamber

Related references

Superseded by
J. Button et al., Phys. Rev. **121** (1961) 1788;

Reactions

$\overline{p}\, p \to \Lambda\, \overline{\Sigma}^0$	1.99 GeV(P_{lab})

Particles studied $\overline{\Sigma}^0$

ALSTON 1960 — Nobel prize

■ First evidence for the $\Sigma(1385\, P_{13})$. Nobel prize to L. W. Alvarez awarded in 1968 "for his decisive contribution to elementary particle physics, in particular the discovery of a large number of resonance states, made possible through his development of the hydrogen bubble chamber technique and data analysis" ■

Resonance in the $\Lambda\pi$ System

M.H. Alston et al.
Phys. Rev. Lett. **5** (1960) 520;

Reprinted in
R. N. Cahn and G. Goldhaber, *The Experimental Foundations of Particle Physics*, Cambridge Univ. Press (1991) 122.
The Physical Review — the First Hundred Years, AIP Press (1995) 873.

Abstract A study, based on 141 events, of the energy distribution in the c.m. system of the pions emitted in the reaction

$$K^-\, p \to \Lambda^0\, \pi^+\, \pi^-$$

using a beam of 1.15 BeV/c K^-. Peaks were revealed in both the π^+ and π^- spectra at around 285 MeV, such as would be expected from the production of the pions through a quasi-two-body reaction

$$K^- \, p \to Y^{*\pm} \, \pi^{\mp}$$

where Y^* has a mass spectrum peaking at about 1380 MeV. The shape of the calculated mass spectrum for the hypothetical "particle" Y^* shows striking resemblance to the well-known $(\frac{3}{2}\frac{3}{2})$ resonance of the πp system. Attempts to determine the spin of the Y^* from study of various possible anisotropies were inconclusive with the limited statistics. *(Science Abstracts, 1961, 3240. S. J. Goldsack)*

Accelerator LBL-Bevatron

Detectors Hydrogen bubble chamber

Related references

See also
L. W. Alvarez et al., Phys. Rev. Lett. **2** (1959) 215;
R. K. Adair, Phys. Rev. **100** (1955) 1540;
E. Eisler and R. G. Sachs, Phys. Rev. **72** (1947) 680;
M. Gell-Mann, Phys. Rev. **106** (1957) 1296;
R. H. Capps, Phys. Rev. **119** (1960) 1753;
R. H. Capps and M. Nauenberg, Phys. Rev. **118** (1960) 593;
R. H. Dalitz and S. F. Tuan, Ann.Phys. **10** (1960) 307;
M. Nauenberg, Phys. Rev. Lett. **2** (1959) 351;
A. Komatsuzawa, R. Sugano, and Y. Nogami, Progr. of Theor. Phys. **21** (1959) 151;
Y. Nogami, Progr. of Theor. Phys. **22** (1959) 25;
D. Amati, A. Stanghellini, and B. Vitale, Nuovo Cim. **13** (1959) 1143;
L. F. Landovitz and B. Margolis, Phys. Rev. Lett. **2** (1959) 318;

Reactions

$K^- \, p \to p \, \overline{K}^0 \, \pi^-$	1.15 GeV(P_{lab})
$K^- \, p \to \Lambda \, \pi^+ \, \pi^0 \, \pi^-$	1.15 GeV(P_{lab})
$K^- \, p \to \Sigma^0 \, \pi^+ \, \pi^0 \, \pi^-$	1.15 GeV(P_{lab})
$K^- \, p \to \Lambda \, \pi^+ \, \pi^-$	1.15 GeV(P_{lab})
$K^- \, p \to \Sigma^0 \, \pi^+ \, \pi^-$	1.15 GeV(P_{lab})
$K^- \, p \to \Lambda \, \pi^+ \, \pi^-$	1.15 GeV(P_{lab})
$K^- \, p \to \Sigma(1385 \, P_{13})^+ \, \pi^-$	1.15 GeV(P_{lab})
$K^- \, p \to \Sigma(1385 \, P_{13})^- \, \pi^+$	1.15 GeV(P_{lab})

Particles studied $\Sigma(1385 \, P_{13})$

KANG-CHANG 1960

■ First evidence for the $\overline{\Sigma}^-$ ■

$\overline{\Sigma}^-$-Hyperon Production by 8.3 BeV/c π^--Mesons

W. Kang-Chang et al.
Zh. Eksp. Teor. Fiz. **38** (1960) 1356;

Excerpt One event of production and decay of a $\overline{\Sigma}^-$ hyperon was found out of 40,000 pictures obtained by a beam of negative 8.3 ± 0.6 BeV/c pions in a propane bubble chamber with a constant magnetic field of 13,700 oe. A photograph and diagram of this event are shown.

...The probability of a chance coincidence in one picture of various events which may have imitated the phenomenon under consideration is, according to our estimate, $\sim 10^{-9}$.
We consider the most probable reaction in the primary star to be

$$\pi^- \, C \to \overline{\Sigma} \, K^0 \, \overline{K}^0 \, K^- \, p \, \pi^+ \, \pi^- + \text{recoil nucleus}$$

For the lifetime of the $\overline{\Sigma}^-$ we obtained the value $(1.18 \pm 0.07) \times 10^{-10}$ sec.
Hence, the data presented is evidence of the fact that we have observed a new type of particle, the charged antihyperon. *(Extracted from the introductory part of the paper.)*

Accelerator Dubna proton synchrotron

Detectors Heavy-liquid bubble chamber

Reactions

π^- nucleus $\to \overline{\Sigma}^-$ X	8.3 GeV(T_{lab})

Particles studied $\overline{\Sigma}^-$

GOLDHABER 1960

■ Interpretation of the enhancement in the production of like charge pairs of pions with similar momenta as an influence of Bose-Einstein correlations ■

Influence of Bose-Einstein Statistics on the Antiproton-Proton Annihilation Process

G. Goldhaber, S. Goldhaber, W. Lee, A. Pais
Phys. Rev. **120** (1960) 300;

Abstract Recent observations of angular distributions of π mesons in $\overline{p}$-p annihilation indicate a deviation from the predictions of the usual Fermi statistical model. In order to shed light on these phenomena, a modification of the statistical model is studied. We retain the assumption that the transition rate into a given final state is proportional to the probability of finding N free π mesons in the reaction volume, but express this probability in terms of wave functions symmetrized with respect to particles of like-charge. The justification of this assumption is discussed. The model reproduces the experimental results qualitatively, provided the radius of the interaction volume is between one-half and three-fourths of the pion Compton wavelength; the dependence of angular correlation effects on the value of the radius is rather sensitive. Quantitatively, there seems to remain some discrepancy, but we cannot say whether this is due to experimental uncertainties or to some other dynamic effects. In the absence of information on π-π interactions and of a fully satisfactory explanation of the mean pion multiplicity for annihilation, we wish to emphasize the preliminary nature of our results, We consider them, however, as an indication that the symmetrization effects discussed here may well play a major role in the analysis of angular distributions. It is pointed out that in this respect the energy dependence of the angular correlations may provide valuable clues for the validity of our model.

Accelerator LBL-Bevatron

Detectors Heavy-liquid bubble chamber

Related references

Analyse information from
G. Goldhaber et al., Phys. Rev. Lett. **3** (1959) 181;

Reactions

$\overline{p}\,p \rightarrow 2\pi^+\,2\pi^-$	1.05 GeV(P_{lab})
$\overline{p}\,p \rightarrow 3\pi^+\,3\pi^-$	1.05 GeV(P_{lab})

REGGE 1960

■ Introduction of Regge poles ■

Bound States, Shadow States and Mandelstam Representation

T. Regge
Nuovo Cim. **18** (1960) 947;

Excerpt In a previous paper a technique involving complex angular momenta was used in order to prove the Mandelstam representation for potential scattering. One of the results was that the number of subtractions in the transmitted momentum depends critically on the location of the poles (shadow states) of the scattering matrix as a function of the complex orbital momentum. In this paper the study of the position of the shadow states is carried out in much greater detail. We give also related inequalities concerning bound states and resonances. The physical interpretation of the shadow states is then discussed.

Related references

See also
T. Regge, Nuovo Cim. **14** (1959) 951;
R. Blankenbecler, M. L. Goldberger, M. N. Khuri, and S. B. Treiman, Ann.Phys. **10** (1960) 62;
A. Klein, Jour. Math. Phys. **1** (1960) 41;

GOLDANSKY 1960

■ First measurement of electrical polarizability of the proton ■

The Elastic $\gamma\,p$ Scattering at Energies of 40-70 MeV and the Polarizability of the Proton

V.I. Goldansky, O.A. Karpukhin, A.V. Kutsenko, V.V. Pavlovskaya
Zh. Eksp. Teor. Fiz. **38** (1960) 1693;

Abstract The elastic $\gamma\,p$ scattering at 40 to 70 MeV energy is investigated and differential cross-sections at 45, 75, 90, 120, 135 and 150° are determined. The results obtained are compared with the theory which takes into account the anomalous proton magnetic moment as well as the effects of mesonic cloud polarization. The cross-section of the $\gamma\,p$ scattering at 90° corresponds to a proton electric polarizability $\alpha_E = (11{\pm}4)\cdot 10^{-43}$ cm^3. As an addition to the experimental results the dispersion relations are used, which allow one to obtain from the π- photoproduction data the value of the sum of electric (α_E) and magnetic (α_M) polarizabilities of proton: $\alpha_E + \alpha_M = 11\cdot 10^{-43}$ cm^3. By taking this value into account the obtained experimental data correspond to the following values of the proton polarizabilities: $\alpha_E = (9{\pm}2)\cdot 10^{-43}$ cm^3, $\alpha_M = (2\pm 2)\cdot 10^{-43}$ cm^3. A mean square fluctuation of the proton electric dipole length $(\overline{r^2})^{1/2} = (3.5 \div 5)\cdot 10^{-14}$ cm corresponds to the electric polarizability value obtained.

Accelerator LEBD-FIAN

Detectors Photon spectrometer

Related references

More (earlier) information
R. G. Vasilkov, B. B. Govorkov and V. I. Goldansky, Zh. Eksp. Teor. Fiz. **37** (1959) 11;
R. G. Vasilkov, B. B. Govorkov and V. I. Goldansky, Nucl. Phys. **12** (1959) 337;

See also
J. L. Powell, Phys. Rev. **75** (1949) 32;
M. Gell-Mann and M. Goldberger, Phys. Rev. **96** (1954) 1433;
F. E. Low, Phys. Rev. **96** (1954) 1428;
A. Klein, Phys. Rev. **99** (1955) 998;
R. Capps, Phys. Rev. **106** (1957) 1031;
E. C. Park, Jour. Sci. Instr. **33** (1956) 257;
W. Barber, W. George, and D. Reagen, Phys. Rev. **98** (1955) 73;
G. Breit and M. L. Rustgi, Phys. Rev. **114** (1959) 830;
R. M. Thaler, Phys. Rev. **114** (1959) 827;
D. I. Blokhintsev, V. S. Barashenkov, and B. M. Barbashov, Usp. Fiz. Nauk **68** (1959) 417;
V. S. Barashenkov and B. M. Barbashov, Nucl. Phys. **9** (1958) 426;
Y. A. Aleksandrov, Zh. Eksp. Teor. Fiz. **33** (1957) 294;
Y. A. Aleksandrov, Zh. Eksp. Teor. Fiz. **32** (1957) 561;
V. S. Barashenkov, I. P. Stakhanov, and Y. A. Aleksandrov, Zh. Eksp. Teor. Fiz. **32** (1957) 1546;
Y. A. Aleksandrov and I. I. Bondarenko, Zh. Eksp. Teor. Fiz. **31** (1956) 726;
F. Curtis Michel, Phys. Rev. **133B** (1964) 329;
S. Wahlborn, Phys. Rev. **138B** (1965) 530;
H. W. Koch, J. W. Motz, Rev. of Mod. Phys. **31** (1959) 920;
L. L. Foldy, Phys. Rev. Lett. **3** (1959) 105;
L. Euges, Phys. Rev. **81** (1951) 982;
P. V. C. Hough, Phys. Rev. **74** (1948) 80;
R. C. Miller, Phys. Rev. **95** (1954) 796;
C. Oxley and V. L. Telegdi, Phys. Rev. **100** (1955) 435;
C. Oxley, Phys. Rev. **110** (1958) 733;
G. Pugh et al., Phys. Rev. **105** (1957) 982;
M. Cini and R. Stroffolini, Nucl. Phys. **5** (1958) 684;

Reactions

$\gamma\,p \rightarrow p\,\gamma$	<75 MeV(E_{lab})

Particles studied p

BOWCOCK 1961

■ Prediction of the spin=1, isospin=1 resonance in the two-pion system ■

The Effect of a Pion-Pion Interaction on Low-Energy Meson-Nucleon Scattering — II.

J. Bowcock, W.N. Cottingham, D. Lurie
Nuovo Cim. **19** (1961) 142;

Excerpt Using the formalism developed in a previous paper we investigated the consequences of a pion-pion interaction on low energy pion-nucleon phase shifts. By confining our considerations to the isotopic spin flip combination of waves we are able to isolate the effects due to a pion-pion interaction in the $T = 1$ state. We find that such an interaction is definitely necessary to give agreement with experiment and that a simple resonance in the $J = 1$, $T = 1$ state gives a good fit to the data.

Related references

More (earlier) information
J. Bowcock, N. Gottingham, and D. Lurie, Nuovo Cim. **16** (1960) 918;

See also
W. R. Frazer and J. R. Fulco, Phys. Rev. **117** (1960) 1603;
G. Chew, M. Golderger, F. Low, and Y. Nambu, Phys. Rev. **106** (1957) 1337;
A. Stanhellini, Nuovo Cim. **10** (1958) 398;

Analyse information from
M. Cini and S. P. Fubini, Nuovo Cim. **10** (1960) 398;
W. D. Walker, J. Davis, and W. D. Shephard, Phys. Rev. **118** (1960) 1612;

Reactions

$\pi\, p \rightarrow p\, \pi$ <0.6 GeV(T_{lab})
$\rho \rightarrow 2\pi$

Particles studied ρ

GOLDSTONE 1961

■ Prediction of unavoidable massless bosons if global symmetry of the Lagrangian is spontaneously broken ■

Field Theories with "Superconductor" Solutions

J. Goldstone
Nuovo Cim. **19** (1961) 154;

Excerpt The conclusion for the existence of non-perturbative type "superconductor" solutions of field theories are examined. A non-covariant canonical transformation method is used to find such solutions for a theory of a fermion interacting with a pseudoscalar boson. A covariant renormalizable method using Feynman integrals is then given. A "superconductor" solution is found whenever in the normal perturbative-type solution the boson mass squared is negative and the coupling constants satisfy certain inequalities. The symmetry properties of such solutions are examined with the aid of a simple model of self-interacting boson field. The solutions have lower symmetry than the Lagrangian and contain mass zero bosons.

Related references

See also
N. N. Bogolyubov, Zh. Eksp. Teor. Fiz. **34** (1958) 73;
N. N. Bogolyubov, JETP **7** (1958) 51;

Particles studied goldstone

SALAM 1961

■ Invention of the gauge principle as basis to construct quantum theories of interacting fundamental fields ■

On a Gauge Theory of Elementary Interactions

A. Salam, J.C. Ward
Nuovo Cim. **19** (1961) 165;

Excerpt A theory of strong as well as weak interactions is proposed using the idea of having only such interactions which arise from generalized gauge transformation.

Related references

More (earlier) information
A. Salam and J. C. Ward, Nuovo Cim. **11** (1959) 568;

See also
J. J. Sakurai, Ann.Phys. **11** (1960) 1;
J. Schwinger, Ann.Phys. **2** (1957) 407;
M. Gell-Mann and M. Levi, Nuovo Cim. **16** (1960) 705;
J. Tiomno, Nuovo Cim. **6** (1957) 69;

BUTTON 1961

■ Confirmation of $\overline{\Lambda}$ production ■

The Reaction $\overline{p}\, p \rightarrow \overline{\Lambda}\, \Lambda$

J. Button et al.
Phys. Rev. **121** (1961) 1788;

Abstract The study of the interaction $\overline{p}\, p \rightarrow \overline{\Lambda}\, \Lambda$, performed with the 72-inch hydrogen bubble chamber, has yielded 11 of these events in a total of 2100 antiproton interactions at 1.61 BeV/c. The cross section for $\overline{\Lambda}\, \Lambda$ production was estimated as 57 ± 18 μb. Eight of the 11 antilambda particles went forward in the c.m. system. At the higher momentum of 1.99 BeV/c, one single-V and one double-V event fitting $\overline{\Lambda}\, \Lambda$ production unambiguously, and one single-V and one double-V event fitting $\overline{\Sigma}^0\, \Lambda$ or $\Sigma^0\, \overline{\Lambda}$ were observed in 4920 antiproton interactions. These events yield a $\overline{\Lambda}$-Λ production cross section of 55 ± 40 μb; this value is consistent with that predicted by the ratio of phase space on the basis of the

1.61 BeV/c data. No charged antisigma events were observed at the higher momentum. Three stages of particle separation utilizing velocity-selecting spectrometers were employed. At the lower momentum, background pions were one-third as numerous as antiprotons at the bubble chamber and the flux of antiprotons was about one per picture. At the higher momentum, the background pion to antiproton ratio was 1.8, and the flux of antiprotons was one every 6 pulses. Delta rays on incident interacting tracks were used to determine beam composition.

Accelerator LBL-Bevatron

Detectors Hydrogen bubble chamber

Related references

More (later) information
G. R. Lynch, Rev. of Mod. Phys. **33** (1961) 395;
See also
J. Button et al., Phys. Rev. Lett. **4** (1960) 530;

Reactions

$\bar{p}\, p \rightarrow \Lambda\, \bar{\Lambda}$ 1.59-1.63 GeV(P_{lab})

Particles studied $\bar{\Lambda}$

NAMBU 1961

■ Nambu-Jona-Lasinio nonlinear model of hadrons ■

Dynamical Model of Elementary Particles Based on an Analogy with Superconductivity. I

Y. Nambu, G. Jona-Lasinio
Phys. Rev. **122** (1961) 345;

Reprinted in
The Physical Review — the First Hundred Years, AIP Press (1995) CD-ROM.

Abstract It is suggested that the nucleon mass arises largely as a self-energy of some primary fermion field through the same mechanism as the appearance of energy gap in the theory of superconductivity. The idea can be put into a mathematical formulation utilizing a generalized Hartree-Fock approximation which regards real nucleons as quasi-particle excitations. We consider a simplified model of nonlinear four-fermion interaction which allows a γ_5-gauge group. An interesting consequence of the symmetry is that there arise automatically pseudoscalar zero-mass bound states of nucleon-antinucleon pair which may be regarded as an idealized pion. In addition, massive bound states of nucleon number zero and two are predicted in a simple approximation.
The theory contains two parameters which can be explicitly related to observed nucleon mass and the pion-nucleon coupling constant. Some paradoxical aspects of the theory in connection with the γ_5 transformation are discussed in detail.

Related references

See also
J. Bardeen, L. N. Cooper, and J. R. Schrieffer, Phys. Rev. **106** (1957) 162;
N. N. Bogolyubov, JETP **34** (1958) 41;
N. N. Bogolyubov, JETP **34** (1958) 51;
N. N. Bogolyubov, Usp. Fiz. Nauk **67** (1959) 549;
D. Pines and J. R. Schrieffer, Nuovo Cim. **10** (1958) 496;
P. W. Anderson, Phys. Rev. **110** (1958) 827;
P. W. Anderson, Phys. Rev. **110** (1958) 1900;
P. W. Anderson, Phys. Rev. **114** (1959) 1002;
G. Rickayzen, Phys. Rev. **115** (1959) 795;
T. D. Lee and C. N. Yang, Phys. Rev. **98** (1955) 1501;
W. Heisenberg, Zeit. Naturforschung **14** (1959) 441;
L. van Hove, Physica **18** (1952) 145;
J. Bernstein, M. Gell-Mann, and L. Michel, Nuovo Cim. **16** (1960) 560;

GLASSER 1961

■ First conclusive measurements of the π^0 lifetime ■

Mean Lifetime of the Neutral Pion

R.G. Glasser, N. Seeman, B. Stiller
Phys. Rev. **123** (1961) 1014;

Abstract An estimate of the mean lifetime of the π^0 meson has been obtained from an experiment employing a direct time-of-flight technique first attempted by Harris et al. in 1957. This method is based upon the observation in nuclear emulsion of the decay of the $K_{\pi 2}{}^+$ meson ($K^+ \rightarrow \pi^+\pi^0$) and the subsequent decay of the π^0 via the Dalitz mode, $\pi^0 \rightarrow e^+e^-\gamma$. In the present experiment we were able to utilize a new fine-grained emulsion (Ilford L.4) that yielded markedly improved resolution. The availability of the separated K^+ beam from the Bevatron at Berkeley permitted detection and measurement of 76 Dalitz decays. We obtain, for the mean lifetime of the π^0, $\tau = (1.9 \pm 0.5) \times 10^{-16}$ sec.

Accelerator LBL-Bevatron

Detectors Nuclear emulsion

Comments on experiment 28600 K^+ endings in emulsion, 75 cases of Dalitz decays, one event of double Dalitz decay.

Related references

See also
B. M. Anand, Proc. Roy. Soc. **A220** (1953) 183;
H. Primakoff, Phys. Rev. **81** (1951) 899;
G. A. Snow and M. M. Shapiro, Rev. of Mod. Phys. **33** (1961) 231;
A. V. Tollestrup, S. Berman, R. Gomes, and M. A. Ruderman, in *Proc. of the 1960 Ann. Int. Conf. of the High Energy Physics at Rochester* (1960) 27;
Analyse information from
R. Bjorklund et al., Phys. Rev. **77** (1950) 213;
J. Steinberger, W. K. H. Panofsky, and J. Steller, Phys. Rev. **78** (1950) 802;
W. Chinowsky and J. Steinberger, Phys. Rev. **93** (1954) 586;
W. K. H. Panofsky, R. L. Aamodt, and J. Hadley, Phys. Rev. **81** (1951) 565;
R. P. Haddock, A. Abashian, K. M. Crowe, and J. B. Czirr, Phys. Rev. Lett. **3** (1959) 478;

R. F. Blackie, A. Engler, and J. H. Mulvey, Phys. Rev. Lett. **5** (1960) 384;
M. Schein, J. Fainberg, D. M. Haskin, and R. G. Glasser, Phys. Rev. **91** (1953) 973;
G. Harris, J. Orear, and S. Taylor, Phys. Rev. **106** (1957) 327;
V. Glasser and R. A. Ferrell, Phys. Rev. **121** (1961) 886;

Reactions
$K^+ \to \pi^+ \pi^0$
$\pi^0 \to 2e^- 2e^+$
$\pi^0 \to e^- e^+ \gamma$

Particles studied π^0

FROISSART 1961

■ Froissart upper bound on the total cross sections of hadronic collisions ■

Asymptotic Behavior and Subtractions in the Mandelstam Representation

M. Froissart
Phys. Rev. **123** (1961) 1053;

Abstract It is proved that a two-body reaction amplitude involving scalar particles and satisfying Mandelstam's representation is bounded by expressions of the form $Cs\ln^2 s$ at the forward and backward angles, and $C^{\frac{3}{4}} \ln^{\frac{3}{2}} s$ at any other fixed angle in the physical region, C being a constant, s being the total squared c.m. energy. This corresponds to cross sections increasing at most like $\ln^2 s$. These restrictions limit the freedom of choice of the subtraction terms to six arbitrary single spectral functions and one subtraction constant.

Related references
See also
S. Mandelstam, Phys. Rev. **112** (1958) 1344;
L. Schwartz, *Theorie des distributions*, Hermann and Cie, Paris (1951);

NAMBU 1961B

■ Nambu-Jona-Lasinio nonlinear model of hadrons ■

Dynamical Model of Elementary Particles Based on an Analogy with Superconductivity. II

Y. Nambu, G. Jona-Lasinio
Phys. Rev. **124** (1961) 246;

Abstract Continuing the program developed in a previous paper, a "superconductive" solution describing the proton-neutron doublet is obtained from a nonlinear spinor field Lagrangian. We find the pions of finite mass as nucleon-antinucleon bound states by introducing a small bare mass into the Lagrangian which otherwise possesses a certain type of the γ_5 invariance. In addition, heavier mesons and two-nucleon bound states are obtained in the same approximation. On the basis of numerical mass relations, it is suggested that the bare nucleon field is similar to the electron-neutrino field, and further speculations are made concerning the complete description of the baryons and leptons.

Related references
See also
Y. Nambu and G. Jona-Lasinio, Phys. Rev. **122** (1961) 345;
J. Goldstone, Nuovo Cim. **19** (1961) 154;
V. G. Vaks and A. I. Larkin, *Proc. of the 1960 Ann. Int. Conf. on High Energy Physics at Rochester*, Interscience Publishers, Inc., New York (1960), 871;
F. Gürsey, Nuovo Cim. **16** (1960) 230;

GOOD 1961

■ First evidence for $K_L \to K_S$ regeneration ■

Regeneration of Neutral K Mesons and Their Mass Difference

R.H. Good et al.
Phys. Rev. **124** (1961) 1223;

Reprinted in
Bologna 1984, Proceedings, *Fifty Years of Weak-Interaction Physics* 631-647.

Abstract A beam of K_2 mesons was produced by passing a beam of 1.1 Bev/c negative pions through a liquid hydrogen target and accepting the neutral reaction products in the forward direction after allowing the K_1 component to decay. The resultant beam was observed in a 30-in. propane bubble chamber fitted with lead and iron plates. About 200 regenerated K_1 mesons were identified by their characteristic Q value and decay rate. All three types of regeneration were observed: by transmission in the plates, by nuclear diffraction, and by interaction with single nucleons. The detection of the first two types of regeneration constitutes strong evidence for the correctness of the Gell-Mann and Pais particle mixture theory. Comparison of the transmission and diffraction regeneration effect, using the method of M. L. Good, gives the K_1-K_2 mass difference δ. Two important corrections must be applied to Good's formula: One originates from the nuclear scattering of the transmission component, the other from the multiplicity of scatterings in a thick plate. The independence from nuclear parameters, which was an advantageous property of Good's formula, is no longer rigorously valid; but due to the sharp dependence of the transmission intensity upon the mass difference, the nuclear properties of K^0 and $\overline{K}^0$, as derived from K^+ and K^- data, still allow a measurement of δ. We find that δ is $0.84^{+0.29}_{-0.22}$ in units of $\hbar/\tau_1$, where τ_1 is the K_1 mean lifetime. With 90% confidence level, the difference is between 0.44 and 1.2 $\hbar/\tau_1$. The probability that the transmission peak we observe is due to a statistical fluctuation is one in a million.

Accelerator LBL-Bevatron

Detectors Heavy-liquid bubble chamber

Related references

More (earlier) information
F. Muller et al., Phys. Rev. Lett. **4** (1960) 418;

See also
M. Gell-Mann and A. Pais, Phys. Rev. **97** (1955) 1387;

Reactions

$p\,C \to \pi^-\,X$	5.3 GeV(T_{lab})
$\pi^-\,p \to \Lambda\,K^0$	1.1 GeV(P_{lab})
$\pi^-\,p \to \Sigma^0\,K^0$	1.1 GeV(P_{lab})
K_L Fe $\to$ Fe K_S	
K_L Pb $\to$ Pb K_S	
K_L nucleon $\to$ nucleon K_S	
$K_S \to \pi^+\,\pi^-$	
$K_L \to \pi^+\,\pi^-$	

Particles studied K_L, K_S

TERNOV 1961

■ Prediction of the "radiation self-polarization" effect for electrons moving in magnetic field ■

On the Possibility of Polarization of an Electron Beam due to Relativistic Radiation in a Magnetic Field

I.M. Ternov, Y.M. Loskutov, L.I. Korovina
Zh. Eksp. Teor. Fiz. **41** (1961) 1294;

Abstract Spin flip of electrons due to radiation during movement in a uniform magnetic field is considered. It is demonstrated that an initially unpolarized beam becomes partially polarized, the magnetic moment mainly being directly along the filed.

ALSTON 1961B — Nobel prize

■ First evidence for the $K^*(892)$ resonance. Nobel prize to L. W. Alvarez awarded in 1968 "for his decisive contribution to elementary particle physics, in particular the discovery of a large number of resonance states, made possible through his development of the hydrogen bubble chamber technique and data analysis" ■

Resonance in the K-π System

M.H. Alston et al.
Phys. Rev. Lett. **6** (1961) 300;

Reprinted in
M. L. Good at the meeting of the American Physical Society, Chicago, November, 1960, Bull. Am. Phys. Soc. 5, 414 (1960).
R. N. Cahn and G. Goldhaber, *The Experimental Foundations of Particle Physics*, Cambridge Univ. Press (1991) 127.

Abstract Reactions of the form $K^-p \to \overline{K}^0\pi^-p$ were studied in a 15 in. hydrogen bubble chamber exposed to a 1.15 BeV/c K^- beam. 48 events were observed giving a total cross section of 2.0 ± 0.3 mb. The kinetic energies of protons resulting from the interactions show a strong peaking around a value of 20 MeV and this is interpreted as arising from a final-state $K\pi$ resonance or K^* particle: $K^-p \to K^{*-}\,p$. The mean value of the total resonance energy (or mass) of the state is 885 ± 3 MeV. After subtraction of estimated background, 22 events lie in the region of the peak and the full width of the peak at half-height is found to be 16 MeV, corresponding to a lifetime of 4×10^{-23} sec. The angular distribution of the decay of the K^* is consistent with a spin $J=0$ or $J=1$ but not $J\geq2$. A preliminary investigation of additional data from the reactions $K^-\,p \to K^-\pi^0\,p$ and $K^-\,p \to K^-\,\pi^+\,n$ also indicates resonances and a crude estimate of the branching ratio

$$R = (K^{*-} \to K^-\,\pi^0)/(K^{*-} \to \overline{K}^0\pi^-)$$

is given as 0.75 ± 0.35. This agrees with a ratio of 0.5 which should occur if the K^* has isotopic spin $I=\frac{1}{2}$. *(Science Abstracts, 1961, 8502. J. D. Dowell)*

Accelerator LBL-Bevatron

Detectors Hydrogen bubble chamber

Related references

More (earlier) information
M. L. Good, Bull.Am.Phys.Soc. **5** (1960) 414;
M. H. Alston et al., Phys. Rev. Lett. **5** (1960) 520;

See also
R. Spitzer and H. P. Stapp, Phys. Rev. **109** (1958) 540;
J. Tiomno, A. L. L. Videira, and N. Zagury, Phys. Rev. Lett. **6** (1961) 120;
M. Gell-Mann, in *Proc. of the 1960 Ann. Int. Conf. on High-Energy Physics at Rochester*, Interscience Publishers, New York (1960) 508;

Reactions

$K^-\,p \to p\,\overline{K}^0\,\pi^-$	1.15 GeV(P_{lab})
$K^-\,p \to p\,K^-\,\pi^0$	1.15 GeV(P_{lab})
$K^-\,p \to n\,K^-\,\pi^+$	1.15 GeV(P_{lab})
$K^-\,p \to p\,K^*(892)^-$	1.15 GeV(P_{lab})
$K^-\,p \to n\,\overline{K}^*(892)^0$	1.15 GeV(P_{lab})
$K^*(892)^- \to \overline{K}^0\,\pi^-$	

Particles studied $K^*(892)^-, K^*(892)^0$

GRIBOV 1961

■ Generalization of Regge asymptotics for relativistic scattering amplitudes ■

Partial Waves with Complex Orbital Momenta and the Asymptotic Behavior of the Scattering Amplitude

V.N. Gribov
Zh. Eksp. Teor. Fiz. **41** (1961) 1962;

Abstract It is shown that in relativistic theory the partial wave amplitudes f_l analytical functions of the angular momentum l. The asymptotic behavior of the scattering amplitude as a function of the transferred momentum is determined by the nearest singularities of f_l. An expression for the scattering amplitude at arbitrary transferred momenta is obtained in terms of f_l and satisfies the Mandelstam equation which relates the spectral function and the absorption terms. The behavior of the scattering amplitude at high energies is discussed.

Related references

More (earlier) information
V. N. Gribov, Nucl. Phys. **22** (1961) 249;
V. N. Gribov, Zh. Eksp. Teor. Fiz. **41** (1961) 667;

See also
T. Regge, Nuovo Cim. **14** (1959) 952;
T. Regge, Nuovo Cim. **18** (1960) 947;
G. F. Chew, S. C. Frautschi, Phys. Rev. Lett. **5** (1960) 580;
M. Froissart, Phys. Rev. **123** (1961) 1053;

Reactions

hadron hadron → 2hadron	>10 GeV(P_{lab})
hadron hadron → X	>10 GeV(P_{lab})

STONEHILL 1961

■ Evidence for the ρ meson res onance ■

Pion-Pion Interaction in Pion Production by $\pi^+ p$ Collisions

D.L. Stonehill et al.
Phys. Rev. Lett. **6** (1961) 624;

Excerpt Since the first conjectures that rise in the total $\pi^- p$ cross section between 300 and 600 MeV might be caused by a pion-pion interaction, this subject has received considerable attention. Theoretical analysis of high energy electron scattering on protons and neutrons has predicted a resonance in the pion-pion interaction at a total di-pion energy (ω) of 4 to 5 pion masses, with isotopic spin and angular momentum both equal to one. Several analyses of $\pi^- p$ experiments in the 1 BeV energy range have tended to confirm this prediction, and application of the Chew-Low method has indicated a steep rise in the pion-pion cross section above $\omega = 4$. Recent work with 1.9 BeV $\pi^- p$ collisions shows a peak in the pion-pion interaction at $\omega \sim 5.5$. We report here evidence of pion-pion interaction in $\pi^- p$ collisions at three separate energies, which show striking effects attributable to a pion-pion resonance with ω of about 5.5 pion masses. *(Extracted from the introductory part of the paper.)*

Accelerator BNL-COSMOTRON

Detectors Hydrogen bubble chamber

Related references

See also
H. W. J. Foelsche and H. Kraybill, Phys. Rev. **134** (1964) B1138;
W. R. Frazer and J. R. Fulco, Phys. Rev. Lett. **2** (1959) 365;
J. Bowcock, W. N. Cottingham, and D. Lurie, Phys. Rev. Lett. **5** (1960) 386;
S. Bergia, A. Stanghellini, S. P. Fubini, and C. Villi, Phys. Rev. Lett. **6** (1961) 367;
W. D. Walker et al., Bull.Am.Phys.Soc. **6** (1961) 311;

Reactions

$\pi^+ p \to X$	0.91-1.26 GeV(E_{lab})
$\pi^+ p \to p\,\pi^+$	0.91-1.26 GeV(E_{lab})
$\pi^+ p \to p\,\pi^+\,\pi^0$	0.91-1.26 GeV(E_{lab})
$\pi^+ p \to n\,2\pi^+$	0.91-1.26 GeV(E_{lab})
$\pi^+ p \to$ nucleon 2π (π's)	0.91-1.26 GeV(E_{lab})
$\pi^+ p \to \Sigma^+ K^+$	0.91-1.26 GeV(E_{lab})
$\pi^+ p \to p\,\pi^+\,\pi^0$	0.91-1.26 GeV(E_{lab})
$\pi^+ p \to n\,2\pi^+$	0.91-1.26 GeV(E_{lab})
$\pi^+ p \to p\,\rho^+$	
$\rho^+ \to \pi^+\,\pi^0$	

Particles studied ρ^+, ρ

ERWIN 1961C

■ Another evidence for the ρ meson resonance ■

Evidence for a π-π Resonance in the I=1, J=1 State

A.R. Erwin, R. March, W.D. Walker, E. West
Phys. Rev. Lett. **6** (1961) 628;

Reprinted in
R. N. Cahn and G. Goldhaber, *The Experimental Foundations of Particle Physics*, Cambridge Univ. Press (1991) 130.
The Physical Review — the First Hundred Years, AIP Press (1995) 878.

Excerpt Since the earliest data became available on pion production by pions, certain features have been quite clear. The main feature which is strongly exhibited above energies of 1 BeV is that collisions are preferred in which there is a small momentum transfer to the nucleon. This is shown by the nucleon angular distributions which are sharply peaked in the backward direction. These results suggest that large-impact-

parameter collisions are important in such processes. The simplest process that could give rise to such collisions is a pion-pion collision with the target pion furnished in a virtual state by the nucleon. The quantitative aspects of such collisions have been discussed by a number of authors. Goebel, Chew and Low, and Salzman and Salzman discussed means of extracting from the data the π π cross section.
Holliday and Frazer and Fulco deduced from electromagnetic data that indeed there must be a strong pion-pion interaction. In particular, Frazer and Fulco deduced that there probably was a resonance in the $I = 1$, $J = 1$ state. A qualitative set of π p phase shifts in the 400-600 MeV region were used by Bowcock *et al.* to deduce an energy of about 660 MeV in the π π system for the resonance. The work of Pickup et al. showed an indication of a peak in the π π spectrum at an energy of about 600 MeV. *(Extracted from the introductory part of the paper.)*

Accelerator BNL-COSMOTRON

Detectors Hydrogen bubble chamber

Related references

See also
W. Holliday, Phys. Rev. **101** (1956) 1198;
W. R. Frazer and J. R. Fulco, Phys. Rev. Lett. **2** (1959) 365;
J. Bowcock, W. N. Cottingham, and D. Lurie, Phys. Rev. Lett. **5** (1960) 386;
J. Bowcock, W. N. Cottingham, and D. Lurie, Nuovo Cim. **19** (1961) 142;
E. Pickup, F. Ayer, and E. O. Salant, Phys. Rev. Lett. **5** (1960) 161;

Reactions

$\pi^- p \to p \pi^0 \pi^-$	1.82-1.96 GeV(P_{lab})
$\pi^- p \to n \pi^+ \pi^-$	1.82-1.96 GeV(P_{lab})
$\pi^- \pi^+ \to \pi^+ \pi^-$	0.4-1.0 GeV(E_{cm})
$\pi^- p \to p \rho^-$	
$\pi^- p \to n \rho^0$	
$\rho^- \to \pi^0 \pi^-$	
$\rho^0 \to \pi^+ \pi^-$	

Particles studied ρ^-, ρ^0, ρ

ALSTON 1961E — Nobel prize

■ First evidence for the $\Lambda(1405\, S_{01})$ resonance. Nobel prize to L. W. Alvarez awarded in 1968 "for his decisive contribution to elementary particle physics, in particular the discovery of a large number of resonance states, made possible through his development of the hydrogen bubble chamber technique and data analysis" ■

Study of Resonances in the Σ π System

M.H. Alston et al.
Phys. Rev. Lett. **6** (1961) 698; CERN HEP-62, P. 311 (1962);

Excerpt Recently a $T = 1$ resonance in the Λ π system, called Y_1^*, has been observed with a mass of 1385 MeV. Two types of resonances have been predicted that might relate this observation to other elementary–particle interactions:

(1) $P_{3/2}$ resonances in the $\Lambda\pi$ and Σ π systems predicted by global symmetry, corresponding to the $(\frac{3}{2}, \frac{3}{2})$ resonance of the π N system,

(2) a spin-$\frac{1}{2}Y$ π resonance resulting from a boundstate in the $\overline{K}$ N system. The position and the width of the observed Y resonance agree with both theories, but since the spin and parity have not yet been determined, we cannot distinguish between the two theoretical interpretations.

Global symmetry (including a phase–space factor) predicts a branching ratio $R = (Y_l^{*\pm} \to \Sigma^0\ \pi^\pm)/\ (Y_1^{*\pm} \to \Lambda^0\ \pi^\pm) = (Y_1^{*\pm} \to \Sigma^\pm\ \pi^0)/\ (Y_1^{*\pm} \to \Lambda^0\ \pi^\pm) = \frac{1}{4}(0.225) \sim 5\%$. The $\overline{K}$ N bound-state model suggests values of R considerably larger than 5%. However, when nonzero effective ranges are taken into account, R can become quite small, especially if the Σ Λ parity should be odd.
To investigate these possibilities, we have continued our study of K^- p interactions at 1.15 BeV/c in the Lawrence Radiation Laboratory 15-in. hydrogen bubble chamber by studying events in which a Σ is observed. *(Extracted from the introductory part of the paper.)*

Accelerator LBL-Bevatron

Detectors Hydrogen bubble chamber

Related references

See also
M. H. Alston et al., Phys. Rev. Lett. **5** (1960) 520;
O. Dahl et al., Phys. Rev. Lett. **6** (1961) 142;
J. P. Berge et al., Phys. Rev. Lett. **6** (1961) 557;
H. J. Martin, L. B. Leipuner, and W. Chinowsky, Phys. Rev. Lett. **6** (1961) 283;
P. Bastien, M. Ferro-Luzzi, and A. H. Rosenfeld, Phys. Rev. Lett. **6** (1961) 702;

Reactions

$K^- p \to \Sigma^0 \pi^+ \pi^-$	1.15 GeV(P_{lab})
$K^- p \to \Sigma^+ \pi^-$	1.15 GeV(P_{lab})
$K^- p \to \Sigma^- \pi^+$	1.15 GeV(P_{lab})
$K^- p \to \Sigma^+ \pi^0 \pi^-$	1.15 GeV(P_{lab})
$K^- p \to \Sigma^- \pi^+ \pi^0$	1.15 GeV(P_{lab})
$K^- p \to \Sigma^+ \pi^+ 2\pi^-$	1.15 GeV(P_{lab})
$K^- p \to \Sigma^- 2\pi^+ \pi^-$	1.15 GeV(P_{lab})
$K^- p \to \Sigma^+ 2\pi^0 \pi^-$	1.15 GeV(P_{lab})
$K^- p \to \Sigma^- \pi^+ 2\pi^0$	1.15 GeV(P_{lab})
$K^- p \to \Sigma^0 \pi^+ \pi^0 \pi^-$	1.15 GeV(P_{lab})
$K^- p \to \Sigma^+ \pi^+ 2\pi^-$	1.15 GeV(P_{lab})
$K^- p \to \Sigma^- 2\pi^+ \pi^-$	1.15 GeV(P_{lab})
$K^- p \to \Sigma(1385\, P_{13})^+ \pi^-$	1.15 GeV(P_{lab})
$K^- p \to \Sigma(1385\, P_{13})^- \pi^+$	1.15 GeV(P_{lab})
$K^- p \to \Lambda(1405\, S_{01}) \pi^0$	1.15 GeV(P_{lab})

$K^- \, p \to \Sigma(1385\, P_{13})^0 \, \pi^0$ 1.15 GeV(P_{lab})

Particles studied $\Lambda(1405\, S_{01})$

GELL-MANN 1961 **Nobel prize**

■ Introduction of the $SU(3)$ octet structure of the known mesons and baryons. Nobel prize to M. Gell-Mann awarded in 1969 "for his contributions and discoveries concerning the classification of elementary particles and their reactions" ■

The Eightfold Way: A Theory of Strong Interaction Symmetry

M. Gell-Mann
CTSL-20 (1961);

Reprinted in

M. Gell-Mann and Y. Ne'eman, *The Eightfold Way: A Review – With Collection of Reprints, Frontiers in Physics*, editor D. Pines, W. A. Benjamin, Inc. New York – Amsterdam (1964) 11.
(translation into Russian) *Elementarnye Chastitsy i Kompensiruyushchie Polya*, red. D. D. Iwanenko, Mir, M. (1964) 117.

Abstract We attempt once more, as in the global symmetry scheme, to treat the eight known baryons as a supermultiplet, degenerate in the limit of a certain symmetry but split into isotopic spin multiplets by a symmetry-breaking term. Here we do not try to describe the symmetry violation in detail, but we ascribe it phenomenologically to the mass differences themselves, supposing that there is some analogy to the $\mu - e$ mass difference.
The symmetry is called unitary symmetry and corresponds to the "unitary group" in three dimensions in the same way that charge independence corresponds to the "unitary group" in two dimensions. The eight infinitesimal generators of the group form a simple Lie algebra, just like the three components of isotopic spin. In this important sense, unitary symmetry is the simplest generalization of charge independence. The baryons then correspond naturally to an eight-dimensional irreducible representation of the group; when the mass differences are turned on, the familiar multiplets appear. The pion and K meson fit into a similar set of eight particles, along with a predicted pseudoscalar meson ξ^0 having I= 0. The pattern of Yukawa coupling of π, K, and ξ if then nearly determined, in the limit of unitary symmetry.

The most attractive feature of the scheme is that it permits the description of eight vector mesons by a unified theory of the Yang-Mills type (with a mass term). Like Sakurai, we have a triplet ρ of vector mesons coupled to the isotopic spin current and a singlet vector meson ω^0 coupled to the hypercharge current. We also have a pair of doublets M and $\overline{M}$, strange vector mesons coupled to strangeness-changing currents that are conserved when the mass differences are turned off. There is only one coupling constant, in the symmetric limit, for the system of eight vector mesons. There is some experimental evidence of ω^0 and M, while ρ is presumably the famous I = 1, $J = 1$, $\pi - \pi$ resonance.
A ninth vector meson coupled to the baryon current can be accommodated naturally in the scheme.
The most important prediction is the qualitative one that the eight baryons should all have same spin and parity and that the pseudoscalar and vector mesons should form "octets", with possible additional "singlets".
If the symmetry is not to badly broken in the case of the renormalized coupling constants of the eight vector mesons, then numerous detailed predictions can be made of experimental results.
The mathematics of the unitary group is described by considering three fictitious "leptons", ν e^-, and μ^-, which may or may not have something to do with real leptons. If there is a connection, then it may throw light on the structure of the week interactions.

Related references

See also
M. Gell-Mann, Phys. Rev. **106** (1957) 1296;
J. Schwinger, Ann.Phys. **2** (1957) 407;
J. J. Sakurai, Ann.Phys. **11** (1960) 1;
C. N. Yang and R. L. Mills, Phys. Rev. **96** (1954) 191;
W. R. Frazer and J. R. Fulco, Phys. Rev. **117** (1960) 1609;
J. Bowcock, W. N. Cottingham, and D. Lurie, Phys. Rev. Lett. **5** (1960) 386;
Y. Nambu, Phys. Rev. **106** (1957) 1366;
G. F. Chew, Phys. Rev. Lett. **4** (1960) 142;
T. D. Lee and C. N. Yang, Phys. Rev. **98** (1955) 1501;
S. L. Glashow, Nucl. Phys. **10** (1959) 107;
A. Salam and J. C. Ward, Nuovo Cim. **11** (1959) 568;
E. Fermi and C. N. Yang, Phys. Rev. **76** (1949) 1739;
R. P. Feynman and M. Gell-Mann, Phys. Rev. **109** (1958) 193;
S. A. Bludman and M. A. Ruderman, Nuovo Cim. **9** (1958) 433;
M. Gell-Mann and M. Levi, Nuovo Cim. **16** (1960) 705;
M. H. Alston et al., Phys. Rev. Lett. **5** (1960) 518;

Particles studied meson, baryon

MAGLIC 1961B **Nobel prize**

■ First evidence for the ω meson resonance. Nobel prize to L. W. Alvarez awarded in 1968 "for his decisive contribution to elementary particle physics, in particular the discovery of a large number of resonance states, made possible through his development of the hydrogen bubble chamber technique and data analysis" ■

Evidence for a T=0 Three-Pion Resonance

B.C. Maglic, L.W. Alvarez, A.H. Rosenfeld, M.L. Stevenson
Phys. Rev. Lett. **7** (1961) 178;

Reprinted in

R. N. Cahn and G. Goldhaber, *The Experimental Foundations of Particle Physics*, Cambridge Univ. Press, (1991) 133.

Excerpt The existence of a heavy neutral meson with $T = 0$ and $J = 1^-$ was predicted by Nambu in an attempt to explain the electromagnetic form factors of the proton and neutron. Chew has pointed out that such a vector meson should exist on dynamical grounds as a three-pion resonance or a bound state. Such a particle is also expected in the vector meson theory of Sakurai and, as a member of an octet of mesons, according to the unitary symmetry theory; and for other reasons. We will refer to it as ω. Previous searches for ω have primarily been confirmed to the mass region $m_\omega \leq 3\mu$,with $\mu =$ the pion mass, where only the following radiative decay modes are allowed: $\omega \to \pi^0\gamma$, $\omega \to 2\pi^0\gamma$, and $\omega \to \pi^+\pi^-\gamma$. The ω cannot decay into two pions.
The present search was made assuming $m_\omega \geq 3m_\pi$, where the decay

$$\omega \to \pi^+\pi^-\pi^0$$

is possible. *(Extracted from the introductory part of the paper.)*

Accelerator Bevatron

Detectors Hydrogen bubble chamber

Related references

More (later) information
N. H. Xuong and G. R. Lynch, Phys. Rev. Lett. **7** (1961) 327;

See also
Y. Nambu, Phys. Rev. **106** (1957) 1366;
G. F. Chew, Phys. Rev. Lett. **4** (1960) 142;
V. de Alfaro and B. Vitale, Phys. Rev. Lett. **7** (1961) 72;
J. J. Sakurai, Ann.Phys. **11** (1960) 1;
J. Weis, Nuovo Cim. **15** (1960) 52;
M. Johnson and E. Teller, Phys. Rev. **98** (1955) 783;
H. Durer, Phys. Rev. **103** (1956) 469;
E. Fabri, Nuovo Cim. **11** (1954) 479;
R. H. Dalitz, Phil. Mag. **44** (1953) 1068;
H. Durer and E. Teller, Phys. Rev. **101** (1956) 494;
S. Sawada and M. Yonezawa, Progr. of Theor. Phys. **22** (1959) 610;
M. Ikeda, S. Ogawa, and Y. Ohnuki, Progr. of Theor. Phys. **22** (1959) 715;
G. F. Chew, R. Karplus, S. Gasiorowicz, and F. Zachariazen, Phys. Rev. **110** (1958) 265;
J. J. Sakurai, Nuovo Cim. **16** (1960) 388;

Analyse information from
S. Bergia, A. Stanghellini, S. P. Fubini, and C. Villi, Phys. Rev. Lett. **6** (1961) 367;
E. Pickup, F. Ayer, and E. O. Salant, Phys. Rev. Lett. **5** (1960) 161;
R. Gomes et al., Phys. Rev. Lett. **5** (1960) 170;
A. Abashian, N. E. Booth, and K. M. Crowe, Phys. Rev. Lett. **5** (1960) 258;
J. G. Rushbrooke and D. Radojicic, Phys. Rev. Lett. **5** (1960) 567;
K. Berkelman, G. Cortellessa, and A. Reale, Phys. Rev. Lett. **6** (1961) 234;
J. A. Anderson et al., Phys. Rev. Lett. **6** (1961) 365;
A. R. Erwin et al., Phys. Rev. Lett. **6** (1961) 628;

Superseded by
M. L. Stevenson et al., Phys. Rev. **125** (1962) 687;

Reactions

$\bar{p}\,p \to 2\pi^+\,\pi^0\,2\pi^-$ 1.61 GeV(P_{lab})
$\omega \to \pi^+\,\pi^0\,\pi^-$

Particles studied ω

NE'EMAN 1961

■ Introduction of the $SU(3)$ octet structure of the known mesons and baryons ■

Derivation of Strong Interactions from a Gauge Invariance

Y. Ne'eman
Nucl. Phys. **26** (1961) 222;

Reprinted in
(translation into Russian) *Elementarnye Chastitsy i Kompensiruyushchie Polya*, red. D. D. Iwanenko, Mir, M. (1964) 176.

Abstract A representation for the baryons and bosons is suggested, based on the Lie algebra of the 3-dimensional traceless matrices. This enables us to generate the strong interactions from a gauge invariance principle, involving 8 vector bosons. Some connections with the electromagnetic and weak interactions are further discussed.

Related references

See also
C. N. Yang and R. L. Mills, Phys. Rev. **96** (1954) 191;
J. J. Sakurai, Ann.Phys. **11** (1960) 1;
A. Salam and J. C. Ward, Nuovo Cim. **11** (1959) 167;
A. Salam and J. C. Ward, Nuovo Cim. **19** (1961) 167;
J. Tiomno, Nuovo Cim. **6** (1957) 69;
J. Schwinger, Ann.Phys. **2** (1957) 407;
M. Ikeda, S. Ogawa, and Y. Ohnuki, Progr. of Theor. Phys. **22** (1959) 715;
R. Utiyama, Phys. Rev. **101** (1956) 1597;

Particles studied meson, baryon

XUONG 1961C

■ Confirmation of the existence of the ω meson ■

Evidence Confirming the $T = 0$ Three-Pion Resonance

N.H. Xuong, G.R. Lynch
Phys. Rev. Lett. **7** (1961) 327;

Excerpt We present here some evidence for a $T = 0$, three-pion resonance with data from events of the type

$$\bar{p}\,p \to 3\pi^+\,3\pi^-\pi^0$$

produced by antiprotons of 1.61 BeV/c in the 72-in. hydrogen bubble chamber. The energy and the width of the resonance agree very well with the ω meson found by Maglić et al. who analyzed the $2\pi^+\,2\pi^-\pi^0$ annihilations in the same experiment. *(Extracted from the introductory part of the paper.)*

Accelerator LBL-Bevatron

Detectors Hydrogen bubble chamber

Related references

More (earlier) information
J. Button et al., Phys. Rev. **121** (1961) 1788;

See also
Y. Nambu, Phys. Rev. **106** (1957) 1366;
J. J. Sakurai, Ann.Phys. **11** (1960) 1;
M. Ikeda, S. Ogawa, and Y. Ohnuki, Progr. of Theor. Phys. **22** (1959) 715;
J. Weis, Nuovo Cim. **15** (1960) 52;
Y. Fujii, Progr. of Theor. Phys. **21** (1959) 232;
G. F. Chew, Phys. Rev. Lett. **4** (1960) 142;
A. Salam and J. C. Ward, Nuovo Cim. **20** (1961) 419;
Y. Yamaguchi, Prog. Theor. Phys. Suppl. **11** (1959) 37;
J. J. Sakurai, Nuovo Cim. **16** (1960) 388;

Analyse information from
B. C. Maglic et al., Phys. Rev. Lett. **7** (1961) 178;

Reactions
$\overline{p}\, p \rightarrow 3\pi^+ \pi^0 3\pi^-$ 1.61 GeV(P_{lab})
$\omega \rightarrow \pi^+ \pi^0 \pi^-$

Particles studied ω

CHEW 1961

■ Invention of equivalence of elementary hadrons and hadronic resonances on the basis of Regge trajectories. Invention of the Chew-Frautschi plot to classify hadrons. Invention of the vacuum pomeron trajectory ■

Principle of Equivalence for all Strongly Interacting Particles within the S-Matrix Framework

G.F. Chew, S.C. Frautschi
Phys. Rev. Lett. **7** (1961) 394;

Excerpt The notion, inherent in Lagrangian field theory, that certain particles are fundamental while others are complex, is becoming less and less palatable for baryons and mesons as the number of candidates for elementary status continues to increase. Sakata has proposed that only neutron, proton, and Λ are elementary, but this choice is rather arbitrary, and strong-interaction consequences of the Sakata model merely reflect the established symmetries. Heisenberg some years ago proposed an underlying spinor field that corresponds to no particular particle but which is supposed to generate all the observed particles on an equivalent basis. The spirit of this approach satisfies Feynman's criterion that the correct theory should not allow a decision as to which particles are elementary, but it has proved difficult to find a convincing mathematical framework in which to fit the fundamental spinor field. On the other hand, the analytically continued S-matrix with only those singularities required by unitarity has progressively, over past half decade, appeared more and more promising as a basis for describing the strongly interacting particles. Our purpose here is to propose a formulation of the Feynman principle within the S-matrix framework. *(Extracted from the introductory part of the paper.)*

Related references

More (earlier) information
G. F. Chew and S. C. Frautschi, Phys. Rev. **124** (1961) 264;
G. F. Chew and S. C. Frautschi, Phys. Rev. **123** (1961) 1478;

See also
S. Sakata, Progr. of Theor. Phys. **16** (1953) 686;
W. Heisenberg, Rev. of Mod. Phys. **29** (1957) 269;
G. F. Chew, *The S-Matrix Theory of Strong Interactions* W. A. Benjamin and Company, New York (1961);
L. Castillejo, R. H. Dalitz, and F. J. Dyson, Phys. Rev. **101** (1956) 453;
R. H. Dalitz and S. F. Tuan, Ann.Phys. **10** (1960) 307;
T. Regge, Nuovo Cim. **14** (1959) 951;
T. Regge, Nuovo Cim. **18** (1960) 947;
M. Froissart, Phys. Rev. **123** (1961) 1053;
V. N. Gribov, Nucl. Phys. **22** (1961) 249;

PEVSNER 1961

■ First evidence for the η meson. Confirmation of the ω meson ■

Evidence for a Three-Pion Resonance Near 550 MeV

A. Pevsner et al.
Phys. Rev. Lett. **7** (1961) 421;

Reprinted in
R. N. Cahn and G. Goldhaber, *The Experimental Foundations of Particle Physics*, Cambridge Univ. Press (1991) 138.

Excerpt A study has been under way of multipion resonances in $\pi^+ d$ reactions observed in the Lawrence Radiation Laboratory 72-in. bubble chamber exposed to a 1.23 BeV/c pion beam from the Bevatron. A preliminary report on this research was given at the Aix-en-Provence Conference on Elementary Particles where the existence of the ω^0 meson reported by the Berkeley group was confirmed. Since then these data have been substantially increased, although the experiment is still in progress. The existence of a second neutral 3-pion resonance with a mass of approximately 550 MeV is indicated by this larger sample of events. *(Extracted from the introductory part of the paper.)*

Accelerator LBL-Bevatron

Detectors Deuterium bubble chamber

Related references

See also
J. J. Sakurai, Ann.Phys. **11** (1960) 1;
Y. Nambu, Phys. Rev. **106** (1957) 1366;
G. F. Chew, Phys. Rev. Lett. **4** (1960) 142;
S. Bergia, A. Stanghellini, S. P. Fubini, and C. Villi, Phys. Rev. Lett. **6** (1961) 367;
P. Strivastava and E. C. G. Sudarshan, Phys. Rev. **110** (1958) 765;
R. Hofstadter and R. Hernan, Phys. Rev. Lett. **6** (1961) 293;
M. M. Block, Phys. Rev. **101** (1956) 796;

Analyse information from
B. C. Maglic et al., Phys. Rev. Lett. **7** (1961) 178;
J. J. Sakurai, Phys. Rev. Lett. **7** (1961) 355;

Reactions

π^+ deuteron $\rightarrow 2p\,\pi^+\,\pi^0\,\pi^-$ 1.23 GeV(P_{lab})
$\pi^+ n \rightarrow p\,\pi^+\,\pi^0\,\pi^-$ 1.23 GeV(P_{lab})
$\omega \rightarrow \pi^+\,\pi^0\,\pi^-$
$\eta \rightarrow \pi^+\,\pi^0\,\pi^-$

Particles studied ω, η

GLASHOW 1961 — Nobel prize

■ First introduction of the neutral intermediate boson. Nobel prize to S. L. Glashow awarded in 1979. Co-winners S. Weinberg and A. Salam "for their contribution to the theory of the unified weak and electromagnetic interaction between elementary particles, including inter alia the prediction of the weak neutral current" ■

Partial-Symmetries of Weak Interactions

S.L. Glashow
Nucl. Phys. **22** (1961) 579;

Abstract Weak and electromagnetic interactions of the leptons are examined under the hypothesis that the weak interactions are mediated by vector bosons. With only an isotopic triplet of leptons coupled to a triplet of vector bosons (two charged decay-intermediaries and the photon) the theory possesses no partial-symmetries. Such symmetries may be established if additional vector bosons or additional leptons are introduced. Since the latter possibility yields a theory disagreeing with experiment, the simplest partially-symmetric model reproducing the observed electromagnetic and weak interactions of leptons requires the existence of at least four vector-boson fields (including the photon). Corresponding partially-conserved quantities suggest leptonic analogues to the conserved quantities associated with strong interactions: strangeness and isobaric spin.

Related references

See also
C. Fronsdal and S. L. Glashow, Phys. Rev. Lett. **3** (1959) 570;
J. Schwinger, Ann.Phys. **2** (1957) 407;
S. L. Glashow, Nucl. Phys. **10** (1959) 107;
J. Bernstein, M. Gell-Mann, and L. Michel, Nuovo Cim. **16** (1960) 560;
M. Gell-Mann and M. Levi, Nuovo Cim. **16** (1960) 705;
G. Feinberg, Phys. Rev. **110** (1958) 1482;
R. P. Feynman and M. Gell-Mann, Phys. Rev. **109** (1958) 193;

Particles studied W^+, W^-, Z^0

AGS STAFF 1961

■ AGS (30 GeV) at BNL — the first strong focusing proton synchrotron ■

Some Operating Characteristics of the Brookhaven Alternating Gradient Synchrotron

AGS Staff;
HEACC-61, 39 (1961);

Reprinted in
The AGS Staff, *1961 International Conference on High Energy Accelerators*, (September 6-12, 1961) 33.

Excerpt Beam was first injected into the AGS May 19, 1960, and was observed on a fluorescent flag after one turn. On May 26, the inflector was pulsed and a spiralling beam obtained. The radiofrequency accelerating system was not completed until July; a trial of the starting oscillator with 5 rf stations, on July 17, accelerated beam for about 2 msec (the duration of the oscillator program). The phase-lock system was switched on and adjusted on July 21 and 22, when occasional beam survival to transition was observed. Although the scarcely completed rf system had not been adequately tested an attempt to accelerate to full energy was made on July 25. It failed. After a few days of electronic measurements and adjustments, acceleration to 31 BeV was achieved on July 29, 1960. The running-in was satisfyingly straightforward, with no difficulties other than the usual defects of untried apparatus. *(Extracted from the introductory part of the paper.)*

DUNAITSEV 1962

■ First observation of the π^+ beta decay. First direct experimental evidence for the validity of the CVC hypothesis ■

Check of the Conserved Vector Current Hypothesis

A.F. Dunaitsev, V.I. Petrukhin, Y.D. Prokoshkin, V.I. Rykalin
Phys. Lett. **1** (1962) 138;

Excerpt The investigation of the pion β decay

$$\pi^+ \rightarrow \pi^0 e^+ \nu_e$$

makes it possible to check directly whether the conserved vector current hypothesis suggested by Gershtein and Zeldovich as earlier as 1955 is testifiable. This mode of decay predicted by Zeldovich is of great interest at present as far as checking one of the fundamental assumptions of the modern weak interaction theory is concerned. *(Extracted from the introductory part of the paper.)*

Accelerator Dubna synchrocyclotron

Detectors Spectrometer

Data comments Four events.

Related references

More (earlier) information
A. F. Dunaitsev et al., Zh. Eksp. Teor. Fiz. **42** (1962) 632;
A. F. Dunaitsev et al., Zh. Eksp. Teor. Fiz. **42** (1962) 1421;

More (later) information
A. F. Dunaitsev et al., Zh. Eksp. Teor. Fiz. **47** (1964) 84;

See also
S. S. Gershtein and Y. B. Zeldovich, Zh. Eksp. Teor. Fiz. **29** (1955) 698;
R. P. Feynman and M. Gell-Mann, Phys. Rev. **109** (1958) 193;
Y. B. Zeldovich, Doklady Akad. Nauk SSSR **97** (1954) 421;

Reactions
$\pi^+ \to \pi^0 \, e^+ \, \nu_e$

Particles studied π^+

TRIPP 1962

■ Determination of the Σ parity ■

Determination of the Σ Parity

R.D. Tripp, M.B. Watson, M. Ferro-Luzzi
Phys. Rev. Lett. **8** (1962) 175;

Excerpt In a previous Letter we reported a hydrogen-bubble-chamber experiment on the $K^- p$ interaction in the vicinity of 400 MeV/c incident K^- momentum. The existence of an excited hyperon of 1520 MeV mass and 16 MeV full width was established; the state was found to have isotopic spin 0, spin 3/2, even parity with respect to $K^- p$, and a $\overline{K}N : \Sigma\pi : \Lambda 2\pi$ branching ratio of 3:5:1. In this Letter we report the study of the angular distributions and polarizations of the different $\Sigma\pi$ charge states, from which we conclude that the $Kp\Sigma$ parity is odd. *(Extracted from the introductory part of the paper.)*

Accelerator LBL-Bevatron

Detectors Hydrogen bubble chamber

Related references

See also
M. Ferro-Luzzi, R. D. Tripp, and M. B. Watson, Phys. Rev. Lett. **8** (1962) 28;
S. Minami, Progr. of Theor. Phys. **11** (1954) 213;
M. Nauenberg and A. Pais, Phys. Rev. **123** (1961) 1058;
E. P. Wigner, Phys. Rev. **98** (1955) 145;
E. F. Beall et al., Phys. Rev. Lett. **7** (1961) 285;
R. W. Birge and W. B. Fowler, Phys. Rev. Lett. **5** (1960) 254;
M. M. Block et al., Phys. Rev. Lett. **3** (1959) 291;
M. Gell-Mann, Phys. Rev. **106** (1957) 1296;
J. Schwinger, Ann.Phys. **2** (1957) 407;
A. Pais, Phys. Rev. **110** (1958) 574;
R. H. Capps, Phys. Rev. Lett. **6** (1961) 375;

Reactions
$K^- \, p \to \Sigma^- \, \pi^0$

Particles studied Σ

BROWN 1962

■ First evidence for $\Xi^- \, \overline{\Xi}^+$ pair production ■

Observation of Production of a $\Xi^- \, \overline{\Xi}^+$ Pair

H.N. Brown et al.
Phys. Rev. Lett. **8** (1962) 255;

Excerpt The reaction $\bar{p}p \to \Xi^- \, \overline{\Xi}^+$ has been observed in a 20-in. liquid hydrogen bubble chamber exposed to a separated antiproton beam of 3.3 BeV/c momentum produced in a tungsten target in the Brookhaven alternating gradient synchrotron. While expected to exist, the $\overline{\Xi}^+$ has so far not been observed, partially because the cross section for Ξ^- production is small in comparison with the production of other hyperons, and partially because beams of high enough energy have been available only recently. Both cascade hyperons decay in the visible region of the chamber, as does the antilambda from the $\overline{\Xi}^+$ decay. *(Extracted from the introductory part of the paper.)*

Accelerator Brookhaven (AGS) proton synchrotron

Detectors Hydrogen bubble chamber

Comments on experiment 34000 pictures.

Related references

Superseded by
C. Baltay et al., Phys. Rev. **140** (1965) B1027;

Reactions
$\bar{p} \, p \to \Xi^- \, \overline{\Xi}^+$ 3.3 GeV(P_{lab})

Particles studied $\overline{\Xi}^+$

CERN 1962

■ Confirmation of the existence of the $\overline{\Xi}^+$ ■

Example of Anticascade $\overline{\Xi}^+$ Particle Production in $\bar{p}$ p Interactions at 3.0 GeV/c.

CERN-Ecole Polytechnique-Saclay collaboration;
Phys. Rev. Lett. **8** (1962) 257;

Excerpt An experiment is in progress at the CERN proton synchrotron to study the interactions of fast antiprotons with protons. A high-energy separated beam has been installed and optimized to provide, in the first instance, a high-purity

beam of 3.0 GeV/c antiprotons. The interactions are being produced and observed in the Saclay 81-cm hydrogen bubble chamber.
In the methodical scanning of the first ten thousand photographs (with an average of seven antiprotons per photograph) an event has been found showing the production of an anticascade particle $\overline{\Xi}^+$. The object of this Letter is to present the data and the analysis leading to this conclusion. *(Extracted from the introductory part of the paper.)*

Accelerator CERN-PS proton synchrotron

Detectors Hydrogen bubble chamber

Related references

More (earlier) information
S. Van der Meer, preprint CERN-60-22 (1960);

Reactions

$\overline{p}\,p \rightarrow \overline{\Xi}^+ X$	3.0 GeV(P_{lab})

Particles studied $\overline{\Xi}^+$

DANBY 1962 — Nobel prize

■ First evidence for the ν_μ. Evidence for more than one kind of neutrinos. Nobel prize to L. M. Lederman, M. Schwartz, and J. Steinberger awarded in 1988 "for the neutrino beam method and the demonstration of the doublet structure of the leptons through the discovery of the muon neutrino" ■

Observation of High-Energy Neutrino Reactions and the Existence of Two Kinds of Neutrinos

G.T. Danby et al.
Phys. Rev. Lett. **9** (1962) 36;

Reprinted in

R. N. Cahn and G. Goldhaber, *The Experimental Foundations of Particle Physics*, Cambridge Univ. Press (1991) 184.
The Physical Review — the First Hundred Years, AIP Press (1995) 881.

Excerpt In the course of an experiment at the Brookhaven AGS, we have observed the interaction of high-energy neutrinos with matter. These neutrinos were produced primarily as the result of the decay of the pion:

$$\pi^\pm \rightarrow \mu^\pm\ (\nu/\overline{\nu})$$

It is the purpose of this Letter to report some of the results of this experiment including (1) demonstration that the neutrinos we have used produce μ mesons but do not produce electrons, and hence are very likely different from the neutrinos involved in β decay and (2) approximate cross sections. *(Extracted from the introductory part of the paper.)*

Accelerator Brookhaven (AGS) proton synchrotron

Detectors Spark chamber, Counters

Related references

See also
T. D. Lee and C. D. Yang, Phys. Rev. Lett. **4** (1960) 307;
Y. Yamaguchi, Progr. of Theor. Phys. **23** (1960) 1117;
N. Cabibbo and R. Gatto, Nuovo Cim. **15** (1960) 304;
G. Feinberg, Phys. Rev. **110** (1958) 1482;
E. J. Konopinski and H. M. Mahmoud, Phys. Rev. **92** (1953) 1045;
J. Schwinger, Ann.Phys. **2** (1957) 407;
I. Kawakami, Progr. of Theor. Phys. **19** (1957) 459;
M. Konuma, Nucl. Phys. **5** (1958) 504;
S. A. Bludman, Bull.Am.Phys.Soc. **4** (1959) 80;
S. Oneda and J. C. Pati, Phys. Rev. Lett. **2** (1959) 125;
K. Nishijima, Phys. Rev. **108** (1957) 907;
D. Bartlett, S. Devons, and A. Sachs, Phys. Rev. Lett. **8** (1962) 120;
S. Frankel et al., Phys. Rev. Lett. **8** (1962) 123;
B. Pontecorvo, JETP **10** (1960) 1236;
M. Schwartz, Phys. Rev. Lett. **4** (1960) 306;
W. F. Baker et al., Phys. Rev. Lett. **7** (1961) 101;
H. L. Anderson et al., Phys. Rev. **119** (1960) 2050;
G. Culligan et al., Phys. Rev. Lett. **7** (1961) 458;
R. H. Hildebrand, Phys. Rev. Lett. **8** (1962) 34;
E. Bleser et al., Phys. Rev. Lett. **8** (1962) 288;
R. P. Feynman and M. Gell-Mann, Phys. Rev. **109** (1958) 193;
E. C. G. Sudarshan and R. E. Marshak, Phys. Rev. **109** (1958) 1860;
R. Plano, Phys. Rev. **119** (1960) 1400;
T. D. Lee, P.Markstein, and C. N. Yang, Phys. Rev. Lett. **7** (1961) 429;
G. Feinberg, F. Gürsey, and A. Pais, Phys. Rev. Lett. **7** (1961) 208;

Reactions

p Be $\rightarrow \nu$ X	15 GeV(P_{lab})
ν Al $\rightarrow \mu^\pm$ X	<3 GeV(P_{lab})

Particles studied ν_μ

GELL-MANN 1962 — Nobel prize

■ Introduction of the $SU(3)$ singlet-octet structure of the known mesons and octet-decuplet structure for the baryons. Prediction of the Ω^- hyperon. Nobel prize to M. Gell-Mann awarded in 1969 "for his contributions and discoveries concerning the classification of elementary particles and their reactions" ■

Symmetries of Baryons and Mesons

M. Gell-Mann
Phys. Rev. **125** (1962) 1067;

Reprinted in

M. Gell-Mann and Y. Ne'eman, *The Eightfold Way: A Review – With Collection of Reprints, Frontiers in Physics*, ed. D. Pines, W. A. Benjamin, Inc. New York – Amsterdam (1964) 278.
The Physical Review — the First Hundred Years, AIP Press (1995) 1026.

Abstract The system of strongly interacting particles is discussed, with electromagnetism, weak interactions, and gravitation considered as perturbations. The electric current j_a, the weak current J_a, and the gravitational tensor $\theta_{\alpha\beta}$ are all well-defined operators, with finite matrix elements obeying dispersion relations. To the extent that the dispersion relations for matrix elements of these operators between the vacuum and other states are highly convergent and dominated by contributions from intermediate one-meson states, we have relations like the Goldberger-Treiman formula and universality principles like that of Sakurai according to which the ρ meson is coupled approximately to the isotopic spin. Homogeneous linear dispersion relations, even without subtractions, do not suffice to fix the scale of these matrix elements; in particular, for the nonconserved currents, the renormalization factors cannot be calculated, and the universality of strength of the weak interactions is undefined. More information than just the dispersion relations must be supplied, for example, by field-theoretic models; we consider, in fact, the equal-time commutation relations of the various parts of j_4 and J_4. These nonlinear relations define an algebraic system (or a group) that underlies the structure of baryons and mesons. It is suggested that the group is in fact $U(3) \times U(3)$, exemplified by the symmetrical Sakata model. The Hamiltonian density θ_{44} is not completely invariant under the group; the noninvariant part transforms according to a particular representation of the group; it is possible that this information also is given correctly by the symmetrical Sakata model. Various exact relations among form factors follow from the algebraic structure. In addition, it may be worthwhile to consider the approximate situation in which the strangeness-changing vector currents are conserved and the Hamiltonian is invariant under $U(3)$; we refer to this limiting case as "unitary symmetry." In the limit, the baryons and mesons form degenerate supermultiplets, which break up into isotopic multiplets when the symmetry breaking term in the Hamiltonian is "turned on." The mesons are expected to form unitary singlets and octets; each octet breaks up into a triplet, a singlet, and a pair of strange doublets. The known pseudoscalar and vector mesons fit this pattern if there exists also an isotopic singlet pseudoscalar meson χ^0. If we consider unitary symmetry in the abstract rather than in connection with a field theory, then we find, as an attractive alternative to the Sakata model, the scheme of Ne'eman and Gell-Mann, which we call the "eightfold way"; the baryons N, Λ, Σ, and Ξ form an octet, like the vector and pseudoscalar meson octets, in the limit of unitary symmetry. Although the violations of unitary symmetry must be quite large, there is some hope of relating certain violations to others. As an example of the methods advocated, we present a rough calculation of the rate of $K^+ \to \mu^+\nu$ in terms of that of $\pi^+ \to \mu^+\nu$.

Related references

See also
M. Gell-Mann, Phys. Rev. **106** (1957) 1296;
J. Schwinger, Ann.Phys. **2** (1957) 407;

Particles studied meson, baryon, Ω^-

PJERROU 1962

■ First evidence for the $\Xi(1530\,P_{13})$ resonance ■

Resonance in the (Ξ π) System at 1.53 GeV

G.M. Pjerrou et al.
Phys. Rev. Lett. **9** (1962) 114;

Reprinted in
R. N. Cahn and G. Goldhaber, *The Experimental Foundations of Particle Physics*, Cambridge Univ. Press (1991) 147.

Excerpt We wish to report the existence of a narrow resonance in the ($\Xi\pi$) system which we observed in the study of the interactions of negative K mesons in the LRL 72-in. hydrogen bubble chamber. The separated incident K^- beam, originating in the Bevatron, had a momentum of 1.80 ± 0.08 GeV/c; the uncertainty includes both the 6% momentum spread of the beam and the momentum loss in the chamber. The film was scanned for events with the following topology: "one positive secondary, one negative secondary with a decay, and one or two associated V's." *(Extracted from the introductory part of the paper.)*

Accelerator LBL-Bevatron

Detectors Hydrogen bubble chamber

Related references

See also
R. K. Adair, Phys. Rev. **100** (1955) 1540;

Reactions

$K^- \; p \to \Xi^- \; K^+$	1.8 GeV(P_{lab})
$K^- \; p \to \Xi^- \; K^+ \; \pi^0$	1.8 GeV(P_{lab})
$K^- \; p \to \Xi^- \; K^0 \; \pi^+$	1.8 GeV(P_{lab})
$K^- \; p \to \Xi(1530\,P_{13})^- \; K^+$	
$K^- \; p \to \Xi(1530\,P_{13})^0 \; K^0$	
$\Xi(1530\,P_{13}) \to \Xi \; \pi$	

Particles studied $\Xi(1530\,P_{13})^0, \Xi(1530\,P_{13})^-, \Xi(1530\,P_{13})$

CRETIEN 1962

■ Confirmation of the existence and evidence for spin zero of the η meson ■

Evidence for Spin Zero of the η from the Two Gamma-Ray Decay Mode

M. Cretien et al.
Phys. Rev. Lett. **9** (1962) 127;

Excerpt Recent experiments by Pevsner et al. (Phys. Rev. Lett. **7** (1961) 421) and others (Bastien et al., Phys. Rev. Lett **8** (1962) 114) report the existence of a mass 546 MeV

particle (called the η) which has both charged and neutral modes of decay. The charged mode of decay is $\eta^0 \to \pi^+\pi^-\pi^0$. Present experimental evidence assigns isotopic-spin zero to this particle.
The original work was done in the reaction

$$\pi^+ \, d \to p \, p \, \eta^0. \quad (1)$$

The proton in the deuteron plays a spectator role, hence the primary reaction is $\pi^+ n \to p \, \eta^0$. The cross section reported for the charged mode of decay of the η is 0.22 ± 0.06 mb and the cross section measured for the neutral modes of decay is 0.8 ± 0.2 mb. The experiment of reference 4 reports a flat plateau for the cross section in the region from 1700-1850 MeV, total energy in the center-of-mass system. The experiment reported here makes use of the reaction

$$\pi^- \, p \to n \eta^0. \quad (2)$$

From charge symmetry, reaction (2) will have the same cross section as reaction (1) at the same energy. This experiment covers an energy band from 1707-1740 MeV and thus one expects a production cross section for the η of about 1 mb. *(Extracted from the introductory part of the paper.)*

Accelerator BNL-COSMOTRON

Detectors Heavy-liquid bubble chamber

Related references

Analyse information from
A. Pevsner et al., Phys. Rev. Lett. **7** (1961) 421;
P. Bastien et al., Phys. Rev. Lett. **8** (1962) 114;
D. D. Carmony, A. H. Rosenfeld, and R. T. Van de Walle, Phys. Rev. Lett. **8** (1962) 117;

Reactions

$\pi^- \, p \to n \, 2\gamma$	1140 MeV(P_{lab})
$\pi^- \, p \to n \, \eta$	1140 MeV(P_{lab})
$\eta \to 2\gamma$	

Particles studied η

BERTANZA 1962D

■ Confirmation of the existence of the $\Xi(1530 \, P_{13})$ resonance. First evidence for the ϕ resonance ■

Possible Resonances in the $\Xi \, \pi$ and $K \, \overline{K}$ Systems

L. Bertanza et al.
Phys. Rev. Lett. **9** (1962) 180;

Reprinted in
The Physical Review — the First Hundred Years, AIP Press (1995) CD-ROM.

Excerpt The purpose of this note is to report the existence of marked departures from phase space in the effective-mass distributions for the $\Xi\pi$ and $K\overline{K}$ states. We present evidence that, in about 25% of the events observed, the $\Xi\pi$ state results from the decay of a resonant state (Ξ^*) with a mass of 1535 MeV and a full width of < 35 MeV. The observed anomaly in the $K\overline{K}$ effective-mass distribution is possibly open to different interpretations. If we assume it to be due to the decay of a resonant state K^*, we find that $M_{K^*} = 1020$ MeV, and that it has a full width of 20 MeV. However, it may also be possible to explain the effect as due to S-wave $K\overline{K}$ scattering. These results, as well as preliminary evidence concerning the properties of the Ξ^* and K^*, are discussed below. *(Extracted from the introductory part of the paper.)*

Accelerator Brookhaven (AGS) proton synchrotron

Detectors Hydrogen bubble chamber

Comments on experiment 79 (Ξ,π) combinations, 37 (K,$\overline{K}$) combinations.

Reactions

$K^- \, p \to \Xi^- \, K^+ \, \pi^0$	2.24,2.5 GeV(P_{lab})
$K^- \, p \to \Xi^- \, K^0 \, \pi^+$	2.24,2.5 GeV(P_{lab})
$K^- \, p \to \Xi^0 \, K^+ \, \pi^-$	2.24,2.5 GeV(P_{lab})
$K^- \, p \to \Xi^0 \, K^0 \, \pi^0$	2.24,2.5 GeV(P_{lab})
$K^- \, p \to \Lambda \, K^+ \, K^-$	2.24,2.5 GeV(P_{lab})
$K^- \, p \to \Lambda \, K^0 \, \overline{K}^0$	2.24,2.5 GeV(P_{lab})
$\Xi(1530 \, P_{13})^0 \to \Xi^- \, \pi^+$	
$\Xi(1530 \, P_{13})^0 \to \Xi^0 \, \pi^0$	
$\Xi(1530 \, P_{13})^- \to \Xi^- \, \pi^0$	
$\Xi(1530 \, P_{13})^- \to \Xi^0 \, \pi^-$	
$\phi \to K^+ \, K^-$	

Particles studied $\Xi(1530 \, P_{13})^0, \Xi(1530 \, P_{13})^-, \Xi(1530 \, P_{13}), \phi$

CHINOWSKY 1962

■ Determination of the spin of the $K^*(892)$ resonance to be 1 ■

On the Spin of the K* Resonance

W. Chinowsky et al.
Phys. Rev. Lett. **9** (1962) 330;

Abstract Investigation of the $K^+ \, p \to K^* \, N^*_{33}$ at 1.96 BeV/c with the 20 inch Brookhaven bubble chamber was achieved by looking at $K^+ \, p \to K^+ \, \pi^- \, p \, \pi^+$ events. 201 events out of 310 were in the double-resonant $K^+\pi^+$ and $p\pi^+$ states as shown by the effective mass distribution. From examination of the distribution of the angle α between the outgoing K^+ meson in the K^* centre–of–mass system, and the incident K^+ direction, the anisotropy can be fitted with a $\cos^2 \alpha$ term. Combining this result with those of Alston et al. allows a spin and parity assignment of 1^- to be given to the K^*. An "Adair analysis" of the data suggests the type of alignment of the K^* spin which is expected if one-pion exchange is the dominant production mode. *(Science Abstracts, 1963, 4457. R. J. Griffiths)*

Accelerator Brookhaven (AGS) proton synchrotron

Detectors Hydrogen bubble chamber

Related references

More (earlier) information
M. H. Alston et al., Phys. Rev. Lett. **6** (1961) 300;
See also
R. K. Adair, Phys. Rev. **100** (1955) 1540;
M. M. Block et al., Phys. Rev. Lett. **3** (1959) 291;
J. P. Berge, F. T. Solmitz, and H. D. Taft, Rev. Sci. Inst. **32** (1961) 538;
A. H. Rosenfeld and J. M. Snyder, Rev. Sci. Inst. **33** (1962) 181;

Reactions

$K^+ \, p \; \to \; p \, K^+ \, \pi^+ \, \pi^-$ 1.96 GeV(P_{lab})
$K^+ \, p \; \to \; \Delta(1232 \, P_{33})^{++} \, K^*(892)^0$ 1.96 GeV(P_{lab})
$K^*(892)^0 \; \to \; K^+ \, \pi^-$

Particles studied $K^*(892)^0$

FRAUTSCHI 1962

■ Application of Regge poles to resonances and particles ■

Experimental Consequences of the Hypothesis of Regge Poles

S.C. Frautschi, M. Gell-Mann, F. Zachariazen
Phys. Rev. **126** (1962) 2204;

Abstract In the nonrelativistic case of the Schrödinger equation, composite particles correspond to Regge poles in scattering amplitudes (poles in the complex plane of angular momentum). It has been suggested that the same may be true in relativistic theory. In that case, the scattering amplitude in which such a particle is exchanged behaves at high energies like $s^{\alpha(t)}[\sin \pi\alpha(t)]^{-1}$, where s is the energy variable and t the momentum transfer variable. When $t = t_R$, the mass squared of the particle, then α equals an integer is related to the spin of the particle. In contrast, we may consider the case of a field theory in which the exchanged particle is treated as elementary and we examine each order of perturbation theory. When $n > 1$, we can usually not renormalize successfully; when $n \leq 1$ and the theory is renormalizable, then the high-energy, behavior is typically $s^n(t - t_R)^{-1}\phi(t)$. Thus an experimental distinction is possible between the two situations. That is particularly interesting in view of the conjecture of Blankenbecler and Goldberger that the nucleon may be composite and that of Chew and Frautschi that all strongly interacting particles may be composite dynamical combinations of one another. We suggest a set of rules for finding the high-energy behavior of scattering cross sections according to the Regge pole hypothesis and apply them to π-π, π-N, and N-N scattering. We show how these cross sections differ from those expected when there are "elementary" nucleons and mesons treated in renormalized perturbation theory. For the case of $N - N$ scattering, we analyze some preliminary experimental data and find indications that an "elementary" neutral vector meson is probably not present. Various reactions are proposed to test the "elementary" or "composite" nature of other baryons and mesons. Higher energies may be needed than are available at present.

Related references

See also
T. Regge, Nuovo Cim. **14** (1959) 951;
T. Regge, Nuovo Cim. **18** (1960) 947;
G. F. Chew and S. C. Frautschi, Phys. Rev. Lett. **7** (1961) 394;
G. F. Chew and S. C. Frautschi, Phys. Rev. **124** (1961) 264;
M. Froissart, Phys. Rev. **123** (1961) 1053;
M. Gell-Mann and F. Zachariazen, Phys. Rev. **123** (1961) 1065;
M. Gell-Mann and F. Zachariazen, Phys. Rev. **124** (1961) 953;
S. C. Frautschi and J. D. Walecka, Phys. Rev. **120** (1960) 1486;
R. Blankenbecler and M. L. Goldberger, Phys. Rev. **126** (1962) 766;
B. Cork, W. A. Wenzel, and C. W. Causey, Phys. Rev. **107** (1957) 859;
G. A. Smith et al., Phys. Rev. **123** (1961) 2160;
W. M. Preston, R. Watson, and J. C. Street, Phys. Rev. **118** (1960) 579;
G. Cocconi et al., Phys. Rev. Lett. **7** (1961) 450;
S. D. Drell and Z. Hiida, Phys. Rev. Lett. **7** (1961) 199;
B. M. Udgaonkar and M. Gell-Mann, Phys. Rev. Lett. **8** (1962) 346;
R. J. Eden, P. V. Landshoff, J. C. Polkinghorne, and J. C. Taylor, Jour. Math. Phys. **2** (1961) 656;
S. Mandelstam, Phys. Rev. **115** (1959) 1741;
M. Gell-Mann, Phys. Rev. **125** (1962) 1067;
G. Salzman, *Proc. of the 1960 Ann. Int. Conf. on High Energy Physics at Rochester*, Interscience Pub. Inc., New York, (1960);
S. D. Drell, Rev. of Mod. Phys. **33** (1961) 458;
W. R. Frazer and J. R. Fulco, Phys. Rev. **117** (1960) 1603;
R. Blankenbecler, M. Goldberger, N. N. Khuri, and S. Trieman, Ann.Phys. **10** (1960) 62;

SALAM 1962

■ Conditions for renormalizability of general gauge theories of massive vector mesons ■

Renormalizability of Gauge Theories

A. Salam
Phys. Rev. **127** (1962) 331;

Abstract By generalization of methods developed by Kamefuchi, O'Raifeartaigh, and Salam, conditions for renormalizability of general gauge theories of massive vector mesons are derived. These conditions are stated explicitly in Eqs. (39) and (40) of the text. It is shown that all theories based on simple Lie groups (with the one exception of the neutral vector meson theory in interaction with a conserved current) are unrenormalizable.

Related references

See also
A. Komar and A. Salam, Nucl. Phys. **21** (1960) 624;
S. Kamefuchi and H. Umezawa, Nucl. Phys. **23** (1961) 399;
S. Kamefuchi, O'Raifeartaigh, and A. Salam, (Nucl. Phys. *to be*

published (1961);
Y. Nambu and G. Jona-Lasinio, Phys. Rev. **122** (1961) 345;
A. Salam, Nucl. Phys. **18** (1960) 681;
Y. Chisholm, Nucl. Phys. **26** (1961) 469;
H. I. Borchers, Nuovo Cim. **15** (1960) 784;
S. Glashow and M. Gell-Mann, Ann.Phys. **15** (1961) 439;

AMATI 1962

■ Invention of the multiperipheral model to analyze a few and many body hadronic reactions. Demonstration that multiperipheral model is capable to predict qualitatively the general features of elastic scattering, inelastic particles spectra, and topological cross sections ■

Theory of High Energy Scattering and Multiple Production

D. Amati, A. Stanghelini, S. Fubini
Nuovo Cim. **24** (1962) 896;

Excerpt In this paper we propose a theoretical model for high-energy interaction, the basic idea of which is that the high-energy processes are reducible to low energy ones, through a peripheral mechanism. The asymptotic properties of this model are studied by means of a linear homogeneous integral equation, whose kernel depends on the low-energy amplitudes. It is shown that many general predictions can be derived which are independent of the detailed form of the low-energy input. The results refer both to high energy elastic scattering and multiple production. For the inelastic processes we obtained simple general predictions for measurable quantities such as multiplicities, inelasticity, and spectra of secondaries. For the elastic scattering we find the characteristic Regge pole behaviour for all scattering amplitudes. The relation between the bound state problem and diffraction can therefore be understood in a relativistic model which contains in itself both phenomena. We finally discuss the possible correction of the model by using approximately the unitarity condition and we find indications of the possible existence of continuous power distribution, or equivalently of cuts in the complex angular momentum variable.

Related references

See also
S. Drell, Rev. of Mod. Phys. **33** (1961) 458;
E. Ferrari and F. Selleri, Nuovo Cim. Suppl. **24** (1962) 453;
I. M. Dremin and D. S. Chernyavski, Zh. Eksp. Teor. Fiz. **40** (1961) 1333;
D. Amati, S. Fubini, A. Stanghellini, and M. Tonin, Nuovo Cim. **22** (1961) 569;
D. Amati et al., Phys. Lett. **1** (1962) 29;
V. B. Berestetski and Y. Pomeranchuk, Nucl. Phys. **22** (1961) 629;
T. Regge, Nuovo Cim. **14** (1959) 951;
T. Regge, Nuovo Cim. **18** (1960) 947;
G. F. Chew and F. Low, Phys. Rev. **113** (1959) 1640;
C. Goebel, Phys. Rev. Lett. **1** (1958) 337;
S. Drell, Phys. Rev. Lett. **5** (1960) 342;
F. Salzman and G. Salzman, Phys. Rev. Lett. **5** (1960) 377;
E. Ferrari and F. Selleri, Phys. Rev. Lett. **7** (1961) 387;
S. Drell and K. Hiida, Phys. Rev. Lett. **7** (1961) 199;
D. H. Perkins, Progr. in Elem. Part. and Cosmic Ray Phys. **5** (1960) 328;
L. Bertocchi, S. Fubini, and M. Tonin, Nuovo Cim. **24** (1962) 626;
E. Predazzi and T. Regge, Nuovo Cim. **24** (1962) 518;
G. C. Wick, Phys. Rev. **96** (1954) 1124;
R. E. Cutkowski, Phys. Rev. **96** (1954) 1135;
M. Gell-Mann, Phys. Rev. Lett. **8** (1962) 263;
I. Y. Pomeranchuk, Zh. Eksp. Teor. Fiz. **34** (1958) 725;
I. Y. Pomeranchuk, JETP **7** (1958) 499;
G. F. Chew, M. Goldberger, F. Low, and Y. Nambu, Phys. Rev. **106** (1957) 1337;
M. Froissart, Phys. Rev. **123** (1961) 1053;
W. D. B. Greenberg and F. Low, Phys. Rev. **124** (1961) 2047;

Reactions
hadron hadron → X
hadron hadron → hadron X
hadron hadron → 2hadron >10 GeV(P_{lab})

GOLDSTONE 1961B

■ Perturbative and general proofs of the Goldstone theorem ■

Broken Symmetries

J. Goldstone, A. Salam, S. Weinberg
Phys. Rev. **127** (1962) 965;

Abstract Some proofs are presented of Goldstone's conjecture, that if there is continuous symmetry transformation under which the Lagrangian is invariant, then either the vacuum state is also invariant under the transformation, or there must exist spinless particles of zero mass.

Related references

More (earlier) information
J. Goldstone, Nuovo Cim. **19** (1961) 154;
Analyse information from
Y. Nambu and G. Jona-Lasinio, Phys. Rev. **122** (1961) 345;
W. Heisenberg, Zeit. Naturforschung **14** (1959) 441;
A. Salam and J. C. Ward, Phys. Rev. Lett. **5** (1960) 512;
H. Lehmann, Nuovo Cim. **11** (1954) 342;

Particles studied goldstone

SCHLEIN 1963

■ Confirmation of the ϕ meson ■

Quantum Numbers of a 1020 MeV $K\,\overline{K}$ Resonance

P. Schlein et al.
Phys. Rev. Lett. **10** (1963) 368;

Abstract The effective mass distribution of the kaons from the reaction $K^- p \to \Lambda\,\overline{K}\,K$ was investigated and a resonance found at 1019 ± 2 MeV, with width $\Gamma < 5$ MeV. The quantum numbers assigned are $G = -I$, $I = 1$, $I = 0$. *(Science Abstracts. 1963, 17336. C. Wilkin)*

Accelerator LBL-Bevatron

Detectors Hydrogen bubble chamber

Related references

See also
A. R. Erwin et al., Phys. Rev. Lett. **9** (1962) 34;
L. Bertanza et al., Phys. Rev. Lett. **9** (1962) 180;
G. Alexander et al., Phys. Rev. Lett. **9** (1962) 460;
M. Goldhaber, T. D. Lee, and C. N. Yang, Phys. Rev. **112** (1958) 1796;
S. B. Treiman, Phys. Rev. **128** (1962) 1342;
S. L. Glashow, Phys. Lett. **2** (1962) 251;
J. J. Sakurai, Phys. Rev. Lett. **9** (1962) 472;

Reactions

$K^- \, p \to \Lambda \, K^+ \, K^-$	1.8,1.95 GeV(P_{lab})
$K^- \, p \to \Lambda \, K^0 \, \overline{K}^0$	1.8,1.95 GeV(P_{lab})

Particles studied ϕ

CONNOLLY 1963

■ Firm establishment of the ϕ meson ■

Existence and Properties of the ϕ Meson

P.L. Connolly et al.
Phys. Rev. Lett. **10** (1963) 371;

Reprinted in
R. N. Cahn and G. Goldhaber, *The Experimental Foundations of Particle Physics*, Cambridge Univ. Press (1991) 141.

Excerpt In a previous publication (L. Bertanza et al., Phys. Rev. Lett. **9** (1962) 180) we reported evidence for the existence of a resonance in the $K\,\overline{K}$ (which we shall call the ϕ meson) with a mass of ~ 1020 and a width ≤ 20 MeV. The purpose of this Letter is to report additional data, the analysis of which confirms the existence of the resonance and provides a conclusive determination of the mass, width, parity, spin, isospin, and branching ratios of this resonance. In particular we find $M = 1019 \pm 1$ MeV, $\Gamma = 1^{+2}_{-1}$ (with $\Gamma > 0$), parity $(P) = -1$, spin $(J) = 1$, isotopic spin $(I) = 0$, and charge conjugation $(C) = -1$. The vector nature of the ϕ clearly establishes that it is not related to the mass enhancement in the $K_1 K_1$ system observed by other groups. *(Extracted from the introductory part of the paper.)*

Accelerator Brookhaven (AGS) proton synchrotron

Detectors Hydrogen bubble chamber

Related references

More (earlier) information
L. Bertanza et al., Phys. Rev. Lett. **9** (1962) 180;

See also
J. J. Sakurai, Ann.Phys. **11** (1960) 1;
J. J. Sakurai, Phys. Rev. Lett. **7** (1961) 335;
A. R. Erwin et al., Phys. Rev. Lett. **9** (1962) 34;
M. Goldhaber, T. D. Lee, and C. N. Yang, Phys. Rev. **112** (1958) 1796;
M. Gell-Mann, Phys. Rev. **125** (1962) 1067;
G. F. Chew and S. C. Frautschi, Phys. Rev. Lett. **8** (1962) 41;
S. Okubo, Progr. of Theor. Phys. **27** (1962) 949;
L. Bertanza et al., Phys. Rev. Lett. **10** (1963) 176;
G. Alexander et al., Phys. Rev. Lett. **9** (1962) 460;

Analyse information from
P. Schlein et al., Phys. Rev. Lett. **10** (1963) 368;
J. J. Sakurai, Phys. Rev. Lett. **9** (1962) 472;

Reactions

$K^- \, p \to \Lambda \, \pi^+ \, \pi^0 \, \pi^-$	2.23 GeV(P_{lab})
$K^- \, p \to \Lambda \, \pi^+ \, \pi^-$	2.23 GeV(P_{lab})
$K^- \, p \to \Lambda \, K^+ \, K^-$	2.23 GeV(P_{lab})
$K^- \, p \to \Lambda \, K^0 \, \overline{K}^0$	2.23 GeV(P_{lab})
$K^- \, p \to \Lambda \, K_S \, K_L$	2.23 GeV(P_{lab})
$K^- \, p \to \Lambda \, \phi$	2.23 GeV(P_{lab})
$\phi \to K^+ \, K^-$	
$\phi \to K_S \, K_L$	
$\phi \to \rho \, \pi$	
$\phi \to$ nonres $< \pi^+ \, \pi^0 \, \pi^- >$	

Particles studied ϕ

CABIBBO 1963

■ $SU(3)$ and hadronic weak currents. Introduction of the Cabibbo angle; predictions for the leptonic decay rates of hyperons ■

Unitary Symmetry and Leptonic Decays

N. Cabibbo
Phys. Rev. Lett. **10** (1963) 531;

Reprinted in
M. Gell-Mann and Y. Ne'eman, *The Eightfold Way: A Review – With Collection of Reprints, Frontiers in Physics*, ed. D. Pines, W. A. Benjamin, Inc . New York – Amsterdam (1964) 207.
The Physical Review — the First Hundred Years, AIP Press (1995) CD-ROM.

Excerpt We present here an analysis of leptonic decays based on the unitary symmetry for strong interactions, in the version known as "eightfold way", and the V−A theory for weak interactions. *(Extracted from the introductory part of the paper.)*

Related references

See also
M. Gell-Mann, preprint CTSL-20 (1961);
Y. Ne'eman, Nucl. Phys. **26** (1961) 222;
R. P. Feynman and M. Gell-Mann, Phys. Rev. **109** (1958) 193;
E. C. G. Sudarshan and R. E. Marshak, Phys. Rev. **109** (1958) 1860;
N. Cabibbo and R. Gatto, Nuovo Cim. **21** (1962) 872;

Analyse information from
B. P. Roe et al., Phys. Rev. Lett. **7** (1961) 346;
W. J. Willis et al., Bull.Am.Phys.Soc. **8** (1963) 349;

Reactions

$\Lambda \rightarrow p\, e^- \, \overline{\nu}_e$
$\Sigma^- \rightarrow n\, e^- \, \overline{\nu}_e$
$\Xi^- \rightarrow \Lambda\, e^- \, \overline{\nu}_e$
$\Xi^- \rightarrow \Sigma^0\, e^- \, \overline{\nu}_e$
$\Xi^0 \rightarrow \Sigma^+\, e^- \, \overline{\nu}_e$

Particles studied $\Lambda, \Sigma^-, \Xi^-, \Xi^0$

DANYSZ 1963

■ First evidence for a double hypernucleus ■

Observation of a Double Hyperfragment

M. Danysz et al.
Phys. Rev. Lett. **11** (1963) 29;

Abstract This event was discovered through a study of the interactions of 1.5 GeV/c K^--mesons in emulsions. The interpretation given to the subsequent events is as follows: A Ξ^- emitted following the interaction of a K^- meson produces a star one track which is a double hyperfragment which subsequently decays into a π^-, a singly charged particle and an ordinary hyperfragment. This latter hyperfragment subsequently decays into a π^- and three other charged particles. From a study of the angle of emission and ranges of all the charged particles involved in these processes it is concluded that the only reasonable interpretation of the event is as suggested. *(Science Abstracts. 1963, 22276. G. I. Crawford.)*

Accelerator CERN-PS proton synchrotron

Detectors Nuclear emulsion

Data comments $^{10}\mathrm{Be}_{SS}$ – double hyperfragment $_{\Lambda\Lambda}\mathrm{Be}^{11}$.

Related references

See also
G. Amato et al., Nucl. Instr. and Meth. **20** (1963) 47;
R. H. Dalitz and B. W. Downs, Phys. Rev. **111** (1958) 967;
S. Iwao, Nuovo Cim. **26** (1962) 1;

Analyse information from
E. R. Fletcher et al., Phys. Lett. **3** (1963) 280;

Reactions

K^- nucleus $\rightarrow \Xi^-$ X	1.3-1.5 GeV(P_{lab})
Ξ^- nucleus $\rightarrow$ nucleus $^{10}\mathrm{Be}_{SS}$	0. GeV(P_{lab})
Ξ^- nucleus $\rightarrow$ nucleus $^{11}\mathrm{Be}_{S}$	0. GeV(P_{lab})

Particles studied hypernucleus, $^{10}\mathrm{Be}_{SS}$, $^{11}\mathrm{Be}_{S}$

LOGUNOV 1963

■ Derivation of the relativistic generalization of the Lippmann-Schwinger equation for the two-body problem in quantum field theory ■

Quasi-Optical Approach in Quantum Field Theory

A.A. Logunov, A.N. Tavkhelidze
Nuovo Cim. **29** (1963) 380;

Excerpt A quasi-optical approach in the quantum field theory is developed.

Related references

See also
T. M. Charp and S. P. Fubini, Nuovo Cim. **14** (1959) 540;
G. F. Chew and S. C. Frautschi, Phys. Rev. **124** (1961) 264;
D. I. Blokhintsev, V. S. Barashenkov, and B. M. Barbashov, Usp. Fiz. Nauk **68** (1956) 1;
A. A. Logunov and A. N. Tavkhelidze, Proc. Georgian Acad. Sc. **18** (1957);
V. L. Bonch-Bruevich, S. V. Tyablikov, *The Green Function Model in Statistical Mechanics*, Moscow, (1961);
B. A. Arbuzov, A. A. Logunov, A. N. Tavkhelidze, and R. N. Faustov, Phys. Lett. **2** (1962) 150;
A. A. Logunov, Nguen Van Hieu, A. N. Tavkhelidze and
O. A. Khrustalev, preprint JINR-1122 (1962);
B. A. Arbuzov et al., Phys. Lett. **2** (1962) 305;
B. A. Arbuzov et al., Phys. Lett. **4** (1963) 272;

SOKOLOV 1963

■ Further investigation of the "radiation self-polarization" effect for electrons moving in magnetic field ■

On the Polarization and Spin Effects in the Synchrotron Radiation Theory

A.A. Sokolov, I.M. Ternov
Doklady Akad. Nauk SSSR **153** (1963) 1052;

Excerpt In this note we want to investigate the influence of electron spin orientation on polarization of synchrotron radiation when electron moves in constant homogeneous magnetic field. *(Extracted from the introductory part of the paper.)*

Related references

More (earlier) information
A. A. Sokolov and I. M. Ternov, Zh. Eksp. Teor. Fiz. **31** (1956) 473;
A. A. Sokolov and I. M. Ternov, Zh. Eksp. Teor. Fiz. **25** (1953) 698;
A. A. Sokolov, Ann.Phys. **8** (1961) 237;
A. A. Sokolov and M. M. Kolesnikova, Zh. Eksp. Teor. Fiz. **38** (1960) 1778;
I. M. Ternov and V. S. Tumanov, Doklady Akad. Nauk SSSR **124** (1959) 1038;
A. A. Sokolov, A. N. Matveev, and I. M. Ternov, Doklady Akad. Nauk SSSR **102** (1956) 65;
I. M. Ternov, Y. M. Loskutov, and L. I. Korovina, Zh. Eksp. Teor. Fiz. **41** (1961) 1294;

See also
J. Hiegevoord and S. A. Wouthuysen, Nucl. Phys. **40** (1963) 1;

F. A. Korolyov et al., Doklady Akad. Nauk SSSR **110** (1956) 542;
D. M. Fradkin, R. H. Good, Rev. of Mod. Phys. **33** (1961) 343;

Reactions
$\gamma^* e^- \to e^- \gamma$

DOLGOSHEIN 1964

■ Invention of the streamer chamber ■

New Gas-discharged Track Detector — The Streamer Chamber

B.A. Dolgoshein, B.I. Luchkov
Zh. Eksp. Teor. Fiz. **46** (1964) 392; JETP **19** (1964) 266;

Excerpt A new device for recording the tracks of charged particles — the spark chamber — has recently been developed and successfully applied in the experimental physics of elementary particles. The principal advantage of the spark chamber over other track detectors is its speed. However, the spark chamber exhibits anisotropy in the degree of localization of particles travelling at various angles ϕ with the direction of the electric field **E**. The spark follows the particle trajectory only up to the angles $\phi \sim 30 - 40°$. If a particle travels making a greater angle ϕ, a spark "shower" is formed along the particle path parallel to **E**. The trajectory of such a particle is well defined in the electronic plane but the third coordinate parallel to **E** is known only to within the value of the interelectrode distance. Another disadvantage of the spark chamber is the difficulty of recording the interaction and decay of particles.
We shall describe a new type of gas-discharge chamber — the "streamer" chamber — which records equally efficiently the particle tracks along any direction, reproducing the space configuration of the event.
The essential feature of the streamer chamber is the use of an incomplete spark discharge. The point of passage of the particle is indicated not by a spark but by a streamer, or more exactly by the initial portions of all the streamers which form from the electron avalanches along the particle path. The gas discharge in the chamber is stopped artificially at that stage when the electron avalanches grow into streamers and the latter begin to travel to the electrodes at a velocity of $\sim 10^3$ cm/sec.
The radiation of the gas in the streamer plasma makes the track visible. The particle track consists of a series of luminous streaks which are the initial portions of the positive and negative streamers. The length of the streaks depends on the duration of the electric field pulse and can be made sufficiently short. It is clear, from the mechanism of the track formation, that a large number of particles, irrespective of their direction, can be recorded in the streamer chamber. *(Extracted from the introductory part of the paper.)*

Accelerator COSMIC

Detectors Streamer chamber

Related references

See also
Y. Kogan and Y. Iosilevsky, Zh. Eksp. Teor. Fiz. **45** (1963) 819;

CHIKOVANI 1964

■ Another invention of the streamer chamber ■

Investigation of the Mechanism of Operation of Track Spark Chambers

G.E. Chikovani, V.N. Roinishvili, V.A. Mikhailov
Zh. Eksp. Teor. Fiz. **46** (1964) 1228; JETP **19** (1964) 833;

Abstract The principle of operation of a new detector of charged particles, the track spark chamber, is based on termination of the streamer discharges induced by primary electrons at an early stage of development. In this case the particle tracks can be discerned by the glowing centers along the particle trajectory. In contrast to the familiar spark chambers, the track spark chambers possesses isotropic properties in the sense that it records in space particles moving in any direction with respect to the electric field. The characteristics of a $100 \times 60 \times 19$ cm^3 operating track chamber where studied. A statistical model of development of the glowing centers is proposed for explaining the operation of the track chamber. The conclusion based on this model are in satisfactory agreement with the experimental results.

Accelerator COSMIC

Detectors Streamer chamber

Related references

More (earlier) information
V. A. Mikhailov, V. N. Roinishvili, and G. E. Chikovani, Zh. Eksp. Teor. Fiz. **45** (1963) 818;
G. E. Chikovani, V. A. Mikhailov, and V. N. Roinishvili, Phys. Lett. **6** (1963) 254;

See also
B. A. Dolgoshein and B. I. Luchkov, Zh. Eksp. Teor. Fiz. **46** (1964) 392;

GELL-MANN 1964 Nobel prize

■ Introduction of quarks as fundamental building blocks for hadrons. Nobel prize to M. Gell-Mann awarded in 1969 "for his contributions and discoveries concerning the classification of elementary particles and their reactions" ■

A Schematic Model of Baryons and Mesons

M. Gell-Mann
Phys. Lett. **8** (1964) 214;

Reprinted in
M. Gell-Mann and Y. Ne'eman, *The Eightfold Way: A Review – With Collection of Reprints, Frontiers in Physics*, ed. D. Pines, W. A. Benjamin, Inc. New York – Amsterdam (1964) 168.

Abstract It is suggested that an $SU(3)$ scheme ought to have unitary triplets as its fundamental objects. The simplest possibility consists of the triplet with spin $\frac{1}{2}$, baryon number $\frac{1}{3}$, and containing an isotopic doublet with charges $\frac{2}{3}$ and $-\frac{1}{3}$, and a singlet with charge $-\frac{1}{3}$. The consequences of assuming that interaction Lagrangians are made up of such objects ("Quarks") are discussed. One of the quarks should be stable, so a search is suggested. *(Science Abstracts. 1964, 15007. E. J. Squires.)*

Related references

See also
Y. Ne'eman, Nucl. Phys. **26** (1961) 222;
M. Gell-Mann, Phys. Rev. **125** (1962) 1067;
R. H. Capps, Phys. Rev. Lett. **10** (1963) 312;
R. E. Cutkosky, J. Kalckar, and P. Tarjanne, Phys. Lett. **1** (1962) 93;
E. Abers, F. Zachariazen, and A. C. Zemach, Phys. Rev. **132** (1963) 1831;
S. L. Glashow, Phys. Rev. **130** (1963) 2132;
R. E. Cutkosky and P. Tarjanne, Phys. Rev. **132** (1963) 1354;
P. Tarjanne and V. L. Teplitz, Phys. Rev. Lett. **11** (1963) 447;
J. Joyce, *Finnegan's Wake* Viking Press, New York (1939) 383;
M. Gell-Mann and M. Levi, Nuovo Cim. **16** (1960) 705;
N. Cabibbo, Phys. Rev. Lett. **10** (1963) 531;

Particles studied $q, \overline{q}, u, d, s$

ZWEIG 1964

■ Introduction of aces (quarks) as fundamental building blocks for hadrons. I ■

An $SU(3)$ Model for Strong Interaction Symmetry and its Breaking I

G. Zweig
CERN-8182-TH-401 (1964);

Related references

More (later) information
G. Zweig, preprint CALT-68-805 (1980), RR = MORE /G. Zweig, preprint CERN-8419-TH-412 (1964);

Particles studied $q, \overline{q}$

ZWEIG 1964B

■ Introduction of aces (quarks) as fundamental building blocks for hadrons. II ■

An $SU(3)$ Model for Strong Interaction Symmetry and its Breaking II

G. Zweig
CERN-8419-TH-412 (1964);

Reprinted in
Development in the Quark Theory of Hadrons, A Reprint Collection V.1: 1964-1978, ed. D. B. Lichtenberg and S. P. Rosen, Hadronic Press, Inc., Monamtum, Mass. (1980) 22.

Abstract Both mesons and baryons are constructed from a set of three fundamental particles called aces. The aces break up into an isospin doublet and singlet. Each ace carries baryon number 1/3 and is fractionally charged. $SU(3)$ (but not the Eightfold Way) is adopted as a higher symmetry for the strong interactions. The breaking of this symmetry is assumed to be universal, being due to mass differences among the aces. Extensive space-time and group theoretic structure is then predicted for both mesons and baryons, in agreement with existing experimental information. Quantitative speculations are presented concerning resonances that have not as yet been definitively classified into representation of $SU(3)$. A weak interaction theory based on right and left handed aces is used to predict rates for $\Delta S = 1$ baryon leptonic decays. An experimental search for the aces is suggested.

Related references

More (earlier) information
G. Zweig, preprint CERN-8419-TH-401 (1964);
More (later) information
G. Zweig, preprint CALT-68-805 (1980);

Particles studied $q, \overline{q}$

BJORKEN 1964

■ Proposal for the existence of a charmed fundamental fermion ■

Elementary Particles and $SU(4)$

J.D. Bjorken, S.L. Glashow
Phys. Lett. **11** (1964) 255;

Reprinted in
J. L. Rosner, *New Particles.* Selected Reprints, Stony Brook, Am. Ass. of Physics Teachers (1981) 15.

Excerpt Recently, models of strong interaction symmetry have been proposed involving four fundamental Fermion fields

ψ_i and approximate symmetry under $SU(4)$. Mesons are identified with bound states $\overline{\psi}^i\psi_i$ a nd baryons with bound states $\overline{\psi}^i\psi_j\psi_k$. In this note we examine a model of this kind whose principal achievements are these: a mass formula relating the masses of the nine vector mesons and predicting a ninth pseudoscalar meson at 950 MeV, a description of weak interactions including all selection rules except the nonleptonic $\Delta I = \frac{1}{2}$ rule, and a significant "baryon"-lepton symmetry. A new quantum number "charm" is violated only by the weak interactions, and the model predicts the existence of many "charmed" particles whose discovery is the crucial test of the idea. *(Extracted from the introductory part of the paper.)*

Related references

See also

M. Gell-Mann, Phys. Lett. **8** (1964) 214;
M. Gell-Mann, Phys. Rev. **125** (1962) 1067;
P. Tarjanne and V. L. Teplitz, Phys. Rev. Lett. **11** (1963) 447;
Y. Hara, Phys. Rev. **134** (1964) B701;
S. Okubo, Progr. of Theor. Phys. **27** (1962) 949;
G. R. Kalbfleisch et al., Phys. Rev. Lett. **12** (1964) 527;
M. Goldberg et al., Phys. Rev. Lett. **12** (1964) 546;
S. Coleman and S. L. Glashow, Phys. Rev. **134** (1964) B671;
N. Cabibbo, Phys. Rev. Lett. **10** (1963) 531;
A. Salam, Phys. Lett. **8** (1964) 216;
J. Schwinger, Phys. Rev. Lett. **12** (1964) 630;

Particles studied charm

HIGGS 1964

■ Further example of a field theory with spontaneous symmetry breakdown, no massless goldstone boson, and massive vector bosons ■

Broken Symmetries, Massless Particles and Gauge Fields

P.W. Higgs
Phys. Lett. **12** (1964) 132;

Excerpt Recently a number of people have discussed the Goldstone theorem (J. Goldstone, Nuovo Cimento **19** (1961) 154; J. Goldstone, A. Salam, and S. Weinberg, Phys. Rev. Letters **12** (1964) 266): that any solution of a Lorentz-invariant theory which violates an internal symmetry operation of that theory must contain a massless scalar particle. Klein and Lee (Phys. Rev. Letters **12** (1964) 266) showed that this theorem does not necessarily apply in non-relativistic theories and implied that their considerations would apply equally well to Lorentz-invariant field theories. Gilbert (Phys. Rev. Letters **12** (1964) 713), however, gave a proof that the failure of the Goldstone theorem in the nonrelativistic case is of a type which cannot exist when Lorentz invariance is imposed on a theory. The purpose of this note is to show that Gilbert's argument fails for an important class of field theories, that in which the conserved currents are coupled to gauge fields. *(Extracted from the introductory part of the paper.)*

Related references

See also

J. Goldstone, Nuovo Cim. **19** (1961) 154;
J. Goldstone, A. Salam, and S. Weinberg, Phys. Rev. **127** (1962) 965;
A. Klein and B. W. Lee, Phys. Rev. Lett. **12** (1964) 266;
W. Gilbert, Phys. Rev. Lett. **12** (1964) 713;
J. Schwinger, Phys. Rev. **127** (1962) 324;

BARNES 1964B

■ First evidence for a hyperon with strangeness −3, the Ω^- ■

Observation of a Hyperon with Strangeness Minus Three

V.E. Barnes et al.
Phys. Rev. Lett. **12** (1964) 204;

Reprinted in

R. N. Cahn and G. Goldhaber, *The Experimental Foundations of Particle Physics*, Cambridge Univ. Press (1991) 151.
M. Gell-Mann and Y. Ne'eman, *The Eightfold Way: A Review - With Collection of Reprints, Frontiers in Physics*, ed. D. Pines, W. A. Benjamin, Inc. New York - Amsterdam, (1964) 88.
The Physical Review — the First Hundred Years, AIP Press (1995) 890.
(translation into Russian) Usp. Fiz. Nauk **85** (1965) 523.

Abstract A particular event in a bubble chamber is interpreted as involving the initial reaction $K^- \ p \rightarrow \Omega^- \ K^+ \ K^0$, where the Ω^- is a negatively charged particle with strangeness minus three and a mass (1686 ± 12) MeV. This particle appears to be the remaining member of the 10 dimensional representation of $SU(3)$ with spin $\frac{3}{2}^+$. The mass agrees extremely well with the Gell-Mann-Okubo mass formula. *(Science Abstracts, 1964, 16969. E. J. Squires.)*

Accelerator Brookhaven (AGS) proton synchrotron

Detectors Hydrogen bubble chamber

Comments on experiment 100,000 pictures. Partially analyzed. One event with characteristic Ω^- decay found.

Related references

See also

M. Gell-Mann, Phys. Rev. **125** (1962) 1067;
Y. Ne'eman, Nucl. Phys. **26** (1961) 222;
S. Okubo, Progr. of Theor. Phys. **27** (1962) 949;
S. L. Glashow and J. J. Sakurai, Nuovo Cim. **25** (1962) 337;
G. Racah, Nucl. Phys. **1** (1956) 302;
H. J. Lipkin, Phys. Lett. **1** (1962) 68;
R. J. Oakes and C. N. Yang, Phys. Rev. Lett. **11** (1963) 174;
R. E. Behrends, J. Dreitlein, C. Fronsdal, and W. Lee, Rev. of Mod. Phys. **34** (1962) 1;

Analyse information from

Y. Eisenberg, Phys. Rev. **96** (1954) 541;

Reactions

$K^- \, p \rightarrow \Omega^- \, K^+ \, K^0$ 5 GeV(P_{lab})
$\pi^0 \rightarrow 2\gamma$
$\Lambda \rightarrow p \, \pi^-$
$\Omega^- \rightarrow \Xi^0 \, \pi^-$
$\Xi^0 \rightarrow \Lambda \, \pi^0$

Particles studied Ω^-

BARNES 1964E

■ Confirmation of the Ω^- hyperon ■

Confirmation of the Existence of the Ω^- Hyperon

V.E. Barnes et al.
Phys. Lett. **12** (1964) 134;

Reprinted in
(translation into Russian) Usp. Fiz. Nauk **85** (1965) 523.

Abstract In a recent publication we have reported evidence for the existence of a new negatively charged hyperon with strangeness S = -3, and mass $M = 1686 \pm 12$ MeV/c^2 which decayed weakly into a Ξ^0 and π^-. This particle was identified with the Ω^- predicted by the Unitary Symmetry scheme of strong interactions proposed by Gell-Mann and Ne'eman. In this model, based on the 8 dimensional representation of $SU(3)$, the strongly interacting particles are identified with the basis vectors of the irreducible representations in the direct product 8×8. Each representation consists of states with the same spin-parity; the hypercharge and I_3 (third component of isospin) quantum numbers serve as labels of the states within the representation. One such representation is the 10 dimensional representation, nine of whose members are associated with the well known resonances $N^*_{\frac{3}{2}}$, Y^*_1, $\Xi^*_{\frac{1}{2}}$ (all with spin–parity $J^P = \frac{3}{2}^+$). The tenth member, the Ω^-, is predicted to be a negatively charged isosinglet with strangeness minus three and the same spin-parity as the other numbers of the multiplet, namely $\frac{3}{2}^+$. Also, the Ω^- mass is predicted to be 1676 MeV/c^2 by the Gell-Mann-Okubo mass formula. In this letter we wish to report what we believe to be a second example of the production and decay of an Ω^-, in this case the Ω^- decaying into a Λ^0 and K^-. *(Extracted from the introductory part of the paper.)*

Accelerator Brookhaven (AGS) proton synchrotron

Detectors Hydrogen bubble chamber

Comments on experiment 120000 pictures. Second event with characteristic Ω^- decay found.

Related references

More (earlier) information
V. E. Barnes et al., Phys. Rev. Lett. **12** (1964) 204;
See also
M. Gell-Mann, Phys. Rev. **125** (1962) 1067;
Y. Ne'eman, Nucl. Phys. **26** (1961) 222;
S. Okubo, Progr. of Theor. Phys. **27** (1962) 949;
Analyse information from
Y. Eisenberg, Phys. Rev. **96** (1954) 541;

Reactions

$K^- \, p \rightarrow \Omega^- \, K^+ \, K^0$ 5 GeV(P_{lab})
$\Lambda \rightarrow p \, \pi^-$
$K^0 \rightarrow \pi^+ \, \pi^-$
$\Omega^- \rightarrow \Lambda \, K^-$
$K^- \rightarrow \pi^+ \, 2\pi^-$

Particles studied Ω^-

KALBFLEISCH 1964 Nobel prize

■ First evidence for the η' meson. Nobel prize to L. W. Alvarez awarded in 1968 "for his decisive contribution to elementary particle physics, in particular the discovery of a large number of resonance states, made possible through his development of the hydrogen bubble chamber technique and data analysis" ■

Observation of a Nonstrange Meson of Mass 959 MeV

G.R. Kalbfleisch et al.
Phys. Rev. Lett. **12** (1964) 527;

Excerpt We present here evidence showing the existence of a nonstrange meson of mass 959 MeV.
In the current experiment, the 72-in. hydrogen bubble chamber was exposed to a separated beam of 2.45, 2.63, and 2.70 BeV/c K^- mesons from the Bevatron. Approximately 370000 pictures were taken to date; approximately 300000 have been scanned . . . Figure 1(a) shows clearly the existence of the 959 MeV meson as an enhancement in the $\pi^+\pi^+\pi^0\pi^-\pi^-$ spectrum from reaction ($K^- \, p \rightarrow \Lambda\pi^+\pi^+\pi^0\pi^-\pi^-$). The mass is 959 ± 2 MeV and the full width is $\Gamma \leq 12$ MeV. We observe 35 events in the interval $0.86 \leq M^2(5\pi) \leq 0.98$ BeV^2. The background is estimated to be less than 10% of the peak . . . *(Extracted from the introductory part of the paper.)*

Accelerator LBL-Bevatron

Detectors Hydrogen bubble chamber

Related references

See also
M. Goldberg et al., Bull.Am.Phys.Soc. **9** (1964) 23;
M. Gell-Mann and J. Schwinger, Phys. Rev. Lett. **12** (1964) 237;

Reactions

$K^- \, p \rightarrow \Lambda \, \pi^+ \, \pi^0 \, \pi^-$ 2.45,2.63,2.7 GeV(P_{lab})
$K^- \, p \rightarrow \Lambda \, \pi^+ \, \pi^-$ 2.45,2.63,2.7 GeV(P_{lab})

$K^- \ p \ \rightarrow \ \Lambda$ neutral (neutrals)	2.45,2.63,2.7 GeV(P_{lab})
$K^- \ p \ \rightarrow \ \Lambda \ \pi^+ \ \pi^-$ (neutrals)	2.45,2.63,2.7 GeV(P_{lab})
$K^- \ p \ \rightarrow \ \Lambda \ 2\pi^+ \ 2\pi^-$	2.45,2.63,2.7 GeV(P_{lab})
$K^- \ p \ \rightarrow \ \Lambda \ 2\pi^+ \ \pi^0 \ 2\pi^-$	2.45,2.63,2.7 GeV(P_{lab})
$K^- \ p \ \rightarrow \ \Lambda \ 2\pi^+ \ 2\pi^-$ (neutrals)	2.45,2.63,2.7 GeV(P_{lab})

Particles studied meson^0, η'

GOLDBERG 1964C

■ Confirmation of the η' meson ■

Existence of a New Meson of Mass 960 MeV

M. Goldberg et al.
Phys. Rev. Lett. **12** (1964) 546;

Excerpt In this note we present evidence for the existence of a new meson of strangeness-zero, which we temporarily call the "X^0". Its mass and full width are $M \sim 960$ MeV and $\Gamma \leq 25$ MeV; its isospin is either 0 or 1. Data relevant to determination of its other quantum numbers are now being analyzed and the results will be reported in a subsequent paper.
The data discussed here come from a bubble chamber study of the K^-p interactions at 2.3 BeV/c. The general features of this study have been described elsewhere. The evidence for the existence of X^0 comes primarily from effective-mass studies in the following reaction channels:

$$K^-p \rightarrow \Lambda + \text{neutrals}, \quad (1)$$

$$K^-p \rightarrow \Lambda\pi^+\pi^-\pi^+\pi^-\pi^0, \quad (2)$$

$$K^-p \rightarrow \Lambda\pi^+\pi^-(\text{ neutrals with mass} > m_{\pi^0}) \quad (3)$$

The number of events in each channel are 1277, 43, and 415, respectively, all coming from a complete sample occurring within a suitable fiducial region. *(Extracted from the introductory part of the paper.)*

Accelerator Brookhaven (AGS) proton synchrotron

Detectors Hydrogen bubble chamber

Related references

More (earlier) information
M. Goldberg et al., Bull.Am.Phys.Soc. **9** (1964) 23;

Reactions

$K^- \ p \ \rightarrow \ \Lambda$ neutral (neutrals)	2.33 GeV(P_{lab})
$K^- \ p \ \rightarrow \ \Lambda \ \pi^+ \ \pi^-$ (neutrals)	2.33 GeV(P_{lab})
$K^- \ p \ \rightarrow \ \Lambda \ 2\pi^+ \ \pi^0 \ 2\pi^-$	2.33 GeV(P_{lab})
$K^- \ p \ \rightarrow \ \Lambda$ meson^0	2.33 GeV(P_{lab})

Particles studied η'

CHRISTENSON 1964 Nobel prize

■ First evidence for CP violation. Nobel prize to J. W. Cronin and V. L. Fitch awarded in 1980 "for their discovery of violations of fundamental symmetry principles in the decay of neutral K mesons" ■

Evidence for the 2π Decay of the K_L Meson

J.H. Christenson, J.W. Cronin, V.L. Fitch, R. Turlay
Phys. Rev. Lett. **13** (1964) 138;

Reprinted in
R. N. Cahn and G. Goldhaber, *The Experimental Foundations of Particle Physics*, Cambridge Univ. Press (1991) 214.
The Physical Review — the First Hundred Years, AIP Press (1995) 893.

Excerpt This Letter reports the results of experimental studies designed to search for the 2π decay of the K_2^0 meson. Several previous experiments have served[1,2] to set an upper limit of 1/300 for the fraction of K_2^0's which decay into two charged pions. The present experiment, using spark chamber techniques, proposed to extend this limit. *(Extracted from the introductory part of the paper.)*

Accelerator Brookhaven (AGS) proton synchrotron

Detectors Double arm spectrometer

Related references

See also
M. Bardon, K. Lande, L. M. Lederman, and W. Chinowsky, Ann.Phys. **5** (1958) 156;
D. Neagu et al., Phys. Rev. Lett. **6** (1961) 552;
D. Luers, I. S. Mittra, W. J. Willis, and S. S. Yamamoto, Phys. Rev. **133** (1964) B1276;
R. K. Adair et al., Phys. Rev. **132** (1963) 2285;

Reactions

p Be $\rightarrow \ \pi^+ \ \pi^-$ X	30 GeV(P_{lab})
$K_L \ \rightarrow \ \pi^+ \ \pi^-$	

Particles studied K^0, K_L

GÜRSEY 1964

■ Introduction of the $SU(6)$ classification of hadrons ■

Spin and Unitary Spin Independence of Strong Interactions

F. Gürsey, L.A. Radicati
Phys. Rev. Lett. **13** (1964) 173;

Excerpt The purpose of this Letter is twofold. We want first to point out that the group $SU(4)$ introduced by Wigner (E. P. Wigner, Phys.Rev. 51 (1937) 105) to classify nu-

clear states can be extended to relativistic domain and it is, therefore, relevant for particle physics. We will next show that when strangeness is taken into account the group $SU(4)$ becomes enlarged to $SU(6)$ which contains, as a subgroup, $SU(3) \times [SU(2)]_q$. $[SU(2)]_q$ is the unitary subgroup (little group) of the Lorentz group that leaves invariant the momentum four-vector q.
The group we consider here embodies $SU(3)$ and the ordinary spin in the same way as Wigner's $SU(4)$ embodies isotopic spin and ordinary spin. Preliminary results on the classification of particles based on $SU(6)$ seem encouraging enough to motivate study of this group. *(Extracted from introductory part of the paper.)*

Related references

See also
P. Franzini and L. A. Radicati, Phys. Lett. **6** (1963) 322;
A. Pais, Phys. Rev. Lett. **13** (1964) 175;
F. Gürsey, T. D. Lee, and M. Nauenberg, Phys. Rev. **135** (1964) B467;
Y. Nambu and P. G. O. Freund, Phys. Rev. Lett. **12** (1964) 714;

ENGLERT 1964

■ Example of a field theory with spontaneous symmetry breakdown, no massless goldstone boson, and massive vector bosons ■

Broken Symmetry and the Mass of Gauge Vector Mesons

F. Englert, R. Brout
Phys. Rev. Lett. **13** (1964) 321;

Abstract It is of interest to inquire whether gauge vector mesons acquire mass through interaction; by a gauge vector meson we mean a Yang-Mills field associated with the extension of a Lie group from global to local symmetry. The importance of this problem resides in the possibility that strong-interaction physics originates from massive gauge fields related to a system of conserved currents. In this note, we shall show that in certain cases vector mesons do indeed acquire mass when the vacuum is degenerate with respect to a compact Lie group.

Theories with degenerate vacuum (broken symmetry) have been the subject of intensive study since their inception by Namby. A characteristic feature of such theories is the possible existence of zero-mass bosons which tend to restore the symmetry. We shall show that it is precisely these singularities which maintain the gauge invariance of the theory, despite the fact that the vector meson acquires mass.

We shall first treat the case where the original fields are a set of bosons ψ_A which transform as a basis for representation of a compact Lie group. This example should be considered as a rather general phenomenological model. As such, we shall not study the particular mechanism by which the symmetry is broken but simply assume that such a mechanism exists. A calculation performed in lowest order perturbation theory indicates that those vector mesons are coupled to currents that "rotate" the original vacuum are th^e ones which acquire mass.

We shall then examine a particular model based on chirality invariance which may have a more fundamental significance. Here we begin with a chirality-invariant Lagrangian and introduce both vector and pseudoscalar gauge fields, thereby guaranteeing invariance under both local phase and local γ_5-phase transformations. In this model the gauge fields themselves may break the γ_5 invariance leading to a mass for the original Fermi field. We shall show in this case that the pseudovector field acquires mass.

In the last paragraph we sketch a simple argument which renders these results reasonable. *(Extracted from the introductory part of the paper.)*

Related references

See also
J. Schwinger, Phys. Rev. **125** (1962) 397;
C. N. Yang and R. L. Mills, Phys. Rev. **96** (1954) 191;
J. J. Sacurai, Ann.Phys. **11** (1960) 1;
Y. Nambu, Phys. Rev. Lett. **4** (1960) 380;
Y. Nambu and G. Jona-Lasinio, Phys. Rev. **122** (1961) 345;
J. Goldstone, A. Salam, and S. Weinberg, Phys. Rev. **127** (1962) 965;
S. A. Bludman and A. Klein, Phys. Rev. **131** (1963) 2364;
R. Utiyama, Phys. Rev. **101** (1956) 1597;

HIGGS 1964B

■ Higgs mechanism of mass generation for vector gauge fields ■

Broken Symmetries and Masses of Gauge Bosons

P.W. Higgs
Phys. Rev. Lett. **13** (1964) 508;

Reprinted in
The Physical Review — the First Hundred Years, AIP Press (1995) CD-ROM.

Excerpt In a recent note it was shown that the Goldstone theorem (Nuovo Cimento **19** (1961) 154), that Lorentz-covariant field theories in which spontaneous breakdown of symmetry under an internal Lie group occurs contain zero-mass particles, fails if and only if the conserved currents associated with the internal group are coupled to gauge fields. The purpose of the present note is to report that, as a consequence of this coupling, the spin-one quanta of some of the gauge fields acquire mass; the longitudinal degrees of freedom of these particles (which would be absent if their mass were zero) go over into the Goldstone bosons when the coupling tends to zero. *(Extracted from the introductory part of the paper.)*

Related references

More (earlier) information
P. W. Higgs, Phys. Lett. **12** (1964) 132;

See also
J. Goldstone, Nuovo Cim. **19** (1961) 154;
J. Goldstone, A. Salam, and S. Weinberg, Phys. Rev. **127** (1962) 965;
P. W. Anderson, Phys. Rev. **130** (1963) 439;

GURALNIK 1964

■ Example of a field theory with spontaneous symmetry breakdown, no massless goldstone boson, and massive vector bosons ■

Global Conservation Laws and Massless Particles

G.S. Guralnik, C.R. Hagen, T.W.B. Kibble
Phys. Rev. Lett. **13** (1964) 585;

Abstract ... In summary then, we have established that it may be possible consistently to break a symmetry by requiring that the vacuum expectation value of a field operator be non-vanishing without generating zero-mass particles. If the theory lacks manifest covariance it may happen that what should be the generator of the theory fail to be time-independent, despite the existence of a local conservation law. Thus the absence of massless bosons is a consequence of the inapplicability of Goldstone's theorem rather than a contradiction of it. Preliminary investigations indicate that superconductivity displays an analogous behavior. *(Extracted from introductory part of the paper.)*

Related references

See also
J. Goldstone, Nuovo Cim. **19** (1961) 154;
J. Goldstone, A.Salam, and S. Weinberg, Phys. Rev. **157** (1962) 965;
S. A. Bludman and A. Klein, Phys. Rev. **131** (1963) 2364;
W. Gilbert, Phys. Rev. Lett. **12** (1964) 713;
P. W. Higgs, Phys. Lett. **12** (1964) 132;
B. Zumino, Phys. Lett. **10** (1964) 224;
F. Englert and R. Brout, Phys. Rev. Lett. **13** (1964) 321;
D. G. Boulware and W. Gilbert, Phys. Rev. **126** (1962) 1563;

WOLFENSTEIN 1964

■ Invention of the superweak theory for CP-violation in weak interactions ■

Violation of *CP* Invariance and the Possibility of Very Weak Interactions

L. Wolfenstein
Phys. Rev. Lett. **13** (1964) 562;

Excerpt The observation (J. H. Christenson et al. Phys. Rev. Lett. 13 (1964) 138) of the decay $K_L \to 2\pi$ provides evidence that the weak interactions are not invariant with respect to the operation CP. The required magnitude of this CP-violating term in the weak-interaction Hamiltonian and the implications for other possible observations depend upon the theoretical model for explaining the $K_L \to 2\pi$ decay. Sachs has provided such a model (R. G. Sachs, Phys. Rev. Lett. **13** (1964) 286) in which

(1) the CP-violating term is maximal in the sense that the $\Delta Q = -\Delta S$ decay interaction is 90° out of phase with $\Delta Q = +\Delta S$ interaction,

(2) CP violation would not occur in nonleptonic decays or in leptonic decays other than that of K^0, and

(3) the strength of the CP-conserving term is comparable to the CP-violating term.

In this note we wish to raise some possible objections to the model of Sachs and to suggest an alternative which essentially satisfies conditions (1) and (2) above but requires that CP-violating weak interactions are weaker than the CP-conserving ones by a factor 10^{-8} to 10^{-7}. *(Extracted from the introductory part of the paper.)*

Related references

See also
R. G. Sachs, Phys. Rev. Lett. **13** (1964) 286;
W. J. Willis et al., Phys. Rev. Lett. **13** (1964) 291;
B. L. Ioffe, Zh. Eksp. Teor. Fiz. **42** (1962) 1411;
B. L. Ioffe, JETP **15** (1962) 978;

Analyse information from
J. H. Christenson, J. W. Cronin, V. L. Fitch, and R. Turlay, Phys. Rev. Lett. **13** (1964) 138;
R. W. Birge et al., Phys. Rev. Lett. **11** (1963) 35;

GREENBERG 1964

■ Introduction of the color quantum number, and colored quarks and gluons ■

Spin and Unitary-Spin Independence in a Paraquark Model of Baryons and Mesons

O.W. Greenberg
Phys. Rev. Lett. **13** (1964) 598;

Reprinted in
The Physical Review — the First Hundred Years, AIP Press (1995) CD-ROM.

Abstract A model of baryons and mesons in which the basic particles are quarks is discussed. *(Science Abstracts. 1965, 8218. W. W. Bell.)*

Related references

See also
E. P. Wigner, Phys. Rev. **51** (1937) 106;
E. P. Wigner and E. L. Feinberg, Rep.Prog.Phys. **8** (1941) 274;
F. Gürsey and L. A. Radicati, Phys. Rev. Lett. **13** (1964) 173;
A. Pais, Phys. Rev. Lett. **13** (1964) 175;
T. K. Kuo and T. Yao, Phys. Rev. Lett. **13** (1964) 415;
M. A. B. Bég, B. W. Lee, and A. Pais, Phys. Rev. Lett.

13 (1964) 514;
J. P. Elliot and T. H. R. Skyrme, Proc. Roy. Soc. **A232** (1955) 561;
E. Baranger and C. W. Lee, Nucl. Phys. **22** (1961) 157;
A. M. L. Messiah and O. W. Greenberg, Phys. Rev. **136** (1964) B248;
M. A. B. Bég and V. Singh, Phys. Rev. Lett. **13** (1964) 418;
F. Gürsey, A. Pais, and L. A. Radicati, Phys. Rev. Lett. **13** (1964) 299;

Analyse information from
W. Blum et al., Phys. Rev. Lett. **13** (1964) 353A;
V. Hagopian et al., Phys. Rev. Lett. **13** (1964) 280;

Particles studied q

ABRAMS 1964

■ Confirmation of the existence of the Ω^- hyperon ■

Example of Decay $\Omega^- \to \Xi^- \pi^0$

G.S. Abrams et al.
Phys. Rev. Lett. **13** (1964) 670;

Excerpt The existence of the Ω^-, a baryon of mass 1676 MeV and strangeness -3, has been predicted from $SU(3)$ invariance of the strong interactions. This prediction was beautifully confirmed by three examples found in the 80-in. hydrogen bubble chamber in a separated 5.0 BeV/c K^- beam at the AGS accelerator at Brookhaven National Laboratory. We have found another example of an Ω^-, which decays into a Ξ^- and π^0. The three previous examples have exhibited the decay modes $\Omega^- \to \Xi^-\pi^-$ and $\Omega^- \to \Lambda K^-$ (two examples). The exposure for this experiment was carried out at Brookhaven National Laboratory using the 80-in. hydrogen bubble chamber. A mass-separated beam of 4.2 BeV/c K^- mesons from the AGS was incident on the bubble chamber. We have scanned 25000 pictures for events initiated by an incident-beam track and containing at least two (charged or neutral) V particles. *(Extracted from the introductory part of the paper.)*

Accelerator Brookhaven (AGS) proton synchrotron

Detectors Hydrogen bubble chamber

Related references

Analyse information from
V. E. Barnes et al., Phys. Rev. Lett. **12** (1964) 204;
V. E. Barnes et al., Phys. Lett. **12** (1964) 134;

Reactions

$K^- \, p \to \Omega^- \, K^+ \, K^0$ 4.2 GeV(P_{lab})
$\Lambda \to p \, \pi^-$
$K^0 \to \pi^+ \, \pi^-$
$\Xi^- \to \Lambda \, \pi^-$
$\Omega^- \to \Xi^- \, \pi^0$

Particles studied Ω^-

SAKITA 1964

■ $SU(6)$ classification of hadrons ■

Supermultiplets of Elementary Particles

B. Sakita
Phys. Rev. **136B** (1964) 1756;

Abstract The notion of supermultiplets first developed by Wigner for the theory of nuclear structure is applied to the structure of elementary particles. The group structure is assumed to be $SU(6)$. The quark model is assumed for the entire discussion, although some of these results can be obtained from other models. It is found that the octet of pseudoscalar mesons along with the octet and the singlet of vector mesons form a supermultiplet. Okubo's speculated mass form for the vector mesons is derived. It is also found that the octet of baryons along with a singlet particle of spin $\frac{3}{2}$ form a supermultiplet. The type of baryonic coupling for the electromagnetic and weak current is derived.

Related references

More (later) information
F. Gürsey and L. A. Radicati, Phys. Rev. Lett. **13** (1964) 173;

See also
E. P. Wigner, Phys. Rev. **51** (1937) 106;
S. Sakata, Progr. of Theor. Phys. **16** (1956) 686;
M. Ikeda, S. Ogawa, and Y. Ohnuki, Progr. of Theor. Phys. **22** (1959) 715;
M. Ikeda, S. Ogawa, and Y. Ohnuki, Progr. of Theor. Phys. **23** (1960) 1073;
M. Gell-Mann, Phys. Rev. **125** (1962) 1067;
M. Gell-Mann, Phys. Lett. **8** (1964) 214;
Y. Ne'eman, Nucl. Phys. **26** (1961) 222;
G. Zweig, preprint CERN-8419-TH-412 (1964);
Y. Hara, Phys. Rev. **134** (1964) B701;
Z. Maki, Progr. of Theor. Phys. **31** (1964) 331;
S. Okubo, Progr. of Theor. Phys. **27** (1962) 949;
G. R. Kalbfleisch et al., Phys. Rev. Lett. **12** (1964) 527;
M. Goldberg et al., Phys. Rev. Lett. **12** (1964) 546;
J. Schwinger, Phys. Rev. **135** (1964) B816;
F. Gürsey, T. D. Lee, and M. Nauenberg, Phys. Rev. **135** (1964) B467;
S. Okubo, Phys. Lett. **5** (1963) 165;
S. Coleman and S. L. Glashow, Phys. Rev. **134** (1964) B671;
S. Coleman and S. L. Glashow, Phys. Rev. Lett. **6** (1961) 423;
N. Cabibbo, Phys. Rev. Lett. **10** (1963) 531;
A. W. Martin and K. C. Wali, Phys. Rev. **130** (1963) 2455;
R. E. Cutkosky, Ann.Phys. **23** (1963) 415;

SALAM 1964	Nobel prize

■ Lagrangian for the electroweak synthesis, first estimations of the W mass. Salam-Ward version. Nobel prize to A. Salam awarded in 1979. Co-winners S. L. Glashow and S. Weinberg "for their contribution to the theory of the unified weak and electromagnetic interaction between elementary particles, including, inter alia, the prediction of the weak neutral current" ■

Electromagnetic and Weak interactions

A. Salam, J.C. Ward
Phys. Lett. **13** (1964) 168;

Excerpt One of the recurrent dreams in elementary particles physics is that of a possible fundamental synthesis between electromagnetism and weak interactions. The idea has its origin in the following shared characteristics:

1) Both forces affect equally all forms of matter-leptons as well as hadrons.
2) Both are vector in character.
3) Both (individually) possess universal coupling strengths. Since universality and vector character are features of a gauge-theory these shared characteristics suggest that weak forces just like the electromagnetic forces arise from a gauge principle.

There of course also are profound differences:

1) Electromagnetic coupling strength is vastly different from the weak. Quantitatively one may state it thus: if weak forces are assumed to have been mediated by intermediate bosons (W), the boson mass would have to equal 137 M_p, in order that the (dimensionless) weak coupling constant $g_W^2/4\pi$ equals $e^2/4\pi$

In the sequel we assume just this. For the outrageous mass value itself ($M_W \approx 137 M_p$) we can offer no explanation. We seek however for a synthesis in terms of a group structure such that the remaining differences, viz:

2) Contrasting space-time behaviour (V for electromagnetic versus V and A for weak).
3) And contrasting ΔS and ΔI behaviours both appear as aspects of the same fundamental symmetry. Naturally for hadrons at least the group structure must be compatible with $SU(3)$.

Related references

More (earlier) information
A. Salam and J. C. Ward, Nuovo Cim. **11** (1959) 568;

See also
A. Salam and J. C. Ward, Nuovo Cim. **27** (1961) 922;
S. L. Glashow, Nucl. Phys. **10** (1959) 103;
J. Schwinger, Ann.Phys. **2** (1957) 407;
R. Gatto, Nuovo Cim. **28** (1963) 567;
Y. Ne'eman, Nuovo Cim. **27** (1963) 567;
Y. Nambu and P. G. O. Freund, Phys. Rev. Lett. **12** (1964) 714;
M. Gell-Mann, Phys. Lett. **12** (1964) 63;
M. Gell-Mann, Phys. Rev. **125** (1962) 1067;

Particles studied W^+, W^-

HAN 1965

■ Introduction of the color quantum number, and colored quarks and gluons ■

Three-Triplet Model with Double SU(3) Symmetry

M.Y. Han, Y. Nambu
Phys. Rev. **139** (1965) B1006;

Reprinted in
The Physical Review — the First Hundred Years, AIP Press (1995) CD-ROM.

Abstract With a view to avoiding some of the kinematical and dynamical difficulties involved in the single-triplet quark model, a model for the low-lying baryons and mesons based on three triplets with integral charges is proposed, somewhat similar to the two-triplet model introduced earlier by one of us (Y. N.). It is shown that in a $U(3)$ scheme of triplets with integral charges, one is naturally led to three triplets located symmetrically about the origin of $I_3 - Y$ diagram under the constraint that the Nishijima–Gell-Mann relation remains intact. A double $SU(3)$ symmetry scheme is proposed in which the large mass splittings between different representations are ascribed to one of the $SU(3)$, while the other $SU(3)$ is the usual one for the mass splittings within a representation of the first $SU(3)$.

Related references

See also
M. Gell-Mann, Phys. Lett. **8** (1964) 214;
G. Zweig, preprint CERN-8419-TH-412 (1964);
Y. Nambu, *Proceedings of the Second Coral Gables Conference on Symmetry Principles at High Energy*, W. H. Freeman and Company, San Francisco (1965);
H. Becry, J. Nuyts, and L. van Hove, Phys. Lett. **9** (1964) 279;
J. D. Bjorken and S. L. Glashow, Phys. Lett. **11** (1964) 255;
C. R. Hagen and A. J. Macfarlane, Phys. Rev. **135** (1964) B432;
I. S. Gerstein and M. L. Whippmann, Phys. Rev. **137** (1965) B1522;
S. Okubo, C. Ryan, and R. E. Marshak, Nuovo Cim. **34** (1964) 759;
I. S. Gerstein and K. T. Mahanthappa, Phys. Rev. Lett. **12** (1964) 570;
I. S. Gerstein and K. T. Mahanthappa, Phys. Rev. Lett. **12** (1964) 656;
S. Okubo, Phys. Lett. **4** (1963) 14;
C. A. Levinson, H. J. Lipkin, and S. Meshkov, Nuovo Cim. **23** (1961) 236;
C. A. Levinson, H. J. Lipkin, and S. Meshkov, Phys. Lett. **1** (1962) 44;
C. A. Levinson, H. J. Lipkin, and S. Meshkov, Phys. Rev. Lett. **10** (1963) 361;
A. J. Macfarlane, E. C. G. Sudarshan, and C. Dullemond, Nuovo

Cim. **30** (1963) 845;

Particles studied color, gluon

BOGOLYUBOV 1965

■ Introduction of an additional quantum number (the color) to resolve conflict with Fermi statistics. Explanation of the relations between magnetic moments of baryons ■

On the Composite Models in Theories of Elementary Particles

N.N. Bogolyubov, B.V. Struminsky, A.N. Tavkhelidze
JINR-D-1968 (1965);

Excerpt The aim of this note is the discussions of a simple dynamical model for hadrons and the study in this model baryonic magnetic moments and formfactors. It is shown that it is possible to reproduce magnetic moments of known baryons with small magnetic moment of the quark.

Related references

More (earlier) information
B. V. Struminsky, preprint JINR-1939 (1965);

See also
G. Zweig, preprint CERN-8419-TH-412 (1964);
M. Gell-Mann, Phys. Lett. **8** (1964) 214;
J. Schwinger, Phys. Rev. Lett. **12** (1964) 237;

Particles studied q

DORFAN 1965D

■ First evidence of the antideuteron ■

Observation of Antideuterons

D.E. Dorfan et al.
Phys. Rev. Lett. **14** (1965) 1003;

Excerpt Using a high-transmission mass analyzer designed to search for unitary-symmetry triplets, we have observed the production of antideuterons in 30 BeV proton-beryllium collisions.
The search was made in a $4\frac{1}{2}$-deg beam described in the preceding Letter. This communication also described the logic designed to suppress $\beta = 1$ particles and their residue while recording beam-defined particles of $0.81 < \beta < 0.96$. To bring antideuterons into the sensitive velocity band, the beam was tuned to momenta between 4.5 and 6 BeV/c. We also report on a search for antitritons in the momentum interval near 9.0 BeV/c. *(Extracted from the introductory part of the paper.)*

Accelerator Brookhaven (AGS) proton synchrotron

Detectors Counters

Related references

See also
D. E. Dorfan et al., Phys. Rev. Lett. **14** (1965) 999;

Reactions

p Be $\rightarrow$ $\overline{\text{deuteron}}$ X	30 GeV(P_{lab})
p Be $\rightarrow$ $\overline{\text{tritium}}$ X	30 GeV(P_{lab})

Particles studied $\overline{\text{deuteron}}, \overline{\text{tritium}}$

FITCH 1965B

■ Confirmation of $K_L \rightarrow K_S$ regeneration phenomenon ■

Evidence for Constructive Interference Between Coherently Regenerated and CP-nonconserving Amplitudes

V.L. Fitch, R.F. Roth, J.S. Russ, W. Vernon
Phys. Rev. Lett. **15** (1965) 73;

Excerpt We report here some preliminary results from an experiment designed to measure the phase of the CP-nonconserving component in a long-lived neutral K-meson beam. While the results are preliminary in that only a small fraction of the total data have been analyzed, we believe the main conclusion drawn at this time is clear. *(Extracted from the introductory part of the paper.)*

Accelerator Brookhaven (AGS) proton synchrotron

Detectors Spectrometer

Related references

Analyse information from
J. H. Christenson, J. W. Cronin, V. L. Fitch, and R. Turlay, Phys. Rev. Lett. **13** (1964) 138;
T. Fujii et al., Phys. Rev. Lett. **13** (1964) 253;

Reactions

p Be $\rightarrow$ neutral X	28 GeV(T_{lab})
K_L Be $\rightarrow$ Be K_S	
$K_S \rightarrow \pi^+ \pi^-$	1.25-1.85 GeV(P_{lab})
$K_L \rightarrow \pi^+ \pi^-$	1.25-1.85 GeV(P_{lab})

Particles studied K_L, K_S

ABOV 1965

■ First evidence of the spatial-parity non-conservation in weak nuclear interactions ■

On the Existence of a Spatial Parity Non-Conserving Internucleon Potential

Y.G. Abov, P.A. Krupchitsky, Y.A. Oratovsky
Yad. Phys. **1** (1965) 479;

Abstract The universal four-fermion weak interaction theory predicts the existence of a space-parity non-conserving nucleon-nucleon potential. All the previous attempts to detect this potential have been unsuccessful. In our experiment we measured the asymmetry coefficient in the angular distribution of γ-quanta, emitted by the ^{114}Cd* nucleus, following the capture of polarized neutrons. The γ-quanta were registered by two identical scintillation spectrometers with NaI crystalls. They picked out γ-quanta of the ^{114}Cd* ground-state transition under the angles $\theta_1 = 0°$ and $\theta_2 = 180°$ with respect to the neutron spin. The influence of the instrumental unstability was minimized by rapid comparisons of effects induced by polarized and unpolarized neutrons, using a rotating depolarizator. Instrumental asymmetries were excluded by performing experiments with two opposite neutron spin directions and also by measurement with interchanged spectrometers. A number of control experiments was performed and their results show, that there is no neglected instrumental asymmetry. The asymmetry coefficient proved to be equal to $a = -(3.7 \pm 0.9)10^{-4}$ and hence $F \approx 4 \cdot 10^{-7}$.

Accelerator Nuclear reactor

Detectors Photon spectrometer

Related references

More (earlier) information
Y. G. Abov, P. A. Krupchitsky and Y. A. Oratovsky, Phys. Lett. **12** (1964) 25;

See also
R. P. Feynmann and M. Gell-Mann, Phys. Rev. **109** (1958) 193;
E. C. G. Sudarshan and R. E. Marshak, Phys. Rev. **109** (1958) 1860;
R. J. Blin-Stoyle, Phys. Rev. **118** (1960) 1605;
R. J. Blin-Stoyle, Phys. Rev. **120** (1961) 181;
N. Tanner, Phys. Rev. **107** (1957) 1203;
R. E. Segel, J. V. Kane, and D. H. Wilkinson, Phil. Mag. **3** (1958) 204;
D. H. Wilkinson, Phys. Rev. **109** (1958) 1603;
D. H. Wilkinson, Phys. Rev. **109** (1958) 1610;
D. H. Wilkinson, Phys. Rev. **109** (1958) 1614;
F. Boehm and U. Hauser, Bull.Am.Phys.Soc. **4** (1959) 460;
F. Boehm and U. Hauser, Nucl. Phys. **14** (1960) 615;
D. A. Bromley et al., Phys. Rev. **114** (1959) 758;
R. Haas and L B. Leipuner, R. K. Adair, Phys. Rev. **116** (1959) 1221;
T. Mayer-Kuckuk, S. A. A. Zaidi, Z. Phys. **159** (1960) 369;
W. Kaufmann, H. Waffler, Nucl. Phys. **24** (1961) 62;
D. E. Alburger et al., Phil. Mag. **6** (1961) 171;
R. E. Segel, J. W. Olness, and E. L. Sprenkel, Phil. Mag. **6** (1961) 163;
R. E. Segel, J. W. Olness, and E. L. Sprenkel, Phys. Rev. **123** (1961) 1382;
L. Grodzins and F. Genovese, Phys. Rev. **121** (1961) 228;
P. Bassi et al., Nuovo Cim. **24** (1962) 560;
P. Bassi et al., Nuovo Cim. **28** (1963) 1049;
A. D. Gulko, Y. A. Oratovsky and S. S. Trostin, preprint ITEP-61 (1962);
T. Vervier, Nucl. Phys. **26** (1961) 10;
A. D. Gulko, Instr. Exp. Tech. **4** (1961) 40;

Reactions
$n\, ^{113}\text{Cd} \rightarrow\, ^{114}\text{Cd}\, \gamma$

KIRILLOVA 1965

■ Evidence for large real part of the nuclear scattering amplitude ■

Elastic $p\,p$ and p deuteron Small Angle Scattering in the Energy Region $2 - 10$ GeV

L.F. Kirillova et al.
Yad. Phys. **1** (1965) 533; Sov. J. Nucl. Phys. **1** (1965) 379;

Abstract Results of differential cross-section measurements for the elastic scattering of protons on protons and deuterons are given for energies 2, 4, 6, 8, 10 GeV. The investigated region of momentum transfers is $3 \cdot 10^3$ GeV$^2/c^2 \leq |t| \leq$GeV$^2/c^2$. Data on the magnitude of the real part of the elastic scattering amplitude are obtained from measurements of the Coulomb and nuclear scattering interference: we have ReA/ImA = -0.25 ± 0.07 for E_{kin} = 10 GeV and ReA/ImA = -0.17 ± 0.07 for E_{kin} = 2 GeV. The experimental results are compared with predictions following from dispersion relations.

Accelerator Dubna proton synchrotron

Detectors Nuclear emulsion

Comments on experiment Region of momentum transfer is $0.003 - 0.2$ (GeV/c)^2.

Related references

More (earlier) information
V. A. Nikitin et al., preprint JINR-1084 (1962);
V. A. Nikitin et al., Instr. Exp. Tech. **6** (1963) 18;
B. Bekker et al., Zh. Eksp. Teor. Fiz. **45** (1963) 1269;
L. F. Kirillova et al., preprint JINR-D-1329 (1963);
L. F. Kirillova, V. A. Nikitin, M. G. Shafranova, preprint JINR-P-1674 (1964);
L. F. Kirillova et al., Zh. Eksp. Teor. Fiz. **45** (1963) 1261;
B. Bekker et al., Zh. Eksp. Teor. Fiz. **46** (1964) 813;

See also
J. B. Cumming, Ann. Rev. Nucl. Sci. **13** (1963) 261;
I. I. Levintov, G. M. Adelson-Velsky, preprint ITEP-257 (1964);
P. Soding, Phys. Lett. **8** (1963) 286;
K. J. Foley et al., Phys. Rev. Lett. **10** (1963) 376;
H. Bethe, Ann.Phys. **3** (1958) 190;
S. J. Lindenbaum et al., Phys. Rev. Lett. **7** (1961) 185;
A. N. Diddens et al., Phys. Rev. Lett. **9** (1962) 32;
M. J. Longo et al., Phys. Rev. Lett. **3** (1959) 569;
A. Ashmore et al., Phys. Rev. Lett. **5** (1960) 576;
G. Von Dardel et al., Phys. Rev. Lett. **5** (1960) 333;

Reactions

$p\,p \rightarrow 2p$	2.78-10.9 GeV(P_{lab})
p deuteron $\rightarrow$ deuteron p	2.78-10.9 GeV(P_{lab})

LOBASHOV 1966

■ Confirmation of the spatial parity nonconservation in weak nuclear interactions ■

Search for Parity Non-conservation Effects in Nuclear γ–Transitions

B.M. Lobashov, V.A. Nazarenko, L.F. Saenco, L.M. Smotritskii

JETP Lett. **3** (1966) 76;

Abstract Recently evidences were found for the effects caused by the weak nucleon nucleon interaction. These effects are the asymmetry in directions of outgoing γ after polarized neutron capture by the ^{113}Cd (the asymmetry $3 \cdot 10^{-4}$ and circular polarization of the outgoing γ quanta from ^{181}Ta. It is desirable to obtain more precise data on these effects because of the great difficulties of such measurements. *(Extracted from the introductory part of the paper.)*

Accelerator Radioactive source

Detectors Photon spectrometer

Related references

See also
Y. G. Abov, P. A. Krupchitskii, and Y. A. Oratovsky, Yad. Phys. **1** (1965) 479;
F. Boehm and E. Kankeleit, Phys. Rev. Lett. **14** (1965) 312;
H. A. Tolhoek, Rev. of Mod. Phys. **28** (1956) 277;
H. W. Koch, J. W. Motz, Rev. of Mod. Phys. **31** (1959) 920;
S. Wahlborn, Phys. Rev. **138B** (1965) 530;
F. Curtis Michel, Phys. Rev. **133B** (1964) 329;

Reactions
$^{181}\mathrm{Hf} \to e^- \, \overline{\nu}_e \, ^{181}\mathrm{Ta}^*$
$^{181}\mathrm{Ta}^* \to \, ^{181}\mathrm{Ta}\gamma$

HIGGS 1966

■ Higgs mechanism of mass generation for vector gauge fields ■

Spontaneous Symmetry Breakdown without Massless Bosons

P.W. Higgs

Phys. Rev. **145** (1966) 1156;

Reprinted in
The Physical Review — the First Hundred Years, AIP Press (1995) 1044.

Abstract We examine a simple relativistic theory of two scalar fields, first discussed by Goldstone, in which as a result of spontaneous breakdown of $U(1)$ symmetry one of the scalar bosons is massless, in conformity with the Goldstone theorem. When the symmetry group of the Lagrangian is extended from global to local $U(1)$ transformations by the introduction of coupling with a vector gauge field, the Goldstone boson becomes the longitudinal state of a massive vector boson whose transverse states are the quanta of the transverse gauge field. A perturbative treatment of the model is developed in which the major features of these phenomena are present in zero order, transition amplitudes for decay and scattering processes are evaluated in lowest order, and it is shown that they may be obtained more directly from an equivalent Lagrangian in which the original symmetry is no longer manifest. When the system is coupled to other systems in a $U(1)$ invariant Lagrangian, the other systems display an induced symmetry breakdown, associated with a partially conserved current which interacts with itself via the massive vector boson.

Related references

See also
J. Schwinger, Phys. Rev. **104** (1956) 1164;
J. Schwinger, Ann.Phys. **2** (1957) 407;
J. Schwinger, Phys. Rev. Lett. **3** (1959) 296;
J. Schwinger, Phys. Rev. **125** (1962) 397;
J. Schwinger, Phys. Rev. **128** (1962) 2425;
A. Salam and J. C. Ward, Nuovo Cim. **11** (1959) 568;
A. Salam and J. C. Ward, Phys. Rev. Lett. **5** (1960) 390;
A. Salam and J. C. Ward, Nuovo Cim. **19** (1961) 167;
A. Salam and J. C. Ward, Phys. Rev. **136** (1964) B763;
S. Coleman and S. L. Glashow, Phys. Rev. **134** (1964) B671;
Y. Nambu and G. Jona-Lasinio, Phys. Rev. **122** (1961) 345;
Y. Nambu and G. Jona-Lasinio, Phys. Rev. **124** (1961) 246;
Y. Nambu and P. Pascual, Nuovo Cim. **30** (1963) 354;
Y. Nambu, Phys. Rev. **117** (1960) 648;
Y. Nambu, Phys. Rev. Lett. **4** (1960) 380;
J. Bardeen, L. N. Cooper, and J. R. Schrieffer, Phys. Rev. **106** (1957) 162;
M. Baker and S. L. Glashow, Phys. Rev. **128** (1962) 2462;
S. L. Glashow, Nucl. Phys. **10** (1959) 107;
S. L. Glashow, Nucl. Phys. **22** (1961) 579;
S. L. Glashow, Phys. Rev. **130** (1963) 2132;
M. Suzuki, Progr. of Theor. Phys. **30** (1963) 138;
M. Suzuki, Progr. of Theor. Phys. **30** (1963) 627;
N. Byrne, C. Iddings, and E. Shrauner, Phys. Rev. **139** (1965) B918;
N. Byrne, C. Iddings, and E. Shrauner, Phys. Rev. **139** (1965) B933;
P. G. O. Freund and Y. Nambu, Phys. Rev. Lett. **13** (1964) 221;
J. Goldstone, Nuovo Cim. **19** (1961) 154;
J. Goldstone, A. Salam, and S. Weinberg, Phys. Rev. **127** (1962) 965;
P. W. Anderson, Phys. Rev. **112** (1958) 1900;
A. Klein and B. W. Lee, Phys. Rev. Lett. **12** (1964) 266;
W. Gilbert, Phys. Rev. Lett. **12** (1964) 713;
P. W. Higgs, Phys. Lett. **12** (1964) 132;
G. S. Guralnik, C. R. Hagen, and T. W. B. Kibble, Phys. Rev. Lett. **13** (1964) 585;
R. V. Lange, Phys. Rev. Lett. **14** (1965) 3;
R. F. Streater, Proc. Roy. Soc. **A287** (1965) 510;
R. F. Streater, Phys. Rev. Lett. **15** (1965) 475;
N. Fuchs, Phys. Rev. Lett. **15** (1965) 911;
S. A. Bludman, Phys. Rev. **100** (1955) 372;
S. A. Bludman and M. A. Ruderman, Nuovo Cim. **9** (1958) 433;
J. J. Sakurai, Ann.Phys. **11** (1960) 1;
C. N. Yang and R. L. Mills, Phys. Rev. **96** (1954) 191;
R. Utiyama, Phys. Rev. **101** (1956) 1597;
M. Gell-Mann and S. L. Glashow, Ann.Phys. **15** (1961) 437;
S. Weinberg, Phys. Rev. Lett. **13** (1964) 495;
F. Englert and R. Brout, Phys. Rev. Lett. **13** (1964) 321;
P. W. Higgs, Phys. Rev. Lett. **13** (1964) 508;
D. Boulware and W. Gilbert, Phys. Rev. **126** (1962) 1563;
P. T. Matthews, Phys. Rev. **76** (1949) 684;
K. Nishijima, Progr. of Theor. Phys. **5** (1950) 405;
C. S. Lam, Nuovo Cim. **38** (1965) 1755;

K. Johnson, Nucl. Phys. **25** (1961) 431;
Y. Nambu and D. Lurie, Phys. Rev. **125** (1962) 1429;
Y. Nambu and E. Shrauner, Phys. Rev. **128** (1962) 862;
E. Shrauner, Phys. Rev. **131** (1963) 1847;
M. Gell-Mann and M. Levy, Nuovo Cim. **16** (1960) 705;

GERSHTEIN 1966

■ First cosmological upper bound on the stable neutrino masses sum ■

Rest Mass of Muonic Neutrino and Cosmology

S.S. Gershtein, Y.B. Zeldovich
Pisma Zh. Eksp. Teor. Fiz. **4** (1966) 174;

Reprinted in
(translation into English) *Unity of Forces in the Universe*, A. Zee, **v.II** World Scientific, (1982) 748.

Excerpt Low-accuracy experimental estimates of the rest mass of the neutrino (A. H. Rosenfeld et al., Rev. Mod. Phys. 37 (1965) 633) yield $m(\nu_e) \leq 200$ eV/c^2 for the electronic neutrino and $m(\nu_\mu) \leq 2.5 \times 10^6$ eV/c^2 for the muonic neutrino.
Cosmological considerations connected with the hot model of the Universe make it possible to strengthen greatly the second inequality. *(Extracted from the introductory part of the paper.)*

Related references

See also
G. Gamow, Phys. Rev. **70** (1946) 572;
G. Gamow, Phys. Rev. **74** (1948) 505;
R. Dicke et al., Astr. Jour. **142** (1965) 414;
Y. B. Zeldovich, Usp. Fiz. Nauk **89** (1966) 647;
Y. B. Zeldovich, Y. A. Smorodinskii, Zh. Eksp. Teor. Fiz. **41** (1961) 907;
A. A. Penzias and R. W. Wilson, Astr. Jour. **142** (1965) 419;
G. Gamow, Vistas in Astronomy **2** (1956) 1726;
P. J. E. Peebles, Phys. Rev. Lett. **16** (1966) 410;

Particles studied $\nu_\mu, \overline{\nu}_\mu$

SOLOVIEV 1966

■ Invention of the dispersion sum rules for hadronic binary amplitudes ■

Dispersion Sum Rules and $SU(6)$-Symmetry

L.D. Soloviev
Yad. Phys. **3** (1966) 188;

Abstract Much attention is payed now to the dispersion relations for the equal time commutators, which allow one to write down sum rules connecting magnetic moments and other coupling constants with multiparticle amplitudes. In this note we will show how such sum rules can easily be obtained from ordinary one-dimensional dispersion relations with some additional assumptions on the high energy behavior of the physical amplitudes. *(Extracted from the introductory part of the paper.)*

Related references

See also
S. Fubini, G. Furlan, C. Rosetti, *A Disperson theory of symmetry breaking*, CERN (1965);
A. A. Logunov and L. D. Soloviev, Nucl. Phys. **10** (1959) 60;
A. A. Logunov, Nucl. Phys. **10** (1959) 71;
G. Chew, M. Goldberger, F. Low, Y. Nambu, Phys. Rev. **106** (1957) 1345;
N. N.Bogolyubov, et al., preprint D-2075 (Dubna, 1965);

NAMBU 1966

■ Invention of the idea of a vector gluon theory for strong interactions ■

A Systematics of Hadrons in Subnuclear Physics

Y. Nambu
Preludes in Theoretical Physics in Honor of V. F. Weisskopf, ed. by A. De-Shalit, H. Feshbach, and L. Van Hove, North-Holland, Amsterdam (1966) 133 (1966);

Excerpt With the recognition that the $SU(3)$ symmetry is the dominant feature of the strong interactions, the main concern of the elementary particle theory has naturally become directed at the understanding of the internal symmetry of particles at a deeper level. An immediate question that arises in this regard is whether there are fundamental objects (such as triplets or quartets) of which all the known baryons and mesons are composed. These fundamental objects would be to the baryons and mesons what the nucleons are to the nuclei, and the electrons and nuclei are to the atoms. If that was really the case, it would certainly precipitate a new revolution in our conceptual image of the world. At the moment we can only hope that the question will be answered within the next ten to twenty years when the 100 GeV to 1000 GeV range accelerators will have been realized.

Even now, the amusing and rather embarrassing success of the $SU(6)$ theory ends support to the existence of those fundamental objects. It is embarassing because this is basically a non-relativistic and static theory, and we do not know exactly how this can cover the realm of high energy relativistic phenomena.

Putting aside those theoretical difficulties mainly associated with relativity, let us make the working hypothesis that there are fundamental objects which are heavy ($>> 1$ GeV), though not necessarily stable, and that inside each baryon or meson they are combined with a large binding energy, yet moving with non-relativistic velocities. Though this might look like a contradiction, at least it does not violate the uncertainty principle in non-relativistic quantum mechanics since the range of the binding forces ($10^{-14} - 10^{-13}$ cm) is large compared to the Compton wave lengths of those constituents, and the strength of the forces can be arbitrarily adjusted. In other words, we have a model very similar to the atomic nuclei except for large binding energies. Theoretical justification of such a hypothesis must await future investigation. ...

...The $SU(6)$ symmetry can be brought in, with the Pauli principle taken into account, since the constituent particles are non-relativistic. In another paper, we also considered a three-triplet model, in which t_1, t_2 and t_3 have charge assignments (1,0,0), (1,0,0) and (0, −1, −1) respectively. This has the advantage that the baryon states (the 56-dimensional representation of $SU(6)$) may be realized with s-state triplets as $\sim t_1 t_2 t_3$.

The reasoning that has gone into the above stability problem is similar to the one used in nuclear physics in deriving the semi-empirical formula of Wiezsäcker. The purpose of the present paper is to put this idea into a more precise form, even though the outcome should still be called at best semi-quantitative. *(Extracted from the introductory part of the paper.)*

Related references

See also
F. Gürsey and L. A. Radicati, Phys. Rev. Lett. **13** (1964) 173;
A. Pais, Phys. Rev. Lett. **13** (1964) 175;
B. Sakata, Phys. Rev. **136** (1964) B1765;
Y. Nambu, *Proc. of the Second Coral Gables Conference on Symmetry Principles at High Energy*, University of Miami, (Jan. 1965);
M. Y. Han and Y. Nambu, Syracuse University preprint 1206-SU-31;
R. Capps, Phys. Rev. Lett. **14** (1965) 31;
J. G. Belinfante and R. E. Cutkosky, Phys. Rev. Lett. **14** (1965) 33;

GREENBERG 1966

■ Three triplet model for hadrons. Beginnings of the quantum chromodynamics — QCD ■

Saturation in Triplet Models of Hadrons

O.W. Greenberg, D. Zwanziger
Phys. Rev. **150** (1966) 1177;

Abstract Triplet models of hadrons are studied according to the criterion of saturation, namely, that the lowest-lying baryons contain exactly three triplets. Two main types of saturation are discussed: Pauli saturation, which depends on antisymmetrization of wave functions, and Coulomb saturation, which relies on the scheme of forces among the particles. The quark, quark-plus-singlet-core, two-triplet, three-triplet, and paraquark models are surveyed, and, using the saturation mechanism discussed in the text, all of the models are made to satisfy the saturation criterion with the sole notable exception of the quark model, which fails.

Related references

More (earlier) information
O. W. Greenberg, Phys. Rev. Lett. **13** (1964) 598;
O. W. Greenberg, Phys. Rev. **147** (1966) 1077;
Y. Nambu, in *Preludes in Theoretical Physics*, ed. A. De-Shalit et al., North-Holland, Amsterdam (1966) 133;
M. Y. Han and Y. Nambu, Phys. Rev. **139** (1965) B1006;
N. N. Bogolyubov et al., preprint JINR-D1968, JINR-D2075, JINR-P2141 (1965);
M. Gell-Mann, Phys. Lett. **8** (1964) 214;
G. Zweig, preprint CERN-8182-TH-401, CERN-8419-TH-412 (1964);
See also
F. Gürsey and L. A. Radicati, Phys. Rev. Lett. **13** (1964) 173;
B. Sakita, Phys. Rev. **136** (1964) B1756;
B. Sakita, Phys. Rev. Lett. **13** (1964) 643;
A. Pais, Rev. of Mod. Phys. **38** (1966) 215;
F. J. Dyson, *Symmetry Groups in Nuclear and Particle Physics* W. A. Benjamin, Inc., New York (1966);
F. Gürsey, A. Pais, and L. A. Radicati, Phys. Rev. Lett. **13** (1964) 299;
M. A. B. Bég, B. W. Lee, and A. Pais, Phys. Rev. Lett. **13** (1964) 514;
W. E. Thirring, Phys. Lett. **16** (1965) 335;
H. J. Lipkin and F. Scheck, Phys. Rev. Lett. **16** (1966) 71;
Y. Nambu, in *Symmetry Principles at High Energy*, ed. B. Kursunoglu et al., W. H. Freeman and Company, San Francisco, (1965) 274;
H. Becry, J. Nuyts, and L. Van Hove, Phys. Lett. **9** (1964) 279;
H. Becry, J. Nuyts, and L. Van Hove, Phys. Lett. **12** (1964) 285;
H. Becry, J. Nuyts, and L. Van Hove, Nuovo Cim. **35** (1965) 510;
T. S. Kuo and L. A. Radicati, Phys. Rev. **139** (1965) B746;
A. N. Mitra, Phys. Rev. **142** (1966) 1119;
F. Gürsey, T. D. Lee, and M. Nauenberg, Phys. Rev. **135** (1964) B467;
M. A. B. Bég and A. Pais, Phys. Rev. **137** (1965) B1514;
P. G. O. Freund and B. W. Lee, Phys. Lett. **13** (1964) 592;

SAKHAROV 1967

■ First attempt to explain baryonic asymmetry of the observable universe ■

CP Symmetry Violation, *C*-Asymmetry and Baryonic Asymmetry of the Universe.

A.D. Sakharov
Pisma Zh. Eksp. Teor. Fiz. **5** (1967) 32;

Related references

See also
Y. B. Zeldovich, Usp. Fiz. Nauk **89** (1966) 647;
L. B. Okun', Usp. Fiz. Nauk **89** (1966) 603;
M. A. Markov, Zh. Eksp. Teor. Fiz. **51** (1966) 878;
A. D. Sakharov, Pisma Zh. Eksp. Teor. Fiz. **3** (1966) 439;
S. S. Gershtein and Y. B. Zeldovich, Pisma Zh. Eksp. Teor. Fiz. **4** (1966) 174;

BUDKER 1967

■ Proposal for electron cooling of the proton and antiproton bunches in storage rings ■

Effective Method of Oscillation Damping in Proton and Antiproton Accumulators

G.I. Budker

Atomic Energy. USSR. **22** (1967) 346;

Abstract A damping method of the synchrotron and betatron oscillations of the heavy particles is proposed. Method is based oh the effect of the sharp rise of the cross sections of heavy particles interactions with electrons at small relative velocities. It is shown that this method can be used to compress and accumulate proton and antiproton bunches.

Related references

See also

E. A. Abramyan et al., in *Proc. Int. Accelerator Conference*, (Dubna 1963), Moscow, Atomizdat (1964) 284;

Analyse information from

S. T. Belyaev, G. I. Budker, Doklady Akad. Nauk SSSR **107** (1956) 807;

DORFAN 1967

■ First evidence for CP violation in $K_L \to \pi^{\pm}\ \mu^{\mp}\ \nu_\mu$ decays ■

Charge Asymmetry in the Muonic Decay of the K_2^0

D.E. Dorfan et al.

Phys. Rev. Lett. **19** (1967) 987;

Excerpt We report herewith the observation and measurement of a charge asymmetry in the muonic decay of the long-lived neutral K meson (K_2^0). In particular we find the decay rate R_{μ^+} into $\pi^-\mu^+\nu$ to be larger than the decay rate R_{μ^-} into $\pi^+\mu^-\overline{\nu}$ with the ratio determined as

$$R \equiv R_{\mu^+}/R_{\mu^-} = 1.0081 \pm 0.0027$$

This result, obtained at the Stanford Linear Accelerator Center (SLAC), is a *prime facile* demonstration of CP noninvariance in K_2^0 decay. It is consistent with theoretical expectations based upon an analysis of experimental data which have been obtained since the first such demonstration by Christenson, Cronin, Fitch, and Turlay. *(Extracted from the introductory part of the paper.)*

Accelerator SLAC

Detectors Spectrometer

Related references

See also

T. D. Lee, R. Oehme, and C. N. Yang, Phys. Rev. **106** (1957) 340;
S. Weinberg, Phys. Rev. **110** (1958) 782;
D. M. Kaplan, Phys. Rev. **139** (1965) B1065;
R. G. Sachs and S. B. Treiman, Phys. Rev. Lett. **8** (1962) 137;

Analyse information from

J. H. Christenson, J. W. Cronin, V. L. Fitch, and R. Turlay, Phys. Rev. Lett. **13** (1964) 138;
T. T. Wu and C. N. Yang, Phys. Rev. Lett. **13** (1964) 380;
T. D. Lee and C. S. Wu, Ann. Rev. Nucl. Sci. **16** (1966) 511;
V. L. Fitch et al., Phys. Rev. Lett. **15** (1965) 73;
X. De Bouard et al., Phys. Rev. Lett. **15** (1965) 58;
W. Galbraith et al., Phys. Rev. Lett. **14** (1965) 383;
J. M. Gaillard et al., Phys. Rev. Lett. **18** (1967) 20;
J. W. Cronin et al., Phys. Rev. Lett. **18** (1967) 25;
C. Rubbia and J. Steinberger, Phys. Lett. **24B** (1967) 531;
M. Bott-Bodenhausen et al., Phys. Lett. **20** (1966) 212;

Reactions

$K_L \to \pi^+\ \mu^-\ \overline{\nu}_\mu$
$K_L \to \pi^-\ \mu^+\ \nu_\mu$

Particles studied K_L

BENNETT 1967

■ Evidence for CP violation in semileptonic decays of the K_L ■

Measurement of the Charge Asymmetry in the Decay $K_L \to \pi^{\pm}\ e^{\mp}\ \nu$

S. Bennett et al.

Phys. Rev. Lett. **19** (1967) 993;

Abstract The charge asymmetry $\delta = (\Gamma_+ - \Gamma_-)/(\Gamma_+ + \Gamma_-)$ for the electron in the decay of the longlived neutral kaon to pion, and neutrino has been measured. We find asymmetry $\delta = (+2.24 \pm 0.36) \times 10^{-3}$. The result shows that CP symmetry is not conserved in this process.

Accelerator Brookhaven (AGS) proton synchrotron

Detectors Spectrometer

Related references

See also

P. A. M. Dirac, Proc. Camb. Phil. Soc. **26** (1930) 361;
J. A. Helland et al., Phys. Rev. **134** (1964) B1062;
J. A. Helland et al., Phys. Rev. **134** (1964) B1079;

Reactions

$K_L \to \pi^+\ e^-\ \overline{\nu}_e$
$K_L \to \pi^-\ e^+\ \nu_e$

Particles studied K_L

WEINBERG 1967 — Nobel prize

■ Lagrangian for the electroweak synthesis, first estimations of the W and Z masses. Nobel prize to S. Weinberg awarded in 1979. Co-winners S. L. Glashow and A. Salam "for their contribution to the theory of the unified weak and electromagnetic interaction between elementary particles, including, inter alia, the prediction of the weak neutral current" ■

A Model of Leptons

S. Weinberg
Phys. Rev. Lett. **19** (1967) 1264;

Reprinted in
The Physical Review — the First Hundred Years, AIP Press (1995) 1052.

Excerpt Leptons interact only with photons, and with the intermediate bosons that presumably mediate weak interactions. What could be more natural than to unite (E. Fermi, Z. Phys. **88** (1934) 161) these spin-one bosons into a multiplet of gauge fields? Standing in the way of this synthesis are the obvious differences in the masses of the photon and intermediate meson, and in their couplings. We might hope to understand these differences by imagining that the symmetries relating the weak and electromagnetic interactions are exact symmetries of the Lagrangian but are broken by the vacuum. However, this raises the specter of unwanted massless Goldstone bosons. (J. Goldstone, Nuovo Cim. **19** (1961) 154) This note will describe a model in which the symmetry between the electromagnetic and weak interactions is spontaneously broken, but in which the Goldstone bosons are avoided by introducing the photon and the intermediate-boson fields as gauge fields. (J. Goldstone, A. Salam, and S. Weinberg, Phys. Rev. **127** (1962) 965) The model may be renormalizable. *(Extracted from the introductory part of the paper.)*

Related references

See also
E. Fermi, Z. Phys. **88** (1934) 161;
S. L. Glashow, Nucl. Phys. **22** (1961) 579;
J. Goldstone, Nuovo Cim. **19** (1961) 154;
J. Goldstone, A. Salam, and S. Weinberg, Phys. Rev. **127** (1962) 965;
P. W. Higgs, Phys. Lett. **12** (1964) 132;
P. W. Higgs, Phys. Rev. Lett. **13** (1964) 508;
P. W. Higgs, Phys. Rev. **145** (1966) 1156;
F. Englert and R. Brout, Phys. Rev. Lett. **13** (1964) 321;
G. S. Guralnik, C. R. Hagen, and T. W. B. Kibble, Phys. Rev. Lett. **13** (1964) 585;
T. W. B. Kibble, Phys. Rev. **155** (1967) 1554;
S. Weinberg, Phys. Rev. Lett. **18** (1967) 188;
S. Weinberg, Phys. Rev. Lett. **18** (1967) 507;
J. Schwinger, Phys. Lett. **24B** (1967) 473;
S. L. Glashow, H. Schnitzer, and S. Weinberg, Phys. Rev. Lett. **19** (1967) 139;
T. D. Lee and C. N. Yang, Phys. Rev. **98** (1955) 101;
R. P. Feynman and M. Gell-Mann, Phys. Rev. **109** (1958) 193;

Particles studied W^+, W^-, Z^0

KIBBLE 1967

■ Extension of the Higgs mechanism of mass generation for non-Abelian gauge field theories. Higgs-Kibble mechanism ■

Symmetry Breaking in Non-Abelian Gauge Theories

T.W.B. Kibble
Phys. Rev. **155** (1967) 1554;

Abstract According to the Goldstone theorem, any manifestly covariant broken-symmetry theory must exhibit massless particles. However, it is known from previous work that such particles need not appear in a relativistic theory such as radiation-gauge electrodynamics, which lacks manifest covariance. Higgs has shown how the massless Goldstone particles may be eliminated from a theory with broken $U(1)$ symmetry by coupling in the electromagnetic field. The primary purpose of this paper is to discuss the analogous problem for the case of broken non-Abelian gauge symmetries. In particular, a model is exhibited which shows how the number of massless particles in a theory of this type is determined, and the possibility of having a broken non-Abelian gauge symmetry with no massless particles whatever is established. A secondary purpose is to investigate the relationship between the radiation-gauge and Lorentz-gauge formalisms. The Abelian-gauge case is reexamined, in order to show that, contrary to some previous assertions, the Lorentz-gauge formalism, properly handled, is perfectly consistent, and leads to physical conclusions identical with those reached using the radiation gauge.

Related references

See also
Y. Nambu, Phys. Rev. Lett. **4** (1960) 380;
Y. Nambu and G. Jona-Lasinio, Phys. Rev. **122** (1961) 345;
M. Baker and S. L. Glashow, Phys. Rev. **128** (1962) 2462;
S. L. Glashow, Phys. Rev. **130** (1963) 2132;
J. Goldstone, Nuovo Cim. **19** (1961) 154;
J. Goldstone, A. Salam, and S. Weinberg, Phys. Rev. **127** (1962) 965;
P. W. Anderson, Phys. Rev. **130** (1963) 439;
P. W. Higgs, Phys. Lett. **12** (1964) 132;
G. S. Guralnik, C. R. Hagen, and T. W. B. Kibble, Phys. Rev. Lett. **13** (1964) 585;
R. F. Streater, Phys. Rev. Lett. **15** (1965) 475;
P. W. Higgs, Phys. Rev. **145** (1966) 1156;
N. Fuchs, Phys. Rev. **140** (1965) B911;
J. Schwinger, Phys. Rev. **125** (1962) 1043;
J. Schwinger, Phys. Rev. **130** (1963) 402;
S. A. Bludman and A. Klein, Phys. Rev. **131** (1963) 2364;
J. Garding and S. Lions, Nuovo Cim. Suppl. **14** (1959) 9;

FADDEEV 1967

■ Faddeev-Popov method for construction of Feynman rules for Yang-Mills type of gauge theories ■

Feynman Diagrams for the Yang-Mills Field

L.D. Faddeev, V.N. Popov
Phys. Lett. **25B** (1967) 29;

Abstract Feynman and De Witt showed that the rules must be changed for the calculation of contributions from diagrams with closed loops in the theory of gauge invariant fields. They suggested also a specific recipe for the case of one loop. In this letter we propose a simple method for calculation of the contribution from arbitrary diagrams. The method of Feynman functional integration is used.

Related references

See also
C. N. Yang and R. L. Mills, Phys. Rev. **96** (1954) 191;
R. Utiyama, Phys. Rev. **101** (1956) 1597;
M. Gell-Mann and S. L. Glashow, Ann.Phys. **15** (1961) 437;
R. P. Feynman, Phys. Rev. **80** (1950) 440;
R. P. Feynman, Acta Phys. Polon. **24** (1963) 697;
B. S. de Witt, *Relativity, groups and topology.* Blackie and Son Ltd (1964) 587;

LOGUNOV 1967

■ Generalization of the dispersion sum rules to non-decreasing hadronic amplitudes ■

Dispersion Sum Rules and High Energy Scattering

A.A. Logunov, L.D. Soloviev, A.N. Tavkhelidze
Phys. Lett. **25B** (1967) 181;

Abstract Dispersion sum rules which connect the high-energy scattering parameters with the integrals of low-energy cross sections are obtained.

Related references

More (earlier) information
L. D. Soloviev, Yad. Phys. **3** (1966) 188;
I. G. Aznauryan and L. D. Soloviev, Yad. Phys. **4** (1966) 615;
See also
A. A. Logunov and L. D. Soloviev, Nucl. Phys. **10** (1959) 60;
V. de Alfaro et al., Phys. Lett. **21** (1966) 576;
V. A. Matveev, B. V. Struminsky, and A. N. Tavkhelidze, Phys. Lett. **23** (1966) 146;
V. A. Matveev, V. G. Pisarenko, and V. B. Struminsky, preprint JINR-E-2822 (Dubna, 1966);
R. E. Kallosh, R. N. Faustov, and V. G. Pisarenko, preprint JINR-E-2865 (Dubna, 1966);
V. A. Matveev, preprint JINR-P-2879 (Dubna, 1966);
V. G. Pisarenko, preprint JINR-E-2-2931 (Dubna, 1966);
G. Chew, M. Goldberger, F. Low, and Y. Nambu, Phys. Rev. **106** (1957) 1345;
K. Ter-Martirosyan, preprint ITEP-417 (Moscow, 1966);
L. Van Hove, preprint CERN-th.714 (1966);

LOGUNOV 1967B

■ First derivation of asymptotic bounds on the behavior of the one particle inclusive differential cross sections from general principles ■

High Energy Behaviour of Inelastic Cross Sections

A.A. Logunov, M.A. Mestvirishvili, guen Van Hieu
Phys. Lett. **25B** (1967) 611;

Abstract Analytic properties of the differential inelastic scattering cross sections in $\cos\theta$ are considered and some bounds for their growth in the high energy region are established.

Related references

See also
R. Ascoli and A. Minguzzi, Phys. Rev. (1435) 18;
M. Froissart, Phys. Rev. **123** (1961) 1053;
O. W. Geenberg and F. E. Low, Phys. Rev. **124** (1961) 2047;
A. Martin, Nuovo Cim. **42A** (1966) 930;
A. Martin, Nuovo Cim. **44A** (1966) 1219;
A. O. Barut and Y. C. Leung, Phys. Rev. **138** (1965) B1128;
L. G. Ratner, K. W. Edwards et al., Phys. Rev. Lett. **18** (1967) 1218;
N. N. Biswas, N. H. Cason et al., Phys. Rev. Lett. **18** (1967) 273;
K. Böckmann, B. Nellen et al., Nuovo Cim. **42** (1966) 954;
H. B. Crawley, R. A. Leacock, and W. J. Kernan, Phys. Rev. **154** (1967) 1264;
E. W. Anderson, E. J. Bleser et al., Phys. Rev. Lett. **19** (1967) 198;
J. D. Bessis, Nuovo Cim. **45** (1966) 974;

SALAM 1968 Nobel prize

■ Lagrangian for the electroweak synthesis. Salam-Ward version. Nobel prize to A. Salam awarded in 1979. Co-winners S. L. Glashow and S. Weinberg "for their contribution to the theory of the unified weak and electromagnetic interaction between elementary particles, including, inter alia, the prediction of the weak neutral current" ■

Weak and Electromagnetic Interactions

A. Salam
Elementary Particle Theory (1968);

Reprinted in

J. L. Rosner, *New Particles.* Selected Reprints, Stony Brook, Am. Ass. of Physics Teachers (1981) 29.

Excerpt One of the recurrent dreams in elementary particles physics is that of a possible fundamental synthesis between electromagnetism and weak interactions. The idea has its origin in the following shared characteristics:

1) Both forces affect equally all forms of matter-leptons as well as hadrons.
2) Both are vector in character.

3) Both (individually) possess universal coupling strengths

Since universality and vector character are features of a gauge-theory those shared characteristics suggest that weak forces just like the electromagnetic forces arise from a gauge principle. There of course also are profound differences: Electromagnetic coupling strength is vastly different from the weak. Quantitatively one may state it thus: if weak forces are assumed to have been mediated by intermediate bosons W, the boson mass would have to equal 137 M_p, in order that the (dimensionless) weak coupling constant $g_W^2/4\pi$ equals $e^2/4\pi$ I shall approach this synthesis from the point of view of renormalization theory. I had hoped that would be able to report on weak and electromagnetic interactions throughout physics, but the only piece of work that is complete is that referring to leptonic interactions, so I will present only that today. Ward and I worked with these ideas intermittently, particularly the last section on renormalization of Yang-Mills theories. The material I shall present today, incorporating some ideas of Higgs-Kibble, was given in lectures (unpublished) at Imperial College. Subsequently I discovered that an almost identical development had been made by Weinberg who apparently was also unaware of Ward's and my work. *(Extracted from the introductory part of the paper.)*

Related references

More (earlier) information
A. Salam and J. C. Ward, Nuovo Cim. **11** (1959) 568;
A. Salam and J. C. Ward, Phys. Lett. **13** (1964) 168;
A. Salam and J. C. Ward, Phys. Rev. **136** (1964) B763;
A. Salam, Phys. Rev. **127** (1962) 331;

See also
S. Weinberg, Phys. Rev. Lett. **19** (1967) 1264;

Particles studied W^+, W^-

CHARPAK 1968 — Nobel prize

■ Invention of multiwire proporional chambers. Nobel prize to G. Charpak awarded in 1992 "for his invention and development of particle detectors, in particular the multiwire proportional chamber" ■

The Use of Multiwire Proportional Counters to Select and Localize Charged Particles

G. Charpak et al.
Nucl. Instr. and Meth. **62** (1968) 262;

Abstract Properties of chambers made of planes of independent wires placed between two plane electrodes have been investigated. A direct voltage is applied to the wires. It has been checked that each wire works as an independent proportional counter down to separations of 0.1 cm between wires. Counting rates of 10^5/wire are easily reached; time resolutions of the order of 100 nsec have been obtained in some gases; it is possible to measure the position of the tracks between the wires using the time delay of the pulses; energy resolution comparable to the one obtained with the best cylindrical chambers is observed; the chambers operate in strong magnetic fields.

Related references

See also
S. C. Curran and J. D. Craggs, *Counting tubes*, Butterworths, London (1949) 32;

ASTVACATUROV 1968

■ Observation of the $\phi \to e^- \; e^+$ decay ■

Observation of the $\phi \to e^+e^-$ Decay

R.G. Astvacaturov et al.
Phys. Lett. **27B** (1968) 45;

Abstract Five events of the $\phi \to e^+e^-$ decay have been identified using spark chambers and Čerenkov total absorption gamma spectrometers. The probability for these events to be simulated by the background is smaller than 0.1%. New data on the e^+e^--decays of ρ and ω-mesons have been obtained.

Accelerator Dubna proton synchrotron

Detectors HBC-50CM, Spark chamber

Related references

More (earlier) information
M. N. Khachaturyan et al., Phys. Lett. **24B** (1967) 349;

See also
R. A. Zdanis et al., Phys. Rev. Lett. **14** (1965) 721;
D. M. Binnie et al., Phys. Lett. **18** (1965) 348;
A. Wehmann et al., Phys. Rev. Lett. **17** (1966) 1113;
J. E. Auqustein et al., preprint Orsay LAL. (1966) 1167;
V. L. Auslander et al., Phys. Lett. **25B** (1967) 433;
G. Lütjens et al., preprint CERN 67/6415;
J. K. de Pagter et al., Phys. Rev. Lett. **16** (1966) 35;
R. C. Chase et al., Phys. Rev. Lett. **18** (1967) 710;
J. G. Asbury et al., Phys. Rev. Lett. **19** (1967) 869;
ABBBHLM Collaboration, Nuovo Cim. **31** (1964) 720;
R. I. Hess et al., Phys. Rev. Lett. **17** (1966) 1109;
H. O. Cohn, W. U. Bugg, and A. T. Condo, Phys. Lett. **15** (1965) 344;
A. H. Rosenfeld et al., *Data on Particles and Resonant States* preprint UCRL-8030. (Sep. 1967);

Reactions

$\pi^- \; p \to n \; e^- \; e^+$	4. GeV(P_{lab})
$\pi^- \; p \to n \; \rho^0$	4. GeV(P_{lab})
$\pi^- \; p \to n \; \omega$	4. GeV(P_{lab})
$\pi^- \; p \to n \; \phi$	4. GeV(P_{lab})
$\rho^0 \to e^- \; e^+$	
$\omega \to e^- \; e^+$	
$\phi \to e^- \; e^+$	

Particles studied ρ^0, ω, ϕ

BJORKEN 1969

■ Invention of Bjorken scaling behavior ■

Asymptotic Sum Rules at Infinite Momentum

J.D. Bjorken
Phys. Rev. **179** (1969) 1547;

Abstract By combining the $q_0 \to i\infty$ method for asymptotic sum rules with $P \to \infty$ method of Fubini and Furlan, we relate the structure functions W_2 and W_1 in inelastic lepton-nucleon scattering to matrix elements of commutators of currents at almost equal times at infinite momentum. We argue that the infinite-momentum limit for these commutators does not diverge, but may vanish. If the limit is nonvanishing, we predict $\nu W_2(\nu, q^2) \to f_2(\nu/q^2)$ and $W_1(\nu, q^2) \to f_1(\nu/q^2)$ as ν and q^2 tend to ∞. From similar analysis for neutrino process, we conclude that at high energies the total neutrino-nucleon cross sections rise linearly with neutrino laboratory energy until nonlocality of the weak current-current coupling sets in. The sum of νp and $\tilde{\nu} p$ cross sections is determined by the equal-time commutator of the Cabibbo current with its time derivative, taken between proton states at infinite momentum.

Related references

See also
J. D. Bjorken, Phys. Rev. **163** (1967) 1767;
J. M. Cornwall and R. E. Norton, Phys. Rev. **177** (1969) 2584;
C. G. Callan, Jr., and D. Gross, Phys. Rev. Lett. **21** (1968) 311;
T. de Forest and J. Walecka, Advances in Phys. **15** (1966) 1;
R. Brandt and J. Sucher, Phys. Rev. Lett. **20** (1968) 1131;
H. Harari, Phys. Rev. Lett. **17** (1963) 1303;
M. Bander and J. D. Bjorken, Phys. Rev. **174** (1968) 1704;
T. D. Lee and C. N. Yang, Phys. Rev. **126** (1962) 2239;
S. Adler, Phys. Rev. **143** (1966) 1144;
M. Menon et al., Can. J. Phys. **46** (1968) S344;
F. Reines et al., Can. J. Phys. **46** (1968) S350;

BLOOM 1969B — Nobel prize

■ First evidence for Bjorken scaling behavior. Nobel prize to J. I. Friedman, H. W. Kendall, and R. E. Taylor awarded in 1990 "for their pioneering investigations concerning deep inelastic scattering of electrons on protons and bound neutrons, which have been of essential importance for the development of the quark model in particle physics" ■

High-Energy Inelastic $e^- p$ Scattering at 6° and 10°

E.D. Bloom et al.
Phys. Rev. Lett. **23** (1969) 930;

Reprinted in
R. N. Cahn and G. Goldhaber, *The Experimental Foundations of Particle Physics*, Cambridge Univ. Press (1991) 240.

Abstract Cross sections for inelastic scattering of electrons from hydrogen were measured for incident energies from 7 to 17 GeV at scattering angles of 6° to 10° covering a range of squared four-momentum transfers up to 7.4 $(\mathrm{GeV}/c)^2$. For low center-of-mass energies of the final hadronic system the cross section shows prominent resonances at low momentum transfer and diminishes markedly at higher momentum transfer. For high excitations the cross section shows only a weak momentum-transfer dependence.

Accelerator SLAC

Detectors Spectrometer

Related references

See also
J. D. Bjorken, Ann.Phys. **24** (1963) 201;
F. J. Gilman, Phys. Rev. **167** (1968) 1365;
A. A. Cone et al., Phys. Rev. **156** (1967) 1490;
A. A. Cone et al., Phys. Rev. **163** (1967) 1854;
F. W. Brasse, J. Engler, E. Ganssauge, and M. Schweizer, Nuovo Cim. **55A** (1968) 679;
W. Bartel, B. Dudelzak, H. Krehbiel, and J. McElroy, Phys. Lett. **28B** (1968) 148;
A. M. Boyarski, F. Bulos, W. Busza, Phys. Rev. Lett. **20** (1968) 300;
R. Anderson, D. Gustavson, R. Prepos, and D. M. Ritson, Nucl. Instr. and Meth. **66** (1968) 328;
L. W. Mo and Y. S. Tsai, Rev. of Mod. Phys. **41** (1969) 205;
H. W. Kendall and D. Isabelle, Bull.Am.Phys.Soc. **9** (1964) 94;

Reactions

$e^- p \to e^- X$ — 7-17.7 GeV(E_{lab})

BREIDENBACH 1969 — Nobel prize

■ Confirmation of Bjorken scaling behavior. Nobel prize to J. I. Friedman, H. W. Kendall, and R. E. Taylor awarded in 1990 "for their pioneering investigations concerning deep inelastic scattering of electrons on protons and bound neutrons, which have been of essential importance for the development of the quark model in particle physics" ■

Observed Behavior of Highly Inelastic Electron-Proton Scattering

M. Breidenbach et al.
Phys. Rev. Lett. **23** (1969) 935;

Reprinted in
R. N. Cahn and G. Goldhaber, *The Experimental Foundations of Particle Physics*, Cambridge Univ. Press (1991) 245.
The Physical Review — the First Hundred Years, AIP Press (1995) 896.

Abstract Results of electron-proton inelastic scattering at 6° and 10° are discussed, and values of the structure function W_2 are estimated. If the interaction is dominated by transverse virtual photons, νW_2 can be expressed as a function of $\omega = 2M\nu/q^2$ within experimental errors for $q^2 > 1$ $(\mathrm{GeV}/c)^2$ and $\omega > 4$, where ν is the invariant energy transfer and q^2 is the invariant momentum transfer of the electron. Various theoretical models and sum rules are briefly discussed.

Accelerator SLAC

Detectors Spectrometer

Related references

See also
J. D. Bjorken, Phys. Rev. **179** (1969) 1547;
J. D. Bjorken, Phys. Rev. Lett. **16** (1966) 408;
R. von Gehlen, Phys. Rev. **118** (1960) 1455;
S. D. Drell, D. J. Levy, and T. M. Yan, Phys. Rev. Lett. **22** (1969) 744;
H. Harari, Phys. Rev. Lett. **22** (1969) 1078;
J. M. Cornwall and R. E. Norton, Phys. Rev. **177** (1969) 2584;
R. Jackiw and G. Preparata, Phys. Rev. Lett. **22** (1969) 975;
H. Cheng and T. T. Wu, Phys. Rev. Lett. **22** (1969) 1409;
K. Gottfried, Phys. Rev. Lett. **18** (1967) 1174;
S. L. Adler and W. K. Tug, Phys. Rev. Lett. **22** (1969) 978;
C. G. Callan and D. J. Gross, Phys. Rev. Lett. **22** (1969) 156;
J. J. Sakurai, Phys. Rev. Lett. **22** (1969) 981;
H. D. Abarbanel and M. L. Goldberger, Phys. Rev. Lett. **22** (1969) 500;
M. Gourdin, Nuovo Cim. **21** (1961) 1094;

Analyse information from
E. D. Bloom et al., Phys. Rev. Lett. **23** (1969) 930;

Reactions

$e^- \, p \to e^- \, X$	7-17.7 GeV(E_{lab})

FEYNMAN 1969

■ Proposal for scaling behavior of the inclusive spectra of produced hadrons. Birth of the partonic picture of hadron collisions. Precise formulation of exclusive and inclusive experiments dichotomy ■

Very High-Energy Collisions of Hadrons

R.P. Feynman
Phys. Rev. Lett. **23** (1969) 1415;

Reprinted in
The Physical Review — the First Hundred Years, AIP Press (1995) CD-ROM.

Abstract Proposals are made predicting the character of longitudinal-momentum distributions in hadron collisions of extreme energies.

Related references

See also
R. P. Feynman, in *Proc. of the Third Topical Conf. on High Energy Collisions of Hadrons*, Stony Brook, N.Y. (1969);

Reactions

hadron $p \to$ hadron X

BUSHNIN 1969

■ First conclusive evidence for scale invariance in hadronic inclusive experiments ■

Production of Negative Particles by 70 GeV Protons

Y.B. Bushnin et al.
Phys. Lett. **29B** (1969) 48; Yad. Phys. **10** (1969) 585; Sov. J. Nucl. Phys. **10** (1969) 337;

Abstract Data are presented on the yields of π^-, K^-, and $\overline{p}$ produced at small angles by 70 GeV proton-aluminium interactions. The momenta of the secondaries studied were greater than 40 GeV/c. Upper limits have been obtained on the production of antideuterons and possible new particles up to a mass of 2.2 GeV in this new energy region.

Accelerator Serpukhov proton synchrotron

Detectors Counters

Related references

See also
W. F. Baker et al., Phys. Rev. Lett. **7** (1961) 101;
D. Dekkers et al., Phys. Rev. **137** (1965) 962;
R. A. Lundy, T. B. Novey, D. D. Yovanovitch, and V. L. Telegdi, Phys. Rev. Lett. **14** (1965) 504;
J. V. Allaby et al., *Proc. XVI Intern. Conf. on High Energy Physics*, Vienna (1968);

Reactions

p Al $\to \pi^-$ X	70 GeV(P_{lab})
p Al $\to K^-$ X	70 GeV(P_{lab})
p Al $\to \overline{p}$ X	70 GeV(P_{lab})
p Al $\to \overline{\text{deuteron}}$ X	70 GeV(P_{lab})

Particles studied longlived

BEZNOGIKH 1969

■ Experimental evidence for the increasing diffraction slope parameter ■

The Slope Parameter of the Differential Cross-Section of Elastic $p\,p$ Scattering in Energy Range 12-70 GeV

G.G. Beznogikh et al.
Phys. Lett. **30B** (1969) 274; Yad. Phys. **10** (1969) 1212; Sov. J. Nucl. Phys. **10** (1969) 687;

Abstract The measurements of the differential cross section of elastic $p\,p$ scattering in relative units were performed in the energy range of 12-70 GeV. The values of the slope parameter were obtained from this data. It was shown that the slope parameter of the differential $p\,p$ scattering is monotonously increasing when the proton energy rises in the range 12-70

GeV. We have obtained the slope Pomeranchuk's pole trajectory from this data: $\alpha'_P = 0.40 \pm 0.09$.

Accelerator Serpukhov proton synchrotron

Detectors Counters

Related references

See also
H. Bethe, Ann.Phys. **3** (1958) 190;
K. A. Ter-Martirosyan, preprint ITEP-417 (1967);
J. D. Bessis, Nuovo Cim. **45** (1966) 974;

Analyse information from
L. F. Kirillova et al., Yad. Phys. **1** (1965) 533;
K. J. Foley et al., Phys. Rev. Lett. **11** (1963) 425;
G. Bellettini et al., Phys. Lett. **14** (1965) 164;

Reactions

$p\,p \to 2p$ 12.-70. GeV(E_{lab})

BINON 1969

■ Confirmation of the antimatter production in hadron nucleus collisions ■

Production of Antideuterons by 43 GeV, 52 GeV, and 70 GeV Protons

IHEP-CERN Collaboration; F. Binon et al.
Phys. Lett. **30B** (1969) 510;

Abstract The production of antideuterons on aluminium nuclei by 43 GeV, 52 GeV, and 70 GeV protons has been measured.

Accelerator Serpukhov proton synchrotron

Detectors Counters

Related references

See also
Y. B. Bushnin et al., Phys. Lett. **29B** (1969) 48;
J. V. Allaby et al., Phys. Lett. **30B** (1969) 500;

Reactions

p Al $\to \overline{\text{deuteron}}$ X 43,52,70 GeV(T_{lab})

Particles studied $\overline{\text{deuteron}}$

BJORKEN 1969B

■ Explanation of Bjorken scaling with use of the parton model ■

Inelastic $e^-\,p$ and $\gamma\,p$ Scattering and the Structure of the Nucleon

J.D. Bjorken, E.A. Paschos
Phys. Rev. **185** (1969) 1975;

Abstract A model for highly inelastic electron-nucleon scattering at high energies is studied and compared with existing data. This model envisages the proton to be composed of pointlike constituents ("partons") from which the electron scatters incoherently. We propose that the model be tested by observing γ rays scattered inelastically in a similar way from the nucleon. The magnitude of this inelastic Compton-scattering cross section can be predicted from existing electron-scattering data, indicating that the experiment is feasible, but difficult, at presently available energies.

Related references

See also
S. Drell and J. Walecka, Ann.Phys. **28** (1964) 18;
J. D. Bjorken, Phys. Rev. **179** (1969) 1547;
M. Gell-Mann, Phys. Lett. **8** (1964) 214;
G. Zweig, preprint CERN-8419-TH-412, CERN-8482-TH-401 (1964);
Y. S. Tsai and Van Whitis, Phys. Rev. **149** (1966) 1948;
R. H. Dalitz, Proc. Roy. Soc. **A64** (1951) 667;

GLASHOW 1970

■ Introduction of lepton-quark symmetry, proposal of a fourth (charmed) quark ■

Weak Interaction with Lepton-Hadron Symmetry

S.L. Glashow, J. Iliopoulos, I. Maiani
Phys. Rev. **D2** (1970) 1285;

Reprinted in

J. L. Rosner, *New Particles.* Selected Reprints, Stony Brook, Am. Ass. of Physics Teachers (1981) 18.
The Physical Review — the First Hundred Years, AIP Press (1995) 1055.

Abstract We propose a model of weak interactions in which the currents are constructed out of four basic quark fields and interact with a charged massive vector boson. We show, to all orders in perturbation theory, that the leading divergencies don't violate any strong-interaction symmetry and the next to the leading divergencies respect all observed weak-interaction selection rules. The model features a remarkable symmetry between leptons and quarks. The extension of our model to a complete Yang-Mills theory is discussed.

Related references

See also
T. D. Lee, Nuovo Cim. **59A** (1969) 579;
C. Bouchiat, J. Iliopoulos, and J. Prentki, Nuovo Cim. **56A** (1968) 1150;
C. Bouchiat, J. Iliopoulos, and J. Prentki, Nuovo Cim. **62A** (1969) 209;
R. Gatto, G. Sartori, and M. Tonin, Phys. Lett. **28B** (1968) 128;
R. Gatto, G. Sartori, and M. Tonin, Nuovo Cim. Lett. **1** (1969) 1;
N. Cabibbo and I. Maiani, Phys. Lett. **28B** (1968) 131;
N. Cabibbo, Phys. Rev. Lett. **10** (1963) 531;
J. D. Bjorken and S. L. Glashow, Phys. Lett. **11** (1964) 255;

Analyse information from
P. Wanderer et al., Phys. Rev. Lett. **23** (1969) 729;
H. E. Bergeson et al., Phys. Rev. Lett. **21** (1968) 1089;
M. G. K. Menon et al., Proc. Roy. Soc. **A301** (1967) 137;

Particles studied $c, \bar{c}$

BINON 1970B

■ Confirmation of scale invariance phenomena in hadronic inclusive experiments ■

Further Investigations of π^-, K^-, and Antiprotons Production at the 70 GeV IHEP Accelerator

IHEP-CERN collaboration; F. Binon et al.
Yad. Phys. **11** (1970) 636; Sov. J. Nucl. Phys. **11** (1970) 357;

Abstract We measured the yields of the π^- and K^- mesons and antiprotons in the central part of the momentum spectrum, produced by bombarding an aluminium target with protons having an energy E_0 equal to 35, 43, 52, and 70 GeV.

Accelerator Serpukhov proton synchrotron

Detectors Counters

Related references

More (earlier) information
F. Binon et al., Phys. Lett. **30B** (1969) 506;

See also
Y. B. Bushnin et al., Phys. Lett. **29B** (1969) 48;
F. Binon et al., *Proc. Intern. Conf. on Elementary Particles*, Lund (June 1969);
J. V. Allaby et al., *Proc. XVI Intern. Conf. on High Energy Physics*, Vienna (1968);

Reactions

p Al $\rightarrow \pi^-$ X	35-70 GeV(P_{lab})
p Al $\rightarrow K^-$ X	35-70 GeV(P_{lab})
p Al $\rightarrow \bar{p}$ X	35-70 GeV(P_{lab})

ANTIPOV 1970

■ First evidence for $^3\overline{\text{He}}$ production ■

Observation of $^3\overline{\text{He}}$

Y.M. Antipov et al.
Yad. Phys. **12** (1970) 311; Sov. J. Nucl. Phys. **12** (1971) 171;

Abstract Observation is reported of $^3\overline{\text{He}}$ nuclei in the beam of negative particles produced by 70 BeV protons in an aluminum target. Among 2.4×10^{11} particles which passed through the apparatus, five $^3\overline{\text{He}}$ nuclei have been identified on the basis of their electric charge and velocity by means of scintillation and Cerenkov counters. The $^3\overline{\text{He}}$ mass was found to be $M_{^3\overline{\text{He}}} = (1.00 \pm 0.03)3m_p$, and the charge Z = (0.99 ± 0.03)2e. The ratio of the differential cross sections for production of $^3\overline{\text{He}}$ nuclei (p = 20 BeV/c) and π^- mesons (p = 10 BeV/c) is 2×10^{-11}, which corresponds to a production cross section for antihelium 3 of $d^2\sigma_{^3\overline{\text{He}}}/d\Omega dp = 2.0 \times 10^{-35}$ cm^2/sr-(BeV/c) per Al nucleus and 2.2×10^{-36} cm^2/sr-(BeV/c) per nucleon.

Accelerator Serpukhov proton synchrotron

Detectors Counters

Related references

More (later) information
Y. M. Antipov et al., Nucl. Phys. **B31** (1971) 235;

See also
O. Chamberlain, E. Segrè, C. Wiegand, and T. Ypsilantis, Phys. Rev. **109** (1958) 947;
B. Cork et al., Phys. Rev. **104** (1956) 1193;
M. Baldo-Ceolin and D. J. Prowse, Bull.Am.Phys.Soc. **3** (1958) 163;
J. Button et al., Phys. Rev. Lett. **4** (1960) 530;
F. Binon et al., Phys. Lett. **30B** (1969) 510;
F. Binon et al., Phys. Lett. **30B** (1969) 506;
V. M. Maksimenko, I. N. Sisakian, E. L. Feinberg, and D. S. Chernavsky, Pisma Zh. Eksp. Teor. Fiz. **3** (1966) 340;
D. E. Dorfan et al., Phys. Rev. Lett. **14** (1965) 1003;
Y. M. Antipov et al., Yad. Phys. **10** (1969) 346;
S. V. Donskov et al., Prib.Tech.Exp. **3** (1969) 60;
A. F. Dunaitsev, V. I. Petrukhin, Y. D. Prokoshkin, and V. I. Rykalin, Prib.Tech.Exp. **2** (1965) 114;
Y. B. Bushnin et al., Yad. Phys. **10** (1969) 585;
A. Liland and N. P. Pilkuhn, Phys. Lett. **29B** (1969) 663;

Reactions

p Al $\rightarrow \pi^-$ X	70 GeV(P_{lab})
p Al $\rightarrow \overline{\text{deuteron}}$ X	70 GeV(P_{lab})
p Al $\rightarrow {}^3\overline{\text{He}}$ X	70 GeV(P_{lab})

Particles studied $^3\overline{\text{He}}$

CHRISTENSON 1970

■ First observation of the high mass muon pairs in hadron collisions — prototype of the experiments which lead to the discovery of the J (Ting) and Υ (Herb) as well as "Drell-Yan" analyses of quark structure functions ■

Observation of Massive Muon Pairs in Hadron Collisions

J.C. Christenson et al.
Phys. Rev. Lett. **25** (1970) 1523;

Abstract Muon pairs in the mass range $1 \leq m_{\mu\mu} \leq 6.7$ GeV/c^2 have been observed in collision of high-energy protons with uranium nuclei. At an incident energy of 29 GeV, the cross section varies smoothly as $d\sigma/dm_{\mu\mu} \approx 10^{-32}/m_{\mu\mu}{}^5$ cm^2 (GeV/c)$^{-2}$ and exhibits no resonant structure. The total cross section increases by a factor of 5 as the proton energy rises from 22 to 29.5 GeV.

Accelerator Brookhaven (AGS) proton synchrotron

Detectors Spectrometer

Related references

See also
T. D. Lee and G. C. Wick, Nucl. Phys. **B9** (1969) 209;
T. D. Lee and G. C. Wick, Phys. Rev. **D2** (1970) 1033;
V. A. Matveev, R. M. Muradyan, and A. N. Tavkhelidze, preprint JINR-E2-4966 (1970);
S. D. Drell and T. M. Yan, Phys. Rev. Lett. **25** (1970) 316;
S. M. Berman, D. J. Levy, and T. L. Neff, Phys. Rev. Lett. **23** (1969) 1363;
J. J. Sakurai, Phys. Rev. Lett. **24** (1970) 968;

Analyse information from
A. Wehmann et al., Phys. Rev. Lett. **17** (1966) 1113;
B. D. Hyams et al., Phys. Lett. **24B** (1967) 634;
J. E. Augustin et al., Phys. Lett. **28B** (1968) 508;
D. R. Earles et al., Phys. Rev. Lett. **25** (1970) 129;
E. W. Anderson et al., Phys. Rev. Lett. **19** (1967) 198;

Reactions

$p\,\mathrm{U} \to \mu^- \mu^+ X$	22-29.5 GeV(P_{lab})

DENISOV 1971F

■ First experimental indication of the rising total hadronic cross sections ■

Total Cross Sections of π^+, K^+ and p on Protons and Deuterons in the Momentum Range 15-60 GeV/c

S.P. Denisov et al.
Phys. Lett. **36B** (1971) 415;

Abstract Total cross section data are presented for protons, positive pions and positive kaons on protons and deuterons in the momentum range 15 GeV/c to 60 GeV/c in 5 GeV/c steps.

Accelerator Serpukhov proton synchrotron

Detectors Counters

Related references

More (later) information
S. P. Denisov et al., Nucl. Phys. **B65** (1973) 1;

See also
J. V. Allaby et al., Phys. Lett. **30B** (1969) 500;
J. V. Allaby et al., Yad. Phys. **12** (1970) 538;
S. P. Denisov et al., *15th Intern. Conf. on High Energy Physics*, Kiev (1970);
S. P. Denisov et al., Nucl. Instr. and Meth. **92** (1971) 77;
S. P. Denisov et al., Nucl. Instr. and Meth. **85** (1970) 101;
R. J. Glauber, Phys. Rev. **100** (1955) 242;
V. Franco and R. J. Glauber, Phys. Rev. **142** (1966) 1195;
V. N. Gribov, JETP **56** (1969) 892;
G. Alberi, L. Bertocchi, and P. J. R. Soper, Phys. Lett. **32B** (1970) 367;
K. J. Foley et al., Phys. Rev. Lett. **19** (1967) 330;
K. J. Foley et al., Phys. Rev. Lett. **19** (1967) 857;
W. Galbraith et al., Phys. Rev. **138** (1965) B913;
L. B. Okun' and I. Y. Pomeranchuk, Zh. Eksp. Teor. Fiz. **30** (1956) 307;
I. Y. Pomeranchuk, Zh. Eksp. Teor. Fiz. **34** (1958) 725;

Reactions

$p\,p \to X$	15.-60. GeV(P_{lab})
$\pi^+ p \to X$	15.-60. GeV(P_{lab})
$K^+ p \to X$	15.-60. GeV(P_{lab})
π^+ deuteron $\to X$	15.-60. GeV(P_{lab})
K^+ deuteron $\to X$	15.-60. GeV(P_{lab})
p deuteron $\to X$	15.-60. GeV(P_{lab})
$K^+ n \to X$	15-60 GeV(P_{lab})
$p\,n \to X$	15-60 GeV(P_{lab})
$\pi^+ n \to X$	15-60 GeV(P_{lab})

FIRESTONE 1971C

■ First evidence for the $\overline{\Omega}^+$ ■

Observation of the $\overline{\Omega}^+$

I. Firestone et al.
Phys. Rev. Lett. **26** (1971) 410;

Abstract We have observed an $\overline{\Omega}$ event. The $\overline{\Omega}$ is produced in the reaction $K^+d \to \overline{\Omega}\Lambda\Lambda p\pi^+\pi^-$ at 12 GeV/c, and decays via the mode $\overline{\Omega} \to \overline{\Lambda}K^+$. The fitted mass for this particle is $M_{\overline{\Omega}} = 1673.1 \pm 1.0$ MeV. The $\overline{\Omega}$-production cross section is of the order of 0.1 μb.

Accelerator SLAC

Detectors Deuterium bubble chamber

Related references

See also
V. E. Barnes et al., Phys. Rev. Lett. **12** (1964) 204;
B. Rossi, Nuovo Cim. Suppl. **2** (1955) 163;
M. Ross et al., Phys. Lett. **33B** (1970) 1;

Reactions

K^+ deuteron $\to \overline{\Omega}^+ X$	12.0 GeV(P_{lab})

Particles studied $\overline{\Omega}^+$

'T HOOFT 1971

■ Rigorous proofs of renormalizability of the massless Yang-Mills quantum fields theory ■

Renormalization of Massless Yang-Mills Fields

G. 't Hooft
Nucl. Phys. **B33** (1971) 173;

Abstract The problem of renormalization of gauge fields is studied. It is observed that the use of non-gauge invariant regulator fields is not excluded provided that in the limit of high regulator mass gauge invariance can be restored by means of a finite number of counter-terms in the Lagrangian. Massless Yang-Mills fields can be treated in this manner, and appear to be renormalizable in the usual sense.
Consistency of the method is proved for diagrams with non-overlapping divergencies by means of gauge invariant regulators, which however, cannot be interpreted in terms of regu-

lator fields. Assuming consistency the S-matrix is shown to be unitary in any order of the coupling constant. A restriction must be made: no local, parity-changing transformations must be contained in the underlying gauge group. The interactions must conserve parity.

Related references

See also
R. P. Feynman, Acta Phys. Polon. **24** (1963) 697;
S. Mandelstam, Phys. Rev. **175** (1968) 1580;
S. Mandelstam, Phys. Rev. **175** (1968) 1604;
B. S. de Witt, Phys. Rev. **162** (1967) 1195;
B. S. de Witt, Phys. Rev. **162** (1967) 1239;
L. D. Faddeev and V. N. Popov, Phys. Lett. **25B** (1967) 29;
E. S. Fradkin and I. V. Tyutin, Phys. Rev. **D2** (1970) 2841;
W. Pauli and F. Villars, Rev. of Mod. Phys. **21** (1949) 434;
J. S. Bell and R. Jackiw, Nuovo Cim. **60A** (1969) 47;
S. L. Adler, Phys. Rev. **177** (1969) 2426;
C. N. Yang and R. L. Mills, Phys. Rev. **96** (1954) 191;
K. Hepp, Comm. Math. Phys. **2** (1966) 301;
R. E. Cutkosky, Jour. Math. Phys. **1** (1960) 429;
M. Veltman, Physica **29** (1963) 186;
M. Veltman, Nucl. Phys. **B7** (1968) 637;
M. Veltman, Nucl. Phys. **B21** (1970) 288;
D. Boulware, Ann.Phys. **56** (1970) 140;
A. A. Slavnow, L. D. Faddeev, *Massless and massive Yang-Mills field*, Moscow preprint 1970;
E. S. Fradkin, U. Esposito and S. Termini, Rivista del Nuovo Cim. **2** (1970) 498;

'T HOOFT 1971B

■ Rigorous proof of renormalizability of massive Yang-Mills quantum fields theory with spontaneously broken gauge invariance ■

Renormalizable Lagrangians for Massive Yang-Mills Fields

G. 't Hooft

Nucl. Phys. **B35** (1971) 167;

Abstract Renormalizable models are constructed in which local gauge invariance is broken spontaneously. Feynman rules and Ward identities can be found by means of a path integral method, and they can be checked by algebra. In one of these models, which is studied in more detail, local $SU(2)$ is broken in such a way that local $U(1)$ remains as a symmetry. A renormalizable and unitary theory results, with photons, charged massive vector particles, and additional neutral scalar particles. It has three independent parameters.
Another model has local $SU(2) \otimes U(1)$ as a symmetry and may serve as a renormalizable theory for ρ-mesons and photons.
In such models electromagnetic mass-differences are finite and can be calculated in perturbation theory.

Related references

More (earlier) information
G. 't Hooft, Nucl. Phys. **B33** (1971) 173;
See also
J. S. Bell and R. Jackiw, Nuovo Cim. **60A** (1969) 47;
S. L. Adler, Phys. Rev. **177** (1969) 2426;
I. S. Gerstein and R. Jackiw, Phys. Rev. **181** (1969) 1955;
S. Weinberg, Phys. Rev. **B140** (1965) 516;
S. Weinberg, Phys. Rev. Lett. **19** (1967) 1264;
T. W. B. Kibble, Phys. Rev. **155** (1967) 1554;
B. W. Lee, Nucl. Phys. **B9** (1969) 649;
J. L. Gerwais and B. W. Lee, Nucl. Phys. **B12** (1969) 627;

GRIBOV 1972

■ Invention of Gribov-Lipatov evolution equations for perturbative parton distribution functions in scalar and vector theories. Scaling violation prediction ■

Deep Inelastic e^- p Scattering in a Perturbative Theory

V.N. Gribov, L.N. Lipatov

Yad. Phys. **15** (1972) 781; Sov. J. Nucl. Phys. **15** (1972) 438;

Abstract Deep inelastic e^- p scattering is considered in vector and pseudoscalar theories at conditions $g^2 \ll 1, g^2 \ln q^2 \sim 1$. In the vector theory deep inelastic scattering amplitudes are shown to increase with increasing of q^2 at fixed ω in spite of the fast decreasing of the electromagnetic form factors.

Related references

More (later) information
V. N. Gribov and L. N. Lipatov, Yad. Phys. **15** (1972) 1218;
See also
J. D. Bjorken, Phys. Rev. **179** (1969) 1547;
J. D. Bjorken and E. Paschos, Phys. Rev. **185** (1969) 1975;
J. D. Bjorken and E. Paschos, Phys. Rev. **D1** (1970) 3151;
S. D. Drell, D. J. Levy, and T. M. Jan, Phys. Rev. **187** (1969) 2159;
S. D. Drell, D. J. Levy, and T. M. Jan, Phys. Rev. **D1** (1970) 1035;
S. D. Drell, D. J. Levy, and T. M. Jan, Phys. Rev. **D1** (1970) 1617;
S. D. Drell, D. J. Levy, and T. M. Jan, Phys. Rev. **D1** (1970) 2402;
P. M. Fishbane and S. J. Chang, Phys. Rev. **D2** (1970) 1084;
L. D. Landau, A. A. Abrikosov, and M. I. Khalatnikov, Doklady Akad. Nauk SSSR **95** (1953) 497;
L. D. Landau, A. A. Abrikosov, and M. I. Khalatnikov, Doklady Akad. Nauk SSSR **95** (1953) 773;
L. D. Landau, A. A. Abrikosov, and M. I. Khalatnikov, Doklady Akad. Nauk SSSR **95** (1953) 1177;
A. D. Galanin, B. L. Ioffe and I. Y. Pomeranchuk, Zh. Eksp. Teor. Fiz. **29** (1955) 51;
C. Calan and D. J. Gross, Phys. Rev. Lett. **22** (1969) 156;
V. V. Sudakov, Zh. Eksp. Teor. Fiz. **30** (1956) 187;
A. A. Abrikosov, Zh. Eksp. Teor. Fiz. **30** (1956) 386;
K. G. Wilson, Phys. Rev. **179** (1969) 1499;
V. N. Gribov, L. N. Lipatov, and G. V. Frolov, Yad. Phys. **12** (1970) 994;
M. Froissart, Phys. Rev. **123** (1961) 1053;
V. G. Gorshkov and L. N. Lipatov, Yad. Phys. **9** (1969) 818;
V. G. Gorshkov, V. N. Gribov, L. N. Lipatov, and G. V. Frolov, Yad. Phys. **6** (1967) 361;
V. G. Gorshkov, V. N. Gribov, L. N. Lipatov, and G. V. Frolov, Yad. Phys. **6** (1967) 129;
A. M. Polyakov, Zh. Eksp. Teor. Fiz. **60** (1971) 1572;
V. N. Gribov and L. N. Lipatov, Yad. Phys. **15** (1972) 6;

MILLER 1972B — Nobel prize

■ Firm establishment of Bjorken scaling behavior. Nobel prize to J. I. Friedman, H. W. Kendall, and R. E. Taylor awarded in 1990 "for their pioneering investigations concerning deep inelastic scattering of electrons on protons and bound neutrons, which have been of essential importance for the development of the quark model in particle physics" ■

Inelastic $e^- \, p$ Scattering at Large Momentum Transfers and the Inelastic Structure Functions of the Proton

G. Miller et al.
Phys. Rev. **D5** (1972) 528;

Abstract Differential cross section for electrons scattered inelastically from hydrogen have been measured at 18°, 26°, and 34°. The range of incident energy was 4.5 to 18 GeV, and the range of four-momentum transfer squared was 1.5 to 21 GeV/c^2. With the use of these data in conjunction with previously measured data at 68° and 10°, the contributions from the longitudinal and transverse components of the exchanged photon have been separately determined. The values of the ratio of the photoabsorbtion cross sections σ_S/σ_T are found to lie in the range 0 to 0.5. The question of scaling of $2M_pW_1$ and νW_2 as a function of ω is discussed, and scaling is verified for a large kinematic range. Also, a new scaling variable which reduced to ω in the Bjorken limit is introduced which extends the scaling region. The behaviour of σ_T and σ_S is also discussed as a function of ν and q^2. Various weighted sum rules of νW_2 are evaluated.

Accelerator SLAC

Detectors Spectrometer

Related references

More (earlier) information
G. Miller. Ph. D. dissertation preprint SLAC-129 (1970);
D. H. Coward et al., Phys. Rev. Lett. **20** (1968) 292;

More (later) information
E. D. Bloom et al., preprint SLAC-PUB-796 (1970);
J. S. Poucher et al., Phys. Rev. Lett. **32** (1974) 118;
A. Bodek et al., Phys. Rev. Lett. **30** (1973) 1087;
A. Bodek et al., Phys. Lett. **51B** (1974) 417;
A. Bodek et al., Phys. Lett. **52B** (1974) 249;
A. Bodek et al., Phys. Rev. **D20** (1979) 1471;
E. M. Riordan et al., Phys. Rev. Lett. **33** (1974) 561;

Reactions
$e^- \, p \rightarrow e^- \, X$ 4.5-18 GeV(P_{lab})

'T HOOFT 1972

■ Universal regularization and renormalization method for gauge fields theories. I ■

Regularization and Renormalization of Gauge Fields

G. 't Hooft, M. Veltman
Nucl. Phys. **B44** (1972) 189;

Abstract A new regularization and renormalization procedure is presented. It is particularly well suited for the treatment of gauge theories. The method works for theories that were known to be renormalizable as well as for Yang-Mills type theories. Overlapping divergencies are disentangled. The procedure respects unitarity, causality and allows shifts of integration variables. In non-anomalous cases also Ward identities are satisfied at all stages. It is transparent when anomalies, such as the Bell-Jackiw-Adler anomaly, may occur.

Related references

More (earlier) information
G. 't Hooft, Nucl. Phys. **B35** (1971) 167;
G. 't Hooft, Nucl. Phys. **B33** (1971) 173;

See also
B. W. Lee, Phys. Rev. **D5** (1972) 823;
P. W. Higgs, Phys. Lett. **12** (1964) 132;
P. W. Higgs, Phys. Rev. Lett. **13** (1964) 508;
P. W. Higgs, Phys. Rev. **145** (1966) 1156;
F. Englert and R. Brout, Phys. Rev. Lett. **13** (1964) 321;
G. S. Guralnik, C. R. Hagen, and T. W. B. Kibble, Phys. Rev. Lett. **13** (1964) 585;
T. W. B. Kibble, Phys. Rev. **155** (1967) 627;
J. S. Bell and R. Jackiw, Nuovo Cim. **51** (1969) 47;
S. L. Adler, Phys. Rev. **177** (1969) 2426;
S. Weinberg, Phys. Rev. Lett. **27** (1971) 1688;
R. E. Cutkosky, Jour. Math. Phys. **1** (1960) 429;
M. Veltman, Physica **29** (1963) 186;

'T HOOFT 1972B

■ Universal regularization and renormalization method for gauge fields theories. II ■

Combinatorics of Gauge Fields

G. 't Hooft, M. Veltman
Nucl. Phys. **B50** (1972) 318;

Abstract Gauge field theories can be described by many different sets of Feynman rules, depending on the particular gauge chosen. In this paper a prescription for obtaining the Feynman rules in different gauges is given. A rigorous combinatorial proof of the independence of the S-matrix of the chosen gauge is presented. The proof is general and applies to Yang-Mills type theories as well as to gravitation. For renormalizable Yang-Mills type theories it is shown that the renormalized theory is invariant with respect to renormalized gauge transformations.

Related references

See also
C. N. Yang and R. L. Mills, Phys. Rev. **96** (1954) 191;
G. 't Hooft, Nucl. Phys. **B33** (1971) 173;
G. 't Hooft, Nucl. Phys. **B35** (1971) 167;
P. W. Higgs, Phys. Lett. **12** (1964) 132;
P. W. Higgs, Phys. Rev. Lett. **13** (1964) 508;
P. W. Higgs, Phys. Rev. **145** (1966) 1156;
F. Englert and R. Brout, Phys. Rev. Lett. **13** (1964) 321;
G. S. Guralnik, C. R. Hagen, and T. W. B. Kibble, Phys. Rev. Lett. **13** (1964) 585;
T. W. B. Kibble, Phys. Rev. **155** (1967) 627;
B. W. Lee, Phys. Rev. **D5** (1972) 823;
B. W. Lee and J. Zinn Justin, Phys. Rev. **D5** (1972) 3121;
B. W. Lee and J. Zinn Justin, Phys. Rev. **D5** (1972) 3137;
B. W. Lee and J. Zinn Justin, Phys. Rev. **D5** (1972) 3155;
S. Weinberg, Phys. Rev. Lett. **27** (1971) 1688;
L. D. Faddeev and V. N. Popov, Phys. Lett. **25B** (1967) 29;
G. 't Hooft, M. Veltman, Nucl. Phys. **B44** (1972) 189;
I. Bialynicki-Birula, Phys. Rev. **D2** (1970) 2877;
M. Veltman, Nucl. Phys. **B21** (1970) 288;

BALDIN 1973

■ Evidence for cumulative effect ■

Cumulative Mesoproduction

A.M. Baldin et al.
Yad. Phys. **18** (1973) 79;

Abstract Probability of production of mesons induced by accelerated deuterium nuclei is measured. The produced pion energy is more than the energy of a nucleon in the deuteron beam. The ratio of the cross section of the pion production induced by deuterium nuclei to the cross section of the pion production induced by nucleons at an equal energy release dose not depend on the ratio of the pion momentum to the maximum value allowed by the kinematics, nor on the primary deuteron energy. The ratio equals to 0.06. This quantity and its energy dependence cannot be explained by means of the Fermi motion.

Accelerator Dubna proton synchrotron

Detectors Spectrometer

Related references

More (earlier) information
A. M. Baldin, preprint JINR-P1-5819 (1971);
A. M. Baldin, preprint JINR-P7-5769 (1971);

See also
H. H. Heckman et al., Phys. Rev. Lett. **28** (1972) 926;
A. M. Baldin, Kr. Soob. Fiz. **1** (1971) 35;
G. W. Akerlof et al., Phys. Rev. **D3** (1971) 645;
D. Dekkers et al., Phys. Rev. **137B** (1965) 962;
J. V. Allaby et al., preprint CERN 70-12 (1970);
A. W. Anderson et al., Phys. Rev. Lett. **19** (1957) 198;
J. Hamberstone and J. S. Wollare, Nucl. Phys. **A141** (1970) 362;
I. J. McGee, Phys. Rev. **151** (1966) 772;

Reactions

deuteron Cu $\rightarrow \pi^-$ X	7.6-8.52 GeV(T_{lab})

KOBAYASHI 1973B

■ Observation that CP violation can be accommodated in the standard electroweak model only if there are at least six quark flavours ■

CP-Violation in the Renormalizable Theory of Weak Interaction

M. Kobayashi, T. Maskawa
Progr. of Theor. Phys. **49** (1973) 652;

Abstract In a framework of the renormalizable theory of weak interaction, problems of CP-violation are studied. It is concluded that no realistic model of CP-violation exist in the quartet scheme without introducing any other new fields. Some possible models of CP-violation are also discussed.

Related references

See also
S. Weinberg, Phys. Rev. Lett. **19** (1967) 1264;
P. W. Higgs, Phys. Lett. **12** (1964) 132;
P. W. Higgs, Phys. Rev. Lett. **13** (1964) 508;
G. S. Guralnik, C. R. Hagen, and T. W. B. Kibble, Phys. Rev. Lett. **13** (1964) 585;
H. Georgi and S. L. Glashow, Phys. Rev. Lett. **28** (1972) 1494;

Particles studied q

AMALDI 1973E

■ Confirmation of rising total hadronic cross sections ■

Energy Dependence of the Proton-Proton Total Cross Section for Center-of-Mass Energies Between 23 and 53 GeV

U. Amaldi et al.
Phys. Lett. **44B** (1973) 112;

Abstract Measurements of proton-proton elastic scattering at angles around 6 mrad have been made at centre-of-mass energies of 23, 31, 45 and 53 GeV using the CERN Intersecting Storage Rings. The absolute scale of the cross section was established by determination of the effective density of the colliding beams in their overlap region. Proton-proton total cross sections were deduced by extrapolation of the elastic differential cross section to the forward direction and by application of the optical theorem. The results indicate that over the energy range studied the proton-proton total cross section increases from about 39 to about 43 mb.

Accelerator CERN-ISR $p\ p$ collider

Detectors Combination

Related references

See also

W. Heisenberg, *Kosmische Strahjung*, Springer Verlag (1953) 148;
M. Froissart, Phys. Rev. **123** (1961) 1053;
A. Martin, Phys. Rev. **129** (1963) 1432;
A. Martin, Nuovo Cim. **42** (1966) 930;
U. Amaldi et al., Phys. Lett. **43B** (1973) 231;
M. Holder et al., Phys. Lett. **35B** (1971) 361;
Analyse information from
M. Holder et al., Phys. Lett. **35B** (1971) 355;
M. Holder et al., Phys. Lett. **36B** (1971) 400;
S. P. Denisov et al., Phys. Lett. **36B** (1971) 415;
U. Amaldi, R. Biancastelli, and C. Bosio, Phys. Lett. **36B** (1971) 504;
G. Barbiellini et al., Phys. Lett. **39B** (1972) 663;
G. B. Yodh, Y. Pal, and J. S. Trefil, Phys. Rev. Lett. **28** (1972) 1005;
F. T. Dao et al., Phys. Rev. Lett. **29** (1972) 1627;
J. W. Chapman et al., Phys. Rev. Lett. **29** (1972) 1686;
G. G. Beznogikh et al., Phys. Lett. **39B** (1972) 411;
S. R. Amendolia et al., Phys. Lett. **44B** (1973) 119;

Reactions
$p\,p \rightarrow X$ 23-53 GeV(E_{cm})

AMENDOLIA 1973B

■ Further confirmation of rising total hadronic cross sections ■

Measurement of the Total Proton-Proton Cross Section at the ISR

S.R. Amendolia et al.
Phys. Lett. **44B** (1973) 119;

Abstract We present the first result of measurement of the total cross section σ_T in proton-proton collisions at equivalent laboratory momenta between 291 and 1480 GeV/c at the CERN Intersecting Storage Ring (ISR). The method is based on the measurement of the ratio of the total interaction rate and the machine luminosity. The data show an increase of about 10% in σ_T in this energy interval.

Accelerator CERN-ISR $p\,p$ collider

Detectors Counters

Related references

See also
W. Heisenberg, *Kosmische Strahjung*, Springer Verlag (1953) 148;
H. Cheng and T. T. Wu, Phys. Rev. Lett. **24** (1970) 1456;
Analyse information from
S. P. Denisov et al., Phys. Lett. **36B** (1971) 415;
G. B. Yodh, Y. Pal and J. S. Trefil, Phys. Rev. Lett. **28** (1972) 1005;
G. Charlton et al., Phys. Rev. Lett. **29** (1972) 515;
F. T. Dao et al., Phys. Rev. Lett. **29** (1972) 1627;
J. W. Chapman et al., Phys. Rev. Lett. **29** (1972) 1686;
G. Barbiellini et al., Phys. Lett. **39B** (1972) 663;
U. Amaldi et al., Phys. Lett. **44B** (1973) 112;

Reactions
$p\,p \rightarrow X$ 291-1480 GeV(P_{lab})

HASERT 1973

■ First experimental indication of the existence of weak neutral currents in pure leptonic interactions ■

Search for Elastic Muon-neutrino Electron Scattering

F.J. Hasert et al.
Phys. Lett. **46B** (1973) 121;

Abstract One possible event of the process $\overline{\nu}_\mu\, e^- \rightarrow \overline{\nu}_\mu\, e^-$ has been observed. The various background processes are discussed and the event interpreted in terms of Weinberg theory. The 90% confidence limits on the Weinberg parameter are $0.1 < \sin^2\Theta_W < 0.6$.

Accelerator CERN-PS proton synchrotron

Detectors Bubble chamber GARGAMELLE

Data comments Upper bounds on the cross sections are given.

Related references

More (later) information
U. Nguyen-Khac, in: *X Moriond Meeting, Meribel-les-Allues*, France **2** (1975) 307;

Reactions
$\overline{\nu}_\mu\, e^- \rightarrow e^-\, \overline{\nu}_\mu$ 1-14 GeV(E_{lab})

HASERT 1973B

■ First experimental evidence for weak neutral currents ■

Observation of Neutrino-like Interactions Without Muon or Electron in the Gargamelle Neutrino Experiment

F.J. Hasert et al.
Phys. Lett. **46B** (1973) 138; Nucl. Phys. **B73** (1974) 1;

Reprinted in
Development in the Quark Theory of Hadrons, A Reprint Collection V.1: 1964-1978, ed. D. B. Lichtenberg and S. P. Rosen, Hadronic Press, Inc., Monamtum, Mass. (1980) 201.
R. N. Cahn and G. Goldhaber, *The Experimental Foundations of Particle Physics*, Cambridge Univ. Press (1991) 394.

Abstract Events induced by neutral particles and producing hadrons, but no muon or electron, have been observed in the CERN neutrino experiment. These events behave as expected if they arise from neutral current induced processes. The rates relative to the corresponding charged current processes are evaluated.

Accelerator CERN-PS proton synchrotron

Detectors Bubble chamber GARGAMELLE

Related references

More (later) information
J. Blietschau et al., Nucl. Phys. **B114** (1976) 189;
U. Nguen Khac, *Internat. Colloquium of High Energy Neutrino Physics*, Paris (1975) 173;
W. F. Fry and D. Haidt, preprint CERN-75-1;
A. Lagarrighue, *Intern. School of Subnuclear Physics*, Erice (1974) 543;
A. Pullia et al, *17 Intern. Conf. on High Energy Physics* London, UK (1974) 114;

Superseded by
J. Blietschau et al., Nucl. Phys. **B118** (1977) 218;

Reactions

ν_μ nucleus $\rightarrow$ ν_μ X	1-10 GeV(P_{lab})
ν_μ nucleus $\rightarrow$ μ^- X	1-10 GeV(P_{lab})
$\overline{\nu}_\mu$ nucleus $\rightarrow$ $\overline{\nu}_\mu$ X	1-10 GeV(P_{lab})
$\overline{\nu}_\mu$ nucleus $\rightarrow$ μ^+ X	1-10 GeV(P_{lab})

GROSS 1973

■ Discovery of the "asymptotic freedom" property of interacting Yang-Mills field theories ■

Ultraviolet Behaviour of Non-Abelian Gauge Theory

D.J. Gross, F. Wilczek
Phys. Rev. Lett. **30** (1973) 1343;

Reprinted in
The Physical Review — the First Hundred Years, AIP Press (1995) 1076.

Abstract It is shown that a wide class of non-Abelian gauge theories have, up to calculable logarithmic corrections, free-field-theory asymptotic behavior. It is suggested that Bjorken scaling may be obtained from strong-interaction dynamics based on non-Abelian gauge symmetry.

Related references

See also
S. Weinberg, Phys. Rev. Lett. **19** (1967) 1264;
S. Weinberg, Phys. Rev. **D5** (1972) 1962;
M. Gell-Mann and F. E. Low, Phys. Rev. **95** (1954) 1300;
C. G. Callan, Phys. Rev. **D2** (1970) 1541;
K. Symanzik, Comm. Math. Phys. **18** (1970) 227;
K. Wilson, Phys. Rev. **D3** (1971) 1818;
H. D. Politzer, Phys. Rev. Lett. **30** (1973) 1346;
Y. Nambu and G. Jona-Lasinio, Phys. Rev. **122** (1961) 345;
S. Coleman and E. Weinberg, Phys. Rev. **D7** (1973) 1888;
H. Georgi and S. L. Glashow, Phys. Rev. Lett. **28** (1972) 1494;

POLITZER 1973

■ Discovery of the "asymptotic freedom" property of interacting Yang-Mills field theories ■

Reliable Perturbative Results for Strong Interactions?

H.D. Politzer
Phys. Rev. Lett. **30** (1973) 1346;

Reprinted in
The Physical Review — the First Hundred Years, AIP Press (1995) CD-ROM.

Abstract An explicit calculation shows perturbation theory to be arbitrarily good for the deep Euclidean Green's functions of any Yang-Mills theory and many Yang-Mills theories with fermions. Under the hypothesis that spontaneous symmetry breakdown is of dynamical origin, these symmetric Green's functions are the asymptotic forms of the physically significant spontaneously broken solution, whose coupling could be strong.

Related references

See also
S. L. Adler, Phys. Rev. **D5** (1972) 3021;
K. Symanzik, preprint DESY-72-73 (1972);
N. N. Bogolyubov and D. V. Shirkov, *Introduction to the Theory of Quantized Fields*, New York, (1959);
S. Coleman and S. Weinberg, Phys. Rev. **D7** (1973) 1888;
B. W. Lee and J. Zinn Justin, Phys. Rev. **D5** (1972) 3121;
G. 't Hooft, Nucl. Phys. **B33** (1971) 173;
D. J. Gross and F. Wilczek, Phys. Rev. Lett. **30** (1973) 1343;
Y. Nambu and G. Jona-Lasinio, Phys. Rev. **122** (1961) 345;

BUESSER 1973

■ First observation of high transverse momentum hadrons at the CERN Intersecting Storage Rings ■

Observation of π^0 Mesons with Large Transverse Momentum in High Energy Proton-Proton Collisions

F.W. Buesser et al.
Phys. Lett. **46B** (1973) 471;

Abstract Invariant cross-sections are presented for the inclusive reaction $p\,p \rightarrow \pi^0 +$ anything. Measurements of large transverse momentum π^0's (2.5 GeV/c $< p_\perp <$ 9 GeV/c) were made near 90° at the CERN ISR at five center-of-mass energies ($\sqrt{s} = 23.5, 30.6, 44.8, 52.7$ and 62.4). At large $p_\perp$, the invariant cross-sections are seen to vary with s and $p_\perp$, in good agreement with fit of the form $Ap_\perp^{-n}F(p_\perp/\sqrt{s})$, with $n \approx 8$ and $F(p_\perp/\sqrt{s}) \approx \exp(-26p_\perp/\sqrt{s})$.

Accelerator CERN-ISR $p\,p$ collider

Detectors Spectrometer

Related references

More (later) information
F. W. Buesser et al., Phys. Lett. **48B** (1974) 371;
F. W. Buesser et al., Phys. Lett. **51B** (1974) 306;
F. W. Buesser et al., Nucl. Phys. **B106** (1976) 1;

Reactions

$p\,p \rightarrow \pi^0\,X$ 23.5-62.4 GeV(E_{cm})

WEINBERG 1973

■ Final formulation of QCD and the Standard Model Lagrangian ■

Non-Abelian Gauge Theories of the Strong Interactions

S. Weinberg
Phys. Rev. Lett. **31** (1973) 494;

Abstract A class of non-Abelian gauge theories of strong interactions is described, for which parity and strangeness are automatically conserved, and for which the nonconservations of parity and strangeness produced by weak interactions are automatically of order α/m_W^2 rather than of order α. When such theories are "asymptotically free," the order-α weak correction to natural zeroth-order symmetries may be calculated ignoring all effects of strong interactions. Speculations are offered on a possible theory of quarks.

Related references

More (earlier) information
D. J. Gross and F. Wilczek, Phys. Rev. Lett. **30** (1973) 1343;
H. D. Politzer, Phys. Rev. Lett. **30** (1973) 1346;

See also
I. Bars, M. B. Halpern, and M. Yoshimura, Phys. Rev. Lett. **29** (1972) 969;
S. Weinberg, Phys. Rev. Lett. **29** (1972) 338;
S. Weinberg, Phys. Rev. Lett. **29** (1972) 1698;
H. Gregori and S. L. Glashow, Phys. Rev. **D6** (1972) 2977;
H. Gregori and S. L. Glashow, Phys. Rev. **D8** (1973) 2457;
K. Wilson, Phys. Rev. **179** (1969) 1499;
C. G. Callan, Phys. Rev. **D5** (1972) 3202;
S. L. Glashow and S. Weinberg, Phys. Rev. Lett. **20** (1968) 224;
M. Gell-Mann, R. J. Oakes, and B. Renner, Phys. Rev. **175** (1968) 2195;
S. L. Glashow, R. Jackiw, and S. S. Shei, Phys. Rev. **187** (1969) 1416;

FRITZSCH 1973

■ Invention of the QCD Lagrangian of Yang–Mills type ■

Advantages of the Color Octet Gluon Picture

H. Fritzsch, M. Gell-Mann, H. Leutwyler
Phys. Lett. **47B** (1973) 365;

Abstract It is pointed out that there are several advantages in abstracting properties of hadrons and their currents from a Yang–Mills gauge model based on colored quarks and color octet gluons.

Related references

More (earlier) information
H. Fritzsch and M. Gell-Mann, in *Proc. XVI Intern. Conf. on High Energy Physics*, Chicago **V.2** (1972) 135;
D. J. Gross and F. Wilczek, Phys. Rev. Lett. **30** (1973) 1343;
H. D. Politzer, Phys. Rev. Lett. **30** (1973) 1346;

See also
H. J. Lipkin, Phys. Lett. **45B** (1973) 267;
J. Schwinger, Phys. Rev. **82** (1951) 664;
S. L. Adler, Phys. Rev. **177** (1969) 2426;
J. S. Bell and R. Jackiw, Nuovo Cim. **60A** (1969) 47;
S. L. Adler and W. A. Bardeen, Phys. Rev. **182** (1969) 1517;
K. Wilson, Phys. Rev. **179** (1969) 1499;

GROSS 1973B

■ Final formulation of QCD theory ■

Asymptotically Free Gauge Theories. I

D.J. Gross, F. Wilczek
Phys. Rev. **D8** (1973) 3633;

Reprinted in

The Physical Review — the First Hundred Years, AIP Press (1995) 1076.

Abstract Asymptotically free gauge theories of the strong interactions are constructed and analyzed. The reason for doing this are recounted, including a review of renormalization-group techniques and their application to scaling phenomena. The renormalization-group equations are derived for Yang-Mills theories. The parameters that enter into the equations are calculated to lowest order and it is shown that these theories are asymptotically free. More specifically the effective coupling constant, which determines the ultraviolet behavior of the theory, vanishes for large spacelike momenta. Fermions are incorporated and the construction of realistic models is discussed. We propose that the strong interactions be mediated by a "color" gauge group which commutes with $SU(3) \times SU(3)$. The problem of symmetry breaking is discussed. It appears likely that this would have a dynamical origin. It is suggested that the gauge symmetry might not be broken and that the severe infrared singularities prevent the occurrence of noncolor singlet physical states. The deep-inelastic structure functions, as well as electron-positron total annihilation cross section are analysed. Scaling obtains up to calculable logarithmic corrections, and the naive light-cone or parton-model results follow. The problems of incorporating scalar mesons and breaking the symmetry by the Higgs mechanism are explained in detail.

Related references

More (earlier) information
D. J. Gross and F. Wilczek, Phys. Rev. Lett. **30** (1973) 1343;

H. D. Politzer, Phys. Rev. Lett. **30** (1973) 1346;

See also
M. Gell-Mann and F. E. Low, Phys. Rev. **95** (1954) 1300;
N. N. Bogolyubov and D. V. Shirkov, *Introduction to the Theory of Quantized Fields*, Interscience, New York (1959);
S. Weinberg, Phys. Rev. **118** (1960) 838;
S. Weinberg, Phys. Rev. Lett. **19** (1967) 1264;
S. Weinberg, Phys. Rev. **D5** (1972) 1962;
K. Johnson, M. Baker, and R. Willey, Phys. Rev. **136** (1964) B1111;
K. Johnson, R. Willey, and M. Baker, Phys. Rev. **163** (1967) 1699;
M. Baker and K. Johnson, Phys. Rev. **183** (1969) 1292;
M. Baker and K. Johnson, Phys. Rev. **D3** (1971) 2516;
M. Baker and K. Johnson, Phys. Rev. **D3** (1971) 2541;
S. L. Adler, Phys. Rev. **D5** (1972) 3021;
J. D. Bjorken, Phys. Rev. **148** (1966) 1467;
J. D. Bjorken, Phys. Rev. **179** (1969) 1547;
K. Wilson, Phys. Rev. **179** (1969) 1499;
K. Wilson, Phys. Rev. **D3** (1971) 1818;
R. P. Feynman, *Photon-Hadron Interactions*, Benjamin, New York (1972) 166;
C. G. Callan and D. J. Gross, Phys. Rev. Lett. **22** (1969) 156;
C. G. Callan, Phys. Rev. **D2** (1970) 1541;
C. G. Callan, Phys. Rev. **D5** (1972) 3202;
L. V. Osviannikov, Doklady Akad. Nauk SSSR **109** (1956) 1112;
N. Christ, B. Hasslacher, and A. Mueller, Phys. Rev. **D6** (1972) 3543;
S. Ferrara et al., Phys. Lett. **38B** (1972) 333;
A. Zee, Phys. Rev. **D7** (1973) 3630;
S. Coleman and D. J. Gross, Phys. Rev. Lett. **31** (1973) 851;
R. P. Feynman, Acta Phys. Polon. **24** (1963) 697;
L. D. Faddeev and V. N. Popov, Phys. Lett. **25B** (1967) 29;
G. 't Hooft and M. Veltman, Nucl. Phys. **B44** (1972) 189;
P. W. Higgs, Phys. Lett. **12** (1964) 132;
Y. Nambu and G. Jona-Lasinio, Phys. Rev. **122** (1961) 345;
K. Symanzik, Comm. Math. Phys. **18** (1970) 227;
H. Pagels, Phys. Rev. **D7** (1973) 3689;
R. Jackiw and K. Johnson, Phys. Rev. **D8** (1973) 2386;
J. M. Cornwall and R. Norton, Phys. Rev. **D8** (1973) 3338;
S. Coleman and S. Weinberg, Phys. Rev. **D7** (1973) 1888;
T. W. B. Kibble, Phys. Rev. **155** (1967) 1554;
B. W. Lee, in *Proc. of the XVI Intern. Conf. on High Energy Physics* Chicago, (Ref. 14) **V.4** (1972) 249;
H. Georgi and S. L. Glashow, Phys. Rev. Lett. **28** (1972) 1494;

MATVEEV 1973

■ Quark counting rules for asymptotic energy power low behavior of the binary hadronic amplitudes at large fixed angles ■

Automodelism in the Large-Angle Elastic Scattering and Structure of Hadrons

V.A. Matveev, R.M. Muradyan, A.N. Tavkhelidze
Lett. Nuovo Cim. **7** (1973) 719;

Abstract In this paper we should like to point to the possibility of revealing a local or pointlike structure of hadrons in pure hadronic reactions of elastic scattering at large momentum transfers. In the framework of the quark model by using the principle of automodelism it is shown that in the limit $-t,\ s \to \infty$ at t/s fixed the automodel asymptotic relations for the differential cross-sections of elastic hadron-hadron and electron-hadron reactions of **Table I** hold.
In the general case of two-particle collisions the differential cross-section of large-angle scattering has the asymptotics

$$\frac{d\sigma}{dt}(ab \to ab) \xrightarrow[t/s \text{ fixed}]{s \to \infty} \frac{1}{s^{2(n_a+n_b-1)}} f_{ab}(t/s). \qquad (1)$$

Table I.

Reaction $a\ b \to a\ b$	$d\sigma/dt$
$pp \to pp$	$(1/s^{10}) f_{pp}(t/s)$
$\pi p \to \pi p$	$(1/s^{8}) f_{\pi p}(t/s)$
$\pi\pi \to \pi\pi$	$(1/s^{6}) f_{\pi\pi}(t/s)$
$ep \to ep$	$(1/s^{6}) f_{ep}(t/s)$
$e\pi \to e\pi$	$(1/s^{4}) f_{e\pi}(t/s)$
$ee \to ee$	$(1/s^{2}) f_{ee}(t/s)$

Here n_a and n_b are the numbers of constituent quarks of a and b particles, *e.g.* for proton $n_p = 3$ and for pion $n_\pi = 2$. Considering an electron as a structureless particle we should put $n_e = 1$ in formula (1) for reactions with electrons. Attempts to the derivation of the power character of the asymptotic behaviour of differential cross-sections at large angles have been done in a number of recent works from the various model assumptions. *(Extracted from the introductory part of the paper.)*

Related references

See also
J. V. Allaby et al., Phys. Lett. **25B** (1967) 156;
J. V. Allaby et al., Phys. Lett. **34B** (1971) 431;
V. A. Matveev, R. M. Muradyan, and A. N. Tavkhelidze, Lett. Nuovo Cim. **5** (1972) 907;
D. Horn and M. Moshe, Nucl. Phys. **48B** (1972) 557;
W. R. Theis, DESY preprint 72/35 (1972);
A. V. Efremov, JINR preprint E2-6612 (1972);
F. Cerulus and A. Martin, Phys. Lett. **8** (1964) 80;
A. A. Logunov and M. A. Mestvirishvili, Phys. Lett. **24B** (1967) 583;
A. M. Baldin, JINR preprint P7-5808 (1971);
D. V. Shirkov, JINR preprint P2-6938 (1973);
A. A. Migdal, Phys. Lett. **37B** (1971) 98;

BRODSKY 1973

■ Quark counting rules for asymptotic energy power low behavior of the binary hadronic amplitudes at large fixed angles ■

Scaling Laws at Large Transverse Momentum

S. Brodsky, G.R. Farrar
Phys. Rev. Lett. **31** (1973) 1153;

Abstract The application of simple dimensional counting to bound states of pointlike particles enables us to derive scaling laws for the asymptotic energy dependence of electromagnetic and hadronic scattering at fixed c.m. angle which only depend on the number of constituent fields of the hadrons. Assuming quark constituents, some of the $s \to \infty$, fixed-t/s predictions are $(d\sigma/dt)_{\pi p \to \pi p} \sim s^{-8}$, $(d\sigma/dt)_{pp \to pp} \sim s^{-10}$,

$(d\sigma/dt)_{\gamma p \to \pi p} \sim s^{-7}$, $(d\sigma/dt)_{\gamma p \to \gamma p} \sim s^{-6}$, $F_\pi(q^2) \sim (q^2)^{-1}$, and $F_{1p}(q^2) \sim (q^2)^{-2}$. We show that such scaling laws are characteristic of renormalizable field theories satisfying certain conditions.

Related references

More (earlier) information
V. Matveev, R. Muradyan, and A. Tavkhelidze, JINR preprint D2-7110 (1973);

See also
S. Mandelstam, Proc. Roy. Soc. **A233** (1953) 248;
S. D. Drell and T. D. Lee, Phys. Rev. **D5** (1972) 1738;
R. Anderson et al., preprint SLAC-PUB-1178;
R. Blankenbeckler, S. Brodsky, and J. Gunion, Phys. Lett. **39B** (1972) 649;
R. Blankenbeckler, S. Brodsky, and J. Gunion, Phys. Rev. **D8** (1973) 187;
R. Blankenbeckler, S. Brodsky, and J. Gunion, Phys. Rev. **D6** (1972) 2652;
R. Blankenbeckler, S. Brodsky, and J. Gunion, Phys. Lett. **42B** (1972) 461;
S. Brodsky, F. Close, and J. Gunion, Phys. Rev. **D6** (1972) 177;
P. Kirk et al., Phys. Rev. **D8** (1973) 63;
E. Bloom and F. Gilman, Phys. Rev. Lett. **25** (1970) 1140;
J. D. Bjorken and J. Kogut, preprint SLAC-PUB-1213;
P. V. Landshoff and J. C. Polkinghorne, Phys. Rev. **D8** (1973) 927;
S. Berman, J. D. Bjorken, and J. Kogut, Phys. Rev. **D4** (1971) 3388;
S. Berman and M. Jacob, Phys. Rev. Lett. **25** (1970) 1683;
D. Horn and F. Moshe, Nucl. Phys. **48B** (1972) 557;

BENVENUTI 1974F

■ Confirmation of the existence of weak neutral currents ■

Observation of Muonless Neutrino-Induced Inelastic Interactions

A. Benvenuti et al.
Phys. Rev. Lett. **32** (1974) 800;

Abstract We report the observation of inelastic interactions induced by high-energy neutrinos and antineutrinos in which no muon is observed in the final state. A possible, but by no means unique, interpretation of this effect is the existence of a neutral weak current.

Accelerator FNAL proton synchrotron

Detectors Magnetic spectrometer HPWF

Related references

More (earlier) information
A. Benvenuti et al., Phys. Rev. Lett. **30** (1973) 1084;

Superseded by
P. Wanderer et al., Phys. Rev. **D17** (1978) 1679;

Reactions

ν_μ nucleon $\to$ ν_μ X	6-150 GeV(P_{lab})
$\overline{\nu}_\mu$ nucleon $\to$ $\overline{\nu}_\mu$ X	6-150 GeV(P_{lab})
ν_μ nucleon $\to$ μ^- X	6-150 GeV(P_{lab})
$\overline{\nu}_\mu$ nucleon $\to$ μ^+ X	6-150 GeV(P_{lab})

AUBERT 1974D — Nobel prize

■ Evidence for the $J/\psi(1S)$. Nobel prize to S. C. C. Ting awarded in 1976. Co-winner B. Richter "for pioneering work in the discovery of a heavy elementary particle of a new kind" ■

Experimental Observation of a Heavy Particle J

J.J. Aubert et al.
Phys. Rev. Lett. **33** (1974) 1404;

Reprinted in
R. N. Cahn and G. Goldhaber, *The Experimental Foundations of Particle Physics*, Cambridge Univ. Press (1991) 279.
The Physical Review — the First Hundred Years, AIP Press (1995) 901.

Abstract We report the observation of a heavy particle J, with mass m =3.1 GeV and width approximately zero. The observation was made from the reaction p Be $\to e^+ e^-$ X by measuring the e^+e^- mass spectrum with a precise pair spectrometer at the Brookhaven National Laboratory's 30 GeV alternating-gradient synchrotron.

Accelerator Brookhaven (AGS) proton synchrotron

Detectors Double arm spectrometer

Related references

See also
T. D. Lee, Phys. Rev. Lett. **26** (1971) 801;
S. Weinberg, Phys. Rev. Lett. **19** (1967) 1264;
S. Weinberg, Phys. Rev. Lett. **27** (1971) 1688;
S. Weinberg, Phys. Rev. **D5** (1972) 1962;
S. D. Drell and T. M. Yan, Phys. Rev. Lett. **25** (1970) 316;
J. E. Augustin et al., Phys. Rev. Lett. **33** (1974) 1406;
S. Weinberg, Phys. Rev. **D5** (1972) 1412;

Analyse information from
J. C. Christenson et al., Phys. Rev. Lett. **25** (1970) 1523;

Reactions

p Be $\to$ e^- e^+ X	19-20 GeV(P_{lab})
p Be $\to$ $J/\psi(1S)$ X	19-20 GeV(P_{lab})

Particles studied $J/\psi(1S)$

AUGUSTIN 1974B — Nobel prize

■ Another evidence for the $J/\psi(1S)$. Nobel prize to B. Richter awarded in 1976. Co-winner S. C. C. Ting "for pioneering work in the discovery of a heavy elementary particle of a new kind" ■

Discovery of a Narrow Resonance in $e^+ e^-$ Annihilation

J.E. Augustin et al.
Phys. Rev. Lett. **33** (1974) 1406;

Reprinted in
R. N. Cahn and G. Goldhaber, *The Experimental Foundations of Particle Physics*, Cambridge Univ. Press (1991) 281.
The Physical Review — the First Hundred Years, AIP Press (1995) 904.

Abstract We have observed a very sharp peak in the cross section for $e^+e^- \to$ hadrons, e^+e^-, and possibly $\mu^+\mu^-$ at a center-of-mass energy of 3.105 ± 0.003 GeV. The upper limit to the full width at half-maximum is 1.3 MeV.

Accelerator SLAC-SPEAR $e^- e^+$ 8.4 GeV ring:

Detectors Mark-I

Related references

More (earlier) information
J. E. Augustin et al., Phys. Rev. Lett. **34** (1975) 233;

More (later) information
A. M. Boyarski et al., Phys. Rev. Lett. **34** (1975) 762;
J. E. Augustin et al., Phys. Rev. Lett. **34** (1975) 764;
J. E. Zipse, preprint LBL-4281 (1975);
G. S. Abrams et al., Phys. Rev. Lett. **33** (1974) 1453;

See also
G. Bonneau and F. Martin, Nucl. Phys. **B27** (1971) 381;

Analyse information from
J. J. Aubert et al., Phys. Rev. Lett. **33** (1974) 1404;
C. Bacci et al., Phys. Rev. Lett. **33** (1974) 1408;

Reactions

$e^+ e^- \to$ 2hadron (hadrons)	3.1-3.14 GeV(E_{cm})
$e^+ e^- \to e^- e^+$	3.1-3.14 GeV(E_{cm})
$e^+ e^- \to K^+ K^-$	3.1-3.14 GeV(E_{cm})
$e^+ e^- \to \pi^+ \pi^-$	3.1-3.14 GeV(E_{cm})
$e^+ e^- \to \mu^- \mu^+$	3.1-3.14 GeV(E_{cm})
$J/\psi(1S) \to$ 2hadron (hadrons)	
$J/\psi(1S) \to e^- e^+$	
$J/\psi(1S) \to K^+ K^-$	
$J/\psi(1S) \to \pi^+ \pi^-$	
$J/\psi(1S) \to \mu^- \mu^+$	

Particles studied $J/\psi(1S)$

BACCI 1974

■ Confirmation of the existence of the $J/\psi(1S)$ ■

Preliminary Result of Frascati (ADONE) on the Nature of a New 3.1 GeV Particle Produced in e^+ e^- Annihilation

C. Bacci et al.
Phys. Rev. Lett. **33** (1974) 1408; Phys. Rev. Lett. **33** (1974) 1649;

Abstract We report on the results at ADONE to study the properties of the newly found 3.1 BeV particle.

Accelerator ADONE Frascati $e^- e^+$ ring:

Detectors Spark chamber ADONE-2GAMMA

Related references

More (later) information
R. Baldini-Celio, M. Berna-Rodini, and G. Capon, Lett. Nuovo Cim. **11** (1974) 711;

See also
J. J. Aubert et al., Phys. Rev. Lett. **33** (1974) 1404;
J. E. Augustin et al., Phys. Rev. Lett. **33** (1974) 1406;

Reactions

$e^+ e^- \to J/\psi(1S)$	3.090-3.112 GeV(E_{cm})
$e^+ e^- \to$ 3hadron (hadrons)	3.090-3.112 GeV(E_{cm})

Particles studied $J/\psi(1S)$

Accelerator ADONE Frascati $e^- e^+$ ring:

Detectors Magnetic detector MEA

Reactions

$e^+ e^- \to$ 2hadron (hadrons)	3.090-3.112 GeV(E_{cm})
$e^+ e^- \to e^- e^+$	3.090-3.112 GeV(E_{cm})
$e^+ e^- \to \mu^- \mu^+$	3.090-3.112 GeV(E_{cm})
$e^+ e^- \to J/\psi(1S)$	3.090-3.112 GeV(E_{cm})

Particles studied $J/\psi(1S)$

Accelerator ADONE Frascati $e^- e^+$ ring:

Detectors PBARP

Reactions

$e^+ e^- \to$ 2hadron (hadrons)	3.090-3.112 GeV(E_{cm})
$e^+ e^- \to e^- e^+$	3.090-3.112 GeV(E_{cm})
$e^+ e^- \to \mu^- \mu^+$	3.090-3.112 GeV(E_{cm})
$e^+ e^- \to J/\psi(1S)$	3.090-3.112 GeV(E_{cm})
$J/\psi(1S) \to \mu^- \mu^+$	

Particles studied $J/\psi(1S)$

ABRAMS 1974

■ First evidence for the $\psi(2S)$ ■

The Discovery of a Second Narrow Resonance in e^+ e^- Annihilation

G.S. Abrams et al.
Phys. Rev. Lett. **33** (1974) 1453;

Reprinted in
R. N. Cahn and G. Goldhaber, *The Experimental Foundations of Particle Physics*, Cambridge Univ. Press (1991) 284.
The Physical Review — the First Hundred Years, AIP Press (1995) CD-ROM.

Abstract We have observed a second sharp peak in the cross section for $e^+e^- \to$ hadrons at a center-of-mass energy of 3.695 ± 0.004 GeV. The upper limit of the full width at half-maximum is 2.7 MeV.

Accelerator SLAC-SPEAR e^- e^+ 8.4 GeV ring:

Detectors Mark-I

Related references

More (earlier) information
J. E. Augustin et al., Phys. Rev. Lett. **33** (1974) 1406;
More (later) information
J. E. Augustin et al., Phys. Rev. Lett. **34** (1975) 764;
A. M. Boyarski et al., Phys. Rev. Lett. **34** (1975) 762;
J. E. Zipse, preprint LBL-4281 (1975);
See also
J. J. Aubert et al., Phys. Rev. Lett. **33** (1974) 1404;
C. Bacci et al., Phys. Rev. Lett. **33** (1974) 1408;
G. Bonneau and F. Martin, Nucl. Phys. **B27** (1971) 321;

Reactions

$e^+ e^- \to X$	3.600-3.720 GeV(E_{cm})
$e^+ e^- \to \psi(2S)$	3.600-3.720 GeV(E_{cm})
$\psi(2S) \to$ 2hadron (hadrons)	

Particles studied $\psi(2S)$

CAZZOLI 1975E

■ First evidence for the charmed baryon $\Sigma_c(2455)^{++}$. First indication of the production of the Λ_c^+ charmed baryon ■

Evidence for $\Delta S = -\Delta Q$ Current or Charmed-Baryon Production by Neutrinos

E.G. Cazzoli et al.
Phys. Rev. Lett. **34** (1975) 1125;

Reprinted in
J. L. Rosner, *New Particles.* Selected Reprints, Stony Brook, Am. Ass. of Physics Teachers (1981) 79.
The Physical Review — the First Hundred Years, AIP Press (1995) CD-ROM.

Abstract We report on the production by neutrinos of an event with negative strangeness. In an exposure of the Brookhaven National Laboratory 7-ft cryogenic bubble chamber to a broad-band neutrino beam 335 events were observed, one of which fits the reaction $\nu p \to \mu^-\Lambda^0\pi^+\pi^+\pi^+\pi^-$. Alternative explanations are examined and none found with a probability greater than 3×10^{-5}. The event thus represents a large violation of the $\Delta S = \Delta Q$ rule or alternatively the production and decay of a charmed baryon state. The most plausible mass for this state is found to be 2426 ± 12 MeV.

Accelerator Brookhaven (AGS) proton synchrotron

Detectors Hydrogen bubble chamber, Deuterium bubble chamber

Comments on experiment One event with charmed-baryon candidate from 100 neutrino events from 62000 pictures with hydrogen. 235 neutrino events from 68000 pictures with deuterium.

Related references

More (later) information
M. K. Gaillard, B. W. Lee, and J. L. Rosner, Rev. of Mod. Phys. **47** (1975) 277;
See also
J. J. Aubert et al., Phys. Rev. Lett. **33** (1974) 1404;
J. E. Augustin et al., Phys. Rev. Lett. **33** (1974) 1406;
C. Bacci et al., Phys. Rev. Lett. **33** (1974) 1408;
C. S. Abrams et al., Phys. Rev. Lett. **33** (1974) 1453;
J. D. Bjorken and S. L. Glashow, Phys. Rev. Lett. **11** (1964) 255;
S. L. Glashow, J. Illiopoulos, and L. Maiani, Phys. Rev. **D2** (1970) 1285;
A. S. Goldhaber and M. Goldhaber, Phys. Rev. Lett. **34** (1975) 37;
J. Schwinger, Phys. Rev. Lett. **34** (1975) 37;
S. Borchardt, V. S. Mathur, and S. Okubo, Phys. Rev. Lett. **34** (1975) 38;
R. M. Barnett, Phys. Rev. Lett. **34** (1975) 41;
T. Appelquist and H. D. Politzer, Phys. Rev. Lett. **34** (1975) 43;
A. De Rujula and S. L. Glashow, Phys. Rev. Lett. **34** (1975) 46;
H. T. Nieh, T. T. Wu, and C. N. Yang, Phys. Rev. Lett. **34** (1975) 49;
C. G. Callan et al., Phys. Rev. Lett. **34** (1975) 52;

Reactions

$\nu_\mu\ p \to \Lambda\ 3\pi^+\ \pi^-\ \mu^-$	13.5 GeV(P_{lab})
$\Sigma_c(2455)^{++} \to \Lambda_c^+\ \pi^+$	
$\Sigma_c(2455)^{++} \to \Lambda\ 3\pi^+\ \pi^-$	
$\Lambda_c^+ \to \Lambda\ 2\pi^+\ \pi^-$	

Particles studied charmed-baryon, Λ_c^+, $\Sigma_c(2455)^{++}$

SCHWITTERS 1975

■ Evidence of azimuthal asymmetry in inclusive hadron production in polarized $e^+ e^-$ collisions. Confirmation of the quark-parton picture of hadron production ■

Azimuthal Asymmetry in Inclusive Hadron Production by e^+e^- Annihilation

R.F. Schwitters et al.
Phys. Rev. Lett. **35** (1975) 1320;

Abstract We have observed an azimuthal asymmetry in inclusive hadron production by e^+e^- annihilation at the center-of-mass energy $\sqrt{s} = 7.4$ GeV. The asymmetry is caused by the polarization of the circulating beams in the storage ring and allows separate determination of the transverse and longitudinal structure functions. We find that transverse production dominates for $x > 0.2$ where x is the scaling variable $2p/\sqrt{s}$.

Accelerator SLAC-SPEAR e^- e^+ 8.4 GeV ring:

Detectors Spectrometer

Related references

See also
J. E. Augustin et al., Phys. Rev. Lett. **34** (1975) 764;
J. E. Augustin et al., Phys. Rev. Lett. **34** (1975) 233;
N. Cabibbo and R. Gatto, Phys. Rev. **124** (1961) 1577;
R. P. Feynman, *Photon-Hadron Interactions*, Benjamin, New York (1972) 166;

Reactions
$e^+ e^- \to$ hadron X   7.4 GeV(E_{cm})

Particles studied q

PERL 1975C   Nobel prize

■ First indication of the τ lepton. Nobel prize to M. Perl awarded in 1995 "for the discovery of the τ lepton".
Co-winner F. Reines "for the detection of the neutrino" ■

Evidence for Anomalous Lepton Production in $e^+ e^-$ Annihilation

M.L. Perl et al.
Phys. Rev. Lett. **35** (1975) 1489;

Reprinted in
J. L. Rosner, *New Particles.* Selected Reprints, Stony Brook, Am. Ass. of Physics Teachers (1981) 89.
Development in the Quark Theory of Hadrons, A Reprint Collection V.1: 1964-1978, ed. D. B. Lichtenberg and S. P. Rosen, Hadronic Press, Inc., Monamtum, Mass. (1980) 210.
R. N. Cahn and G. Goldhaber, *The Experimental Foundations of Particle Physics*, Cambridge Univ. Press (1991) 297.
The Physical Review — the First Hundred Years, AIP Press (1995) 907.

Abstract We have found events of the form $e^+ e^- \to e^\pm \mu^\mp +$ missing energy, in which no other charged particles or photons are detected. Most of these events are detected at or above a center-of-mass energy of 4 GeV. The missing-energy and missing-momentum spectra require that at least two additional particles be produced in each event. We have no conventional explanation for these events.

Accelerator SLAC-SPEAR e^- e^+ 8.4 GeV ring:

Detectors Mark-I

Comments on experiment 64 events.

Related references

More (earlier) information
J. E. Augustin et al., Phys. Rev. Lett. **34** (1975) 233;
See also
M. Perl and P. A. Rapides, preprint SLAC-PUB-1496 (1974);
Y. S. Tsai, Phys. Rev. **D4** (1971) 2821;
J. D. Bjorken and C. H. Llewellyn Smith, Phys. Rev. **D7** (1973) 887;
M. A. B. Bég and A. Sirlin, Ann. Rev. Nucl. Sci. **24** (1974) 379;
B. C. Barish et al., Phys. Rev. Lett. **31** (1973) 180;
V. M. Budnev et al., Phys. Rept. **15C** (1975) 182;
H. Terazawa, Rev. of Mod. Phys. **45** (1973) 615;
G. J. Feldman and M. L. Perl, Phys. Rept. **19C** (1975) 233;
M. K. Gaillard, B. W. Lee, and J. L. Rosner, Rev. of Mod. Phys. **47** (1975) 277;
Analyse information from
V. Alles-Borelli et al., Nuovo Cim. Lett. **4** (1970) 1156;
M. Berna-Rodini et al., Nuovo Cim. **17A** (1973) 383;
S. Orito et al., Phys. Lett. **48B** (1974) 165;

Reactions
$e^+ e^- \to \mu^- e^+ 0\gamma$ (neutrals)
$e^+ e^- \to \mu^+ e^- 0\gamma$ (neutrals) 3-7.5 GeV(E_{cm})

Particles studied heavy-lepton$^+$, heavy-lepton$^-$

HANSON 1975

■ First evidence for quark jets in $e^+ e^-$ annihilation ■

Evidence for Jet Structure in Hadron Production by $e^+ e^-$ Annihilation

G. Hanson et al.
Phys. Rev. Lett. **35** (1975) 1609;

Reprinted in
R. N. Cahn and G. Goldhaber, *The Experimental Foundations of Particle Physics*, Cambridge Univ. Press (1991) 318.
Summer Institute on Particle Physics, Stanford (1975) 237.
The Physical Review — the First Hundred Years, AIP Press (1995) 911.

Abstract We have found evidence for jet structure in $e^+e^- \to$ hadrons at center-of-mass energies of 6.2 and 7.4 GeV. At

7.4 GeV the jet-axis angular distribution integrated over azimuthal angle was determined to be proportional to $1+(0.78\pm 0.12)\cos^2\theta$.

Accelerator SLAC-SPEAR $e^- e^+$ 8.4 GeV ring:

Detectors Mark-I

Related references

More (earlier) information
J. E. Augustin et al., Phys. Rev. Lett. **34** (1975) 233;
J. E. Augustin et al., Phys. Rev. Lett. **34** (1975) 764;

See also
S. D. Drell, D. J. Levy, and T. M. Yan, Phys. Rev. **187** (1969) 2159;
S. D. Drell, D. J. Levy, and T. M. Yan, Phys. Rev. **D1** (1970) 1617;
N. Cabibbo, G. Parisi, and M. Testa, Nuovo Cim. Lett. **4** (1970) 35;
J. D. Bjorken and S. J. Brodsky, Phys. Rev. **D1** (1970) 1416;
R. F. Schwitters et al., Phys. Rev. Lett. **35** (1975) 1320;

Reactions

$e^+ e^- \rightarrow$ 2hadron (hadrons)	6.2,7.4 GeV(E_{cm})
$e^+ e^- \rightarrow$ 2jet	6.2,7.4 GeV(E_{cm})
$e^+ e^- \rightarrow q\,\bar{q}$	6.2,7.4 GeV(E_{cm})
$q \rightarrow$ jet	
$\bar{q} \rightarrow$ jet	

APEL 1975B

■ First evidence for the spin 4 $f_4(2050)$ resonance. Confirmation of linearity of Regge trajectories for the spin ≤ 4 resonances ■

Observation of a Spin 4 Neutral Meson with 2 GeV Mass Decaying in $\pi^0\ \pi^0$

Serpukhov-CERN collaboration; W.D. Apel et al.
Phys. Lett. **57B** (1975) 398;

Abstract The invariant mass spectrum of neutral meson states from π^- interactions at 40 GeV/c incident momentum has been investigated in a high statistics experiment performed at the 70 GeV IHEP accelerator. To detect the high energy photons coming from the produced neutral states, a hodoscope spectrometer with a computer on-line was used. A clear structure on the mass spectrum of dipions produced in the reaction $\pi^-\ p \rightarrow n\ \pi^0\ \pi^0$ is observed at 2 GeV. The decay angular distribution show in this mass region the variation with mass typical for of a state with spin $J = 4$. The mass of the observed meson is found to be M = (2020 ± 30) MeV and the estimate of the full width is (180 ± 60) MeV.

Accelerator Serpukhov proton synchrotron

Detectors Spectrometer NICE

Comments on experiment 10^7 events.

Related references

More (earlier) information
W. D. Apel et al., Phys. Lett. **56B** (1975) 190;

Reactions

$\pi^-\ p \rightarrow n\ 2\pi^0$	40 GeV(P_{lab})
$\pi^-\ p \rightarrow n\ \pi^0\ \gamma$	40 GeV(P_{lab})
$\pi^-\ p \rightarrow n\ 2\pi^0$	40 GeV(P_{lab})
$\pi^-\ p \rightarrow n\ f_4(2050)$	40 GeV(P_{lab})
$f_4(2050) \rightarrow 2\pi^0$	

Particles studied $f_4(2050)$

BLUM 1975

■ Evidence for a Spin 4 Boson Resonance at 2050 MeV ■

Evidence for a Spin 4 Boson Resonance at 2050 MeV

CERN-MPI (Munich) collaboration; W. Blum et al.
Phys. Lett. **57B** (1975) 403;

Abstract We report the observation of a spin 4 resonance from the analysis of the reaction $\pi^-\ p \rightarrow K^+\ K^-\ n$. The mass and width of the h-meson were determined to be 2050 MeV and 225 MeV. The quantum numbers are $J^P = 4^+, C = +1$ and very probably $I^G = 0^+$.

Accelerator CERN-PS proton synchrotron

Detectors Spectrometer CERN-MUNICH

Comments on experiment 27000 events.

Related references

More (earlier) information
W. Manner, *Invited talk at the 4th Int. Conf. on Exp. Meson Spectroscopy*, Boston (1974);
G. Grayer et al., Nucl. Phys. **B75** (1974) 189;

Reactions

$\pi^-\ p \rightarrow n\ K^+\ K^-$	18.4 GeV(P_{lab})
$\pi^-\ p \rightarrow K^+\ K^-$ (neutrals)	18.4 GeV(P_{lab})
$\pi^-\ p \rightarrow n\ f_4(2050)$	18.4 GeV(P_{lab})
$f_4(2050) \rightarrow K^+\ K^-$	

Particles studied $f_4(2050)$

BELAVIN 1975

■ Invention of the BPST-instanton – the pseudoparticle solution of the Yang-Mills equation ■

Pseudoparticle Solutions of the Yang-Mills Equations

A.A. Belavin, A.M. Polyakov, A.S. Schwartz, Y.S. Tyupkin
Phys. Lett. **59B** (1975) 85;

Abstract We find regular solutions of the four-dimensional euclidean Yang-Mills equations. The solutions minimize locally the action integral which is finite in this case. The topological nature of the solutions is discussed.

Related references

More (earlier) information
A. M. Polyakov, Phys. Lett. **59B** (1975) 82;

PERL 1976C — Nobel prize

■ Evidence for the production of the τ lepton. Nobel prize to M. Perl awarded in 1995 "for the discovery of the τ lepton". Co-winner F. Reines "for the detection of the neutrino" ■

Properties of Anomalous e μ Events produced in e^+ e^- Annihilation

M.L. Perl et al.
Phys. Lett. **63B** (1976) 466;

Reprinted in
J. L. Rosner, *New Particles.* Selected Reprints, Stony Brook, Am. Ass. of Physics Teachers (1981) 93.

Abstract We present the properties of 105 events of the form $e^+e^- \to e^\pm\ \mu^\pm$ + missing energy, in which no other charged particles or photons are detected. The simplest hypothesis compatible with all the data is that these events come from the production of a pair of heavy leptons, the mass of the lepton being in the range 1.6 to 2.0 GeV/c^2.

Accelerator SLAC-SPEAR e^- e^+ 8.4 GeV ring:

Detectors Mark-I

Related references

More (earlier) information
M. L. Perl et al., Phys. Rev. Lett. **35** (1975) 1489;

Reactions

$e^+ e^- \to \mu^- e^+$ (neutrals)	3-7.5 GeV(E_{cm})
$e^+ e^- \to \mu^+ e^-$ (neutrals)	3-7.5 GeV(E_{cm})
$e^+ e^- \to K_S\ \mu^- e^+$ (neutrals)	3-7.5 GeV(E_{cm})
$e^+ e^- \to$ heavy-lepton$^-$ X	3-7.5 GeV(E_{cm})
$e^+ e^- \to K_S\ \mu^+ e^-$ (neutrals)	3-7.5 GeV(E_{cm})
$e^+ e^- \to$ heavy-lepton$^+$ X	3-7.5 GeV(E_{cm})

Particles studied heavy-lepton$^+$, heavy-lepton$^-$, τ^+, τ^-

GOLDHABER 1976D

■ First evidence for the D^0 charmed meson ■

Observation in e^+ e^- Annihilation of a Narrow State at 1865 MeV/c^2 Decaying to $K\pi$ and $K\pi\pi\pi$

G. Goldhaber et al.
Phys. Rev. Lett. **37** (1976) 255;

Reprinted in
J. L. Rosner, *New Particles.* Selected Reprints, Stony Brook, Am. Ass. of Physics Teachers (1981) 66.
Development in the Quark Theory of Hadrons, A Reprint Collection V.1: 1964-1978, ed. D. B. Lichtenberg and S. P. Rosen, Hadronic Press, Inc., Monamtum, Mass. (1980) 500.
R. N. Cahn and G. Goldhaber, *The Experimental Foundations of Particle Physics*, Cambridge Univ. Press (1991) 301.
The Physical Review — the First Hundred Years, AIP Press (1995) 915.

Abstract We present evidence, from a study of multihadronic final states produced in e^+e^- annihilation at center-of-mass energies between 3.90 and 4.60 GeV, for the production of a new neutral state with mass 1865 ± 15 MeV/c^2 and decay width less than 40 MeV/c^2 that decays to $K^\pm\pi^\mp$ and $K^\pm\pi^\mp\pi^\pm\pi^\mp$. The recoil-mass spectrum for this state suggests that it is produced only in association with systems of comparable or larger mass.

Accelerator SLAC-SPEAR e^- e^+ 8.4 GeV ring:

Detectors Mark-I

Comments on experiment Based on 29000 hadronic events. 110 D^0 events.

Related references

More (earlier) information
J. E. Augustin et al., Phys. Rev. Lett. **34** (1975) 233;
J. E. Augustin et al., Phys. Rev. Lett. **34** (1975) 764;
A. M. Boyarski et al., Phys. Rev. Lett. **34** (1975) 1357;
A. M. Boyarski et al., Phys. Rev. Lett. **35** (1975) 196;

See also
G. S. Abrams et al., Phys. Rev. Lett. **34** (1975) 1181;
J. Siegrist et al., Phys. Rev. Lett. **36** (1976) 700;
S. L. Glashow, J. Iliopoulos, and L. Maiani, Phys. Rev. **D2** (1970) 1285;

Analyse information from
A. Benvenuti et al., Phys. Rev. Lett. **34** (1975) 419;
E. G. Cazzoli et al., Phys. Rev. Lett. **34** (1975) 1125;
J. Blietschau et al., Phys. Lett. **60B** (1976) 207;
J. Von Krogh et al., Phys. Rev. Lett. **36** (1976) 710;
B. C. Barish et al., Phys. Rev. Lett. **36** (1976) 939;

Reactions

$e^+ e^- \to K^+\ \pi^-$ X	3.9-4.6 GeV(E_{cm})
$e^+ e^- \to K^+\ \pi^+\ 2\pi^-$ X	3.9-4.6 GeV(E_{cm})
$e^+ e^- \to K^-\ \pi^+$ X	3.9-4.6 GeV(E_{cm})
$e^+ e^- \to K^-\ 2\pi^+\ \pi^-$ X	3.9-4.6 GeV(E_{cm})

$e^+ e^- \rightarrow D^0$ X	3.9-4.6 GeV(E_{cm})
$e^+ e^- \rightarrow D^0 \overline{D}^0$	3.9-4.6 GeV(E_{cm})
$D^0 \rightarrow K^- \pi^+$	
$D^0 \rightarrow K^- 2\pi^+ \pi^-$	
$\overline{D}^0 \rightarrow K^+ \pi^-$	
$\overline{D}^0 \rightarrow K^+ \pi^+ 2\pi^-$	

Particles studied $D^0, \overline{D}^0$

BUNCE 1976

■ Evidence for large polarization of produced hyperons in p Be collisions ■

Λ Hyperon Polarization in Inclusive Production by 300 GeV Protons on Beryllium

G. Bunce et al.

Phys. Rev. Lett. **36** (1976) 1113;

Abstract Λ polarization has been observed in p Be $\rightarrow \Lambda$ X at 300 GeV. A total of 1.2×10^6 Λ decays were recorded at fixed lab angles between 0 and 9.5 mrad, covering a range of kinematic variables $0.3 \leq x \leq 0.7$ and $0 \leq p_\perp \leq 1.5$ GeV/c. The observed polarization was consistent with parity conservation and increased monotonically with increasing $p_\perp$, independently of x, reaching $P_\Lambda = 0.28 \pm 0.08$ at 1.5 GeV/c.

Accelerator FNAL proton synchrotron

Detectors Spectrometer

Related references

See also

R. G. Roberts, in *Phenomenology of Particles at High Energies*, ed. R. L. Crawford and R. Jennings, Academic, New York (1974);
G. Giacomelli, Phys. Rept. **23C** (1976) 123;
L. Dick et al., Phys. Lett. **57B** (1975) 93;
A. Lesnik et al., Phys. Rev. Lett. **35** (1975) 770;
O. Overseth and R. Roth, Phys. Rev. Lett. **19** (1967) 319;
D. Hill et al., Phys. Rev. **D4** (1971) 1979;
E. Dahl-Jensen et al., Nuovo Cim. **3A** (1971) 1;

Reactions

p Be $\rightarrow \Lambda$ X	300 GeV(P_{lab})
p Be $\rightarrow \Lambda$ X	300 GeV(P_{lab})

PERUZZI 1976C

■ First evidence for the production of D^+ and D^- charmed mesons ■

Observation of a Narrow Charged State at 1876 MeV/c^2 Decaying to an Exotic Combination of $K\pi\pi$

I. Peruzzi et al.

Phys. Rev. Lett. **37** (1976) 569;

Reprinted in

J. L. Rosner, *New Particles.* Selected Reprints, Stony Brook, Am. Ass. of Physics Teachers (1981) 71.

Abstract We report evidence for the production of a new narrow charged state in e^+e^- annihilation at a center-of-mass energy of 4.03 GeV. This state, which has a mass of 1876 ± 15 MeV/c^2, is observed in the exotic channel $K^\mp\pi^\pm\pi^\pm$, but not in the nonexotic channel $K^\mp\pi^+\pi^-$. It is produced primarily in association with a system of mass 2.01 ± 0.02 GeV/c^2. These characteristics are just those expected of a charged charmed meson.

Accelerator SLAC-SPEAR e^- e^+ 8.4 GeV ring:

Detectors Mark-I

Related references

More (earlier) information

G. Goldhaber et al., Phys. Rev. Lett. **37** (1976) 255;

Reactions

$e^+ e^- \rightarrow K^+ 2\pi^-$ X	4.03 GeV(E_{cm})
$e^+ e^- \rightarrow K^+ \pi^+ \pi^-$ X	4.03 GeV(E_{cm})
$e^+ e^- \rightarrow K^- 2\pi^+$ X	4.03 GeV(E_{cm})
$e^+ e^- \rightarrow K^- \pi^+ \pi^-$ X	4.03 GeV(E_{cm})
$e^+ e^- \rightarrow D^+$ X	4.03 GeV(E_{cm})
$e^+ e^- \rightarrow D^-$ X	4.03 GeV(E_{cm})
$e^+ e^- \rightarrow$ 2hadron (hadrons)	4.03 GeV(E_{cm})
$D^+ \rightarrow K^- 2\pi^+$	
$D^- \rightarrow K^+ 2\pi^-$	

Particles studied D^+, D^-

KNAPP 1976B

■ First evidence for the charmed antibaryon $\overline{\Lambda}_c^-$ ■

Observation of a Narrow Anti-Baryon State at 2.26 GeV/c^2

B. Knapp et al.

Phys. Rev. Lett. **37** (1976) 882;

Abstract We report evidence, from a study of multihadron final states produced in the wide-band photon beam at Fermilab, for the production of a new antibaryon state which decays into $\overline{\Lambda}\pi^-\pi^-\pi^+$ The mass of this state is 2.26 ± 0.01 GeV/c^2 and its decay width is less than 75 MeV/c^2. We also report evidence of a state of higher mass (~ 2.5 GeV/c^2) which decays into the state at 2.26 GeV/c^2.

Accelerator FNAL proton synchrotron

Detectors Spectrometer

Related references

More (earlier) information

B. Knapp et al., Phys. Rev. Lett. **34** (1975) 1040;

Reactions

γ Be $\rightarrow \overline{\Lambda}$ mult[π] X <220 GeV(P_{lab})

$\overline{\Lambda}_c^- \rightarrow \overline{\Lambda}\, \pi^+\, 2\pi^-$

Particles studied $\overline{\Lambda}_c^-$

BLIETSCHAU 1976

■ Confirmation of muon-antineutrino scattering off electrons. Confirmation of the weak neutral current ■

Evidence for the Leptonic Neutral Current Reaction $\overline{\nu}_\mu\, e^- \rightarrow \overline{\nu}_\mu\, e^-$

Aachen-Brussels-CERN-Paris-Milan-Orsay-London Collaboration; J. Blietschau et al.
Nucl. Phys. **B114** (1976) 189;

Abstract In the Gargamelle neutrino experiment, three unambiguous candidates for the reaction $\overline{\nu}_\mu\, e^- \rightarrow \overline{\nu}_\mu\, e^-$ have been observed corresponding to a cross section for a recoil electron energy within the range $0.3 < E_{e^-} < 2.0$ GeV of $0.06 \times 10^{-41}\, E_{\overline{\nu}}$ (GeV) cm^2/electron. The calculated background is 0.44 ± 0.13 events and the probability that all three candidates could be due to this background is 1%.

Accelerator CERN-PS proton synchrotron

Detectors Bubble chamber GARGAMELLE

Related references

More (earlier) information
F. J. Hasert et al., Phys. Lett. **46B** (1973) 121;

Reactions

$\overline{\nu}_\mu\, e^- \rightarrow e^-\, \overline{\nu}_\mu$ 1-14 GeV(P_{lab})

BUDKER 1976

■ Evidence for electron cooling ■

Experimental Studies of Electron Cooling

G.I. Budker et al.
PAAC 7, 197 (1976);

Abstract Input of a beam of "cool" electrons into a straight-section orbit of a heavy charged-particle beam circulating in a storage ring introduces an effective friction when the velocities of heavy particles and electrons coincidence magnitude and direction. This friction causes the phase-space volume of the heavy-particle beam to decrease —"electron cooling".
In the work presented here, an experimental study was made of electron cooling of a beam of 35-80 MeV protons. A study was made of the electron-cooling effect on the proton lifetime in the storage ring. At a proton energy of 65 MeV and electron current of 100 mA, the betatron-oscillation damping time was obtained and the equilibrium proton-beam dimensions were measured to be: diameter ≤ 0.8 mm, angular spread $\leq 4 \times 10^{-5}$. Damping-time dependence on the parameters was also studied. The problem of the equilibrium value of the proton momentum spread is considered, and its experimental value is measured to be: $\delta(p)/p \leq 1 \times 10^{-5}$.
A detailed description is given of methods for measurement of the proton-beam apertures. Some possible applications of the electron-cooling methods are described.

Related references

More (earlier) information
G. I. Budker et al., Atomic Energy. USSR. **40** (1976) 49;
G. I. Budker et al., IEEE Transactions on Nucl. Sci. **NS-20** (1975) 2093;
G. I. Budker et al., Atomic Energy. USSR. **40** (1976) 1;

See also
G. I. Budker, *Proc. Intern. Symp. on Electron and Positron Storage Rings*, Saclay (1966) 11-1-1;
C. van der Meer, preprint CERN/ISR/70-5 (1970), CERN/ISR/72-31 (1975);
P. Bramhan et al., Nucl. Instr. and Meth. **125** (1975) 201;
B. Vosicki and K. Zankel, IEEE Transactions on Nucl. Sci. **NS-22** (1975) 1475;
A. A. Kolomensky and A. N. Lebedev, *Cyclic Accelerator Theory*, Fizmatgiz (1962);
B. A. Trubnikov, *Problems of Plasma Theory*, Ed. I.M. (1963);

SEREDNYAKOV 1976

■ Experimental confirmation of the "radiation self-polarization" effect for electrons moving in magnetic field ■

Radiative Polarization of Beams in the VEPP-2M Storage Ring

S.I. Serednyakov, A.N. Skrinsky, G.M. Tumaikin, Y.M. Shatunov
Zh. Eksp. Teor. Fiz. **71** (1976) 2025;

Abstract Radiative polarization of the beams in the VEPP-2M electron-positron storage ring is studied experimentally. Polarization of a beam was measured on basis of the variation of the counting rate of particles scattered within a cluster on resonant depolarization by an external electromagnetic field. The values of the time and degree of polarization are in good agreement with the theoretical predictions. It is shown that the beam energy region between 450 and 670 MeV can be passed without destruction of polarization. The conservation of polarization in the presence of an intense counter-beam is demonstrated.

Accelerator NOVO-VEPP-2M

Detectors Counters

Related references

See also

A. A. Sokolov and I. M. Ternov, Doklady Akad. Nauk SSSR **153** (1963) 1052;
V. N. Baier and V. M. Katkov, Zh. Eksp. Teor. Fiz. **52** (1967) 1422;
V. N. Baier, Usp. Fiz. Nauk **105** (1971) 3;
I. LeDuff et al., preprint Orsay-4-73 (1973);
V. Bargman, L. Mishel, and V. Telegdi, Phys. Rev. Lett. **2** (1959) 435;
Y. S. Derbenev, A. M. Kondratenko, and A. N. Skrinsky, Zh. Eksp. Teor. Fiz. **60** (1971) 1260;
Y. S. Derbenev and A. M. Kondratenko, Zh. Eksp. Teor. Fiz. **62** (1972) 430;
V. N. Baier and Y. F. Orlov, Doklady Akad. Nauk SSSR **165** (1965) 783;
V. N. Baier and V. A. Khoze, Usp. Fiz. Nauk **105** (1971) 3;
V. M. Aulchenko and G. I. Budker et al., preprint NOVO-75-65 (1975);
S. E. Baru et al., preprint NOVO-75-86 (1975);
A. D. Bukin et al., preprint NOVO-75-64 (1975);
A. M. Kondratenko, Zh. Eksp. Teor. Fiz. **66** (1974) 1211;
L. M. Kurdadze et al., preprint NOVO-75-66 (1975);

BLIETSCHAU 1977B

■ Confirmation of the existence of the weak neutral current ■

Determination of the Neutral to Charged Current Inclusive Cross Section Ratio for Neutrino and Anti-Neutrino Interactions in the "Gargamelle" Experiment

GARGAMELLE-neutrino collaboration; J. Blietschau et al.
Nucl. Phys. **B118** (1977) 218;

Abstract The ratio of the neutral current to charged current cross sections has been measured for events with an energy of the hadron system greater than 1 GeV. These ratios are found to be $R_\nu = 0.25 \pm 0.04$ for ν and $R_{\overline{\nu}} = 0.56 \pm 0.08$ for $\overline{\nu}$. Assuming the validity of scaling and a V and/or A type structure for the neutral current, the ratios have been corrected for the cut on the hadronic energy. Our results differ from the predictions of parity conserving models by more than 3 standard deviations. Within the framework of the Weinberg model, $\sin^2\theta_W$ is found to be 0.31 ± 0.06 from R_ν and $0.33^{+0.05}_{-0.08}$ from $R_{\overline{\nu}}$.

Accelerator CERN-PS proton synchrotron

Detectors Bubble chamber GARGAMELLE

Related references

See also
F. J. Hasert et al., Phys. Lett. **46B** (1973) 138;
F. J. Hasert et al., Nucl. Phys. **B73** (1974) 1;

Reactions

ν_μ nucleon $\to$ ν_μ X	1-10 GeV(P_{lab})
$\overline{\nu}_\mu$ nucleon $\to$ $\overline{\nu}_\mu$ X	1-10 GeV(P_{lab})
ν_μ nucleon $\to$ μ^- X	1-10 GeV(P_{lab})
$\overline{\nu}_\mu$ nucleon $\to$ μ^+ X	1-10 GeV(P_{lab})

BRANDELIK 1977D

■ First evidence for $D_S^\pm$, D_S^{*+}, and D_S^{*-} strange charmed mesons ■

Evidence for the F Meson

DASP Collaboration; R. Brandelik et al.
Phys. Lett. **70B** (1977) 132;

Reprinted in
J. L. Rosner, *New Particles.* Selected Reprints, Stony Brook, Am. Ass. of Physics Teachers (1981) 74.

Abstract Inclusive η production by e^+e^- annihilation for cm energies between 4 and 5.2 GeV was studied. A strong η signal was observed at 4.4 GeV, produced predominantly in conjunction with a low energy photon suggestive of $F\overline{F}^*$ or $F^*\overline{F}^*$ production. From events containing a $\pi^\pm$ and η, and a low energy photon the F and F^* masses were found to be 2.03 ± 0.06 and 2.14 ± 0.06 GeV, respectively.

Accelerator DESY-DORIS

Detectors Spectrometer DASP

Related references

See also
S. L. Glashow, J. Iliopoulos, and L. Maini, Phys. Rev. **D2** (1970) 1285;
E. G. Cazzoli et al., Phys. Rev. Lett. **34** (1975) 1125;
B. Knapp et al., Phys. Rev. Lett. **37** (1976) 882;
M. K. Gaillard, B. W. Lee, and J. L. Rosner, Rev. of Mod. Phys. **47** (1975) 277;
G. Goldhaber et al., Phys. Rev. Lett. **37** (1976) 255;
I. Peruzzi et al., Phys. Rev. Lett. **37** (1976) 569;
J. E. Wiss et al., Phys. Rev. Lett. **37** (1976) 1531;
DASP Coll., W. Braunschweig et al., Phys. Lett. **63B** (1976) 471;
PLUTO Coll., J. Burmester et al., Phys. Lett. **64B** (1976) 369;
PLUTO Coll., J. Burmester et al., Phys. Lett. **66B** (1977) 395;
DASP Coll., R. Brandelik et al., Phys. Lett. **67B** (1977) 243;
DASP Coll., R. Brandelik et al., Phys. Lett. **67B** (1977) 358;
J. W. Tukey, *Exploratory data analysis*, Vol. III, ch.24, Addison Wesley, Reading, Mass. (1971);
J. I. Friedman, SLAC-preprint 176 (1974);
J. Siegrist et al., Phys. Rev. Lett. **36** (1976) 700;

Reactions

$e^+ e^- \to 2\gamma$ X	4-5.1 GeV(E_{cm})
$e^+ e^- \to \eta$ X	4-5.1 GeV(E_{cm})
$e^+ e^- \to D_S^+ \eta \pi^- \gamma$	4-5.1 GeV(E_{cm})
$e^+ e^- \to D_S^- \eta \pi^+ \gamma$	4-5.1 GeV(E_{cm})

Particles studied $D_S^+, D_S^-, D_S^{*+}, D_S^{*-}$

ALTARELLI 1977

■ Invention of Altarelli-Parisi evolution equations for quark and gluon densities in colliding hadrons ■

Asymptotic Freedom in Parton Language

G. Altarelli, G. Parisi
Nucl. Phys. **B126** (1977) 298;

Abstract A novel derivation of the Q^2 dependence of quark and gluon densities (of given helicity) as predicted by quantum chromodynamics is presented. The main body of predictions of the theory for deep-inelastic scattering on either unpolarized or polarized targets is re-obtained by a method which only makes use of the simplest tree diagrams and is entirely phrased in parton language with no reference to the conventional operator formalism.

Related references

More (earlier) information
R. P. Feynman, *Photon-Hadron Interactions*, Benjamin, New York (1972) 166;

See also
J. Kogut and L. Susskind, Phys. Rept. **8** (1973) 76;
J. Kogut and L. Susskind, Phys. Rev. **D9** (1974) 697;
J. Kogut and L. Susskind, Phys. Rev. **D9** (1974) 3391;
G. Altarelli, Rivista del Nuovo Cim. **4** (1974) 335;
J. Ellis, *Lectures at Les Houches Summer School* (1976);
H. D. Politzer, Phys. Rept. **14** (1974) 129;
H. Georgi and H. D. Politzer, Phys. Rev. **D9** (1974) 416;
D. J. Gross and F. Wilczek, Phys. Rev. **D9** (1974) 980;
M. A. Ahmed and G. G. Ross, Phys. Lett. **56B** (1975) 385;
M. A. Ahmed, Nucl. Phys. **B111** (1976) 441;
K. Sasaki, Kyoto preprint KUNS 318;
C. F. Weizsäcker, Z. Phys. **88** (1934) 612;
E. J. Williams, Phys. Rev. **45** (1934) 729;
P. Kessler, Nuovo Cim. **16** (1960) 809;
V. N. Baier, V. S. Fadin, and V. A. Khoze, Nucl. Phys. **B65** (1973) 381;
G. Parisi, *Proc. 11th Rencontre de Moriond on weak interactions and neutrino physics*, ed. J. Tran Thanh Van (1976);
A. J. G. Hey and J. E. Mandula, Phys. Rev. **D5** (1972) 2610;
R. L. Heimann, Nucl. Phys. **B64** (1973) 429;
G. Parisi, Phys. Lett. **43B** (1973) 207;
G. Parisi, Phys. Lett. **50B** (1974) 367;
T. Hematsu, Kyoto preprint RIFP-251 (1976);

PERL 1977F — Nobel prize

■ Firm establishment of the τ lepton properties. Nobel prize to M. Perl awarded in 1995 "for the discovery of the τ lepton". Co-winner F. Reines "for the detection of the neutrino" ■

Properties of the Proposed τ Charged Lepton

M.L. Perl et al.
Phys. Lett. **70B** (1977) 487;

Abstract The anomalous $e\mu$ and 2-prong μx events produced in e^+e^- annihilation are used to determine the properties of the proposed τ charged lepton. We find the τ mass is 1.90 ± 0.10 GeV/c^2; the mass of the associated neutrino, ν_τ, is less than 0.6 GeV/c^2 with 95% confidence; V−A coupling is favored over V+A coupling for the $\tau - \nu_\tau$ current and the leptonic branching ratios are $0.186 \pm 0.010 \pm 0.028$ from the $e\mu$ events and $0.175 \pm 0.027 \pm 0.030$ from the μx events where the first error is statistical and the second is systematic.

Accelerator SLAC-SPEAR e^- e^+ 8.4 GeV ring:

Detectors Mark-I

Related references

More (earlier) information
M. L. Perl et al., Phys. Rev. Lett. **35** (1975) 1489;
M. L. Perl et al., Phys. Lett. **63B** (1976) 466;

Reactions

$e^+ e^- \to \mu^- e^+$ (neutrals)	3.8-7.8 GeV(E_{cm})
$e^+ e^- \to \mu^+ e^-$ (neutrals)	3.8-7.8 GeV(E_{cm})
$e^+ e^- \to \tau^- X$	3.8-7.8 GeV(E_{cm})
$e^+ e^- \to \tau^+ X$	3.8-7.8 GeV(E_{cm})

Particles studied τ^+, τ^-, ν_τ

PECCEI 1977

■ Proposal of a Peccei-Quinn spontaneously broken symmetry to explain CP conservation of strong interactions ■

CP Conservation in the Presence of Pseudoparticles

R.D. Peccei, H.R. Quinn
Phys. Rev. Lett. **38** (1977) 1440;

Reprinted in
The Physical Review — the First Hundred Years, AIP Press (1995) CD-ROM.

Abstract We give an explanation of the CP conservation of strong interactions which includes the effects of pseudoparticles. We find it is a natural result for any theory where at least one flavor of fermion acquires its mass through a Yukawa coupling to a scalar field which has nonvanishing vacuum expectation value.

Related references

More (later) information
R. D. Peccei and H. R. Quinn, Phys. Rev. **D16** (1977) 1791;

See also
G. 't Hooft, Phys. Rev. Lett. **37** (1976) 8;
G. 't Hooft, Phys. Rev. **D14** (1976) 3432;
L. Schulman, Phys. Rev. **176** (1968) 1558;
R. Jackiw and C. Rebbi, Phys. Rev. Lett. **37** (1976) 172;
C. G. Callan, R. F. Dashen, and D. J. Gross, Phys. Rev. Lett. **36** (1976) 334;
J. Kiskis, Phys. Rev. **D15** (1977) 2329;
S. Weinberg, Phys. Rev. Lett. **31** (1973) 494;

Particles studied axion

HERB 1977

■ First evidence of the $\Upsilon(1S)$ meson interpreted as a bound state of the new quarks $b\,\bar{b}$. Further indication on the existence of the third quark–lepton family ■

Observation of a Dimuon Resonance at 9.5 GeV in 400 GeV Proton-Nucleus Collisions

S.W. Herb et al.

Phys. Rev. Lett. **39** (1977) 252;

Reprinted in

Development in the Quark Theory of Hadrons, ed. D. B. Lichtenberg and S. P. Rosen, A Reprint Collection **V.1: 1964-1978** Hadronic Press, Inc., Monamtum, Mass. (1980) 214.
J. L. Rosner, *New Particles.* Selected Reprints, Stony Brook, American Association of Physics Teachers (1981) 98.
R. N. Cahn and G. Goldhaber, *The Experimental Foundations of Particle Physics*, Cambridge Univ. Press, (1991) 350.
The Physical Review — the First Hundred Years, AIP Press (1995) 920.

Abstract Dimuon production is studied in 400 GeV proton-nucleus collisions. A strong enhancement is observed at 9.5 GeV mass in a sample of 9000 dimuon events with a mass $m_{\mu^+\mu^-}$.

Accelerator FNAL proton synchrotron

Detectors Double arm spectrometer

Related references

More (earlier) information
D. C. Hom et al., Phys. Rev. Lett. **36** (1976) 1236;
D. C. Hom et al., Phys. Rev. Lett. **37** (1976) 1374;
H. D. Snyder et al., Phys. Rev. Lett. **36** (1976) 1415;

Reactions

p nucleus $\rightarrow \mu^- \mu^+$ X	400 GeV(P_{lab})
p Cu $\rightarrow \mu^- \mu^+$ X	400 GeV(P_{lab})
p Pt $\rightarrow \mu^- \mu^+$ X	400 GeV(P_{lab})
p nucleus $\rightarrow \Upsilon(1S)$ X	400 GeV(P_{lab})
$\Upsilon(1S) \rightarrow \mu^- \mu^+$	

Particles studied $\Upsilon(1S)$

INNES 1977

■ Evidence for the $\Upsilon(2S)$ resonance ■

Observation of Structure in the Υ Region

W.R. Innes et al.

Phys. Rev. Lett. **39** (1977) 1240; Phys. Rev. Lett. **39** (1977) 1640;

Reprinted in

J. L. Rosner, *New Particles.* Selected Reprints, Stony Brook, American Association of Physics Teachers (1981) 102.

Abstract The properties of the dimuon enhancement seen in 400 GeV proton-nucleus collisions have been clarified by a threefold increase in data. We find two peaks whose widths are consistent with our resolution:
$M_1 = 9.4$ GeV with $B\ d\sigma/dy\,|_{y=0} = 1.8 \times 10^{-37}$ cm^2/nucleon and
$M_2 = 10.0$ GeV with $B\ d\sigma/dy\,|_{y=0} = 0.7 \times 10^{-37}$ cm^2/nucleon.
Evidence for the possible existence of a third peak near 10.4 GeV is discuss ed as are the comparisons with the properties of a $q\bar{q}$ system, where q is a new heavy quark.

Accelerator FNAL proton synchrotron

Detectors Spectrometer

Comments on experiment High mass peak may be split.

Related references

See also
S. W. Herb et al., Phys. Rev. Lett. **39** (1977) 252;
M. Barnett, in *Proceedings of the European Conference on Particle Physics*, Budapesht (July 1977);
T. K. Gaisser, F. Halzen, and E. A. Paschos, Phys. Rev. **D15** (1977) 2572;
Analyse information from
E. Eichten and K. Gottfried, Phys. Lett. **66B** (1977) 286;

Reactions

p Cu $\rightarrow \mu^- \mu^+$ X	400 GeV(P_{lab})
p Pt $\rightarrow \mu^- \mu^+$ X	400 GeV(P_{lab})
p nucleus $\rightarrow \Upsilon(1S)$ X	400 GeV(P_{lab})
p nucleus $\rightarrow \Upsilon(2S)$ X	400 GeV(P_{lab})
p nucleus $\rightarrow \mu^- \mu^+$ X	400 GeV(P_{lab})
p nucleus $\rightarrow \Upsilon(1S)$ X	400 GeV(P_{lab})
p nucleus $\rightarrow \Upsilon(2S)$ X	400 GeV(P_{lab})

Particles studied $\Upsilon(1S), \Upsilon(2S)$

LIN 1978

■ Evidence of strong energy dependence of spin-spin correlation parameters in large angle elastic $p\,p$ scattering. "Argonne spin-effect" ■

Energy Dependence of Spin-Spin Forces in $90^\circ_{c.m.}$ Elastic $p\,p$ Scattering

A. Lin et al.
Phys. Lett. **74B** (1978) 273;

Abstract We measured $d\sigma/dt$ ($90^\circ_{c.m.}$) for $p_\uparrow\,p_\uparrow \to p\,p$ from 1.75 to 5.5 GeV/c, using the Argonne zero-gradient synchrotron 70% polarized proton beam and a 70% polarized proton target. We found that the spin–spin correlation parameter, $A_{\rm nn}$ equals 60% at low energy, then drops sharply to about 10% near 3.5 Gev/c, and remains constant up to 5.5 Gev/c.

Accelerator ANL

Detectors Double arm spectrometer

Related references

See also
R. C. Fernow et al., Phys. Lett. **52B** (1974) 243;
L. G. Ratner et al., Phys. Rev. **D15** (1977) 604;
K. Abe et al., Phys. Lett. **63B** (1976) 239;
J. R. O'Fallon et al., Phys. Rev. Lett. **39** (1977) 733;
T. Khoe et al., Particle Accelerators **6** (1975) 213;
H. E. Miettinen et al., Phys. Rev. **D16** (1977) 549;

Analyse information from
C. W. Akerlof et al., Phys. Rev. **159** (1967) 1138;
R. C. Kammerund et al., Phys. Rev. **D4** (1971) 1309;
P. Grannis et al., Phys. Rev. **148** (1966) 1297;
M. Borghini et al., Phys. Lett. **24B** (1967) 77;
M. Borghini et al., Phys. Lett. **31B** (1970) 405;
M. Borghini et al., Phys. Lett. **36B** (1970) 405;
M. G. Albrow et al., Nucl. Phys. **B23** (1970) 445;
N. E. Booth et al., Phys. Rev. Lett. **21** (1968) 651;
N. E. Booth et al., Phys. Rev. Lett. **23** (1969) 192;
D. J. Sherden et al., Phys. Rev. Lett. **25** (1970) 898;
J. H. Parry et al., Phys. Rev. **D8** (1973) 45;
E. F. Parker et al., Phys. Rev. Lett. **31** (1973) 783;
W. de Boer et al., Phys. Rev. Lett. **34** (1975) 558;
G. Hicks et al., Phys. Rev. **D12** (1975) 2594;
D. H. Miller et al., Phys. Rev. Lett. **36** (1976) 1727;
I. P. Auer et al., Phys. Rev. Lett. **37** (1976) 1727;

Reactions
$p\,p \to 2p$ 1.75-5.5 GeV($P_{\rm lab}$)

BERGER 1978

■ Confirmation of the $\Upsilon(1S)$ resonance ■

Observation of a Narrow Resonance Formed in e^+e^- Annihilation at 9.46 GeV

PLUTO collaboration; C. Berger et al.
Phys. Lett. **76B** (1978) 243;

Reprinted in
J. L. Rosner, *New Particles.* Selected Reprints, Stony Brook, Am. Ass. of Physics Teachers (1981) 106.
R. N. Cahn and G. Goldhaber, *The Experimental Foundations of Particle Physics*, Cambridge Univ. Press (1991) 354.

Abstract An experiment using the PLUTO detector has observed the formation of a narrow, high mass, resonance in e^+e^- annihilations at the DORIS storage ring. The mass is determined to be 9.46 ± 0.01 GeV which is consistent with that of the Upsilon. The gaussian width σ is observed as 8 ± 1 MeV and is equal to the DORIS energy resolution. This suggests that the resonance is a bound state of a new heavy quark-antiquark pair. An electronic width $\Gamma_{ee} = 1.3 \pm 0.4$ KeV was obtained. In standard theoretical models, this favors a quark charge assignment of $-1/3$.

Accelerator DESY-DORIS-II

Detectors Spectrometer PLUTO

Related references

See also
S. W. Herb et al., Phys. Rev. Lett. **39** (1977) 252;
W. R. Innes et al., Phys. Rev. Lett. **39** (1977) 1240;
T. Appelquist and H. D. Politzer, Phys. Rev. Lett. **34** (1975) 43;
PLUTO Coll., J. Burmester et al., Phys. Lett. **66B** (1977) 395;
K. Gottfried, *Proc. Int. Symp. on Lepton and Photon Interactions at High Energies*, Hamburg (1977) 667;

Reactions
$e^+\,e^- \to \Upsilon(1S)$
$\Upsilon(1S) \to$ 2hadron (hadrons)

Particles studied $\Upsilon(1S)$

DARDEN 1978

■ Another confirmation of the $\Upsilon(1S)$ resonance ■

Observation of a Narrow Resonance at 9.46 GeV in Electron Positron Annihilations

C.W. Darden et al.
Phys. Lett. **76B** (1978) 246;

Reprinted in
J. L. Rosner, *New Particles.* Selected Reprints, Stony Brook, American Association of Physics Teachers (1981) 109.

Abstract We observe a narrow resonance in the reaction $e^+e^- \to$ hadrons using the DASP detector at the DORIS storage ring. The mass is found to be (9.46 ± 0.01) GeV and the observed width is compatible with the storage ring resolution of ± 8 MeV. The energy-integrated cross section results in an electronic width $\Gamma_{ee} = (1.3 \pm 0.4)$ keV.

Accelerator DESY-DORIS-II

Detectors Spectrometer DASP

Related references

See also
S. W. Herb et al., Phys. Rev. Lett. **39** (1977) 252;
J. Ellis, M. K. Gaillard, D. V. Nanopoulos, and S. Rudaz, Nucl. Phys. **B131** (1977) 285;
DASP Coll., R. Brandelik et al., Phys. Lett. **56B** (1975) 491;
DASP Coll., R. Brandelik et al., Phys. Lett. **67B** (1977) 243;
J. D. Jackson and D. L. Scharre, Nucl. Instr. and Meth. **128** (1975) 13;
E. Eichten and K. Gottfried, Phys. Lett. **66B** (1977) 286;

Reactions

$e^+ e^- \to$ 2hadron (hadrons)	9.2-9.48 GeV(E_{cm})
$e^+ e^- \to \Upsilon(1S)$	9.2-9.48 GeV(E_{cm})
$e^+ e^- \to$ mult[γ]	9.2-9.48 GeV(E_{cm})
$e^+ e^- \to$ mult[charged]	9.2-9.48 GeV(E_{cm})

Particles studied $\Upsilon(1S)$

BARKOV 1978

■ First evidence of the weak neutral current in atomic transitions ■

Observation of Parity Nonconservation in Atomic Transitions

L.M. Barkov, M.S. Zolotorev
Pisma Zh. Eksp. Teor. Fiz. **27** (1978) 379; JETP Lett. **27** (1978) 357;

Reprinted in
Neutrinos 78, West Lafayette (1978) 423.
19th Int. Conf. on High Energy Physics, Tokyo (1979) 425.

Abstract Parity nonconservation in atomic transitions has been observed. The rotation of the plane of polarization of light was measured on the components of the hyperfine splitting of the 6477 Å line of bismuth.

Accelerator Laser

Detectors OPTICAL ROTATION

Reactions
γ Bi(atom) $\to$ Bi(atom) γ

PRESCOTT 1978C

■ Confirmation of weak neutral currents ■

Parity Non-conservation in Inelastic Electron Scattering

C.Y. Prescott et al.
Phys. Lett. **77B** (1978) 347;

Reprinted in
A Festschrift in Honor of Vernon W. Hughes, New Haven (1991) 246.
R. N. Cahn and G. Goldhaber, *The Experimental Foundations of Particle Physics*, Cambridge Univ. Press (1991) 397.
Development in the Quark Theory of Hadrons, A Reprint Collection V.1: 1964-1978, ed. D. B. Lichtenberg and S. P. Rosen, Hadronic Press, Inc., Monamtum, Mass. (1980) 478.

Abstract We have measured parity violating asymmetries in the inelastic scattering of longitudinally polarized electrons from deuterium and hydrogen. For deuterium $Q^2 = 1.6$ $(\mathrm{GeV}/c)^2$ the asymmetry is (-9.5×10^5) Q^2 with statistical and systematic uncertainties each about 10%.

Accelerator SLAC

Detectors Spectrometer SSF

Related references

See also
A. Love et al., Nucl. Phys. **B49** (1972) 513;
S. M. Berman and J. R. Primack, Phys. Rev. **D9** (1974) 2171;
S. M. Berman and J. R. Primack, Phys. Rev. **D10** (1974) 3895;
E. Dermon, Phys. Rev. **D7** (1973) 2755;
W. W. Wilson, Phys. Rev. **D10** (1974) 218;
S. M. Bilenkii et al., Sov. J. Nucl. Phys. **21** (1975) 657;
M. A. B. Bég and G. Feinberg, Phys. Rev. Lett. **33** (1974) 606;
S. Weinberg, Phys. Rev. Lett. **19** (1967) 1264;
Z. D. Farkas, preprint SLAC-PUB-1823;
P. S. Cooper et al., Phys. Rev. Lett. **34** (1975) 1589;
D. T. Pierce et al., Phys. Lett. **51B** (1974) 465;
L. M. Barkov and M. S. Zolotorev, Pisma Zh. Eksp. Teor. Fiz. **26** (1978) 379;
P. E. G. Baird et al., Phys. Rev. Lett. **39** (1977) 798;
L. L. Lewis et al., Phys. Rev. Lett. **39** (1977) 795;
W. B. Atwood et al., preprint SLAC-PUB-2123;
M. J. Alguard et al., Phys. Rev. Lett. **41** (1978) 70;
M. J. Alguard et al., Phys. Rev. Lett. **37** (1976) 1261;
M. J. Alguard et al., Phys. Rev. Lett. **37** (1976) 1258;
Y. B. Bushnin et al., Sov. J. Nucl. Phys. **24** (1976) 279;
R. N. Cahn and F. J. Gilman, Phys. Rev. **D17** (1978) 1313;

Reactions
e^- deuteron $\to e^-$ X   16.18-22.20 GeV(T_{lab})

CARRON 1978 Nobel prize

■ Evidence for stochastic cooling. Nobel prize to S. van der Meer awarded in 1984. Co-winner C. Rubbia "for decisive contribution to the large project, which led to the discovery of the field particles W and Z^0, communicators of weak interactions" ■

Stochastic Cooling Tests in ICE

G. Carron et al.
Phys. Lett. **77B** (1978) 353;

Excerpt ICE (for Initial Cooling Experiment) is a storage ring built for testing both stochastic and electron cooling of proton beams in view of future application in a scheme for collecting antiprotons. *(Extracted from the introductory part of the paper.)*

Related references

See also
W. Schnell, *Proc. 1st Course Intern. School of Particle Accelerators*, Erice, 1976, CERN 77-13 (1977) 302;
G. Carron and L. Thorndahl, preprint CERN/ISR-RF/78-12 (1978);
H. Hereward and W. Schnell, *Proc. 1st Course Intern. School of Particle Accelerators*, Erice, 1976, preprint CERN 77-13 (1977) 281;
H. Hereward, *Proc. 1st Course Intern. School of Particle Accelerators*, Erice, 1976, CERN 77-13 (1977) 284;
F. Sacherer, preprint CERN/ISR-TH/78-11 (1978);

FAISSNER 1978D

■ First evidence for elastic muon-neutrino scattering off electrons. Confirmation of weak neutral currents ■

Measurement of Muon-Neutrino and -Antineutrino Scattering off Electrons

H. Faissner et al.
Phys. Rev. Lett. **41** (1978) 213; Phys. Rev. Lett. **41** (1978) 1083;

Abstract Muon-neutrino and -antineutrino scattering off electrons was detected in a 19-ton Al spark chamber, exposed to the wide-band ν ($\overline{\nu}$) beam from the CERN proton synchrotron. The background was determined experimentally. 11 (10) genuine ν_μ- ($\overline{\nu}_\mu$-) e scattering events were found. The respective cross sections are $(1.1 \pm 0.6) \times 10^{-42} (E_\nu/\text{GeV})$ cm^2 and $(2.2 \pm 1.0) \times 10^{-42} (E_\nu/\text{GeV})$ cm^2. The analysis excludes a pure $V - A$ interaction, and makes a pure V or A theory improbable. The data agree well with the Salam-Weinberg model and $\sin^2 \theta_W = 0.35 \pm 0.08$.

Accelerator CERN-PS proton synchrotron

Detectors Optical spark chamber

Related references

See also
F. Reines, H. S. Curr, and H. W. Sobel, Phys. Rev. Lett. **37** (1976) 315;
R. P. Feynman and M. Gell-Mann, Phys. Rev. **109** (1958) 193;
E. C. G. Sudarshan and R. E. Marshak, Phys. Rev. **109** (1958) 1860;
F. J. Hasert et al., Phys. Lett. **46B** (1973) 121;
J. Blietschau et al., Nucl. Phys. **B114** (1976) 189;
G. 't Hooft, Phys. Lett. **37B** (1971) 195;
A. Salam and J. C. Ward, Phys. Lett. **13** (1964) 168;
S. Wienberg, Phys. Rev. Lett. **19** (1967) 1264;
S. Wienberg, Phys. Rev. Lett. **27** (1971) 1688;
S. Wienberg, Phys. Rev. **D5** (1972) 1412;

Reactions

$\nu_\mu e^- \to e^- \nu_\mu$	2.2 GeV(E_{lab})
$\overline{\nu}_\mu e^- \to e^- \overline{\nu}_\mu$	2.0 GeV(E_{lab})

BIENLEIN 1978

■ Confirmation of the $\Upsilon(2S)$ resonance ■

Observation of a Narrow Resonance at 10.02 GeV in $e^+ e^-$ Annihilations

J.K. Bienlein et al.
Phys. Lett. **78B** (1978) 360;

Reprinted in
J. L. Rosner, *New Particles.* Selected Reprints, Stony Brook, American Association of Physics Teachers (1981) 112.
R. N. Cahn and G. Goldhaber, *The Experimental Foundations of Particle Physics*, Cambridge Univ. Press (1991) 357.

Abstract The Υ' state has been observed as a narrow resonance at $M(\Upsilon') = 10.02 \pm 0.02$ GeV in e^+e^- annihilations, using a NaI and lead-glass detector in the DORIS storage ring at DESY. The ratio $\Gamma_{ee}\Gamma_{had}/\Gamma_{tot}$ of electronic, hadronic, and total widths has been measured to be 0.32 ± 0.13 keV. The parameters of Υ particle have also been determined to be $M(\Upsilon) = 9.56 \pm 0.01$ and $\Gamma_{ee}\Gamma_{had}/\Gamma_{tot} = 1.04 \pm 0.28$ keV. The mass difference is $M(\Upsilon') - M(\Upsilon) = 0.56 \pm 0.01$ GeV.

Accelerator DESY-DORIS-II

Detectors Wire chamber

Related references

See also
S. W. Herb et al., Phys. Rev. Lett. **39** (1977) 252;
W. R. Innes et al., Phys. Rev. Lett. **39** (1977) 1240;
K. Gottfried, *Proc. Int. Symp. on Lepton and Photon Interactions at High Energies*, Hamburg (1977) 667;
C. Berger et al., Phys. Lett. **76B** (1978) 243;
C. W. Darden et al., Phys. Lett. **76B** (1978) 246;
W. Bartel et al., Phys. Lett. **66B** (1977) 483;
W. Bartel et al., Phys. Lett. **77B** (1978) 331;
J. D. Jackson and D. L. Scharre, Nucl. Instr. and Meth. **128** (1975) 13;
J. L. Rosner, C. Quigg, and H. B. Thacker, Phys. Lett. **74B** (1978) 350;

Reactions

$e^+ e^- \to \Upsilon(1S)$	9.4-10.1 GeV(E_{cm})
$e^+ e^- \to \Upsilon(2S)$	9.4-10.1 GeV(E_{cm})
$\Upsilon(1S) \to$ 2hadron (hadrons)	
$\Upsilon(2S) \to$ 2hadron (hadrons)	

Particles studied $\Upsilon(1S), \Upsilon(2S)$

DARDEN 1978B

■ Another confirmation of the $\Upsilon(2S)$ resonance ■

Evidence for a Narrow Resonance at 10.01 GeV in Electron Positron Annihilations

C.W. Darden et al.
Phys. Lett. **78B** (1978) 364;

Reprinted in
J. L. Rosner, *New Particles.* Selected Reprints, Stony Brook, American Association of Physics Teachers (1981) 116.

Abstract We observe evidence for a second narrow resonance in the reaction $e^+e^- \to$ hadrons at $\sqrt{s}$ around 10 GeV using the DASP detector at the DORIS storage ring. The mass of the resonance is (10.01 ± 0.02) GeV; its width is in agreement with the storage ring resolution of 9 MeV. From the integrated cross section, an electronic width of $\Gamma_{ee} = (0.35 \pm 0.14)$ keV is derived.

Accelerator DESY-DORIS-II

Detectors Spectrometer DASP

Related references

Analyse information from
C. W. Darden et al., Phys. Lett. **76B** (1978) 246;
C. W. Darden et al., Internal report DESY F15-78/01;
C. Berger et al., Phys. Lett. **76B** (1978) 243;
S. W. Herb et al., Phys. Rev. Lett. **39** (1977) 252;
T. Appelquist and H. D. Politzer, Phys. Rev. Lett. **34** (1975) 43;
E. Eichten and K. Gottfried, Phys. Lett. **66B** (1977) 286;
DASP Coll., R. Brandelik et al., Phys. Lett. **56B** (1975) 491;
DASP Coll., R. Brandelik et al., Phys. Lett. **67B** (1977) 243;
W. R. Innes et al., Phys. Rev. Lett. **39** (1977) 1240;
J. D. Jackson and D. L. Scharre, Nucl. Instr. and Meth. **128** (1975) 13;
J. L. Rosner, C. Quigg, and H. B. Thacker, Phys. Lett. **74B** (1978) 350;

Reactions

$e^+ e^- \to \Upsilon(2S)$	9.98-10.10 GeV(E_{cm})
$\Upsilon(2S) \to$ 2hadron (hadrons)	

Particles studied $\Upsilon(2S)$

DERBENEV 1978

■ Invention of the method of the acceleration of polarized particles to high energies - Sibirian snakes ■

Radiative Polarization: Obtaining, Control, Using

Y.S. Derbenev et al.
Particle Accelerators **8** (1978) 115;

Abstract Theoretical and experimental studies in the Institute (Novosibirsk) on the behavior of particle polarization in storage rings are reviewed. In theoretical works the motion of particle spins in arbitrary inhomogeneous fields was investigated. Methods are described for obtaining beams with required polarization direction in storage rings and accelerators. It is shown that variable-direction fields in some parts of the orbit may be used to avoid resonance depolarization during acceleration of polarized particles to high energies. The conditions for the existence of radiative polarization of electrons and positrons are discussed. Methods for measuring the polarization of the single and colliding beams are described. The results of measuring the time and degree of radiative polarization are presented. The action of spin resonances was studied. The use of polarized beams to measure the absolute energy of particles in a storage ring and to compare precisely the anomalous magnetic momenta of electrons and positrons is described.

Related references

See also
A. A. Sokolov and I. M. Ternov, Doklady Akad. Nauk SSSR **153** (1963) 1052;
Y. S. Derbenev and A. M. Kondratenko, Zh. Eksp. Teor. Fiz. **64** (1973) 1918;
V. L. Lyuboshitz, V.L., Yad. Phys. **4** (1966) 269;
V. N. Baier and V. M. Katkov, Yad. Phys. **3** (1966) 81;
V. N. Baier and V. M. Katkov, Zh. Eksp. Teor. Fiz. **52** (1967) 1422;
Y. S. Derbenev and A. M. Kondratenko, Zh. Eksp. Teor. Fiz. **62** (1972) 430;
Y. A. Pliss and L. M. Soroko, Usp. Fiz. Nauk **107** (1972) 281;
M. Froissart and R. Stora, Nucl. Instr. and Meth. **7** (1960) 297;
V. N. Baier and V. A. Khoze, Yad. Phys. **9** (1969) 409;

FIDECARO 1978

■ Evidence for the nonnegligible spin effects in the strong interactions at high energies ■

Evidence for Spin Effects in $p\,p$ Elastic Scattering at 150 GeV/c

G. Fidecaro et al.
Phys. Lett. **76B** (1978) 369;

Abstract Proton elastic scattering off a polarized proton target has been measured at 150 GeV/c, in the $|t|$-range $0.2-3.0$ GeV^2. The results on polarization and differential cross section are presented.

Accelerator CERN-SPS super proton synchrotron:

Detectors Spectrometer

Related references

See also
A. Gaidot et el., Phys. Lett. **61B** (1976) 103;
D. S. Aytes et el., Phys. Rev. **D15** (1977) 3105;

Reactions

$p\,p \to 2p$	150 GeV(P_{lab})

SHIFMAN 1979

■ Invention of the quantum chromodynamic sum rules ■

Resonance Properties in Quantum Chromodynamics

M.A. Shifman, A.I. Vainstein, V.I. Zakharov
Phys. Rev. Lett. **42** (1979) 297;

Abstract Quantum-chromodynamic sum rules are derived which are sensitive to resonance contributions and in fact allow one to compute leptonic decay widths and masses of low-lying resonances. The crucial point is the inclusion of power corrections which are related to the vacuum structure of quantum chromodynamics.

Related references

More (earlier) information
A. Vainshtein, V. Zakharov, and M. Shifman, Pisma Zh. Eksp. Teor. Fiz. **27** (1978) 60;
A. Vainshtein, V. Zakharov, and M. Shifman, JETP Lett. **27** (1978) 59;
M. A. Shifman, A. I. Vainshtein, and V. I. Zakharov, Phys. Lett. **76B** (1978) 471;

See also
A. Belavin et al., Phys. Lett. **59B** (1975) 85;
V. N. Gribov, Nucl. Phys. **B139** (1978) 1;
D. J. Gross and F. Wilczek, Phys. Rev. Lett. **30** (1973) 1343;
H. D. Politzer, Phys. Rev. Lett. **30** (1973) 1346;
S. Weinberg, Phys. Rev. Lett. **18** (1967) 507;
C. G. Callan, R. Dashen, and D. J. Gross, Phys. Rev. **D17** (1978) 2717;
T. Appelquist and D. H. Politzer, Phys. Rev. Lett. **34** (1975) 43;
M. Gell-Mann, R. Oakes, and B. Renner, Phys. Rev. **175** (1968) 2195;
H. Leutwyller, Phys. Lett. **48B** (1974) 45;

UENO 1979

■ First evidence for the $\Upsilon(3S)$ state. Confirmation of the $\Upsilon(1S)$ and $\Upsilon(2S)$ states ■

Evidence for the Υ'' and a Search for New Narrow Resonances

K. Ueno et al.
Phys. Rev. Lett. **42** (1979) 486;

Reprinted in
J. L. Rosner, *New Particles.* Selected Reprints, Stony Brook, American Association of Physics Teachers, (1981) 118.

Abstract The production of the Υ family in proton-nucleus collisions is clarified by a sixfold increase in statistics. Constraining Υ, Υ' masses to those observed at DORIS we find the statistical significance of the Υ'' to be 11 standard deviations. The dependence of Υ production on p_t, y, and s is presented. Limits for other resonance production in the mass range 4-18 GeV are determined.

Accelerator FNAL proton synchrotron

Detectors Double arm spectrometer

Related references

More (earlier) information
S. W. Herb et al., Phys. Rev. Lett. **39** (1977) 252;
W. R. Innes et al., Phys. Rev. Lett. **39** (1977) 1240;
D. M. Kaplan et al., Phys. Rev. Lett. **40** (1978) 435;
J. K. Yoh et al., Phys. Rev. Lett. **41** (1978) 684;

Reactions

p nucleus $\rightarrow \mu^- \mu^+ X$	400 GeV(P_{lab})
p nucleus $\rightarrow \Upsilon(1S) X$	400 GeV(P_{lab})
p nucleus $\rightarrow \Upsilon(2S) X$	400 GeV(P_{lab})
p nucleus $\rightarrow \Upsilon(3S) X$	400 GeV(P_{lab})
p nucleus $\rightarrow$ longlived X	400 GeV(P_{lab})
$\Upsilon(1S) \rightarrow \mu^- \mu^+$	
$\Upsilon(2S) \rightarrow \mu^- \mu^+$	
$\Upsilon(3S) \rightarrow \mu^- \mu^+$	
longlived $\rightarrow \mu^- \mu^+$	

Particles studied $\Upsilon(1S), \Upsilon(2S), \Upsilon(3S)$, longlived

BALTAY 1979E

■ Confirmation of the $\Sigma_c(2455)^{++}$ and the Λ_c^+ ■

Confirmation of the Existence of the $\Sigma_c(2455)^{++}$ and Λ_c^+ Charmed Baryons and Observation of the Decay $\Lambda_c^+ \rightarrow \Lambda \pi^+$ and $\Lambda_c^+ \rightarrow p \overline{K}^0$

C. Baltay et al.
Phys. Rev. Lett. **42** (1979) 1721;

Abstract In a broadband neutrino exposure of the Fermilab 15-ft bubble chamber, we observe the production of the $\Sigma_c^{++}(2426)$ charmed baryon followed by its decay to $\Lambda_c^+(2260)$ and π^+. We find the mass of the Λ_c^+ to be 2257 ± 10 MeV and the $m(\Sigma_c^{++}) - m(\Lambda_c^+)$ mass difference to be 168 ± 3 MeV. Previously unseen two-body decay modes of the $\Lambda_c^+(2260)$ are observed.

Accelerator FNAL proton synchrotron

Detectors HLBC-15FT-HYB

Related references

Analyse information from
E. G. Cazzoli et al., Phys. Rev. Lett. **34** (1975) 1125;
B. Knapp et al., Phys. Rev. Lett. **37** (1976) 882;
A. M. Cnops et al., Phys. Rev. Lett. **42** (1979) 197;
G. Goldhaber et al., Phys. Rev. Lett. **37** (1976) 255;
C. Baltay et al., Phys. Rev. Lett. **41** (1978) 73;

Reactions

ν_μ Ne $\rightarrow \Lambda \pi^+ \mu^- X$	20-200 GeV(P_{lab})
ν_μ Ne $\rightarrow \Sigma_c(2455)^{++} \mu^- X$	20-200 GeV(P_{lab})
ν_μ Ne $\rightarrow \Lambda_c^+ \mu^- X$	20-200 GeV(P_{lab})
$\Sigma_c(2455)^{++} \rightarrow \Lambda_c^+ \pi^+$	
$\Lambda_c^+ \rightarrow \Lambda \pi^+$	
$\Lambda_c^+ \rightarrow p \overline{K}^0$	

Particles studied $\Lambda_c^+, \Sigma_c(2455)^{++}$

BARBER 1979F

■ First evidence for the gluon jet in $e^+e^- \to 3$jet annihilations ■

Discovery of Three Jet Events and a Test of Quantum Chromodynamics at PETRA Energies

D.P. Barber et al.
Phys. Rev. Lett. **43** (1979) 830;

Abstract We report the analysis of the spatial energy distribution of data for $e^+e^- \to$ hadrons obtained with the MARK-J detector at PETRA. We define the quantity "oblateness" to describe the flat shape of the energy configuration and the three-jet structure which is unambiguously observed for the first time. Our data can be explained by quantum chromodynamic predictions for the production of quark-antiquark pairs accompanied by hard noncollinear gluons.

Accelerator DESY-PETRA

Detectors Mark-J

Related references

More (earlier) information
D. P. Barber et al., Phys. Rev. Lett. **42** (1979) 1110;
See also
J. Ellis, M. K. Gaillard, and G. C. Ross, Nucl. Phys. **B111** (1976) 253;
T. A. de Grand, Y. J. Ng, and S. H. H. Tye, Phys. Rev. **D16** (1977) 3251;
G. Kramer et al., Phys. Lett. **79B** (1978) 249;

Reactions

$e^+ e^- \to$ 2hadron (hadrons)	27.4,30,31.6 GeV(E_{cm})
$e^+ e^- \to$ 3jet	27.4,30,31.6 GeV(E_{cm})
$e^+ e^- \to q\ \bar{q}$ gluon	27.4,30,31.6 GeV(E_{cm})
$q \to$ jet	
$\bar{q} \to$ jet	
gluon $\to$ jet	

Particles studied gluon

PRESCOTT 1979C

■ Confirmation of weak neutral currents ■

Further Measurements of Parity Non-conservation in Inelastic Electron Scattering

C.Y. Prescott et al.
Phys. Lett. **84B** (1979) 524;

Reprinted in
Lepton and Photon Interactions at High Energies, Batavia (1979) 271.

Abstract We have extended our earlier measurements of parity violating asymmetries in the inelastic scattering of longitudinally polarized electrons from deuterium to cover the range $0.15 \le y \le 0.36$. The observed asymmetry shows only slight y dependence over this range. Our results are consistent with the expectations of the Weinberg-Salam model for a value of $\sin^2\theta_W = 0.224 \pm 0.020$.

Accelerator SLAC

Detectors Spectrometer SSF

Related references

More (earlier) information
C. Y. Prescott et al., Phys. Lett. **77B** (1978) 347;

Reactions

e^- deuteron $\to e^-$ X	19.4 GeV(T_{lab})

BARKOV 1979C

■ Confirmation of parity nonconservation effects in atomic transitions ■

Parity Violation in Atomic Bismuth

L.M. Barkov, M.S. Zolotorev
Phys. Lett. **85B** (1979) 308; Zh. Eksp. Teor. Fiz. **79** (1980) 713; JETP **52** (1980) 360;

Abstract The results of a new run of measurements of parity violation in atomic bismuth in the $^4S_{3/2} \to ^2 D_{5/2}$ M1-transition at Λ = 648 nm are presented. The value $R = \mathrm{Im}(E1/M1)$ measured for the F=6$\to$ F'=7 and $F = 6 \to F' = 6$ hyperfine structure components is found to be $(-20.6 \pm 3.2) \cdot 10^{-8}$. The average value for all our measurements $\langle R \rangle = (-20.2 \pm 2.7) \cdot 10^{-8}$ is in agreement with the theoretical prediction obtained in the framework of the standard gauge model with $\sin^2\Theta = 0.25$.

Accelerator Laser

Detectors OPTICAL ROTATION

Related references

More (earlier) information
L. M. Barkov and M. S. Zolotorev, Pisma Zh. Eksp. Teor. Fiz. **27** (1978) 379;
L. M. Barkov and M. S. Zolotorev, JETP Lett. **27** (1978) 357;
L. M. Barkov and M. S. Zolotorev, Pisma Zh. Eksp. Teor. Fiz. **28** (1978) 544;

Reactions

γ Bi(atom) $\to$ Bi(atom) γ

BRANDELIK 1979H

■ Confirmation of the production of gluon jets ■

Evidence for Planar Events in $e^+ e^-$ Annihilation at High Energies

TASSO collaboration; R. Brandelik et al.
Phys. Lett. **86B** (1979) 243;

Reprinted in
R. N. Cahn and G. Goldhaber, *The Experimental Foundations of Particle Physics*, Cambridge Univ. Press (1991) 322.

Abstract Hadron jets produced in e^+e^- annihilation between 13 GeV and 31.6 GeV in c.m. at PETRA are analyzed. The transverse momentum of the jets is found to increase strongly with c.m. energy. The broadening of the jets is not uniform in azimuthal angle around the quark direction but tends to yield planar events with large and growing transverse momenta in the plane and smaller transverse momenta normal to the plane. The simple $q\overline{q}$ collinear jet picture is ruled out. The observation of planar events shows that there are three basic particles in the final state. Indeed several events with three well-separated jets of hadrons are observed at the highest energies. This occurs naturally when the outgoing quark radiates a hard noncollinear gluon, i.e., $e^+e^- \to q\overline{q}g$ with the quarks and the gluons fragmenting into hadrons with the limited transverse momenta.

Accelerator DESY-PETRA

Detectors Spectrometer TASSO

Related references

See also
S. D. Drell, D. J. Levy, and T. M. Yan, Phys. Rev. **187** (1969) 2159;
S. D. Drell, D. J. Levy, and T. M. Yan, Phys. Rev. **D1** (1970) 1617;
N. Cabibbo, G. Parisi, and M. Testa, Nuovo Cim. Lett. **4** (1970) 35;
J. D. Bjorken and S. J. Brodsky, Phys. Rev. **D1** (1970) 1416;
R. P. Feynman, *Photon-Hadron Interactions*, Benjamin, New York (1972) 166;
G. Hanson et al., Phys. Rev. Lett. **35** (1975) 1609;
B. H. Wiik, *Proc. Intern. Neutrino Conf.* Bergen, Norway, June (1979);
D. J. Fox et al., Phys. Rev. Lett. **37** (1974) 1504;
H. L. Anderson et al., Phys. Rev. Lett. **37** (1976) 4;
P. C. Bosetti et al., Nucl. Phys. **B142** (1978) 1;
J. G. H. Groot et al., Z. Phys. **C1** (1979) 143;
J. Kogut and L. Susskind, Phys. Rev. **D9** (1974) 697;
J. Kogut and L. Susskind, Phys. Rev. **D9** (1974) 3391;
A. M. Polyakov, *Proc. Intern. Symp. on Lepton and Photon Interactions at High Energies* ed. W. T. Kirk, Stanford (1975) 855;
PLUTO Coll., C. Berger et al., Phys. Lett. **78B** (1978) 176;
PLUTO Coll., C. Berger et al., Phys. Lett. **82B** (1979) 449;
TASSO Coll., R. Brandelik et al., Phys. Lett. **83B** (1979) 201;
J. Ellis, M. K. Gaillard, and G. C. Ross, Nucl. Phys. **B111** (1976) 253;
A. de Rujula et al., Nucl. Phys. **B138** (1978) 387;
T. A. de Grand, Y. J. Ng, and S. H. H. Tye, Phys. Rev. **D16** (1977) 3251;
G. Kramer and G. Schierholz, Phys. Lett. **82B** (1979) 102;
E. Fahri, Phys. Rev. Lett. **39** (1977) 1587;
S. Brandt and H. D. Dahmen, Z. Phys. **C1** (1978) 273;
R. D. Field and R. P. Feynman, Nucl. Phys. **B136** (1978) 1;
A. Ali, J. G. Korner, J. Willrodt, and C. Kramer, Z. Phys. **C1** (1979) 269;
A. Ali, J. G. Korner, J. Willrodt, and C. Kramer, Z. Phys. **C2** (1979) 33;
S. L. Wu and G. Zobernig, Z. Phys. **C2** (1979) 107;

Reactions

$e^+ e^- \to$ 2hadron (hadrons)	13,17-31.6 GeV(E_{cm})
$e^+ e^- \to$ (jets) 2jet	13,17-31.6 GeV(E_{cm})
$e^+ e^- \to q\ \overline{q}$ gluon	
$q \to$ jet	
$\overline{q} \to$ jet	
gluon $\to$ jet	

Particles studied gluon

BERGER 1979C

■ Another confirmation of the production of gluon jets ■

Evidence for Gluon Bremsstrahlung in $e^+ e^-$ Annihilations at High Energies

PLUTO collaboration; C. Berger et al.
Phys. Lett. **86B** (1979) 418;

Abstract We report our results on the reactions $e^+e^- \to$ hadrons in the energy range $13 \le E_{cm} \le 31.6$ GeV and compare them with $q\overline{q}$ jets described in the quark–parton model and with first-order QCD predictions including gluon emission ($q\overline{q}g$jets). At high energies the observed features of one-sided jet broadening, sea-gull effect and planar events with a three-jet structure represent a gross violation of the parton model without gluons and find a most natural interpretation if gluon bremsstrahlung is included.

Accelerator DESY-PETRA

Detectors Spectrometer PLUTO

Related references

More (earlier) information
C. Berger et al., Phys. Lett. **81B** (1979) 410;
C. Berger et al., Phys. Lett. **86B** (1979) 413;

Reactions

$e^+ e^- \to$ 2hadron (hadrons)	13,17,31.6 GeV(E_{cm})
$e^+ e^- \to$ 3jet	13,17,31.6 GeV(E_{cm})
$e^+ e^- \to q\ \overline{q}$ gluon	
$q \to$ jet	
$\overline{q} \to$ jet	
gluon $\to$ jet	

Particles studied gluon

BARTEL 1980D

■ Confirmation of gluon jets ■

Observation of Planar Three Jet Events in $e^+ e^-$ Annihilation and Evidence for Gluon Bremsstrahlung

JADE collaboration; W. Bartel et al.
Phys. Lett. **91B** (1980) 142;

Abstract Topological distributions of charged and neutral hadrons from the reaction $e^+e^- \to$ multihadrons are studied at $\sqrt{s}$ of about 30 GeV. An excess of planar events is observed at a rate which cannot be explained by statistical fluctuations in the standard two-jet process. The planar events, mostly consisting of a slim jet on one side and a broader jet on the other, are shown actually to possess three-jet structure by demonstrating that the broader jet itself consists of two collinear jets in its own rest system. Detailed agreement between data and predictions is obtained if the process $e^+e^- \to q\bar{q}g$ is taken into account. This strongly suggests gluon bremsstrahlung as the origin of the planar three-jet events. By comparison of the data with the $q\bar{q}g$-model we obtain a value for the strong coupling constant of $\alpha_s(q^2) = 0.17 \pm 0.04$.

Accelerator DESY-PETRA

Detectors Specrtometer JADE

Related references

See also
JADE Coll., W. Bartel et al., Phys. Lett. **89B** (!979) 136;
JADE Coll., W. Bartel et al., Phys. Lett. **88B** (1979) 171;
J. Ellis, M. K. Gaillard, and G. C. Ross, Nucl. Phys. **B111** (1976) 253;
T. A. de Grand, Y. J. Ng, and S. H. H. Tye, Phys. Rev. **D16** (1977) 3251;
TASSO Coll., R. Brandelik et al., Phys. Lett. **86B** (1979) 243;
MARK J.Coll., D. P. Barber et al., Phys. Rev. Lett. **43** (1979) 830;
MARK J.Coll., D. P. Barber et al., Phys. Rev. Lett. **42** (1979) 1113;
PLUTO Coll., C. Berger et al., Phys. Lett. **86B** (1979) 418;
PLUTO Coll., C. Berger et al., Phys. Lett. **78B** (1978) 176;
R. D. Field and R. P. Feynman, Nucl. Phys. **B136** (1978) 1;
G. Hanson et al., Phys. Rev. Lett. **35** (1975) 1609;
F. H. Heimlich et al., Phys. Lett. **86B** (1979) 399;
D. J. Gross and F. Wilczek, Phys. Rev. **D8** (1973) 3633;
H. D. Politzer, Phys. Rev. Lett. **30** (1973) 1346;
H. D. Politzer, Phys. Rept. **14C** (1974) 129;
H. D. Politzer, Phys. Lett. **70B** (1977) 430;
S. Brand et al., Phys. Lett. **12** (1964) 57;
E. Fahri, Phys. Rev. Lett. **39** (1977) 1587;
A. de Rujula et al., Nucl. Phys. **B138** (1978) 387;
J. D. Bjorken and S. J. Brodsky, Phys. Rev. **D1** (1970) 1416;
J. Ellis and I. Karliner, Nucl. Phys. **B148** (1979) 141;
S. Brand and H. D. Dahmen, Z. Phys. **C1** (1979) 61;

Reactions
$e^+ e^- \to$ 2hadron (hadrons) 30 GeV(E_{cm})
$e^+ e^- \to$ (jets) 2jet 30 GeV(E_{cm})
$e^+ e^- \to q\ \bar{q}$ gluon
$q \to$ jet
$\bar{q} \to$ jet
gluon $\to$ jet

Particles studied gluon

DERBENEV 1980

■ Invention of the resonance depolarization method of the beams energy calibration in an electron-positron storage ring ■

Accurate Calibration of the Beam Energy in a Storage Ring Based on Measurement of Spin Precession Frequency of Polarized Particles

Y.S. Derbenev et al.
Particle Accelerators **10** (1980) 177;

Abstract A method is described for measuring the particle energy in an electron-positron storage ring by means of resonance depolarization by a high frequency field. The measurement accuracy is discussed taking into account energy spread and synchrotron oscillations. It is found that in practice the limitation in accuracy is due to the irregular pulsations of the magnetic guide field. As a result, the electron beam energy in the storage ring VEPP-2M has been measured with an accuracy of $\pm 2 \cdot 10^{-3}$..

Related references

See also
A. A. Sokolov and I.M.Ternov, Doklady Akad. Nauk SSSR **153** (1963) 1052;
V. N. Baier, Usp. Fiz. Nauk **105** (1971) 441;
S. I. Serednyakov et al., Zh. Eksp. Teor. Fiz. **71** (1976) 2025;
Y. S. Derbenev and A. M. Kondratenko, Zh. Eksp. Teor. Fiz. **62** (1972) 430;
V. N. Baier and V. A. Khose, Atomic Energy. USSR. **25** (1968) 440;
R. S. Van Dyck, P. B. Schwinberg, and H. G. Dehmelt, Phys. Rev. Lett. **38** (1977) 310;
Y. S. Derbenev and A. M. Kondratenko, Doklady Akad. Nauk SSSR **19** (1975) 438;
Analyse information from
U. Camerini et al., Phys. Rev. **D12** (1975) 1855;
A. D. Bukin et al., Yad. Phys. **27** (1978) 976;
L. M. Barkov et al., Nucl. Phys. **B148** (1979) 53;
S. I. Serednyakov et al., Phys. Lett. **66B** (1977) 102;

BRANDELIK 1980G

■ First experimental determination of the gluon spin ■

Evidence for a Spin-1 Gluon in Three Jet Events

TASSO collaboration; R. Brandelik et al.
Phys. Lett. **97B** (1980) 453;

Abstract High-energy e^+e^--annihilation events obtained in the TASSO detector at PETRA have been used to determine the spin of the gluon in the reaction $e^+e^- \to q\bar{q}g$. We analysed angular correlations between the three jet axes. While vector gluons are consistent with the data (55% confidence limit), scalar gluons are disfavoured by 3.8 standard deviations, corresponding to a confidence level of about 10^{-4}. Our conclusion is free of possible biases due to uncertainties in the fragmentation process or in determining the $q\bar{q}g$ kinematics from the observed hadron.

Accelerator DESY-PETRA

Detectors Spectrometer TASSO

Data comments 2229 hadronic events.

Related references

Analyse information from
TASSO Coll., R. Brandelik et al., Phys. Lett. **83B** (1979) 261;
TASSO Coll., R. Brandelik et al., Z. Phys. **C4** (1980) 87;

Reactions

$e^+ e^- \rightarrow$ 2hadron (hadrons) 27.4-36.6 GeV(E_{cm})
$e^+ e^- \rightarrow$ 3jet 27.4-36.6 GeV(E_{cm})
$e^+ e^- \rightarrow q\,\bar{q}$ gluon

Particles studied gluon

BERGER 1980L

■ Confirmation of the gluon spin = 1 ■

A Study of Multi-Jet Events in $e^+ e^-$ Annihilation

PLUTO collaboration; C. Berger et al.
Phys. Lett. **97B** (1980) 459;

Abstract A multi-jet analysis of hadronic final states from e^+e^- annihilation in the energy range $27 < E_{cm} < 32$ GeV is presented. The analysis uses a cluster method to identify the jets in a hadronic event. The distribution of the number of jets per event is compared with several models. From the number of identified coplanar three-jet events the strong coupling constant is determined to be $\alpha_S = 0.15 \pm 0.03$(stat. error)± 0.02(syst. error). The inferred energy distribution of the most energetic parton is in good agreement with the first-order QCD prediction. A scalar-gluon model is strongly disflavoured. Higher-twist contributions to the three-jet sample are found to be small.

Accelerator DESY-PETRA

Detectors Spectrometer PLUTO

Reactions

$e^+ e^- \rightarrow$ mult[jet] 27.-32. GeV(E_{cm})
$e^+ e^- \rightarrow q\,\bar{q}$ gluon 27.-32. GeV(E_{cm})
gluon $\rightarrow$ mult[charged] (neutrals)

Particles studied gluon

ANDREWS 1980

■ Confirmation of the $\Upsilon(3S)$ ■

Observation of Three Upsilon States

CLEO Collaboration; D. Andrews et al.
Phys. Rev. Lett. **44** (1980) 1108;

Abstract Three narrow resonances have been observed in e^+e^- annihilation into hadrons at total energies between 9.4 and 10.4 GeV. Measurements of mass spacings and ratios of lepton pair widths support the interpretation of these "Υ" states as the lowest triplet-S levels of the $b\bar{b}$ quark-antiquark system.

Accelerator Cornell $e^+ e^-$ storage ring

Detectors CLEO

Related references

See also
J. L. Rosner, C. Quigg, and H. B. Thacker, Phys. Lett. **74B** (1978) 350;
J. L. Richardson, Phys. Lett. **82B** (1979) 272;
G. Bhanot and S. Rudaz, Phys. Lett. **78B** (1978) 119;
E. Eichten et al., Phys. Rev. **D21** (1980) 203;

Analyse information from
S. W. Herb et al., Phys. Rev. Lett. **39** (1977) 252;
C. Berger et al., Phys. Lett. **76B** (1978) 243;
C. Berger et al., Phys. Lett. **78B** (1978) 176;
C. W. Darden et al., Phys. Lett. **76B** (1978) 246;
C. W. Darden et al., Phys. Lett. **78B** (1978) 364;
J. K. Bienlein et al., Phys. Lett. **78B** (1978) 360;
F. H. Heimlich et al., Phys. Lett. **86B** (1979) 399;
K. Ueno et al., Phys. Rev. Lett. **42** (1979) 486;
T. Bohringer et al., Phys. Rev. Lett. **44** (1980) 1111;

Reactions

$e^+ e^- \rightarrow$ 2hadron (hadrons) 9.4-10.4 GeV(E_{cm})
$e^+ e^- \rightarrow \Upsilon(1S)$
$e^+ e^- \rightarrow \Upsilon(2S)$
$e^+ e^- \rightarrow \Upsilon(3S)$
$\Upsilon(1S) \rightarrow e^- e^+$
$\Upsilon(2S) \rightarrow e^- e^+$
$\Upsilon(3S) \rightarrow e^- e^+$

Particles studied $\Upsilon(1S), \Upsilon(2S), \Upsilon(3S), b$

BOHRINGER 1980B

■ Another confirmation of the $\Upsilon(3S)$ ■

Observation of Υ, Υ' and Υ'' at the Cornell Electron Storage Ring

T. Bohringer et al.
Phys. Rev. Lett. **44** (1980) 1111;

Reprinted in
The Physical Review — the First Hundred Years, AIP Press

(1995) 924.

Abstract The Υ, Υ', and Υ'' states have been observed at the Cornell Electron Storage Ring as narrow peaks in $\sigma(e^+e^- \to$ hadrons) versus beam energy. Data were collected during a run with integrated luminosity of 1000 nb^{-1}, using the Columbia University-Stony Brook segmented NaI detector. The measured mass differences are $M(\Upsilon') - M(\Upsilon) = 559 \pm 1(\pm 3)$ MeV and $M(\Upsilon'') - M(\Upsilon) = 889 \pm 1(\pm 5)$ MeV, where the errors in parentheses represent systematic uncertainties. Preliminary values for the leptonic width ratios were also obtained.

Accelerator Cornell $e^+ e^-$ storage ring

Detectors Calorimeter CUSB

Related references

See also
C. Quigg and J. L. Rosner, Phys. Lett. **71B** (1977) 153;
E. Eichten et al., Phys. Rev. **D17** (1978) 3090;
G. Bhanot and S. Rudaz, Phys. Lett. **78B** (1978) 119;
E. Eichten et al., Phys. Rev. **D21** (1980) 203;
Analyse information from
S. W. Herb et al., Phys. Rev. Lett. **39** (1977) 252;
W. R. Innes et al., Phys. Rev. Lett. **39** (1977) 1240;
C. Berger et al., Phys. Lett. **76B** (1978) 243;
C. Berger et al., Phys. Lett. **78B** (1978) 176;
C. W. Darden et al., Phys. Lett. **76B** (1978) 246;
C. W. Darden et al., Phys. Lett. **78B** (1978) 364;
J. K. Bienlein et al., Phys. Lett. **78B** (1978) 360;
F. H. Heimlich et al., Phys. Lett. **86B** (1979) 399;
K. Ueno et al., Phys. Rev. Lett. **42** (1979) 486;
D. Andrews et al., Phys. Rev. Lett. **44** (1980) 1108;

Reactions
$e^+ e^- \to$ hadron (hadrons) 9.4-10.4 GeV(E_{cm})
$e^+ e^- \to \Upsilon(1S)$
$e^+ e^- \to \Upsilon(2S)$
$e^+ e^- \to \Upsilon(3S)$

Particles studied $\Upsilon(1S), \Upsilon(2S), \Upsilon(3S), b$

CALICCHIO 1980

■ First evidence for the $\Sigma_c(2455)^+$ ■

First Observation of the Production and Decay of the $\Sigma_c(2455)^+$

BEBC TST neutrino collaboration; M. Calicchio et al.
Phys. Lett. **93B** (1980) 521;

Abstract An event with the decay chain $\Sigma_c^+ \to \Lambda_c^+ \pi^0$, $\Lambda_c^+ \to K^- p \pi^+$, has been observed in an exposure of BEBC, equipped with a track sensitive target, to the wide band neutrino beam from the SPS at CERN. The event has a unique three constraint kinematic fit to the $\Delta S = -\Delta Q$ reaction $\nu p \to \bar{\mu} p K^- \pi^+ \pi^+ \pi^0$ with both gammas from the π^0 decay detected. The proton and other final state particles are identified. The masses are $M(\Lambda_c^+) = 2290 \pm 3$ MeV/c^2, $M(\Sigma_c^+) = 2457 \pm 4$ MeV/c^2 and $M(\Sigma_c^+) - M(\Lambda_c^+) = 168 \pm 3$ MeV/c^2. Including other data one obtains $M(\Sigma_c^{++}) - M(\Sigma_c^+) = 0 \pm 4$ MeV/c^2.

Accelerator CERN-SPS super proton synchrotron:

Detectors HBC-BEBC-TST-HYB

Related references

See also
E. G. Cazzoli et al., Phys. Rev. Lett. **34** (1975) 1125;
B. Knapp et al., Phys. Rev. Lett. **37** (1976) 882;
A. M. Cnops et al., Phys. Rev. Lett. **42** (1979) 197;
W. Lockman et al., Phys. Lett. **85B** (1979) 443;
D. Drijard et al., Phys. Lett. **85B** (1979) 452;
G. S. Abrams et al., Phys. Rev. Lett. **44** (1980) 10;
C. Baltay et al., Phys. Rev. Lett. **42** (1979) 1721;
K. L. Giboni et al., Phys. Lett. **85B** (1979) 437;
R. Beuselinck et al., Nucl. Instr. and Meth. **154** (1978) 445;
V. Flaminio et al., *Compilation of cross-sections*, preprint CERN-HERA 79-01, 79-02 (1979);

Reactions
$\nu_\mu p \to p K^- 2\pi^+ \pi^0 \mu^-$ 30-300 GeV(P_{lab})
$\Lambda_c^+ \to p K^- \pi^+$
$\Sigma_c(2455)^+ \to p K^- \pi^+ \pi^0$

Particles studied $\Sigma_c(2455)^+, \Lambda_c^+$

ANDREWS 1980B

■ Evidence for the $\Upsilon(4S)$ ■

Observation of a Fourth Upsilon State in $e^+ e^-$ Annihilations

CLEO Collaboration; D. Andrews et al.
Phys. Rev. Lett. **45** (1980) 219;

Abstract A fourth state in the upsilon energy region has been seen in e^+e^- collisions at the Cornell Electron Storage Ring. A resonance is observed with a mass 1112 ± 5 MeV above the lowest upsilon state. The 9.6 MeV rms width is greater than the 4.6 MeV energy resolution of the e^+e^- beams. The observed characteristics of the new state make it a likely candidate for the 4^3S state of the $b\bar{b}$ system, lying above the threshold for the production of B mesons.

Accelerator Cornell $e^+ e^-$ storage ring

Detectors CLEO

Related references

More (earlier) information
D. Andrews et al., Phys. Rev. Lett. **44** (1980) 1108;
Analyse information from
T. Bohringer et al., Phys. Rev. Lett. **44** (1980) 1111;
S. W. Herb et al., Phys. Rev. Lett. **39** (1977) 252;
W. R. Innes et al., Phys. Rev. Lett. **39** (1977) 1240;
C. Berger et al., Phys. Lett. **76B** (1978) 243;
C. W. Darden et al., Phys. Lett. **76B** (1978) 246;
C. W. Darden et al., Phys. Lett. **78B** (1978) 364;
J. K. Bienlein et al., Phys. Lett. **78B** (1978) 360;

G. Finocchiaro et al., Phys. Rev. Lett. **45** (1980) 222;

Reactions

$e^+ e^- \to \Upsilon(4S)$ 10.46-10.64 GeV(E_{cm})

$\Upsilon(4S) \to$ 2hadron (hadrons)

Particles studied $\Upsilon(4S)$

FINOCCHIARO 1980

■ Another evidence for the $\Upsilon(4S)$ ■

Observation of the Υ''' at CESR

G. Finocchiaro et al.
Phys. Rev. Lett. **45** (1980) 222;

Abstract During an energy scan at the Cornell Electron Storage Ring, with use of the Columbia University-Stony Brook NaI detector, an enhancement in $\sigma(e^+e^- \to \text{hadrons})$ is observed at center-of-mass energy ~ 10.55 GeV. The mass and leptonic width of this state (Υ''') suggest that it is the 4^3S_1 bound state of the b quark and its antiquark. After applying to the data a cut in a (pseudo) thrust variable, the natural width is measured to be $\Gamma = 12.6 \pm 6.0$ MeV, indicating that the Υ''' is above threshold for $B\overline{B}$ production.

Accelerator Cornell $e^+ e^-$ storage ring

Detectors Calorimeter CUSB

Related references

More (earlier) information
T. Bohringer et al., Phys. Rev. Lett. **44** (1980) 1111;

Analyse information from
D. Andrews et al., Phys. Rev. Lett. **44** (1980) 1108;
D. Andrews et al., Phys. Rev. Lett. **45** (1980) 219;
S. W. Herb et al., Phys. Rev. Lett. **39** (1977) 252;
W. R. Innes et al., Phys. Rev. Lett. **39** (1977) 1240;
C. Berger et al., Phys. Lett. **76B** (1978) 243;
C. Berger et al., Phys. Lett. **78B** (1978) 176;
C. Berger et al., Phys. Lett. **82B** (1979) 449;
J. K. Bienlein et al., Phys. Lett. **78B** (1978) 360;
C. W. Darden et al., Phys. Lett. **78B** (1978) 364;
C. W. Darden et al., Phys. Lett. **76B** (1978) 246;
F. H. Heimlich et al., Phys. Lett. **86B** (1979) 399;

Reactions

$e^+ e^- \to$ 2hadron (hadrons) 9.390-10.60 GeV(E_{cm})

$\Upsilon(4S) \to$ 2hadron (hadrons)

Particles studied $\Upsilon(4S)$

BEBEK 1981

■ First evidence for the B meson ■

Evidence for New Flavor Production at the $\Upsilon(4S)$

CLEO Collaboration; C. Bebek et al.
Phys. Rev. Lett. **46** (1981) 84;

Abstract An enhancement has been observed in the inclusive cross section for direct single electrons produced in e^+e^- annihilations at the $\Upsilon(4S)$. This is interpreted as evidence for a new weakly decaying particle, the B meson. A branching ratio for $B \to Xe\nu$ of $[13 \pm 3(\pm 3)]\%$, is inferred, where the first set of errors is statistical and the estimated systematic error is enclosed in parentheses.

Accelerator Cornell $e^+ e^-$ storage ring

Detectors CLEO

Related references

See also
D. Andrews et al., Phys. Rev. Lett. **45** (1980) 219;
G. Finocchiaro et al., Phys. Rev. Lett. **45** (1980) 222;
K. Chadwick et al., Phys. Rev. Lett. **46** (1981) 88;
M. Kobayashi and T. Maskawa, Progr. of Theor. Phys. **49** (1973) 652;

Reactions

$e^+ e^- \to$ 5charged X 10.378-10.598 GeV(E_{cm})

$e^+ e^- \to e^\pm$ X 10.378-10.598 GeV(E_{cm})

$e^+ e^- \to \Upsilon(4S)$ 10.378-10.598 GeV(E_{cm})

$\Upsilon(4S) \to B$(unspec) $\overline{B}$(unspec)

B(unspec) $\to e^-$ X

Particles studied $\Upsilon(4S), B$(unspec)

BASILE 1981I

■ First indication of the existence of the bottom baryon Λ_b ■

Evidence for a New Particle with Naked "Beauty" and for its Associated Production in High-Energy (pp) Interactions

M. Basile et al.
Lett. Nuovo Cim. **31** (1981) 97;

Excerpt Evidence is reported for the existence, with six-standard-deviation significance, of a new particle: the heaviest baryon observed so far, whose mass is measured to be $m = (5425^{+175}_{-75})$ MeV/c^2. Its width is compatible with zero. It is electrically neutral and it decays into a proton, a D^0 and a π^0. This particle is produced in association with a hadronic state whose semi-leptonic decay contains a positive electron. The partial cross section for the observed effect is

$\Delta\sigma = (3.8 \pm 1.2) \cdot 10^{-35}$ cm^2. The interpretation of these results is in terms of the associated production of naked-"beauty" states in (pp) interactions at $\sqrt{s} = 62$ GeV c.m. energy. The new particle is identified as the first "beauty"-flavoured baryon with quark composition (udb), *i.e.* the Λ_b^0. The discrimination power of the experiment against "known" physics is of the order of 10^{-10}.

Accelerator CERN-ISR $p\ p$ collider

Detectors Split field magnet SFM

Reactions

$p\ p \rightarrow$ bottom X	62 GeV(E_{cm})
$p\ p \rightarrow p\ K^-\ \pi^+\ \pi^-\ e^+$ X	62 GeV(E_{cm})

Particles studied Λ_b

CHADWICK 1981

■ Confirmation of the production of the B meson ■

Decay of B-flavored Hadrons to Single-Muon and Dimuon Final States

CLEO Collaboration; K. Chadwick et al.

Phys. Rev. Lett. **46** (1981) 88;

Abstract An enhancement in the inclusive cross section for single muons produced in e^+e^- annihilation at the $\Upsilon(4S)$ is observed, confirming the interpretation that a new bare flavor (B mesons) is produced at the $\Upsilon(4S)$. A branching ratio for $B \rightarrow X\mu\nu$ of $(9.4 \pm 3.6)\%$ is obtained. The two-muon decay, $B \rightarrow X\mu^+\mu^-$ is not observed, providing a 90%-confidence level upper limit for the branching ratio for that decay of 1.7%. Combining this with our previously reported limit of 5% for $B \rightarrow Xe^+e^-$, we obtain 1.3% as an upper limit for $B \rightarrow Xl^+l^-$.

Accelerator Cornell $e^+\ e^-$ storage ring

Detectors CLEO

Related references

See also

D. Andrews et al., Phys. Rev. Lett. **44** (1980) 1108;
D. Andrews et al., Phys. Rev. Lett. **45** (1980) 219;
G. Finocchiaro et al., Phys. Rev. Lett. **45** (1980) 222;
V. Barger and S. Pakvasa, Phys. Lett. **81B** (1979) 195;
H. Georgi and A. Pais, Phys. Rev. **D19** (1979) 2746;
H. Georgi and S. L. Glashow, Nucl. Phys. **B167** (1980) 173;
W. Bacino et al., Phys. Rev. Lett. **45** (1980) 329;

Reactions

$e^+\ e^- \rightarrow \mu^\pm$ X	10.55 GeV(E_{cm})
$e^+\ e^- \rightarrow e^-\ e^+$ X	10.55 GeV(E_{cm})
$e^+\ e^- \rightarrow \mu^-\ \mu^+$ X	10.55 GeV(E_{cm})
$e^+\ e^- \rightarrow \Upsilon(4S)$	10.55 GeV(E_{cm})
$\Upsilon(4S) \rightarrow B$(unspec) $\overline{B}$(unspec)	
B(unspec) $\rightarrow \mu^-$ X	
B(unspec) $\rightarrow \mu^-\ \mu^+$ X	
B(unspec) $\rightarrow e^-\ e^+$ X	

Particles studied $\Upsilon(4S), B$(unspec)

KIM 1981

■ First global comparison of data on weak neutral currents with minimal electroweak theory ■

A Theoretical and Experimental Review of the Weak Neutral Current: a Determination of its Structure and Limits on Deviations from the Minimal $SU(2)_L \times U(1)$ Electroweak Theory

J.E. Kim, P. Langacker, M. Levine, H.H. Williams

Rev. of Mod. Phys. **53** (1981) 211;

Abstract A detailed analysis of existing neutral-current data has been performed in order (a) to determine as fully as possible the structure of the hadronic and leptonic neutral currents without recourse to a specific weak-interaction model; (b) to search for the effects of small deviations from the Weinberg-Salam (WS-GIM) model; and (c) to determine the value of $\sin^2\Theta_W$ as accurately as possible. The authors attempt to incorporate the best possible theoretical expressions in the treatment of each of the reactions. For deep-inelastic scattering, for example, the effects of quantum chromodynamics, including the contributions of the s and c quarks, have been included. The sensitivity of the results both to systematic uncertainties in the data and to theoretical uncertainties in the treatment of deep-inelastic scattering, semi-inclusive pion production, ν elastic scattering from protons, and asymmetry in polarized eD scattering have been considered; the systematic errors are generally found to be smaller than the statistical uncertainties. In the model-independent analyses the authors find that the hadronic neutral current parameters are uniquely determined to lie within small domain consistent with the WS-GIM model. The leptonic couplings are determined to within a twofold ambiguity; one solution, the axial-vector-dominant, is consistent with WS-GIM model. If factorization is assumed then the axial-dominant solution is uniquely determined and null atomic parity violation experiments are inconsistent with other neutral-current experiments. Within generalized $SU(2) \times U(1)$ models we find the following limits on mixing between right-handed singlets and doublets: $\sin^2\alpha_u \leq 0.103$, $\sin^2\alpha_d \leq 0.348$, and $\sin^2\alpha_e \leq 0.064$. Assuming these mixing angles to be zero, a fit to the most accurate data (deep-inelastic and the polarized eD asymmetry) yields $\rho = 0.992 \pm 0.017(\pm 0.011)$ and $\sin^2\Theta_W = 0.224 \pm 0.015(\pm 0.012)$, where $\rho = M_W^2/M_Z^2\cos^2\Theta_W$ and the numbers in parentheses are the theoretical uncertainties. The value of ρ is remarkably close to 1.0 and strongly suggests that the Higgs meson occur only as doublets and singlets. If one makes this assumption, then the limit on ρ implies $m_L \leq 500$

GeV, where m_L is the mass of any heavy lepton with a massless partner. In addition, for $\rho = 1.0$, the authors determine $\sin^2 \Theta_W = 0.220 \pm 0.009(\pm 0.005)$. Fits which also include the semi-inclusive, elastic, and leptonic data yield very similar results. A two-parameter fit gives $\rho = 1.002 \pm 0.015(\pm 0.011)$ and $\sin^2 \Theta_W = 0.234 \pm 0.013(\pm 0.009)$, while a one-parameter fit to $\sin^2 \Theta_W$ gives $\sin^2 \Theta_W = 0.223 \pm 0.009(\pm 0.005)$. Finally, the authors have found no evidence for a violation of factorization or for the existence of additional Z bosons. Fits to two explicit two-boson model yield the lower limits $M_{Z_2}/M_{Z_1} \geq 1.61$ and 3.44 for the mass of the second Z boson. The desirability of a complete analysis of radiative and higher-order weak corrections, which have not been included in the authors' theoretical uncertainties, is emphasized.

Related references

See also
S. A. Bludman and M. A. Ruderman, Nuovo Cim. **9** (1958) 433;
S. L. Glashow, Nucl. Phys. **22** (1961) 579;
S. L. Glashow, J. Iliopoulos, and L. Maiani, Phys. Rev. **D2** (1970) 1285;
A. Salam and J. C. Ward, Phys. Lett. **13** (1964) 168;
A. Salam, *Elementary Particle Theory* edited by D. B. Cline and F. E. Mills, Harwood Academic, London (1977) 153;
J. Schwinger, Ann.Phys. **2** (1957) 407;
S. Weinberg, Phys. Rev. Lett. **19** (1967) 1264;
S. Weinberg, Phys. Rev. Lett. **27** (1971) 1688;

ARNISON 1983C — Nobel prize

■ First evidence for the charged intermediate bosons W^+ and W^-. Nobel prize to C. Rubbia awarded in 1984. Co-winner S. van der Meer "for decisive contribution to the large project, which led to the discovery of the field particles W and Z^0, communicators of weak interactions" ■

Experimental Observation of Isolated Large Transverse Energy Electrons with Associated Missing Energy at $\sqrt{s} = 540$ GeV

UA1 collaboration; G. Arnison et al.
Phys. Lett. **122B** (1983) 103;

Reprinted in
Gluons and Heavy Flavours, La Plagne (1983) 611.
Proton Antiproton Collider Physics, Rome (1983) 123.
Fifty Years of Weak-Interaction Physics, Bologna (1984) 753.
R. N. Cahn and G. Goldhaber, *The Experimental Foundations of Particle Physics*, Cambridge Univ. Press (1991) 403.

Abstract We report the results of two searches made at the CERN SPS Proton-Antiproton Collider: one for isolated large-E_T electrons, the other for large-E_T neutrinos using the technique of missing transverse energy. Both searches converge to the same events, which have the signature of two-body decay of a particle of mass ~ 80 GeV/c^2. The topology as well as the number of events fits well the hypothesis that they are produced by the process $\bar{p}\,p \to W^\pm$ X, with $W^\pm \to e^\pm \nu_e$; where $W^\pm$ is the Intermediate Vector Boson postulated by the unified theory of weak and electromagnetic interactions.

Accelerator CERN-PBAR/P $\bar{p}\,p$ collider

Detectors UA1

Related references

See also
C. Rubbia, P. McIntyre, and D. Cline, *Proc. Intern. Neutrino Conf.*, Aachen, (1976);
S. Weinberg, Phys. Rev. Lett. **19** (1967) 1264;
A. Salam, *Elementary particle theory*, ed. N. Svartholm, Stockholm: Almqvist and Wiksell (1968) 367;
A. Sirlin, Phys. Rev. **D22** (1980) 971;
W. J. Marciano and A. Sirlin, Phys. Rev. **D22** (1980) 2695;
C. H. Llewellyn Smith and J. A. Wheater, Phys. Lett. **105B** (1981) 486;
L. B. Okun' and M. B. Voloshin, Nucl. Phys. **B120** (1977) 459;
C. Quigg, Rev. of Mod. Phys. **49** (1977) 297;
J. Kogut and J. Shigemitsu, Nucl. Phys. **B129** (1977) 461;
R. F. Peierls, T. Trueman, and L. L. Wang, Phys. Rev. **D16** (1977) 1397;
M. Calvetti et al., Nucl. Instr. and Meth. **176** (1980) 255;
K. Eggert et al., Nucl. Instr. and Meth. **176** (1980) 217;
H. M. Kroll and W. Wada, Phys. Rev. **98** (1955) 1355;
F. Halzen et al., Phys. Lett. **106B** (1981) 147;
M. Chaichian, M. Hayashi, and K. Yamagishi, Phys. Rev. **D25** (1982) 130;
F. Halzen, A. D. Martin, and D. M. Scott, Phys. Rev. **D25** (1982) 754;

Reactions

$\bar{p}\,p \to e^-$ X	540 GeV(E_{cm})

Particles studied W^+, W^-

BIAGI 1983C

■ First evidence for the $\Xi_c(2460)^+$ ■

Observation of a Narrow State at 2.46 GeV/c^2 — A Candidate for the Charmed Strange Baryon A^+

S.F. Biagi et al.
Phys. Lett. **122B** (1983) 455;

Reprinted in
Gluons and Heavy Flavours, La Plagne (1983) 341.

Abstract A narrow state has been observed in the reaction $\Sigma^- Be \to (\Lambda K^- \pi^+ \pi^+) + X$ in an experiment at the CERN SPS hyperon beam. At 2.46 GeV/c^2 the effective $(\Lambda K^- \pi^+ \pi^+)$ mass distribution shows an excess of 82 events above a background estimated to be 147, corresponding to a statistical significance of more than 6 standard deviations. The positive charge of the observed final state, which has strangeness-2, suggests the interpretation as a Cabibbo favored decay of the charmed strange baryon, A^+ [quark content (csu)]. The cross section times branching ratio is measured to be $\sigma \cdot B = (5.3 \pm 2.0)\ \mu b$/(Be nucleus) for $x > 0.6$. The invariant production cross section is described by $E\ d^3\sigma/d\mathbf{p}^3 \propto (1-x)^{(1.7\pm0.7)} \exp[-(1.1^{+0.7}_{-0.4})p_T^2]$.

Accelerator CERN-SPS super proton synchrotron:

Detectors Spectrometer

Related references

See also
M. Bourquin et al., Nucl. Phys. **B153** (1979) 13;
H. J. Burckhart, Ph.D.Thesis, Universitat Heidelberg (1983);

Reactions

$\Sigma^- \text{Be} \to \Lambda K^- 2\pi^+ X$	135 GeV(P_{lab})
$\Sigma^- \text{Be} \to \Xi_c(2460)^+ X$	135 GeV(P_{lab})
$\Xi_c(2460)^+ \to \Lambda K^- 2\pi^+$	

Particles studied $\Xi_c(2460)^+$

BANNER 1983B

■ First evidence for the production of the charged intermediate bosons W^+ and W^- ■

Observation of Single Isolated Electrons of High Transverse Momentum in Events with Missing Transverse Energy at the CERN $\overline{p}\, p$ Collider

UA2 collaboration; M. Banner et al.
Phys. Lett. **122B** (1983) 476;

Abstract We report the results of a search for single isolated electrons of high transverse momentum at the CERN $\overline{p}p$ collider. Above 15 GeV/c, four events are found having large missing transverse energy along a direction opposite in azimuth to that of the high-p_T electron. Both the configuration of the events and their number are consistent with the expectations from the process $\overline{p}\, p \to W^\pm$ anything, with $W \to e\,\nu$, where $W^\pm$ is the charged Intermediate Vector Boson postulated by the unified electroweak theory.

Accelerator CERN-PBAR/P $\overline{p}\, p$ collider

Detectors UA2

Related references

See also
L. B. Okun' and M. B. Voloshin, Nucl. Phys. **B120** (1977) 459;
C. Quigg, Rev. of Mod. Phys. **49** (1977) 297;
J. Kogut and J. Shigemitsu, Nucl. Phys. **B129** (1977) 461;
R. F. Peierls, T. Trueman, and L. L. Wang, Phys. Rev. **D16** (1977) 1397;
F. Rapuano, Lett. Nuovo Cim. **26** (1979) 219;
S. L. Glashow, Nucl. Phys. **22** (1961) 579;
S. Weinberg, Phys. Rev. Lett. **19** (1967) 1264;
A. Salam, *Elementary particle theory*, ed. N. Svartholm, Stockholm: Almqvist and Wiksell (1968) 367;
C. Rubbia, P. McIntyre, and D. Cline, *Proc. Intern. Neutrino Conf.*, Aachen, (1976);
S. van der Meer, *Stochastic damping of betatron oscillations*, preprint CERN/ISR-PO/72-31 (1972);
UA1 Coll., G. Arnison et al., Phys. Lett. **122B** (1983) 103;
UA2 Coll., M. Banner et al., Phys. Lett. **115B** (1982) 59;
UA2 Coll., M. Banner et al., Phys. Lett. **121B** (1983) 187;
UA2 Coll., M. Banner et al., Phys. Lett. **122B** (1983) 322;
UA2 Coll., M. Banner et al., Phys. Lett. **118B** (1982) 203;
R. Battiston et al., Phys. Lett. **117B** (1982) 126;
E. Calva-Tellez et al., Lett. Nuovo Cim. **4** (1971) 619;

Reactions

$\overline{p}\, p \to$ shower X	540 GeV(E_{cm})
$\overline{p}\, p \to e^+ \nu_e X$	540 GeV(E_{cm})
$\overline{p}\, p \to e^- \overline{\nu}_e X$	540 GeV(E_{cm})
$\overline{p}\, p \to W^+ X$	540 GeV(E_{cm})
$\overline{p}\, p \to W^- X$	540 GeV(E_{cm})
$W^+ \to e^+ \nu_e$	
$W^- \to e^- \overline{\nu}_e$	

Particles studied W^+, W^-

AUBERT 1983

■ Evidence for difference between structure functions of bound and free nucleons — EMC effect ■

The Ratio of the Nucleon Structure Function F_2^N for Iron and Deuterium

The european collaboration; J.J. Aubert et al.
Phys. Lett. **123B** (1983) 275;

Abstract Using the data on deep inelastic muon scattering on iron and deuterium the ratio of the nucleon structure functions $F_2^N(Fe)/F_2^N(D)$ is presented. The observed x-dependence of this ratio is in disagreement with existing theoretical predictions.

Accelerator CERN-SPS super proton synchrotron:

Detectors EMC

Data comments The kinematical range covered by the data is $0.03 < x < 0.65$.

Related references

More (earlier) information
EMC, J. J. Aubert et al., Phys. Lett. **105B** (1981) 315;
EMC, J. J. Aubert et al., Phys. Lett. **105B** (1981) 322;
EMC, J. J. Aubert et al., Phys. Lett. **123B** (1983) 123;

See also
G. Alterelli and G. Parisi, Nucl. Phys. **B126** (1977) 298;
L. L. Frankfurt and M. I. Strikman, Nucl. Phys. **B148** (1979) 107;
L. L. Frankfurt and M. I. Strikman, Nucl. Phys. **B181** (1981) 22;
L. L. Frankfurt and M. I. Strikman, Phys. Lett. **114B** (1982) 345;
L. L. Frankfurt and M. I. Strikman, Phys. Rept. **76C** (1981) 215;
W. B. Atwood and G. B. West, Phys. Rev. **D7** (1973) 773;
A. Bodek and J. L. Ritchie, Phys. Rev. **D23** (1981) 1070;
G. Berlad et al., Phys. Rev. **D22** (1980) 1547;
J. V. Noble, Phys. Rev. Lett. **46** (1981) 412;
A. M. Green, Rep.Prog.Phys. **39** (1976) 1109;
H. J. Weber and H. Arenhövel, Phys. Rept. **36C** (1978) 277;
E. Lehman, Phys. Lett. **62B** (1976) 296;
V. A. Matveev and P. Sorba, Nuovo Cim. Lett. **20** (1977) 435;
H. Högaasen et al., Z. Phys. **C4** (1980) 131;
L. Bergström and S. Fredriksson, Rev. of Mod. Phys. **52** (1980) 675;
H. J. Priner and J. P. Vary, Phys. Rev. Lett. **46** (1981) 1376;
M. Namiki et al., Phys. Rev. **C25** (1982) 2157;
R. M. Godbole and K. V. L. Sarma, Phys. Rev. **D25** (1982) 120;

Reactions

μ^+ deuteron $\to$ μ^+ X	9-170 GEV2 (Q2)
μ^+ Fe $\to$ μ^+ X	9-170 GEV2 (Q2)

ARNISON 1983D — Nobel prize

■ First evidence for the neutral intermediate boson Z^0. Nobel prize to C. Rubbia awarded in 1984. Co-winner S. van der Meer "for decisive contribution to the large project, which led to the discovery of the field particles W and Z^0, communicators of weak interactions" ■

Experimental Observation of Lepton Pairs of Invariant Mass Around 95 GeV at the CERN SPS Collider

UA1 collaboration; G. Arnison et al.
Phys. Lett. **126B** (1983) 398;

Reprinted in
1st Asia-Pacific Physics Conf., **V.1** Singapore (1983) 101.
Lepton and Photon Inter. at High Energies, Ithaca (1983) 27.
Fifty Years of Weak-Interaction Phys., Bologna (1984) 779.

Abstract We report the observation of four electron-positron pairs and one muon pair which have the signature of a two-body decay of a particle of mass $\sim$ 95 GeV/c^2. These events fit well the hypothesis that they are produced by the process $\bar{p}p \to Z^0$X (with $Z^0 \to e^+e^-$), where Z^0 is the Intermediate Vector Boson postulated by the electroweak theories as the mediator of weak neutral currents.

Accelerator CERN-PBAR/P $\bar{p}\,p$ collider

Detectors UA1

Related references

See also
UA1 Coll., G. Arnison et al., Phys. Lett. **122B** (1983) 103;
UA2 Coll., M. Banner et al., Phys. Lett. **122B** (1983) 476;
S. Weinberg, Phys. Rev. Lett. **19** (1967) 1264;
A. Salam, *Elementary particle theory*, ed. N. Svartholm, Stockholm: Almqvist and Wiksell (1968) 367;
C. Rubbia, P. McIntyre and D. Cline, *Proc. Intern. Neutrino Conf.*, Aachen, (1976);
M. Barranco Luque et al., Nucl. Instr. and Meth. **176** (1980) 175;
M. Calvetti et al., Nucl. Instr. and Meth. **176** (1980) 255;
K. Eggert et al., Nucl. Instr. and Meth. **176** (1980) 217;
K. Eggert et al., Nucl. Instr. and Meth. **176** (1980) 233;
J. Kim et al., Rev. of Mod. Phys. **53** (1981) 211;
I. Liede and M. Roos, Nucl. Phys. **B167** (1980) 397;
F. A. Berends et al., Nucl. Phys. **B202** (1982) 63;
G. Passarino and M. Veltman, Nucl. Phys. **B160** (1979) 151;
M. Greco et al., Nucl. Phys. **B171** (1980) 118;
M. Greco et al., Nucl. Phys. **B197** (1982) 543;
V. N. Baier et al., Phys. Rept. **78** (1981) 293;
S. D. Drell and T. M. Yan, Phys. Rev. Lett. **25** (1970) 316;
F. Halzen and D. H. Scott, Phys. Rev. **D18** (1978) 3378;
R. W. Brown and K. O. Mikaelian, Phys. Rev. **D19** (1979) 922;
R. W. Brown, D. Sahdev, and K. O. Mikaelian, Phys. Rev. **D20** (1979) 1164;
T. K. Gaisser, F. Halzen, and E. A. Paschos, Phys. Rev. **D15** (1977) 2572;
V. N. Baier and R. Ruckl, Phys. Lett. **102B** (1981) 364;
P. Aurenche and J. Lindfors, Nucl. Phys. **B185** (1981) 274;
P. Chapetta and M. Greco, Nucl. Phys. **B199** (1982) 77;

Reactions

$\bar{p}\,p \to Z^0$ X	540. GeV(E_{cm})
$Z^0 \to e^-\, e^+$	
$Z^0 \to \mu^-\, \mu^+$	

Particles studied Z^0

BAGNAIA 1983B

■ Confirmation of Z^0 boson production. Observation of $Z^0 \to e^+e^-$ decay ■

Evidence for $Z^0 \to e^+\,e^-$at the CERN $\bar{p}\,p$ Collider

UA2 collaboration; P. Bagnaia et al.
Phys. Lett. **129B** (1983) 130;

Abstract From a search for electron pairs produced in $\bar{p}p$ collisions at $\sqrt{s}$ = 550 GeV we report the observation of eight events which we interpret as resulting from the process $\bar{p}\,p \to Z^0+$ anything, followed by the decay $Z^0 \to e^+\,e^-$ or $Z^0 \to e^+\,e^-\,\gamma$, where Z^0 is the neutral Intermediate Vector Boson postulated by the unified electroweak theory. We used four of these events to measure the Z^0 mass M_Z = $91.9 \pm 1.3 \pm 1.4$(systematic) GeV/c^2.

Accelerator CERN-PBAR/P $\bar{p}\,p$ collider

Detectors UA2

Related references

See also
S. L. Glashow, Nuovo Cim. **22** (1961) 579;
S. Weinberg, Phys. Rev. Lett. **19** (1967) 1264;
A. Salam, *Elementary particle theory*, ed. N. Svartholm, Stockholm: Almqvist and Wiksell (1968) 367;
UA1 Coll., G. Arnison et al., Phys. Lett. **122B** (1983) 103;
UA1 Coll., G. Arnison et al., Phys. Lett. **126B** (1983) 398;
UA2 Coll., M. Banner et al., Phys. Lett. **115B** (1982) 59;
UA2 Coll., M. Banner et al., Phys. Lett. **118B** (1982) 203;
UA2 Coll., M. Banner et al., Phys. Lett. **122B** (1983) 322;
UA2 Coll., M. Banner et al., Phys. Lett. **122B** (1983) 476;

Reactions

$\bar{p}\,p \to e^-\,e^+$ X	550 GeV(E_{cm})
$\bar{p}\,p \to e^-\,e^+\,\gamma$ X	550 GeV(E_{cm})
$Z^0 \to e^-\,e^+$	
$Z^0 \to e^-\,e^+\,\gamma$	

Particles studied Z^0

BODEK 1983

■ Confirmation for difference between structure functions of bound and free nucleons — EMC effect ■

Electron Scattering from Nuclear Targets and Quark Distributions in Nuclei

A. Bodek et al.
Phys. Rev. Lett. **50** (1983) 1431;

Abstract The deep-inelastic electromagnetic structure functions of steel, deuterium, and hydrogen nuclei have been measured with use of the high-energy electron beam beam at the Stanford Linear Accelerator Center. The ratio of the structure functions of steel and deuterium cannot be understood simply by corrections due to Fermi-motion effects. The data indicate that the quark momentum distribution in the nucleon become distorted in the nucleus. The present results are consistent with recent measurements with high-energy muon beams.

Accelerator SLAC

Detectors Spectrometer SSF

Related references

More (earlier) information
A. Bodek et al., Phys. Lett. **51B** (1974) 417;
A. Bodek et al., Phys. Rev. **D20** (1979) 1471;
A. Bodek et al., Phys. Rev. Lett. **30** (1973) 1087;
A. Bodek, Nucl. Instr. and Meth. **109** (1973) 603;
A. Bodek, Nucl. Instr. and Meth. **117** (1974) 613;
A. Bodek, Nucl. Instr. and Meth. **150** (1978) 367;
See also
J. J. Aubert et al., Phys. Lett. **105B** (1981) 315;
J. J. Aubert et al., Phys. Lett. **105B** (1981) 322;
J. J. Aubert et al., Phys. Lett. **123B** (1983) 275;
J. C. Poucher et al., Phys. Rev. Lett. **32** (1974) 118;
W. B. Atwood et al., Phys. Lett. **64B** (1976) 479;
S. Stein et al., Phys. Rev. **D12** (1975) 1884;
W. R. Ditzler et al., Phys. Lett. **57B** (1975) 201;
J. Eickmeyer et al., Phys. Rev. Lett. **36** (1976) 289;
J. Bailey et al., Nucl. Phys. **B151** (1979) 367;
G. Huber et al., Z. Phys. **C2** (1979) 279;
J. Franz et al., Z. Phys. **C10** (1979) 105;
M. May et al., Phys. Rev. Lett. **35** (1975) 407;
M. Miller et al., Phys. Rev. **D24** (1981) 1;
G. Grammer and J.D. Sullivan, in *Electromegnetic Interactions of Hadrons*, ed. A. Donnachie and G. Shaw **2** (1978) 195;
A. Bodek and J. L. Ritchie, Phys. Rev. **D23** (1981) 1070;
A. Bodek and J. L. Ritchie, Phys. Rev. **D24** (1981) 1400;
R. L. Jaffe, Phys. Rev. Lett. **50** (1983) 228;
O. Nachtmann, Nucl. Phys. **B63** (1973) 237;
H. Georgy and H. D. Politzer, Phys. Rev. **D14** (1976) 1829;
L. W. Mo and Y. S. Tsai, Rev. of Mod. Phys. **41** (1969) 205;
L. C. Maximon, Rev. of Mod. Phys. **41** (1969) 193;
G. B. West, Ann.Phys. **74** (1972) 464;
W. B. Atwood and G. B. West, Phys. Rev. **D7** (1973) 773;
G. R. Farrar et al., Phys. Lett. **69B** (1977) 112;
P. Allen et al., Phys. Lett. **103B** (1981) 71;
J. Hanlon et al., Phys. Rev. Lett. **45** (1980) 1817;

Reactions

$e^- p \to e^- X$	8.7-20 GeV(E_{lab})
e^- deuteron $\to e^- X$	8.7-20 GeV(E_{lab})
e^- Fe $\to e^- X$	8.7-20 GeV(E_{lab})

ARNISON 1984C

■ Confirmation of W^+ and W^- production. First observation of $W^\pm \to \mu^\pm \nu$ decays ■

Observation of the Muonic Decay of the Charged Intermediate Vector Boson

UA1 collaboration; G. Arnison et al.
Phys. Lett. **134B** (1984) 469;

Abstract Muons of high transverse momentum p_T^μ have been observed in the large drift chambers surrounding the UA1 detector at the CERN 540 GeV $p\bar{p}$ collider. For an integrated luminosity of 108 nb^{-1}, 14 isolated muons have been found with $p_T > 15$ GeV/c. They are correlated with a large imbalance in total transverse energy, and show a kinematic behaviour consistent with the muonic decay of the Intermediate Vector Boson $W^\pm$ of weak interactions. The partial cross section is in agreement with previous measurements for electronic decay and muon-electron universality. The W mass is determined to be $m_W = 81^{+6}_{-7}$ GeV/c^2.

Accelerator CERN-PBAR/P $\bar{p}\,p$ collider

Detectors UA1

Related references

See also
UA1 Coll., G. Arnison et al., Phys. Lett. **118B** (1982) 167;
UA1 Coll., G. Arnison et al., Phys. Lett. **122B** (1983) 103;
UA1 Coll., G. Arnison et al., Phys. Lett. **132B** (1983) 214;
UA1 Coll., G. Arnison et al., Phys. Lett. **129B** (1983) 273;
UA2 Coll., M. Banner et al., Phys. Lett. **122B** (1983) 322;
UA2 Coll., M. Banner et al., Phys. Lett. **122B** (1983) 476;
S. L. Glashow, Nucl. Phys. **22** (1961) 579;
S. Weinberg, Phys. Rev. Lett. **19** (1967) 1264;
A. Salam, *Elementary particle theory*, ed. N. Svartholm, Stockholm: Almqvist and Wiksell (1968) 367;
A. Astbury, Phys. Scr. **23** (1981) 397;
M. J. Corden et al., Phys. Scr. **25** (1982) 5;
M. J. Corden et al., Phys. Scr. **25** (1982) 11;
B. Aubert et al., Nucl. Instr. and Meth. **176** (1980) 195;
M. Barranco Luque et al., Nucl. Instr. and Meth. **176** (1980) 175;
M. Calvetti et al., Nucl. Instr. and Meth. **176** (1980) 255;
K. Eggert et al., Nucl. Instr. and Meth. **176** (1980) 217;
K. Eggert et al., Nucl. Instr. and Meth. **176** (1980) 233;

Reactions

$\bar{p}\,p \to \mu^+ X$	540 GeV(E_{cm})
$\bar{p}\,p \to \mu^- X$	540 GeV(E_{cm})
$\bar{p}\,p \to \mu^+$ jet X	540 GeV(E_{cm})
$\bar{p}\,p \to \mu^-$ jet X	540 GeV(E_{cm})
$\bar{p}\,p \to W^+ X$	540 GeV(E_{cm})
$\bar{p}\,p \to W^- X$	540 GeV(E_{cm})
$W^+ \to \mu^+$ jet	
$W^- \to \mu^-$ jet	

Particles studied W^+, W^-

ARTAMONOV 1984B

■ Implementation of the resonance depolarization method of the beams energy calibration to the high precision measurements of heavy $e^+ e^-$ resonances ■

A High Precision Measurement of the $\Upsilon(1S)$, $\Upsilon(2S)$ and $\Upsilon(3S)$ Meson Masses

A.S. Artamonov et al.
Phys. Lett. **137B** (1984) 272; NOVO-83-84 (1983);

Abstract An experiment has been performed at the storage ring VEPP-4 using MD-1 detector. The resonance depolarization method has been employed for the absolute calibration of the beam energy. This has allowed the accuracy of the mass measurement to be improved by a factor of ten. The following mass values of $\Upsilon(1S)$, $\Upsilon(2S)$ and $\Upsilon(3S)$ mesons have been obtained: $M(\Upsilon(1S)) = 9460.0 \pm 0.4$ MeV, $M(\Upsilon(2S)) = 10023.8 \pm 0.5$ MeV, $M(\Upsilon(3S)) = 10355.5 \pm 0.5$ MeV.

Accelerator NOVO-VEPP-4

Detectors MD-1

Related references

More (earlier) information
A. D. Bukin et al., Yad. Phys. **27** (1978) 976;
L. M. Barkov et al., Nucl. Phys. **B148** (1979) 53;
A. A. Zholentz et al., Phys. Lett. **96B** (1980) 214;
A. A. Zholentz et al., Yad. Phys. **34** (1981) 1471;
A. S. Artamonov et al., Phys. Lett. **118B** (1982) 225;

See also
S. I. Serednyakov et al., Zh. Eksp. Teor. Fiz. **71** (1976) 2025;
J. D. Jackson and D. L. Scharre, Nucl. Instr. and Meth. **128** (1975) 13;
Particle Data Group, *Review of Particle Properties*, M. Roos et al., Phys. Lett. **111B** (1982) 1;
C. Bonneau and F. Martin, Nucl. Phys. **B27** (1971) 381;
A. E. Blinov et al., Phys. Lett. **113B** (1982) 423;
V. N. Baer and V. A. Khose, Yad. Phys. **9** (1969) 409;
D. R. Yennie et al., Ann.Phys. **13** (1961) 379;
Y. S. Derbenev et al., Particle Accelerators **10** (1980) 177;

Reactions

$e^+ e^- \to$ 2hadron (hadrons) 9.4-10.4 GeV(E_{cm})

Particles studied $\Upsilon(1S), \Upsilon(2S), \Upsilon(3S)$

ARNISON 1984G

■ Confirmation of Z^0 boson production. Observation of $Z^0 \to \mu^+\mu^-$ decay ■

Observation of Muonic Z^0 Decay at the $\bar{p}\, p$ Collider

UA1 collaboration; G. Arnison et al.
Phys. Lett. **147B** (1984) 241;

Abstract We report the observation of five muonic Z^0 decays. The mass and cross section times branching ratio is consistent with the previous measurements of $Z^0 \to e^+e^-$. Three of the muonic decays have unexpected features. One event is of the type $Z^0 \to e^+e^-\gamma$. Two of the $Z^0 \to e^+e^-$ decays are accompanied by several (> 4) energetic ($E_T > 10$ GeV) jets which are difficult to explain within the framework of standard QCD corrections.

Accelerator CERN-PBAR/P $\bar{p}\, p$ collider

Detectors UA1

Related references

See also
UA1 Coll., G. Arnison et al., Phys. Lett. **132B** (1983) 214;
UA1 Coll., G. Arnison et al., Phys. Lett. **126B** (1983) 398;
UA1 Coll., G. Arnison et al., Phys. Lett. **139B** (1984) 115;
UA1 Coll., G. Arnison et al., Phys. Lett. **135B** (1984) 250;
UA1 Coll., G. Arnison et al., Phys. Lett. **134B** (1984) 469;
M. Barranco Luque et al., Nucl. Instr. and Meth. **176** (1980) 175;
M. Calvetti et al., Nucl. Instr. and Meth. **176** (1980) 255;
B. Humpert, Phys. Lett. **85B** (1979) 293;
S. L. Glashow, Nucl. Phys. **22** (1961) 579;
S. Weinberg, Phys. Rev. Lett. **19** (1967) 1264;
A. Salam, *Elementary particle theory*, ed. N. Svartholm, Stockholm: Almqvist and Wiksell (1968) 367;

Reactions

$\bar{p}\, p \to \mu^- \mu^+$ X 540 GeV(E_{cm})
$\bar{p}\, p \to \mu^- \mu^+ \gamma$ X 540 GeV(E_{cm})
$\bar{p}\, p \to Z^0$ X 540 GeV(E_{cm})
$\bar{p}\, p \to Z^0$ mult[jet] 540 GeV(E_{cm})
$Z^0 \to \mu^- \mu^+$
$Z^0 \to \mu^- \mu^+ \gamma$

Particles studied Z^0

BIAGI 1985C

■ First evidence for the Ω_c ■

Properties of the Charmed Strange Baryon $\Xi_c(2460)^+$ and Evidence for the Charmed Doubly Strange Baryon Ω_c at 2.74 GeV/c^2

S.F. Biagi et al.
Z. Phys. **C28** (1985) 175;

Abstract Results are presented from experiment WA62, which searched for charmed strange baryon states produced in Σ^--nucleus interactions in the SPS charged hyperon beam at CERN. Properties of the A^+ (csu) baryon at 2.46 GeV/c^2 are summarized and upper limits are given for decay branching ratios into various channels. Three events observed at 2.74 GeV/c^2 in the $\Xi^- K^- \pi^+ \pi^+$ mass spectrum are interpreted as the first evidence for the T^0 baryon with quark content css. Results of a search for the A^0 (csd), the isospin partner of the A^+, are presented. The results are discussed in the context of current theoretical understanding, and a comparison with other experiments on hadroproduction of charmed baryons is made.

Accelerator CERN-SPS super proton synchrotron:

Detectors Spectrometer

Related references

More (earlier) information

S. F. Biagi et al., Phys. Lett. **122B** (1983) 455;
S. F. Biagi et al., Phys. Lett. **150B** (1985) 230;

Reactions

Σ^- Be $\to \Lambda K^- 2\pi^+$ X	135 GeV(P_{lab})
Σ^- Be $\to p \overline{K}^0 K^- \pi^+$ X	135 GeV(P_{lab})
Σ^- Be $\to p\, 2K^- 2\pi^+$ X	135 GeV(P_{lab})
Σ^- Be $\to \Omega^- K^+ \pi^+$ X	135 GeV(P_{lab})
Σ^- Be $\to \Lambda \overline{K}^*(892)^0 \pi^+$ X	135 GeV(P_{lab})
Σ^- Be $\to \Lambda K^- \pi^+ \pi^0$ X	135 GeV(P_{lab})
Σ^- Be $\to \Omega^- K^+$ X	135 GeV(P_{lab})
Σ^- Be $\to p\, 2K^- \pi^+$ X	135 GeV(P_{lab})
Σ^- Be $\to p \overline{K}^0 K^-$ X	135 GeV(P_{lab})
Σ^- Be $\to \Xi^- K^- 2\pi^+$ X	135 GeV(P_{lab})
Σ^- Be $\to \Xi_c(2460)^+$ X	135 GeV(P_{lab})
Σ^- Be $\to \Xi_c(2460)^0$ X	135 GeV(P_{lab})
Σ^- Be $\to \Omega_c$ X	135 GeV(P_{lab})
$\Xi_c(2460)^+ \to \Lambda K^- 2\pi^+$	
$\Xi_c(2460)^+ \to p \overline{K}^0 K^- \pi^+$	
$\Xi_c(2460)^+ \to p\, 2K^- 2\pi^+$	
$\Xi_c(2460)^+ \to \Omega^- K^+ \pi^+$	
$\Xi_c(2460)^+ \to \Lambda \overline{K}^*(892)^0 \pi^+$	
$\Xi_c(2460)^0 \to \Lambda K^- \pi^+ \pi^0$	
$\Xi_c(2460)^0 \to \Omega^- K^+$	
$\Xi_c(2460)^0 \to p\, 2K^- \pi^+$	
$\Xi_c(2460)^0 \to p \overline{K}^0 K^-$	
$\Omega_c \to \Xi^- K^- 2\pi^+$	

Particles studied $\Xi_c(2460)^+, \Omega_c$

ARNISON 1985D

■ Firm establishment of the properties of W^+, W^- and Z^0 bosons ■

Recent Results on Intermediate Vector Boson Properties at the CERN Super Proton Synchrotron Collider

UA1 collaboration; G. Arnison et al.

Phys. Lett. **166B** (1986) 484; Eur. Lett. **1** (1986) 327;

Abstract The properties of a sample of 172 charged intermediate vector bosons decaying in the ($e\nu_e$) channel and 16 neutral intermediate vector bosons decaying in the (e^+e^-) channel are summarized. Masses, decay widths, decay angular distributions, and production cross sections are given, and a limit is put on the number of light neutrino types $N_\nu \leq 10$ at 90% CL.

Accelerator CERN-PBAR/P $\overline{p}\, p$ collider

Detectors UA1

Related references

See also

UA1 Coll., G. Arnison et al., Phys. Lett. **122B** (1983) 103;
UA1 Coll., G. Arnison et al., Phys. Lett. **132B** (1983) 214;
UA1 Coll., G. Arnison et al., Phys. Lett. **122B** (1983) 273;
UA1 Coll., G. Arnison et al., Phys. Lett. **126B** (1983) 398;
UA1 Coll., G. Arnison et al., Phys. Lett. **135B** (1984) 250;
UA1 Coll., G. Arnison et al., Phys. Lett. **147B** (1984) 241;
UA1 Coll., G. Arnison et al., Phys. Lett. **134B** (1984) 469;
UA1 Coll., G. Arnison et al., Nuovo Cim. Lett. **44** (1985) 1;
M. Barranco Luque et al., Nucl. Instr. and Meth. **176** (1980) 175;
M. Calvetti et al., Nucl. Instr. and Meth. **176** (1980) 255;
B. Humpert, Phys. Lett. **85B** (1979) 293;
UA2 Coll., M. Banner et al., Phys. Lett. **122B** (1983) 476;
UA2 Coll., P. Bagnaia et al., Phys. Lett. **129B** (1983) 130;
UA2 Coll., P. Bagnaia et al., Z. Phys. **C24** (1984) 1;
S. D. Drell and T. M. Yan, Phys. Rev. Lett. **25** (1970) 316;
S. D. Drell and T. M. Yan, Ann.Phys. **66** (1971) 578;
S. Weinberg, Phys. Rev. Lett. **19** (1967) 1264;
A. Salam, *Elementary particle theory*, ed. N. Svartholm, Stockholm: Almqvist and Wiksell (1968) 367;
S. Geer and W. J. Stirling, Phys. Lett. **152B** (1985) 373;
G. Altarelli, R. K. Ellis, M. Greco, and G. Martinelli, Nucl. Phys. **B246** (1984) 12;
G. Altarelli, R. K. Ellis, and G. Martinelli, Z. Phys. **C27** (1985) 617;
N. G. Deshpande et al., Phys. Rev. Lett. **54** (1985) 1757;
F. Halzen and K. Mursula, Phys. Rev. Lett. **51** (1983) 857;
K. Hikasa, Phys. Rev. **D29** (1984) 1939;
W. J. Marciano and A. Sirlin, Phys. Rev. **D29** (1984) 945;

Reactions

$\overline{p}\, p \to W^+$ X	546,630 GeV(E_{cm})
$\overline{p}\, p \to W^-$ X	546,630 GeV(E_{cm})
$\overline{p}\, p \to Z^0$ X	546,630 GeV(E_{cm})
$W^+ \to e^+ \nu_e$	
$W^- \to e^- \overline{\nu}_e$	
$Z^0 \to e^- e^+$	
$Z^0 \to \mu^- \mu^+$	
$W^+ \to \mu^+ \nu_\mu$	

$W^- \to \mu^- \bar{\nu}_\mu$

Particles studied W^+, W^-, Z^0

VOYVODIC 1986B

■ First evidence for the $\Sigma_c(2455)^0$ ■

Observation in Nuclear Emulsion of a Charmed $\Sigma_c(2455)^0$ Baryon Decay to Λ_c^+ π^- with a Subsequent Λ_c^+ Decay to Σ^+ π^- π^+

L. Voyvodic et al.
Pisma Zh. Eksp. Teor. Fiz. **43** (1986) 401; JETP Lett. **43** (1986) 515;

Abstract An event in which the decay of a charmed Σ_c^0 baryon into Λ_c^+ and π^- has been detected in a neutrino experiment with a nuclear emulsion is described. The decay of the baryon occurs in the channel $\Lambda_c^+ \to \Sigma^+\pi^-\pi^+$ with an observable subsequent decay of $\Sigma^+ \to \pi^+ n$. The mass of the Λ_c^+ is 2.300 ± 0.025 GeV/c^2, and it decays in a time of $(3.13 \pm 0.02) \times 10^{-13}$ s. The mass of the Σ_c^0 baryon and the difference between the Σ_c^0 and Λ_c^+ masses are 2.462 ± 0.026 GeV/c^2 and 0.163 ± 0.002 GeV/c^2, respectively. *(Science Abstracts. 1986, 15150.)*

Accelerator FNAL proton synchrotron

Detectors HLBC-15FT-HYB

Comments on experiment Wide band beam.

Related references

More (earlier) information
W. M. Smart et al., Acta Phys. Polon. **B17** (1986) 41;

Reactions

ν_μ nucleus $\to n\ 2\pi^+\ 2\pi^-\ \mu^-$ X	10-200 GeV(P_{lab})
ν_μ nucleus $\to \Sigma^+\ \pi^+\ 2\pi^-\ \mu^-$ X	10-200 GeV(P_{lab})
ν_μ nucleus $\to \Lambda_c^+\ \pi^-\ \mu^-$ X	10-200 GeV(P_{lab})
ν_μ nucleus $\to \Sigma_c(2455)^0\ \mu^-$ X	10-200 GeV(P_{lab})
$\Sigma^+ \to n\ \pi^+$	
$\Sigma_c(2455)^0 \to \Lambda_c^+\ \pi^-$	
$\Lambda_c^+ \to \Sigma^+\ \pi^+\ \pi^-$	

Particles studied $\Sigma_c(2455)^0, \Lambda_c^+$

VAN DYCK 1986 — Nobel prize

■ High precision measurement of the electron g−2 factor. Nobel prize to H. G. Dehmelt and W. Paul awarded in 1989 "for the development of the ion trap technique". Co-winner N. F. Ramsey "for the invention of the separated oscillatory fields method and its use in the hydrogen maser and other atomic clocks" ■

Electron Magnetic Moment from Geonium Spectra: Early Experiments and Background Concepts

R.S. Van Dyck, P.B. Schwinberg, H.G. Dehmelt
Phys. Rev. **D34** (1986) 722;

Abstract The magnetic moment of a free electron has been measured by observing both its low-energy spin and cyclotron resonances (at $\nu_s = \omega_s/2\pi$ and $\nu_c = \omega_c/2\pi$, respectively) by means of a sensitive frequency-shift technique. Using radiation and tuned-circuit damping of a single electron, isolated in a special anharmonicity-compensated Penning trap, also cooled to 4 K, the electron's motion is brought nearly to rest, thus preparing it in a cold quasipermanent state of the geonium "atom." The magnetic-coupling scheme, described as a continuous Stern-Gerlach effect, is made possible through a weak Lawrence magnetic bottle which causes the very narrow axial resonance, at $\nu_z = \omega_z/2\pi$ for the harmonically bound electron, to change in frequency by a small fixed amount δ per unit change in magnetic quantum number. Spin flips are indirectly induced by a scheme which weakly drives the axial motion at the $\nu_a = \omega_a/2\pi$ spin-cyclotron difference frequency within the inhomogeneous magnetic field, thus yielding a measure of $\omega_a = \omega_s - \omega_c$. The magnetic moment μ_s in terms of the Bohr magneton μ_B equals $\frac{1}{2}$ the spin's g factor, which in turn is described by ω_s and ω_c: $g = 2\mu_s/\mu_B = 2\omega_s/\omega_c$. In a Penning trap, however, these resonance frequencies are obtained from the observed cyclotron frequency at $\omega_c' = \omega_c - \delta_e$ and the observed anomaly frequency at $\omega_a' = \omega_s - \omega_c'$, which are related by the small electric shift δ_e computed using the measured axial frequency and $2\delta_e\omega_c' = \omega_z{}^2$. This last expression, derived for a perfectly axially symmetric trap, happens to be practically invariant against small imperfections in the electric quadrupole field (error in $\omega_c < 10^{-16}$). The magnetic-bottle-determined line shapes are analyzed and found to have sharp low-frequency edge features which correspond to the electron being temporarily at the trap center and at the bottom of the magnetic well. Relativistic shifts are considered and found to be $< 10^{-11}$. Our result at the time of submission, $g/2 = 1.001159652200(40)$, is the most accurately determined parameter of any elementary charged particle which in addition can be directly compared with theory.

Accelerator Non accelerator experiment

Detectors Geonium "atom"

Related references

See also
H. Louisel, R. W. Pidd, and H. R. Crane, Phys. Rev. **109** (1958) 381;
H. G. Dehmelt, Science **124** (1956) 1039;
H. G. Dehmelt and F. L. Walls, Phys. Rev. Lett. **21** (1968) 127;
H. G. Dehmelt and F. G. Major, Phys. Rev. Lett. **8** (1962) 213;
G. Gräff et al., Phys. Rev. Lett. **21** (1968) 340;
G. Gräff, E. Klempt, and G. Werth, Z. Phys. **222** (1969) 201;
A. Rich and J. C. Wesley, Rev. of Mod. Phys. **44** (1972) 250;
J. Mehra and H. Rechenberg, *The Historic Development of Quantum Theory*, Springer, New York **V.1, Pt.2** (1982) 696;
W. Pauli, *Exclusion Principle in Quantum Mechanics*, ed. by R. Kronig and V. F. Weisskopf, Interscience-Wiley, N.Y. **2** (1964) 1082;
P. A. M. Dirac, *The Principles of Quantum Mechanics*, 4th ed. Clarendon, Oxford, (1958) 262;
H. G. Dehmelt and P. Ekstrom, Bull.Am.Phys.Soc. **18** (1973) 727;
J. H. Field, E. Picasso, and F. Combley, Usp. Fiz. Nauk **127** (1979) 553;
T. Kinoshita and J. Sapirstein, *Atomic Physics 9*, ed. by R. S. van Dyck and E. N. Fortson, World Scientific, Singapore (1984) 38;
R. S. van Dyck, P. Ekstrom, and H. G. Dehmelt, Nature **262** (1976) 776;
R. S. van Dyck, P. B. Schwinberg, and H. G. Dehmelt, Phys. Rev. Lett. **38** (1977) 310;
P. B. Schwinberg, R. S. van Dyck, and H. G. Dehmelt, Phys. Lett. **81A** (1981) 119;
T. Kinoshita and W. B. Lindquist, Phys. Rev. Lett. **47** (1981) 1573;
T. Kinoshita and W. B. Lindquist, Phys. Rev. **D27** (1983) 853;
T. Kinoshita and W. B. Lindquist, Phys. Rev. **D27** (1983) 867;
T. Kinoshita and W. B. Lindquist, Phys. Rev. **D27** (1983) 877;
T. Kinoshita and W. B. Lindquist, Phys. Rev. **D27** (1983) 886;
D. Wineland, P. Ekstrom, and H. G. Dehmelt, Phys. Rev. Lett. **31** (1973) 1279;
S. Liebes and P. Franken, Phys. Rev. **116** (1959) 633;
I. I. Rabi, Z. Phys. **49** (1928) 507;
W. Tsai, Phys. Rev. **D7** (1973) 1945;
H. Mendlowitz and K. M. Case, Phys. Rev. **97** (1955) 33;

Particles studied e^-

ALBAJAR 1987C

■ First indication of B_S-$\overline{B}_S$ mixing ■

Search for B^0 $\overline{B}^0$ Oscillations at the CERN Proton Antiproton Collider. Paper 2

UA1 collaboration; C. Albajar et al.
Phys. Lett. **186B** (1987) 247; Phys. Lett. **197B** (1987) 565;

Abstract We report a search for $B^0 \leftrightarrow \overline{B}^0$ oscillations using events with two identified muons from data collected at the CERN $\overline{p}p$ collider. In the absence of $B^0 \leftrightarrow \overline{B}^0$ oscillations, dimuons coming directly from decays of beauty-antibeauty pairs must have *opposite* signs. Like-sign dimuons are expected from events where one muon arises from beauty decay and the other from the charm decay of the associated beauty-charm cascade. Taking these processes into account, together with the contribution from charm production, the predicted ratio of like-sign to unlike-sign muon pairs is 0.26 ± 0.03. Experimentally we measure $0.42 \pm 0.07 \pm 0.03$. A natural explanation for the excess of like-sign events is the existence of a significant amount of $B^0 \leftrightarrow \overline{B}^0$ transitions. The fraction of beauty particles that produce first-generation decay muons with the opposite electric charge from that expected with out mixing is deduced to be: $\chi = 0.121 \pm 0.047$. Combined with the null result from searching $B^0 \leftrightarrow \overline{B}^0$ oscillations at e^+e^- colliders, our results are consistent with transitions in the B_S system, as favored theoretically.

Accelerator CERN-PBAR/P $\overline{p}\,p$ collider

Detectors UA1

Related references

See also
T. D. Lee and C. S. Wu, Ann. Rev. Nucl. Sci. **16** (1966) 511;
J. Ellis, M. K. Gaillard, D. V. Nanopoulos, and S. Rudaz, Nucl. Phys. **B131** (1977) 285;
A. Ali and Z. Aydin, Nucl. Phys. **B148** (1979) 165;
W. C. Louis et al., Phys. Rev. Lett. **56** (1986) 1027;
M. Kobayashi and T. Maskawa, Progr. of Theor. Phys. **49** (1973) 652;
N. Cabibbo, Phys. Rev. Lett. **10** (1963) 531;
L. Wolfenstein, Phys. Rev. Lett. **51** (1983) 1945;
A. Ali and C. Jarlskog, Phys. Lett. **144B** (1984) 266;
V. Barger and R. J. N. Philips, Phys. Rev. Lett. **55** (1985) 2752;
V. Barger and R. J. N. Philips, Phys. Lett. **143B** (1984) 259;
I. I. Bigi and A. I. Sanda, Nucl. Phys. **B193** (1981) 85;
I. I. Bigi and A. I. Sanda, Phys. Rev. **D29** (1984) 1393;
Mark II Coll., T. Schaad et al., Phys. Lett. **160B** (1985) 188;
UA1 Coll., C. Albajar et al., Phys. Lett. **186B** (1987) 237;
E. Eichten et al., Rev. of Mod. Phys. **56** (1984) 579;
A. Chen et al., Phys. Rev. Lett. **52** (1984) 1084;
J. Green et al., Phys. Rev. Lett. **51** (1983) 347;
S. E. Csorna et al., Phys. Rev. Lett. **54** (1985) 1894;
W. Bacino et al., Phys. Rev. Lett. **43** (1979) 1073;
TASSO Coll., M. Althoff et al., Phys. Lett. **138B** (1984) 441;
ABCDHW Coll., A. Breakstone et al., Phys. Lett. **135B** (1984) 510;
Particle Data Group, *Review of Particle Properties*, M. Aguilar-Benitez, Phys. Lett. **170B** (1986) 1;

Reactions

$\overline{p}\,p \to \mu^-\,\mu^+$ X	540,630 GeV(E_{cm})
$\overline{p}\,p \to 2\mu^+$ X	540,630 GeV(E_{cm})
$\overline{p}\,p \to 2\mu^-$ X	540,630 GeV(E_{cm})
$\overline{p}\,p \to$ bottom $\overline{\text{bottom}}$ X	540,630 GeV(E_{cm})
$B^0 \to \overline{B}^0$	
$B_S \to \overline{B}_S$	
$B_S \to \mu^-$ X	
$\overline{B}_S \to \mu^+$ X	
$B^0 \to \mu^-$ X	
$\overline{B}^0 \to \mu^-$ X	

ANSARI 1987E

■ Firm establishment of properties of the W^+, W^-, and Z^0 bosons ■

Measurement of the Standard Model Parameters from a Study of W and Z Bosons

UA2 collaboration; R. Ansari et al.
Phys. Lett. **186B** (1987) 440;

Abstract A study has been made of the decays $W \to e\nu$ and $Z \to e^+e^-$, using the UA2 detector at the CERN $\bar{p}p$ Collider. The data correspond to an integrated luminosity of 142 nb^{-1} at a centre-of-mass collision energy $\sqrt{s} = 546$ GeV, and 768 nb^{-1} at $\sqrt{s} = 630$ GeV. Measurements of the standard model parameters from samples of 251 W decay and 39 Z decay candidates are compared with expectations of the standard electroweak model.

Accelerator CERN-PBAR/P $\bar{p}\ p$ collider

Detectors UA2

Related references

See also
UA2 Coll., M. Banner et al., Phys. Lett. **122B** (1983) 476;
UA2 Coll., P. Bagnaia et al., Phys. Lett. **129B** (1983) 130;
J. A. Appel et al., Z. Phys. **C30** (1986) 1;

Reactions

$\bar{p}\,p \to W^+$ X	546,630 GeV(E$_{cm}$)
$\bar{p}\,p \to W^-$ X	546,630 GeV(E$_{cm}$)
$\bar{p}\,p \to Z^0$ X	546,630 GeV(E$_{cm}$)
$W^+ \to e^+\ \nu_e$	
$W^- \to e^-\ \bar{\nu}_e$	
$Z^0 \to e^-\ e^+$	

Particles studied W^+, W^-, Z^0

ALBRECHT 1987P

■ First evidence for the B^0-$\overline{B}^0$ mixing ■

Observation of B^0-$\overline{B}^0$ Mixing

ARGUS collaborations; H. Albrecht et al.
Phys. Lett. **192B** (1987) 245;

Reprinted in
Heavy Flavours, San Miniato (1987) 145.
Hadrons, Quarks and Gluons, Les Arcs (1987) 51.
R. N. Cahn and G. Goldhaber, *The Experimental Foundations of Particle Physics*, Cambridge Univ. Press (1991) 369.

Abstract Using the ARGUS detector at the DORIS II storage ring we have searched in three different ways for B^0-$\overline{B}^0$ mixing in $\Upsilon(4S)$ decays. One explicitly mixed event, a decay $\Upsilon(4S) \to B^0 B^0$, has been completely reconstructed. Futhermore, we observe a 4.0 standard deviation signal of 24.8 events with like-sign lepton pairs and 3.0 standard deviation signal of 4.1 events containing one reconstructed B^0-$\overline{B}^0$ and an additional fast $l^+(l)^-$. This leads to the conclusion that B^0-$\overline{B}^0$ mixing is substantial. For the mixing parameter we obtain $r = 0.21 \pm 0.08$.

Accelerator DESY-DORIS-II

Detectors ARGUS

Related references

See also
M. Kobayashi and T. Maskawa, Progr. of Theor. Phys. **49** (1973) 652;
M. K. Gaillard and B. W. Lee, Phys. Rev. **D10** (1974) 897;
J. Ellis, M. K. Gaillard, D. V. Nanopoulos, and S. Rudaz, Nucl. Phys. **B131** (1977) 285;
A. Ali and Z. Aydin, Nucl. Phys. **B148** (1979) 165;
E. A. Paschos and U. Tuerke, Nucl. Phys. **B243** (1984) 29;
M. B. Gavela et al., Phys. Lett. **154B** (1985) 147;
A. A. Anselm et al., Phys. Lett. **156B** (1985) 102;
M. Gronau and J. Schechter, Phys. Rev. **D31** (1985) 1668;
G. Ecker and W. Grimus, Z. Phys. **C30** (1986) 293;
L. B. Okun', V. I. Zakharov, and B. Pontecorvo, Nuovo Cim. Lett. **13** (1986) 218;
A. Pais and S. B. Treiman, Phys. Rev. **D12** (1975) 2744;
I. I. Bigi and A. I. Sanda, Nucl. Phys. **B193** (1981) 123;
I. I. Bigi and A. I. Sanda, Phys. Rev. **D29** (1984) 1393;
I. I. Bigi and A. I. Sanda, Phys. Lett. **171B** (1986) 320;
CLEO Coll., A. Bean et al., Phys. Rev. Lett. **58** (1987) 183;
Mark II Coll., T. Schaad et al., Phys. Lett. **160B** (1985) 188;
ARGUS Coll., H. Albrecht et al., Phys. Lett. **134B** (1984) 137;
ARGUS Coll., H. Albrecht et al., Phys. Lett. **150B** (1985) 235;
ARGUS Coll., H. Albrecht et al., Phys. Lett. **185B** (1987) 218;
G. C. Fox and S. Wolfram, Phys. Lett. **82B** (1979) 134;
G. Altarelli, N. Cabibbo, and L. Maiani, Nucl. Phys. **B100** (1975) 313;
B. Anderson et al., Phys. Rept. **97** (1983) 31;
MARK III Coll., R. M. Baltrusaitis et al., Phys. Rev. Lett. **54** (1985) 1976;
A. J. Buras, W. Slonimski, and H. Steger, Nucl. Phys. **B245** (1984) 369;
Particle Data Group, *Review of Particle Properties*, M. Aguilar-Benitez, Phys. Lett. **170B** (1986) 1;

Reactions

$e^+\ e^- \to \Upsilon(4S)$
$\Upsilon(4S) \to B^0\ \overline{B}^0$
$\Upsilon(4S) \to B^+\ B^-$
$\Upsilon(4S) \to B^0\ e^+$ X
$\Upsilon(4S) \to B^0\ \mu^+$ X
$\Upsilon(4S) \to \overline{B}^0\ e^-$ X
$\Upsilon(4S) \to 2e^-$ X
$\Upsilon(4S) \to \mu^-\ e^-$ X
$\Upsilon(4S) \to 2\mu^-$ X
$\Upsilon(4S) \to 2e^+$ X
$\Upsilon(4S) \to \mu^+\ e^+$ X
$\Upsilon(4S) \to 2\mu^+$ X
$\Upsilon(4S) \to e^-\ e^+$ X
$\Upsilon(4S) \to \mu^-\ \mu^+$ X
$\Upsilon(4S) \to \mu^-\ e^+$ X
$\Upsilon(4S) \to \mu^+\ e^-$ X
$B^0 \to \overline{B}^0$
$D^0 \to K^-\ \pi^+$
$D^0 \to K^-\ 2\pi^+\ \pi^-$
$\overline{D}^0 \to K^+\ \pi^-$
$\overline{D}^0 \to K^+\ \pi^+\ 2\pi^-$
$B^0 \to D^*(2010)^-\ \pi^+$
$B^0 \to D^*(2010)^-\ \pi^+\ \pi^0$
$B^0 \to D^*(2010)^-\ 2\pi^+\ \pi^-$
$B^0 \to D^*(2010)^-\ e^+\ \nu_e$
$B^0 \to D^*(2010)^-\ \mu^+\ \nu_\mu$
$\overline{B}^0 \to D^*(2010)^+\ \pi^-$

$\overline{B}^0 \rightarrow D^*(2010)^+ \pi^0 \pi^-$
$\overline{B}^0 \rightarrow D^*(2010)^+ \pi^+ 2\pi^-$
$\overline{B}^0 \rightarrow D^*(2010)^+ e^- \overline{\nu}_e$
$\overline{B}^0 \rightarrow D^*(2010)^+ \mu^- \overline{\nu}_\mu$
$D^*(2010)^- \rightarrow \overline{D}^0 \pi^-$
$D^*(2010)^+ \rightarrow D^0 \pi^+$
$D^0 \rightarrow K^- \pi^+ \pi^0$
$D^0 \rightarrow \overline{K}^0 \pi^+ \pi^-$
$\overline{D}^0 \rightarrow K^+ \pi^0 \pi^-$
$\overline{D}^0 \rightarrow K^0 \pi^+ \pi^-$

Particles studied $B^0, \overline{B}^0$

HIRATA 1987C

■ First observation of the neutrino burst from supernova SN1987A ■

Observation of a Neutrino Burst from the Supernova $SN1987A$

K. Hirata et al.
Phys. Rev. Lett. **58** (1987) 1490;

Abstract A neutrino burst was observed in the Kamikande II detector on 23 February 87, 7:33:35 UT($\pm$ 1 min) during a time interval of 13 sec. The signal consisted of eleven electron events of energy 7.5 to 36 MeV, of which the first two point back to the Large Magellanic Cloud with angles $18° \pm 18°$ and $15° \pm 27°$.

Accelerator Local cosmic rays source SN1987A

Detectors KAMIOKANDE-II

Data comments Reported neutrino burst of Mont Blanc Neutrino Observatory not confirmed.

Related references

See also
S. A. Colgate and R. H. White, Astr. Jour. **143** (1966) 626;
S. E. Woosley, J. Wilson, and R. Mayle, Astr. Jour. **302** (1986) 19;
A. Burrows and J. M. Lattimer, Astr. Jour. **307** (1986) 178;
R. Mayle, *Ph.D. thesis, University of California*, Berkeley (1985);

Reactions

$\overline{\nu}_e\, p \rightarrow n\, e^+$	5.-50. MeV(P_{lab})
$\nu_e\, e^- \rightarrow e^-\, \nu_e$	5.-50. MeV(P_{lab})
ν_e	
$\overline{\nu}_e$	

Particles studied $\nu_e, \overline{\nu}_e$

BIONTA 1987C

■ First observation of the neutrino burst from supernova SN1987A. Birth of neutrino astronomy ■

Observation of a Neutrino Burst in Coincidence with Supernova 1987A in the Large Magellanic Cloud

R.M. Bionta et al.
Phys. Rev. Lett. **58** (1987) 1494;

Abstract A burst of eight neutrino events preceding the optical detection of the supernova in the Large Magellanic Cloud has been observed in a large underground water Cerenkov detector. The events span an interval of 6 s and have visible energies in the range 20-40 MeV.

Accelerator Local cosmic rays source SN1987A

Detectors Nucleon decay detector IMB

Data comments Reported neutrino burst of Mont Blanc Neutrino Observatory not confirmed.

Related references

See also
K. Hirata et al., Phys. Rev. Lett. **58** (1987) 1490;
S. A. Colgate and R. H. White, Astr. Jour. **143** (1966) 626;
J. R. Wilson et al., Ann. N.Y. Acad. Sci. **470** (1986) 267;

Reactions

ν_e	20.-50. MeV(P_{lab})

VAN DYCK 1987 Nobel prize

■ High precision measurements of electron and positron g−2 factors. High precision test of QED and CPT symmetry. Nobel prize to H. G. Dehmelt and W. Paul awarded in 1989 "for the development of the ion trap technique". Co-winner N. F. Ramsey "for the invention of the separated oscillatory fields method and its use in the hydrogen maser and other atomic clocks" ■

New High Precision Comparison of Electron and Positron g Factors

R.S. Van Dyck, P.B. Schwinberg, H.G. Dehmelt
Phys. Rev. Lett. **59** (1987) 26;

Abstract Single electrons and positrons have been alternately isolated in the same compensated Penning trap in order to form the geonium pseudoatom under nearly identical conditions. For each, the g-factor anomaly is obtained by measurement of both the spin-cyclotron difference frequency and the cyclotron frequency. A search for systematic effects uncovered a small (but common) residual shift due to the cyclotron excitation field. Extrapolation to zero power yields

e^+ and e^- g factors with a smaller statistical error and a new particle-antiparticle comparison: $g(e^-)/g(e^+) = 1 + (0.5 \pm 2.1) \times 10^{-12}$.

Accelerator Non accelerator experiment

Detectors Geonium "atom"

Related references

More (earlier) information
P. B. Schwinberg, R. S. van Dyck, and H. G. Dehmelt, Phys. Rev. Lett. **47** (1981) 1679;
P. B. Schwinberg, R. S. van Dyck, and H. G. Dehmelt, Phys. Lett. **81A** (1981) 119;
R. S. van Dyck, P. B. Schwinberg, and H. G. Dehmelt, Phys. Rev. **D34** (1986) 722;

Particles studied e^+, e^-

ASHMAN 1988B

■ Evidence for complex spin structure of the proton, "proton spin crisis" ■

A Measurement of the Spin Asymmetry and Determination of the Structure Function g_1 in Deep Inelastic Muon-Proton Scattering.

EMC Collaboration; J.G. Ashman et al.
Phys. Lett. **206B** (1988) 364;

Abstract The spin asymmetry in deep inelastic scattering of longitudinally polarized muons by longitudinally polarized protons has been measured over a large $(0.01 < x < 0.7)$. The spin-dependent structure function $g_1(x)$ for the proton has been determined and its integral over x found to be $0.114 \pm 0.012 \pm 0.026$, in disagreement with the Ellis-Jaffe sum rule. Assuming the validity of the Bjorken sum rule, this result implies a significant negative value for the integral of g_1 for the neutron. These values for the integrals of g_1 lead to the conclusion that the total quark spin constitutes a rather small fraction of the spin of the nucleon.

Accelerator CERN-SPS super proton synchrotron:

Detectors EMC

Related references

See also
J. D. Bjorken, Phys. Rev. **148** (1966) 1467;
J. Kuti and V. F. Weisskopf, Phys. Rev. **D4** (1971) 3418;
A. J. G. Hey and J. E. Mandula, Phys. Rev. **D5** (1972) 2610;
N. S. Craigie et al., Phys. Rept. **99** (1983) 69b;
V. W. Hughes and J. Kuti, Ann. Rev. Nucl. Part. Sci. **33** (1983) 611;
E. Gabathuler, *Proc. 6th Intern. Symp. on High Energy Spin Physics* (Marseille, 1984), ed. J. Soffer;
F. Sciulli, *Proc. Intern. Symp. on Lepton and Photon Interactions at High Energies* (Kyoto, 1985), eds. M. Konuma and K. Takahashi;
EM Coll., J. J. Aubert et al., Nucl. Phys. **B259** (1985) 189;
M. G. Doncel and E. de Rafael, Nuovo Cim. **4A** (1971) 363;
R. P. Feynman, *Photon-Hadron Interactions*, Benjamin, New York (1972) 166;
J. D. Bjorken, Phys. Rev. **D1** (1970) 1376;
J. Kodaira et al., Phys. Rev. **D20** (1979) 627;
J. Kodaira et al., Nucl. Phys. **B159** (1979) 99;
J. Kodaira et al., Nucl. Phys. **B165** (1983) 129;
J. Ellis and R. L. Jaffe, Phys. Rev. **D9** (1974) 1444;
J. Ellis and R. L. Jaffe, Phys. Rev. **D10** (1974) 1669(E);
M. Bourquin et al., Z. Phys. **C21** (1983) 27;
D. Bollini et al., Nuovo Cim. **63A** (1981) 441;
EM Coll., O. C. Alkofer et al., Nucl. Instr. and Meth. **179** (1981) 445;
G. R. Court and W. G. Heyes, Nucl. Instr. and Meth. **A243** (1986) 37;
L. W. Mo and Y. S. Tsai, Rev. of Mod. Phys. **41** (1965) 205;
Y. S. Tsai, preprint SLAC-PUB-848 (1971);
T. V. Kukhto and N. M. Shumeiko, Yad. Phys. **36** (1982) 707;
G. Altarelli and G. Martinelli, Phys. Lett. **B76** (1978) 89;
M. Glück and E. Reya, Nucl. Phys. **B145** (1978) 24;
M. J. Alguard et al., Phys. Rev. Lett. **37** (1976) 1261;
M. J. Alguard et al., Phys. Rev. Lett. **41** (1978) 70;
G. Baum et al., Phys. Rev. Lett. **51** (1983) 1135;
R. Carlitz and J. Kaur, Phys. Rev. Lett. **38** (1977) 673;
J. Kaur, Nucl. Phys. **B128** (1977) 219;
O. Darrigol and F. Hayot, Nucl. Phys. **B141** (1978) 391;
B. L. Ioffe, V. A. Khoze, and L. N. Lipatov, *Hard processes* North-Holland, Amsterdam, (1984) 61;
V. M. Belyaev, B. L. Ioffe, and Y. I. Kogan, Phys. Lett. **B151** (1985) 290;
F. E. Close and D. Sivers, Phys. Rev. Lett. **39** (1977) 1116;
R. L. Jaffe, Phys. Lett. **B193** (1987) 101;
S. L. Adler, Phys. Rev. **177** (1969) 2426;
J. S. Bell and R. Jackiw, Nuovo Cim. **60A** (1969) 47;
L. M. Sehgal, Phys. Rev. **D10** (1974) 1663;
P. G. Ratcliffe, Phys. Lett. **B192** (1987) 180;

Reactions

$\mu^+ p \to \mu^+ X$	100,120,200 GeV(P_{lab})

BAND 1988

■ Confirmation of B^0-$\overline{B}^0$ mixing ■

Additional Evidence for B^0-$\overline{B}^0$ Mixing

MAC collaboration; H. Band et al.
Phys. Lett. **200B** (1988) 221;

Abstract The rate of like-charge dimuons has been measured with the MAC detector in hadronic events produced in e^+e^- annihilation at $\sqrt{s} = 29$ GeV. If the observed excess is attributed to B^0-$\overline{B}^0$ mixing, the corresponding value of the mixing parameter $\chi = \Gamma(B \to \mu^- X)/\Gamma(B \to \mu^\pm X)$ is $\chi = 0.21^{+0.29}_{-0.15}$ and $\chi > 0.02$ at 90% CL.

Accelerator SLAC-PEP e^- e^+ 30 GeV ring

Detectors Magnetic calorimeter

Comments on experiment 310 pb^{-1} integrated luminosity.

Related references

See also
S. W. Herb et al., Phys. Rev. Lett. **39** (1977) 252;
J. Ellis, M. K. Gaillard, D. V. Nanopoulos, and S. Rudaz, Nucl. Phys. **B131** (1977) 285;
UA1 Coll., C. Albajar et al., Phys. Lett. **186B** (1987) 247;
ARGUS Coll., H. Albrecht et al., Phys. Lett. **192B** (1987) 245;
MAC Coll., E. Fernandez et al., Phys. Rev. **D31** (1985) 1537;

E. Fahri, Phys. Rev. Lett. **39** (1977) 1587;
C. Peterson et al., Phys. Rev. **D27** (1983) 105;
MAC Coll., E. Fernandez et al., Phys. Rev. Lett. **50** (1983) 2054;
A. Pais and S. B. Treiman, Phys. Rev. **D12** (1975) 2744;
Mark II Coll., T. Schaad et al., Phys. Lett. **160B** (1985) 188;
JADE Coll., W. Bartel et al., Phys. Lett. **146B** (1984) 437;
CLEO Coll., A. Bean et al., Phys. Rev. Lett. **58** (1987) 183;

Reactions

$e^+ e^- \to \mu^- \mu^+ X$ — 29 GeV(E_{cm})
$e^+ e^- \to \mu^+ X$ — 29 GeV(E_{cm})
$e^+ e^- \to \mu^- X$ — 29 GeV(E_{cm})
$e^+ e^- \to 2\mu^+ X$ — 29 GeV(E_{cm})
$e^+ e^- \to B^0 \overline{B}^0 X$ — 29 GeV(E_{cm})
$B^0 \to \overline{B}^0$
$B^0 \to \mu^- X$
$\overline{B}^0 \to \mu^+ X$
$\overline{B}^0 \to B^0$

Particles studied $B^0, \overline{B}^0$

AVERY 1989C

■ First evidence for the $\Xi_c(2460)^0$ and $\overline{\Xi}_c(2460)^0$ ■

Observation of the Charmed Strange Baryon $\Xi_c(2460)^0$

CLEO collaboration; P. Avery et al.
Phys. Rev. Lett. **62** (1989) 863;

Abstract We present evidence from the CLEO detector for the charmed strange baryon Ξ_c^0. It is seen in nonresonant e^+e^- annihilations at $\sqrt{s}$ of 10.5 GeV through its decay to $\Xi^-\pi^+$. The measured Ξ_c^0 mass is $2471 \pm 3 \pm 4$ MeV/c^2.

Accelerator Cornell $e^+ e^-$ storage ring

Detectors CLEO

Comments on experiment The sample consists of 101 pb^{-1} of integrated luminosity at energies just below $\Upsilon(4S)$ resonances, 212 pb^{-1} at the $\Upsilon(4S)$ and 117 pb^{-1} at the Υ(5S).

Related references

See also
S. F. Biagi et al., Phys. Lett. **122B** (1983) 455;
S. F. Biagi et al., Phys. Lett. **150B** (1985) 230;
S. F. Biagi et al., Z. Phys. **C28** (1985) 175;
P. Cotes et al., Phys. Rev. Lett. **59** (1987) 1530;
D. Andrews et al., Nucl. Instr. and Meth. **47** (1983) 211;
C. Bebek et al., Phys. Rev. **D36** (1987) 690;
S. Behrends et al., Phys. Rev. **D31** (1985) 2161;
D. Cassel et al., Nucl. Instr. and Meth. **A252** (1986) 325;
W. Kwong, J. L. Rosner, and C. Quigg, Ann. Rev. Nucl. Part. Sci. **37** (1987) 325;
Particle Data Group, *Review of Particle Properties*, G. P. Yost et al., Phys. Lett. **204B** (1988) 1;

Reactions

$e^+ e^- \to p \pi^+ 2\pi^- X$ — 10.5 GeV(E_{cm})
$e^+ e^- \to \overline{p} 2\pi^+ \pi^- X$ — 10.5 GeV(E_{cm})
$e^+ e^- \to \Lambda \pi^+ \pi^- X$ — 10.5 GeV(E_{cm})
$e^+ e^- \to \overline{\Lambda} \pi^+ \pi^- X$ — 10.5 GeV(E_{cm})
$e^+ e^- \to \Xi^- \pi^+ X$ — 10.5 GeV(E_{cm})
$e^+ e^- \to \overline{\Xi}^+ \pi^- X$ — 10.5 GeV(E_{cm})
$e^+ e^- \to \Xi_c(2460)^0 X$ — 10.5 GeV(E_{cm})
$e^+ e^- \to \overline{\Xi}_c(2460)^0 X$ — 10.5 GeV(E_{cm})
$\Lambda \to p \pi^-$
$\Xi^- \to \Lambda \pi^-$
$\overline{\Lambda} \to \overline{p} \pi^+$
$\Xi_c(2460)^0 \to \Xi^- \pi^+$
$\overline{\Xi}_c(2460)^0 \to \overline{\Xi}^+ \pi^-$
$\overline{\Xi}^+ \to \overline{\Lambda} \pi^+$

Particles studied $\Xi_c(2460)^0, \overline{\Xi}_c(2460)^0$

ARTUSO 1989C

■ Confirmation of B^0-$\overline{B}^0$ mixing ■

B^0 $\overline{B}^0$ Mixing at the $\Upsilon(4S)$

CLEO collaboration; M. Artuso et al.
Phys. Rev. Lett. **62** (1989) 2233;

Abstract We have measured B^0 $\overline{B}^0$ mixing by observing like-sign dilepton events in $\Upsilon(4S)$ decay. Assuming that the semileptonic branching fraction of the charged and neutral B mesons are equal and that the $\Upsilon(4S)$ decays to B^+B^- 55% of the time and to B^0 $\overline{B}^0$ 45% of the time, we measure the mixing parameter r to be $0.19 \pm 0.06 \pm 0.06$, where the first error is statistical and the second is systematic.

Accelerator Cornell $e^+ e^-$ storage ring

Detectors CLEO

Related references

See also
ARGUS Coll., H. Albrecht et al., Phys. Lett. (1985) 62B 395;
ARGUS Coll., H. Albrecht et al., Phys. Lett. **192B** (1987) 245;
CLEO Coll., A. Bean et al., Phys. Rev. Lett. **58** (1987) 183;
I. I. Bigi and A. I. Sanda, Phys. Rev. **D29** (1984) 1393;
K. Chadwick et al., Phys. Rev. **D27** (1983) 475;
D. Andrews et al., Nucl. Instr. and Meth. **211** (1983) 47;
G. Altarelli et al., Nucl. Phys. **B208** (1982) 365;
P. Haas et al., Phys. Rev. Lett. **55** (1985) 1248;
C. Bebek et al., Phys. Rev. **D36** (1987) 1289;
Particle Data Group, *Review of Particle Properties*, G. P. Yost et al., Phys. Lett. **204B** (1988) 1;

Reactions

$e^+ e^- \to \Upsilon(4S)$ — 10.58 GeV(E_{cm})
$\Upsilon(4S) \to B^0 \overline{B}^0$
$\Upsilon(4S) \to B^+ B^-$
$B \to \ell^+ \nu X$
$\overline{B} \to \ell^- \nu X$
$B^0 \to \overline{B}^0$

Particles studied $B^0, \overline{B}^0$

ABRAMS 1989B

■ First evidence that the number of light neutrinos = 3. MARK-II Collaboration ■

Measurements of Z^0-Boson Resonance Parameters in $e^+ e^-$ Annihilation

MARK-II Collaboration; G.S. Abrams et al.
Phys. Rev. Lett. **63** (1989) 2173;

Abstract We have measured the mass of the Z boson to be 91.14 ± 0.12 GeV/c^2, and its width to be $2.42^{+0.45}_{-0.35}$ GeV. If we constrain the visible width to its standard-model value, we find the partial width to invisible decay modes to be 0.46 ± 0.10 GeV, corresponding to 2.8 ± 0.6 neutrino species, with a 95% confidence-level upper limit of 3.9.

Accelerator SLAC-SLC linear e^- e^+ collider

Detectors Mark-II

Comments on experiment Total integrated luminosity of 19 nb−1.

Related references

More (earlier) information
G. S. Abrams et al., Phys. Rev. Lett. **63** (1989) 724;

Reactions

$e^+ e^- \to Z^0$	91.3 GeV(E_{cm})
$Z^0 \to \mu^- \mu^+$	
$Z^0 \to \tau^- \tau^+$	
$Z^0 \to$ 2hadron (hadrons)	

Particles studied Z^0, ν

ALAM 1989B

■ Confirmation of the $\Xi_c(2460)^+$, $\Xi_c(2460)^0$, $\overline{\Xi}_c(2460)^0$, and $\overline{\Xi}_c(2460)^-$ states ■

Measurement of the Isospin Mass Splitting between $\Xi_c(2460)^+$ and $\Xi_c(2460)^0$

CLEO collaboration; M.S. Alam et al.
Phys. Lett. **226B** (1989) 401;

Abstract Using the CLEO detector, we present a measurement of the mass of the charmed strange baryon Ξ_c^+, and its neutral counterpart the Ξ_c^0. The Ξ_c^+ is observed through its decay to $\Xi^- \pi^+ \pi^+$, and its mass is measured to be $(2467 \pm 3 \pm 4)$ MeV. The isospin mass splitting of the Ξ_c system is found to be $M(\Xi_c^+) - M(\Xi_c^0) = (-5 \pm 4 \pm 1)$ MeV.

Accelerator Cornell e^+ e^- storage ring

Detectors CLEO

Comments on experiment The sample consists of 102 pb^{-1} of integrated luminosity at energies just below $\Upsilon(4S)$ resonances, 212 pb^{-1} at the $\Upsilon(4S)$ and 114 pb^{-1} at the $\Upsilon(5S)$.

Related references

More (earlier) information
P. Avery et al., Phys. Rev. Lett. **62** (1989) 863;

Reactions

$e^+ e^- \to p\, 4\pi^+ \pi^-$ X	10.5-10.7 GeV(E_{cm})
$e^+ e^- \to \Lambda\, 3\pi^+$ X	10.5-10.7 GeV(E_{cm})
$e^+ e^- \to \Xi^-\, 2\pi^+$ X	10.5-10.7 GeV(E_{cm})
$e^+ e^- \to \overline{p}\, \pi^+ 4\pi^-$ X	10.5-10.7 GeV(E_{cm})
$e^+ e^- \to \overline{\Lambda}\, 3\pi^-$ X	10.5-10.7 GeV(E_{cm})
$e^+ e^- \to \overline{\Xi}^+\, 2\pi^-$ X	10.5-10.7 GeV(E_{cm})
$e^+ e^- \to \Xi_c(2460)^0$ X	10.5-10.7 GeV(E_{cm})
$e^+ e^- \to \overline{\Xi}_c(2460)^0$ X	10.5-10.7 GeV(E_{cm})
$e^+ e^- \to \Xi_c(2460)^+$ X	10.5-10.7 GeV(E_{cm})
$e^+ e^- \to \overline{\Xi}_c(2460)^-$ X	10.5-10.7 GeV(E_{cm})

Particles studied $\Xi_c(2460)^+, \overline{\Xi}_c(2460)^-, \Xi_c(2460)^0, \overline{\Xi}_c(2460)^0$

WEIR 1989B

■ Confirmation of B_S-$\overline{B}_S$ mixing ■

A Reanalysis of B^0-$\overline{B}^0$ Mixing in $e^+ e^-$ Annihilation at 29 GeV

MARK-II Collaboration; A.J. Weir et al.
Phys. Lett. **240B** (1990) 289;

Abstract Data taken by the Mark II detector at the PEP storage ring was used to measure the rate of dilepton production in multihadronic events produced by e^+e^- annihilation at $\sqrt{s} = 29$ GeV. We determine the probability that a hadron initially containing a $b(\overline{b})$ quark decays to a positive (negative) lepton to be $0.17^{+0.15}_{-0.08}$, with 90% confidence level limits of 0.06 and 0.38.

Accelerator SLAC-PEP e^- e^+ 30 GeV ring

Detectors Mark-II

Related references

More (earlier) information
Mark II Coll., T. Schaad et al., Phys. Lett. **160B** (1985) 188;

Reactions

$e^+ e^- \to B^0\, \overline{B}^0$ X	29 GeV(E_{cm})
$e^+ e^- \to B_S\, \overline{B}_S$ X	29 GeV(E_{cm})
$e^+ e^- \to \mu^+$ X	29 GeV(E_{cm})
$e^+ e^- \to \mu^-$ X	29 GeV(E_{cm})
$e^+ e^- \to 2\mu^+$ X	29 GeV(E_{cm})
$e^+ e^- \to e^+$ X	29 GeV(E_{cm})
$e^+ e^- \to e^-$ X	29 GeV(E_{cm})
$e^+ e^- \to 2\mu^-$ X	29 GeV(E_{cm})
$e^+ e^- \to \mu^+ e^+$ X	29 GeV(E_{cm})

$e^+ e^- \to \mu^- e^- X$	29 GeV(E_{cm})
$e^+ e^- \to 2e^+ X$	29 GeV(E_{cm})
$e^+ e^- \to 2e^- X$	29 GeV(E_{cm})
$\overline{B}^0 \to \mu^- X$	
$B^0 \to e^+ X$	
$\overline{B}^0 \to e^- X$	
$B^+ \to e^+ X$	
$B^- \to e^- X$	
$B^0 \to \mu^+ X$	
$B^+ \to \mu^+ X$	
$B^- \to \mu^- X$	
$B^0 \to \overline{B}^0$	
$B_S \to \overline{B}_S$	
$\overline{B}^0 \to B^0$	
$\overline{B}_S \to B_S$	

Particles studied $B^0, \overline{B}^0, B_S, \overline{B}_S$

ADEVA 1990R

■ Another confirmation of B_S-$\overline{B}_S$ mixing ■

A Measurement of B^0 $\overline{B}^0$ Mixing in Z^0 Decays

L3 collaboration; B. Adeva et al.
Phys. Lett. **252B** (1990) 703;

Abstract We have observed B^0 $\overline{B}^0$ mixing in $Z^0 \to b\overline{b}$ decays using hadronic events containing dileptons. The data sample corresponds to 118200 hadron events at $\sqrt{s} = M_Z$. From a fit to the dilepton p and $p_\perp$ spectra, we determine the mixing parameter to be $\chi_B = 0.178^{+0.049}_{-0.040}$.

Accelerator CERN-LEP $e^+ e^-$ collider

Detectors L3

Comments on experiment Total integrated luminosity 5.5 pb^{-1}.

Related references

More (earlier) information
L3 Collab., B. Adeva et al., Phys. Lett. **241B** (1990) 416;

Reactions

$e^+ e^- \to e^- e^+ X$	88.2-94.2 GeV(E_{cm})
$e^+ e^- \to \mu^- \mu^+ X$	88.2-94.2 GeV(E_{cm})
$e^+ e^- \to 2\mu^+ X$	88.2-94.2 GeV(E_{cm})
$e^+ e^- \to 2\mu^- X$	88.2-94.2 GeV(E_{cm})
$e^+ e^- \to \mu^+ e^+ X$	88.2-94.2 GeV(E_{cm})
$e^+ e^- \to \mu^- e^- X$	88.2-94.2 GeV(E_{cm})
$e^+ e^- \to 2e^+ X$	88.2-94.2 GeV(E_{cm})
$e^+ e^- \to 2e^- X$	88.2-94.2 GeV(E_{cm})
$e^+ e^- \to \mu^+ e^- X$	88.2-94.2 GeV(E_{cm})
$e^+ e^- \to \mu^- e^+ X$	88.2-94.2 GeV(E_{cm})
$e^+ e^- \to Z^0$	88.2-94.2 GeV(E_{cm})
$Z^0 \to B^0 \overline{B}^0$	
$Z^0 \to B_S \overline{B}_S$	
$B^0 \to \overline{B}^0$	
$B_S \to \overline{B}_S$	
$\overline{B}^0 \to B^0$	
$\overline{B}_S \to B_S$	
$B^0 \to e^+ X$	
$B^0 \to \mu^+ X$	
$B_S \to \mu^+ X$	
$B_S \to e^+ X$	

Particles studied $B^0, \overline{B}^0, B_S, \overline{B}_S$

DECAMP 1991D

■ Further confirmation of B_S-$\overline{B}_S$ mixing ■

Measurement of B-$\overline{B}$ Mixing at the Z^0

ALEPH collaboration; D. Decamp et al.
Phys. Lett. **258B** (1991) 236;

Abstract From more than 175000 hadronic Z^0 decays observed with the ALEPH detector at LEP, we select 823 events with pairs of leptons in the final state. From these we measure χ, the probability that a b hadron which is observed to decay originated as an anti-b hadron. We find $\chi = 0.132^{+0.027}_{-0.026}$.

Accelerator CERN-LEP $e^+ e^-$ collider

Detectors ALEPH

Related references

See also
ARGUS Coll., H. Albrecht et al., Phys. Lett. **192B** (1987) 245;
CLEO Coll., M. Artuso et al., Phys. Rev. Lett. **62** (1989) 2233;
ALEPH Coll., D. Decamp et al., Nucl. Instr. and Meth. **A294** (1990) 121;
ALEPH Coll., D. Decamp et al., Phys. Lett. **244B** (1990) 551;
JADE Coll., W. Bartel et al., Z. Phys. **C33** (1986) 23;
JADE Coll., W. Bartel et al., Z. Phys. **C33** (1987) 339;
JADE Coll., S. Bethke et al., Phys. Lett. **213B** (1988) 235;
MARK J Coll., B. Adeva et al., Phys. Rev. Lett. **51** (1983) 443;
CELLO Coll., H. J. Berend et al., Z. Phys. **C19** (1983) 291;
TASSO Coll., M. Althoff et al., Z. Phys. **C22** (1984) 219;
TASSO Coll., M. Althoff et al., Phys. Lett. **146B** (1984) 443;
TPC Coll., H. Aihara et al., Phys. Rev. **D31** (1985) 2719;
TPC Coll., H. Aihara et al., Z. Phys. **C27** (1985) 39;
DELCO Coll., T. Pal et al., Phys. Rev. **D33** (1986) 2708;
Particle Data Group, *Review of Particle Properties*, J. J. Hernandez et al., Phys. Lett. **239B** (1990) 1;
A. Pais and S. B. Treiman, Phys. Rev. **D12** (1975) 2744;
L. B. Okun', V. I. Zakharov, and B. Pontecorvo, Lett. Nuovo Cim. **15** (1975) 218;
J. Ellis, M. K. Gaillard, and D. V. Nanopoulos, Nucl. Phys. **B109** (1976) 213;
P. J. Franzini, Phys. Rept. **173** (1989) 1;
UA1 Coll., C. Albajar et al., Phys. Lett. **186B** (1987) 247;
UA1 Coll., C. Albajar et al., Phys. Lett. **187B** (1987) 565(E);
MAC Coll., H. R. Band et al., Phys. Lett. **200B** (1988) 221;
MAC Coll., E. Fernandez et al., Phys. Rev. Lett. **50** (1983) 2054;
Mark II Coll., A. J. Weir et al., Phys. Lett. **192B** (1987) 289;
Mark II Coll., R. A. Ong et al., Phys. Rev. Lett. **60** (1988) 2587;

Reactions

$e^+ e^- \to Z^0$	91.3 GeV(E_{cm})

$Z^0 \rightarrow 2e^+ X$
$Z^0 \rightarrow \mu^+ e^+ X$
$Z^0 \rightarrow 2\mu^+ X$
$Z^0 \rightarrow 2e^- X$
$Z^0 \rightarrow \mu^- e^- X$
$Z^0 \rightarrow 2\mu^- X$
$Z^0 \rightarrow e^- e^+ X$
$Z^0 \rightarrow \mu^- e^+ X$
$Z^0 \rightarrow \mu^+ e^- X$
$Z^0 \rightarrow \mu^- \mu^+ X$
$Z^0 \rightarrow$ 2jet
$Z^0 \rightarrow 2\ell^+ X$
$Z^0 \rightarrow 2\ell^- X$
$Z^0 \rightarrow \ell^+ \ell^- X$
$Z^0 \rightarrow \bar{b}\, b$
$Z^0 \rightarrow$ 2bottom
$Z^0 \rightarrow 2\overline{\text{bottom}}$
$B^0 \rightarrow \overline{B}^0$
$B_S \rightarrow \overline{B}_S$
$\overline{B}^0 \rightarrow B^0$
$\overline{B}_S \rightarrow B_S$
bottom $\rightarrow \overline{\text{bottom}}$
$\overline{\text{bottom}} \rightarrow$ bottom

ADEVA 1991D

■ Confirmation of the number of light neutrinos = 3. L3 Collaboration ■

Measurement of Electroweak Parameters from Hadronic and Leptonic Decays of the Z^0

L3 collaboration; B. Adeva et al.
Z. Phys. **C51** (1991) 179;

Abstract We have studied the reactions $e^+e^- \rightarrow$ hadrons, e^+e^-, $\mu^+\mu^-$ and $\tau^+\tau^-$, in the energy range $88.2 \leq \sqrt{s} \leq 94.2$ GeV. A total luminosity of 5.5 pb^{-1}, corresponding to approximately 115000 hadronic and 10000 leptonic Z^0 decays, has been recorded with the L3 detector. From a simultaneous fit to all of our measured cross section data, we obtain assuming lepton universality:

$$M_Z = 91.181 \pm 0.010 \pm 0.02(\text{LEP})\text{GeV},$$
$$\Gamma_Z = 2501 \pm 17\text{MeV},$$
$$\Gamma_{had} = 1742 \pm 19\text{MeV},$$
$$\Gamma_l = 83.6 \pm 0.8\text{MeV}.$$

If we do not assume lepton universality, we obtain for the partial decay widths of the Z^0 into e^+e^-, $\mu^+\mu^-$ and $\tau^+\tau^-$:

$$\Gamma_e = 83.3 \pm 1.1\text{MeV},$$
$$\Gamma_\mu = 84.5 \pm 2.0\text{MeV},$$
$$\Gamma_\tau = 84.0 \pm 2.7 MeV.$$

From the measured ratio of the invisible and the leptonic decay widths of the Z^0, we determine the number of light neutrino species to be $N_\nu = 3.05 \pm 0.10$. We include our measurements of the forward-backward asymmetry for the leptonic channels in a fit to determine the vector and axial-vector neutral current coupling constants of charged leptons to the Z^0. We obtain $\tilde{g}_V = -0.046^{+0.015}_{-0.012}$ and $\tilde{g}_A = -0.500 \pm 0.003$. In the framework of the Standard Model, we estimate the top quark mass to be $m_t = 193^{+52}_{-69} \pm 16$ (Higgs) GeV, and we derive a value for the weak mixing angle of $\sin^2\theta_W \equiv 1-(M_W/M_Z)^2 = 0.222 \pm 0.008$, corresponding to an effective weak mixing angle of $\sin^2\tilde{\theta}_W = 0.2315 \pm 0.0025$.

Accelerator CERN-LEP $e^+ e^-$ collider

Detectors L3

Related references

More (earlier) information
L3 Coll., B. Adeva et al., Phys. Lett. **231B** (1989) 509;
L3 Coll., B. Adeva et al., Phys. Lett. **236B** (1990) 109;
L3 Coll., B. Adeva et al., Phys. Lett. **237B** (1990) 136;
L3 Coll., B. Adeva et al., Phys. Lett. **238B** (1990) 122;
L3 Coll., B. Adeva et al., Phys. Lett. **247B** (1990) 473;
L3 Coll., B. Adeva et al., Phys. Lett. **248B** (1990) 464;
L3 Coll., B. Adeva et al., Phys. Lett. **249B** (1990) 341;
L3 Coll., B. Adeva et al., Phys. Lett. **250B** (1990) 183;
L3 Coll., B. Adeva et al., Phys. Lett. **250B** (1990) 199;
L3 Coll., B. Adeva et al., Phys. Lett. **252B** (1990) 713;
L3 Coll., B. Adeva et al., Phys. Lett. **257B** (1991) 469;
L3 Coll., B. Adeva et al., Nucl. Instr. and Meth. **A289** (1990) 35;
See also
S. L. Glashow, Nucl. Phys. **22** (1961) 579;
S. Weinberg, Phys. Rev. Lett. **19** (1967) 1264;
A. Salam, *Elementary particle theory*, ed. N. Svartholm, Stockholm: Almqvist and Wiksell (1968) 367;
ALEPH Coll., D. Decamp et al., Phys. Lett. **231B** (1989) 519;
ALEPH Coll., D. Decamp et al., Phys. Lett. **234B** (1990) 399;
ALEPH Coll., D. Decamp et al., Phys. Lett. **235B** (1990) 399;
ALEPH Coll., D. Decamp et al., Z. Phys. **C48** (1990) 365;
DELPHI Coll., P. Aarnio et al., Phys. Lett. **231B** (1989) 539;
DELPHI Coll., P. Aarnio et al., Phys. Lett. **241B** (1990) 425;
DELPHI Coll., P. Abreu et al., Phys. Lett. **241B** (1990) 435;
OPAL Coll., M. Z. Akrawy et al., Phys. Lett. **231B** (1989) 530;
OPAL Coll., M. Z. Akrawy et al., Phys. Lett. **235B** (1990) 379;
OPAL Coll., M. Z. Akrawy et al., Phys. Lett. **240B** (1990) 497;
OPAL Coll., M. Z. Akrawy et al., Phys. Lett. **247B** (1990) 458;
M. Böhm, A. Denner, and W. Hollik, Nucl. Phys. **B304** (1988) 687;
F. A. Berends, R. Kleiss, and W. Hollik, Nucl. Phys. **B304** (1988) 712;
S. Jadach, B. F. L. Ward, Phys. Rev. **D40** (1989) 3582;
F. A. Berends and R. Kleiss, Nucl. Phys. **B186** (1981) 22;
G. Marchesini and B. Webber, Nucl. Phys. **B310** (1988) 461;
F. A. Berends, P. H. Daverveldt, and R. Kleiss, Nucl. Phys. **B253** (1985) 441;
O. Adriani et al., Nucl. Instr. and Meth. **A302** (1991) 53;
D. Bardin et al., Nucl. Phys. **B351** (1991) 1;
D. Bardin et al., Z. Phys. **C44** (1989) 493;
D. Bardin et al., Phys. Lett. **255B** (1991) 290;
M. Greco, Phys. Lett. **177B** (1986) 97;
A. Borrelli et al., Nucl. Phys. **B333** (1990) 357;
K. G. Chetyrkin, A. L. Kataev, and F. V. Tkachov, Phys. Lett. **85B** (1979) 277;
M. Dine and J. Sapirsten, Phys. Rev. Lett. **43** (1979) 668;
W. Celmaster and R. J. Gonsalves, Phys. Rev. Lett. **44** (1980) 560;
W. Celmaster and R. J. Gonsalves, Phys. Rev. **D21** (1980) 3112;
CHARM Coll., J. Dorenbosch et al., Z. Phys. **C44** (1989) 567;
K. Abe et al., Phys. Rev. Lett. **62** (1989) 1709;
CHARM II Coll., D. Geiregat et al., Phys. Lett. **232B** (1989) 539;

F. Avignone et al., Phys. Rev. **D16** (1977) 2383;
U. Amaldi et al., Phys. Rev. **D36** (1987) 1385;
CDF Coll., F. Abe et al., Phys. Rev. Lett. **65** (1990) 2243;
UA2 Coll., J. Alitti et al., Phys. Lett. **241B** (1990) 150;
CHARM Coll., J. V. Allaby et al., Z. Phys. **C36** (1987) 611;
CDHSW Coll., H. Abramovicz et al., Phys. Rev. Lett. **57** (1986) 298;
CDHSW Coll., A. Blondel et al., Z. Phys. **C45** (1990) 361;
CCFR Coll., P. G. Reutens et al., Z. Phys. **C45** (1990) 539;
L. A. Ahrens et al., Phys. Rev. **D41** (1990) 3297;
Mark II Coll., G. S. Abrams et al., Phys. Rev. Lett. **63** (1989) 724;
Mark II Coll., G. S. Abrams et al., Phys. Rev. Lett. **63** (1989) 2173;

Reactions

$e^+ e^- \to$ 2hadron (hadrons)	88.2-94.2 GeV(E_{cm})
$e^+ e^- \to e^- e^+$	88.2-94.2 GeV(E_{cm})
$e^+ e^- \to \mu^- \mu^+$	88.2-94.2 GeV(E_{cm})
$e^+ e^- \to \tau^- \tau^+$	88.2-94.2 GeV(E_{cm})
$e^+ e^- \to Z^0$	88.2-94.2 GeV(E_{cm})
$Z^0 \to e^- e^+$	
$Z^0 \to \mu^- \mu^+$	
$Z^0 \to \tau^- \tau^+$	
$Z^0 \to$ 2hadron (hadrons)	
$Z^0 \to \nu \overline{\nu}$	
$Z^0 \to$ X	

Particles studied Z^0, ν

ALEXANDER 1991C

■ Confirmation of the number of light neutrinos = 3. OPAL Collaboration ■

Measurement of the Z^0 Line Shape Parameters and the Electroweak Couplings of Charged Leptons

OPAL collaboration; G. Alexander et al.
Z. Phys. **C52** (1991) 175;

Abstract We report on an improved measurement of the mass of the Z^0 boson, its total width and its partial decay width into hadrons and leptons, as well as the effective axial vector and vector couplings to charged leptons. These measurements are based on a data set of approximately 160000 hadronic Z^0 decays and 18000 decays into electrons, muons and taus, recorded by the OPAL experiment at centre of mass energies near the mass of Z^0. The total width and the partial width to visible final states, derived from the measured cross sections, are used to extract the invisible width. The effective couplings of the Z^0 to charged leptons are studied using measurements of the lepton pair cross sections and forward-backward asymmetries at the different centre of mass energy points of the Z^0 scan. The implications of our results in the context of the Standard Model are discussed.

Accelerator CERN-LEP $e^+ e^-$ collider

Detectors OPAL

Related references

More (earlier) information
OPAL Coll., M. Z. Akrawy et al., Phys. Lett. **240B** (1990) 497;
OPAL Coll., M. Z. Akrawy et al., Phys. Lett. **247B** (1990) 458;

Reactions

$e^+ e^- \to$ 2hadron (hadrons)	88.233-95.036 GeV(E_{cm})
$e^+ e^- \to e^- e^+$	88.233-95.036 GeV(E_{cm})
$e^+ e^- \to \mu^- \mu^+$	88.233-95.036 GeV(E_{cm})
$e^+ e^- \to \tau^- \tau^+$	88.233-95.036 GeV(E_{cm})
$e^+ e^- \to e^- e^+$ 2jet	88.233-95.036 GeV(E_{cm})
$e^+ e^- \to \mu^- \mu^+$ 2jet	88.233-95.036 GeV(E_{cm})
$Z^0 \to e^- e^+$	
$Z^0 \to \mu^- \mu^+$	
$Z^0 \to \tau^- \tau^+$	
$Z^0 \to$ 2hadron (hadrons)	

Particles studied Z^0, ν

ABREU 1991B

■ Confirmation of the number of light neutrinos = 3. DELPHI Collaboration ■

Determination of Z^0 Resonance Parameters and Couplings from its Hadronic and Leptonic Decays

DELPHI collaboration; P. Abreu et al.
Nucl. Phys. **B367** (1991) 511;

Abstract From measurements of the cross section for $e^+e^- \to$ hadrons and the cross sections and forward-backward charge asymmetries for $e^+e^- \to e^+e^-$, $\mu^+\mu^-$ and $\tau^+\tau^-$ at several centre-of-mass energies around the Z^0 pole with the DELPHI apparatus, using approximately 150000 hadronic and leptonic events from 1989 and 1990, one determines the following Z^0 parameters: the mass and the total width $M_Z = 91.177 \pm 0.022$ GeV, $\Gamma_Z = 2.465 \pm 0.020$ GeV, the hadronic and leptonic partial widths $\Gamma_h = 1.726 \pm 0.019$ GeV, $\Gamma_l = 83.4 \pm 0.8$ MeV, the invisible width $\Gamma_{inv} = 488 \pm 17$ MeV, the ratio of hadronic over leptonic partial widths $R_z = 20.70 \pm 0.29$ and the Born level hadronic peak cross section $\sigma_0 = 41.84 \pm 0.45$ nb. A flavour-independent measurement of the leptonic cross section gives very consistent results to those presented above ($\Gamma_l = 83.7 \pm 0.8$ MeV). From these results the number of light neutrinos species is determined to be $N_\nu = 2.94 \pm 0.10$. The individual leptonic width obtained is are: $\Gamma_e = 82.4 \pm 1.2$ MeV, $\Gamma_\mu = 86.9 \pm 2.1$ MeV and $\Gamma_\tau = 82.7 \pm 2.4$ MeV. Assuming universality, the squared vector and axial-vector couplings of the Z_0 to charged leptons are: $V_l^2 = 0.0003 \pm 0.0010$ and $A_l^2 = 0.2508 \pm 0.0027$. These values correspond to the electroweak parameters: $\rho_{eff} = 1.003 \pm 0.011$ and $\sin^2\theta_{W eff} = 0.241 \pm 0.009$. Within the Minimal Standard Model (MSM), the results can be expressed in terms of a single parameter: $\sin^2\theta_W{}^{\overline{MS}} = 0.2338 \pm 0.0027$. All these values are in good agreement with the predictions of the MSM. Fits yield

$43 < m_t < 215$ GeV at the 95% level. Finally, the measured values of Γ_Z and Γ_{inv} are used to derive lower mass bounds for possible new particles.

Accelerator CERN-LEP $e^+ e^-$ collider

Detectors DELPHI

Related references

More (earlier) information
DELPHI Coll., P. Aarnio et al., Phys. Lett. **241B** (1990) 425;
DELPHI Coll., P. Abreu et al., Phys. Lett. **241B** (1990) 435;
S. Jadach et al., Phys. Lett. **260B** (1991) 240;

Reactions

$e^+ e^- \to$ 2hadron (hadrons)	88.22-95.35 GeV(E_{cm})
$e^+ e^- \to \ell^+ \ell^-$	88.22-94.23 GeV(E_{cm})
$e^+ e^- \to e^- e^+$	88.22-94.22 GeV(E_{cm})
$e^+ e^- \to \mu^- \mu^+$	88.22-94.22 GeV(E_{cm})
$e^+ e^- \to \tau^- \tau^+$	88.22-94.22 GeV(E_{cm})
$Z^0 \to e^- e^+$	
$Z^0 \to \mu^- \mu^+$	
$Z^0 \to \tau^- \tau^+$	
$Z^0 \to$ 2hadron (hadrons)	
$Z^0 \to \ell^+ \ell^-$	
$Z^0 \to \nu \overline{\nu}$	

Particles studied Z^0, ν

ALBAJAR 1991C

■ Confirmation of B_S-$\overline{B}_S$ mixing ■

Measurement of B^0-$\overline{B}^0$ Mixing at the CERN $Sp\overline{p}S$ Collider

UA1 collaboration; C. Albajar et al.
Phys. Lett. **262B** (1991) 171;

Abstract We report on a new measurement of B_S-$\overline{B}_S$ mixing at the CERN $Sp\overline{p}S$ Collider. Mixing is measured in the non-isolated high mass dimuon sample using data from the 1988-1989 collider runs. The measured value of the mixing parameter, χ, is 0.145 ± 0.135 (*stat.*) ± 0.014 (*syst.*). The average of this measurement and that from our 1984-1985 data is $\chi = 0.148 \pm 0.029$ (*stat.*) ± 0.017 (*syst.*) assuming fully correlated systematic errors. Using the measurements of χ_d from ARGUS and CLEO, we obtain $\chi_s = 0.50 \pm 0.20$, which gives a limit of $\chi_s > 0.17(0.12)$ at 90%(95%) CL. Including the measurement of χ from the ALEPH and L3 experiments gives $\chi_s = 0.53 \pm 0.15$, and a limit of $\chi_s > 0.27(0.23)$ at 90%(95%) CL.

Accelerator CERN-PBAR/P $\overline{p}\, p$ collider

Detectors UA1

Comments on experiment Total integrated luminosity 4.7 pb^{-1}.

Related references

See also
UA1 Coll., C. Albajar et al., Phys. Lett. **186B** (1987) 237;
UA1 Coll., C. Albajar et al., Phys. Lett. **186B** (1987) 247;
UA1 Coll., C. Albajar et al., Z. Phys. **C48** (1990) 1;
UA1 Coll., C. Albajar et al., Phys. Lett. **256B** (1991) 121;
ARGUS Coll., H. Albrecht et al., Phys. Lett. **192B** (1987) 245;
CLEO Coll., A. Bean et al., Phys. Rev. Lett. **58** (1987) 183;
P. J. Franzini, Phys. Rev. **173** (1989) 1;
Mark II Coll. R. A. Ong et al., Phys. Rev. Lett. **60** (1988) 2587;
UA2 Coll., M. Banner et al., Phys. Lett. **122B** (1983) 322;
ABCDHW Coll., A. Breakstone et al., Phys. Lett. **135B** (1984) 510;
UA5 Coll., G. J. Alner et al., Nucl. Phys. **B258** (1985) 505;
K. Kleinknecht and B. Renk, Z. Phys. **C34** (1987) 209;

Reactions

$\overline{p}\, p \to \mu^- \mu^+ X$	630 GeV(E_{cm})
$\overline{p}\, p \to 2\mu^+ X$	630 GeV(E_{cm})
$\overline{p}\, p \to 2\mu^- X$	630 GeV(E_{cm})
$\overline{p}\, p \to$ bottom $\overline{\text{bottom}}$ X	630 GeV(E_{cm})
$B^0 \to \overline{B}^0$	
$B_S \to \overline{B}_S$	
$B_S \to \mu^- X$	
$\overline{B}_S \to \mu^+ X$	
$B^0 \to \mu^- X$	
$\overline{B}^0 \to \mu^+ X$	

ALBAJAR 1991D

■ Confirmation of the beauty baryon Λ_b ■

First Observation of the Beauty Baryon Λ_b in the Decay Channel $\Lambda_b \to J/\psi(1S)\ \Lambda$ at the CERN Proton-Antiproton Collider

UA1 collaboration; C. Albajar et al.
Phys. Lett. **273B** (1991) 540;

Abstract We report on the first observation of the beauty baryon Λ_b in an exclusive decay channel at the CERN $p\overline{p}$ collider. Using 4.7 pb^{-1} of muon data collected in the 1988/89 collider runs we reconstruct 16 ± 5 Λ_b's in the decay mode $\Lambda_b \to J/\psi\Lambda$ above a background of 9 ± 1 events, corresponding to a significance of about five standard deviations. We measure the Λ_b mass to be $m_{\Lambda_b} = 5640 \pm 50 \pm 30$ MeV/c^2. Using the beauty cross section measured by UA1 we deduce for the product of the production fraction and branching ratio $f_{\Lambda_b} Br(\Lambda_b \to J/\psi\Lambda) = (1.8 \pm 1.0) \times 10^{-3}$. Our sample contains a three-muon event in which the beauty particle opposite to $\overline{\Lambda}_b$ is tagged by the third muon. We also observe an indication of a signal in the decay channel $B^0 \to J/\psi K^{*0}$ with a significance of three standard deviations.

Accelerator CERN-PBAR/P $\overline{p}\, p$ collider

Detectors UA1

Related references

Analyse information from

M. Basile et al., Lett. Nuovo Cim. **31** (1981) 97;
D. Drijard et al., Phys. Lett. **108B** (1982) 361;
M. W. Arenton et al., Nucl. Phys. **B274** (1986) 707;

Reactions
$\overline{p}\,p \rightarrow \Lambda_b\, B^+\, X$ 630. GeV(E_{cm})
$\overline{p}\,p \rightarrow \Lambda_b\, B^0\, X$ 630. GeV(E_{cm})
$\Lambda_b \rightarrow \Lambda\, J/\psi(1S)$
$B^+ \rightarrow K^+\, J/\psi(1S)$
$B^+ \rightarrow K^*(892)^+\, J/\psi(1S)$
$B^0 \rightarrow K^0\, J/\psi(1S)$
$B^0 \rightarrow K^*(892)^0\, J/\psi(1S)$
$B^0 \rightarrow J/\psi(1S)\, \phi$
$\Lambda \rightarrow p\, \pi^-$
$K^*(892)^0 \rightarrow K^+\, \pi^-$
$J/\psi(1S) \rightarrow \mu^-\, \mu^+$

Particles studied Λ_b

JUNG 1992

■ First direct observation of the β decay into a bound electron state ■

First Observation of Bound-State β^- Decay

S. Jung et al.
Phys. Rev. Lett. **69** (1992) 2164;

Abstract Bound-state β^- decay was observed for the first time by storing bare $^{163}_{66}Dy^{66+}$ ions in a heavy-ion storage ring. From the number of $^{163}_{67}Ho^{66+}$ daughter ions, measured as a function of the storage time, a half-life of 47^{+5}_{-4} d was derived. By comparing this result with reported half-lives for electron capture (EC) from the M_1 and M_2 shells of neutral $^{163}_{67}Ho$, bounds for both the Q_{EC} value of neutral $^{163}_{67}Ho$ and for the electron-neutrino mass were set.

Accelerator Darmstadt heavy ion facility

Detectors Nonelectronic detectors

Related references

See also
R. Daudell, M. Jean, and M. Lecoin, J. Phys. Radium **8** (1947) 238;
J. N. Bahcall, Phys. Rev. **124** (1961) 495;

Analyse information from
S. Yasumi et al., Phys. Lett. **181B** (1986) 169;

Reactions
$\text{ion}(^{163}Dy)^+ \rightarrow \text{ion}(^{163}Ho)^+\, \overline{\nu}_e$

Particles studied $\nu_e, \overline{\nu}_e$

DECAMP 1991N

■ Confirmation of the number of light neutrinos = 3. ALEPH Collaboration ■

Improved Measurements of Electroweak Parameters from Z^0 Decays into Fermion Pairs

ALEPH collaboration; D. Decamp et al.
Z. Phys. **C53** (1992) 1;

Abstract The properties of the Z resonance are measured on the basis of 190000 Z decays into fermion pairs collected with the ALEPH detector at LEP. Assuming lepton universality, $M_Z = (91.182 \pm 0.009_{exp} \pm 0.020_{LEP})$ GeV, $\Gamma_Z = (2484 \pm 17)$ MeV, $\sigma^0_{had} = (41.44 \pm 0.36)$ nb, and $\Gamma_{had}/\Gamma_{ll} = 21.00 \pm 0.20$. The corresponding number of light neutrino species is 2.97 ± 0.07. The forward-backward asymmetry in leptonic decays is used to determine the ratio of vector to axial-vector coupling constants of leptons: $g^2_V(M^2_Z)/g^2_A(M^2_Z) = 0.0072 \pm 0.0027$. Combining these results with ALEPH results on quark charge and $b\overline{b}$ asymmetries, and τ polarization, $\sin^2\Theta_W(M^2_Z) = 0.2312 \pm 0.0018$. In the context of the Minimal Standard Model, limits are placed on the top-quark mass.

Accelerator CERN-LEP $e^+\, e^-$ collider

Detectors ALEPH

Data comments Improved measurements of the parameters of the Z resonance based on 8 pb^{-1} of data collected during 1989 and 1990 are presented.The data include 165000 hadronic and 25000 leptonic Z decays.

Related references

More (earlier) information
ALEPH Coll., D. Decamp et al., Z. Phys. **C48** (1990) 365;

Reactions
$e^+\, e^- \rightarrow$ 2hadron (hadrons) 88.22-94.28 GeV(E_{cm})
$e^+\, e^- \rightarrow e^-\, e^+$ 88.22-94.28 GeV(E_{cm})
$e^+\, e^- \rightarrow \mu^-\, \mu^+$ 88.22-94.28 GeV(E_{cm})
$e^+\, e^- \rightarrow \tau^-\, \tau^+$ 88.22-94.28 GeV(E_{cm})
$e^+\, e^- \rightarrow \ell^+\, \ell^-$ 88.22-94.28 GeV(E_{cm})
$Z^0 \rightarrow e^-\, e^+$
$Z^0 \rightarrow \mu^-\, \mu^+$
$Z^0 \rightarrow \tau^-\, \tau^+$
$Z^0 \rightarrow$ 2hadron (hadrons)
$Z^0 \rightarrow \ell^+\, \ell^-$
$Z^0 \rightarrow \nu\, \overline{\nu}$

Particles studied Z^0, ν

ACTON 1992B

■ Confirmation of B_S-$\overline{B}_S$ mixing ■

Measurement of B_S-$\overline{B}_S$ in Hadronic Z^0 Decays

OPAL collaboration; D.P. Acton et al.
Phys. Lett. **276B** (1992) 379;

Reprinted in
High Energy Physics, Dallas **v.1** (1992) 453.

Abstract From a sample of approximately 135 000 hadronic Z^0 decays recorded with the OPAL detector, 1 536 events were selected with two lepton candidates, either electrons or muons. A signal for B_S-$\overline{B}_S$ mixing was observed using the sign of the lepton charge to tag the charge of the b-quark in decaying b-flavoured hadrons. A flavour discrimination variable was constructed from the lepton momentum and its component perpendicular to the jet axis. By fitting the fraction of events in which the two lepton charges are of the same sign, as a function of this variable, the average mixing parameter was measured to be $\chi = 0.145^{+0.041}_{-0.035} \pm 0.018$, where the first error is statistical and the second is systematic.

Accelerator CERN-LEP e^+ e^- collider

Detectors OPAL

Reactions

$e^+ e^- \to Z^0$ 91.3 GeV(E_{cm})
$Z^0 \to \overline{b}\, b$
$Z^0 \to$ bottom $\overline{\text{bottom}}$
$Z^0 \to B^0\, \overline{B}^0$ X
$Z^0 \to B_S\, \overline{B}_S$ X
$Z^0 \to 2B^0$ X
$Z^0 \to 2\overline{B}^0$ X
$Z^0 \to 2B_S$ X
$Z^0 \to 2\overline{B}_S$ X
$Z^0 \to B^0\, B_S$ X
$Z^0 \to \overline{B}^0\, \overline{B}_S$ X
$Z^0 \to 2e^+$ X
$Z^0 \to \mu^+\, e^+$ X
$Z^0 \to 2\mu^+$ X
$Z^0 \to 2e^-$ X
$Z^0 \to \mu^-\, e^-$ X
$Z^0 \to 2\mu^-$ X
$Z^0 \to e^-\, e^+$ X
$Z^0 \to \mu^-\, e^+$ X
$Z^0 \to \mu^+\, e^-$ X
$Z^0 \to \mu^-\, \mu^+$ X
$B^0 \to \overline{B}^0$
$B_S \to \overline{B}_S$
$\overline{B}^0 \to B^0$
$\overline{B}_S \to B_S$

DECAMP 1991S

■ Precise determination of the Z^0 parameters. Confirmation of the number of light neutrinos = 3 ■

Electroweak Parameters of the Z^0 Resonance and the Standard Model

ALEPH collaboration; LEP Collaborations; DELPHI collaboration; L3 collaboration; OPAL collaboration; D. Decamp et al.
Phys. Lett. **276B** (1992) 247;

Abstract The four LEP experiments have each performed precision measurements of Z parameters. A method is described for combining the results of the four experiments, which takes into account the experimental and theoretical systematic errors and their correlations. We apply this method to the 1989 and 1990 LEP data, corresponding to approximately 650,000 Z decays into hadrons and charged leptons, to obtain precision values for the Z parameters. We use these results to test the Standard Model and to constrain its parameters.

Accelerator CERN-LEP e^+ e^- collider

Detectors ALEPH

Related references

Analyse information from
D. Decamp et al., Z. Phys. **C53** (1992) 1;

Reactions

$e^+ e^- \to e^-\, e^+$ 88.2-94.2 GeV(E_{cm})
$e^+ e^- \to \mu^-\, \mu^+$ 88.2-94.2 GeV(E_{cm})
$e^+ e^- \to \tau^-\, \tau^+$ 88.2-94.2 GeV(E_{cm})
$e^+ e^- \to \ell^+\, \ell^-$ 88.2-94.2 GeV(E_{cm})
$e^+ e^- \to$ 2hadron (hadrons) 88.2-94.2 GeV(E_{cm})
$e^+ e^- \to Z^0$ 88.2-94.2 GeV(E_{cm})
$Z^0 \to e^-\, e^+$
$Z^0 \to \mu^-\, \mu^+$
$Z^0 \to \tau^-\, \tau^+$
$Z^0 \to$ 2hadron (hadrons)
$Z^0 \to \ell^+\, \ell^-$
$Z^0 \to \nu\, \overline{\nu}$

Particles studied Z^0, ν

Accelerator CERN-LEP e^+ e^- collider

Detectors DELPHI

Related references

Analyse information from
P. Abreu et al., Nucl. Phys. **B367** (1991) 511;

Reactions

$e^+ e^- \to e^-\, e^+$ 88.2-94.2 GeV(E_{cm})
$e^+ e^- \to \mu^-\, \mu^+$ 88.2-94.2 GeV(E_{cm})
$e^+ e^- \to \tau^-\, \tau^+$ 88.2-94.2 GeV(E_{cm})

$e^+ e^- \to \ell^+ \ell^-$	88.2-94.2 GeV(E_{cm})
$e^+ e^- \to$ 2hadron (hadrons)	88.2-94.2 GeV(E_{cm})
$e^+ e^- \to Z^0$	88.2-94.2 GeV(E_{cm})
$Z^0 \to e^- e^+$	
$Z^0 \to \mu^- \mu^+$	
$Z^0 \to \tau^- \tau^+$	
$Z^0 \to$ 2hadron (hadrons)	
$Z^0 \to \ell^+ \ell^-$	
$Z^0 \to \nu \overline{\nu}$	

Particles studied Z^0, ν

Accelerator CERN-LEP $e^+ e^-$ collider

Detectors L3

Related references

Analyse information from
B. Adeva et al., Z. Phys. **C51** (1991) 179;

Reactions

$e^+ e^- \to e^- e^+$	88.2-94.2 GeV(E_{cm})
$e^+ e^- \to \mu^- \mu^+$	88.2-94.2 GeV(E_{cm})
$e^+ e^- \to \tau^- \tau^+$	88.2-94.2 GeV(E_{cm})
$e^+ e^- \to \ell^+ \ell^-$	88.2-94.2 GeV(E_{cm})
$e^+ e^- \to$ 2hadron (hadrons)	88.2-94.2 GeV(E_{cm})
$e^+ e^- \to Z^0$	88.2-94.2 GeV(E_{cm})
$Z^0 \to e^- e^+$	
$Z^0 \to \mu^- \mu^+$	
$Z^0 \to \tau^- \tau^+$	
$Z^0 \to$ 2hadron (hadrons)	
$Z^0 \to \ell^+ \ell^-$	
$Z^0 \to \nu \overline{\nu}$	

Particles studied Z^0, ν

Accelerator CERN-LEP $e^+ e^-$ collider

Detectors OPAL

Related references

Analyse information from
G. Alexander et al., Z. Phys. **C52** (1991) 175;

Reactions

$e^+ e^- \to e^- e^+$	88.2-94.2 GeV(E_{cm})
$e^+ e^- \to \mu^- \mu^+$	88.2-94.2 GeV(E_{cm})
$e^+ e^- \to \tau^- \tau^+$	88.2-94.2 GeV(E_{cm})
$e^+ e^- \to \ell^+ \ell^-$	88.2-94.2 GeV(E_{cm})
$e^+ e^- \to$ 2hadron (hadrons)	88.2-94.2 GeV(E_{cm})
$e^+ e^- \to Z^0$	88.2-94.2 GeV(E_{cm})
$Z^0 \to e^- e^+$	
$Z^0 \to \mu^- \mu^+$	
$Z^0 \to \tau^- \tau^+$	
$Z^0 \to$ 2hadron (hadrons)	
$Z^0 \to \ell^+ \ell^-$	
$Z^0 \to \nu \overline{\nu}$	

Particles studied Z^0, ν

BUSKULIC 1993H

■ First direct and precise measurement of the B_S meson mass ■

First Measurement of the B_S Meson Mass

ALEPH collaboration; D. Buskulic et al.
Phys. Lett. **311B** (1993) 425; Phys. Lett. **316B** (1993) 631;

Abstract In a sample of about 1.1 million hadronic Z decays recorded with the ALEPH detector during the 1990-1992 running of LEP, two unambiguous B_S meson candidates were observed. From these events the mass of the B_S meson has been measured to be 5.3686 ± 0.0056 (stat.)±0.0015 (syst.) GeV.

Accelerator CERN-LEP $e^+ e^-$ collider

Detectors ALEPH

Reactions

$e^+ e^- \to B_S$ X	91.2 GeV(E_{cm})
$e^+ e^- \to \overline{B}_S$ X	91.2 GeV(E_{cm})
$B_S \to \pi^+ \pi^- \mu^- \mu^+$ neutral	
$\overline{B}_S \to K^+ K^- \pi^+ \pi^-$	
$K_S \to \pi^+ \pi^-$	
$\phi \to K_S K_L$	
$B_S \to \psi(2S) \phi$	
$\psi(2S) \to \mu^- \mu^+$	
$\phi \to K^+ K^-$	
$\overline{B}_S \to D_S^+ \pi^-$	
$D_S^+ \to \phi \pi^+$	

Particles studied $B_S, \overline{B}_S$

ABE 1994ZE

■ First direct evidence of top quark production ■

Evidence for Top Quark Production in $\overline{p}\, p$ Collisions at $\sqrt{s} = 1.8$ TeV

CDF collaboration; F. Abe et al.
Phys. Rev. **D50** (1994) 2966;

Abstract We present the result of a search for the top quark in 19.3 pb^{-1} of $\overline{p}\, p$ collisions at $\sqrt{s} = 1.8$ TeV. The data were collected at the Fermilab Tevatron collider using the Collider Detector at Fermilab (CDF). The search includes standard model $t\bar{t}$ decays to final states $ee\nu\overline{\nu}$, $e\mu\nu\overline{\nu}$, and $\mu\mu\nu\overline{\nu}$ as well as $e\,\nu$ jets or $\mu\,\nu$ jets. In the (e,, $\mu\,\nu$ jets channel we search for b quarks from t decays via secondary vertex identification and via semileptonic decays of the b and cascade c quarks. In the dilepton final states we find two events with a background of $0.56^{+0.25}_{-0.13}$ events. In the $e, \mu\,\nu$ jets channel with a b identified via a secondary vertex, we find six events with a background of 2.3 ± 0.3. With a b identified via a semileptonic decay, we find seven events with a background of 3.1 ± 0.3. The secondary vertex and semileptonic decay samples have three events in common. The probability that the observed yield is consistent with the background is estimated to be 0.26%. The statistics are too limited to firmly establish the existence of the top quark; however, a natural interpretation of the excess is that it is due to $t\bar{t}$ production. We present several cross-checks. Some support this hypothesis; other do not. Under the assumption that the excess yield over background is due to $t\bar{t}$, constrained fitting on a subset of the events yields a mass of $174 \pm 10^{+13}_{-12}$ GeV/c^2 for the top quark. The $t\bar{t}$ cross-section, using this top quark mass to compute the acceptance, is measured to be $13.9^{+6.1}_{-4.8}$ pb.

Accelerator FNAL-COLLIDER

Detectors CDF

Related references

More (later) information
F. Abe et al., Phys. Rev. Lett. **73** (1994) 225;

Reactions

$\bar{p}\,p \to \mu^+\, e^-$ 3jet X	1.8 TeV(E_{cm})
$\bar{p}\,p \to \mu^-\, e^+$ 3jet X	1.8 TeV(E_{cm})
$\bar{p}\,p \to e^+$ 4jet X	1.8 TeV(E_{cm})
$\bar{p}\,p \to e^-$ 4jet X	1.8 TeV(E_{cm})
$\bar{p}\,p \to \mu^-$ 4jet X	1.8 TeV(E_{cm})
$\bar{p}\,p \to \mu^+$ 4jet X	1.8 TeV(E_{cm})
$\bar{p}\,p \to \bar{t}\,t$ X	1.8 TeV(E_{cm})
$W^+ \to e^+\,\nu_e$	
$W^- \to e^-\,\bar{\nu}_e$	
$W^+ \to \mu^+\,\nu_\mu$	
$W^- \to \mu^-\,\bar{\nu}_\mu$	
$t \to W^+\,b$	
$\bar{t} \to W^-\,\bar{b}$	
$W^+ \to$ 2jet	
$W^- \to$ 2jet	
$b \to$ jet	
$\bar{b} \to$ jet	

Particles studied $t, \bar{t}$

ABE 1995

■ Observation of the top quark. CDF Collaboration ■

Observation of Top Quark Production in $\bar{p}\,p$ Collisions

CDF collaboration; F. Abe et al.
Phys. Rev. Lett. **74** (1995) 2626;

Abstract We establish the existence of the top quark using a 67 pb^{-1} data sample of $\bar{p}p$ collisions at $\sqrt{s} = 1.8$ TeV collected with the Collider Detector at Fermilab (CDF). Employing techniques similar to those we previously published, we observe a signal consistent with $t\bar{t}$ decay to $WWb\bar{b}$, but inconsistent with the background prediction by 4.8σ. Additional evidence for the top quark is provided by a peak in the reconstructed mass distribution. We measure the top quark mass to be 176 ± 8(stat) ± 10(syst) GeV/c^2, and the $t\bar{t}$ production cross section to be $6.8^{+3.6}_{-2.4}$ pb.

Accelerator FNAL-COLLIDER

Detectors CDF

Related references

More (earlier) information
F. Abe et al., Phys. Rev. **D50** (1994) 2966;
F. Abe et al., Phys. Rev. Lett. **73** (1994) 225;

See also
F. Abe et al., preprint HEP-EX-9412009 (1994);
S. Abachi et al., Phys. Rev. Lett. **72** (1994) 2138;

Reactions

$\bar{p}\,p \to \mu^+\, e^-$ 3jet X	1.8 TeV(E_{cm})
$\bar{p}\,p \to \mu^-\, e^+$ 3jet X	1.8 TeV(E_{cm})
$\bar{p}\,p \to e^+$ 4jet X	1.8 TeV(E_{cm})
$\bar{p}\,p \to e^-$ 4jet X	1.8 TeV(E_{cm})
$\bar{p}\,p \to \mu^-$ 4jet X	1.8 TeV(E_{cm})
$\bar{p}\,p \to \mu^+$ 4jet X	1.8 TeV(E_{cm})
$\bar{p}\,p \to \mu^+\, e^-$ 2jet X	1.8 TeV(E_{cm})
$\bar{p}\,p \to \mu^-\, e^+$ 2jet X	1.8 TeV(E_{cm})
$\bar{p}\,p \to \mu^-\, \mu^+$ 2jet X	1.8 TeV(E_{cm})
$\bar{p}\,p \to e^-\, e^+$ 2jet X	1.8 TeV(E_{cm})
$\bar{p}\,p \to \mu^+\, e^+$ 3jet X	1.8 TeV(E_{cm})
$\bar{p}\,p \to 2\mu^+$ 3jet X	1.8 TeV(E_{cm})
$\bar{p}\,p \to \mu^-\, e^-$ 3jet X	1.8 TeV(E_{cm})
$\bar{p}\,p \to \mu^-\, \mu^+$ 3jet X	1.8 TeV(E_{cm})
$\bar{p}\,p \to \bar{t}\,t$ X	1.8 TeV(E_{cm})
$W^+ \to e^+\,\nu_e$	
$W^- \to e^-\,\bar{\nu}_e$	
$W^+ \to \mu^+\,\nu_\mu$	
$W^- \to \mu^-\,\bar{\nu}_\mu$	
$t \to W^+\,b$	
$\bar{t} \to W^-\,\bar{b}$	
$W^+ \to$ 2jet	
$W^- \to$ 2jet	
$b \to$ jet	
$\bar{b} \to$ jet	
$b \to \mu^-$ X	
$\bar{b} \to \mu^+$ X	

Particles studied $t, \bar{t}$

ABACHI 1995G

■ Observation of the top quark. D0 Collaboration ■

Observation of the Top Quark

D0 collaboration; S. Abachi et al.
Phys. Rev. Lett. **74** (1995) 2632;

Abstract The D0 Collaboration reports on a search for the standard model top quark in $p\,\overline{p}$ collisions at $\sqrt{s} = 1.8$ TeV at the Fermilab Tevatron with an integrated luminosity of approximately 50 pb^{-1}. We have searched for $t\overline{t}$ production in the dilepton and single-lepton decay channels with and without tagging of b-quark jets. We observed 17 events with an expected background of 3.8 ± 0.6 events. The probability for an upward fluctuation of the background to produce the observed signal is 2×10^{-6} (equivalent to 4.6 standard deviations). The kinematic properties of the excess events are consistent with top quark decay. We conclude that we have observed the top quark and measured its mass to be $199^{+19}_{-21}(stat) \pm 22(syst)$ GeV/c^2 and its cross section to be 6.4 ± 2.2 pb.

Accelerator FNAL-COLLIDER

Detectors D0

Related references

More (earlier) information
S. Abachi et al., Phys. Rev. Lett. **72** (1994) 2138;

See also
F. Abe et al., Phys. Rev. **D50** (1994) 2966;
F. Abe et al., Phys. Rev. Lett. **73** (1994) 225;

Reactions

$\overline{p}\,p \to \mu^+\, e^-$ 3jet X	1.8 TeV($\mathrm{E_{cm}}$)
$\overline{p}\,p \to \mu^-\, e^+$ 3jet X	1.8 TeV($\mathrm{E_{cm}}$)
$\overline{p}\,p \to e^+$ 4jet X	1.8 TeV($\mathrm{E_{cm}}$)
$\overline{p}\,p \to e^-$ 4jet X	1.8 TeV($\mathrm{E_{cm}}$)
$\overline{p}\,p \to \mu^-$ 4jet X	1.8 TeV($\mathrm{E_{cm}}$)
$\overline{p}\,p \to \mu^+$ 4jet X	1.8 TeV($\mathrm{E_{cm}}$)
$\overline{p}\,p \to \mu^+\, e^-$ 2jet X	1.8 TeV($\mathrm{E_{cm}}$)
$\overline{p}\,p \to \mu^-\, e^+$ 2jet X	1.8 TeV($\mathrm{E_{cm}}$)
$\overline{p}\,p \to \mu^-\, \mu^+$ 2jet X	1.8 TeV($\mathrm{E_{cm}}$)
$\overline{p}\,p \to e^-\, e^+$ 2jet X	1.8 TeV($\mathrm{E_{cm}}$)
$\overline{p}\,p \to \mu^+\, e^+$ 3jet X	1.8 TeV($\mathrm{E_{cm}}$)
$\overline{p}\,p \to 2\mu^+$ 3jet X	1.8 TeV($\mathrm{E_{cm}}$)
$\overline{p}\,p \to \mu^-\, e^-$ 3jet X	1.8 TeV($\mathrm{E_{cm}}$)
$\overline{p}\,p \to \mu^-\, \mu^+$ 3jet X	1.8 TeV($\mathrm{E_{cm}}$)
$\overline{p}\,p \to \overline{t}\,t$ X	1.8 TeV($\mathrm{E_{cm}}$)

$W^+ \to e^+\,\nu_e$
$W^- \to e^-\,\overline{\nu}_e$
$W^+ \to \mu^+\,\nu_\mu$
$W^- \to \mu^-\,\overline{\nu}_\mu$
$t \to W^+\,b$
$\overline{t} \to W^-\,\overline{b}$
$W^+ \to$ 2jet
$W^- \to$ 2jet
$b \to$ jet
$\overline{b} \to$ jet
$b \to \mu^-$ X
$\overline{b} \to \mu^+$ X

Particles studied $t, \overline{t}$

For reference purposes we include the following:

BARNETT 1996

■ Summary of Current Status of Particle Physics ■

Review of Particle Physics

Particle Data Group collaboration; *R. M. Barnett et al.*
Phys. Rev. D54, 1 (1996);

Abstract This biennial review summarizes much of Particle Physics. Using data from previous editions, plus 2300 new measurements from 700 papers, we list, evaluate, and average measured properties of gauge bosons, leptons, quarks, monopoles, and supersymmetric particles. All the particle properties and search limits are listed in Summary Tables. We also give numerous tables, figures, formulae, and reviews of topics such as the Standard Model, particle detectors, probability, and statistics. A booklet is available containing the Summary Tables and abbreviated version of some of the other sections of this full *Review*.
The Summary Tables of Particle Properties include: gauge and Higgs bosons, leptons and quarks, mesons, baryons, searches, and tests of conservation laws.
The Reviews, Tables, and Plots section covers many topics such as: Physical constants, Atomic and nuclear properties of materials, Passage of particles through matter, Particle detectors, Probability, Statistics, Kinematics, Cross-section formulae for specific processes, Quantum chromodynamics, Standard Model of electroweak interactions, Quark model, Astrophysics, and Plots of cross sections and related quantities.

Related references

More (earlier) information
Particle Data Group, *Review of Particle Properties*, L. Montanet et al., Phys. Rev. **D50** (1994) 1173;
Particle Data Group, *Review of Particle Properties*, K. Hikasa et al., Phys. Rev. **D45** (1992) 1;
Particle Data Group, *Review of Particle Properties*, J. J. Hernandez et al., Phys. Lett. **239B** (1990) 1;
Particle Data Group, *Review of Particle Properties*, G. P. Yost et al., Phys. Lett. **204B** (1988) 1;
Particle Data Group, *Review of Particle Properties*, M. Aguilar-Benitez, Phys. Lett. **170B** (1986) 1;
Particle Data Group, *Review of Particle Properties*, C. G. Wohl et al., Rev. of Mod. Phys. **56** (1984) S1;
Particle Data Group, *Review of Particle Properties*, M. Roos et al., Phys. Lett. **111B** (1982) 1;
Particle Data Group, *Review of Particle Properties*, C. Bricman et

al., Rev. of Mod. Phys. **52** (1980) 1;
Particle Data Group, *Review of Particle Properties*, C. Brickman et al., Phys. Lett. **75B** (1978) 1;
Particle Data Group, *Review of Particle Properties*, T. G. Trippe et al., Phys. Lett. **68B** (1977) 1;
Particle Data Group, *Review of Particle Properties*, T. G. Trippe et al., Rev. of Mod. Phys. **48** (1976) 1;
Particle Data Group, *Review of Particle Properties*, V. Chaloupka et al., Phys. Lett. **50B** (1974) 1;
Particle Data Group, *Review of Particle Properties*,
N. Barash-Schmidt et al., Rev. of Mod. Phys. **41** (1969) 109;
A. H. Rosenfeld, Ann. Rev. Nucl. Sci. **25** (1975) 555;
W. H. Barkas and A. H. Rosenfeld, *Data for Elementary Particle Physics*, preprint UCRL-8030 (1958);
F. G. Dunnington, *A Re-evaluation of the Atomic Constants*, Phys. Rev. **55** (1933) 683;
R. T. Birge, *Probable Values of the General Physical Constants*, Rev. of Mod. Phys. **1** (1929) 1;
M. Gell-Mann and A. H. Rosenfeld, *Hyperons and Heavy Mesons. Systematics and Decay*, Ann. Rev. Nucl. Sci. **7** (1957) 407;

IV. Author Index

Aamodt R.L. PANOFSKY 1951
Abachi S. ABACHI 1995G
Abbaneo D. DECAMP 1991N, DECAMP 1991S, BUSKULIC 1993H
Abbott B. ABACHI 1995G
Abe F. ABE 1994ZE, ABE 1995
Abe K. LIN 1978
Abolins M.A. ABACHI 1995G
Abov Y.G. ABOV 1965
Abrams G.S. ABRAMS 1964, AUGUSTIN 1974B, ABRAMS 1974, SCHWITTERS 1975, PERL 1975C, HANSON 1975, PERL 1976C, GOLDHABER 1976D, PERUZZI 1976C, PERL 1977F, ABRAMS 1989B, WEIR 1989B
Abreu P. ABREU 1991B, DECAMP 1991S
Acharya B.S. ABACHI 1995G
Achterberg O. BERGER 1978, BERGER 1979C, BERGER 1980L
Ackermann H. BERGER 1979C, BERGER 1980L
Acton D.P. ACTON 1992B
Adam I. ABACHI 1995G
Adam W. ABREU 1991B, DECAMP 1991S
Adami F. ABREU 1991B, DECAMP 1991S
Adams D.L. ABACHI 1995G
Adams M.R. ABACHI 1995G
Adeva B. ADEVA 1990R, ADEVA 1991D, DECAMP 1991S
Adlung S. BUSKULIC 1993H
Adolphsen C.E. ABRAMS 1989B, WEIR 1989B
Adriani O. ADEVA 1990R, ADEVA 1991D, DECAMP 1991S
Adye T. ABREU 1991B, DECAMP 1991S
Agnew L. AGNEW 1958
Aguilar-Benitez M. ADEVA 1990R, ADEVA 1991D, DECAMP 1991S, BARNETT 1996
Ahn S. ABACHI 1995G
Aihara H. ABACHI 1995G
Aizu H. MINAKAWA 1959
Ajaltouni Z. DECAMP 1991D, DECAMP 1991N, DECAMP 1991S, BUSKULIC 1993H
Akbari H. ADEVA 1990R, ADEVA 1991D, DECAMP 1991S
Akesson T. ABREU 1991B, DECAMP 1991S
Akimoto H. ABE 1995
Akopian A. ABE 1995
Al-Agil I. BRANDELIK 1980G
Alam M.S. PERL 1976C, GOLDHABER 1976D, PERUZZI 1976C, PERL 1977F, ANDREWS 1980, ANDREWS 1980B, BEBEK 1981, CHADWICK 1981, AVERY 1989C, ARTUSO 1989C, ALAM 1989B
Albajar C. ALBAJAR 1987C, ALBAJAR 1991C, ALBAJAR 1991D
Albanese J.P. DECAMP 1991D, DECAMP 1991N, DECAMP 1991S
Albrecht H. DARDEN 1978, DARDEN 1978B, ALBRECHT 1987P, DECAMP 1991D
Albrow M.G. ARNISON 1985D, ALBAJAR 1987C, ALBAJAR 1991C, ALBAJAR 1991D, ABE 1994ZE, ABE 1995
Alcaraz J. ADEVA 1990R, ADEVA 1991D, DECAMP 1991S
Aldrovandi A. BLIETSCHAU 1977B
Alekseev G.D. ABREU 1991B, DECAMP 1991S
Alemany R. DECAMP 1991D, DECAMP 1991N, DECAMP 1991S, BUSKULIC 1993H
Alexander G. BERGER 1978, BERGER 1979C, ALEXANDER 1991C, ACTON 1992B, DECAMP 1991S
Alexander J. ALAM 1989B, WEIR 1989B
Alford W.L. LEIGHTON 1951
Alitti J. CALICCHIO 1980, ABACHI 1995G
Allaby J.V. BUSHNIN 1969, DENISOV 1971F, AMALDI 1973E
Allen J.E. DANYSZ 1963
Allen P. DANYSZ 1963, ABREU 1991B, DECAMP 1991S
Allison J. BARTEL 1980D, ALEXANDER 1991C, ACTON 1992B, DECAMP 1991S
Allkofer O.C. ARNISON 1984G, ARNISON 1985D, ALBAJAR 1987C, ALBAJAR 1991C, ALBAJAR 1991D
Allport P.P. ALEXANDER 1991C, ACTON 1992B, DECAMP 1991S
Almehed S. ABREU 1991B, DECAMP 1991S
Aloisio A. ADEVA 1990R, ADEVA 1991D, DECAMP 1991S
Alston M.H. ALSTON 1960, ALSTON 1961B, ALSTON 1961E
Altarelli G. ALTARELLI 1977
Alted F. ABREU 1991B, DECAMP 1991S
Althauss E.J. SARD 1948
Altoon B. DECAMP 1991D, DECAMP 1991N, DECAMP 1991S
Altshuler C.A. ALTSHULER 1934
Alvarez G. ABACHI 1995G
Alvarez L.W. ALVAREZ 1938, LAWRENCE 1939, ALVAREZ 1940, ALVAREZ 1955, ALVAREZ 1956, ALVAREZ 1959, ALSTON 1960, ALSTON 1961B, ALSTON 1961E, MAGLIC 1961B, KALBFLEISCH 1964
Alvarez M. WEIR 1989B
Alverson G. ADEVA 1990R, ADEVA 1991D, DECAMP 1991S
Alves G.A. ABACHI 1995G
Alvial C. DAVIES 1955
Alviggi M.G. ADEVA 1990R, ADEVA 1991D, DECAMP 1991S
Alvsvaag S.J. ABREU 1991B, DECAMP 1991S
Amaldi E. CHAMBERLAIN 1956E, CHAMBERLAIN 1956F, BARKAS 1957
Amaldi U. AMALDI 1973E, ABREU 1991B, DECAMP 1991S
Amati D. AMATI 1962
Ambler E. WU 1957
Amendolia S.R. AMENDOLIA 1973B, DECAMP 1991D, DECAMP 1991N, DECAMP 1991S, ABE 1995
Amidei D. WEIR 1989B, ABE 1994ZE, ABE 1995
Amidi E. ABACHI 1995G
Ammar R. VOYVODIC 1986B, ALBRECHT 1987P, ALAM 1989B
Ammosov V.V. VOYVODIC 1986B
Amos N. ABACHI 1995G
An Q. ADEVA 1990R, ADEVA 1991D, DECAMP 1991S
Anassontzis E. ABREU 1991B, DECAMP 1991S
Andam A.A. ALBRECHT 1987P
Anderhub H. ADEVA 1990R, ADEVA 1991D, DECAMP 1991S

Anderson A.L. ADEVA 1990R, ADEVA 1991D, DECAMP 1991S
Anderson C.D. ANDERSON 1932, ANDERSON 1933, NEDDERMEYER 1937, LEIGHTON 1949, SERIFF 1950, ANDERSON 1953B, COWAN 1953, SORRELS 1955
Anderson F. DAVIES 1955
Anderson H.L. ANDERSON 1952B, ANDERSON 1952E
Anderson K.J. ALEXANDER 1991C, ACTON 1992B, DECAMP 1991S
Anderson W. ABACHI 1995G
Andreev V.P. ADEVA 1990R, ADEVA 1991D, DECAMP 1991S
Andrews D. ANDREWS 1980, ANDREWS 1980B, BEBEK 1981, CHADWICK 1981
Angelov T. ADEVA 1990R, ADEVA 1991D, DECAMP 1991S
Ankoviak K. ALBAJAR 1991C, ALBAJAR 1991D
Annis M. BRIDGE 1951, ANNIS 1952
Ansari R. ANSARI 1987E
Antilogus P. ABREU 1991B, DECAMP 1991S
Antipov Y.M. ANTIPOV 1970
Antonelli A. DECAMP 1991D, DECAMP 1991N, DECAMP 1991S, BUSKULIC 1993H
Antonov L. ADEVA 1990R, ADEVA 1991D, DECAMP 1991S
Antos J. ABE 1994ZE, ABE 1995
Antreasyan D. ADEVA 1990R, ADEVA 1991D, DECAMP 1991S
Anway-Weise C. ABE 1994ZE, ABE 1995
Aota S. ABE 1995
Apel W.D. APEL 1975B, ABREU 1991B, DECAMP 1991S
Apollinari G. DECAMP 1991D, DECAMP 1991N, DECAMP 1991S, ABE 1994ZE, ABE 1995
Appel J.A. HERB 1977, INNES 1977
Apsimon R. ALBAJAR 1991C, ALBAJAR 1991D
Arce P. ADEVA 1990R, ADEVA 1991D, DECAMP 1991S
Arcelli S. ALEXANDER 1991C, ACTON 1992B, DECAMP 1991S
Arefev A.V. ADEVA 1990R, ADEVA 1991D, DECAMP 1991S
Areti H. ABE 1994ZE
Ariztizabal F. BUSKULIC 1993H
Armenteros R. ARMENTEROS 1951, ARMENTEROS 1951B, ARMENTEROS 1952, ARMENTEROS 1955
Armitage J.C. BARTEL 1980D, ALEXANDER 1991C, DECAMP 1991S
Armstrong B. BARNETT 1996
Arnison G. ARNISON 1983C, ARNISON 1983D, ARNISON 1984C, ARNISON 1984G, ARNISON 1985D, ALBAJAR 1987C
Arnold W.H. HODSON 1954
Aronson S.H. ABACHI 1995G
Artamonov A.S. ARTAMONOV 1984B
Artuso M. AVERY 1989C, ARTUSO 1989C, ALAM 1989B
Asakawa T. ABE 1995
Ashman J.G. ASHMAN 1988B, DECAMP 1991D, DECAMP 1991N, DECAMP 1991S, BUSKULIC 1993H
Ashmanskas W. ABE 1995
Ashton P. ALEXANDER 1991C, ACTON 1992B, DECAMP 1991S
Asman B. ABREU 1991B, DECAMP 1991S
Assmann R. BUSKULIC 1993H
Astbury A. ARNISON 1983C, ARNISON 1983D, ARNISON 1984C, ARNISON 1984G, ARNISON 1985D, ALBAJAR 1987C, ALEXANDER 1991C, ACTON 1992B, DECAMP 1991S
Astier P. ABREU 1991B, DECAMP 1991S
Astur R. ABACHI 1995G
Astvacaturov R.G. ASTVACATUROV 1968
Atac M. ABE 1995
Atwood W.B. PRESCOTT 1978C, PRESCOTT 1979C, BODEK 1983, DECAMP 1991D, DECAMP 1991N, DECAMP 1991S
Aubert B. HASERT 1973, HASERT 1973B, BLIETSCHAU 1976, BLIETSCHAU 1977B, ARNISON 1983C, ARNISON 1983D, ARNISON 1984C, ARNISON 1984G, ARNISON 1985D, ALBAJAR 1987C, ALBAJAR 1991C, ALBAJAR 1991D
Aubert J.J. AUBERT 1974D, AUBERT 1983, DECAMP 1991D, DECAMP 1991N, DECAMP 1991S
Auchincloss P. ABE 1994ZE, ABE 1995
Augenstein K. APEL 1975B
Augustin J.E. AUGUSTIN 1974B, ABRAMS 1974, ABREU 1991B, DECAMP 1991S
Augustinus A. ABREU 1991B, DECAMP 1991S
Austern M. ABE 1994ZE
Averill D. ABRAMS 1989B
Averill J.A. CRETIEN 1962
Avery P. AVERY 1989C, ARTUSO 1989C, ALAM 1989B
Avery R.E. ABACHI 1995G
Axen D. ALEXANDER 1991C, ACTON 1992B, DECAMP 1991S
Axon T. ALBAJAR 1987C
Azemoon T. BERGER 1979C, CALICCHIO 1980, ADEVA 1990R, ADEVA 1991D, DECAMP 1991S
Azfar F. ABE 1994ZE, ABE 1995
Azimov M.A. ASTVACATUROV 1968
Aziz T. ADEVA 1990R, ADEVA 1991D, DECAMP 1991S
Azuelos G. ALEXANDER 1991C, ACTON 1992B, DECAMP 1991S
Azzi P. ABE 1994ZE
Azzi-Bacchetta P. ABE 1995
Baba P.V.K.S. ADEVA 1990R, ADEVA 1991D, DECAMP 1991S
Babaev A.I. ALBRECHT 1987P
Babbage W. BUSKULIC 1993H
Bacchetta N. ABE 1994ZE, ABE 1995
Bacci C. BACCI 1974, ARNISON 1983C, ARNISON 1983D, ARNISON 1984C, ARNISON 1984G, ARNISON 1985D, ALBAJAR 1987C, ALBAJAR 1991C, ALBAJAR 1991D
Backer A. BERGER 1978, BERGER 1979C, BERGER 1980L
Bacon T. ALBAJAR 1987C
Badaud F. BUSKULIC 1993H
Badelek B. ASHMAN 1988B
Baden A.R. WEIR 1989B, ABACHI 1995G
Badgett W. ABE 1994ZE, ABE 1995
Badier J. DECAMP 1991D, DECAMP 1991N, DECAMP 1991S, BUSKULIC 1993H
Bagdasarov S. ABE 1995
Bagliesi G. DECAMP 1991D, DECAMP 1991N, DECAMP 1991S, BUSKULIC 1993H
Bagnaia P. BANNER 1983B, BAGNAIA 1983B, ANSARI 1987E, ADEVA 1990R, ADEVA 1991D, DECAMP 1991S
Bahan G.A. ALEXANDER 1991C, ACTON 1992B, DECAMP 1991S
Bailey M.W. ABE 1994ZE, ABE 1995
Baillon P. ABREU 1991B, DECAMP 1991S
Baines J.T.M. ALEXANDER 1991C, ACTON 1992B, DECAMP 1991S
Bains N. ALBAJAR 1987C
Baker N.J. CALICCHIO 1980
Bakich A. VOYVODIC 1986B
Bakken J.A. ADEVA 1990R, ADEVA 1991D, DECAMP 1991S
Baksay L. ADEVA 1990R, ADEVA 1991D, DECAMP 1991S
Balamurali V. ABACHI 1995G
Balderston J. ABACHI 1995G
Baldi R. HASERT 1973B
Baldin A.M. BALDIN 1960, ASTVACATUROV 1968, BALDIN 1973
Baldin B.Y. ABACHI 1995G

Baldini R. DECAMP 1991D, DECAMP 1991N, DECAMP 1991S, BUSKULIC 1993H
Baldini-Celio R. BACCI 1974
Baldo-Ceolin M. PROWSE 1958, FAISSNER 1978D
Ball A.H. ALEXANDER 1991C, ACTON 1992B, DECAMP 1991S
Ball R.C. ADEVA 1990R, ADEVA 1991D, DECAMP 1991S
Ballam J. HODSON 1954, ABRAMS 1989B
Baltay C. STONEHILL 1961, BROWN 1962, HASERT 1973, BALTAY 1979E
Bambade P. ABREU 1991B, DECAMP 1991S
Bamberger A. ABE 1994ZE
Band H. BAND 1988
Banerjee S. ADEVA 1990R, ADEVA 1991D, DECAMP 1991S
Banks J. ALEXANDER 1991C, ACTON 1992B, DECAMP 1991S
Banner M. BANNER 1983B, BAGNAIA 1983B, ANSARI 1987E
Bantly J. ABACHI 1995G
Bao J. ADEVA 1990R, ADEVA 1991D, DECAMP 1991S, ABE 1994ZE, ABE 1995
Baranov V.I. VOYVODIC 1986B
Barao F. ABREU 1991B, DECAMP 1991S
Barbarino G. BACCI 1974
Barbaro-Galtieri A. KALBFLEISCH 1964, ABE 1994ZE, ABE 1995
Barber D.P. BARBER 1979F
Barberio E. DECAMP 1991N, DECAMP 1991S, BUSKULIC 1993H
Barbiellini G. BACCI 1974, ABREU 1991B, DECAMP 1991S
Barczewski T. DECAMP 1991D, DECAMP 1991N, DECAMP 1991S
Bardadin-Otwinowska M. DECAMP 1991D, DECAMP 1991N, DECAMP 1991S, BUSKULIC 1993H
Bardin D.Y. ABREU 1991B, DECAMP 1991S
Baringer P. AVERY 1989C, ARTUSO 1989C, ALAM 1989B
Barish B.C. ABRAMS 1989B, WEIR 1989B
Barkas W.H. BARKAS 1957, BARKAS 1958
Barker G.J. ALEXANDER 1991C, ACTON 1992B, DECAMP 1991S
Barker K.H. ARMENTEROS 1951, ARMENTEROS 1951B, ARMENTEROS 1952
Barklow T. BRANDELIK 1980G, ABRAMS 1989B, WEIR 1989B
Barkov L.M. BARKOV 1978, BARKOV 1979C
Barlow R.J. BRANDELIK 1979H, BRANDELIK 1980G, ALEXANDER 1991C, ACTON 1992B, DECAMP 1991S
Barnes V.E. BARNES 1964B, BARNES 1964E, ABE 1994ZE, ABE 1995
Barnett B.A. ABRAMS 1989B, WEIR 1989B, ABE 1994ZE, ABE 1995
Barnett R.M. BARNETT 1996
Baroncelli A. ABREU 1991B, DECAMP 1991S
Barone L. ADEVA 1990R, ADEVA 1991D, DECAMP 1991S
Baroni G. CHAMBERLAIN 1956E, CHAMBERLAIN 1956F, BARKAS 1957
Barreiro F. BERGER 1979C, BERGER 1980L
Barring O. ABREU 1991B, DECAMP 1991S
Bartalini P. ABE 1994ZE, ABE 1995
Bartel W. AMALDI 1973E, BARTEL 1980D
Bartelt J. ABRAMS 1989B, WEIR 1989B
Bartha S. ALBAJAR 1991C, ALBAJAR 1991D
Bartl W. FIDECARO 1978, ABREU 1991B, DECAMP 1991S
Bartlett J.F. ABACHI 1995G
Bartley J.H. CALICCHIO 1980
Bartoli B. BACCI 1974
Baru S.E. ARTAMONOV 1984B
Basile M. BASILE 1981I
Bassompierre G. AUBERT 1983
Bates M.J. ABREU 1991B, DECAMP 1991S
Batignani G. DECAMP 1991D, DECAMP 1991N, DECAMP 1991S, BUSKULIC 1993H
Batley J.R. ARNISON 1985D, ALBAJAR 1987C, ALEXANDER 1991C, ACTON 1992B, DECAMP 1991S
Baton J.P. CALICCHIO 1980
Battaglia M. ABREU 1991B, DECAMP 1991S
Battiston R. BANNER 1983B, BAGNAIA 1983B, ANSARI 1987E
Batusov Y.A. VOYVODIC 1986B
Baubillier M. ABREU 1991B, DECAMP 1991S
Bauer C. BUSKULIC 1993H
Bauer G. ARNISON 1983D, ARNISON 1984C, ARNISON 1984G, ARNISON 1985D, ALBAJAR 1987C, ALBAJAR 1991C, ALBAJAR 1991D, ABE 1994ZE, ABE 1995
Bauerdick L.A.T. DECAMP 1991D, DECAMP 1991N, DECAMP 1991S
Baum G. ASHMAN 1988B
Baumann T. ABE 1994ZE, ABE 1995
Bay A. ADEVA 1990R, ADEVA 1991D, DECAMP 1991S
Bazizi K. ABACHI 1995G
Beaudoin G. ALEXANDER 1991C, ACTON 1992B, DECAMP 1991S
Beaufays J. ASHMAN 1988B
Bebek C. ANDREWS 1980, ANDREWS 1980B, BEBEK 1981, CHADWICK 1981, AVERY 1989C, ARTUSO 1989C, ALAM 1989B
Bechis D. ANDREWS 1980, ANDREWS 1980B, BEBEK 1981
Beck A. ALEXANDER 1991C, ACTON 1992B, DECAMP 1991S
Becker H. DECAMP 1991N, DECAMP 1991S
Becker J. ALEXANDER 1991C, ACTON 1992B, DECAMP 1991S
Becker U. AUBERT 1974D, BARBER 1979F, ADEVA 1990R, ADEVA 1991D, DECAMP 1991S
Beckert K. JUNG 1992
Becks K.H. AUBERT 1983, ABREU 1991B, DECAMP 1991S
Becquerel H. BECQUEREL 1896B, BECQUEREL 1896C, BECQUEREL 1896D, BECQUEREL 1896F, BECQUEREL 1900
Bedeschi F. ABE 1994ZE, ABE 1995
Bee C.P. ASHMAN 1988B
Beeston C.J. ABREU 1991B, DECAMP 1991S
Begalli M. ABREU 1991B, DECAMP 1991S
Behnke T. ALEXANDER 1991C, ACTON 1992B, DECAMP 1991S
Behrends S. ABE 1994ZE, ABE 1995
Behrens J. ADEVA 1990R, ADEVA 1991D, DECAMP 1991S
Beier E.W. HIRATA 1987C
Beilliere P. ABREU 1991B, DECAMP 1991S
Beingessner S. ALBAJAR 1987C, ADEVA 1990R, ADEVA 1991D, DECAMP 1991S
Belavin A.A. BELAVIN 1975
Belforte S. ABE 1994ZE, ABE 1995
Belk A.T. DECAMP 1991D, DECAMP 1991N, DECAMP 1991S
Bell K.W. BRANDELIK 1979H, BRANDELIK 1980G, ALEXANDER 1991C, ACTON 1992B, DECAMP 1991S
Bella G. ALEXANDER 1991C, ACTON 1992B, DECAMP 1991S
Bellantoni L. DECAMP 1991D, DECAMP 1991N, DECAMP 1991S, BUSKULIC 1993H
Bellettini G. AMENDOLIA 1973B, ABE 1994ZE, ABE 1995
Bellinger J. ALBAJAR 1987C, ABE 1994ZE, ABE 1995
Bellotti E. HASERT 1973, HASERT 1973B, BLIETSCHAU 1976, BLIETSCHAU 1977B
Belokopytov Y.A. ABREU 1991B, DECAMP 1991S
Belousov A.S. ASTVACATUROV 1968
Beltran P. ABREU 1991B, DECAMP 1991S
Belusevic R. CALICCHIO 1980
Bemporad C. BACCI 1974
Bencheikh A.M. BUSKULIC 1993H

Benchouk C. ASHMAN 1988B, DECAMP 1991D, DECAMP 1991N, DECAMP 1991S, BUSKULIC 1993H

Bencivenni G. DECAMP 1991D, DECAMP 1991N, DECAMP 1991S, BUSKULIC 1993H

Bencze G.L. ADEVA 1990R, ADEVA 1991D, DECAMP 1991S

Benda H. BARBER 1979F

Bendich J. ABACHI 1995G

Benedic D. ABREU 1991B, DECAMP 1991S

Beniston M.J. DANYSZ 1963

Benjamin D. ABE 1994ZE, ABE 1995

Benlloch J.M. ABREU 1991B, DECAMP 1991S, ABE 1994ZE, ABE 1995

Bennett S. BENNETT 1967

Bensinger J. ABE 1994ZE, ABE 1995

Benton D. ABE 1994ZE, ABE 1995

Benvenuti A. BENVENUTI 1974F

Berdugo J. ADEVA 1990R, ADEVA 1991D, DECAMP 1991S

Beretvas A. ABE 1994ZE, ABE 1995

Berge J.P. ABE 1994ZE, ABE 1995

Berger C. BERGER 1978, BERGER 1979C, BERGER 1980L

Berges P. ADEVA 1990R, ADEVA 1991D, DECAMP 1991S

Berggren M. ABREU 1991B, DECAMP 1991S

Beri S.B. ABACHI 1995G

Berkelman K. ANDREWS 1980, ANDREWS 1980B, BEBEK 1981, CHADWICK 1981, AVERY 1989C, ARTUSO 1989C, ALAM 1989B

Berlich P. ACTON 1992B

Berna-Rodini M. BACCI 1974

Bernabei I. ARNISON 1983C

Bernard V. DECAMP 1991D, DECAMP 1991N, DECAMP 1991S

Bernlohr K. ANSARI 1987E

Bertanza L. BERTANZA 1962D

Bertin V. DECAMP 1991D, DECAMP 1991N, DECAMP 1991S

Bertolucci E. APEL 1975B

Bertolucci S. ABE 1994ZE, ABE 1995

Bertram I. ABACHI 1995G

Bertrand D. CALICCHIO 1980, ABREU 1991B, DECAMP 1991S

Bertrand-Coremans G. HASERT 1973, HASERT 1973B, BLIETSCHAU 1976, BLIETSCHAU 1977B

Bertucci B. ADEVA 1990R, ADEVA 1991D, DECAMP 1991S

Besson D. AVERY 1989C, ARTUSO 1989C, ALAM 1989B

Best C. AUBERT 1983

Betev B.L. ADEVA 1990R, ADEVA 1991D, DECAMP 1991S

Bethe H.A. BETHE 1947, SALPETER 1951

Bethke S. ABRAMS 1989B, ALEXANDER 1991C, ACTON 1992B, DECAMP 1991S

Betteridge A.P. BUSKULIC 1993H

Bettini A. ARNISON 1985D, ALBAJAR 1987C, ALBAJAR 1991C, ALBAJAR 1991D

Beuselinck R. DECAMP 1991D, DECAMP 1991N, DECAMP 1991S, BUSKULIC 1993H

Bezaguet A. ARNISON 1983C, ARNISON 1983D, ARNISON 1984C, ARNISON 1984G, ARNISON 1985D, ALBAJAR 1987C, ALBAJAR 1991C, ALBAJAR 1991D

Beznogikh G.G. BEZNOGIKH 1969

Bezzubov V.A. ABACHI 1995G

Bhagavantam S. RAMAN 1932

Bhat P.C. ABACHI 1995G

Bhatnagar V. ABACHI 1995G

Bhattacharjee M. ABACHI 1995G

Bhatti A. ABE 1994ZE, ABE 1995

Biagi S.F. BIAGI 1983C, BIAGI 1985C, ABREU 1991B, DECAMP 1991S

Biancastelli R. AMALDI 1973E, BACCI 1974

Bianchi F. ABREU 1991B, DECAMP 1991S

Bibby J.H. ABREU 1991B, DECAMP 1991S

Biddulph P. ALBAJAR 1991C, ALBAJAR 1991D

Biebel O. ALEXANDER 1991C, ACTON 1992B, DECAMP 1991S

Bieler E.S. CHADWICK 1921

Bienlein J.K. BIENLEIN 1978

Biery K. ABE 1994ZE, ABE 1995

Biggs P.J. AUBERT 1974D

Biland A. ADEVA 1990R, ADEVA 1991D, DECAMP 1991S

Bilenky M.S. ABREU 1991B, DECAMP 1991S

Billing M. ANDREWS 1980

Billoir P. ABREU 1991B, DECAMP 1991S

Binder U. ALBRECHT 1987P, ALEXANDER 1991C, ACTON 1992B, DECAMP 1991S

Bingham H.H. CALICCHIO 1980

Binkley M. KNAPP 1976B, ABE 1994ZE, ABE 1995

Binnie D.M. BRANDELIK 1979H, BRANDELIK 1980G, DECAMP 1991D, DECAMP 1991N, DECAMP 1991S, BUSKULIC 1993H

Binon F. BUSHNIN 1969, BINON 1969, BINON 1970B

Bionta R.M. BIONTA 1987C

Bird F. DECAMP 1991D, DECAMP 1991N, DECAMP 1991S, ABE 1994ZE

Bird I.G. ASHMAN 1988B

Birge R.T. BIRGE 1929

Birge R.W. BIRGE 1955B, BARKAS 1957, GOOD 1961

Birsa R. FIDECARO 1978

Bischoff A. ABACHI 1995G

Bisello D. BACCI 1974, ABE 1994ZE, ABE 1995

Biswas N. ABACHI 1995G

Bizzarri R. ADEVA 1990R, ADEVA 1991D, DECAMP 1991S

Bjarne J. ABREU 1991B, DECAMP 1991S

Bjorken J.D. BJORKEN 1964, BJORKEN 1969, BJORKEN 1969B

Bjorklund R. BJORKLUND 1950

Bjornerud E.K. YORK 1953

Blackett P.M.S. BLACKETT 1932, BLACKETT 1933

Blair R.E. ABE 1994ZE, ABE 1995

Blaising J.J. ADEVA 1990R, ADEVA 1991D, DECAMP 1991S

Blanar G. BIENLEIN 1978

Blazey G.C. ABACHI 1995G

Blessing S. ABACHI 1995G

Blewitt G. BIONTA 1987C

Blietschau J. BLIETSCHAU 1976, BLIETSCHAU 1977B

Blinov A.E. ARTAMONOV 1984B

Blobel V. BERGER 1978, BERGER 1979C, BERGER 1980L

Bloch D. ABREU 1991B, DECAMP 1991S

Bloch F. BLOCH 1937, ALVAREZ 1940

Bloch P. BANNER 1983B, BAGNAIA 1983B

Bloch-Devaux B. DECAMP 1991D, DECAMP 1991N, DECAMP 1991S, BUSKULIC 1993H

Block M.M. PEVSNER 1961

Blocker C. ABE 1994ZE, ABE 1995

Blockus D. ABRAMS 1989B, WEIR 1989B

Blomeke P. ADEVA 1990R, ADEVA 1991D, DECAMP 1991S

Blondel A. DECAMP 1991D, DECAMP 1991N, DECAMP 1991S, BUSKULIC 1993H

Bloodworth I.J. ALEXANDER 1991C, ACTON 1992B, DECAMP 1991S

Bloom E.D. BLOOM 1969B, BREIDENBACH 1969, MILLER 1972B

Blucher E. AVERY 1989C, ARTUSO 1989C, DECAMP 1991D, DECAMP 1991N, DECAMP 1991S, BUSKULIC 1993H

Blum D. HASERT 1973B, BLIETSCHAU 1976, BLIETSCHAU 1977B

Blum W. BLUM 1975, BIENLEIN 1978, DECAMP 1991D, DECAMP 1991N, DECAMP 1991S, BUSKULIC 1993H

Blumenfeld B. ADEVA 1990R, ADEVA 1991D, DECAMP 1991S

Blumenfeld B.J. BUESSER 1973

Bobbink G.J. ADEVA 1990R, ADEVA 1991D, DECAMP 1991S
Bobisut F. FAISSNER 1978D
Bocciolini M. ADEVA 1990R, ADEVA 1991D, DECAMP 1991S
Bock P. BIENLEIN 1978, ALEXANDER 1991C, ACTON 1992B, DECAMP 1991S
Bock R.K. ARNISON 1983C, ARNISON 1983D, ARNISON 1984C, ARNISON 1984G, ARNISON 1985D, ADEVA 1991D, DECAMP 1991S
Bockmann P. DARDEN 1978, DARDEN 1978B, ALBRECHT 1987P
Bodek A. BODEK 1983, ABE 1994ZE, ABE 1995
Boden B. ACTON 1992B
Boehm A. BARBER 1979F
Boehnlein A. ABACHI 1995G
Boerner H. BRANDELIK 1979H, BRANDELIK 1980G
Boesten L. BERGER 1978, BERGER 1979C, BERGER 1980L
Boggild J. DAVIES 1955
Bogolyubov N.N. BOGOLYUBOV 1956, BOGOLYUBOV 1956B, BOGOLYUBOV 1965
Bogolyubov P.N. ABREU 1991B, DECAMP 1991S
Bohlen W. ADEVA 1990R
Bohm A. ADEVA 1990R, ADEVA 1991D, DECAMP 1991S
Böhm E. AUBERT 1983
Bohn H. ALBAJAR 1991C, ALBAJAR 1991D
Bohr N. BOHR 1913, BOHR 1913B, BOHR 1918
Bohrer A. ALBAJAR 1991C, ALBAJAR 1991D, BUSKULIC 1993H
Bohringer T. BOHRINGER 1980B, FINOCCHIARO 1980, ADEVA 1990R
Bokhari W. ABE 1995
Boldt E. BOLDT 1958B
Bologna G. DECAMP 1991D, DECAMP 1991N, DECAMP 1991S, BUSKULIC 1993H
Bolognese T. ABREU 1991B, DECAMP 1991S
Bolognesi V. ABE 1994ZE, ABE 1995
Bonapart M. ABREU 1991B, DECAMP 1991S
Bonaudi F. BANNER 1983B, BAGNAIA 1983B
Bondar A.E. ARTAMONOV 1984B
Bonesini M. ABREU 1991B, DECAMP 1991S
Bonetti A. BONETTI 1953, BONETTI 1953B, DAVIES 1955
Bonetti S. HASERT 1973, HASERT 1973B, BLIETSCHAU 1976, BLIETSCHAU 1977B
Bonino R. ALBAJAR 1991C, ALBAJAR 1991D
Bonissent A. DECAMP 1991D, DECAMP 1991N, DECAMP 1991S, BUSKULIC 1993H
Bonivento W. ABREU 1991B, DECAMP 1991S
Bonneaud G. DECAMP 1991D, DECAMP 1991N, DECAMP 1991S, BUSKULIC 1993H
Bonvicini G. BASILE 1981I, ABRAMS 1989B, WEIR 1989B, DECAMP 1991D, DECAMP 1991N, DECAMP 1991S, BUSKULIC 1993H
Booth C.N. ANSARI 1987E, DECAMP 1991D, DECAMP 1991N, DECAMP 1991S, BUSKULIC 1993H
Booth E.T. LANDE 1956
Booth P.S.L. ABREU 1991B, DECAMP 1991S
Boratav M. ABREU 1991B, DECAMP 1991S
Borcherding F. ABACHI 1995G
Borders J. ABACHI 1995G
Bordner C.A. CRETIEN 1962
Borer K. BANNER 1983B, BAGNAIA 1983B, ANSARI 1987E
Borgeaud P. ABREU 1991B, DECAMP 1991S
Borger J. BERGER 1979C
Borghini M. PRESCOTT 1978C, PRESCOTT 1979C, BANNER 1983B, BAGNAIA 1983B, ANSARI 1987E
Borgia B. ADEVA 1990R, ADEVA 1991D, DECAMP 1991S
Borisov G.V. ABREU 1991B, DECAMP 1991S
Born M. BORN 1925, BORN 1926, BORN 1926B, BORN 1926C, BORN 1926D
Borner H. ABREU 1991B, DECAMP 1991S
Borone R. FRAUENFELDER 1957
Bortoletto D. AVERY 1989C, ARTUSO 1989C, ALAM 1989B, ABE 1994ZE, ABE 1995
Bos K. ALBAJAR 1987C, ALBAJAR 1991C, ALBAJAR 1991D
Bosch F. JUNG 1992
Bosch H.M. ALEXANDER 1991C, ACTON 1992B, DECAMP 1991S
Bose S.N. BOSE 1924
Bosio C. AMALDI 1973E, ABREU 1991B, DECAMP 1991S
Bosisio L. DECAMP 1991D, DECAMP 1991N, DECAMP 1991S, BUSKULIC 1993H
Bossi F. DECAMP 1991D, DECAMP 1991N, DECAMP 1991S, BUSKULIC 1993H
Bostjancic B. ALBRECHT 1987P
Boswell C. ABE 1994ZE, ABACHI 1995G
Bothe W. BOTHE 1929
Botlo M. ALBAJAR 1991C, ALBAJAR 1991D
Botner O. ABREU 1991B, DECAMP 1991S
Botterill D.R. DECAMP 1991D, DECAMP 1991N, DECAMP 1991S, BUSKULIC 1993H
Bottigli U. DECAMP 1991D, DECAMP 1991N, DECAMP 1991S, BUSKULIC 1993H
Bouclier R. CHARPAK 1968
Boucrot J. DECAMP 1991D, DECAMP 1991N, DECAMP 1991S, BUSKULIC 1993H
Boudreau J.F. DECAMP 1991D, DECAMP 1991N, DECAMP 1991N, DECAMP 1991S, DECAMP 1991S, BUSKULIC 1993H, ABE 1995
Bougerolle S. ALEXANDER 1991C, ACTON 1992B, DECAMP 1991S
Boulos T. ABE 1994ZE
Bouquet B. ABREU 1991B, DECAMP 1991S
Bourilkov D. ADEVA 1990R, ADEVA 1991D, DECAMP 1991S
Bourotte J. DECAMP 1991D, DECAMP 1991N, DECAMP 1991S
Bourquin M. BIAGI 1983C, BIAGI 1985C, ADEVA 1990R, ADEVA 1991D, DECAMP 1991S
Boutigny D. ADEVA 1990R, ADEVA 1991D, DECAMP 1991S
Bouwens B. ADEVA 1990R, ADEVA 1991D, DECAMP 1991S
Bowcock J. BOWCOCK 1961
Bowcock T. AVERY 1989C, ARTUSO 1989C, ALAM 1989B
Bowcock T.J.V. ARNISON 1983D, ARNISON 1984C, ARNISON 1984G
Bowdery C.K. DECAMP 1991D, DECAMP 1991N, DECAMP 1991S, BUSKULIC 1993H
Boyarski A.M. AUGUSTIN 1974B, ABRAMS 1974, SCHWITTERS 1975, PERL 1975C, HANSON 1975, PERL 1976C, GOLDHABER 1976D, PERUZZI 1976C, PERL 1977F, ABRAMS 1989B, WEIR 1989B
Boyer J. WEIR 1989B
Boyle O. DECAMP 1991D, DECAMP 1991N, DECAMP 1991S
Bozhko N.I. ABACHI 1995G
Bozzi C. BUSKULIC 1993H
Bozzo M. ABREU 1991B, DECAMP 1991S
Brabant J.M. BRABANT 1956C
Brabson B.B. ABRAMS 1989B, WEIR 1989B, ALEXANDER 1991C, ACTON 1992B, DECAMP 1991S
Braccini P.L. AMENDOLIA 1973B
Bradamante F. FIDECARO 1978
Bradaschia C. AMENDOLIA 1973B, DECAMP 1991D, DECAMP 1991N, DECAMP 1991S
Braems F. DECAMP 1991D, DECAMP 1991N, DECAMP 1991S
Braibant S. ABREU 1991B, DECAMP 1991S
Branchini P. ABREU 1991B, DECAMP 1991S
Brand K.D. ABREU 1991B, DECAMP 1991S
Brandelik R. BRANDELIK 1977D, BRANDELIK 1979H, BRANDELIK 1980G

Brandenburg G. ABE 1994ZE, ABE 1995
Brandl B. DECAMP 1991D, DECAMP 1991N, DECAMP 1991S, BUSKULIC 1993H
Brandt A. ABACHI 1995G
Brandt S. BERGER 1979C, BERGER 1980L, DECAMP 1991D, DECAMP 1991N, DECAMP 1991S, BUSKULIC 1993H
Branson J.G. BARBER 1979F, ADEVA 1990R, ADEVA 1991D, DECAMP 1991S
Brasse F.W. AUBERT 1983
Bratton C.B. BIONTA 1987C
Braun O. DECAMP 1991D, DECAMP 1991N, DECAMP 1991S, BUSKULIC 1993H
Braunschweig W. BRANDELIK 1977D, BRANDELIK 1979H, BRANDELIK 1980G
Breakstone A. ABRAMS 1989B, WEIR 1989B
Breccia L. ABE 1995
Breidenbach M. BLOOM 1969B, BREIDENBACH 1969, AUGUSTIN 1974B, ABRAMS 1974, SCHWITTERS 1975, PERL 1975C, HANSON 1975, PERL 1976C, GOLDHABER 1976D, PERUZZI 1976C, PERL 1977F
Breit G. BREIT 1936, BREIT 1936B
Brena C. ARNISON 1985D
Brene N. DAVIES 1955
Brenner A.E. CRETIEN 1962
Brenner R.A. ABREU 1991B, DECAMP 1991S
Bressani T. CHARPAK 1968
Breuker H. ALEXANDER 1991C, ACTON 1992B, DECAMP 1991S
Brickwedde F.G. UREY 1932, UREY 1932B
Bricman C. ABREU 1991B, DECAMP 1991S
Bridge H.S. BRIDGE 1951, ANNIS 1952, BRIDGE 1955
Bridges D.L. ANDREWS 1980, ANDREWS 1980B, BEBEK 1981, CHADWICK 1981
Brient J.C. DECAMP 1991D, DECAMP 1991N, DECAMP 1991S, BUSKULIC 1993H
Briggs D. AUGUSTIN 1974B, ABRAMS 1974
Briggs D.D. PERL 1975C
Brisson V. BERTANZA 1962D, HASERT 1973, HASERT 1973B, BLIETSCHAU 1976, BLIETSCHAU 1977B, CALICCHIO 1980
Britten A.J. BIAGI 1983C, BIAGI 1985C
Brobeck W.M. LAWRENCE 1939, BROBECK 1947
Brock I.C. BRANDELIK 1980G, AVERY 1989C, ARTUSO 1989C, ALAM 1989B, ADEVA 1990R, ADEVA 1991D, DECAMP 1991S
Brock R. ABACHI 1995G
Brockhausen D. ALBAJAR 1991C, ALBAJAR 1991D
Brodbeck T.J. DECAMP 1991D, DECAMP 1991N, DECAMP 1991S, BUSKULIC 1993H
Brodsky S. BRODSKY 1973
Brody A. ANDREWS 1980, ANDREWS 1980B, BEBEK 1981, CHADWICK 1981
Broll C. AUBERT 1983
Brom J.M. ABRAMS 1989B, WEIR 1989B
Bromberg C. ABE 1995
Bron J. BARBER 1979F
Bronstein J. KNAPP 1976B
Bross A.D. ABACHI 1995G
Brout R. ENGLERT 1964
Brown B.C. HERB 1977, INNES 1977, UENO 1979
Brown C.N. HERB 1977, INNES 1977, UENO 1979
Brown D. DECAMP 1991D, DECAMP 1991N, DECAMP 1991S, BUSKULIC 1993H
Brown H.N. BROWN 1962
Brown R.C.A. ABREU 1991B, DECAMP 1991S
Brown R.H. BROWN 1949B
Brown R.M. BIAGI 1983C, BIAGI 1985C, ALEXANDER 1991C, ACTON 1992B, DECAMP 1991S
Brown S.C. AUBERT 1983, ASHMAN 1988B
Brueckner K.A. BRUECKNER 1951
Brummer N. ABREU 1991B, DECAMP 1991S
Brun R. ALEXANDER 1991C, ACTON 1992B, DECAMP 1991S
Brunet J.M. ABREU 1991B, DECAMP 1991S
Bruyant F. ADEVA 1990R, ADEVA 1991D, DECAMP 1991S
Buchanan C. ALBAJAR 1991C, ALBAJAR 1991D
Buchholz D. ABACHI 1995G
Buckley E.J. ARNISON 1985D, ALBAJAR 1987C
Buckley-Geer E. ABE 1994ZE, ABE 1995
Budd H.S. ABE 1994ZE, ABE 1995
Budker G.I. BUDKER 1967, BUDKER 1976
Buesser F.W. BUESSER 1973
Bugge L. ABREU 1991B, DECAMP 1991S
Buhring R. BRANDELIK 1979H
Buijs A. ALEXANDER 1991C, ACTON 1992B, DECAMP 1991S
Buikman D. BARBER 1979F
Buisson C. ADEVA 1990R, ADEVA 1991D, DECAMP 1991S
Bujak A. ADEVA 1990R, ADEVA 1991D, DECAMP 1991S
Bukin A.D. ARTAMONOV 1984B
Bull V.A. DANYSZ 1963
Bullock F.W. HASERT 1973, HASERT 1973B, BLIETSCHAU 1976, BLIETSCHAU 1977B, CALICCHIO 1980
Bulos F. CRETIEN 1962, AUGUSTIN 1974B, ABRAMS 1974, SCHWITTERS 1975, PERL 1975C, HANSON 1975, PERL 1976C, ABRAMS 1989B, WEIR 1989B
Bunce G. BUNCE 1976, ALBAJAR 1987C
Bunn J. ARNISON 1985D
Bunyatov S.A. VOYVODIC 1986B
Buran T. ABREU 1991B, DECAMP 1991S
Burchat P.R. ABRAMS 1989B, WEIR 1989B
Burckhart H.J. BIAGI 1983C, BIAGI 1985C, ALEXANDER 1991C, ACTON 1992B, DECAMP 1991S
Burfening J. BURFENING 1949
Burger J. AUBERT 1974D, BERGER 1978, BARBER 1979F, BERGER 1980L
Burger J.D. ADEVA 1990R, ADEVA 1991D, DECAMP 1991S
Burke D.L. ABRAMS 1989B, WEIR 1989B
Burkett K. ABE 1994ZE, ABE 1995
Burkhardt H. BRANDELIK 1980G, DECAMP 1991D, DECAMP 1991N, DECAMP 1991S
Burmeister H. ABREU 1991B, DECAMP 1991S
Burnett T.H. VOYVODIC 1986B, DECAMP 1991D, DECAMP 1991N, DECAMP 1991S
Burnstein R.A. ABRAMS 1964
Burq J.P. ADEVA 1990R, ADEVA 1991D, DECAMP 1991S
Burtovoy V.S. ABACHI 1995G
Buschbeck B. ALBAJAR 1991C, ALBAJAR 1991D
Buschhorn G. MILLER 1972B, BRANDELIK 1977D
Busenitz J. ADEVA 1990R, ADEVA 1991D, DECAMP 1991S
Busetto G. ARNISON 1985D, ALBAJAR 1987C, ALBAJAR 1991C, ALBAJAR 1991D, ABE 1994ZE, ABE 1995
Bushnin Y.B. BUSHNIN 1969
Buskirk A.V. THOMPSON 1953, THOMPSON 1953B
Buskulic D. BUSKULIC 1993H
Butler C.C. ROCHESTER 1947, ARMENTEROS 1951, ARMENTEROS 1951B, ARMENTEROS 1952
Butler F. WEIR 1989B
Butler J.M. ABACHI 1995G
Buttar C. DECAMP 1991D, DECAMP 1991N, DECAMP 1991S, BUSKULIC 1993H
Button J. BUTTON 1960, BUTTON 1961
Buyak A. BEZNOGIKH 1969
Buytaert J.A.M.A. ABREU 1991B, DECAMP 1991S
Byon-Wagner A. ABE 1994ZE, ABE 1995
Byrd J. AVERY 1989C, ARTUSO 1989C
Byrum K.L. ABE 1994ZE, ABE 1995

Cabenda R. ANDREWS 1980, ANDREWS 1980B, BEBEK 1981, CHADWICK 1981
Cabibbo N. CABIBBO 1963
Caccia M. ABREU 1991B, DECAMP 1991S
Cachon A. ARMENTEROS 1951, ARMENTEROS 1951B, ARMENTEROS 1952
Cai X.D. ADEVA 1990R, ADEVA 1991D, DECAMP 1991S
Calderini G. BUSKULIC 1993H
Caldwell D.O. BOLDT 1958B
Calicchio M. CALICCHIO 1980
Callot O. DECAMP 1991D, DECAMP 1991N, DECAMP 1991S, BUSKULIC 1993H
Calvetti M. ARNISON 1983C, ARNISON 1983D, ARNISON 1984C, ARNISON 1984G
Calvi M. ABREU 1991B, DECAMP 1991S
Calvino F. WEIR 1989B
Camacho Rozas A.J. ABREU 1991B, DECAMP 1991S
Camerini U. BROWN 1949B, HASERT 1973B
Cameron W. DECAMP 1991D, DECAMP 1991N, DECAMP 1991S, BUSKULIC 1993H
Camilleri L. BUESSER 1973
Cammerata J. ABE 1995
Campagnari C. ABE 1994ZE, ABE 1995
Campagne J.E. ABREU 1991B, DECAMP 1991S
Campana P. DECAMP 1991D, DECAMP 1991N, DECAMP 1991S, BUSKULIC 1993H
Campbell M. ABE 1994ZE, ABE 1995
Campion A. ABREU 1991B, DECAMP 1991S
Camporesi T. BAND 1988, ABREU 1991B, DECAMP 1991S
Camps C. ADEVA 1990R
Canale V. ABREU 1991B, DECAMP 1991S
Candlin D.J. DECAMP 1991D, DECAMP 1991N, DECAMP 1991S, BUSKULIC 1993H
Caner A. ALBAJAR 1991C, ALBAJAR 1991D, ABE 1994ZE, ABE 1995
Canzler T. BARTEL 1980D
Cao F. ABREU 1991B, DECAMP 1991S
Capell M. ADEVA 1990R, ADEVA 1991D, DECAMP 1991S
Capiluppi P. ALEXANDER 1991C, ACTON 1992B, DECAMP 1991S
Capon G. BACCI 1974, DECAMP 1991D, DECAMP 1991N, DECAMP 1991S, BUSKULIC 1993H
Caputo M.C. ASHMAN 1988B
Cara Romeo G. BASILE 1981I
Carbonara F. ADEVA 1990R, ADEVA 1991D, DECAMP 1991S
Carboni G. ANSARI 1987E
Cardenal P. ADEVA 1991D, DECAMP 1991S
Carithers W. GOLDHABER 1976D, PERUZZI 1976C, ABE 1994ZE, ABE 1995
Carlsmith D. ABE 1994ZE, ABE 1995
Carlson A.G. CARLSON 1950
Carminati F. ADEVA 1990R, ADEVA 1991D, DECAMP 1991S
Carnegie R.K. ALEXANDER 1991C, ACTON 1992B, DECAMP 1991S
Carney R. DECAMP 1991D, DECAMP 1991N, DECAMP 1991S
Caroumbalis D. BALTAY 1979E
Carpinelli M. DECAMP 1991D, DECAMP 1991N, DECAMP 1991S, BUSKULIC 1993H
Carr J. AUBERT 1983, BUSKULIC 1993H
Carroll L. ABREU 1991B, DECAMP 1991S
Carroll T. ARNISON 1983C
Carron G. CARRON 1978
Cartacci A.M. ADEVA 1990R, ADEVA 1991D, DECAMP 1991S
Carter A.A. BIAGI 1983C, BIAGI 1985C, ALEXANDER 1991C, ACTON 1992B, DECAMP 1991S
Carter J.M. DECAMP 1991D, DECAMP 1991N, DECAMP 1991S, BUSKULIC 1993H
Carter J.R. BIAGI 1983C, ALEXANDER 1991C, ACTON 1992B, DECAMP 1991S
Cartwright S.L. DECAMP 1991D, DECAMP 1991N, DECAMP 1991S, BUSKULIC 1993H
Casey D. ABACHI 1995G
Cashmore R. BRANDELIK 1979H, BRANDELIK 1980G, ABE 1994ZE
Caso C. ABREU 1991B, DECAMP 1991S, BARNETT 1996
Casoli P. ALBAJAR 1991C, ALBAJAR 1991D
Casper D. BIONTA 1987C, BUSKULIC 1993H
Cassel D.G. BRANDELIK 1980G, ANDREWS 1980, ANDREWS 1980B, BEBEK 1981, CHADWICK 1981, AVERY 1989C, ARTUSO 1989C, ALAM 1989B
Cassen B. CASSEN 1936
Castagnoli C. CHAMBERLAIN 1956E, CHAMBERLAIN 1956F, BARKAS 1957
Castaldi R. AMENDOLIA 1973B
Castelli E. ABREU 1991B, DECAMP 1991S
Castilla-Valdez H. ALBAJAR 1991C, ALBAJAR 1991D, ABACHI 1995G
Castillo Gimenez M.V. ABREU 1991B, DECAMP 1991S
Castro A. ABE 1994ZE, ABE 1995
Catanesi M. DECAMP 1991D, DECAMP 1991N, DECAMP 1991S
Cattai A. ABREU 1991B, DECAMP 1991S
Cattaneo M.C. DECAMP 1991D, DECAMP 1991N, DECAMP 1991S, BUSKULIC 1993H
Cattaneo P. DECAMP 1991D, DECAMP 1991N, DECAMP 1991S, BUSKULIC 1993H
Catz P. ARNISON 1983C, ARNISON 1983D, ARNISON 1984C, ARNISON 1984G, ARNISON 1985D, ALBAJAR 1987C
Cauz G. ABE 1995
Cavalli D. HASERT 1973, HASERT 1973B, BLIETSCHAU 1976, BLIETSCHAU 1977B
Cavallo F.R. ABREU 1991B, DECAMP 1991S
Cavanna F. ALBAJAR 1991C, ALBAJAR 1991D
Cavasinni V. AMENDOLIA 1973B, ANSARI 1987E
Cazzoli E.G. CAZZOLI 1975E
Ceccarelli M. DAVIES 1955
Celvetti M. BACCI 1974
Cen Y. ABE 1994ZE, ABE 1995
Cence R.J. ABRAMS 1989B, WEIR 1989B
Cenci P. ANSARI 1987E
Cennini P. ARNISON 1983D, ARNISON 1984C, ARNISON 1984G, ARNISON 1985D, ALBAJAR 1987C, ALBAJAR 1991C, ALBAJAR 1991D
Centro S. ARNISON 1983C, ARNISON 1983D, ARNISON 1984C, ARNISON 1984G, ARNISON 1985D, ALBAJAR 1987C, ALBAJAR 1991C, ALBAJAR 1991D
Ceradini F. ARNISON 1983C, ARNISON 1983D, ARNISON 1984C, ARNISON 1984G, ARNISON 1985D, ALBAJAR 1987C, ALBAJAR 1991C, ALBAJAR 1991D
Cerenkov P.A. CERENKOV 1934, CERENKOV 1937
Cerrada M. ADEVA 1990R, ADEVA 1991D, DECAMP 1991S
Cerri C. AMENDOLIA 1973B
Cerrito L. ABREU 1991B, DECAMP 1991S
Cerutti F. DECAMP 1991N, DECAMP 1991S, BUSKULIC 1993H
Cervelli F. ABE 1994ZE, ABE 1995
Cesaroni F. ADEVA 1990R, ADEVA 1991D, DECAMP 1991S
Cevenini F. BACCI 1974
Chabaud V. BLUM 1975
Chadwick G.B. BAND 1988
Chadwick J. CHADWICK 1914, CHADWICK 1921, CHADWICK 1932, CHADWICK 1932B, CHADWICK 1934
Chadwick K. ANDREWS 1980, ANDREWS 1980B, BEBEK 1981, CHADWICK 1981
Chai Y. BUSKULIC 1993H
Chakraborty D. ABACHI 1995G

Chamberlain O. CHAMBERLAIN 1955, CHAMBERLAIN 1956E, CHAMBERLAIN 1956F
Chang C.C. BARBER 1979F
Chang C.Y. BERGER 1979C, BERGER 1980L, ALEXANDER 1991C, ACTON 1992B, DECAMP 1991S
Chang G.K. ANDREWS 1980
Chang S.M. ABACHI 1995G
Chang Y.H. ADEVA 1990R, ADEVA 1991D, DECAMP 1991S
Chao H.Y. ABE 1995
Chapkin M. ABREU 1991B, DECAMP 1991S
Chapman A.H. ARMENTEROS 1951
Chapman J. ABRAMS 1989B, WEIR 1989B, ABE 1994ZE, ABE 1995
Charlton D.G. ALBAJAR 1987C, ALEXANDER 1991C, ACTON 1992B, DECAMP 1991S
Charlton G.R. ABRAMS 1964
Charpak G. CHARPAK 1968
Charpentier P. ABREU 1991B, DECAMP 1991S
Chaturvedi U.K. ADEVA 1990R, ADEVA 1991D, DECAMP 1991S
Checchia P. ABREU 1991B, DECAMP 1991S
Chekulaev S.V. ABACHI 1995G
Chelkov G.A. ABREU 1991B, DECAMP 1991S
Chemarin M. ADEVA 1990R, ADEVA 1991D, DECAMP 1991S
Chen A. ANDREWS 1980, ANDREWS 1980B, BEBEK 1981, CHADWICK 1981, ADEVA 1990R, ADEVA 1991D, DECAMP 1991S
Chen C. ADEVA 1990R, ADEVA 1991D, DECAMP 1991S
Chen D. AVERY 1989C
Chen G.M. ADEVA 1990R, ADEVA 1991D, DECAMP 1991S
Chen H.F. ADEVA 1990R, ADEVA 1991D, DECAMP 1991S
Chen H.S. BARBER 1979F, ADEVA 1990R, ADEVA 1991D, DECAMP 1991S
Chen L.P. ABACHI 1995G
Chen M. AUBERT 1974D, BARBER 1979F, ADEVA 1990R, ADEVA 1991D, DECAMP 1991S
Chen M.C. ADEVA 1990R, DECAMP 1991S
Chen M.L. ADEVA 1990R, ADEVA 1991D, DECAMP 1991S
Chen W. BUSKULIC 1993H, ABACHI 1995G
Chen W.Y. ARTUSO 1989C, ALAM 1989B, ADEVA 1991D
Chen X. DECAMP 1991D, DECAMP 1991N, DECAMP 1991S
Cheng C.P. BARBER 1979F
Cheng D.C. BENVENUTI 1974F
Cheng M.T. ABE 1994ZE, ABE 1995
Chernev K. KIRILLOVA 1965
Chernyaev K. VOYVODIC 1986B
Cheu E. AVERY 1989C, ARTUSO 1989C, ALAM 1989B
Cheung H.W.K. ASHMAN 1988B
Chevalier L. ABREU 1991B, DECAMP 1991S, ABACHI 1995G
Chew G.F. CHEW 1956, CHEW 1959, CHEW 1961
Chiarella V. DECAMP 1991D, DECAMP 1991N, DECAMP 1991S, BUSKULIC 1993H
Chiarelli G. ABE 1994ZE, ABE 1995
Chiefari G. ADEVA 1990R, ADEVA 1991D, DECAMP 1991S
Chien C.Y. ADEVA 1990R, ADEVA 1991D, DECAMP 1991S
Chikamatsu T. ABE 1994ZE, ABE 1995
Chikovani G.E. CHIKOVANI 1964
Childers R. ALBRECHT 1987P
Chima J.S. ASHMAN 1988B
Chinowsky W. CHINOWSKY 1955, LANDE 1956, LANDE 1957, CHINOWSKY 1962, AUGUSTIN 1974B, ABRAMS 1974, SCHWITTERS 1975, PERL 1975C, HANSON 1975, PERL 1976C, GOLDHABER 1976D, PERUZZI 1976C, PERL 1977F, BRANDELIK 1979H
Chiou C.N. ABE 1995
Chliapnikov P.V. ABREU 1991B, DECAMP 1991S
Chmeissani M. ABRAMS 1989B
Chodorow M. CHODOROW 1955
Chollet F. ADEVA 1990R, ADEVA 1991D, DECAMP 1991S
Chollet J.C. BANNER 1983B, BAGNAIA 1983B, ANSARI 1987E
Chopra S. ABACHI 1995G
Chorowicz V. ABREU 1991B, DECAMP 1991S
Choudhary B.C. ABACHI 1995G
Chounet L.M. HASERT 1973, HASERT 1973B, BLIETSCHAU 1976, BLIETSCHAU 1977B
Chrin J.T.M. ALEXANDER 1991C, DECAMP 1991S
Christenson J.C. CHRISTENSON 1970
Christenson J.H. CHRISTENSON 1964, ABACHI 1995G
Christofek L. ABE 1995
Chu Y.S. BARBER 1979F
Chu-Chien W. KANG-CHANG 1960
Chung M. ABACHI 1995G
Chupp W.W. CHUPP 1955, ILOFF 1955, CHAMBERLAIN 1956E, CHAMBERLAIN 1956F, BARKAS 1957
Chuvilo I.V. ASTVACATUROV 1968
Ciapetti G. ARNISON 1985D, ALBAJAR 1987C, ALBAJAR 1991C, ALBAJAR 1991D
Ciborowski J. ASHMAN 1988B
Cifarelli L. BASILE 1981I
Cihangir S. ABE 1994ZE, ABE 1995
Cinabro D. DECAMP 1991D, DECAMP 1991N, DECAMP 1991S, BUSKULIC 1993H
Ciocci M.A. DECAMP 1991D, DECAMP 1991N, DECAMP 1991S, BUSKULIC 1993H
Ciocio A. BIONTA 1987C
Cirio R. ABREU 1991B, DECAMP 1991S
Cittolin S. ARNISON 1983C, ARNISON 1983D, ARNISON 1984C, ARNISON 1984G, ARNISON 1985D, ALBAJAR 1987C, ALBAJAR 1991C, ALBAJAR 1991D
Civinini C. ADEVA 1990R, ADEVA 1991D, DECAMP 1991S
Claes D. ABACHI 1995G
Clara M.P. ABREU 1991B, DECAMP 1991S
Clare I. ADEVA 1990R, ADEVA 1991D, DECAMP 1991S
Clare R. BARBER 1979F, ADEVA 1990R, ADEVA 1991D, DECAMP 1991S
Clark A.G. BANNER 1983B, BAGNAIA 1983B, ABE 1994ZE, ABE 1995
Clark A.R. ABACHI 1995G
Clark D. CLARK 1951
Clarke D. BARTEL 1980D, ALBAJAR 1987C
Clarke P.E.L. ALEXANDER 1991C, ACTON 1992B, DECAMP 1991S
Claus R. BIONTA 1987C
Clayton E. ALBAJAR 1991C, ALBAJAR 1991D
Clendenin J.E. PRESCOTT 1978C, PRESCOTT 1979C
Clifft R.W. AUBERT 1983, ASHMAN 1988B, DECAMP 1991D, DECAMP 1991N, DECAMP 1991S, BUSKULIC 1993H
Cline D. BENVENUTI 1974F, ARNISON 1983D, ARNISON 1984C, ARNISON 1984G, ARNISON 1985D, ALBAJAR 1987C, ALBAJAR 1991C, ALBAJAR 1991D
Cnops A.M. CAZZOLI 1975E, BALTAY 1979E, ARNISON 1983C
Cobal M. ABE 1994ZE, ABE 1995
Cobau W.G. ABACHI 1995G
Cobb J.H. AUBERT 1983

Cocconi G. AMALDI 1973E
Cochet C. ARNISON 1983C, ARNISON 1983D, ARNISON 1984C, ARNISON 1984G, ARNISON 1985D, ALBAJAR 1987C
Cochran J. ABACHI 1995G
Cockcroft J.D. COCKCROFT 1932
Coffman D.M. AVERY 1989C, ARTUSO 1989C, ALAM 1989B
Cohen I. ALEXANDER 1991C, ACTON 1992B, DECAMP 1991S
Cohn H.O. ADEVA 1990R, ADEVA 1991D, DECAMP 1991S
Coignet G. AUBERT 1983, ASHMAN 1988B, ADEVA 1990R, ADEVA 1991D, DECAMP 1991S
Colas J. ARNISON 1983C, ARNISON 1983D, ARNISON 1984C, ARNISON 1984G, ARNISON 1985D, ALBAJAR 1987C, ALBAJAR 1991C, ALBAJAR 1991D
Colas P. ARNISON 1985D, ALBAJAR 1987C, DECAMP 1991D, DECAMP 1991N, DECAMP 1991S, BUSKULIC 1993H
Cole F.T. KERST 1956
Coleman R. KNAPP 1976B
Colino N. ADEVA 1990R, ADEVA 1991D, DECAMP 1991S
Colley D.C. CALICCHIO 1980
Colling D.J. BUSKULIC 1993H
Collins P. ABREU 1991B, DECAMP 1991S
Collins W.J. ALEXANDER 1991C, ACTON 1992B, DECAMP 1991S
Colrain P. BUSKULIC 1993H
Coltman J.W. MARSHALL 1947
Comas P. BUSKULIC 1993H
Combley F. AUBERT 1983, ASHMAN 1988B, DECAMP 1991D, DECAMP 1991N, DECAMP 1991S, BUSKULIC 1993H
Commichau V. ADEVA 1990R, ADEVA 1991D, DECAMP 1991S
Compton A.H. COMPTON 1923
Compton K.T. VAN DE GRAAFF 1933
Conboy J.E. ALEXANDER 1991C, ACTON 1992B, DECAMP 1991S
Condon E.U. BREIT 1936B, CASSEN 1936
Conforto G. ADEVA 1990R, ADEVA 1991D, DECAMP 1991S, BARNETT 1996
Connolly P.L. BERTANZA 1962D, CONNOLLY 1963, BARNES 1964B, BARNES 1964E, GOLDBERG 1964C, CAZZOLI 1975E, BALTAY 1979E
Conta C. HASERT 1973, HASERT 1973B, BANNER 1983B, BAGNAIA 1983B, ANSARI 1987E
Conte R. ALBAJAR 1991C, ALBAJAR 1991D
Contin A. BASILE 1981I, ADEVA 1990R, ADEVA 1991D, DECAMP 1991S
Contreras J.L. ABREU 1991B, DECAMP 1991S
Contreras M. ABE 1994ZE, ABE 1995
Contri R. ABREU 1991B, DECAMP 1991S
Conversi M. CONVERSI 1947, CONVERSI 1955, EISLER 1958
Conway J. ABE 1994ZE, ABE 1995
Conway J.S. DECAMP 1991D, DECAMP 1991N, DECAMP 1991S, BUSKULIC 1993H
Cooksey D. LAWRENCE 1939
Cool R.L. BUESSER 1973
Cooper A.M. CALICCHIO 1980
Cooper J. ABE 1994ZE, ABE 1995
Cooper M. ALEXANDER 1991C, ACTON 1992B, DECAMP 1991S
Cooper S. GOLDHABER 1976D
Cooper W.E. ABACHI 1995G
Copie T. AVERY 1989C, ARTUSO 1989C
Coppage D. ALBRECHT 1987P
Cordelli M. ABE 1994ZE, ABE 1995
Corden M. ARNISON 1983C, ARNISON 1983D, ARNISON 1984C, ARNISON 1984G, ARNISON 1985D, ALBAJAR 1987C, DECAMP 1991D, DECAMP 1991N, DECAMP 1991S, BUSKULIC 1993H
Cordier A. DECAMP 1991D, DECAMP 1991N, DECAMP 1991S, BUSKULIC 1993H
Cords D. BRANDELIK 1977D, BARTEL 1980D, ABRAMS 1989B, WEIR 1989B
Cork B. BRABANT 1956C, CORK 1956
Corson D.R. LAWRENCE 1939
Cortez B.G. HIRATA 1987C, BIONTA 1987C
Cosme G. ABREU 1991B, DECAMP 1991S
Costantini F. BACCI 1974, BOHRINGER 1980B, FINOCCHIARO 1980, ANSARI 1987E
Cottingham W.N. BOWCOCK 1961
Cottrell R.L. PRESCOTT 1978C, PRESCOTT 1979C
Couch M. ALEXANDER 1991C, ACTON 1992B, DECAMP 1991S
Couchot P. ABREU 1991B, DECAMP 1991S
Coughlan J.A. ALBAJAR 1987C, ALBAJAR 1991C, ALBAJAR 1991D
Coupal D.P. ABRAMS 1989B, WEIR 1989B, ABE 1994ZE
Coupland M. ALEXANDER 1991C, ACTON 1992B, DECAMP 1991S
Courant E.D. COURANT 1952
Courant H. ANNIS 1952, STONEHILL 1961
Court G.R. AUBERT 1983, ASHMAN 1988B
Courvoisier D. DECAMP 1991D, DECAMP 1991N, DECAMP 1991S
Couyoumtzelis C. ABE 1995
Cowan B. DECAMP 1991D, DECAMP 1991N, DECAMP 1991S
Cowan C.L. REINES 1953B, COWAN 1956, REINES 1956, REINES 1959
Cowan E.W. SERIFF 1950, ANDERSON 1953B, COWAN 1953, COWAN 1954
Cowan G. BUSKULIC 1993H
Coward D.H. BLOOM 1969B, BREIDENBACH 1969, MILLER 1972B, BODEK 1983
Cowen D.F. DECAMP 1991D, DECAMP 1991N, DECAMP 1991S
Cox G. ARNISON 1985D, ALBAJAR 1987C
Coyle P. BUSKULIC 1993H
Crabb D.G. LIN 1978
Crandall W.E. BJORKLUND 1950
Crane D. ABE 1994ZE, ABE 1995
Crane H.R. KERST 1956
Crane L. DAVIES 1955
Crawford F.S. ALVAREZ 1956, CRAWFORD 1958
Crawford G. AVERY 1989C, ARTUSO 1989C, ALAM 1989B
Crawford R.L. BARNETT 1996
Crawley H.B. ABREU 1991B, DECAMP 1991S
Creanza D. DECAMP 1991D, DECAMP 1991N, DECAMP 1991S, BUSKULIC 1993H
Crennell D. BARNES 1964E, ABREU 1991B, DECAMP 1991S
Crennell D.J. BARNES 1964B
Crespo J.M. DECAMP 1991D, DECAMP 1991N, DECAMP 1991S, BUSKULIC 1993H
Cresti M. CRAWFORD 1958
Cretien M. CRETIEN 1962
Cretsinger C. ABACHI 1995G
Criegee L. BERGER 1978, BERGER 1979C, BERGER 1980L
Crijns F. ADEVA 1990R, ADEVA 1991D, DECAMP 1991S
Criscuolo P. ADEVA 1991D, DECAMP 1991S
Cronin J.W. CHRISTENSON 1964
Cronin-Hennessy D. ABE 1995
Crosetti G. ABREU 1991B, DECAMP 1991S
Crosland N. ABREU 1991B, DECAMP 1991S
Crouch H.R. CRETIEN 1962
Crouch M. BIONTA 1987C
Crozon M. ABREU 1991B, DECAMP 1991S

Csorna S.E. ANDREWS 1980, ANDREWS 1980B, BEBEK 1981, CHADWICK 1981, AVERY 1989C, ARTUSO 1989C, ALAM 1989B
Cuevas Maestro J. ABREU 1991B, DECAMP 1991S
Cuffiani M. ALEXANDER 1991C, ACTON 1992B, DECAMP 1991S
Cui X.Y. ADEVA 1990R, ADEVA 1991D, DECAMP 1991S
Culbertson R. ABE 1995
Cullen-Vidal D. ABACHI 1995G
Culwick B.B. BROWN 1962, BARNES 1964B, BARNES 1964E
Cummings M. ABACHI 1995G
Cundy D.C. HASERT 1973, HASERT 1973B, BLIETSCHAU 1976, BLIETSCHAU 1977B
Cunningham J.D. ABE 1994ZE, ABE 1995
Cutts D. ABACHI 1995G
Czellar S. ABREU 1991B, DECAMP 1991S
Dado S. ALEXANDER 1991C, ACTON 1992B, DECAMP 1991S
Dagoret S. ABREU 1991B, DECAMP 1991S
D'Agostini G.D. AUBERT 1983, ASHMAN 1988B
Dahl O.I. KALBFLEISCH 1964, ABACHI 1995G
Dahl-Jensen E. ABREU 1991B, DECAMP 1991S
Dai T.S. ADEVA 1990R, ADEVA 1991D, DECAMP 1991S
Daion M.I. VOYVODIC 1986B
Dakin J.T. AUGUSTIN 1974B, PERL 1975C
D'Alessandro R. ADEVA 1990R, ADEVA 1991D, DECAMP 1991S
D'Ali G. BASILE 1981I
Dalitz R.H. DALITZ 1953, DALITZ 1954
Dalkhazhav N. KIRILLOVA 1965
Dallavalle G.M. ALEXANDER 1991C, ACTON 1992B, DECAMP 1991S
Dallman D. ARNISON 1983C, ARNISON 1983D, ARNISON 1984C, ARNISON 1984G, ARNISON 1985D
Dalmagne B. ABREU 1991B, DECAMP 1991S
Dam M. ABREU 1991B, DECAMP 1991S
Damgaard G. ABREU 1991B, DECAMP 1991S
Damyanov S. KIRILLOVA 1965
Danby G.T. DANBY 1962
Dangelo S. ARNISON 1983C
Daniels T. ABE 1994ZE, ABE 1995
Danilchenko I.A. HASERT 1973B, BLIETSCHAU 1976
Danilov M.V. ALBRECHT 1987P
Danysz M. DANYSZ 1953, DANYSZ 1963
Darbo G. ABREU 1991B, DECAMP 1991S
Darden C.W. DARDEN 1978, DARDEN 1978B, ALBRECHT 1987P
Darriulat P. BANNER 1983B, BAGNAIA 1983B, ANSARI 1987E
Darvill D.C. BARTEL 1980D
Dau D. ARNISON 1983D, ARNISON 1984C, ARNISON 1984G, ARNISON 1985D, ALBAJAR 1987C, ALBAJAR 1991C, ALBAJAR 1991D
Dau W.D. AUBERT 1983
Daubie E. ABREU 1991B, DECAMP 1991S
Daum C. ALBAJAR 1991C, ALBAJAR 1991D
Daum H.J. BERGER 1978, BERGER 1979C, BERGER 1980L
Daumann H. BERGER 1978
Dauncey P.D. ABRAMS 1989B, ABREU 1991B, DECAMP 1991S
Davenport M. ABREU 1991B, DECAMP 1991S
David P. ABREU 1991B, DECAMP 1991S
Davier M. DECAMP 1991D, DECAMP 1991N, DECAMP 1991S, BUSKULIC 1993H
Davies J.H. DAVIES 1955
Davies J.K. AUBERT 1983
Davies R. DAVIES 1955B
Davis D.H. DANYSZ 1963
Davis R. VOYVODIC 1986B, ALBRECHT 1987P
Davisson C.J. DAVISSON 1927, DAVISSON 1927B
Dawson I. BUSKULIC 1993H
Day T.B. ABRAMS 1964
de Angelis A. ABREU 1991B, DECAMP 1991S
de Asmundis R. ADEVA 1990R, ADEVA 1991D, DECAMP 1991S
de Barbaro P. ABE 1994ZE, ABE 1995
de Beer M. ARNISON 1983C, ARNISON 1983D, ARNISON 1984C, ARNISON 1984G, ARNISON 1985D, ALBAJAR 1987C, ABREU 1991B, DECAMP 1991S
de Boeck H. ABREU 1991B, DECAMP 1991S
de Boer W. BRANDELIK 1977D, ABREU 1991B, DECAMP 1991S
de Bonis I. BUSKULIC 1993H
de Bouard G. DECAMP 1991D
de Bouard X. AUBERT 1983
de Brion J.P. ALBAJAR 1987C
de Broglie L. DE BROGLIE 1923, DE BROGLIE 1923B
de Cesare P. BASILE 1981I
de Clercq C. ABREU 1991B, DECAMP 1991S
de Fez Laso M.D.M. ABREU 1991B, DECAMP 1991S
de Giorgi M. ARNISON 1985D, ALBAJAR 1987C
de Groot N. ABREU 1991B, DECAMP 1991S
de Jong S. ALEXANDER 1991C, ACTON 1992B, DECAMP 1991S
de Jongh F. ABE 1994ZE, ABE 1995
de La Vaissiere C. ABREU 1991B, DECAMP 1991S
de Lotto B. ANSARI 1987E, ABREU 1991B, DECAMP 1991S
de Mello Neto J.R.T. ABACHI 1995G
de Miranda J.M. ABACHI 1995G
de Mortier L. ABE 1994ZE, ABE 1995
de Notaristefani F. ADEVA 1990R, ADEVA 1991D, DECAMP 1991S
de Palma M. DECAMP 1991D, DECAMP 1991N, DECAMP 1991S, BUSKULIC 1993H
de Pasquali G. FRAUENFELDER 1957
de Salvo R. ARNISON 1983C, AVERY 1989C, ARTUSO 1989C
de Sangro R. BAND 1988
de Staebler H.C. BRIDGE 1955, BLOOM 1969B, BREIDENBACH 1969, MILLER 1972B, PRESCOTT 1978C, PRESCOTT 1979C, ABRAMS 1989B, WEIR 1989B
de Troconiz J.F. ABE 1994ZE, ABE 1995
de Voe R. GOLDHABER 1976D, PERUZZI 1976C
de Weerd A.J. DECAMP 1991D
de Wire J.W. ANDREWS 1980, ANDREWS 1980B, BEBEK 1981, CHADWICK 1981, AVERY 1989C, ARTUSO 1989C, ALAM 1989B
de Wit M. BLIETSCHAU 1976, BLIETSCHAU 1977B
de Witt H. FAISSNER 1978D
De K. ABACHI 1995G
Debu P. ALEXANDER 1991C, ACTON 1992B, DECAMP 1991S
Decamp D. DECAMP 1991D, DECAMP 1991N, DECAMP 1991S, BUSKULIC 1993H
Deden H. BLIETSCHAU 1977B
Defoix C. ABREU 1991B, DECAMP 1991S
Degrange B. HASERT 1973, HASERT 1973B, BLIETSCHAU 1976, BLIETSCHAU 1977B
Degre A. ADEVA 1990R, ADEVA 1991D, DECAMP 1991S
Dehmelt H.G. VAN DYCK 1986, VAN DYCK 1987
Dehne H.C. BERGER 1978, BERGER 1979C, BERGER 1980L
Dehning B. BUSKULIC 1993H
Dehning F. DECAMP 1991D, DECAMP 1991N, DECAMP 1991S
Deiters K. ADEVA 1990R, ADEVA 1991D, DECAMP 1991S
del Fabbro R. BACCI 1974
del Pozo L. ACTON 1992B
del Prete T. AMENDOLIA 1973B, ANSARI 1987E
Delaney W.C. BARNES 1964B, BARNES 1964E
Delchamps S. ABE 1994ZE, ABE 1995

Delfino M. BAND 1988, DECAMP 1991D, DECAMP 1991N, DECAMP 1991S, BUSKULIC 1993H
Delikaris D. ABREU 1991B, DECAMP 1991S
Della Marina R. BUSKULIC 1993H
Della Negra M. ARNISON 1983C, ARNISON 1983D, ARNISON 1984C, ARNISON 1984G, ARNISON 1985D, ALBAJAR 1987C, ALBAJAR 1991C, ALBAJAR 1991D
Dell'Agnello S. ABE 1994ZE, ABE 1995
Dell'Orso M. ABE 1994ZE, ABE 1995
Dell'Orso R. DECAMP 1991D, DECAMP 1991N, DECAMP 1991S, BUSKULIC 1993H
Delorme S. ABREU 1991B, DECAMP 1991S
Delpierre P. ABREU 1991B, DECAMP 1991S
Demaria N. ABREU 1991B, DECAMP 1991S
Demarteau M. ABACHI 1995G
Demin A. ABREU 1991B, DECAMP 1991S
Demina R.Y. ABACHI 1995G
Demoulin M. ARNISON 1983C, ARNISON 1983D, ARNISON 1984C, ARNISON 1984G, ARNISON 1985D, ALBAJAR 1987C, ALBAJAR 1991C, ALBAJAR 1991D
Denby B. ARNISON 1985D, ALBAJAR 1987C, ABE 1994ZE, ABE 1995
Denegri D. ARNISON 1983C, ARNISON 1983D, ARNISON 1984C, ARNISON 1984G, ARNISON 1985D, ALBAJAR 1987C, ALBAJAR 1991C, ALBAJAR 1991D
Denes E. ADEVA 1990R, ADEVA 1991D, DECAMP 1991S
Denes P. ADEVA 1990R, ADEVA 1991D, DECAMP 1991S
Deninno M.M. ALEXANDER 1991C, ACTON 1992B, DECAMP 1991S, ABE 1994ZE, ABE 1995
Denisenko K.G. ABACHI 1995G
Denisenko N.L. ABACHI 1995G
Denisov D.S. ABACHI 1995G
Denisov S.P. BUSHNIN 1969, BINON 1970B, ANTIPOV 1970, DENISOV 1971F, ABACHI 1995G
Derbenev Y.S. DERBENEV 1978, DERBENEV 1980
Derikum K. BERGER 1978, BERGER 1979C, BERGER 1980L
Derkaoui J. ABREU 1991B, DECAMP 1991S
Derwent P.F. ABE 1994ZE, ABE 1995
Deschizeaux B. DECAMP 1991D, DECAMP 1991N, DECAMP 1991S
D'Ettore-Piazzoli B. DECAMP 1991D, DECAMP 1991N, DECAMP 1991S, BUSKULIC 1993H
Devenish R. BERGER 1978, BERGER 1979C, BRANDELIK 1980G
Devlin T. BUNCE 1976, ABE 1994ZE, ABE 1995
Dharmaratna W. ABACHI 1995G
Dhina M. ADEVA 1990R, ADEVA 1991D, DECAMP 1991S
di Bitonto D. ARNISON 1983C, ARNISON 1983D, ARNISON 1984C, ARNISON 1984G, ADEVA 1990R, ADEVA 1991D, DECAMP 1991S
di Ciaccio A. ARNISON 1983D, ARNISON 1984C, ARNISON 1984G, ARNISON 1985D, ALBAJAR 1987C, ALBAJAR 1991C, ALBAJAR 1991D
di Ciaccio L. ABREU 1991B, DECAMP 1991S
di Corato M. DAVIES 1955
di Lella L. BUESSER 1973, BANNER 1983B, BAGNAIA 1983B, ANSARI 1987E
Dibon H. ALBAJAR 1991C, ALBAJAR 1991D
Dickson M. ABE 1994ZE, ABE 1995
Diddens A.N. BUSHNIN 1969, AMALDI 1973E
Dieckmann A. ALEXANDER 1991C, ACTON 1992B, DECAMP 1991S
Diehl H.T. ABACHI 1995G
Diekmann A. BERGER 1979C
Diemoz M. ADEVA 1990R, ADEVA 1991D, DECAMP 1991S
Diesburg M. ABACHI 1995G
Dietl H. BLUM 1975, BIENLEIN 1978, DECAMP 1991D, DECAMP 1991N, DECAMP 1991S, BUSKULIC 1993H
Diez-Hedo F.J. ADEVA 1990R, ADEVA 1991D, ALBAJAR 1991C, ALBAJAR 1991D, DECAMP 1991S
Dijkstra H. ABREU 1991B, DECAMP 1991S
Dikansky N.S. BUDKER 1976
Diloreto G. ABACHI 1995G
Dilworth C. DAVIES 1955
Dimitrov H.R. ADEVA 1990R, ADEVA 1991D, DECAMP 1991S
Dines-Hansen J. BANNER 1983B, BAGNAIA 1983B, ANSARI 1987E
Dinsdale M. DECAMP 1991D, DECAMP 1991N, DECAMP 1991S
Dionisi C. ADEVA 1990R, ADEVA 1991D, DECAMP 1991S
Dirac P.A.M. DIRAC 1926, DIRAC 1927, DIRAC 1927B, DIRAC 1928, DIRAC 1928B, DIRAC 1930, DIRAC 1931
Dittmann J.R. ABE 1995
Dittmann P. BARTEL 1980D
Dittmar M. ALEXANDER 1991C, ACTON 1992B, DECAMP 1991S
Dittus F. ADEVA 1990R
Divia R. ADEVA 1991D, DECAMP 1991S
Dixit M.S. ALEXANDER 1991C, ACTON 1992B, DECAMP 1991S
Dixon R. ABACHI 1995G
Djama F. ABREU 1991B, DECAMP 1991S
Dobbins J. BOHRINGER 1980B, FINOCCHIARO 1980
Dobinson R.W. AMALDI 1973E, AUBERT 1983
Dobrzynski L. ARNISON 1983C, ARNISON 1983D, ARNISON 1984C, ARNISON 1984G, ARNISON 1985D, ALBAJAR 1987C, ALBAJAR 1991C, ALBAJAR 1991D
Dogru M. DECAMP 1991D, DECAMP 1991N, DECAMP 1991S
Dolbeau J. ABREU 1991B, DECAMP 1991S
Dolgoshein B.A. DOLGOSHEIN 1964
Dolin R. ADEVA 1990R
Dominick J. ALAM 1989B
Donati S. ABE 1994ZE, ABE 1995
Donker J.P. ALBRECHT 1987P
Donskov S.V. ANTIPOV 1970, DENISOV 1971F, APEL 1975B
Donszelmann M. ABREU 1991B, DECAMP 1991S
Dore C. BIAGI 1983C, BIAGI 1985C
Dorenbosch J. ARNISON 1985D, ALBAJAR 1987C, ALBAJAR 1991C, ALBAJAR 1991D
Dorfan D.E. DORFAN 1965D, DORFAN 1967, ABRAMS 1989B, WEIR 1989B
Dorfan J.M. PERL 1976C, GOLDHABER 1976D, PERUZZI 1976C, PERL 1977F, ABRAMS 1989B, WEIR 1989B
Dornan P.J. BRANDELIK 1979H, BRANDELIK 1980G, DECAMP 1991D, DECAMP 1991N, DECAMP 1991S, BUSKULIC 1993H
Doroba K. ABREU 1991B, DECAMP 1991S
Dorsaz P.A. BANNER 1983B, BAGNAIA 1983B
Dosselli U. AUBERT 1983
Dova M.T. ADEVA 1991D, DECAMP 1991S
Dowell J.D. ARNISON 1983C, ARNISON 1983D, ARNISON 1984C, ARNISON 1984G, ARNISON 1985D, ALBAJAR 1987C, ALBAJAR 1991C, ALBAJAR 1991D
Downie N.A. BRANDELIK 1979H, BRANDELIK 1980G
Dracos M. ABREU 1991B, DECAMP 1991S
Drago E. ADEVA 1990R, ADEVA 1991D, DECAMP 1991S
Draper P. ABACHI 1995G
Drees J. BLOOM 1969B, BREIDENBACH 1969, MILLER 1972B, AUBERT 1983, ASHMAN 1988B, ABREU 1991B, DECAMP 1991S

Drell P.S. AVERY 1989C, ARTUSO 1989C, ALAM 1989B, WEIR 1989B
Drescher A. ALBRECHT 1987P
Drevermann H. DECAMP 1991D, DECAMP 1991N, DECAMP 1991S, BUSKULIC 1993H
Drewer D. ABRAMS 1989B
Driever T. ADEVA 1990R, ADEVA 1991D, DECAMP 1991S
Drijard D. ALBAJAR 1991C, ALBAJAR 1991D
Drinkard J. BUSKULIC 1993H, ABACHI 1995G
Dris M. ABREU 1991B, DECAMP 1991S
Drucker R.B. ABE 1994ZE, ABE 1995
Drumm H. BARTEL 1980D
Duarte H. BUSKULIC 1993H
Dubin D.L. BODEK 1983
Duchesneau D. ADEVA 1990R, ADEVA 1991D, DECAMP 1991S
Duchovni E. BRANDELIK 1980G, ARNISON 1985D, ALBAJAR 1987C, ALEXANDER 1991C, ACTON 1992B, DECAMP 1991S
Duckeck G. ALEXANDER 1991C, ACTON 1992B, DECAMP 1991S
Ducros Y. ABACHI 1995G
Duerdoth I.P. BARTEL 1980D, ALEXANDER 1991C, ACTON 1992B, DECAMP 1991S
Duflot L. BUSKULIC 1993H
Dufour J.Y. ABREU 1991B, DECAMP 1991S
Dugad S.R. ABACHI 1995G
Dugeay S. DECAMP 1991D, DECAMP 1991N, DECAMP 1991S
Duinker P. BARBER 1979F, ADEVA 1990R, ADEVA 1991D, DECAMP 1991S
Dulinski W. ABREU 1991B, DECAMP 1991S
Dumas D.J.P. ALEXANDER 1991C, ACTON 1992B, DECAMP 1991S
Dunaitsev A.F. DUNAITSEV 1962, BUSHNIN 1969
Dunn A. ABE 1994ZE, ABE 1995
Dunnington F.G. DUNNINGTON 1939
Duran I. ADEVA 1990R, ADEVA 1991D, DECAMP 1991S
Durbin R. DURBIN 1951C
Düren M. ASHMAN 1988B
Durston-Johnson S. ABACHI 1995G
Duteil P. BUSHNIN 1969, BINON 1969, BINON 1970B
Dyce N. ASHMAN 1988B
Dydak F. DECAMP 1991D, DECAMP 1991N, DECAMP 1991S, BUSKULIC 1993H
Dye S.T. BIONTA 1987C
Dyson F.J. DYSON 1949, DYSON 1949B
Dzhelyadin R.I. ABREU 1991B, DECAMP 1991S
Déclais Y. AUBERT 1983
Eades J. DORFAN 1965D
Eberhard P.H. ALVAREZ 1959, BUTTON 1960, ALSTON 1960, BUTTON 1961, ALSTON 1961B, ALSTON 1961E, KALBFLEISCH 1964
Eckerlin G. ALEXANDER 1991C, ACTON 1992B, DECAMP 1991S
Eddy N. ABE 1995
Edelman B. BUNCE 1976
Edgecock R. ARNISON 1985D, ALBAJAR 1987C
Edgecock T.R. DECAMP 1991D, DECAMP 1991N, DECAMP 1991S, BUSKULIC 1993H
Edmunds D. ABACHI 1995G
Edwards A.W. AUBERT 1983, ASHMAN 1988B
Edwards B. EDWARDS 1958
Edwards K.W. ALBRECHT 1987P
Edwards M. ARNISON 1983C, AUBERT 1983, ASHMAN 1988B, DECAMP 1991D, DECAMP 1991N, DECAMP 1991S
Edwards R. BUNCE 1976
Eek L.O. ABREU 1991B, DECAMP 1991S
Eerola P.A.M. ABREU 1991B, DECAMP 1991S
Efimov A. ABACHI 1995G
Efthymiopoulos I. DECAMP 1991D, DECAMP 1991N, DECAMP 1991S, BUSKULIC 1993H
Eggert K. ARNISON 1983C, ARNISON 1983D, ARNISON 1984C, ARNISON 1984G, ARNISON 1985D, ALBAJAR 1987C, ALBAJAR 1991C, ALBAJAR 1991D
Egorov O.K. VOYVODIC 1986B
Ehmann C. ALBRECHT 1987P
Ehrlich R. ANDREWS 1980, ANDREWS 1980B, BEBEK 1981, CHADWICK 1981, AVERY 1989C, ARTUSO 1989C, ALAM 1989B
Eichler R. BARTEL 1980D
Eickhoff H. JUNG 1992
Eidelman S. BARNETT 1996
Eidelman Y.I. ARTAMONOV 1984B
Einstein A. EINSTEIN 1905, EINSTEIN 1905B, EINSTEIN 1905C, EINSTEIN 1906, EINSTEIN 1924, EINSTEIN 1925
Einsweiler K. ANSARI 1987E, ABE 1994ZE, ABE 1995
Eisenberg Y. BRANDELIK 1979H, BRANDELIK 1980G
Eisenhandler E. ARNISON 1983C, ARNISON 1983D, ARNISON 1984C, ARNISON 1984G, ARNISON 1985D, ALBAJAR 1987C, ALBAJAR 1991C, ALBAJAR 1991D
Eisler F. EISLER 1958
Ekelof T. ABREU 1991B, DECAMP 1991S
Ekspong A.G. CHAMBERLAIN 1956F, BARKAS 1957
Ekspong G. ABREU 1991B, DECAMP 1991S
El Fellous R. DECAMP 1991D, DECAMP 1991N, DECAMP 1991S, BUSKULIC 1993H
El Mamouni H. ADEVA 1990R, ADEVA 1991D, DECAMP 1991S
Elcombe P.A. ALEXANDER 1991C, ACTON 1992B, DECAMP 1991S
Elia R. ABRAMS 1989B
Elias J.E. BODEK 1983, ABE 1994ZE, ABE 1995
Elioff T. AGNEW 1958
Elliot Peisert A. ABREU 1991B, DECAMP 1991S
Ellis C.D. ELLIS 1927, ELLIS 1927B
Ellis N. ARNISON 1983C, ARNISON 1983D, ARNISON 1984C, ARNISON 1984G, ARNISON 1985D, ALBAJAR 1987C, ALBAJAR 1991C, ALBAJAR 1991D
Ellison J. ABACHI 1995G
Elsen E. BARTEL 1980D
Elvira V.D. ABACHI 1995G
Ely R. ABE 1994ZE, ABE 1995
Emery S. BUSKULIC 1993H
Engel J.P. ABREU 1991B, DECAMP 1991S
Engelmann R. ABACHI 1995G
Engels E.Jr. ABE 1994ZE, ABE 1995
Engler A. PEVSNER 1961, ADEVA 1990R, ADEVA 1991D, DECAMP 1991S
Englert F. ENGLERT 1964
Eno S. ABE 1994ZE, ABACHI 1995G
Enstrom J. DORFAN 1967
Eppley G. ABACHI 1995G
Eppling F.J. ADEVA 1990R, ADEVA 1991D, DECAMP 1991S
Erhard P. ARNISON 1983C, ARNISON 1983D, ARNISON 1984C, ARNISON 1984G, ARNISON 1985D, ALBAJAR 1987C
Ermolov P.F. ABACHI 1995G
Erne F.C. ADEVA 1990R, ADEVA 1991D, DECAMP 1991S
Ernst T. ASHMAN 1988B
Eroshin O.V. ABACHI 1995G
Errede D. ABE 1994ZE, ABE 1995
Errede S. BIONTA 1987C, ABE 1994ZE, ABE 1995
Erriquez O. CALICCHIO 1980
Erwin A.R. ERWIN 1961C
Eskreys A. BERGER 1979C, BERGER 1980L
Esposito B. BACCI 1974, BASILE 1981I
Estabrooks P.G. ALEXANDER 1991C, ACTON 1992B, DECAMP 1991S
Esten M.J. HASERT 1973, HASERT 1973B, BLIETSCHAU 1976
Esterman I. ESTERMAN 1933, ESTERMAN 1934

Etienne F. DECAMP 1991D, DECAMP 1991N, DECAMP 1991S, BUSKULIC 1993H
Etter L.R. THOMPSON 1953, THOMPSON 1953B
Etzion E. ALEXANDER 1991C, ACTON 1992B, DECAMP 1991S
Evans D. DAVIES 1955
Evans H. ALBAJAR 1991C, ALBAJAR 1991D
Evdokimov V.N. ABACHI 1995G
Everhart G. AUBERT 1974D
Extermann P. BIAGI 1983C, BIAGI 1985C, ADEVA 1990R, ADEVA 1991D, DECAMP 1991S
Fabbretti R. ADEVA 1990R, ADEVA 1991D, DECAMP 1991S
Fabbri F. ALEXANDER 1991C, ACTON 1992B, DECAMP 1991S
Faber G. ADEVA 1990R, ADEVA 1991D, DECAMP 1991S
Fabre M. ADEVA 1990R, ADEVA 1991D, DECAMP 1991S
Fackler O. ADEVA 1990R, ADEVA 1991D, DECAMP 1991S
Faddeev L.D. FADDEEV 1967
Fahey S. ABACHI 1995G
Fahland T. ABACHI 1995G
Fainberg J. LORD 1950
Faissner H. HASERT 1973, BLIETSCHAU 1976, FAISSNER 1978D, ARNISON 1983C, ARNISON 1983D, ARNISON 1984C, ARNISON 1984G, ARNISON 1985D, ALBAJAR 1987C, ALBAJAR 1991C, ALBAJAR 1991D
Falciano S. ADEVA 1990R, ADEVA 1991D, DECAMP 1991S
Falvard A. DECAMP 1991D, DECAMP 1991N, DECAMP 1991S, BUSKULIC 1993H
Fan Q. ADEVA 1990R, ADEVA 1991D, DECAMP 1991S, ABE 1994ZE, ABE 1995
Fan S.J. ADEVA 1990R, ADEVA 1991D, DECAMP 1991S
Fang G.Y. BARBER 1979F
Farhat B. ABE 1994ZE
Farilla A. DECAMP 1991D, DECAMP 1991N, DECAMP 1991S, BUSKULIC 1993H
Farrar G.R. BRODSKY 1973
Fasold H.G. FAISSNER 1978D
Fassouliotis D. ABREU 1991B, DECAMP 1991S
Fatyga M.K. ABACHI 1995G
Fatyga M.N. ABACHI 1995G
Favier J. CHARPAK 1968, AUBERT 1983
Fay J. ADEVA 1990R, ADEVA 1991D, DECAMP 1991S
Fayard L. BANNER 1983B, BAGNAIA 1983B, ANSARI 1987E
Fazzini T. FAZZINI 1958
Fearnley T. BUSKULIC 1993H
Featherly J. ABACHI 1995G
Feher S. ABACHI 1995G
Fehlmann J. ADEVA 1990R, ADEVA 1991D, DECAMP 1991S
Fein D. ABACHI 1995G
Felcini M. ALBAJAR 1991C, ALBAJAR 1991D
Feldman G.J. AUGUSTIN 1974B, ABRAMS 1974, SCHWITTERS 1975, PERL 1975C, HANSON 1975, PERL 1976C, GOLDHABER 1976D, PERUZZI 1976C, PERL 1977F, ABRAMS 1989B, WEIR 1989B
Feldscher L.R. HIRATA 1987C
Felicetti F. BACCI 1974
Felici G. DECAMP 1991D, DECAMP 1991N, DECAMP 1991S, BUSKULIC 1993H
Felst R. BRANDELIK 1977D, BARTEL 1980D
Feng Z. DECAMP 1991D, DECAMP 1991N, DECAMP 1991S, BUSKULIC 1993H
Fenker H. ADEVA 1990R
Fensome I.F. ALBAJAR 1987C, ALBAJAR 1991C, ALBAJAR 1991D
Ferbel T. ABACHI 1995G
Ferguson D.P.S. DECAMP 1991D, DECAMP 1991N, DECAMP 1991S, BUSKULIC 1993H
Ferguson T. ANDREWS 1980, ANDREWS 1980B, BEBEK 1981, CHADWICK 1981, AVERY 1989C, ARTUSO 1989C, ALAM 1989B, ADEVA 1990R, ADEVA 1991D, DECAMP 1991S
Fermi E. FERMI 1926, FERMI 1926B, FERMI 1934, FERMI 1943, FERMI 1949, ANDERSON 1952B, ANDERSON 1952E
Fernandes D. ABRAMS 1989B
Fernandez Alonso M. ABREU 1991B, DECAMP 1991S
Fernandez E. WEIR 1989B, DECAMP 1991D, DECAMP 1991N, DECAMP 1991S, BUSKULIC 1993H
Fernandez G. ADEVA 1990R, ADEVA 1991D, DECAMP 1991S
Fernandez-Bosman M. DECAMP 1991D, DECAMP 1991N, DECAMP 1991S, BUSKULIC 1993H
Fernow R.C. LIN 1978
Ferrando A. ALBAJAR 1991C, ALBAJAR 1991D
Ferrante I. DECAMP 1991D, DECAMP 1991N, DECAMP 1991S, BUSKULIC 1993H
Ferrari R. ANSARI 1987E
Ferrer A. ABREU 1991B, DECAMP 1991S
Ferrero M.I. AUBERT 1983, ASHMAN 1988B
Ferro-Luzzi M. TRIPP 1962
Ferroni F. ADEVA 1990R, ADEVA 1991D, DECAMP 1991S
Fesefeldt H. BARBER 1979F, ADEVA 1990R, ADEVA 1991D, DECAMP 1991S
Feynman R.P. FEYNMAN 1948, FEYNMAN 1948B, FEYNMAN 1948C, FEYNMAN 1949, FEYNMAN 1949B, FEYNMAN 1950, FEYNMAN 1958, FEYNMAN 1969
Fickinger W. STONEHILL 1961
Fidecaro F. DECAMP 1991D, DECAMP 1991N, DECAMP 1991S, BUSKULIC 1993H
Fidecaro G. FAZZINI 1958, FIDECARO 1978
Fidecaro M. FIDECARO 1978
Field J. ADEVA 1990R, ADEVA 1991D, DECAMP 1991S
Field R.C. ABRAMS 1989B, WEIR 1989B
Filippas T. ABREU 1991B, DECAMP 1991S
Filthaut F. ADEVA 1990R, ADEVA 1991D, DECAMP 1991S
Finch A.J. DECAMP 1991D, DECAMP 1991N, DECAMP 1991S, BUSKULIC 1993H
Fincke M. ARNISON 1983D, ARNISON 1984C, ARNISON 1984G
Fincke-Keeler M. ARNISON 1985D, ALBAJAR 1987C, ALEXANDER 1991C, ACTON 1992B, DECAMP 1991S
Finocchiaro G. AMENDOLIA 1973B, BOHRINGER 1980B, FINOCCHIARO 1980, ADEVA 1990R, ADEVA 1991D, DECAMP 1991S, ABACHI 1995G
Fiori I. ABE 1994ZE, ABE 1995
Fiorini E. HASERT 1973, HASERT 1973B, BLIETSCHAU 1976, BLIETSCHAU 1977B
Firestone A. ABREU 1991B, DECAMP 1991S
Firestone I. FIRESTONE 1971C
Firth D.R. CRETIEN 1962
Fischer G.E. AUGUSTIN 1974B, ABRAMS 1974, PERUZZI 1976C
Fischer H.M. BRANDELIK 1979H, BRANDELIK 1980G, ALEXANDER 1991C, ACTON 1992B, DECAMP 1991S
Fisher P.H. ADEVA 1990R, ADEVA 1991D, DECAMP 1991S
Fisher S.M. DECAMP 1991D, DECAMP 1991N, DECAMP 1991S
Fisk H.E. ABACHI 1995G
Fisk R.J. INNES 1977, UENO 1979
Fisyak Y.V. ABACHI 1995G
Fitch V.L. FITCH 1956, CHRISTENSON 1964, FITCH 1965B

Flattum E. ABACHI 1995G
Flauger W. AUBERT 1983
Flaugher B. ABE 1994ZE, ABE 1995
Fletcher E.R. DANYSZ 1963
Flugge G. BERGER 1978, BERGER 1979C
Flynn P. ARNISON 1984G, ARNISON 1985D, ALBAJAR 1987C
Foa L. AMENDOLIA 1973B, DECAMP 1991D, DECAMP 1991N, DECAMP 1991S, BUSKULIC 1993H
Focardi E. DECAMP 1991D, DECAMP 1991N, DECAMP 1991S, BUSKULIC 1993H
Fock V.A. FOCK 1932
Foeth H. ABREU 1991B, DECAMP 1991S
Fogli-Muciaccia M.T. CALICCHIO 1980
Fohrmann R. BRANDELIK 1979H, BRANDELIK 1980G
Fokitis E. ABREU 1991B, DECAMP 1991S
Folegati P. ABREU 1991B, DECAMP 1991S
Foley H.M. KUSCH 1947, FOLEY 1948
Folger H. JUNG 1992
Fominykh B.A. ALBRECHT 1987P
Fong D. BARBER 1979F
Fong D.G. ALEXANDER 1991C, ACTON 1992B, DECAMP 1991S
Fontaine G. ARNISON 1983C, ARNISON 1983D, ARNISON 1984C, ARNISON 1984G, ARNISON 1985D, ALBAJAR 1987C
Fontanelli F. ABREU 1991B, DECAMP 1991S
Forconi G. ADEVA 1990R, ADEVA 1991D, DECAMP 1991S
Ford W.T. BENVENUTI 1974F, BAND 1988, ABRAMS 1989B, WEIR 1989B
Forden G.E. ABACHI 1995G
Fordham C. ABRAMS 1989B, WEIR 1989B
Foreman T. ADEVA 1990R, ADEVA 1991D, DECAMP 1991S
Forsbach H. ABREU 1991B, DECAMP 1991S
Forti F. DECAMP 1991D, DECAMP 1991N, DECAMP 1991S, BUSKULIC 1993H
Fortner M. ABACHI 1995G
Fortson L. ALBAJAR 1991C, ALBAJAR 1991D
Forty R.W. DECAMP 1991D, DECAMP 1991N, DECAMP 1991S, BUSKULIC 1993H
Foster B. BRANDELIK 1979H, BRANDELIK 1980G
Foster F. BARTEL 1980D, DECAMP 1991D, DECAMP 1991N, DECAMP 1991S, BUSKULIC 1993H
Foster G.W. BIONTA 1987C, ABE 1994ZE, ABE 1995
Fouque G. DECAMP 1991D, DECAMP 1991N, DECAMP 1991S, BUSKULIC 1993H
Fournier J.P. ARNISON 1983C
Fowler E.C. STONEHILL 1961, BROWN 1962
Fowler P.H. BROWN 1949B, FOWLER 1951, DAVIES 1955
Fowler W.B. FOWLER 1953C, FOWLER 1953B, FOWLER 1954C, AGNEW 1958, GOLDHABER 1959, GOOD 1961, BROWN 1962, BARNES 1964B, BARNES 1964E
Frame K.C. ABACHI 1995G
Francis D. ASHMAN 1988B
Francois P.E. DAVIES 1955
Francois T. CALICCHIO 1980
Franek B. ABREU 1991B, DECAMP 1991S
Frank I.M. TAMM 1937
Frank M. BUSKULIC 1993H
Franke G. BERGER 1978, BERGER 1979C, BERGER 1980L
Franklin M. ABE 1994ZE, ABE 1995
Franzinetti C. CHAMBERLAIN 1956E, CHAMBERLAIN 1956F, BARKAS 1957
Franzini P. EISLER 1958, BOHRINGER 1980B, FINOCCHIARO 1980, ABACHI 1995G
Franzke B. JUNG 1992
Fraternali M. BANNER 1983B, BAGNAIA 1983B, ANSARI 1987E
Frauenfelder H. FRAUENFELDER 1957
Frautschi M. ABE 1994ZE, ABE 1995
Frautschi S.C. CHEW 1961, FRAUTSCHI 1962
Fredriksen S. ABACHI 1995G
Freeman J. BRANDELIK 1979H, BRANDELIK 1980G, ABE 1994ZE, ABE 1995
French H. BALTAY 1979E
Frenkiel P. ABREU 1991B, DECAMP 1991S
Frenzel E. FAISSNER 1978D
Fretter W.B. FRETTER 1951
Freudenreich K. ADEVA 1990R, ADEVA 1991D, DECAMP 1991S
Frey R. ARNISON 1983C, ARNISON 1983D, ARNISON 1984C, ARNISON 1984G, ARNISON 1985D, ALBAJAR 1987C, ABRAMS 1989B, WEIR 1989B
Friebel W. ADEVA 1990R, ADEVA 1991D, DECAMP 1991S
Friedberg C.E. AUGUSTIN 1974B, ABRAMS 1974, SCHWITTERS 1975, PERL 1975C, HANSON 1975, PERL 1976C, GOLDHABER 1976D, PERUZZI 1976C
Friedlander M.W. DAVIES 1955
Friedman J.I. FRIEDMAN 1957, BLOOM 1969B, BREIDENBACH 1969, MILLER 1972B, BODEK 1983, ABE 1994ZE, ABE 1995
Fries D.C. ABREU 1991B, DECAMP 1991S
Fries R. BRANDELIK 1977D
Frisch H. ABE 1994ZE, ABE 1995
Frisch R. FRISCH 1933
Frisken W.R. ALBRECHT 1987P
Fritzsch H. FRITZSCH 1973
Frodesen A.G. ABREU 1991B, DECAMP 1991S
Froidevaux D. BANNER 1983B, BAGNAIA 1983B, ANSARI 1987E
Froissart M. FROISSART 1961
Fruhwirth R. FIDECARO 1978, ARNISON 1983C, ARNISON 1983D, ARNISON 1984C, ARNISON 1984G, ARNISON 1985D, ALBAJAR 1987C, ABREU 1991B, DECAMP 1991S
Fry A. ABE 1994ZE
Fry W.F. FRY 1956, HASERT 1973B
Fryberger D. AUGUSTIN 1974B, ABRAMS 1974, PERL 1975C, HANSON 1975, GOLDHABER 1976D, PERUZZI 1976C
Fubini S. AMATI 1962
Fues W. BRANDELIK 1977D
Fuess S. ABACHI 1995G
Fuess T.A. ALBAJAR 1991C, ALBAJAR 1991D, ABE 1994ZE, ABE 1995
Fujimoto Y. MINAKAWA 1959
Fujino D. ABRAMS 1989B, WEIR 1989B
Fukui S. FUKUI 1959
Fukui Y. ABE 1994ZE, ABE 1995
Fukunaga C. ALEXANDER 1991C, ACTON 1992B, DECAMP 1991S
Fukushima M. BARBER 1979F, ADEVA 1990R, ADEVA 1991D, DECAMP 1991S
Fulda-Quenzer F. ABREU 1991B, DECAMP 1991S
Fulton R. AVERY 1989C, ARTUSO 1989C, ALAM 1989B
Fumagalli G. BANNER 1983B, BAGNAIA 1983B
Funaki S. ABE 1994ZE, ABE 1995
Fung S. FUNG 1956
Furnival K. ABREU 1991B, DECAMP 1991S
Furstenau H. ABREU 1991B, DECAMP 1991S
Fuster J. ABREU 1991B, DECAMP 1991S
Gabathuler E. AUBERT 1983, ASHMAN 1988B
Gabriel J.C. ALBRECHT 1987P
Gabriel W. BERGER 1979C, BERGER 1980L
Gadermann E. BRANDELIK 1977D
Gagliardi G. ABE 1994ZE, ABE 1995
Gago J.M. ABREU 1991B, DECAMP 1991S
Gaidot A. ALEXANDER 1991C, ACTON 1992B, DECAMP 1991S
Gaillard J.M. DANBY 1962, BANNER 1983B, BAGNAIA 1983B, ANSARI 1987E

Gailloud M. BROWN 1962, BIAGI 1983C, BIAGI 1985C, ADEVA 1990R, ADEVA 1991D, DECAMP 1991S
Gaines I. KNAPP 1976B
Gaitan V. DECAMP 1991D, DECAMP 1991N, DECAMP 1991S, BUSKULIC 1993H
Gajewski J. ASHMAN 1988B
Gajewski W. BIONTA 1987C
Galaktionov Y.V. ADEVA 1990R, ADEVA 1991D, DECAMP 1991S
Galeazzi G. ABREU 1991B, DECAMP 1991S
Galeotti S. ABE 1994ZE, ABE 1995
Galik R.S. AVERY 1989C, ARTUSO 1989C, ALAM 1989B
Galjaev A.N. ABACHI 1995G
Gallas E. ABACHI 1995G
Gallinaro M. ABE 1994ZE, ABE 1995
Gallo E. ADEVA 1990R, ADEVA 1991D, DECAMP 1991S
Gamba D. ABREU 1991B, DECAMP 1991S
Gamess A. DECAMP 1991D, DECAMP 1991N, DECAMP 1991S
Gamet R. AUBERT 1983, ASHMAN 1988B
Gamow G. GAMOW 1928, GAMOW 1936, GAMOW 1946
Gan K.K. ABRAMS 1989B, WEIR 1989B
Ganci P. ANDREWS 1980, ANDREWS 1980B, BEBEK 1981, CHADWICK 1981
Ganel O. ALEXANDER 1991C, ACTON 1992B, DECAMP 1991S
Ganezer K.S. BIONTA 1987C
Ganguli S.N. ADEVA 1990R, ADEVA 1991D, DECAMP 1991S
Ganis G. DECAMP 1991D, DECAMP 1991N, DECAMP 1991S, BUSKULIC 1993H
Gao C.S. ABACHI 1995G
Gao S. ABACHI 1995G
Gao Y. DECAMP 1991D, DECAMP 1991N, DECAMP 1991S, BUSKULIC 1993H
Gao Y.S. DECAMP 1991D, DECAMP 1991N, DECAMP 1991S, BUSKULIC 1993H
Gapienko V.A. VOYVODIC 1986B
Garbovska K. DANYSZ 1963
Garbutt D.A. DANYSZ 1963, BRANDELIK 1979H, BRANDELIK 1980G
Garcia C. ABREU 1991B, DECAMP 1991S
Garcia J. ABREU 1991B, DECAMP 1991S
Garcia-Abia P. ADEVA 1990R, ADEVA 1991D, DECAMP 1991S
Garcia-Sciveres M. ABE 1995
Gardner E. BURFENING 1949
Gardner F.T. HILL 1956
Garelick D. BLUM 1975
Garfinkel A.F. BERGER 1978, ABE 1994ZE, ABE 1995
Garren L. AVERY 1989C, ARTUSO 1989C, ALAM 1989B
Garrido L. DECAMP 1991D, DECAMP 1991N, DECAMP 1991S, BUSKULIC 1993H
Garvey J. ARNISON 1983C, ARNISON 1983D, ARNISON 1984C, ARNISON 1984G, ARNISON 1985D, ALBAJAR 1987C, ALBAJAR 1991C, ALBAJAR 1991D
Garwin E.L. PRESCOTT 1978C, PRESCOTT 1979C
Garwin R.L. GARWIN 1957
Gary J.W. ALEXANDER 1991C, ACTON 1992B, DECAMP 1991S
Gascon J. ALEXANDER 1991C, ACTON 1992B, DECAMP 1991S
Gasparini F. FIDECARO 1978
Gasparini U. ABREU 1991B, DECAMP 1991S
Gately L. ALBAJAR 1987C
Gather K. BRANDELIK 1979H, BRANDELIK 1980G
Gau S.S. ADEVA 1990R, ADEVA 1991D, DECAMP 1991S
Gavillet P. ABREU 1991B, DECAMP 1991S
Gay C. BUSKULIC 1993H, ABE 1995
Gay P. DECAMP 1991D, DECAMP 1991N, DECAMP 1991S, BUSKULIC 1993H
Gayler J. AUBERT 1983
Gazis E.N. ABREU 1991B, DECAMP 1991S
Geddes N.I. ALEXANDER 1991C, ACTON 1992B, DECAMP 1991S
Gee C.N. BIAGI 1983C, BIAGI 1985C
Gee D. ALBAJAR 1987C
Geer S. ARNISON 1983C, ARNISON 1983D, ARNISON 1984C, ARNISON 1984G, ARNISON 1985D, ALBAJAR 1987C, ABE 1994ZE, ABE 1995
Geich-Gimbel C. ALEXANDER 1991C, ACTON 1992B, DECAMP 1991S
Geiges G. DECAMP 1991D, DECAMP 1991N, DECAMP 1991S
Geiser A. ALBAJAR 1987C, ALBAJAR 1991C, ALBAJAR 1991D
Geld T.L. ABACHI 1995G
Gele D. ADEVA 1991D, DECAMP 1991S
Gell-Mann M. GELLMANN 1953, GELL-MANN 1954B, GELL-MANN 1954, GELL-MANN 1955, GELL-MANN 1956, GELL-MANN 1957, FEYNMAN 1958, GELL-MANN 1961, GELL-MANN 1962, FRAUTSCHI 1962, GELL-MANN 1964, FRITZSCH 1973
Genik R.J.II. ABACHI 1995G
Gennow H. ALBRECHT 1987P
Genser K. ABACHI 1995G
Gensler S.W. ALEXANDER 1991C, ACTON 1992B, DECAMP 1991S
Gentile S. ADEVA 1990R, ADEVA 1991D, DECAMP 1991S
Gentile T. ANDREWS 1980, ANDREWS 1980B, BEBEK 1981, CHADWICK 1981
Gentit F.X. ALEXANDER 1991C, ACTON 1992B, DECAMP 1991S
Genzel H. BERGER 1979C, BERGER 1980L
Georgiopoulos C. DECAMP 1991D, DECAMP 1991N, DECAMP 1991S, BUSKULIC 1993H
Gerasimov S.B. BALDIN 1973
Gerber C.E. ABACHI 1995G
Gerber J.P. ABREU 1991B, DECAMP 1991S
Gerbier G. CALICCHIO 1980
Gerdes D.W. ABE 1994ZE, ABE 1995
Gerhardt V. AUBERT 1983
Gerke C. BERGER 1978, BERGER 1979C, BERGER 1980L
Germer L.H. DAVISSON 1927, DAVISSON 1927B
Gero E. ABRAMS 1989B
Gershtein S.S. GERSHTEIN 1955, GERSHTEIN 1966
Gessaroli R. PEVSNER 1961
Gettner M. BAND 1988
Geweniger C. DECAMP 1991D, DECAMP 1991N, DECAMP 1991S, BUSKULIC 1993H
Ghesquiere C. ARNISON 1983C, ARNISON 1983D, ARNISON 1984C, ARNISON 1984G, ARNISON 1985D, ALBAJAR 1987C
Ghez P. ARNISON 1983C, ARNISON 1983D, ARNISON 1984C, ARNISON 1984G, ARNISON 1985D, ALBAJAR 1987C, BUSKULIC 1993H
Ghio F. ARNISON 1985D, ALBAJAR 1987C
Giacomelli G. BUSHNIN 1969, DENISOV 1971F, ALEXANDER 1991C, ACTON 1992B, DECAMP 1991S
Giacomelli P. ABREU 1991B, DECAMP 1991S
Giannetti P. ABE 1994ZE, ABE 1995
Giannini G. BOHRINGER 1980B, FINOCCHIARO 1980, DECAMP 1991D, DECAMP 1991N, DECAMP 1991S, BUSKULIC 1993H
Giassi A. DECAMP 1991D, DECAMP 1991N, DECAMP 1991S, BUSKULIC 1993H
Gibbard B.G. ANDREWS 1980, ABACHI 1995G
Giboni K.L. ARNISON 1983C, ARNISON 1983D, ARNISON 1984C, ARNISON 1984G
Gibson V. ASHMAN 1988B, ALEXANDER 1991C, ACTON 1992B, DECAMP 1991S
Gibson W.M. BIAGI 1983C, BIAGI 1985C

Gibson W.R. ARNISON 1983C, ARNISON 1983D, ARNISON 1984C, ARNISON 1984G, ARNISON 1985D, ALBAJAR 1987C, ALEXANDER 1991C, ACTON 1992B, DECAMP 1991S

Gidal G. ABRAMS 1989B, WEIR 1989B

Gilchriese M.G.D. ANDREWS 1980, ANDREWS 1980B, BEBEK 1981, CHADWICK 1981

Gildemeister O. BANNER 1983B, BAGNAIA 1983B, ANSARI 1987E

Gilkinson D.J. ALBRECHT 1987P

Gillies J. ASHMAN 1988B

Gillies J.D. ALEXANDER 1991C, ACTON 1992B, DECAMP 1991S

Gilly L. AGNEW 1958

Gingrich D.M. ALBRECHT 1987P

Ginzton E.L. CHODOROW 1955

Giokaris N. BODEK 1983, ABE 1994ZE, ABE 1995

Giordenescu N. BALDIN 1973

Giorgi M. FIDECARO 1978

Giorgi M.A. DECAMP 1991D, DECAMP 1991N, DECAMP 1991S, BUSKULIC 1993H

Giraud-Heraud Y. ARNISON 1983C, ARNISON 1983D, ARNISON 1984C, ARNISON 1984G, ARNISON 1985D, ALBAJAR 1987C

Giromini P. AMENDOLIA 1973B, ABE 1994ZE, ABE 1995

Girtler P. DECAMP 1991D, DECAMP 1991N, DECAMP 1991S, BUSKULIC 1993H

Giselman K. BARNETT 1996

Gittelman B. ANDREWS 1980, ANDREWS 1980B, BEBEK 1981, CHADWICK 1981, AVERY 1989C, ARTUSO 1989C, ALAM 1989B

Giusti P. BASILE 1981I

Givernaud A. ARNISON 1983C, ARNISON 1983D, ARNISON 1984C, ARNISON 1984G, ARNISON 1985D, ALBAJAR 1987C, ALBAJAR 1991C, ALBAJAR 1991D

Gladding G. BUESSER 1973, KNAPP 1976B

Gladney L. WEIR 1989B, ABE 1994ZE, ABE 1995

Glanzman T. ABRAMS 1989B, WEIR 1989B

Glaser D.A. GLASER 1953

Glaser R. ALBRECHT 1987P

Glashow S.L. GLASHOW 1961, BJORKEN 1964, GLASHOW 1970

Glasser R.G. GLASSER 1961, ABRAMS 1964, BERGER 1979C, BERGER 1980L

Glaubman M. ADEVA 1990R, ADEVA 1991D, DECAMP 1991S, ABACHI 1995G

Glawe U. BIENLEIN 1978

Glebov V.Y. ABACHI 1995G

Glenn S. ABACHI 1995G

Glenzinski D. ABE 1994ZE, ABE 1995

Glicenstein J.F. ABACHI 1995G

Glitza K.W. ABREU 1991B, DECAMP 1991S

Gobbi B. ABACHI 1995G

Gobbo B. DECAMP 1991D, DECAMP 1991N, DECAMP 1991S, BUSKULIC 1993H

Goddard M.C. BARTEL 1980D

Goderre G.P. BAND 1988

Goebel C. GOEBEL 1958

Goforth M. ABACHI 1995G

Goggi V.G. BANNER 1983B, BAGNAIA 1983B, ANSARI 1987E

Gokieli R. ABREU 1991B, DECAMP 1991S

Gold M. ABE 1994ZE, ABE 1995

Gold M.S. WEIR 1989B

Goldansky V.I. GOLDANSKY 1960

Goldberg J. ALEXANDER 1991C, ACTON 1992B, DECAMP 1991S

Goldberg M. BERTANZA 1962D, CONNOLLY 1963, GOLDBERG 1964C, ANDREWS 1980, ANDREWS 1980B, BEBEK 1981, CHADWICK 1981, AVERY 1989C, ARTUSO 1989C, ALAM 1989B

Goldberger M.L. GELL-MANN 1954, GOLDBERGER 1955, GOLDBERGER 1955B, GOLDBERGER 1958

Goldfarb S. ADEVA 1990R, ADEVA 1991D, DECAMP 1991S

Goldhaber G. CHUPP 1955, ILOFF 1955, CHAMBERLAIN 1956E, CHAMBERLAIN 1956F, BARKAS 1957, GOLDHABER 1959, GOLDHABER 1960, CHINOWSKY 1962, FIRESTONE 1971C, AUGUSTIN 1974B, ABRAMS 1974, SCHWITTERS 1975, PERL 1975C, HANSON 1975, PERL 1976C, GOLDHABER 1976D, PERUZZI 1976C, PERL 1977F, ABRAMS 1989B, WEIR 1989B

Goldhaber M. CHADWICK 1934, GOLDHABER 1958C, BIONTA 1987C

Goldhaber S. CHUPP 1955, ALVAREZ 1955, ILOFF 1955, CHAMBERLAIN 1956F, BARKAS 1957, GOLDHABER 1959, GOLDHABER 1960, CHINOWSKY 1962

Goldhagen P. AUBERT 1974D

Goldschmidt A. ABACHI 1995G

Goldstone J. GOLDSTONE 1961, GOLDSTONE 1961B

Golovatyuk V.M. ABREU 1991B, DECAMP 1991S

Golutvin A.I. ALBRECHT 1987P

Gomez Y Cadenas J.J. ABRAMS 1989B, ABREU 1991B, DECAMP 1991S

Gomez B. ABACHI 1995G

Goncharov P.I. ABACHI 1995G

Gong Z.F. ADEVA 1990R, ADEVA 1991D, DECAMP 1991S

Gonidec A. PRESCOTT 1978C, PRESCOTT 1979C, ARNISON 1983C, ARNISON 1983D, ARNISON 1984C, ARNISON 1984G, ARNISON 1985D, ALBAJAR 1987C, ALBAJAR 1991C, ALBAJAR 1991D

Gonzales B. ALBAJAR 1991C, ALBAJAR 1991D

Gonzalez E. ADEVA 1990R, ADEVA 1991D, DECAMP 1991S

Gonzalez J. ABE 1994ZE, ABE 1995

Goobar A. ABREU 1991B, DECAMP 1991S

Good M.L. ALVAREZ 1956, CRAWFORD 1958, ALVAREZ 1959, ALSTON 1960, ALSTON 1961B, ALSTON 1961E

Good R.H. GOOD 1961

Goodman M. KNAPP 1976B, ARNISON 1985D

Goodrick M.J. ALEXANDER 1991C, ACTON 1992B, DECAMP 1991S

Goos L.T. ABACHI 1995G

Gopal G. ABREU 1991B, DECAMP 1991S

Gordeev A.M. ADEVA 1990R, ADEVA 1991D, DECAMP 1991S

Gordon A. ABE 1994ZE, ABE 1995

Gordon H. ABACHI 1995G

Gordon J.C. BIAGI 1983C, BIAGI 1985C

Gorelov I.V. ALBRECHT 1987P

Gorichev P.A. VOYVODIC 1986B

Gorin Y.P. BUSHNIN 1969, BUSHNIN 1969, ANTIPOV 1970, DENISOV 1971F

Gormley M. KNAPP 1976B

Gorn W. ALEXANDER 1991C, ACTON 1992B, DECAMP 1991S

Gorski M. ABREU 1991B, DECAMP 1991S

Goshaw A.T. ABE 1994ZE, ABE 1995

Gössling C. AUBERT 1983

Gossling C. ANSARI 1987E

Gottfried C. FIDECARO 1978

Gottlicher P. ADEVA 1990R, ADEVA 1991D, DECAMP 1991S

Goudsmit S. UHLENBECK 1925

Goujon D. ADEVA 1990R, ADEVA 1991D, DECAMP 1991S

Goulianos K. DANBY 1962, ABE 1994ZE, ABE 1995

Goy C. DECAMP 1991D, DECAMP 1991N, DECAMP 1991S, BUSKULIC 1993H

Goz B. BARNES 1964E

Gozzini A. CONVERSI 1955

Grab C. DECAMP 1991D, DECAMP 1991N, DECAMP 1991S, BARNETT 1996

Gracco V. ABREU 1991B, DECAMP 1991S
Graf N. ABACHI 1995G
Grafström P. ASHMAN 1988B
Grahl J. DECAMP 1991D, DECAMP 1991N, DECAMP 1991S, BUSKULIC 1993H
Grandi C. ALEXANDER 1991C, ACTON 1992B, DECAMP 1991S
Grannis P.D. AMENDOLIA 1973B, ABACHI 1995G
Grant A. ABREU 1991B, DECAMP 1991S
Grant F.C. ACTON 1992B
Grard F. ABREU 1991B, DECAMP 1991S
Grassmann H. ARNISON 1985D, ALBAJAR 1987C, ABE 1994ZE, ABE 1995
Gratta G. ABRAMS 1989B, ADEVA 1990R, ADEVA 1991D, DECAMP 1991S
Gray L. BERTANZA 1962D
Gray R.J. BIAGI 1983C, BIAGI 1985C
Gray S.W. AVERY 1989C, ARTUSO 1989C, ALAM 1989B
Grayer G. BLUM 1975, ARNISON 1983C, ARNISON 1983D, ARNISON 1984C, ARNISON 1984G, ARNISON 1985D, ALBAJAR 1987C
Graziani E. ABREU 1991B, DECAMP 1991S
Graziano W. ALVAREZ 1959, ALSTON 1960, ALSTON 1961B, ALSTON 1961E
Green A. WEIR 1989B
Green D.R. AMENDOLIA 1973B, ABACHI 1995G
Green J. ABACHI 1995G
Green M.G. DECAMP 1991D, DECAMP 1991N, DECAMP 1991S, BUSKULIC 1993H
Greenberg O.W. GREENBERG 1964, GREENBERG 1966
Greene A.M. DECAMP 1991D, DECAMP 1991N, DECAMP 1991S, BUSKULIC 1993H
Greenlee H. ABACHI 1995G
Gregorio A. BUSKULIC 1993H
Gregory B.P. GREGORY 1954, ARMENTEROS 1955
Gregory J.M. ALBAJAR 1991C, ALBAJAR 1991D
Gress D. VOYVODIC 1986B
Grewal A. ABE 1994ZE
Gribov V.N. GRIBOV 1961, GRIBOV 1972
Grieco G. ABE 1994ZE
Griffin G. ABACHI 1995G
Grigull R. BERGER 1979C, BERGER 1980L
Grilli M. DAVIES 1955, BACCI 1974
Grindhammer G. BRANDELIK 1977D, ABRAMS 1989B
Grinnell C. ADEVA 1990R, ADEVA 1991D, DECAMP 1991S
Grivaz J.F. DECAMP 1991D, DECAMP 1991N, DECAMP 1991S, BUSKULIC 1993H
Grodzins L. GOLDHABER 1958C
Groer L. ABE 1994ZE, ABE 1995
Gronberg J. ALBAJAR 1991C, ALBAJAR 1991D
Groom D.E. BAND 1988, BARNETT 1996
Gros M.H. ABREU 1991B, DECAMP 1991S
Grosdidier G. ABREU 1991B, DECAMP 1991S
Groshev V.R. ARTAMONOV 1984B
Gross D.J. GROSS 1973, GROSS 1973B
Gross E. ALEXANDER 1991C, DECAMP 1991S
Grosse-Wiesmann P. ABRAMS 1989B, WEIR 1989B
Grossetete G. ABREU 1991B, DECAMP 1991S
Grossman N. ABACHI 1995G
Grossmann P. BRANDELIK 1980G
Grosso-Pilcher C. ABE 1994ZE, ABE 1995
Grote H. BAGNAIA 1983B
Gruber A. JUNG 1992
Grudberg P. ABACHI 1995G
Grundzik M. CONNOLLY 1963, GOLDBERG 1964C
Grunendahl S. ABACHI 1995G
Grunewald M. ADEVA 1990R, ADEVA 1991D, DECAMP 1991S
Grupen C. BERGER 1978, BERGER 1979C, BERGER 1980L, DECAMP 1991D, DECAMP 1991N, DECAMP 1991S, BUSKULIC 1993H
Guanziroli M. ADEVA 1990R, ADEVA 1991D, DECAMP 1991S
Guerriero L. CRETIEN 1962
Guicheney C. BUSKULIC 1993H
Guida J.A. ABACHI 1995G
Guida J.M. ABACHI 1995G
Guillian G. ABE 1995
Guirlet R. DECAMP 1991D, DECAMP 1991N, DECAMP 1991S
Gumenyuk S.A. ABREU 1991B, DECAMP 1991S
Gunderson B. BRANDELIK 1977D
Guo J.C. BARBER 1979F
Guo J.K. ADEVA 1991D, DECAMP 1991S
Guo R.S. ABE 1995
Guralnik G.S. GURALNIK 1964
Gürsey F. GÜRSEY 1960, GÜRSEY 1964
Gurtu A. ADEVA 1990R, ADEVA 1991D, DECAMP 1991S, BARNETT 1996
Guryn W. ARNISON 1985D, ABACHI 1995G
Gurzhiev S.N. ABACHI 1995G
Gustafson H.R. ADEVA 1990R, ADEVA 1991D, DECAMP 1991S
Gutay L.J. ADEVA 1990R, ADEVA 1991D, DECAMP 1991S
Gutierrez P. ARNISON 1983C
Gutnikov Y.E. ABACHI 1995G
Guy J. ABREU 1991B, DECAMP 1991S
Guy J.G. CALICCHIO 1980
Haan H. ADEVA 1990R, ADEVA 1991D, DECAMP 1991S
Haas J. AUBERT 1983
Haas P. AVERY 1989C, ARTUSO 1989C, ALAM 1989B
Haber C. ABE 1994ZE, ABE 1995
Hackmack E. BERGER 1978
Hadley J. PANOFSKY 1951
Hadley N.J. ABACHI 1995G
Hagelberg R. DECAMP 1991D, DECAMP 1991N, DECAMP 1991S, BUSKULIC 1993H
Hagemann J. ALEXANDER 1991C, ACTON 1992B, DECAMP 1991S
Hagen C.R. GURALNIK 1964
Hagerty P.E. BARNES 1964B, BARNES 1964E
Haggerty H. ABACHI 1995G
Haggerty J. ANDREWS 1980, ANDREWS 1980B, BEBEK 1981, CHADWICK 1981, WEIR 1989B
Hagopian S. ABACHI 1995G
Hagopian V. ABACHI 1995G
Haguenauer M. HASERT 1973, HASERT 1973B, BLIETSCHAU 1976, BLIETSCHAU 1977B
Hahn B. BANNER 1983B, BAGNAIA 1983B, ANSARI 1987E
Hahn F. ABREU 1991B, DECAMP 1991S
Hahn K.S. ABACHI 1995G
Hahn M. ABREU 1991B, DECAMP 1991S
Hahn S.R. ABE 1994ZE, ABE 1995
Haidan R. ARNISON 1983C
Haider S. ABREU 1991B, DECAMP 1991S
Haidt D. HASERT 1973, HASERT 1973B, BLIETSCHAU 1976, BLIETSCHAU 1977B, BARTEL 1980D
Haines T.J. BIONTA 1987C
Hajduk Z. ABREU 1991B, DECAMP 1991S
Hakansson A. ABREU 1991B, DECAMP 1991S
Hall R.E. ABACHI 1995G
Halley A.W. DECAMP 1991D, DECAMP 1991N, DECAMP 1991S, BUSKULIC 1993H
Hallgren A. ABREU 1991B, DECAMP 1991S
Halling A.M. AVERY 1989C, ARTUSO 1989C, ALAM 1989B
Hamacher K. AUBERT 1983, ASHMAN 1988B, ABREU 1991B, DECAMP 1991S
Hamel deMonchenault G ABREU 1991B, DECAMP 1991S
Hamilton R. ABE 1994ZE, ABE 1995
Han K. BOHRINGER 1980B, FINOCCHIARO 1980
Han M.Y. HAN 1965
Hancke S. ADEVA 1990R
Handler R. BUNCE 1976, ABE 1994ZE, ABE 1995
Hangarter K. ADEVA 1990R

Hanke P. DECAMP 1991D, DECAMP 1991N, DECAMP 1991S, BUSKULIC 1993H
Hanni H. BANNER 1983B, BAGNAIA 1983B, ANSARI 1987E
Hans R.M. ABE 1994ZE, ABE 1995
Hansen J.B. BUSKULIC 1993H
Hansen J.D. DECAMP 1991D, DECAMP 1991N, DECAMP 1991S, BUSKULIC 1993H
Hansen J.R. BANNER 1983B, BAGNAIA 1983B, ANSARI 1987E, DECAMP 1991D, DECAMP 1991N, DECAMP 1991S, BUSKULIC 1993H
Hansen P.H. BANNER 1983B, BAGNAIA 1983B, ANSARI 1987E, DECAMP 1991D, DECAMP 1991N, DECAMP 1991S, BUSKULIC 1993H
Hansen S. ABACHI 1995G
Hansen W.W. CHODOROW 1955
Hansl T. FAISSNER 1978D
Hansl-Kozanecka T. ARNISON 1983C, ARNISON 1983D, ARNISON 1984C, ARNISON 1984G, DECAMP 1991D, DECAMP 1991N, DECAMP 1991S
Hanson A.O. LYMAN 1951
Hanson G. AUGUSTIN 1974B, ABRAMS 1974, SCHWITTERS 1975, PERL 1975C, HANSON 1975, PERL 1976C, GOLDHABER 1976D, PERUZZI 1976C, PERL 1977F, ABRAMS 1989B, WEIR 1989B
Hanson G.G. ALEXANDER 1991C, ACTON 1992B, DECAMP 1991S
Hansroul M. ALEXANDER 1991C, ACTON 1992B, DECAMP 1991S
Hara K. ANSARI 1987E, ABE 1994ZE, ABE 1995
Harder G. ALBRECHT 1987P
Hargrove C.K. ALEXANDER 1991C, ACTON 1992B, DECAMP 1991S
Hariri A. BARBER 1979F
Harms P. BERGER 1978
Harnew N. ANSARI 1987E
Harr R. ABRAMS 1989B, WEIR 1989B
Harral B. ABRAMS 1989B, ABE 1994ZE, ABE 1995
Harris D.R. HODSON 1954
Harris F.A. ABRAMS 1989B, WEIR 1989B
Harris J.F. ABREU 1991B, DECAMP 1991S
Harris M. ADEVA 1990R
Harris R.M. ABE 1994ZE, ABE 1995
Harrison F.B. REYNOLDS 1950, COWAN 1956
Harrison P.F. ALEXANDER 1991C, ACTON 1992B, DECAMP 1991S
Hart E.L. BERTANZA 1962D, CONNOLLY 1963, BARNES 1964B, BARNES 1964E, GOLDBERG 1964C
Hart J.C. BRANDELIK 1979H, BRANDELIK 1980G, ALEXANDER 1991C, ACTON 1992B, DECAMP 1991S
Hartill D. SCHWITTERS 1975
Hartill D.L. HANSON 1975, ANDREWS 1980, ANDREWS 1980B, BEBEK 1981, CHADWICK 1981, AVERY 1989C, ARTUSO 1989C, ALAM 1989B
Hartmann G.C. MILLER 1972B
Hartmann H. BRANDELIK 1979H, BRANDELIK 1980G
Harton J.L. DECAMP 1991D, DECAMP 1991N, DECAMP 1991S, BUSKULIC 1993H
Harvey J. DECAMP 1991D, DECAMP 1991N, DECAMP 1991S, BUSKULIC 1993H
Hasan A. ADEVA 1990R, ADEVA 1991D, DECAMP 1991S
Hasegawa H. MINAKAWA 1959
Hasegawa S. MINAKAWA 1959
Hasemann H. DARDEN 1978, DARDEN 1978B
Hasert F.J. HASERT 1973, HASERT 1973B, BLIETSCHAU 1976, BLIETSCHAU 1977B
Hassard J.F. BARTEL 1980D, DECAMP 1991D, DECAMP 1991N, DECAMP 1991S, BUSKULIC 1993H
Hatcher R. ABACHI 1995G
Hatfield F. DECAMP 1991D, DECAMP 1991N, DECAMP 1991S
Hattersley P.M. ALEXANDER 1991C, ACTON 1992B, DECAMP 1991S
Hauger S.A. ABE 1994ZE, ABE 1995
Hauptman J.M. ABACHI 1995G
Hauschild M. ALEXANDER 1991C, ACTON 1992B, DECAMP 1991S
Hauschildt D. ADEVA 1990R, ADEVA 1991D, DECAMP 1991S
Hauser J. ABE 1994ZE, ABE 1995
Hawk C. ABE 1994ZE, ABE 1995
Hawkes C.M. ABRAMS 1989B, WEIR 1989B, ALEXANDER 1991C, ACTON 1992B, DECAMP 1991S
Hayashi E. ABE 1995
Hayes K. ABRAMS 1989B, WEIR 1989B
Hayes K.G. BARNETT 1996
Hayman P. AUBERT 1983, ASHMAN 1988B
Haynes W.J. ARNISON 1983C, ARNISON 1983D, ARNISON 1984C, ARNISON 1984G, ARNISON 1985D, ALBAJAR 1987C
Hayward R.W. WU 1957
Haywood S. ALBAJAR 1987C, DECAMP 1991D, DECAMP 1991N, DECAMP 1991S, BUSKULIC 1993H
He C.F. ADEVA 1990R, ADEVA 1991D, DECAMP 1991S
Hearns J.A. DECAMP 1991D, DECAMP 1991N, DECAMP 1991S
Hearty C. ABRAMS 1989B
Hebbeker T. ADEVA 1990R, ADEVA 1991D, DECAMP 1991S
Hebert M. ADEVA 1990R, ADEVA 1991D, DECAMP 1991S
Heck B. ABREU 1991B, DECAMP 1991S
Heckman H.H. BARKAS 1957
Hedgecock R. BARTEL 1980D
Hedin D. ABACHI 1995G
Heeran M. DANYSZ 1963
Heflin E. ALEXANDER 1991C, ACTON 1992B, DECAMP 1991S
Heile F.B. PERL 1975C, PERL 1976C
Heimlich F.H. BIENLEIN 1978
Heinrich J. ABE 1994ZE, ABE 1995
Heinson A.P. ABACHI 1995G
Heintz U. ABACHI 1995G
Heintze J. BARTEL 1980D
Heinzelmann G. BIENLEIN 1978, BARTEL 1980D
Heisenberg W. HEISENBERG 1925, BORN 1926, HEISENBERG 1927, HEISENBERG 1932, HEISENBERG 1943, HEISENBERG 1952
Heller K. BUNCE 1976
Helm M. BARTEL 1980D
Heltsley B.K. AVERY 1989C, ARTUSO 1989C, ALAM 1989B
Hemingway R.J. ALEXANDER 1991C, ACTON 1992B, DECAMP 1991S
Hempstead M. AVERY 1989C, ARTUSO 1989C, ALAM 1989B
Henckes M. AUBERT 1983
Hendel A. ARMENTEROS 1955
Hennessy D. ABE 1994ZE
Henrard P. DECAMP 1991D, DECAMP 1991N, DECAMP 1991S, BUSKULIC 1993H
Hentschel G. BLUM 1975
Hepp V. BERGER 1980L, DECAMP 1991D, DECAMP 1991N, DECAMP 1991S, BUSKULIC 1993H
Herb S.W. HERB 1977, INNES 1977, UENO 1979, BOHRINGER 1980B, FINOCCHIARO 1980
Herbst I. ABREU 1991B, DECAMP 1991S
Hernandez J.J. ABREU 1991B, DECAMP 1991S, BARNETT 1996
Hernandez-Montoya R. ABACHI 1995G
Herquet P. ABREU 1991B, DECAMP 1991S
Herr H. ABREU 1991B, DECAMP 1991S
Herr R. CARRON 1978

Herrup D. ANDREWS 1980, ANDREWS 1980B, BEBEK 1981, CHADWICK 1981, WEIR 1989B
Herten G. BARBER 1979F, ADEVA 1990R, ADEVA 1991D, DECAMP 1991S
Herten U. ADEVA 1990R, ADEVA 1991D, DECAMP 1991S
Hertzberger L.O. ARNISON 1983C, ARNISON 1983D, ARNISON 1984C, ARNISON 1984G
Herve A. ADEVA 1990R, ADEVA 1991D, DECAMP 1991S
Herzlinger M. ANDREWS 1980, ANDREWS 1980B, BEBEK 1981, CHADWICK 1981
Hess V.F. HESS 1912, HESS 1913, HESS 1926
Heuer R.D. BARTEL 1980D, ALEXANDER 1991C, ACTON 1992B, DECAMP 1991S
Heuring T. ABACHI 1995G
Heusch C.A. ABRAMS 1989B, WEIR 1989B
Heusse P. HASERT 1973, HASERT 1973B, BLIETSCHAU 1976, BLIETSCHAU 1977B, DECAMP 1991D, DECAMP 1991N, DECAMP 1991S, BUSKULIC 1993H
Heyland D. BRANDELIK 1979H, BRANDELIK 1980G
Hibbs M. BALTAY 1979E
Hicks G.S. CHRISTENSON 1970
Hietanen I. ABREU 1991B, DECAMP 1991S
Higgs P.W. HIGGS 1964, HIGGS 1964B, HIGGS 1966
Higon E. ABREU 1991B, DECAMP 1991S
Hikasa K. BARNETT 1996
Hildreth M. ABRAMS 1989B
Hilgart J. DECAMP 1991D, DECAMP 1991N, DECAMP 1991S, BUSKULIC 1993H
Hilger E. BRANDELIK 1979H, BRANDELIK 1980G
Hilgers K. ADEVA 1990R, ADEVA 1991D, DECAMP 1991S
Hilke H.J. ABREU 1991B, DECAMP 1991S
Hill J.C. ALEXANDER 1991C, ACTON 1992B, DECAMP 1991S
Hill R.O. HILL 1956
Hillen W. BRANDELIK 1979H, BRANDELIK 1980G
Hillier R. DAVIES 1955
Hillier S.J. ALEXANDER 1991C, ACTON 1992B, DECAMP 1991S
Himel T. BANNER 1983B, BAGNAIA 1983B, ABRAMS 1989B, WEIR 1989B
Hincks E.P. HINCKS 1948, HINCKS 1949
Hinshaw D.A. ABRAMS 1989B, ALEXANDER 1991C, ACTON 1992B, DECAMP 1991S
Hirata K. HIRATA 1987C
Hirosky R. ABACHI 1995G
Hladky J. ASTVACATUROV 1968
Ho C. ALEXANDER 1991C, ACTON 1992B, DECAMP 1991S
Ho M.C. BARBER 1979F
Hoag J.B. ROSSI 1939
Hoang T.F. GOLDHABER 1959
Hobbs J.D. ALEXANDER 1991C, ACTON 1992B, DECAMP 1991S, ABACHI 1995G
Hobson P.R. ALEXANDER 1991C, ACTON 1992B, DECAMP 1991S
Hochman D. ALEXANDER 1991C, ACTON 1992B, DECAMP 1991S
Hodges C. ARNISON 1983C, ARNISON 1983D
Hodgson S.D. ABREU 1991B, DECAMP 1991S
Hodson A.L. HODSON 1954
Hoeneisen B. ABACHI 1995G
Hofer H. ADEVA 1990R, ADEVA 1991D, DECAMP 1991S
Hoffmann D. FAISSNER 1978D, ARNISON 1983C, ARNISON 1983D, ARNISON 1984C, ARNISON 1984G
Hoffmann H. ARNISON 1983C, ARNISON 1983D, ARNISON 1984C, ARNISON 1984G, ARNISON 1985D, ALBAJAR 1987C
Hofmann W. DARDEN 1978, DARDEN 1978B
Hofmokl T. ABREU 1991B, DECAMP 1991S
Hofstadter R. MCALLISTER 1956
Hoftun J. ABACHI 1995G
Hohler G. BARNETT 1996
Hohlmann M. ABE 1995
Holck C. ABE 1995
Holder M. BRANDELIK 1979H, BRANDELIK 1980G
Holl B. ALEXANDER 1991C, ACTON 1992B, DECAMP 1991S
Hollebeek R.J. AUGUSTIN 1974B, ABRAMS 1974, WEIR 1989B, ABE 1994ZE, ABE 1995
Holloway L. ABE 1994ZE, ABE 1995
Holmes R. ABREU 1991B, DECAMP 1991S
Holmgren S.O. ABREU 1991B, DECAMP 1991S
Holscher A. ABE 1994ZE, ABE 1995
Holt J.R. ASHMAN 1988B
Holthuizen D.J. ARNISON 1983C, ARNISON 1983D, ARNISON 1984C, ARNISON 1984G, ARNISON 1985D, ALBAJAR 1987C, ABREU 1991B, ALBAJAR 1991C, ALBAJAR 1991D, DECAMP 1991S
Hom D.C. HERB 1977, INNES 1977, UENO 1979
Homer R.J. ARNISON 1983C, ARNISON 1983D, ARNISON 1984C, ARNISON 1984G, ARNISON 1985D, ALBAJAR 1987C, ALEXANDER 1991C, ACTON 1992B, DECAMP 1991S
Hong S. ABRAMS 1989B, ABE 1994ZE, ABE 1995
Honma A. ARNISON 1983C, ARNISON 1983D, ARNISON 1984C, ARNISON 1984G, ARNISON 1985D, ALBAJAR 1987C, ACTON 1992B
Honore P.F. ABREU 1991B, DECAMP 1991S
Hooper J.E. CARLSON 1950, DAVIES 1955, ABREU 1991B, DECAMP 1991S
Hoorani H. ADEVA 1990R, ADEVA 1991D, DECAMP 1991S
Hoppes D.D. WU 1957
Horber E. BIENLEIN 1978
Horlitz G. BERGER 1978
Horwitz N. BRABANT 1956C, BARNES 1964B, BARNES 1964E, ANDREWS 1980, ANDREWS 1980B, BEBEK 1981, CHADWICK 1981, AVERY 1989C, ARTUSO 1989C, ALAM 1989B
Hou S.R. ALEXANDER 1991C, ACTON 1992B, DECAMP 1991S
Hough P.V.C. BARNES 1964B, BARNES 1964E
Houk G. ABE 1994ZE, ABE 1995
Houlden M. ABREU 1991B, DECAMP 1991S
Howarth C.P. ALEXANDER 1991C, ACTON 1992B, DECAMP 1991S
Hrubec J. ABREU 1991B, DECAMP 1991S
Hsiao C. SERIFF 1950
Hsieh F. ABACHI 1995G
Hsu H.K. BARBER 1979F
Hsu L.S. ADEVA 1990R, ADEVA 1991D, DECAMP 1991S
Hsu T.T. BARBER 1979F
Hu G. ADEVA 1990R, ADEVA 1991D, DECAMP 1991S
Hu G.Q. ADEVA 1990R, ADEVA 1991D, DECAMP 1991S
Hu H. DECAMP 1991D, DECAMP 1991N, DECAMP 1991S, BUSKULIC 1993H
Hu P. ABE 1994ZE, ABE 1995
Hu Ting. ABACHI 1995G
Hu Tong. ABACHI 1995G
Huang D. DECAMP 1991D, DECAMP 1991N, DECAMP 1991S, BUSKULIC 1993H
Huang X. DECAMP 1991D, DECAMP 1991N, DECAMP 1991S, BUSKULIC 1993H
Hudson R.P. WU 1957
Huehn T. ABACHI 1995G
Huffman B.T. ABE 1994ZE, ABE 1995
Hugentobler E. ANSARI 1987E
Hughes G. BARTEL 1980D, DECAMP 1991D, DECAMP 1991N, DECAMP 1991S, BUSKULIC 1993H
Hughes R. ABE 1994ZE, ABE 1995

Hughes V.W. PRESCOTT 1978C, PRESCOTT 1979C, ASHMAN 1988B
Hughes-Jones R.E. ALEXANDER 1991C, ACTON 1992B, DECAMP 1991S
Hulth P.O. ABREU 1991B, DECAMP 1991S
Hultqvist K. ABREU 1991B, DECAMP 1991S
Hultschig H. BRANDELIK 1977D, BRANDELIK 1979H, BRANDELIK 1980G
Humbert R. ALEXANDER 1991C, ACTON 1992B, DECAMP 1991S
Humpfrey W.E. KALBFLEISCH 1964
Hungerbuhler V. BANNER 1983B, BAGNAIA 1983B
Hurst P. ABE 1994ZE
Hurst R.B. BAND 1988
Husson D. ABREU 1991B, DECAMP 1991S
Huston J. ABE 1994ZE, ABE 1995
Hutchinson D. ABRAMS 1989B, WEIR 1989B
Huth J. ABE 1994ZE, ABE 1995
Huzita H. FAISSNER 1978D
Hyams B.D. BLUM 1975, ABREU 1991B, DECAMP 1991S
Hylen J. ABRAMS 1989B, WEIR 1989B, ABE 1994ZE, ABE 1995
Hylton R. BALTAY 1979E
Iacopini E. ANSARI 1987E
Iarocci E. BACCI 1974
Iaselli G. DECAMP 1991D, DECAMP 1991N, DECAMP 1991S, BUSKULIC 1993H
Ichimiya T. NISHINA 1937
Iconomidou-Fayard L. ANSARI 1987E
Igarashi S. ABACHI 1995G
Igo-Kemenes P. BIAGI 1983C, BIAGI 1985C, ALEXANDER 1991C, ACTON 1992B, DECAMP 1991S
Ihssen H. ALEXANDER 1991C, ACTON 1992B, DECAMP 1991S
Ikeda H. ABE 1995
Ikeda M. IKEDA 1959, ALBAJAR 1987C, DECAMP 1991D, DECAMP 1991N, DECAMP 1991S, BUSKULIC 1993H
Iliopoulos J. GLASHOW 1970
Ille B. ADEVA 1990R, ADEVA 1991D, DECAMP 1991S
Illingworth J. BRANDELIK 1979H, BRANDELIK 1980G
Iloff E.L. ILOFF 1955
Ilyas M.M. ADEVA 1990R, ADEVA 1991D, DECAMP 1991S
Imaeda K. MINAKAWA 1959
Imlay R. BENVENUTI 1974F, ANDREWS 1980
Imore M. BARTEL 1980D
Impeduglia J. LANDE 1956, IMPEDUGLIA 1958
Imrie D.C. ALEXANDER 1991C, ACTON 1992B, DECAMP 1991S
In K.H. KANG-CHANG 1960
Incagli M. ABE 1994ZE, ABE 1995
Incandela J. ABE 1994ZE, ABE 1995
Innes W.R. HERB 1977, INNES 1977, UENO 1979, ABRAMS 1989B, WEIR 1989B
Innocente V. ADEVA 1990R, ADEVA 1991D, DECAMP 1991S
Inozemtzev N.I. ARTAMONOV 1984B
Inyakin A.V. APEL 1975B
Ioannou P. ABREU 1991B, DECAMP 1991S
Ioffe B.L. IOFFE 1957
Iori M. CALICCHIO 1980
Iovchev K.I. BEZNOGIKH 1969
Iredale P. DAVIES 1955
Isenhower D. ABREU 1991B, DECAMP 1991S
Ishii Y. MINAKAWA 1959
Isiksal E. ADEVA 1990R, ADEVA 1991D, DECAMP 1991S
Iso H. ABE 1994ZE
Ito A.S. HERB 1977, INNES 1977, UENO 1979, ABACHI 1995G
Ivanov V.I. ASTVACATUROV 1968
Ivanova L.K. BALDIN 1973
Ivanova M. VOYVODIC 1986B
Iversen P.S. ABREU 1991B, DECAMP 1991S
Iwai J. ABE 1995
Iwanenko D.D. IWANENKO 1932, IWANENKO 1944
Iwata Y. ABE 1995
Izen J.M. ANDREWS 1980, ANDREWS 1980B, BEBEK 1981, CHADWICK 1981
Jacholkowska A. ASHMAN 1988B
Jackson D. BUSKULIC 1993H
Jackson J.N. ABREU 1991B, DECAMP 1991S
Jacobs K. ANSARI 1987E, BUSKULIC 1993H
Jacobsen J.E. DECAMP 1991D, DECAMP 1991N, DECAMP 1991S
Jacobsen R. ABRAMS 1989B, BUSKULIC 1993H
Jacot-Guillarmod P. BIAGI 1985C
Jaffe D.E. BUSKULIC 1993H
Jaffre M. HASERT 1973, BLIETSCHAU 1976, BLIETSCHAU 1977B, WEIR 1989B
Jagel E. ADEVA 1990R
Jahn A. DECAMP 1991D, DECAMP 1991N, DECAMP 1991S
Jalocha P. ABREU 1991B, DECAMP 1991S
James E. ABACHI 1995G
Janissen L. ALEXANDER 1991C, ACTON 1992B, DECAMP 1991S
Jank W. ARNISON 1983C, ARNISON 1983D, ARNISON 1984C, ARNISON 1984G, ARNISON 1985D, ALBAJAR 1987C, ALBAJAR 1991C, ALBAJAR 1991D
Janot P. DECAMP 1991D, DECAMP 1991N, DECAMP 1991S, BUSKULIC 1993H
Janssen H. ADEVA 1991D, DECAMP 1991S
Jaques J. ABACHI 1995G
Jared R.C. DECAMP 1991D, DECAMP 1991N, DECAMP 1991S
Jarlskog G. ABREU 1991B, DECAMP 1991S
Jaros J.A. PERL 1976C, GOLDHABER 1976D, PERUZZI 1976C, PERL 1977F, ABRAMS 1989B, WEIR 1989B
Jarry P. ABREU 1991B, DECAMP 1991S
Jauneau L. BLIETSCHAU 1976, BLIETSCHAU 1977B
Jawahery A. AVERY 1989C, ARTUSO 1989C, ALAM 1989B, ALEXANDER 1991C, ACTON 1992B, DECAMP 1991S
Jean-Marie B. AUGUSTIN 1974B, ABRAMS 1974, SCHWITTERS 1975, PERL 1975C, HANSON 1975, ABREU 1991B, DECAMP 1991S
Jeffreys P.W. ALEXANDER 1991C, ACTON 1992B, DECAMP 1991S
Jenni P. BANNER 1983B, BAGNAIA 1983B, ANSARI 1987E
Jensen H. ABE 1994ZE, ABE 1995
Jensen J.E. BARNES 1964B, BARNES 1964E
Jensen T. AVERY 1989C, ARTUSO 1989C, ALAM 1989B
Jentschke W. PRESCOTT 1978C, PRESCOTT 1979C
Jeremie H. ALEXANDER 1991C, ACTON 1992B, DECAMP 1991S
Jerger S.A. ABACHI 1995G
Jessop C.P. ABE 1994ZE
Jiang J.Z.Y. ABACHI 1995G
Jimack M. ALBAJAR 1987C, ALEXANDER 1991C, ACTON 1992B, DECAMP 1991S
Jin B.N. ADEVA 1990R, ADEVA 1991D, DECAMP 1991S
Jobes M. ALEXANDER 1991C, ACTON 1992B, DECAMP 1991S
Jobin P. BLIETSCHAU 1977B
Joffe-Minor T. ABACHI 1995G
Johansson E.K. ABREU 1991B, DECAMP 1991S
Johansson S.D. HILL 1956
Johari H. ABACHI 1995G
Johns K. ABACHI 1995G
Johnson A.D. GOLDHABER 1976D, PERUZZI 1976C
Johnson D. ABREU 1991B, DECAMP 1991S
Johnson D.R. AVERY 1989C, ARTUSO 1989C, ALAM 1989B
Johnson J.R. BAND 1988
Johnson M. ABACHI 1995G

Johnson R.P. DECAMP 1991D, DECAMP 1991N, DECAMP 1991S, BUSKULIC 1993H
Johnson W.R. CHUPP 1955
Johnstad H. ABACHI 1995G
Johnston H. MURPHY 1934
Johnston R.H.W. DAVIES 1955
Jona-Lasinio G. NAMBU 1961, NAMBU 1961B
Jonckheere A.M. ABACHI 1995G
Jones C. BLUM 1975
Jones G.T. CALICCHIO 1980
Jones L.W. KERST 1956, ADEVA 1990R, ADEVA 1991D, DECAMP 1991S
Jones R.W.L. ALEXANDER 1991C, ACTON 1992B, DECAMP 1991S
Jones T. ASHMAN 1988B
Jones T.J. DECAMP 1991D, DECAMP 1991N, DECAMP 1991S
Jones T.W. HASERT 1973, HASERT 1973B, BLIETSCHAU 1976, BLIETSCHAU 1977B, BIONTA 1987C
Jones W.G. BRANDELIK 1979H, BRANDELIK 1980G
Jonker M. ABREU 1991B, DECAMP 1991S
Jonsson L. DARDEN 1978, DARDEN 1978B, ALBRECHT 1987P, ABREU 1991B, DECAMP 1991S
Joos P. BRANDELIK 1977D, BRANDELIK 1979H, BRANDELIK 1980G
Jorat G. ARNISON 1983D, ARNISON 1984C, ARNISON 1984G, ARNISON 1985D, ALBAJAR 1987C, ALBAJAR 1991C, ALBAJAR 1991D
Jordan C.L. MILLER 1972B
Jordan P. BORN 1925, BORN 1926
Josa M.I. ALBAJAR 1991C, ALBAJAR 1991D
Joshi U. ABE 1994ZE, ABE 1995
Jost B. DECAMP 1991D, DECAMP 1991N, DECAMP 1991S, BUSKULIC 1993H
Jostlein H. HERB 1977, INNES 1977, UENO 1979, ABACHI 1995G
Journe V. DECAMP 1991D
Jousset J. DECAMP 1991D, DECAMP 1991N, DECAMP 1991S, BUSKULIC 1993H
Jovanovic P. ALEXANDER 1991C, ACTON 1992B, DECAMP 1991S
Juillot P. ABREU 1991B, DECAMP 1991S
Jun S.Y. ABACHI 1995G
Jung C.K. ABRAMS 1989B, ABACHI 1995G
Jung S. JUNG 1992
Juricic I. WEIR 1989B
Kabe S. HASERT 1973B
Kabuss E.M. ASHMAN 1988B
Kachanov V.A. BUSHNIN 1969, BINON 1969, BINON 1970B, ANTIPOV 1970, APEL 1975B
Kadansky V. BRANDELIK 1979H, BRANDELIK 1980G
Kadel R.W. BARBER 1979F, ABE 1994ZE, ABE 1995
Kadyk J.A. AUGUSTIN 1974B, ABRAMS 1974, SCHWITTERS 1975, PERL 1975C, HANSON 1975, PERL 1976C, GOLDHABER 1976D, PERUZZI 1976C, PERL 1977F, ABRAMS 1989B, WEIR 1989B
Kagan H. ANDREWS 1980, ANDREWS 1980B, BEBEK 1981, CHADWICK 1981, AVERY 1989C, ARTUSO 1989C, ALAM 1989B
Kahl T. BERGER 1978
Kahn S.A. BALTAY 1979E, ABACHI 1995G
Kajfasz E. ABE 1994ZE, ABE 1995
Kajita T. HIRATA 1987C
Kalbfleisch G.R. CRAWFORD 1958, BUTTON 1960, BUTTON 1961, KALBFLEISCH 1964
Kalelkar M. BALTAY 1979E
Kalkanis G. ABREU 1991B, DECAMP 1991S
Kallmann H. KALLMANN 1950
Kalmus G. ABREU 1991B, DECAMP 1991S
Kalmus P.I.P. ARNISON 1983C, ARNISON 1983D, ARNISON 1984C, ARNISON 1984G, ARNISON 1985D, ALBAJAR 1987C, ALBAJAR 1991C, ALBAJAR 1991D
Kalogeropoulos T.E. GOLDHABER 1959, BROWN 1962
Kamon T. ABE 1994ZE, ABE 1995
Kamp D. ALBRECHT 1987P
Kamyshkov Y. ADEVA 1990R, ADEVA 1991D, DECAMP 1991S
Kanazirski K. KIRILLOVA 1965
Kandaswamy J. ANDREWS 1980, ANDREWS 1980B, BEBEK 1981, CHADWICK 1981, AVERY 1989C, ARTUSO 1989C, ALAM 1989B
Kanekal S. ALBRECHT 1987P
Kaneko T. ABE 1994ZE, ABE 1995
Kang J.S. ABACHI 1995G
Kang-Chang W. KANG-CHANG 1960
Kantardjian G. ABREU 1991B, DECAMP 1991S
Kapitza H. BERGER 1978, BERGER 1979C, BERGER 1980L, ALBRECHT 1987P
Kaplan D.M. HERB 1977, INNES 1977, UENO 1979, BOHRINGER 1980B, FINOCCHIARO 1980
Kapusta F. ABREU 1991B, DECAMP 1991S
Kapusta P. ABREU 1991B, DECAMP 1991S
Kardelis D.A. ABE 1994ZE
Karimaki V. ARNISON 1983D, ARNISON 1984C, ARNISON 1984G, ARNISON 1985D, ALBAJAR 1987C, ALBAJAR 1991C, ALBAJAR 1991D
Karlen D. WEIR 1989B, ALEXANDER 1991C, ACTON 1992B, DECAMP 1991S
Karplus R. KARPLUS 1955
Karpukhin O.A. GOLDANSKY 1960
Karr K. ABE 1995
Karshon U. BRANDELIK 1979H, BRANDELIK 1980G
Karyotakis Y. ADEVA 1990R, ADEVA 1991D, DECAMP 1991S
Karzmark C.J. THOMPSON 1953, THOMPSON 1953B
Kasemann M. DECAMP 1991D, DECAMP 1991N, DECAMP 1991S
Kasha H. ABE 1994ZE, ABE 1995
Kass R. ANDREWS 1980B, BEBEK 1981, CHADWICK 1981, AVERY 1989C, ARTUSO 1989C, ALAM 1989B
Kasser A. ADEVA 1991D, DECAMP 1991S
Katayama N. AVERY 1989C, ARTUSO 1989C, ALAM 1989B
Kato Y. ABE 1994ZE, ABE 1995
Katsanevas S. ABREU 1991B, DECAMP 1991S
Katsoufis E.C. ABREU 1991B, DECAMP 1991S
Kaur M. ADEVA 1990R, ADEVA 1991D, DECAMP 1991S
Kawabata S. BARTEL 1980D, BARNETT 1996
Kawagoe K. ALEXANDER 1991C, ACTON 1992B, DECAMP 1991S
Kawamoto T. ALEXANDER 1991C, ACTON 1992B, DECAMP 1991S
Kazuno M. MINAKAWA 1959
Keeble L. ABE 1994ZE, ABE 1995
Keefe D. DAVIES 1955
Keeler R. ARNISON 1983C, ARNISON 1983D, ARNISON 1984C, ARNISON 1984G, ARNISON 1985D, ALBAJAR 1987C
Keeler R.K. ALEXANDER 1991C, ACTON 1992B, DECAMP 1991S
Keemer N.R. DECAMP 1991D, DECAMP 1991N, DECAMP 1991S, BUSKULIC 1993H
Kehoe B. ABRAMS 1964
Kehoe R. ABACHI 1995G
Kelley K. ABE 1995
Kellner G. DECAMP 1991D
Kellogg J.M.B. KELLOGG 1939
Kellogg R.G. BERGER 1979C, BERGER 1980L, ALEXANDER 1991C, ACTON 1992B, DECAMP 1991S
Kelly M. ABACHI 1995G
Kendall H.W. BLOOM 1969B, BREIDENBACH 1969, MILLER 1972B, BODEK 1983
Kennedy B.W. ALEXANDER 1991C, ACTON 1992B, DECAMP 1991S
Kennedy R.D. ABE 1994ZE, ABE 1995
Kent J. ABRAMS 1989B, WEIR 1989B

Kenyon I. ARNISON 1983C, ARNISON 1983D, ARNISON 1984C, ARNISON 1984G, ARNISON 1985D, ALBAJAR 1987C, ALBAJAR 1991C, ALBAJAR 1991D
Kephart R.D. HERB 1977, INNES 1977, UENO 1979, ABE 1994ZE, ABE 1995
Keranen R. ABREU 1991B, DECAMP 1991S
Kernan A. ARNISON 1983C, ARNISON 1983D, ARNISON 1984C, ARNISON 1984G, ARNISON 1985D, ALBAJAR 1987C, ABACHI 1995G
Kernel G. ALBRECHT 1987P
Kerst D.W. KERST 1940, KERST 1941, KERST 1956
Kerth L. ABACHI 1995G
Kesteman J. ABREU 1991B, DECAMP 1991S
Kesten P. ABE 1994ZE, ABE 1995
Kestenbaum D. ABE 1994ZE, ABE 1995
Keuffel J.W. KEUFFEL 1949
Keup R.M. ABE 1994ZE, ABE 1995
Keutelian H. ABE 1994ZE, ABE 1995
Keyvan F. ABE 1994ZE, ABE 1995
Khac U.N. HASERT 1973
Khachaturyan M.N. KIRILLOVA 1965, ASTVACATUROV 1968
Khan A. ALBAJAR 1987C
Khan R.A. ADEVA 1990R, ADEVA 1991D, DECAMP 1991S
Khodyrev Y.S. BUSHNIN 1969
Khokhar S. ADEVA 1990R, ADEVA 1991D, DECAMP 1991S
Khomenko B.A. ABREU 1991B, DECAMP 1991S
Khovanski N.N. ABREU 1991B, DECAMP 1991S
Khoze V.A. ADEVA 1990R, ADEVA 1991D, DECAMP 1991S
Khristov L.G. KIRILLOVA 1965
Khromov V.P. BINON 1969, ANTIPOV 1970
Khvastunov M.S. ASTVACATUROV 1968
Kibble T.W.B. GURALNIK 1964, KIBBLE 1967
Kielczewska D. BIONTA 1987C
Kienle P. JUNG 1992
Kienzle M.N. ADEVA 1990R
Kienzle W. ALBAJAR 1987C
Kienzle-Focacci M.N. ADEVA 1991D, DECAMP 1991S
Kifune T. HIRATA 1987C
Kim B.J. ABE 1995
Kim C.L. ABACHI 1995G
Kim D.H. ABE 1994ZE, ABE 1995
Kim D.W. DECAMP 1991D, DECAMP 1991N, DECAMP 1991S, BUSKULIC 1993H
Kim H.S. ABE 1994ZE, ABE 1995
Kim I.J. AVERY 1989C, ARTUSO 1989C, ALAM 1989B
Kim J.E. KIM 1981
Kim P.C.H. ALBRECHT 1987P
Kim S.B. HIRATA 1987C, ABE 1994ZE, ABE 1995
Kim S.H. ABE 1994ZE, ABE 1995
Kim S.K. ABACHI 1995G
Kim Y.K. ABE 1994ZE, ABE 1995
King B. BARTEL 1980D, ABREU 1991B, DECAMP 1991S
King D.T. CARLSON 1950
King J. ABRAMS 1989B
Kinnison W. ADEVA 1990R, ADEVA 1991D, DECAMP 1991S
Kinnunen R. ARNISON 1983C, ARNISON 1983D, ARNISON 1984C, ARNISON 1984G, ARNISON 1985D, ALBAJAR 1987C, ALBAJAR 1991C, ALBAJAR 1991D
Kinoshita K. AVERY 1989C, ARTUSO 1989C, ALAM 1989B
Kirillov A.D. BALDIN 1973
Kirillova L. BEZNOGIKH 1969
Kirillova L.F. KIRILLOVA 1965
Kirk H.G. BALTAY 1979E
Kirkby D. ADEVA 1990R, ADEVA 1991D, DECAMP 1991S
Kirsch L. ABE 1994ZE, ABE 1995
Kirschfink F.J. BRANDELIK 1980G
Kiselev V.A. ARTAMONOV 1984B
Kittel W. ADEVA 1990R, ADEVA 1991D, DECAMP 1991S
Kittenberger W. APEL 1975B
Kjaer N.J. ABREU 1991B, DECAMP 1991S
Kladnitskaya E.N. KANG-CHANG 1960
Klatchko A. ABACHI 1995G
Klein H. ABREU 1991B, DECAMP 1991S
Klein S.R. ABRAMS 1989B, WEIR 1989B
Kleinknecht K. DECAMP 1991D, DECAMP 1991N, DECAMP 1991S, BUSKULIC 1993H
Kleinwort C. ALEXANDER 1991C, ACTON 1992B, DECAMP 1991S
Klem D.E. ALEXANDER 1991C, ACTON 1992B, DECAMP 1991S
Klempt W. ABREU 1991B, DECAMP 1991S
Klepper O. JUNG 1992
Klima B. ABACHI 1995G
Klimenko S.G. ARTAMONOV 1984B
Klimentov A. ADEVA 1990R, ADEVA 1991D, DECAMP 1991S
Kline R. ANDREWS 1980
Klochkov B.I. ABACHI 1995G
Klopfenstein C. DECAMP 1991D, DECAMP 1991N, DECAMP 1991S, ABACHI 1995G
Klovning A. BERGER 1979C, BERGER 1980L, ABREU 1991B, DECAMP 1991S
Kluberg L. HASERT 1973, HASERT 1973B, BLIETSCHAU 1976, BLIETSCHAU 1977B
Kluge E.E. ANSARI 1987E, DECAMP 1991D, DECAMP 1991N, DECAMP 1991S, BUSKULIC 1993H
Kluit P. ABREU 1991B, DECAMP 1991S
Klyukhin V.I. VOYVODIC 1986B, ABACHI 1995G
Knapp B. KNAPP 1976B
Knauer J. KNAPP 1976B
Knies G. BERGER 1978, BERGER 1979C, BERGER 1980L
Knobloch J. DECAMP 1991D, DECAMP 1991N, DECAMP 1991S, BUSKULIC 1993H
Knop G. BRANDELIK 1979H, BRANDELIK 1980G
Koba Z. KOBA 1947, KOBA 1947B
Kobayashi M. KOBAYASHI 1973B
Kobayashi T. BARTEL 1980D, ALEXANDER 1991C, ACTON 1992B, DECAMP 1991S
Koch W. BRANDELIK 1977D, BRANDELIK 1979H, BRANDELIK 1980G
Kochetkov V.I. ABACHI 1995G
Kochowski C. CALICCHIO 1980
Koehler P. BRANDELIK 1980G
Koehn P. ABE 1994ZE, ABE 1995
Koehne J.H. ABREU 1991B, DECAMP 1991S
Koene B. ABREU 1991B, DECAMP 1991S
Koenig W. JUNG 1992
Koetke D. ABRAMS 1989B
Kofoed-Hansen O. BANNER 1983B, BAGNAIA 1983B, ANSARI 1987E
Kogan E. BRANDELIK 1979H
Kohli J.M. ABACHI 1995G
Kokkinias P. ABREU 1991B, DECAMP 1991S
Kokott T.P. ALEXANDER 1991C, ACTON 1992B, DECAMP 1991S
Kolanoski H. BRANDELIK 1980G, ALBRECHT 1987P
Kolganova E.D. VOYVODIC 1986B
Kolhorster W. BOTHE 1929
Koltick D. ABACHI 1995G
Komamiya S. BARTEL 1980D, ABRAMS 1989B, ALEXANDER 1991C, ACTON 1992B, DECAMP 1991S
Kondo K. ABE 1994ZE, ABE 1995
Kondratenko A.M. DERBENEV 1978, DERBENEV 1980
Konig A.C. ADEVA 1990R, ADEVA 1991D, DECAMP 1991S
Konigsberg J. ABE 1994ZE, ABE 1995
Konopinski E.J. KONOPINSKI 1953
Kooy H. ANDREWS 1980, ANDREWS 1980B, BEBEK 1981, CHADWICK 1981
Kopf M. ABREU 1991B, DECAMP 1991S

Kopke L. BRANDELIK 1980G, ALEXANDER 1991C, ACTON 1992B, DECAMP 1991S
Kopp J.K. BROWN 1962, BARNES 1964B, BARNES 1964E
Kopp S. ABE 1994ZE, ABE 1995
Koppadzh D. VOYVODIC 1986B
Koppitz B. BERGER 1978, BERGER 1979C, BERGER 1980L
Koratzinos M. ABREU 1991B, DECAMP 1991S
Korbach W. BRANDELIK 1979H
Korbel V. AUBERT 1983
Korbel Z. KIRILLOVA 1965
Korcyl K. ABREU 1991B, DECAMP 1991S
Kordas K. ABE 1994ZE, ABE 1995
Koreshev V.I. VOYVODIC 1986B
Kornadt O. ADEVA 1990R, ADEVA 1991D, DECAMP 1991S
Korovina L.I. TERNOV 1961
Korytov A.V. ABREU 1991B, DECAMP 1991S
Korzen B. ASHMAN 1988B, ABREU 1991B, DECAMP 1991S
Koshiba M. BARTEL 1980D, HIRATA 1987C
Koska W. ABRAMS 1989B, WEIR 1989B, ABE 1994ZE, ABE 1995
Kostritskii A.V. ABACHI 1995G
Kostyukhin V.V. ABREU 1991B, DECAMP 1991S
Kotcher J. ABACHI 1995G
Kotov V.I. BUSHNIN 1969
Kotthaus R. BRANDELIK 1977D
Kotz U. BRANDELIK 1977D, BRANDELIK 1979H, BRANDELIK 1980G
Kourkoumelis C. ABREU 1991B, DECAMP 1991S
Kourlas J. ABACHI 1995G
Koutsenko V. ADEVA 1990R, ADEVA 1991D, DECAMP 1991S
Kovacs E. ABE 1994ZE, ABE 1995
Kowald W. ABE 1994ZE, ABE 1995
Kowalewski R. AVERY 1989C, ARTUSO 1989C, ALAM 1989B, ALEXANDER 1991C, ACTON 1992B, DECAMP 1991S
Kowalski H. BRANDELIK 1979H, BRANDELIK 1980G, ARNISON 1983C
Kowalski L. ABRAMS 1989B
Kozanecki W. ARNISON 1983C, ARNISON 1983D, ARNISON 1984C, ARNISON 1984G, ABRAMS 1989B, WEIR 1989B, DECAMP 1991D, DECAMP 1991N, DECAMP 1991S, BUSKULIC 1993H
Kozelov A. ABACHI 1995G
Kozhuharov C. JUNG 1992
Koziol H. CARRON 1978
Kozlovskii E.A. ABACHI 1995G
Kraemer R.W. PEVSNER 1961, ADEVA 1990R, ADEVA 1991D, DECAMP 1991S
Kral F. ACTON 1992B
Kral J. ABRAMS 1989B
Kramer T. ADEVA 1990R, ADEVA 1991D, DECAMP 1991S
Krammer M. ALBAJAR 1987C, ALBAJAR 1991C, ALBAJAR 1991D
Krasberg M. ABE 1994ZE, ABE 1995
Krasnokutsky R.N. APEL 1975B
Krastev V.R. ADEVA 1990R, ADEVA 1991D, DECAMP 1991S
Kraybill H. STONEHILL 1961
Krehbiel H. BRANDELIK 1977D, BARTEL 1980D
Kreinick D. BRANDELIK 1977D
Kreinick D.L. ANDREWS 1980, ANDREWS 1980B, BEBEK 1981, CHADWICK 1981, AVERY 1989C, ARTUSO 1989C, ALAM 1989B
Krenz W. HASERT 1973, HASERT 1973B, BLIETSCHAU 1976, BLIETSCHAU 1977B, BARBER 1979F, ADEVA 1990R, ADEVA 1991D, DECAMP 1991S
Kreutzmann H. ALEXANDER 1991C, ACTON 1992B, DECAMP 1991S
Kreuzberger T. ABREU 1991B, DECAMP 1991S
Krienen F. CARRON 1978
Krisch A.D. LIN 1978
Krishnaswamy M.R. ABACHI 1995G
Krizmanic J. ADEVA 1990R, ADEVA 1991D, DECAMP 1991S
Krolikowski J. ABREU 1991B, DECAMP 1991S
Kroll J. ARNISON 1985D, ALBAJAR 1987C, ALEXANDER 1991C, ACTON 1992B, DECAMP 1991S, ABE 1994ZE, ABE 1995
Krolzig A. DARDEN 1978, DARDEN 1978B
Kropp W.R. BIONTA 1987C
Kruener-Marquis U. ABREU 1991B, DECAMP 1991S
Kruger M. APEL 1975B
Krüner U. ASHMAN 1988B
Krupchitsky P.A. ABOV 1965
Krupinski W. ABREU 1991B, DECAMP 1991S
Kruse H.W. COWAN 1956
Kruse M. ABE 1994ZE, ABE 1995
Kryn D. ARNISON 1983C, ARNISON 1983D, ARNISON 1984C, ARNISON 1984G, ARNISON 1985D, ALBAJAR 1987C, ALBAJAR 1991C, ALBAJAR 1991D
Krzywdzinski S. VOYVODIC 1986B, ABACHI 1995G
Kubota Y. AVERY 1989C, ARTUSO 1989C, ALAM 1989B
Kucewicz W. ABREU 1991B, DECAMP 1991S
Kudelainen V.I. BUDKER 1976
Kuhlen M. ABRAMS 1989B
Kuhlmann S.E. ABE 1994ZE, ABE 1995
Kuhn A. ADEVA 1990R
Kuhn D. DECAMP 1991D, DECAMP 1991N, DECAMP 1991S, BUSKULIC 1993H
Kulka K. ABREU 1991B, DECAMP 1991S
Kullander S. ASHMAN 1988B
Kumar K.S. ADEVA 1990R, ADEVA 1991D, DECAMP 1991S
Kumar R.C. DANYSZ 1963
Kumar V. ADEVA 1990R, ADEVA 1991D, DECAMP 1991S
Kunin A. ADEVA 1990R, ADEVA 1991D, DECAMP 1991S
Kunori S. ABACHI 1995G
Kuns E. ABE 1994ZE, ABE 1995
Kuper E.A. ARTAMONOV 1984B
Kurvinen K. ABREU 1991B, DECAMP 1991S
Kusch P. KUSCH 1947, FOLEY 1948
Kutjin V.M. BUSHNIN 1969, BINON 1969, BINON 1970B, ANTIPOV 1970
Kutschke R. ALBRECHT 1987P
Kutsenko A.V. GOLDANSKY 1960
Kuwabara T. ABE 1995
Kuwano M. ALEXANDER 1991C, ACTON 1992B, DECAMP 1991S
Kuznetsov A.A. KANG-CHANG 1960
Kuznetsov O.M. VOYVODIC 1986B
Kuznetsov V.A. BALDIN 1973
Kwak N. VOYVODIC 1986B, ALBRECHT 1987P
Kyberd P. ARNISON 1984G, ARNISON 1985D, ALBAJAR 1987C, ALEXANDER 1991C, ACTON 1992B, DECAMP 1991S
Kyhl R.L. CHODOROW 1955
Kyriakis A. BUSKULIC 1993H
Laasanen A.T. ABE 1994ZE, ABE 1995
Labanca N. ABE 1995
Labarga M. ABRAMS 1989B
Lacasta C. ABREU 1991B, DECAMP 1991S
Lacava F. ARNISON 1983C, ARNISON 1983D, ARNISON 1984C, ARNISON 1984G, ARNISON 1985D, ALBAJAR 1987C, ALBAJAR 1991C, ALBAJAR 1991D
Lackas W. BERGER 1978, BERGER 1979C, BERGER 1980L
Lacourt A. DECAMP 1991D, DECAMP 1991N, DECAMP 1991S
Ladage A. BRANDELIK 1979H, BRANDELIK 1980G
Lafferty G.D. ALEXANDER 1991C, ACTON 1992B, DECAMP 1991S

Lagarrigue A. GREGORY 1954, ARMENTEROS 1955, HASERT 1973, HASERT 1973B
Lai K.W. CONNOLLY 1963, BARNES 1964B, BARNES 1964E, GOLDBERG 1964C
Lalieu V. ADEVA 1990R, ADEVA 1991D, DECAMP 1991S
Lam Ha. ARTUSO 1989C, ALAM 1989B
Lamarche F. ALEXANDER 1991C, ACTON 1992B, DECAMP 1991S
Lamb W.E. LAMB 1947
Lambertson G.R. CORK 1956
Lambropoulos C. ABREU 1991B, DECAMP 1991S
Lami S. ABACHI 1995G
Lammel S. ALBAJAR 1991C, ALBAJAR 1991D, ABE 1994ZE, ABE 1995
Lamoureux J.I. ABE 1994ZE, ABE 1995
Lamsa J.W. ABREU 1991B, DECAMP 1991S
Lanceri L. FIDECARO 1978, ABREU 1991B, DECAMP 1991S
Lancon E. BANNER 1983B, BAGNAIA 1983B, ANSARI 1987E, DECAMP 1991D, DECAMP 1991N, DECAMP 1991S, BUSKULIC 1993H
Landau L.D. LANDAU 1944, LANDAU 1957, LANDAU 1957B, LANDAU 1959
Lande K. LANDE 1956, LANDE 1957
Lander R. AGNEW 1958
Landgraf U. AUBERT 1983, ASHMAN 1988B
Landi G. ADEVA 1990R, ADEVA 1991D, DECAMP 1991S
Landon M. ALBAJAR 1991C, ALBAJAR 1991D
Landsberg G.L. ABACHI 1995G
Landsberg L.G. ANTIPOV 1970
Langacker P. KIM 1981
Lange E. DECAMP 1991D, DECAMP 1991N, DECAMP 1991S
Lanius K. ADEVA 1990R
Lankford A.J. ABRAMS 1989B, WEIR 1989B
Lannutti J.E. CHUPP 1955, ILOFF 1955, BUTTON 1960, BUTTON 1961, DECAMP 1991D, DECAMP 1991N, DECAMP 1991S
Lanou R.E. CRETIEN 1962, ABACHI 1995G
Lanske D. HASERT 1973, HASERT 1973B, BLIETSCHAU 1977B, ASHMAN 1988B, ADEVA 1990R, ADEVA 1991D, DECAMP 1991S
Lantero P BARNETT 1996
Lanzano S. ADEVA 1990R, ADEVA 1991D, DECAMP 1991S
Lapin V. ABREU 1991B, DECAMP 1991S
Lapshin V.G. BINON 1969, ANTIPOV 1970
Lariccia P. BACCI 1974, ANDREWS 1980, ANDREWS 1980B, BEBEK 1981, CHADWICK 1981, ANSARI 1987E
Larsen R.R. AUGUSTIN 1974B, ABRAMS 1974, SCHWITTERS 1975, PERL 1975C, HANSON 1975, PERL 1976C, GOLDHABER 1976D, PERUZZI 1976C, ABRAMS 1989B, WEIR 1989B
Larson D. ANDREWS 1980
Larson W.J. ALEXANDER 1991C, ACTON 1992B, DECAMP 1991S
Laslett L.J. KERST 1956
Lattes C.M.G. LATTES 1947, LATTES 1947B, BURFENING 1949
Lau K.H. BERGER 1979C, BERGER 1980L
Lauber J. DECAMP 1991N, DECAMP 1991S, BUSKULIC 1993H
Laugier J.P. ARNISON 1983C, ARNISON 1983D, ARNISON 1984C, ARNISON 1984G, ARNISON 1985D, ALBAJAR 1987C, ABREU 1991B, DECAMP 1991S
Lauhakangas R. ABREU 1991B, DECAMP 1991S
Laurelli P. AMENDOLIA 1973B, DECAMP 1991D, DECAMP 1991N, DECAMP 1991S, BUSKULIC 1993H
Lavine T.L. BAND 1988
Law M.E. CRETIEN 1962
Lawlor G. DAVIES 1955
Lawrence E.O. LAWRENCE 1931, LAWRENCE 1931B, LAWRENCE 1932, LAWRENCE 1939, BROBECK 1947
Layter J.G. ALEXANDER 1991C, ACTON 1992B, DECAMP 1991S
Le Claire B.W. WEIR 1989B, DECAMP 1991D, DECAMP 1991N, DECAMP 1991S, BUSKULIC 1993H
Le Compte T. ABE 1994ZE, ABE 1995
Le Coultre P. ADEVA 1990R, ADEVA 1991D, DECAMP 1991S
Le Diberder F. ABRAMS 1989B, DECAMP 1991N, DECAMP 1991S, BUSKULIC 1993H
Le Du P. ALEXANDER 1991C, ACTON 1992B, DECAMP 1991S
Le Francois J. DECAMP 1991D, DECAMP 1991N, DECAMP 1991S, BUSKULIC 1993H
Le Goff J.M. ADEVA 1990R, ADEVA 1991D, DECAMP 1991S
Lea R.M. BROWN 1962
Learned J.G. BIONTA 1987C
Lebedev A.A. ANTIPOV 1970
Lebedev P.N. LEBEDEV 1901
Leblanc P. ALEXANDER 1991C, ACTON 1992B, DECAMP 1991S
Lebrat J.F. ABACHI 1995G
Lebrun P. ADEVA 1990R, ADEVA 1991D, DECAMP 1991S
Lecomte P. BRANDELIK 1979H, BRANDELIK 1980G, ADEVA 1990R, ADEVA 1991D, DECAMP 1991S
Lecoq P. ADEVA 1990R, ADEVA 1991D, DECAMP 1991S
Leder G. APEL 1975B, FIDECARO 1978, ABREU 1991B, DECAMP 1991S
Lederman L.M. LANDE 1956, GARWIN 1957, LANDE 1957, DANBY 1962, DORFAN 1965D, CHRISTENSON 1970, BUESSER 1973, HERB 1977, INNES 1977, UENO 1979, BOHRINGER 1980B, FINOCCHIARO 1980
Lednev A.A. APEL 1975B
Ledroit F. ABREU 1991B, DECAMP 1991S
Lee A.M. ALEXANDER 1991C, ACTON 1992B, DECAMP 1991S
Lee D. ADEVA 1990R, ADEVA 1991D, DECAMP 1991S
Lee T.D. LEE 1949, LEE 1956, LEE 1956B, LEE 1957B, LEE 1957, LEE 1957C
Lee W. GOLDHABER 1960, CHINOWSKY 1962, DORFAN 1965D, KNAPP 1976B
Lee Y.Y. AUBERT 1974D
Lee-Franzini J. BOHRINGER 1980B, FINOCCHIARO 1980, ABACHI 1995G
Leedom I. ADEVA 1990R, ADEVA 1991D, DECAMP 1991S
Leedy R.E. BAND 1988
Leenen M. AUBERT 1983
Lees J.P. ARNISON 1983C, ARNISON 1983D, ARNISON 1984C, ARNISON 1984G, ARNISON 1985D, ALBAJAR 1987C, DECAMP 1991D, DECAMP 1991N, DECAMP 1991S, BUSKULIC 1993H
Leflat A. ABACHI 1995G
Lehmann H. LEHMANN 1955, BERGER 1978, BERGER 1979C, BERGER 1980L, ARNISON 1983C, ARNISON 1983D, ARNISON 1984C, ARNISON 1984G
Lehraus I. DECAMP 1991D, DECAMP 1991N, DECAMP 1991S, BUSKULIC 1993H
Lehto M.H. ALEXANDER 1991C, ACTON 1992B, DECAMP 1991S
Leighton R.B. LEIGHTON 1949, SERIFF 1950, LEIGHTON 1951, YORK 1953, ANDERSON 1953B, COWAN 1953, SORRELS 1955
Leissner M. BIENLEIN 1978

Leistam L. ADEVA 1990R, ADEVA 1991D, DECAMP 1991S
Leiste R. ADEVA 1990R, ADEVA 1991D, DECAMP 1991S
Leitner J. BERTANZA 1962D, CONNOLLY 1963, BARNES 1964B, BARNES 1964E, GOLDBERG 1964C
Lellouch D. ALEXANDER 1991C, ACTON 1992B, DECAMP 1991S
Lemaire M.C. BUSKULIC 1993H
Lemoigne Y. ALBAJAR 1991C, ALBAJAR 1991D
Lemonne J. DANYSZ 1963, HASERT 1973, ABREU 1991B, DECAMP 1991S
Lennert P. BARTEL 1980D, ALEXANDER 1991C, ACTON 1992B, DECAMP 1991S
Lenti M. ADEVA 1990R, ADEVA 1991D, DECAMP 1991S
Lenzen G. ABREU 1991B, DECAMP 1991S
Leonardi E. ADEVA 1991D, DECAMP 1991S
Leone S. ABE 1994ZE, ABE 1995
Leong J. AUBERT 1974D
Lepeltier V. ABREU 1991B, DECAMP 1991S
Leprince-Ringuet L. LEPRINCE-RINGUET 1944, GREGORY 1954, ARMENTEROS 1955
Leroy C. ALEXANDER 1991C, ACTON 1992B, DECAMP 1991S
Lessard L. ALEXANDER 1991C, DECAMP 1991S
Letessier-Selvon A. ABREU 1991B, DECAMP 1991S
Letson T. AVERY 1989C, ARTUSO 1989C, ALAM 1989B
Lettenström F. ASHMAN 1988B
Lettry J. ADEVA 1990R, ADEVA 1991D, DECAMP 1991S
Letts J. ACTON 1992B
Leu P. BRANDELIK 1979H, BRANDELIK 1980G
Leuchs R. ARNISON 1983C, ARNISON 1983D, ARNISON 1984C, ARNISON 1984G, ARNISON 1985D, ALBAJAR 1987C
Leung P. KNAPP 1976B
Leutwyler H. FRITZSCH 1973
Leutz H. CALICCHIO 1980
Levchenko P.M. ADEVA 1990R, ADEVA 1991D, DECAMP 1991S
Levegrun S. ARNISON 1985D, ALBAJAR 1987C, ALEXANDER 1991C, ALBAJAR 1991C, ALBAJAR 1991D, ACTON 1992B, DECAMP 1991S
Leveque A. ARNISON 1983C, ARNISON 1983D, ARNISON 1984C, ARNISON 1984G, ARNISON 1985D
Levi M.E. ARNISON 1985D, ALBAJAR 1987C, ABRAMS 1989B, WEIR 1989B
Levi-Setti R. BONETTI 1953, BONETTI 1953B, DAVIES 1955
Levine M. KIM 1981
Levine N. FRAUENFELDER 1957
Levinson L. ALEXANDER 1991C, ACTON 1992B, DECAMP 1991S
Levinthal D. DECAMP 1991D, DECAMP 1991N, DECAMP 1991S, BUSKULIC 1993H
Levman G. FINOCCHIARO 1980
Levy M.M. YENNIE 1957
Lewendel B. BERGER 1980L
Lewis H.R. FRAUENFELDER 1957
Lewis J.D. AVERY 1989C, ARTUSO 1989C, ALAM 1989B, ABE 1994ZE, ABE 1995
Leytens X. ADEVA 1990R, ADEVA 1991D, DECAMP 1991S
Lezoch P. BIENLEIN 1978
Lheritier M. LEPRINCE-RINGUET 1944
Li C. ADEVA 1990R, ADEVA 1991D, DECAMP 1991S
Li H. ABACHI 1995G
Li H.T. ADEVA 1990R, ADEVA 1991D, DECAMP 1991S
Li J. BARBER 1979F, ABACHI 1995G
Li J.F. ADEVA 1990R, ADEVA 1991D, DECAMP 1991S
Li L. ADEVA 1990R, ADEVA 1991D, DECAMP 1991S
Li P.J. ADEVA 1990R, ADEVA 1991D, DECAMP 1991S
Li Q. ADEVA 1990R, ADEVA 1991D, DECAMP 1991S
Li Q.Z. BARBER 1979F
Li S. ALBAJAR 1987C
Li W.C. AVERY 1989C, ARTUSO 1989C, ALAM 1989B
Li X.G. ADEVA 1990R, ADEVA 1991D, DECAMP 1991S
Li Y.K. ABACHI 1995G
Li-Demarteau Q.Z. ABACHI 1995G
Liao J.Y. ADEVA 1990R, ADEVA 1991D, DECAMP 1991S
Lichtman S. BERTANZA 1962D, CONNOLLY 1963, GOLDBERG 1964C
Lieb E. ABREU 1991B, DECAMP 1991S
Liello F. DECAMP 1991D, DECAMP 1991N, DECAMP 1991S
Lierl H. BRANDELIK 1977D
Lieske N.M. DECAMP 1991D, DECAMP 1991N, DECAMP 1991S, BUSKULIC 1993H
Ligabue F. DECAMP 1991D, DECAMP 1991N, DECAMP 1991S, BUSKULIC 1993H
Liko D. ABREU 1991B, DECAMP 1991S
Lillestol E. BERGER 1979C, BERGER 1980L
Lillethun E. BERGER 1979C, ABREU 1991B, DECAMP 1991S
Lima J.G.R. ABACHI 1995G
Limon P. ABE 1994ZE, ABE 1995
Limon P.J. CHRISTENSON 1970
Lin A. LIN 1978
Lin J. DECAMP 1991D, DECAMP 1991N, DECAMP 1991S, BUSKULIC 1993H
Lin Z.Y. ADEVA 1990R, ADEVA 1991D, DECAMP 1991S
Lincoln D. ABACHI 1995G
Linde F.L. ADEVA 1990R, ADEVA 1991D, DECAMP 1991S
Lindemann B. ADEVA 1991D, DECAMP 1991S
Lindenbaum S.J. LINDENBAUM 1953
Lindgren J. ABREU 1991B, DECAMP 1991S
Lindgren M. ABE 1994ZE, ABE 1995
Lindquist T. ASHMAN 1988B
Lindsey J.S. KALBFLEISCH 1964
Ling T.Y. BENVENUTI 1974F
Linglin D. ARNISON 1983C, ARNISON 1983D, ARNISON 1984C, ARNISON 1984G, ARNISON 1985D, ALBAJAR 1987C, ALBAJAR 1991C, ALBAJAR 1991D
Linn S.L. ABACHI 1995G
Linnemann J.T. ABACHI 1995G
Linnhofer D. ADEVA 1990R, ADEVA 1991D, DECAMP 1991S
Lipa P. ALBAJAR 1991C, ALBAJAR 1991D
Lipatov L.N. GRIBOV 1972
Lipniacka A. ABREU 1991B, DECAMP 1991S
Lippi I. ABREU 1991B, DECAMP 1991S
Lippmann B.A. LIPPMANN 1950
Lipton R. ABACHI 1995G
Lishka C. BUSKULIC 1993H
Lisowski B. ANSARI 1987E
Liss T.M. ABE 1994ZE, ABE 1995
Lissauer D. FIRESTONE 1971C
Litke A. ABRAMS 1989B, WEIR 1989B, BUSKULIC 1993H
Litke A.M. ABRAMS 1974, SCHWITTERS 1975, PERL 1975C, HANSON 1975, PERL 1976C
Litt L. BUESSER 1973
Liu R. ADEVA 1990R, ADEVA 1991D, DECAMP 1991S
Liu Y. ADEVA 1990R, ADEVA 1991D, DECAMP 1991S
Liu Y.C. ABACHI 1995G
Livan M. BANNER 1983B, BAGNAIA 1983B, ANSARI 1987E

Livingston M.S. LAWRENCE 1931, LAWRENCE 1931B, LAWRENCE 1932, COURANT 1952
Llosa R. ABREU 1991B, DECAMP 1991S
Lloyd J.L. BARNES 1964B, BARNES 1964E
Lloyd S.L. BRANDELIK 1979H, BRANDELIK 1980G, ALEXANDER 1991C, ACTON 1992B, DECAMP 1991S
Lo Secco J.M. BIONTA 1987C
Loar H. DURBIN 1951C
Lobashov B.M. LOBASHOV 1966
Lobkowicz F. ANDREWS 1980, ANDREWS 1980B, BEBEK 1981, CHADWICK 1981, ABACHI 1995G
Locci E. ARNISON 1983C, ARNISON 1983D, ARNISON 1984C, ARNISON 1984G, ARNISON 1985D, ALBAJAR 1987C, DECAMP 1991D, DECAMP 1991N, DECAMP 1991S, BUSKULIC 1993H
Locci M. BACCI 1974
Lockyer N.S. WEIR 1989B, ABE 1994ZE, ABE 1995
Loebinger F. BARTEL 1980D
Loebinger F.K. ALEXANDER 1991C, ACTON 1992B, DECAMP 1991S
Loerstad B. ABREU 1991B, DECAMP 1991S
Lofgren E.J. CHAMBERLAIN 1956F
Logunov A.A. LOGUNOV 1956, LOGUNOV 1957, LOGUNOV 1963, LOGUNOV 1967, LOGUNOV 1967B
Lohmann W. ADEVA 1990R, ADEVA 1991D, DECAMP 1991S
Lohr B. BRANDELIK 1979H, BRANDELIK 1980G
Lohse T. DECAMP 1991D, DECAMP 1991N, DECAMP 1991S, BUSKULIC 1993H
Lokajicek M. ABREU 1991B, DECAMP 1991S
Loken J. ASHMAN 1988B
Loken J.G. ABREU 1991B, DECAMP 1991S
Loken S.C. ABACHI 1995G
Lokos S. ADEVA 1990R, ABACHI 1995G
London G.W. CONNOLLY 1963, BARNES 1964B, BARNES 1964E, GOLDBERG 1964C
Long E.A. ANDERSON 1952B, ANDERSON 1952E
Long K. ALBAJAR 1987C
Long O. ABE 1994ZE, ABE 1995
Longo E. ADEVA 1990R, ADEVA 1991D, DECAMP 1991S
Longuemare C. BLIETSCHAU 1976, BLIETSCHAU 1977B, BEBEK 1981, CHADWICK 1981
Loomis C. ABE 1995
Loomis W.A. ANDREWS 1980, ANDREWS 1980B, BEBEK 1981, CHADWICK 1981
Lopez Aguera M.A. ABREU 1991B, DECAMP 1991S
Lopez-Fernandez A. ABREU 1991B, DECAMP 1991S
Lorah J.M. ALEXANDER 1991C, ACTON 1992B, DECAMP 1991S
Lorazo B. ALEXANDER 1991C, ACTON 1992B, DECAMP 1991S
Lord J.J. LORD 1950, VOYVODIC 1986B
Lorenz E. BLUM 1975, BIENLEIN 1978
Loret M. ARNISON 1984G
Loreti M. FAISSNER 1978D, ABE 1994ZE, ABE 1995
Los M. ABREU 1991B, DECAMP 1991S
Loskutov Y.M. TERNOV 1961
Losty J. EDWARDS 1958
Losty M.J. ALEXANDER 1991C, ACTON 1992B, DECAMP 1991S
Lou J. DECAMP 1991D, DECAMP 1991N, DECAMP 1991S
Lou X.C. AVERY 1989C, ARTUSO 1989C, ABRAMS 1989B, ALAM 1989B, ALEXANDER 1991C, ACTON 1992B, DECAMP 1991S
Loucatos S. BANNER 1983B, BAGNAIA 1983B, ANSARI 1987E, DECAMP 1991D, DECAMP 1991N, DECAMP 1991S
Louis W.C. BIAGI 1983C, BIAGI 1985C
Loukas D. ABREU 1991B, DECAMP 1991S
Lounis A. ABREU 1991B, DECAMP 1991S
Louttit R.I. BROWN 1962, CAZZOLI 1975E
Low E.H. ABE 1994ZE
Low F.E. GELL-MANN 1954B, CHEW 1956, CHEW 1959
Lozano J.J. ABREU 1991B, DECAMP 1991S
Lu J. ABE 1994ZE, ABE 1995
Lu M. BARBER 1979F
Lu Y.S. ADEVA 1990R, ADEVA 1991D, DECAMP 1991S
Lubbers J.M. ADEVA 1990R, ADEVA 1991D, DECAMP 1991S
Lubelsmeyer K. PRESCOTT 1978C, PRESCOTT 1979C, BRANDELIK 1979H, BRANDELIK 1980G, ADEVA 1990R, ADEVA 1991D, DECAMP 1991S
Lubrano P. AVERY 1989C
Lucchesi D. ABE 1994ZE, ABE 1995
Luchini C.B. ABE 1994ZE
Luchkov B.I. DOLGOSHEIN 1964
Luci C. ADEVA 1990R, ADEVA 1991D, DECAMP 1991S
Luckey D. BARBER 1979F, ADEVA 1990R, ADEVA 1991D, DECAMP 1991S
Lucock R. ABREU 1991B, DECAMP 1991S
Luders G. LUDERS 1954
Ludovici L. ADEVA 1990R, ADEVA 1991D, DECAMP 1991S
Ludwig J. BRANDELIK 1977D, ALEXANDER 1991C, ACTON 1992B, DECAMP 1991S
Lue X. ADEVA 1990R, ADEVA 1991D, DECAMP 1991S
Lueking L. ABACHI 1995G
Luhrsen W. BERGER 1978, BERGER 1979C, BERGER 1980L
Luke D. SCHWITTERS 1975, PERL 1975C, HANSON 1975, PERL 1976C, GOLDHABER 1976D, PERUZZI 1976C, PERL 1977F, BRANDELIK 1979H, BRANDELIK 1980G, DECAMP 1991D
Lukens P. ABE 1994ZE, ABE 1995
Lukov V.V. VOYVODIC 1986B
Lulu B.A. AUGUSTIN 1974B, ABRAMS 1974, SCHWITTERS 1975, PERL 1975C, HANSON 1975, PERL 1976C
Luminari L. ADEVA 1990R, ADEVA 1991D, DECAMP 1991S
Lurie D. BOWCOCK 1961
Lusiani A. DECAMP 1991D, DECAMP 1991N, DECAMP 1991S, BUSKULIC 1993H
Lusin S. ABE 1995
Luth V. AUGUSTIN 1974B, ABRAMS 1974, SCHWITTERS 1975, PERL 1975C, HANSON 1975, PERL 1976C, GOLDHABER 1976D, PERUZZI 1976C, PERL 1977F, ABRAMS 1989B, WEIR 1989B
Lutjens G. BLUM 1975, DECAMP 1991D, DECAMP 1991N, DECAMP 1991S, BUSKULIC 1993H
Lutters G. BUSKULIC 1993H
Lutz A.M. HASERT 1973, HASERT 1973B, BLIETSCHAU 1976, BLIETSCHAU 1977B, DECAMP 1991D, DECAMP 1991N, DECAMP 1991S, BUSKULIC 1993H
Lutz G. BLUM 1975, DECAMP 1991D, DECAMP 1991N, DECAMP 1991S, BUSKULIC 1993H
Lutz P. ABREU 1991B, DECAMP 1991S
Lyman E.M. LYMAN 1951
Lynch G.R. BUTTON 1960, BUTTON 1961, XUONG 1961C
Lynch H.L. AUGUSTIN 1974B, ABRAMS 1974, SCHWITTERS 1975, HANSON 1975, GOLDHABER 1976D, PERUZZI 1976C, BRANDELIK 1977D, BRANDELIK 1979H, BRANDELIK 1980G
Lynch J.G. DECAMP 1991D, DECAMP 1991N, DECAMP 1991S, BUSKULIC 1993H

Lyon A.L. ABACHI 1995G
Lyon D. AUGUSTIN 1974B, ABRAMS 1974, PERL 1975C
Lyons L. ABREU 1991B, DECAMP 1991S
Lys J. ABE 1994ZE, ABE 1995
Lyubimov V.A. ALBRECHT 1987P
Ma C.M. BARBER 1979F
Ma D.A. BARBER 1979F
Ma W.G. ADEVA 1990R, ADEVA 1991D, DECAMP 1991S
Maas P. ABE 1994ZE
Mac Kay W.W. CHADWICK 1981
Macbeth A. BARTEL 1980D
MacDermott M. ADEVA 1990R, ADEVA 1991D, DECAMP 1991S
MacFarlane D.B. ALBRECHT 1987P
Maciel A.K.A. ABACHI 1995G
MacKenzie K.R. BROBECK 1947
Madaras R. PERL 1976C, GOLDHABER 1976D, PERUZZI 1976C, PERL 1977F, ABACHI 1995G
Madden R. ABACHI 1995G
Madsen B. BANNER 1983B, BAGNAIA 1983B, ANSARI 1987E
Maehlum G. ABREU 1991B, DECAMP 1991S
Maeshima K. ABE 1994ZE, ABE 1995
Magahiz R. ADEVA 1990R, ADEVA 1991D, DECAMP 1991S
Mageras G. BOHRINGER 1980B, FINOCCHIARO 1980
Maggi G. DECAMP 1991D, DECAMP 1991N, DECAMP 1991S, BUSKULIC 1993H
Maggi M. DECAMP 1991D, DECAMP 1991N, DECAMP 1991S, BUSKULIC 1993H
Maghakian A. ABE 1994ZE, ABE 1995
Maglic B.C. BUTTON 1960, BUTTON 1961, MAGLIC 1961B
Mahmoud H.M. KONOPINSKI 1953
Maiani I. GLASHOW 1970
Maillard J. ABREU 1991B, DECAMP 1991S
Maire M. AUBERT 1983, ADEVA 1990R, ADEVA 1991D, DECAMP 1991S
Maitland W. BUSKULIC 1993H
Majerotto W. FIDECARO 1978
Majorana E. MAJORANA 1933, MAJORANA 1937
Makhlyueva I.V. VOYVODIC 1986B
Maksimovic P. ABE 1994ZE, ABE 1995
Malhotra P.K. ADEVA 1990R, ADEVA 1991D, DECAMP 1991S
Malik R. ADEVA 1990R, ADEVA 1991D, DECAMP 1991S
Malinin A. ADEVA 1990R, ADEVA 1991D, DECAMP 1991S
Malosse J.J. ARNISON 1983D, ARNISON 1984C
Maltezos A. ABREU 1991B, DECAMP 1991S
Maltezos S. ABREU 1991B, DECAMP 1991S
Mana C. ADEVA 1990R, ADEVA 1991D, DECAMP 1991S
Mandelstam S. MANDELSTAM 1958
Mandl F. ABREU 1991B, DECAMP 1991S
Mandrichenko I.V. ABACHI 1995G
Manelli I. EISLER 1958, APEL 1975B
Manfredini A. CHAMBERLAIN 1956E, CHAMBERLAIN 1956F, BARKAS 1957
Mangano M. ABE 1994ZE, ABE 1995
Mangeot P. ABACHI 1995G
Mani P. BANNER 1983B, BAGNAIA 1983B
Mani S. ABACHI 1995G
Manley D.M. BARNETT 1996
Mann A.K. BENVENUTI 1974F, HIRATA 1987C
Mann R. JUNG 1992
Mannelli E.B. DECAMP 1991D, DECAMP 1991N, DECAMP 1991S, BUSKULIC 1993H
Mannelli M. ALEXANDER 1991C, ACTON 1992B, DECAMP 1991S
Manner W. BLUM 1975, DECAMP 1991D, DECAMP 1991N, DECAMP 1991S, BUSKULIC 1993H
Mannocchi G. DECAMP 1991D, DECAMP 1991N, DECAMP 1991S, BUSKULIC 1993H
Manohar A. BARNETT 1996
Mansoulie B. BANNER 1983B, BAGNAIA 1983B, ANSARI 1987E, ABACHI 1995G
Mansour J. ABE 1994ZE, ABE 1995
Mantovani G.C. BANNER 1983B, BAGNAIA 1983B, ANSARI 1987E
Mao D.N. ADEVA 1990R, ADEVA 1991D, DECAMP 1991S
Mao H.S. ABACHI 1995G
Mao Y.F. ADEVA 1990R, ADEVA 1991D, DECAMP 1991S
Maolinbay M. ADEVA 1990R, ADEVA 1991D, DECAMP 1991S
Mapelli L. BANNER 1983B, BAGNAIA 1983B, ANSARI 1987E
Marage P. CALICCHIO 1980
Marcellini S. ALEXANDER 1991C, ACTON 1992B, DECAMP 1991S
March P.V. DANYSZ 1963, DECAMP 1991D, DECAMP 1991N, DECAMP 1991S, BUSKULIC 1993H
March R. ERWIN 1961C, BUNCE 1976
Marchesini P. ADEVA 1990R, ADEVA 1991D, DECAMP 1991S
Marchionni A. ADEVA 1990R, ADEVA 1991D, DECAMP 1991S
Marchioro A. DECAMP 1991D, DECAMP 1991N, DECAMP 1991S
Marco J. ABREU 1991B, DECAMP 1991S
Margoni J. ABREU 1991B, DECAMP 1991S
Margulies S. ABACHI 1995G
Marin J.C. ABREU 1991B, DECAMP 1991S
Maringer G. ALEXANDER 1991C, ACTON 1992B, DECAMP 1991S
Mariotti M. ABE 1994ZE, ABE 1995
Markees A. DARDEN 1978B
Markeloff R. ABACHI 1995G
Markiewicz T. ARNISON 1983C, ARNISON 1983D, ARNISON 1984C, ARNISON 1984G, ARNISON 1985D, ALBAJAR 1987C
Markosky L. ABACHI 1995G
Markou A. ABREU 1991B, DECAMP 1991S
Markou C. ALBAJAR 1987C, ALBAJAR 1991C, ALBAJAR 1991D, BUSKULIC 1993H
Markov P. KIRILLOVA 1965, BEZNOGIKH 1969
Markytan M. ALBAJAR 1987C, ALBAJAR 1991C, ALBAJAR 1991D
Markytan T. ARNISON 1985D
Marquina M.A. ALBAJAR 1991C, ALBAJAR 1991D
Marriner J.P. ABE 1994ZE, ABE 1995
Marrocchesi P.S. DECAMP 1991D, DECAMP 1991N, DECAMP 1991S, BUSKULIC 1993H
Marshall F. MARSHALL 1947
Marshall R. BARTEL 1980D
Marshall T. ABACHI 1995G
Marti S. ABREU 1991B, DECAMP 1991S
Martin A.J. ALEXANDER 1991C, ACTON 1992B, DECAMP 1991S, ABE 1994ZE, ABE 1995
Martin B. ADEVA 1991D, DECAMP 1991S
Martin J.P. ADEVA 1990R, DECAMP 1991D, ADEVA 1991D, ALEXANDER 1991C, DECAMP 1991N, ACTON 1992B, DECAMP 1991S, DECAMP 1991S, DECAMP 1991S
Martin M.I. ABACHI 1995G
Martin P. BUNCE 1976
Martin R. ANDERSON 1952B
Martin T. ARNISON 1985D
Martinelli R. ALBAJAR 1991C, ALBAJAR 1991D
Martinez L. ADEVA 1990R
Martinez M. DECAMP 1991D, DECAMP 1991N, DECAMP 1991S, BUSKULIC 1993H
Martinez-Laso L. ADEVA 1991D, DECAMP 1991S
Martyn H.U. BRANDELIK 1977D, BRANDELIK 1979H, BRANDELIK 1980G

Maruyama T. BAND 1988
Marx B. BUSKULIC 1993H
Marx M. ABACHI 1995G
Marzano F. ADEVA 1990R, ADEVA 1991D, DECAMP 1991S
Maschuw R. BERGER 1978, BERGER 1979C
Mashimo T. ALEXANDER 1991C, ACTON 1992B, DECAMP 1991S
Maskawa T. KOBAYASHI 1973B
Massam T. BASILE 1981I
Massaro G.G.G. BARBER 1979F, ADEVA 1990R, ADEVA 1991D, DECAMP 1991S
Massimo J.T. CRETIEN 1962
Massimo-Brancacci F. DECAMP 1991D
Mathis L. ABREU 1991B, DECAMP 1991S
Mato P. DECAMP 1991D, DECAMP 1991N, DECAMP 1991S, BUSKULIC 1993H
Matorras F. ABREU 1991B, DECAMP 1991S
Matsen R.P. GOOD 1961
Matsuda T. BARBER 1979F, ADEVA 1990R, ADEVA 1991D, DECAMP 1991S
Matsumura H. BARTEL 1980D
Matteuzzi C. ABREU 1991B, DECAMP 1991S
Matthews J. BIONTA 1987C
Matthews J.A.J. ABRAMS 1989B, WEIR 1989B, ABE 1994ZE, ABE 1995
Matthews M. ASHMAN 1988B
Matthiae G. AMALDI 1973E, ABREU 1991B, DECAMP 1991S
Matthiesen U. ALBRECHT 1987P
Mattig P. BRANDELIK 1979H, BRANDELIK 1980G, ALEXANDER 1991C, ACTON 1992B, DECAMP 1991S
Mattingly R. ABE 1994ZE, ABE 1995
Mattison T. ABRAMS 1989B, BUSKULIC 1993H
Matveev V.A. MATVEEV 1973, ALBRECHT 1987P
Matyushin A.T. ASTVACATUROV 1968
Matyushin V.T. ASTVACATUROV 1968
Maull K. FAISSNER 1978D
Maumary Y. DECAMP 1991D, DECAMP 1991N, DECAMP 1991S, BUSKULIC 1993H
Maur U. ALEXANDER 1991C, ACTON 1992B, DECAMP 1991S
Maurin G. ARNISON 1983C, ARNISON 1983D, ARNISON 1984C, ARNISON 1984G, ARNISON 1985D, ALBAJAR 1987C, ALBAJAR 1991C, ALBAJAR 1991D
May B. ABACHI 1995G
May J. DECAMP 1991D
Mayorov A.A. ABACHI 1995G
Mazumdar K. ADEVA 1990R, ADEVA 1991D, DECAMP 1991S
Mazzucato M. ABREU 1991B, DECAMP 1991S
McAllister R.W. MCALLISTER 1956
McBride P. ADEVA 1990R, ADEVA 1991D, DECAMP 1991S
McCarthy R. ABACHI 1995G
McCord R.V. SNELL 1950
McCorriston T.P. AUBERT 1974D
McCubbin M. ABREU 1991B, DECAMP 1991S
McGowan R.F. ALEXANDER 1991C, ACTON 1992B, DECAMP 1991S
McGuire A.D. COWAN 1956
McIlwain R.L. AVERY 1989C, ARTUSO 1989C, ALAM 1989B
McIntyre P. ABE 1994ZE, ABE 1995
McKay K.G. MCKAY 1949
McKay R. ABREU 1991B, DECAMP 1991S
McKenna J. ALBRECHT 1987P, ABRAMS 1989B, ACTON 1992B
McKenzie J. HASERT 1973, HASERT 1973B
McKibben T. ABACHI 1995G
McKinley J. ABACHI 1995G
McLean K. ALBRECHT 1987P
McMahon S. ALBAJAR 1991C, ALBAJAR 1991D
McMahon T.J. ARNISON 1983C, ARNISON 1983D, ARNISON 1984C, ARNISON 1984G, ARNISON 1985D, ADEVA 1990R, ADEVA 1991D, ALEXANDER 1991C, ACTON 1992B, DECAMP 1991S, DECAMP 1991S
McMillan E.M. LAWRENCE 1939, MCMILLAN 1945, MCMILLAN 1946, BROBECK 1947
McNally D. ADEVA 1990R, ADEVA 1991D, DECAMP 1991S
McNeely W.A. BRANDELIK 1977D
McNicholas J. AUBERT 1983
McNulty R. ABREU 1991B, DECAMP 1991S
McNutt J.R. ALEXANDER 1991C, ACTON 1992B, DECAMP 1991S
Medcalf T. DECAMP 1991D, DECAMP 1991N, DECAMP 1991S, BUSKULIC 1993H
Medvedev B.V. BOGOLYUBOV 1956B
Meier K. ANSARI 1987E
Meijers F. ALEXANDER 1991C, ACTON 1992B, DECAMP 1991S
Meinhard H. DECAMP 1991D, DECAMP 1991N, DECAMP 1991S, BUSKULIC 1993H
Meinholz T. ADEVA 1990R, ADEVA 1991D, DECAMP 1991S
Meissburger J. BLUM 1975
Meitner L. MEITNER 1930
Melese P. ABE 1994ZE, ABE 1995
Melissinos A.C. ANDREWS 1980B, BEBEK 1981, CHADWICK 1981
Meltzer C. PEVSNER 1961
Menary S. DECAMP 1991D, DECAMP 1991N, DECAMP 1991S
Mencuccini C. BACCI 1974
Mendiburu J.P. ARNISON 1983C, ARNISON 1983D, ARNISON 1984C, ARNISON 1984G, ARNISON 1985D, ALBAJAR 1987C, ALBAJAR 1991C, ALBAJAR 1991D
Meneguzzo A. ARNISON 1985D, ALBAJAR 1987C, ALBAJAR 1991C, ALBAJAR 1991D
Menichetti E. ABREU 1991B, DECAMP 1991S
Menon M.G.K. FOWLER 1951, DAVIES 1955
Menszner D. ALEXANDER 1991C, ACTON 1992B, DECAMP 1991S
Menzione A. AMENDOLIA 1973B, ABE 1994ZE, ABE 1995
Meola G. ABREU 1991B, DECAMP 1991S
Merk M. ADEVA 1990R, ADEVA 1991D, DECAMP 1991S
Merkel B. BANNER 1983B, BAGNAIA 1983B, ANSARI 1987E
Merlin M. DAVIES 1955
Merlo J.P. ALBAJAR 1987C, ALBAJAR 1991C, ALBAJAR 1991D
Mermikides M. BAGNAIA 1983B, DECAMP 1991D, DECAMP 1991N, DECAMP 1991S
Merola L. ADEVA 1990R, ADEVA 1991D, DECAMP 1991S
Meroni C. ABREU 1991B, DECAMP 1991S
Merrill D.W. KALBFLEISCH 1964
Merrison A.W. FAZZINI 1958
Merritt F.S. ALEXANDER 1991C, ACTON 1992B, DECAMP 1991S
Merritt K.W. ABACHI 1995G
Mertiens H.D. BERGER 1979C, BERGER 1980L
Mes H. ALEXANDER 1991C, ACTON 1992B, DECAMP 1991S
Meschi E. ABE 1994ZE, ABE 1995
Meschini M. ADEVA 1990R, ADEVA 1991D, DECAMP 1991S
Meshkov I.N. BUDKER 1976
Mess K.H. BRANDELIK 1977D
Messineo A. DECAMP 1991D, DECAMP 1991N, DECAMP 1991S, BUSKULIC 1993H
Messing F. BENVENUTI 1974F
Messner R. KNAPP 1976B
Messner R.L. BAND 1988
Mestayer M.D. AVERY 1989C, ARTUSO 1989C, ALAM 1989B

Mestvirishvili M.A. LOGUNOV 1967B
Metcalf W. ANDREWS 1980
Metzger W.J. ADEVA 1990R, ADEVA 1991D, DECAMP 1991S
Metzler S. ABE 1995
Meunier R. BUSHNIN 1969
Meyer D.I. WEIR 1989B
Meyer H. BERGER 1978, BERGER 1979C, BERGER 1980L
Meyer H.J. BERGER 1978, BERGER 1979C, BERGER 1980L
Meyer O. BERGER 1978, BERGER 1979C, BERGER 1980L, ARNISON 1985D
Meyer T. BRANDELIK 1979H, BRANDELIK 1980G, ARNISON 1985D, ALBAJAR 1987C, ALBAJAR 1991C, ALBAJAR 1991D
Meyer W.T. ABREU 1991B, DECAMP 1991S
Mi Y. ADEVA 1990R, ADEVA 1991D, DECAMP 1991S
Miao C. ABE 1995
Michail G. ABE 1994ZE, ABE 1995
Michel B. DECAMP 1991D, DECAMP 1991N, DECAMP 1991S, BUSKULIC 1993H
Michelini A. ALEXANDER 1991C, ACTON 1992B, DECAMP 1991S
Michelotto M. ABREU 1991B, DECAMP 1991S
Michelsen U. BERGER 1980L
Michette A.G. HASERT 1973, HASERT 1973B, BLIETSCHAU 1976, BLIETSCHAU 1977B, CALICCHIO 1980
Micke M. ADEVA 1990R
Micke U. ADEVA 1990R
Middleton R.P. ALEXANDER 1991C, ACTON 1992B, DECAMP 1991S
Miettinen H. ABACHI 1995G
Mikamo S. ABE 1994ZE, ABE 1995
Mikenberg G. BRANDELIK 1977D, BRANDELIK 1979H, BRANDELIK 1980G, ALEXANDER 1991C, ACTON 1992B, DECAMP 1991S
Mikhailov V.A. CHIKOVANI 1964
Mikhailov Y.V. APEL 1975B
Mikhul A. KANG-CHANG 1960
Mildenberger J. ALEXANDER 1991C, ACTON 1992B, DECAMP 1991S
Milder A. ABACHI 1995G
Milekhin G.A. MILEKHIN 1957
Miller D.H. DORFAN 1967, AVERY 1989C, ARTUSO 1989C, ALAM 1989B
Miller D.J. ALEXANDER 1991C, ACTON 1992B, DECAMP 1991S
Miller G. MILLER 1972B
Miller L.C. SNELL 1948
Miller M. ABE 1994ZE
Miller R. ABE 1994ZE, ABE 1995
Miller R.H. PRESCOTT 1978C, PRESCOTT 1979C, BIONTA 1987C
Millikan R.A. MILLIKAN 1911, MILLIKAN 1913, MILLIKAN 1916
Milliken B.D. ABRAMS 1989B, WEIR 1989B
Mills G.B. ADEVA 1990R, ADEVA 1991D, DECAMP 1991S
Mills H. BARTEL 1980D
Mills R.L. YANG 1954
Milner C. ABACHI 1995G
Milone A. DAVIES 1955
Milstene C. ALEXANDER 1991C, DECAMP 1991S
Mimashi T. ABE 1994ZE
Minakawa O. MINAKAWA 1959
Minard M.N. ARNISON 1983C, ARNISON 1983D, ARNISON 1984C, ARNISON 1984G, ARNISON 1985D, ALBAJAR 1987C, DECAMP 1991D, DECAMP 1991N, DECAMP 1991S, BUSKULIC 1993H
Minato H. ABE 1995
Mincer A. ABACHI 1995G
Minowa M. BARTEL 1980D
Minssieux H. AUBERT 1983
Minten A. DECAMP 1991D, DECAMP 1991N, DECAMP 1991S, BUSKULIC 1993H
Miotto A. DECAMP 1991D, DECAMP 1991N, DECAMP 1991S, BUSKULIC 1993H
Miquel R. DECAMP 1991D, DECAMP 1991N, DECAMP 1991S, BUSKULIC 1993H
Mir L.M. DECAMP 1991D, DECAMP 1991N, DECAMP 1991S, BUSKULIC 1993H
Mir R. ALEXANDER 1991C, ACTON 1992B, DECAMP 1991S
Mir Y. ADEVA 1990R, ADEVA 1991D, DECAMP 1991S
Mirabelli G. ADEVA 1990R, ADEVA 1991D, DECAMP 1991S
Mirabito L. DECAMP 1991N, DECAMP 1991S
Miscetti S. ABE 1994ZE, ABE 1995
Mishina M. ABE 1994ZE, ABE 1995
Mishnev S.I. ARTAMONOV 1984B
Mishra C.S. ABACHI 1995G
Mistry N. DANBY 1962
Mistry N.B. ANDREWS 1980, ANDREWS 1980B, BEBEK 1981, CHADWICK 1981, AVERY 1989C, ARTUSO 1989C, ALAM 1989B
Mitaroff W.A. ABREU 1991B, DECAMP 1991S
Mitselmakher G.V. ABREU 1991B, DECAMP 1991S
Mitsushio H. ABE 1994ZE, ABE 1995
Mittra I.S. BERTANZA 1962D
Miyamoto S. FUKUI 1959
Miyamoto T. ABE 1995
Miyano K. HIRATA 1987C
Miyashita S. ABE 1994ZE, ABE 1995
Miyazawa H. GOLDBERGER 1955B
Mizuno Y. ASHMAN 1988B
Mjoernmark U. ABREU 1991B, DECAMP 1991S
Mnich J. ADEVA 1990R, ADEVA 1991D, DECAMP 1991S
Mo L.W. BLOOM 1969B, BREIDENBACH 1969, MILLER 1972B
Moa T. ABREU 1991B, DECAMP 1991S
Modis T. BIAGI 1983C, BIAGI 1985C
Moeller R. ABREU 1991B, DECAMP 1991S
Moenig K. ABREU 1991B, DECAMP 1991S
Moers T. ALBAJAR 1991C, ALBAJAR 1991D
Moffeit K.C. BRANDELIK 1977D, ABRAMS 1989B, WEIR 1989B
Mohammadi M. ARNISON 1983D, ARNISON 1984C, ARNISON 1984G, ARNISON 1985D, ALBAJAR 1987C, ALBAJAR 1991C, ALBAJAR 1991D
Mohammadi-BarmandM ABACHI 1995G
Mohl D. CARRON 1978
Mohler R. FUNG 1956
Mohr W. AUBERT 1983, ALEXANDER 1991C, ACTON 1992B, DECAMP 1991S
Moisan C. ALEXANDER 1991C, ACTON 1992B, DECAMP 1991S
Mokhov N.V. ABACHI 1995G
Moller M. ADEVA 1990R, ADEVA 1991D, DECAMP 1991S
Mollerud R. BANNER 1983B, BAGNAIA 1983B, ANSARI 1987E, DECAMP 1991D, DECAMP 1991N, DECAMP 1991S, BUSKULIC 1993H
Monacelli P. BACCI 1974
Mondal N.K. ABACHI 1995G
Moneta L. DECAMP 1991D, DECAMP 1991N, DECAMP 1991S, BUSKULIC 1993H
Moneti G.C. BERTANZA 1962D, CONNOLLY 1963, ANDREWS 1980, ANDREWS 1980B, BEBEK 1981, CHADWICK 1981, AVERY 1989C, ARTUSO 1989C, ALAM 1989B
Monge M.R. ABREU 1991B, DECAMP 1991S
Moniez M. ANSARI 1987E
Mönig K. ASHMAN 1988B
Moning R. ANSARI 1987E
Monnier E. DECAMP 1991D, DECAMP 1991N, DECAMP 1991S

Montanari A. ALEXANDER 1991C, ACTON 1992B, DECAMP 1991S
Montanet F. ASHMAN 1988B
Montanet L BARNETT 1996
Monteleoni B. ADEVA 1990R, ADEVA 1991D, DECAMP 1991S
Montgomery H.E. AUBERT 1983, ABACHI 1995G
Montret J.C. DECAMP 1991D, DECAMP 1991N, DECAMP 1991S, BUSKULIC 1993H
Montwill A. DANYSZ 1963
Mooney P. ABACHI 1995G
Morand G. ADEVA 1990R, ADEVA 1991D, DECAMP 1991S
Morand R. ADEVA 1990R, ADEVA 1991D, DECAMP 1991S
Moreels J. CALICCHIO 1980
Morehouse C.C. AUGUSTIN 1974B, ABRAMS 1974, SCHWITTERS 1975, PERL 1975C, HANSON 1975, PERL 1976C, GOLDHABER 1976D, PERUZZI 1976C
Morettini P. ABREU 1991B, DECAMP 1991S
Morfin J. HASERT 1973, HASERT 1973B, BLIETSCHAU 1976, BLIETSCHAU 1977B
Morgan K. ARNISON 1983C, ARNISON 1983D, ARNISON 1984C, ARNISON 1984G, ARNISON 1985D, ALBAJAR 1987C, ALBAJAR 1991C, ALBAJAR 1991D
Morganti M. ANSARI 1987E
Morganti S. ADEVA 1990R, ADEVA 1991D, DECAMP 1991S
Morgunov V.L. ADEVA 1990R
Mori T. ALEXANDER 1991C, ACTON 1992B, DECAMP 1991S
Moricca M. ARNISON 1983C, ARNISON 1983D, ARNISON 1984C, ARNISON 1984G, ARNISON 1985D, ARNISON 1985D, ALBAJAR 1987C
Morita Y. ABE 1994ZE, ABE 1995
Moroz N.S. BALDIN 1973
Morozov A.G. ANTIPOV 1970
Morozov B.A. BEZNOGIKH 1969
Morris T.W. BROWN 1962, BARNES 1964B, BARNES 1964E
Morrison R.J. BARNETT 1996
Morrow F. ARTUSO 1989C, ALAM 1989B
Morsch A. ALBAJAR 1991C, ALBAJAR 1991D
Morton W.T. DECAMP 1991D, DECAMP 1991N, DECAMP 1991S, BUSKULIC 1993H
Moser H.G. ARNISON 1985D, ALBAJAR 1987C, DECAMP 1991D, DECAMP 1991N, DECAMP 1991S, BUSKULIC 1993H
Moser K. AUBERT 1983
Moshammer R. JUNG 1992
Moss L.J. BAND 1988
Moss M.W. ALEXANDER 1991C, ACTON 1992B, DECAMP 1991S
Motley R. FITCH 1956
Moulai N.E. ADEVA 1991D, DECAMP 1991S
Moulding S. ABE 1994ZE
Moulin A. ALBAJAR 1991C, ALBAJAR 1991D
Mount R. ADEVA 1990R, ADEVA 1991D, DECAMP 1991S
Mount R.P. AUBERT 1983
Mours B. ARNISON 1985D, ALBAJAR 1987C, ALBAJAR 1991D
Mouthuy T. ALEXANDER 1991C, ACTON 1992B, DECAMP 1991S
Moutoussi A. BUSKULIC 1993H
Moyer B.J. BJORKLUND 1950, BRABANT 1956C
Mrabito L. DECAMP 1991D
Mudan M. BIONTA 1987C, ABACHI 1995G
Mueller H. APEL 1975B, ABREU 1991B, DECAMP 1991S
Mueller J.J. ANDREWS 1980, ANDREWS 1980B, BEBEK 1981, CHADWICK 1981, AVERY 1989C, ARTUSO 1989C, ALAM 1989B, ABE 1994ZE, ABE 1995
Muhlemann P. BIAGI 1983C
Muirhead H. LATTES 1947, BROWN 1949B, ARNISON 1983D, ARNISON 1984C
Mukherjee A. ABE 1994ZE, ABE 1995
Mulkens H. BLIETSCHAU 1976, BLIETSCHAU 1977B
Muller F. GREGORY 1954, ARMENTEROS 1955, GOOD 1961, ARNISON 1983C, ARNISON 1983D, ARNISON 1984C, ARNISON 1984G, ARNISON 1985D, BAND 1988
Muller H. ABREU 1991B, DECAMP 1991S
Muller L. WEIR 1989B
Muller S. ADEVA 1991D, DECAMP 1991S
Muller T. ALBAJAR 1987C, ALBAJAR 1991C, ALBAJAR 1991D, ABE 1994ZE, ABE 1995
Munger C. ABRAMS 1989B
Munoz R. ALBAJAR 1991C, ALBAJAR 1991D
Muradyan R.M. MATVEEV 1973
Murat P. ABE 1995
Murayama H. BARNETT 1996
Murphy C. ABACHI 1995G
Murphy C.T. ABACHI 1995G
Murphy G.M. UREY 1932, UREY 1932B, MURPHY 1934
Murphy P.G. BARTEL 1980D, ALEXANDER 1991C, DECAMP 1991S
Murray J.J. BRABANT 1956C, KALBFLEISCH 1964
Murray W. ABRAMS 1989B
Murray W.J. ABREU 1991B, DECAMP 1991S
Murtagh M.J. CAZZOLI 1975E, BALTAY 1979E
Murtas F. DECAMP 1991D, DECAMP 1991N, DECAMP 1991S, BUSKULIC 1993H
Murtas G.P. BACCI 1974, DECAMP 1991D, DECAMP 1991N, DECAMP 1991S, BUSKULIC 1993H
Musgrave P. ABE 1994ZE
Musset P. HASERT 1973, HASERT 1973B, BLIETSCHAU 1976, BLIETSCHAU 1977B
Mustard R. AMENDOLIA 1973B
Myatt G. HASERT 1973, HASERT 1973B, BLIETSCHAU 1976, BLIETSCHAU 1977B, ABREU 1991B, DECAMP 1991S
Myklebost K. BLIETSCHAU 1976, BLIETSCHAU 1977B
Nagle D.E. ANDERSON 1952B, ANDERSON 1952E
Nagovitsin N. ALBRECHT 1987P
Nagy E. ADEVA 1990R, ADEVA 1991D, DECAMP 1991S
Nakada H. ABE 1995
Nakae L.F. ABE 1994ZE
Nakahata M. HIRATA 1987C
Nakano I. ABE 1994ZE, ABE 1995
Nambu Y. NAMBU 1961, NAMBU 1961B, HAN 1965, NAMBU 1966
Namjoshi R. AVERY 1989C, ARTUSO 1989C, ALAM 1989B
Nandi A.K. ARNISON 1983C, ARNISON 1983D, ARNISON 1984C, ARNISON 1984G, ARNISON 1985D, ALBAJAR 1987C
Nandi S. AVERY 1989C, ARTUSO 1989C, ALAM 1989B
Nang F. ABACHI 1995G
Nania R. BASILE 1981I
Napolitano M. ADEVA 1990R, ADEVA 1991D, DECAMP 1991S
Naraghi F. ABREU 1991B, DECAMP 1991S
Narain M. ABACHI 1995G
Narasimham V.S. ABACHI 1995G
Narayanan A. ABACHI 1995G
Naroska B. BARTEL 1980D
Nash J. ABRAMS 1989B, WEIR 1989B, DECAMP 1991D, DECAMP 1991N, DECAMP 1991S, BUSKULIC 1993H
Nassalski J. ASHMAN 1988B
Natali S. HASERT 1973, HASERT 1973B, CALICCHIO 1980, DECAMP 1991D, DECAMP 1991N, DECAMP 1991S, BUSKULIC 1993H
Nau-Korzen U. ABREU 1991B, DECAMP 1991S

Naumann L. ARNISON 1983C, ARNISON 1983D, ARNISON 1984C, ARNISON 1984G, ARNISON 1985D, ALBAJAR 1987C, ALBAJAR 1991C, ALBAJAR 1991D
Navarria F.L. ABREU 1991B, DECAMP 1991S
Nazarenko V.A. LOBASHOV 1966
Neal H.A. ABACHI 1995G
Neal R.B. CHODOROW 1955
Neddermeyer S.H. NEDDERMEYER 1937
Nedelec P. ALBAJAR 1987C, ALBAJAR 1991C, ALBAJAR 1991D
Ne'eman Y. NE'EMAN 1961
Negret J.P. ABACHI 1995G
Negri P. ABREU 1991B, DECAMP 1991S
Neis E. ABACHI 1995G
Nellen B. ALEXANDER 1991C, ACTON 1992B, DECAMP 1991S
Nelson B. CRETIEN 1962
Nelson C. ABE 1994ZE, ABE 1995
Nelson H.N. BAND 1988
Nelson M.E. WEIR 1989B
Nemethy P. ABACHI 1995G
Nereson N. ROSSI 1942, NERESON 1942
Nesic D. ABACHI 1995G
Neuberger D. ABE 1994ZE, ABE 1995
Neugebauer E. DECAMP 1991D
Neuhofer G. FIDECARO 1978
Neumann B. BERGER 1978, BERGER 1979C, BERGER 1980L
Neveu M. CALICCHIO 1980
Nevin T.E. DAVIES 1955
Newcomer F.M. HIRATA 1987C
Newman H. BARBER 1979F, ADEVA 1990R, ADEVA 1991D, DECAMP 1991S
Newman-Holmes C. ABE 1994ZE, ABE 1995
Neyer C. ADEVA 1991D, DECAMP 1991S
Ng C.R. AVERY 1989C, ALAM 1989B
Ng R. ARTUSO 1989C
Nguen Van Hieu LOGUNOV 1967B
Nguyen H.H. ALEXANDER 1991C, ACTON 1992B, DECAMP 1991S
Nguyen H.K. PERL 1976C, GOLDHABER 1976D, PERUZZI 1976C, PERL 1977F
Nguyen-Khac U. HASERT 1973B, BLIETSCHAU 1976, BLIETSCHAU 1977B
Niaz M.A. ADEVA 1990R, ADEVA 1991D, DECAMP 1991S
Nicod D. BUSKULIC 1993H
Nicoletti G. DECAMP 1991D, DECAMP 1991N, DECAMP 1991S
Niczyporuk B. BIENLEIN 1978
Nielsen B.S. ABREU 1991B, DECAMP 1991S
Nielsen E.R. DECAMP 1991D, DECAMP 1991N, DECAMP 1991S
Niessen L. ADEVA 1990R, ADEVA 1991D, DECAMP 1991S
Nigro M. BACCI 1974
Niinikoski T. ASHMAN 1988B
Nijjhar B. ABREU 1991B, DECAMP 1991S
Nikitas M. ALBAJAR 1991C, ALBAJAR 1991D
Nikitin A.V. KANG-CHANG 1960, BEZNOGIKH 1969
Nikitin S.A. ARTAMONOV 1984B
Nikitin V.A. KIRILLOVA 1965
Nikolaenko V.I. ABREU 1991B, DECAMP 1991S
Nilsen N.V. KERST 1956
Nilsson A. ALBRECHT 1987P
Nilsson B.S. BAGNAIA 1983B, DECAMP 1991D, DECAMP 1991N, DECAMP 1991S, BUSKULIC 1993H
Ninomiya M. ABE 1995
Nippe A. ALBRECHT 1987P
Nisati A. ALBAJAR 1991C, ALBAJAR 1991D
Nishijima K. NISHIJIMA 1955, NISHIJIMA 1957
Nishikawa K. MINAKAWA 1959
Nishimura J. MINAKAWA 1959
Nishimura Y. MINAKAWA 1959
Nishina Y. NISHINA 1937
Nitz D. WEIR 1989B
Niu K. MINAKAWA 1959
Nodulman L. ABE 1994ZE, ABE 1995
Nolden F. JUNG 1992
Nomokonov P.V. BEZNOGIKH 1969
Nordberg E. ANDREWS 1980, ANDREWS 1980B, BEBEK 1981, CHADWICK 1981, AVERY 1989C, ARTUSO 1989C, ALAM 1989B
Nordin P. NORDIN 1958
Nordsieck A. BLOCH 1937
Norem J. BUNCE 1976
Norman D. ABACHI 1995G
Norton A. ARNISON 1983C, ARNISON 1983D, ARNISON 1984C, ARNISON 1984G, ARNISON 1985D, ALBAJAR 1987C, ALBAJAR 1991C, ALBAJAR 1991D
Norton P.R. AUBERT 1983, ASHMAN 1988B, DECAMP 1991D, DECAMP 1991N, DECAMP 1991S, BUSKULIC 1993H
Notz D. BRANDELIK 1977D, BRANDELIK 1979H, BRANDELIK 1980G
Nowak H. ADEVA 1991D, DECAMP 1991S
Nozaki M. ALEXANDER 1991C, ACTON 1992B, DECAMP 1991S
Nozaki T. BARTEL 1980D
Nurushev S. FIDECARO 1978
Nurushev S.B. BEZNOGIKH 1969
Nussbaum M. PEVSNER 1961
Nuttall M. DECAMP 1991D, DECAMP 1991N, DECAMP 1991S, BUSKULIC 1993H
Nuzzo S. CALICCHIO 1980, DECAMP 1991D, DECAMP 1991N, DECAMP 1991S, BUSKULIC 1993H
Nygren D. BENNETT 1967
Oakham F.G. ASHMAN 1988B, ALEXANDER 1991C, ACTON 1992B, DECAMP 1991S
Oberlack H. BRANDELIK 1977D
Obraztsov V.F. ABREU 1991B, DECAMP 1991S
Occhialini G.P.S. BLACKETT 1932, BLACKETT 1933, OCCHIALINI 1947, LATTES 1947, LATTES 1947B
Occhialini O. DAVIES 1955
O'Ceallaigh C. O'CEALLAIGH 1951, DAVIES 1955
O'Dell V. ALBAJAR 1991C, ALBAJAR 1991D
Odorici F. ALEXANDER 1991C, ACTON 1992B, DECAMP 1991S
Oehme R. GOLDBERGER 1955B, LEE 1957
Oesch L. ABACHI 1995G
O'Fallon J.R. LIN 1978
Ogawa S. IKEDA 1959, ABE 1994ZE, ABE 1995
Ogg M. BRANDELIK 1979H, BRANDELIK 1980G, ALEXANDER 1991C, ACTON 1992B, DECAMP 1991S
O'Grady C. AVERY 1989C, ARTUSO 1989C, ALAM 1989B
Ogren H.O. ABRAMS 1989B, WEIR 1989B, ALEXANDER 1991C, ACTON 1992B, DECAMP 1991S
Oguri V. ABACHI 1995G
Oh H. ALEXANDER 1991C, ACTON 1992B, DECAMP 1991S
Oh S.H. ABE 1994ZE, ABE 1995
O'Halloran T. CHINOWSKY 1962, KNAPP 1976B
Ohkawa T. KERST 1956
Ohl K.E. ABE 1994ZE, ABE 1995
Ohmoto Y. ABE 1995
Ohnuki Y. IKEDA 1959
Ohsugi T. ABE 1995
Oishi R. ABE 1994ZE, ABE 1995
Okabe M. ABE 1995
Oku Y. ALBRECHT 1987P
Okun L.B. IOFFE 1957, OKUN 1958
Okusawa T. ABE 1994ZE, ABE 1995
Olbert S. ANNIS 1952
Olive K. BARNETT 1996
Oliver R. ABE 1995

Olsen J. ABE 1995
Olsen S.L. ANDREWS 1980, ANDREWS 1980B, BEBEK 1981, CHADWICK 1981
Olshevski A.G. VOYVODIC 1986B, ABREU 1991B, DECAMP 1991S
Olsson J. BARTEL 1980D
Oltman E. ABACHI 1995G
O'Neale S. CALICCHIO 1980
O'Neale S.W. ALEXANDER 1991C, ACTON 1992B, DECAMP 1991S
O'Neill B.P. ALEXANDER 1991C, ACTON 1992B, DECAMP 1991S
O'Neill L.H. BARTEL 1980D
Ong R.A. ABRAMS 1989B, WEIR 1989B
Onions C. BANNER 1983B, BAGNAIA 1983B, ANSARI 1987E
Onuchin A.P. ARTAMONOV 1984B
Oppenheim R.F. ASHMAN 1988B
Oppenheimer J.R. OPPENHEIMER 1930
Oram C.J. ALEXANDER 1991C, ACTON 1992B, DECAMP 1991S
Oratovsky Y.A. ABOV 1965
Orava R. ABREU 1991B, DECAMP 1991S
Orear J. NORDIN 1958
Oreglia M.J. ALEXANDER 1991C, ACTON 1992B, DECAMP 1991S
Oren Y. BARNES 1964B, BARNES 1964E
Orito S. BRANDELIK 1977D, BARTEL 1980D, ALEXANDER 1991C, ACTON 1992B, DECAMP 1991S
Orkin-Lecourtois A. HASERT 1973B, ARNISON 1983C, ARNISON 1983D, ARNISON 1984C, ARNISON 1984G
Orr R.S. ALBRECHT 1987P
Ortel W.C.G. DAVIES 1955
Orteu S. DECAMP 1991D, DECAMP 1991N, DECAMP 1991S, BUSKULIC 1993H
Orthmann W. MEITNER 1930
Osborne A.M. AUBERT 1983, ASHMAN 1988B
Osculati B. HASERT 1973B
O'Shaughnessy K.F. ABRAMS 1989B, WEIR 1989B
Oshima N. ABACHI 1995G
Ostankov A. ABREU 1991B, DECAMP 1991S
O'Sullivan D. DANYSZ 1963
Oswald L. AGNEW 1958
Otwinowski S. ALBAJAR 1991C, ALBAJAR 1991D
Ouraou A. ABREU 1991B, DECAMP 1991S
Overseth O. BUNCE 1976
Owen D. ABACHI 1995G
Oyama Y. HIRATA 1987C
Pacheco A. DECAMP 1991D, DECAMP 1991N, DECAMP 1991S, BUSKULIC 1993H
Paciotti M. DORFAN 1967
Padilla C. BUSKULIC 1993H
Padley P. ALBRECHT 1987P, ABACHI 1995G
Pagliarone C. ABE 1994ZE, ABE 1995
Pain R. ABREU 1991B, DECAMP 1991S
Pais A. PAIS 1952, GELL-MANN 1955, PAIS 1955, PAIS 1957, GOLDHABER 1960
Pal T. ANSARI 1987E
Pal Y. BOLDT 1958B
Palazzi P. DECAMP 1991D, DECAMP 1991N, DECAMP 1991S, BUSKULIC 1993H
Palka H. ABREU 1991B, DECAMP 1991S
Palla F. DECAMP 1991D, DECAMP 1991N, DECAMP 1991S, BUSKULIC 1993H
Pallin D. DECAMP 1991D, DECAMP 1991N, DECAMP 1991S, BUSKULIC 1993H
Palmer R.B. BARNES 1964B, BARNES 1964E, HASERT 1973B, CAZZOLI 1975E, BALTAY 1979E
Palmonari F. BASILE 1981I
Pan Y. DECAMP 1991D, DECAMP 1991N, DECAMP 1991S
Pan Y.B. DECAMP 1991D, DECAMP 1991N, DECAMP 1991S, BUSKULIC 1993H
Pancheri G. ALBAJAR 1991C, ALBAJAR 1991D
Pancini E. CONVERSI 1947
Pandoulas D. BRANDELIK 1979H, BRANDELIK 1980G, ADEVA 1990R, ADEVA 1991D, DECAMP 1991S
Panetti M. BONETTI 1953, BONETTI 1953B
Pang M. ABACHI 1995G
Panin V.S. ARTAMONOV 1984B
Panofsky W.K.H. STEINBERGER 1950, PANOFSKY 1951, CHODOROW 1955
Pansart J.P. ALEXANDER 1991C, ACTON 1992B, DECAMP 1991S
Pantuev V.S. KIRILLOVA 1965
Panvini R.S. ANDREWS 1980, ANDREWS 1980B, BEBEK 1981, CHADWICK 1981
Panzer-Steindel B. ALEXANDER 1991C, ACTON 1002B, DECAMP 1991S
Paoletti R. ABE 1994ZE, ABE 1995
Paoluzi L. BACCI 1974, ARNISON 1983C, ARNISON 1983D, ARNISON 1984C, ARNISON 1984G, ARNISON 1985D
Papadimitriou V. ABE 1994ZE, ABE 1995
Papadopoulou T. ABREU 1991B, DECAMP 1991S
Papalexiou S. DECAMP 1991D, DECAMP 1991N, DECAMP 1991S, BUSKULIC 1993H
Papavassiliou V. ASHMAN 1988B
Pape K.H. BERGER 1979C, BERGER 1980L
Pape L. ABREU 1991B, DECAMP 1991S
Pappas S.P. ABE 1995
Para A. ABACHI 1995G
Paradiso J. BARBER 1979F
Parascandalo P. BACCI 1974
Parchomchuk V.V. BUDKER 1976
Parisi G. ALTARELLI 1977
Park C.H. AVERY 1989C, ARTUSO 1989C, ALAM 1989B, ABACHI 1995G
Park H.S. BIONTA 1987C
Park S. ABE 1994ZE, ABE 1995
Park Y.M. ABACHI 1995G
Parker D. DECAMP 1991D, DECAMP 1991N, DECAMP 1991S
Parker M.A. ANSARI 1987E
Parker S.I. ABRAMS 1989B, WEIR 1989B
Parrini G. DECAMP 1991D, DECAMP 1991N, DECAMP 1991S, BUSKULIC 1993H
Parrour G. BANNER 1983B, BAGNAIA 1983B, ANSARI 1987E
Parsons J.A. ALBRECHT 1987P
Parsons M.I. BUSKULIC 1993H
Partridge R. ABACHI 1995G
Parva N. ABACHI 1995G
Pascaud C. BLIETSCHAU 1976, BLIETSCHAU 1977B
Paschevici P. ALEXANDER 1991C, ACTON 1992B, DECAMP 1991S
Paschos E.A. BJORKEN 1969B
Pascoli D. ALBAJAR 1987C
Pascoli L. ARNISON 1985D
Pascual A. BUSKULIC 1993H
Passalacqua L. DECAMP 1991D, DECAMP 1991N, DECAMP 1991S, BUSKULIC 1993H
Passaleva G. ADEVA 1990R, ADEVA 1991D, DECAMP 1991S
Passaseo M. ALBAJAR 1991C, ALBAJAR 1991D
Passeri A. ABREU 1991B, DECAMP 1991S
Pastore F. BANNER 1983B, BAGNAIA 1983B, ANSARI 1987E
Patel A. DECAMP 1991D, DECAMP 1991N, DECAMP 1991S, BUSKULIC 1993H
Patel P.M. ALBRECHT 1987P
Pater J.R. DECAMP 1991D, DECAMP 1991N, DECAMP 1991S, BUSKULIC 1993H
Paterno M. ABACHI 1995G
Paternoster G. ADEVA 1990R, ADEVA 1991D, DECAMP 1991S

Paterson J.M. AUGUSTIN 1974B, ABRAMS 1974, SCHWITTERS 1975, PERL 1975C, HANSON 1975, PERL 1976C, GOLDHABER 1976D, PERUZZI 1976C, PERL 1977F

Patricelli S. ADEVA 1990R, ADEVA 1991D, DECAMP 1991S

Patrick G.N. ALEXANDER 1991C, ACTON 1992B, DECAMP 1991S

Patrick J. ABE 1994ZE, ABE 1995

Pattison J.B.M. HASERT 1973, HASERT 1973B, BLIETSCHAU 1976, BLIETSCHAU 1977B

Patton S.J. DECAMP 1991D, DECAMP 1991N, DECAMP 1991S, BUSKULIC 1993H

Paul H. FAZZINI 1958

Pauletta G. ABE 1994ZE, ABE 1995

Pauli W. PAULI 1925, PAULI 1925B, PAULI 1930, PAULI 1934, PAULI 1940B

Paulini M. ABE 1994ZE, ABE 1995

Pauss F. ARNISON 1983D, ARNISON 1984C, ARNISON 1984G, ARNISON 1985D, ALBAJAR 1987C, ALBAJAR 1991C, ALBAJAR 1991D

Pavel N. ASHMAN 1988B

Pavlovskaya V.V. GOLDANSKY 1960

Pawley S.J. ALEXANDER 1991C, ACTON 1992B, DECAMP 1991S

Payne G.J. DECAMP 1991D, DECAMP 1991N, DECAMP 1991S, BUSKULIC 1993H

Payre P. AUBERT 1983, DECAMP 1991D, DECAMP 1991N, DECAMP 1991S, BUSKULIC 1993H

Peacock R.N. FRAUENFELDER 1957

Pearce G.F. BARTEL 1980D

Peccei R.D. PECCEI 1977

Peck C. ABRAMS 1989B, WEIR 1989B

Peggs S. ANDREWS 1980

Pegoraro M. ABREU 1991B, DECAMP 1991S

Pei Y.J. ADEVA 1990R, ADEVA 1991D, DECAMP 1991S

Peigneux J.P. BUSHNIN 1969, BINON 1969, BINON 1970B

Peise G. BRANDELIK 1979H, BRANDELIK 1980G

Penso G. BACCI 1974

Penzo A. FIDECARO 1978

Peoples J. KNAPP 1976B

Pepe M. ANSARI 1987E

Pepe-Altarelli M. DECAMP 1991D, DECAMP 1991N, DECAMP 1991S, BUSKULIC 1993H

Perault C. ARNISON 1985D, ALBAJAR 1987C

Perchonok R. ANDREWS 1980, ANDREWS 1980B, BEBEK 1981, CHADWICK 1981

Perevozchikov V.M. ABREU 1991B, DECAMP 1991S

Perez P. DECAMP 1991D, DECAMP 1991N, DECAMP 1991S, BUSKULIC 1993H

Perkins D.H. PERKINS 1947, DAVIES 1955, BARKAS 1957, EDWARDS 1958, HASERT 1973, HASERT 1973B, BLIETSCHAU 1976

Perkins J. ABACHI 1995G

Perl M.L. AUGUSTIN 1974B, ABRAMS 1974, SCHWITTERS 1975, PERL 1975C, HANSON 1975, PERL 1976C, GOLDHABER 1976D, PERUZZI 1976C, PERL 1977F, ABRAMS 1989B, WEIR 1989B

Perlas J.A. DECAMP 1991D, DECAMP 1991N, DECAMP 1991S, BUSKULIC 1993H

Pernicka M. FIDECARO 1978, ABREU 1991B, DECAMP 1991S

Peroni C. AUBERT 1983, ASHMAN 1988B

Perret P. DECAMP 1991D, DECAMP 1991N, DECAMP 1991S, BUSKULIC 1993H

Perret-Gallix D. ADEVA 1990R, ADEVA 1991D, DECAMP 1991S

Perrier F. ABRAMS 1989B, DECAMP 1991D, DECAMP 1991N, DECAMP 1991S

Perrier J. BIAGI 1983C, ADEVA 1990R, ADEVA 1991D, DECAMP 1991S

Perrin J.B. PERRIN 1895

Perrotta A. ABREU 1991B, DECAMP 1991S

Perticone D. AVERY 1989C, ARTUSO 1989C, ALAM 1989B

Peruzzi I. BACCI 1974, PERL 1976C, GOLDHABER 1976D, PERUZZI 1976C, PERL 1977F, BAND 1988

Peryshkin A. ABACHI 1995G

Pescara L. ABE 1994ZE, ABE 1995

Peschel H. ASHMAN 1988B

Pessard H. AUBERT 1983

Pestrikov D.V. BUDKER 1976

Petermann A. STÜCKELBERG 1953

Peters M.D. ABE 1994ZE, ABE 1995, ABACHI 1995G

Petersen A. BRANDELIK 1977D, BARTEL 1980D, WEIR 1989B

Peterson D. BOHRINGER 1980B, FINOCCHIARO 1980, AVERY 1989C, ARTUSO 1989C, ALAM 1989B

Peterson J.R. BIRGE 1955B

Petiau P. HASERT 1973, HASERT 1973B, BLIETSCHAU 1976, BLIETSCHAU 1977B, CALICCHIO 1980

Petradza M. ABRAMS 1989B, WEIR 1989B

Petridou C. ANSARI 1987E

Petrolo E. ALBAJAR 1987C, ALBAJAR 1991C, ALBAJAR 1991D

Petrov V.V. ARTAMONOV 1984B

Petrucci C. CARRON 1978

Petrukhin A.I. BUSHNIN 1969, ANTIPOV 1970, DENISOV 1971F

Petrukhin V.I. DUNAITSEV 1962

Petta P. ALBAJAR 1987C

Pevsner A. ILOFF 1955, FUNG 1956, PEVSNER 1961, BRANDELIK 1979H, ADEVA 1990R, ADEVA 1991D, DECAMP 1991S

Peyrou C. GREGORY 1954, ARMENTEROS 1955

Pfister P. ALEXANDER 1991C, ACTON 1992B, DECAMP 1991S

Phillips M.J. DECAMP 1991D, DECAMP 1991N, DECAMP 1991S, BUSKULIC 1993H

Phillips T.J. ABE 1994ZE, ABE 1995

Piacentino A.G. ABE 1994ZE, ABE 1995

Piano Mortari G. BACCI 1974, ARNISON 1983C, ARNISON 1983D, ARNISON 1984C, ARNISON 1984G, ARNISON 1985D, ALBAJAR 1987C, ALBAJAR 1991C, ALBAJAR 1991D

Picchi P. DECAMP 1991D, DECAMP 1991N, DECAMP 1991S, BUSKULIC 1993H

Piccioni O. CONVERSI 1947, PICCIONI 1948, PAIS 1955, CORK 1956, GOOD 1961

Piccioni R.L. BENVENUTI 1974F

Piccolo D. ADEVA 1991D, DECAMP 1991S

Piccolo M. BACCI 1974, PERL 1976C, GOLDHABER 1976D, PERUZZI 1976C, PERL 1977F, BAND 1988

Piegaia R. ASHMAN 1988B

Piekarz H. ABACHI 1995G

Piemontese L. FIDECARO 1978

Pierazzini G.M. APEL 1975B

Pieri M. ADEVA 1990R, ADEVA 1991D, DECAMP 1991S

Pierre F.M. AUGUSTIN 1974B, ABRAMS 1974, SCHWITTERS 1975, PERL 1975C, HANSON 1975, PERL 1976C, GOLDHABER 1976D, PERUZZI 1976C, ABREU 1991B, DECAMP 1991S

Pietarinen E. ARNISON 1983D, ARNISON 1984C, ARNISON 1984G, ARNISON 1985D, ALBAJAR 1987C, ALBAJAR 1991C, ALBAJAR 1991D

Pietrzyk B. BIENLEIN 1978, ASHMAN 1988B, DECAMP 1991D, DECAMP 1991N, DECAMP 1991S, BUSKULIC 1993H

Pietrzyk U. AUBERT 1983, ASHMAN 1988B

Pigot C. ALBAJAR 1987C

Pik L. VOYVODIC 1986B

Pilcher J.E. BENVENUTI 1974F, ALEXANDER 1991C, ACTON 1992B, DECAMP 1991S

Pillai M. ABE 1994ZE, ABE 1995
Pimenta M. ABREU 1991B, DECAMP 1991S
Pimia M. ARNISON 1983C, ARNISON 1983D, ARNISON 1984C, ARNISON 1984G, ARNISON 1985D, ALBAJAR 1987C, ALBAJAR 1991C, ALBAJAR 1991D
Pinfold J.L. BLIETSCHAU 1976, ALEXANDER 1991C, ACTON 1992B, DECAMP 1991S
Pingot O. ABREU 1991B, DECAMP 1991S
Pinkau K. EDWARDS 1958
Pinsent A. ABREU 1991B, DECAMP 1991S
Pipkin F.M. ANDREWS 1980, ANDREWS 1980B, BEBEK 1981, CHADWICK 1981, AVERY 1989C, ARTUSO 1989C, ALAM 1989B
Piroue P.A. ADEVA 1990R, ADEVA 1991D, DECAMP 1991S
Pisharody M. AVERY 1989C, ARTUSO 1989C, ALAM 1989B
Pishchalnikov Y.M. ABACHI 1995G
Pitman D. ARNISON 1984G, ARNISON 1985D, ALBAJAR 1987C, ACTON 1992B
Pitthan R. ABRAMS 1989B
Pitts K.T. ABE 1995
Pittuck D. BLIETSCHAU 1976
Pitukhin P.V. VOYVODIC 1986B
Pizzuto D. ABACHI 1995G
Pjerrou G.M. PJERROU 1962
Placci A. BUESSER 1973, ARNISON 1983C, ARNISON 1983D, ARNISON 1984C, ARNISON 1984G, ARNISON 1985D, ALBAJAR 1987C, ALBAJAR 1991C, ALBAJAR 1991D
Planck M. PLANCK 1900, PLANCK 1900B, PLANCK 1901
Plane D.E. ALEXANDER 1991C, ACTON 1992B, DECAMP 1991S
Plano R. PLANO 1957, IMPEDUGLIA 1958, EISLER 1958, PLANO 1959
Plasil F. ADEVA 1990R, ADEVA 1991D, DECAMP 1991S
Pleasonton F. SNELL 1950
Pleshko M. ALBRECHT 1987P
Pless I.A. CRETIEN 1962
Plothow-Besch H. BANNER 1983B, BAGNAIA 1983B, ANSARI 1987E
Plunkett R.K. ANDREWS 1980, ANDREWS 1980B, BEBEK 1981, CHADWICK 1981, ABE 1994ZE, ABE 1995
Pluquet A. ABACHI 1995G
Plyaskin V. ADEVA 1990R, ADEVA 1991D, DECAMP 1991S
Pniewski J. DANYSZ 1953, DANYSZ 1963
Pniewski T. DANYSZ 1963
Podlyski F. BUSKULIC 1993H
Podstavkov V.M. ABACHI 1995G
Poelz G. BRANDELIK 1977D, BRANDELIK 1979H, BRANDELIK 1980G
Poffenberger P. ALEXANDER 1991C, ACTON 1992B, DECAMP 1991S
Pohl M. BLIETSCHAU 1977B, ADEVA 1990R, ADEVA 1991D, DECAMP 1991S
Poincare H. POINCARE 1905
Pol M.E. ABREU 1991B, DECAMP 1991S
Poli B. ALEXANDER 1991C, ACTON 1992B, DECAMP 1991S
Poling R. ANDREWS 1980, ANDREWS 1980B, BEBEK 1981, CHADWICK 1981, AVERY 1989C, ARTUSO 1989C, ALAM 1989B
Politzer H.D. POLITZER 1973
Polivanov M.K. BOGOLYUBOV 1956B
Polok G. ABREU 1991B, DECAMP 1991S
Polvado R.O. BERGER 1980L
Polverel M. BANNER 1983B, BAGNAIA 1983B, ANSARI 1987E
Polyakov A.M. BELAVIN 1975
Pomeranchuk I.Y. IWANENKO 1944, POMERANCHUK 1958
Pondrom L. BUNCE 1976, ABE 1994ZE, ABE 1995
Pontecorvo B. PONTECORVO 1946, PONTECORVO 1947, HINCKS 1948, HINCKS 1949, PONTECORVO 1958, PONTECORVO 1959
Pontecorvo L. ALBAJAR 1991C, ALBAJAR 1991D
Pope B.G. CHRISTENSON 1970, BUESSER 1973, ABACHI 1995G
Popov V.N. FADDEEV 1967
Poropat P. ABREU 1991B, DECAMP 1991S
Porte J.P. ARNISON 1983D, ARNISON 1984C, ARNISON 1984G, ARNISON 1985D, ALBAJAR 1987C, ALBAJAR 1991C, ALBAJAR 1991D
Porter F.C. ABRAMS 1989B, WEIR 1989B, BARNETT 1996
Poschmann F.P. BARBER 1979F
Posocco M. FIDECARO 1978
Potter D. ANDREWS 1980, ANDREWS 1980B, BEBEK 1981, CHADWICK 1981
Poucher J.S. ANDREWS 1980, ANDREWS 1980B, BEBEK 1981, CHADWICK 1981, BODEK 1983
Pouladdej A. ALEXANDER 1991C, ACTON 1992B, DECAMP 1991S
Povh B. ASHMAN 1988B
Powell C.F. OCCHIALINI 1947, LATTES 1947, LATTES 1947B, BROWN 1949B, FOWLER 1951, DAVIES 1955
Powell W.M. AGNEW 1958, GOLDHABER 1959, GOOD 1961
Poyer C. FIDECARO 1978
Pozharova E.A. VOYVODIC 1986B
Pozhidaev V. ADEVA 1990R, ADEVA 1991D, DECAMP 1991S
Prebys E. ALEXANDER 1991C, ACTON 1992B, DECAMP 1991S
Prentice J.D. ALBRECHT 1987P
Prepost R. BAND 1988
Prescott C.Y. PRESCOTT 1978C, PRESCOTT 1979C
Present R.D. BREIT 1936B
Price L.R. BIONTA 1987C
Priem R. ALBAJAR 1991C
Primer M. GOLDBERG 1964C
Pritchard T.W. ALEXANDER 1991C, ACTON 1992B, DECAMP 1991S
Privitera P. ABREU 1991B, DECAMP 1991S
Procario M. AVERY 1989C, ARTUSO 1989C, ALAM 1989B
Prodell A.G. EISLER 1958, PLANO 1959, BARNES 1964B, BARNES 1964E
Produit N. ADEVA 1990R, ADEVA 1991D, DECAMP 1991S, ABE 1994ZE
Prokoshkin Y.D. DUNAITSEV 1962, BUSHNIN 1969, BINON 1969, BINON 1970B, ANTIPOV 1970, DENISOV 1971F, APEL 1975B
Proriol J. DECAMP 1991D, DECAMP 1991N, DECAMP 1991S, BUSKULIC 1993H
Prosi R. ALBAJAR 1991C, ALBAJAR 1991D
Prosper H. BARTEL 1980D, ABACHI 1995G
Protopopescu S. ABACHI 1995G
Protopopov I.Y. ARTAMONOV 1984B
Proudfoot J. BRANDELIK 1979H, BRANDELIK 1980G, ABE 1994ZE, ABE 1995
Prowse D.J. PROWSE 1958, PJERROU 1962
Prulhiere F. DECAMP 1991D, DECAMP 1991N, DECAMP 1991S, BUSKULIC 1993H
Przysiezniak H. ALEXANDER 1991C, ACTON 1992B, DECAMP 1991S
Ptohos F. ABE 1994ZE, ABE 1995
Puglierin G. FAISSNER 1978D
Pullia A. HASERT 1973, HASERT 1973B, BLIETSCHAU 1976, BLIETSCHAU 1977B, ABREU 1991B, DECAMP 1991S
Pun T.P. SCHWITTERS 1975, PERL 1975C, HANSON 1975, PERL 1976C, GOLDHABER 1976D, PERUZZI 1976C, PERL 1977F
Punzi G. ABE 1994ZE, ABE 1995
Puseljic D. ABACHI 1995G
Pusztaszeri J.F. BUSKULIC 1993H

Putzer A. DECAMP 1991D, DECAMP 1991N, DECAMP 1991S, BUSKULIC 1993H
Pyrlik J. BRANDELIK 1979H, BRANDELIK 1980G, BAND 1988
Pyyhtia J. ABREU 1991B, DECAMP 1991S
Qi N. BAND 1988
Qian J.M. ADEVA 1990R, ADEVA 1991D, DECAMP 1991S, ABACHI 1995G
Qian Z. DECAMP 1991D, DECAMP 1991N, DECAMP 1991S
Qiao C. DECAMP 1991D, DECAMP 1991N, DECAMP 1991S
Quaglia M. APEL 1975B
Quarrie D.R. BRANDELIK 1979H, BRANDELIK 1980G, BRANDELIK 1980G
Quast G. ALEXANDER 1991C, ACTON 1992B, DECAMP 1991S
Quattromini M. DECAMP 1991D, DECAMP 1991N, DECAMP 1991S, BUSKULIC 1993H
Quazi I.S. DECAMP 1991D, DECAMP 1991N, DECAMP 1991S, BUSKULIC 1993H
Quinn H.R. PECCEI 1977
Quintas P.Z. ABACHI 1995G
Qureshi K.N. ADEVA 1990R, ADEVA 1991D, DECAMP 1991S
Raab J. DECAMP 1991N, DECAMP 1991S, BUSKULIC 1993H
Rabany M. ARNISON 1983C
Rabi I.I. KELLOGG 1939
Radermacher E. FAISSNER 1978D, ARNISON 1983C, ARNISON 1983D, ARNISON 1984C, ARNISON 1984G, ARNISON 1985D, ALBAJAR 1987C, ALBAJAR 1991C, ALBAJAR 1991D
Radicati L.A. GÜRSEY 1964
Radojicic D. BARNES 1964B, BARNES 1964E, ABREU 1991B, DECAMP 1991S
Radomanov V.B. BALDIN 1973
Ragan K. ABE 1994ZE, ABE 1995
Ragazzi S. ABREU 1991B, DECAMP 1991S
Raghavan R. ADEVA 1990R, ADEVA 1991D, DECAMP 1991S
Ragusa F. DECAMP 1991N, DECAMP 1991S, BUSKULIC 1993H
Rahal-Callot G. ADEVA 1990R, ADEVA 1991D, DECAMP 1991S
Rahm D.C. BARNES 1964B, BARNES 1964E
Raine C. DECAMP 1991D, DECAMP 1991N, DECAMP 1991S, BUSKULIC 1993H
Raja R. ARNISON 1985D, ABACHI 1995G
Rajagopalan S. ABACHI 1995G
Rajmer R. VOYVODIC 1986B
Raman C.V. RAMAN 1932
Ramirez O. ABACHI 1995G
Ramsey N.F. KELLOGG 1939
Ramzhin V.N. BALDIN 1973
Rander J. DECAMP 1991D, DECAMP 1991N, DECAMP 1991S, BUSKULIC 1993H
Ranga Swamy T.N. BERGER 1979C
Range W.H. ABREU 1991B, DECAMP 1991S
Ranieri A. DECAMP 1991D, DECAMP 1991N, DECAMP 1991S, BUSKULIC 1993H
Ranjard F. DECAMP 1991D, DECAMP 1991N, DECAMP 1991S, BUSKULIC 1993H
Rankin P. ABRAMS 1989B, WEIR 1989B
Ransdell J. ARNISON 1983C, ARNISON 1983D, ARNISON 1984C, ARNISON 1984G, ARNISON 1985D
Rao M.V.S. ABACHI 1995G
Rapidis P.A. AUGUSTIN 1974B, ABRAMS 1974, SCHWITTERS 1975, PERL 1975C, HANSON 1975, PERL 1976C, GOLDHABER 1976D, PERUZZI 1976C, PERL 1977F, ABACHI 1995G
Rappoport V.M. VOYVODIC 1986B
Rasetti F. RASETTI 1941, RASETTI 1941B
Rasmussen L. ANSARI 1987E, ABACHI 1995G
Raso G. DECAMP 1991D, DECAMP 1991N, DECAMP 1991S, BUSKULIC 1993H
Ratner L.G. LIN 1978
Ratoff P.N. ABREU 1991B, DECAMP 1991S
Rau R.R. HODSON 1954, BERTANZA 1962D, CONNOLLY 1963
Raupach F. BERGER 1978, BERGER 1979C, BERGER 1980L
Rauschkolb M. ALBAJAR 1991C, ALBAJAR 1991D
Ravenhall D.G. YENNIE 1957
Raymond D. DORFAN 1967
Razis P. ADEVA 1990R, ADEVA 1991D, DECAMP 1991S
Razuvaev E.A. BUSHNIN 1969, BINON 1969, BINON 1970B, ANTIPOV 1970
Read A.L. BAND 1988, ABREU 1991B, DECAMP 1991S, ABACHI 1995G
Read K. ADEVA 1990R, ADEVA 1991D, DECAMP 1991S
Redaelli N.G. ABREU 1991B, DECAMP 1991S
Redelberger T. ARNISON 1985D, ALBAJAR 1987C
Rediker R.H. THOMPSON 1953, THOMPSON 1953B
Redlinger G. DECAMP 1991D, DECAMP 1991N, DECAMP 1991S, BUSKULIC 1993H
Redmond M.W. ALEXANDER 1991C, ACTON 1992B, DECAMP 1991S
Reed L. NORDIN 1958
Reeder D.D. BENVENUTI 1974F
Rees D.L. ALEXANDER 1991C, ACTON 1992B, DECAMP 1991S
Reeves P. DECAMP 1991D, DECAMP 1991N, DECAMP 1991S, BUSKULIC 1993H
Regge T. REGGE 1959, REGGE 1960
Regler M. FIDECARO 1978, ABREU 1991B, DECAMP 1991S
Reich H.D. BERGER 1979C, BERGER 1980L
Reid D. ABREU 1991B, DECAMP 1991S
Reines F. REINES 1953B, COWAN 1956, REINES 1956, REINES 1959, BIONTA 1987C
Reithler H. FAISSNER 1978D, ARNISON 1983C, ARNISON 1983D, ARNISON 1984C, ARNISON 1984G, ARNISON 1985D, ALBAJAR 1987C, ALBAJAR 1991C, ALBAJAR 1991D
Ren D. ADEVA 1990R, ADEVA 1991D, DECAMP 1991S
Ren Z. ADEVA 1990R, ADEVA 1991D, DECAMP 1991S
Renard P. DANYSZ 1963
Renardy J.F. DECAMP 1991D, DECAMP 1991N, DECAMP 1991S, BUSKULIC 1993H
Renk B. DECAMP 1991D, DECAMP 1991N, DECAMP 1991S, BUSKULIC 1993H
Rensch B. DECAMP 1991D, DECAMP 1991N, DECAMP 1991S, BUSKULIC 1993H
Renton P. ASHMAN 1988B, ABREU 1991B, DECAMP 1991S
Repellin J.P. BANNER 1983B, BAGNAIA 1983B, ANSARI 1987E
Resvanis L.K. ABREU 1991B, DECAMP 1991S
Retherford R.C. LAMB 1947
Reucroft S. ADEVA 1990R, ADEVA 1991D, DECAMP 1991S, ABACHI 1995G
Revel D. BRANDELIK 1979H, BRANDELIK 1980G
Revol J.P. BARBER 1979F, ARNISON 1983D, ARNISON 1984C, ARNISON 1984G, ARNISON 1985D, ALBAJAR 1987C, ALBAJAR 1991C, ALBAJAR 1991D
Reynolds G.T. REYNOLDS 1950, HODSON 1954
Reynolds J. EDWARDS 1958
Rhoades T.G. AUBERT 1974D
Ribon A. ABE 1995
Rice E. BOHRINGER 1980B, FINOCCHIARO 1980
Rich J. ARNISON 1983C, ARNISON 1983D, ARNISON 1984C, ARNISON 1984G
Richard F. ABREU 1991B, DECAMP 1991S
Richardson C.R. BARNES 1964B, BARNES 1964E

Richardson C.S. PEVSNER 1961
Richardson M. ABREU 1991B, DECAMP 1991S
Richman J. ALBAJAR 1987C, ABRAMS 1989B
Richter B. AUGUSTIN 1974B, ABRAMS 1974, SCHWITTERS 1975, PERL 1975C, HANSON 1975, PERL 1976C, GOLDHABER 1976D, PERUZZI 1976C, PERL 1977F, WEIR 1989B
Richter R. BLUM 1975, BIENLEIN 1978, DECAMP 1991D, DECAMP 1991N, DECAMP 1991S, BUSKULIC 1993H
Ricker A. ADEVA 1991D, DECAMP 1991S
Ridky J. ABREU 1991B, DECAMP 1991S
Riemann S. ADEVA 1991D, DECAMP 1991S
Rieseberg H. BARTEL 1980D
Riethmuller R. BRANDELIK 1979H, BRANDELIK 1980G
Rieubland J.M. ASHMAN 1988B
Rijllart A. ASHMAN 1988B
Rijssenbeek M. ARNISON 1983C, ARNISON 1983D, ARNISON 1984C, ARNISON 1984G, ARNISON 1985D, ABACHI 1995G
Riles K. ABRAMS 1989B, WEIR 1989B, ALEXANDER 1991C, ACTON 1992B, DECAMP 1991S
Riley D. AVERY 1989C, ARTUSO 1989C, ALAM 1989B
Rimkus J. BRANDELIK 1979H, BRANDELIK 1980G
Rimoldi A. BANNER 1983B, BAGNAIA 1983B
Rimondi F. ABE 1994ZE, ABE 1995
Rinaudo G. ABREU 1991B, DECAMP 1991S
Rind O. ADEVA 1990R, ADEVA 1991D, DECAMP 1991S
Ringel J. BRANDELIK 1977D, BRANDELIK 1979H
Riordan E.M. BODEK 1983
Rippich C. BIENLEIN 1978, ADEVA 1990R, ADEVA 1991D, DECAMP 1991S
Ristori L. AMENDOLIA 1973B, ABE 1994ZE, ABE 1995
Rith K. AUBERT 1983, ASHMAN 1988B
Ritson D.M. BROWN 1949B, ILOFF 1955, FUNG 1956, BAND 1988
Rittenberg A. KALBFLEISCH 1964
Rivera F. BUSKULIC 1993H
Rizvi H.A. ADEVA 1990R, ADEVA 1991D, DECAMP 1991S
Rizzo G. BUSKULIC 1993H
Roach-Bellino M. ABE 1994ZE
Rob L. KIRILLOVA 1965
Roberts A. CLARK 1951
Roberts C. ARNISON 1983C, ARNISON 1983D, ARNISON 1984C, ARNISON 1984G, ARNISON 1985D
Roberts G.E. WILLIAMS 1940
Robertson W.J. ABE 1994ZE, ABE 1995
Robins S.A. ALEXANDER 1991C, ALBAJAR 1991C, ACTON 1992B, DECAMP 1991S
Robinson D. ALBAJAR 1987C, ALEXANDER 1991C, ALBAJAR 1991C, ALBAJAR 1991D, ACTON 1992B, DECAMP 1991S
Robson J.M. ROBSON 1950, ROBSON 1950B
Rochat O. FOWLER 1951
Rochester G.D. ROCHESTER 1947
Rochester L.S. PRESCOTT 1978C, PRESCOTT 1979C
Rockwell T. ABACHI 1995G
Roditi I. ABREU 1991B, DECAMP 1991S
Rodrigo T. ALBAJAR 1991C, ALBAJAR 1991D, ABE 1994ZE, ABE 1995
Roe B.P. BRANDELIK 1980G, ADEVA 1990R, ADEVA 1991D, DECAMP 1991S
Roe N.A. ABACHI 1995G
Roehn S. DECAMP 1991D, DECAMP 1991N, DECAMP 1991S
Roeser U. ADEVA 1990R, ADEVA 1991D, DECAMP 1991S
Rohde M. AUBERT 1974D, BARBER 1979F
Rohlf J. ANDREWS 1980B, BEBEK 1981, CHADWICK 1981, ARNISON 1983D, ARNISON 1984C, ARNISON 1984G, ARNISON 1985D, ALBAJAR 1987C, ALBAJAR 1991C, ALBAJAR 1991D
Rohner M. ADEVA 1990R, ADEVA 1991D, DECAMP 1991S
Rohner S. ADEVA 1990R, ADEVA 1991D, DECAMP 1991S
Roinishvili V.N. CHIKOVANI 1964
Rolandi L. DECAMP 1991D, DECAMP 1991N, DECAMP 1991S, BUSKULIC 1993H
Roldan J.M.R. ABACHI 1995G
Rollier M. HASERT 1973, HASERT 1973B, BLIETSCHAU 1976, BLIETSCHAU 1977B
Rollnik A. ALEXANDER 1991C, ACTON 1992B, DECAMP 1991S
Romano F. BLIETSCHAU 1976, BLIETSCHAU 1977B, CALICCHIO 1980, DECAMP 1991D, DECAMP 1991N, DECAMP 1991S, BUSKULIC 1993H
Romano J. ABE 1994ZE, ABE 1995
Rombach Th. ADEVA 1990R
Romer O. BRANDELIK 1977D, BRANDELIK 1979H, BRANDELIK 1980G
Romero A. ABREU 1991B, DECAMP 1991S
Romero L. ADEVA 1990R, ADEVA 1991D, DECAMP 1991S
Ronat E.E. CRETIEN 1962, BRANDELIK 1979H, BRANDELIK 1980G
Roncagliolo I. ABREU 1991B, DECAMP 1991S
Ronchese P. ABREU 1991B, DECAMP 1991S
Rondio E. ASHMAN 1988B
Roney J.M. ALEXANDER 1991C, ACTON 1992B, DECAMP 1991S
Ronga F. BACCI 1974
Ronnqvist C. ABREU 1991B, DECAMP 1991S
Röntgen W.C. RÖNTGEN 1895
Roos L. BUSKULIC 1993H
Roos M. BARNETT 1996
Ropelewski L. ASHMAN 1988B
Ros E. ACTON 1992B
Rose J. ADEVA 1990R, ADEVA 1991D, DECAMP 1991S
Rose M.E. ROSE 1948
Rosenberg E.I. ABREU 1991B, DECAMP 1991S
Rosenberg L.J. BAND 1988
Rosenbluth M.N. LEE 1949
Rosenfeld A.H. GELL-MANN 1957, BARKAS 1958, NORDIN 1958, MAGLIC 1961B
Rosenfeld C. ANDREWS 1980, ANDREWS 1980B, BEBEK 1981, CHADWICK 1981
Rosenson L. CRETIEN 1962, ABE 1994ZE, ABE 1995
Roser R. ABE 1995
Rosier-Lees S. ADEVA 1990R, ADEVA 1991D, DECAMP 1991S
Rosmalen R. ADEVA 1990R, ADEVA 1991D, DECAMP 1991S
Rosowsky A. DECAMP 1991D, DECAMP 1991N, DECAMP 1991S, BUSKULIC 1993H
Ross R.R. KALBFLEISCH 1964
Rossberg S. ALEXANDER 1991C, ACTON 1992B, DECAMP 1991S
Rosselet P. BIAGI 1983C, BIAGI 1985C, ADEVA 1990R, ADEVA 1991D, DECAMP 1991S
Rossi A. FRAUENFELDER 1957
Rossi A.M. ALEXANDER 1991C, ACTON 1992B, DECAMP 1991S
Rossi B. ROSSI 1939, ROSSI 1942, NERESON 1942, ANNIS 1952, BRIDGE 1955
Rossi P. ARNISON 1983D, ARNISON 1984C, ARNISON 1984G, ARNISON 1985D, ALBAJAR 1987C
Rossi U. ABREU 1991B, DECAMP 1991S
Rossler M. BERGER 1978, BERGER 1979C, BERGER 1980L
Rosso E. ABREU 1991B, DECAMP 1991S
Rost M. BERGER 1979C, BERGER 1980L

Rosvick M. ACTON 1992B
Roth A. DECAMP 1991D, DECAMP 1991N, DECAMP 1991S
Roth F. BRANDELIK 1979H
Roth R.F. FITCH 1965B
Rothberg J. DECAMP 1991D, DECAMP 1991N, DECAMP 1991S, BUSKULIC 1993H
Rothenberg A. BANNER 1983B, BAGNAIA 1983B
Rotscheidt H. DECAMP 1991D, DECAMP 1991N, DECAMP 1991S
Roudeau P. ABREU 1991B, DECAMP 1991S
Rouge A. DECAMP 1991D, DECAMP 1991N, DECAMP 1991S, BUSKULIC 1993H
Rouse F. ABRAMS 1989B
Roussarie A. BANNER 1983B, BAGNAIA 1983B, ANSARI 1987E, DECAMP 1991D, DECAMP 1991N, DECAMP 1991S, BUSKULIC 1993H
Rousseau D. BUSKULIC 1993H
Rousset A. HASERT 1973, HASERT 1973B
Routenburg P. ALEXANDER 1991C, ACTON 1992B, DECAMP 1991S
Rovelli T. ABREU 1991B, DECAMP 1991S
Rowlingson W.S. DECAMP 1991D, DECAMP 1991N, DECAMP 1991S
Rowson P.C. WEIR 1989B
Rozenblat R. VOYVODIC 1986B
Rozental I.L. MILEKHIN 1957
Ruan T. DECAMP 1991D, DECAMP 1991N, DECAMP 1991S, BUSKULIC 1993H
Rubbia A. ADEVA 1991D, DECAMP 1991S
Rubbia C. BENVENUTI 1974F, CARRON 1978, ARNISON 1983C, ARNISON 1983D, ARNISON 1984C, ARNISON 1984G, ARNISON 1985D, ALBAJAR 1987C, ALBAJAR 1991C, ALBAJAR 1991D
Rubin G. VOYVODIC 1986B
Rubinov P. ABACHI 1995G
Rubio J.A. ADEVA 1990R, ADEVA 1991D, DECAMP 1991S
Rubio M. ADEVA 1991D, DECAMP 1991S
Ruchti R. ABACHI 1995G
Rucinski G. ANDREWS 1980, ANDREWS 1980B, BEBEK 1981, CHADWICK 1981
Ruckstuhl W. ADEVA 1990R, ADEVA 1991D, ABREU 1991B, DECAMP 1991S, DECAMP 1991S
Ruderman M.A. KARPLUS 1955
Rudik A.P. IOFFE 1957
Rudolph G. BRANDELIK 1980G, DECAMP 1991D, DECAMP 1991N, DECAMP 1991S, BUSKULIC 1993H
Ruf T. ALBRECHT 1987P
Ruggieri F. CALICCHIO 1980, DECAMP 1991D, DECAMP 1991N, DECAMP 1991S, BUSKULIC 1993H
Ruhlmann V. ANSARI 1987E, ABREU 1991B, DECAMP 1991S
Ruhm W. ALBAJAR 1987C
Ruhmer W. BRANDELIK 1979H
Ruiz A. ABREU 1991B, DECAMP 1991S
Rumpf M. DECAMP 1991D, DECAMP 1991N, DECAMP 1991S, BUSKULIC 1993H
Runge K. ALEXANDER 1991C, ACTON 1992B, DECAMP 1991S
Runolfsson O. ALEXANDER 1991C, ACTON 1992B, DECAMP 1991S
Rusch R. BRANDELIK 1977D, BRANDELIK 1979H, BRANDELIK 1980G
Rusin S. ABACHI 1995G
Russ J.S. FITCH 1965B
Rust D.R. ABRAMS 1989B, WEIR 1989B, ALEXANDER 1991C, ACTON 1992B, DECAMP 1991S
Rutherford E. RUTHERFORD 1899, RUTHERFORD 1903, RUTHERFORD 1911, RUTHERFORD 1913, RUTHERFORD 1919, RUTHERFORD 1920
Rutherford J. ABACHI 1995G
Rybicki K. BLUM 1975, ABREU 1991B, DECAMP 1991S
Rykaczewski H. BARBER 1979F, ADEVA 1990R, ADEVA 1991D, DECAMP 1991S
Rykalin V.I. DUNAITSEV 1962, BINON 1969, ANTIPOV 1970
Ryltsov V.V. ALBRECHT 1987P
Saadi F. BUSKULIC 1993H
Saadi Y. DECAMP 1991D, DECAMP 1991N, DECAMP 1991S, BUSKULIC 1993H
Saal H. BENNETT 1967
Saarikko H. ABREU 1991B, DECAMP 1991S
Sacherer F. CARRON 1978
Sachwitz M. ADEVA 1990R, ADEVA 1991D, DECAMP 1991S
Sacquin Y. ABREU 1991B, DECAMP 1991S
Sacton J. DANYSZ 1963, HASERT 1973, HASERT 1973B, BLIETSCHAU 1976, BLIETSCHAU 1977B, CALICCHIO 1980
Sadoff A.J. ANDREWS 1980, ANDREWS 1980B, BEBEK 1981, CHADWICK 1981, AVERY 1989C, ARTUSO 1989C, ALAM 1989B
Sadoulet B. ABRAMS 1974, SCHWITTERS 1975, PERL 1975C, HANSON 1975, PERL 1976C, GOLDHABER 1976D, ARNISON 1983C, ARNISON 1983D, ARNISON 1984C, ARNISON 1984G, ARNISON 1985D
Sadoulet L. CARRON 1978
Sadrozinski H.F.W. ABRAMS 1989B, WEIR 1989B
Saenko. L.F. LOBASHOV 1966
Saich M.R. DECAMP 1991D, DECAMP 1991N, DECAMP 1991S, BUSKULIC 1993H
Sajot G. ARNISON 1983C, ARNISON 1983D, ARNISON 1984C, ARNISON 1984G, ARNISON 1985D, ALBAJAR 1987C, ALBAJAR 1991C, ALBAJAR 1991D
Sakata S. SAKATA 1956
Sakharov A.D. SAKHAROV 1967
Sakita B. SAKITA 1964
Sakumoto W.K. ABE 1994ZE, ABE 1995
Sakurai J.J. SAKURAI 1958
Salam A. SALAM 1957, SALAM 1961, SALAM 1962, GOLDSTONE 1961B, SALAM 1964, SALAM 1968
Salandin G.A. DAVIES 1955, CRETIEN 1962
Salicio J. ADEVA 1990R, ADEVA 1991D, DECAMP 1991S
Salicio J.M. ADEVA 1990R, ADEVA 1991D, ALBAJAR 1991C, ALBAJAR 1991D, DECAMP 1991S
Salisbury W.W. LAWRENCE 1939
Salmon D. ASHMAN 1988B
Salmon D.P. DECAMP 1991D, DECAMP 1991N, DECAMP 1991S
Salmon G.L. BRANDELIK 1979H, BRANDELIK 1980G
Salpeter E.E. SALPETER 1951
Salt J. ABREU 1991B, DECAMP 1991S
Salthouse A.J. LIN 1978
Saltzberg D. ABE 1994ZE, ABE 1995
Salvi G. ARNISON 1983C, ARNISON 1983D, ARNISON 1984C
Salvini G. REYNOLDS 1950, ARNISON 1983C, ARNISON 1983D, ARNISON 1984C, ARNISON 1984G, ARNISON 1985D, ALBAJAR 1987C
Samios N.P. PLANO 1957, IMPEDUGLIA 1958, EISLER 1958, PLANO 1959, BERTANZA 1962D, CONNOLLY 1963, BARNES 1964B, BARNES 1964E, GOLDBERG 1964C, CAZZOLI 1975E, BALTAY 1979E
Samyn D. ALBAJAR 1987C, ALBAJAR 1991C, ALBAJAR 1991D
Sanchez E. ABREU 1991B, DECAMP 1991S
Sanchez J. ABREU 1991B, DECAMP 1991S
Sandacz A. ASHMAN 1988B

Sander H.G. BRANDELIK 1977D, BRANDELIK 1979H, BRANDELIK 1980G, DECAMP 1991D, DECAMP 1991N, DECAMP 1991S, BUSKULIC 1993H
Sanders G. ADEVA 1990R, ADEVA 1991D, DECAMP 1991S
Sandler B. LIN 1978
Sandweiss J. BARKAS 1957, STONEHILL 1961, BROWN 1962
Sanford J.R. STONEHILL 1961, BROWN 1962, BARNES 1964B, BARNES 1964E
Sanghera S. ALEXANDER 1991C, ACTON 1992B, DECAMP 1991S
Sanguinetti G. AMENDOLIA 1973B, DECAMP 1991D, DECAMP 1991N, DECAMP 1991S, BUSKULIC 1993H
SanMartin G. BUSKULIC 1993H
Sannes F. ANDREWS 1980, ANDREWS 1980B, BEBEK 1981, CHADWICK 1981
Sannino M. ABREU 1991B, DECAMP 1991S
Sansoni A. ABE 1994ZE, ABE 1995
Santangelo R. EISLER 1958
Santi L. ABE 1995
Santoro A. ABACHI 1995G
Sapper M. AVERY 1989C, ARTUSO 1989C, ALAM 1989B
Sarakinos M.S. ADEVA 1991D, DECAMP 1991S
Sard R.D. SARD 1948
Sarracino J. KNAPP 1976B
Sartorelli G. BASILE 1981I, ADEVA 1991D, DECAMP 1991S
Sartorelli J. ADEVA 1990R
Sasaki M. ALEXANDER 1991C, ACTON 1992B, DECAMP 1991S
Sasao N. PRESCOTT 1978C, PRESCOTT 1978C, PRESCOTT 1979C
Sass J. ARNISON 1983C, ARNISON 1983D, ARNISON 1984C, ARNISON 1984G, ARNISON 1985D, ALBAJAR 1987C
Sassi E. BACCI 1974
Sato A. BARTEL 1980D
Sato H. ABE 1995
Sato N. HIRATA 1987C
Sato T. PRESCOTT 1978C, PRESCOTT 1979C
Saudraix J. ARNISON 1983C, ARNISON 1983D, ARNISON 1984C
Sauerberg K. BRANDELIK 1977D
Saunders B.J. BIAGI 1983C, BIAGI 1985C
Sauvage G. BANNER 1983B, BAGNAIA 1983B, ANSARI 1987E, ADEVA 1990R, ADEVA 1991D, DECAMP 1991S
Savin A. ADEVA 1990R, ADEVA 1991D, DECAMP 1991S
Savoy-Navarro A. ARNISON 1983D, ARNISON 1984C, ARNISON 1984G, ARNISON 1985D, ALBAJAR 1987C
Sawyer L. DECAMP 1991D, DECAMP 1991N, DECAMP 1991S, ABACHI 1995G
Saxon D.H. BRANDELIK 1979H, BRANDELIK 1980G
Scarpine V. ABE 1994ZE, ABE 1995
Scarr J.M. DECAMP 1991D, DECAMP 1991N, DECAMP 1991S, BUSKULIC 1993H
Scarr M. BERGER 1980L
Scarsi L. DAVIES 1955
Schaad M. ABRAMS 1989B
Schaad T. WEIR 1989B
Schaaf U. JUNG 1992
Schacher J. BANNER 1983B, BAGNAIA 1983B, ANSARI 1987E
Schachinger L. BUNCE 1976
Schaeffer M. ABREU 1991B, DECAMP 1991S
Schafer M. ALBRECHT 1987P
Schafer U. DECAMP 1991D, DECAMP 1991N, DECAMP 1991S, BUSKULIC 1993H
Schaile A.D. ALEXANDER 1991C, ACTON 1992B, DECAMP 1991S
Schaile O. ALEXANDER 1991C, ACTON 1992B, DECAMP 1991S
Schalk T.L. WEIR 1989B
Schamberger R.D. BOHRINGER 1980B, FINOCCHIARO 1980, ABACHI 1995G
Schappert W. ALEXANDER 1991C, ACTON 1992B, DECAMP 1991S
Scharff M. DAVIES 1955
Scharff-Hansen P. ALEXANDER 1991C, ACTON 1992B, DECAMP 1991S
Scheck H. ALBRECHT 1987P
Schein M. LORD 1950
Schellman H. WEIR 1989B, ABACHI 1995G
Schenk P. ALEXANDER 1991C, ACTON 1992B, DECAMP 1991S
Schiavon P. FIDECARO 1978
Schindler A. ABE 1994ZE
Schindler R.H. GOLDHABER 1976D, PERUZZI 1976C, BARNETT 1996
Schinzel D. ARNISON 1983C, ARNISON 1983D, ARNISON 1984C, ARNISON 1984G, ARNISON 1985D, ALBAJAR 1987C, ALBAJAR 1991C, ALBAJAR 1991D
Schirato P. BIAGI 1983C, BIAGI 1985C
Schlabach P. ABE 1994ZE, ABE 1995
Schlatter D. DECAMP 1991D, DECAMP 1991N, DECAMP 1991S, BUSKULIC 1993H
Schleichert R. ALBAJAR 1991C, ALBAJAR 1991D
Schlein P. PEVSNER 1961, PJERROU 1962, SCHLEIN 1963
Schliwa M. BRANDELIK 1977D, BRANDELIK 1979H, BRANDELIK 1980G
Schlupmann K. BUSHNIN 1969
Schmelling M. DECAMP 1991D, DECAMP 1991N, DECAMP 1991S, BUSKULIC 1993H
Schmid D. ABACHI 1995G
Schmidke W.B. WEIR 1989B
Schmidt D. BERGER 1979C, BERGER 1980L
Schmidt E.E. ABE 1994ZE, ABE 1995
Schmidt H. DECAMP 1991D, DECAMP 1991N, DECAMP 1991S, BUSKULIC 1993H
Schmidt M.P. ABE 1994ZE, ABE 1995
Schmidt-Parzefall W. DARDEN 1978, DARDEN 1978B, ALBRECHT 1987P
Schmiemann K. ADEVA 1991D, DECAMP 1991S
Schmitt M. BUSKULIC 1993H
Schmitz D. BRANDELIK 1977D, BRANDELIK 1979H, BRANDELIK 1980G, ADEVA 1990R, ADEVA 1991D, DECAMP 1991S
Schmitz M. BIENLEIN 1978
Schmitz P. ADEVA 1990R, ADEVA 1991D, DECAMP 1991S
Schmuser P. BRANDELIK 1977D, BRANDELIK 1979H, BRANDELIK 1980G
Schneegans M. AUBERT 1983, ADEVA 1990R, ADEVA 1991D, DECAMP 1991S
Schneider H. APEL 1975B, ABREU 1991B, DECAMP 1991S
Schneider O. ABE 1994ZE
Schneps J. FRY 1956
Schontag M. ADEVA 1990R
Schopper H. ADEVA 1990R, ADEVA 1991D, DECAMP 1991S
Schotanus D.J. ADEVA 1990R, ADEVA 1991D, DECAMP 1991S
Schreiber H.J. ADEVA 1990R, ADEVA 1991D, DECAMP 1991S
Schreiber S. ALEXANDER 1991C, ACTON 1992B, DECAMP 1991S
Schroder H. DARDEN 1978, DARDEN 1978B, ALBRECHT 1987P
Schroder J. DECAMP 1991D, DECAMP 1991N, DECAMP 1991S, BUSKULIC 1993H
Schroder M. ALBAJAR 1991C, ALBAJAR 1991D
Schröder T. ASHMAN 1988B
Schrödinger E. SCHRÖDINGER 1926, SCHRÖDINGER 1926B, SCHRÖDINGER 1926C, SCHRÖDINGER 1926D, SCHRÖDINGER 1926E

Schubert K.R. DARDEN 1978, DARDEN 1978B, ALBRECHT 1987P
Schuler K.P. PRESCOTT 1978C, PRESCOTT 1978C, PRESCOTT 1979C
Schüler K.P. ASHMAN 1988B
Schuller J.P. DECAMP 1991D, DECAMP 1991N, DECAMP 1991S, BUSKULIC 1993H
Schulte R. ADEVA 1990R, ADEVA 1991D, DECAMP 1991S
Schulte S. ADEVA 1990R, ADEVA 1991D, DECAMP 1991S
Schultz J. BIONTA 1987C
Schultz P.F. LIN 1978
Schultze K. HASERT 1973, HASERT 1973B, BLIETSCHAU 1976, BLIETSCHAU 1977B, ASHMAN 1988B, ADEVA 1990R, ADEVA 1991D, DECAMP 1991S
Schulz H.D. DARDEN 1978, DARDEN 1978B, ALBRECHT 1987P
Schumm B. ABRAMS 1989B
Schune M.H. BUSKULIC 1993H
Schutte J. ADEVA 1990R, ADEVA 1991D, DECAMP 1991S
Schwartz A. ARNISON 1985D, ALBAJAR 1987C
Schwartz A.S. BELAVIN 1975
Schwartz M. PLANO 1957, EISLER 1958, PLANO 1959, SCHWARTZ 1960, DANBY 1962, DORFAN 1967
Schwarz A.S. WEIR 1989B, DECAMP 1991D, DECAMP 1991N, DECAMP 1991S, BUSKULIC 1993H
Schwarz J. ALEXANDER 1991C, DECAMP 1991S
Schwemling P. BUSKULIC 1993H
Schwenke J. ADEVA 1990R, ADEVA 1991D, DECAMP 1991S
Schwering G. ADEVA 1990R, ADEVA 1991D, DECAMP 1991S
Schwiening J. ACTON 1992B
Schwinberg P.B. VAN DYCK 1986, VAN DYCK 1987
Schwindling J. DECAMP 1991D, DECAMP 1991N, DECAMP 1991S, BUSKULIC 1993H
Schwinger J. SCHWINGER 1937, SCHWINGER 1948B, SCHWINGER 1948, SCHWINGER 1949, LIPPMANN 1950
Schwitters R.F. AUGUSTIN 1974B, ABRAMS 1974, SCHWITTERS 1975, PERL 1975C, HANSON 1975, PERL 1976C, GOLDHABER 1976D, PERUZZI 1976C
Sciacca C. ADEVA 1990R, ADEVA 1991D, DECAMP 1991S
Sciacca G.F. ABE 1994ZE, ABE 1995
Scotoni I. FAISSNER 1978D
Scott I. ADEVA 1990R, ADEVA 1991D, DECAMP 1991S
Scott M.B. LYMAN 1951
Scott W. ARNISON 1983C, ARNISON 1983D, ARNISON 1984C, ARNISON 1984G, ARNISON 1985D, ALBAJAR 1987C
Scott W.G. HASERT 1973, HASERT 1973B, ALEXANDER 1991C, ACTON 1992B, DECAMP 1991S
Scribano A. APEL 1975B, ABE 1994ZE, ABE 1995
Sculli J. ABACHI 1995G
Scuri F. ABREU 1991B, DECAMP 1991S
Scwartz M. IMPEDUGLIA 1958
Sebastiani F. BACCI 1974
Sechi B. DAVIES 1955
Sechi-Zorn B. ABRAMS 1964, BERGER 1979C, BERGER 1980L
Sedgbeer J.K. BRANDELIK 1979H, BRANDELIK 1980G, DECAMP 1991D, DECAMP 1991N, DECAMP 1991S, BUSKULIC 1993H
Seeman J. ANDREWS 1980
Seeman N. GLASSER 1961
Seez C. ALBAJAR 1987C, ALBAJAR 1991C, ALBAJAR 1991D
Sefkow F. BUSKULIC 1993H
Segar A.M. ABREU 1991B, DECAMP 1991S
Segler S. ABE 1994ZE, ABE 1995
Segler S.L. BUESSER 1973
Segrè E. CHAMBERLAIN 1955, CHAMBERLAIN 1956E, CHAMBERLAIN 1956F, BARKAS 1957, AGNEW 1958
Sehgal R. ADEVA 1990R, ADEVA 1991D, DECAMP 1991S
Seidel S. BIONTA 1987C, ABE 1994ZE, ABE 1995
Seiden A. ABRAMS 1989B, WEIR 1989B
Seiler P.G. ADEVA 1990R, ADEVA 1991D, DECAMP 1991S
Seiya Y. ABE 1994ZE, ABE 1995
Sekulin R. ABREU 1991B, DECAMP 1991S
Selonke F. DARDEN 1978, DARDEN 1978B
Selvaggi G. DECAMP 1991D, DECAMP 1991N, DECAMP 1991S, BUSKULIC 1993H
Semenov A.V. ALBRECHT 1987P
Sens J.C. HERB 1977, INNES 1977, ADEVA 1990R, ADEVA 1991D, DECAMP 1991S
Serber R. BROBECK 1947, BRUECKNER 1951
Serednyakov S.I. SEREDNYAKOV 1976, DERBENEV 1978, DERBENEV 1980
Sergiampietri F. APEL 1975B
Seriff A.J. LEIGHTON 1949, SERIFF 1950
Sessa M. ABREU 1991B, DECAMP 1991S
Sessler A.M. KERST 1956
Sette G. ABREU 1991B, DECAMP 1991S
Settles M. ALEXANDER 1991C, ACTON 1992B, DECAMP 1991S
Settles R. DECAMP 1991D, DECAMP 1991N, DECAMP 1991S, BUSKULIC 1993H
Seufert R. ABREU 1991B, DECAMP 1991S
Sewell D.C. BROBECK 1947
Sewell S.J. CALICCHIO 1980
Seywerd H.C.J. ALBRECHT 1987P, DECAMP 1991D, DECAMP 1991N, DECAMP 1991S, BUSKULIC 1993H
Sganos G. ABE 1994ZE, ABE 1995
Sgolacchia A. ABE 1994ZE, ABE 1995
Shabalina E. ABACHI 1995G
Shafer J.B. KALBFLEISCH 1964
Shaffer C. ABACHI 1995G
Shafranova M. BEZNOGIKH 1969
Shafranova M.G. KIRILLOVA 1965
Shah T.P. DANYSZ 1963, ARNISON 1983C, ARNISON 1983D, ARNISON 1984C, ARNISON 1984G, ARNISON 1985D, ALBAJAR 1987C, ALBAJAR 1991C, ALBAJAR 1991D
Shamanov V.V. VOYVODIC 1986B
Shamov A.G. ARTAMONOV 1984B
Shankar H.C. ABACHI 1995G
Shapira A. BRANDELIK 1977D, BRANDELIK 1979H, BRANDELIK 1980G
Shapiro A.M. CRETIEN 1962
Shapiro M. ABE 1994ZE, ABE 1995
Sharma V. AVERY 1989C, ARTUSO 1989C, ALAM 1989B, DECAMP 1991D, DECAMP 1991N, DECAMP 1991S, BUSKULIC 1993H
Shastri K. BALTAY 1979E
Shatunov Y.M. SEREDNYAKOV 1976, DERBENEV 1978, DERBENEV 1980
Shaw N.M. ABE 1994ZE, ABE 1995
Shchegelsky V. ADEVA 1990R, ADEVA 1991D, DECAMP 1991S
Sheaff M. BUNCE 1976
Sheer I. ARNISON 1984G, ARNISON 1985D, ALBAJAR 1987C, ADEVA 1990R, ADEVA 1991D, DECAMP 1991S
Sheldon B.M. FIRESTONE 1971C
Sheldon P.D. WEIR 1989B
Shellard R.C. ABREU 1991B, DECAMP 1991S
Shen B.C. ALEXANDER 1991C, ACTON 1992B, DECAMP 1991S
Shen D.Z. ADEVA 1991D, DECAMP 1991S
Shen Q. ABE 1994ZE, ABE 1995

Shephard P.F. ABE 1994ZE, ABE 1995
Sherden D. PRESCOTT 1978C, PRESCOTT 1979C
Sherden D.J. BODEK 1983
Sherwood P. ALEXANDER 1991C, ACTON 1992B, DECAMP 1991S
Shevchenko S.V. ADEVA 1990R, ADEVA 1991D, DECAMP 1991S
Shevchenko V.G. VOYVODIC 1986B, ALBRECHT 1987P, ADEVA 1990R, ADEVA 1991D, DECAMP 1991S
Shi X.R. ADEVA 1990R, ADEVA 1991D, DECAMP 1991S
Shi Z.H. DECAMP 1991N, DECAMP 1991S, BUSKULIC 1993H
Shibata E.I. AVERY 1989C, ARTUSO 1989C, ALAM 1989B
Shibata T.A. ASHMAN 1988B
Shifman M.A. SHIFMAN 1979
Shimojima M. ABE 1994ZE, ABE 1995
Shinsky K.A. ANDREWS 1980, ANDREWS 1980B, BEBEK 1981, CHADWICK 1981
Shipsey I.P.J. AVERY 1989C, ARTUSO 1989C, ALAM 1989B
Shirkov D.V. BOGOLYUBOV 1956
Shively F.T. KALBFLEISCH 1964
Shivpuri R.K. ABACHI 1995G
Shmakov K. ADEVA 1990R, ADEVA 1991D, DECAMP 1991S
Shochet M. ABE 1994ZE, ABE 1995
Shotkin S.M. ADEVA 1991D, DECAMP 1991S
Shoutko V. ADEVA 1990R, ADEVA 1991D, DECAMP 1991S
Shrock R.E. BARNETT 1996
Shtarkov L.N. ASTVACATUROV 1968
Shumard E. BIONTA 1987C
Shumilov E. ADEVA 1990R, ADEVA 1991D, DECAMP 1991S
Shupe M. ABACHI 1995G
Shutt R.P. FOWLER 1953C, FOWLER 1953B, FOWLER 1954C, BROWN 1962, BARNES 1964B, BARNES 1964E
Shuvalov R.S. BUSHNIN 1969, BINON 1969, BINON 1970B, ANTIPOV 1970, APEL 1975B
Shypin R. ALEXANDER 1991C, DECAMP 1991S
Shypit R. ACTON 1992B
Sidorov V.A. ARTAMONOV 1984B
Siebert H.W. BIAGI 1983C, BIAGI 1985C
Siegel D.M. KALBFLEISCH 1964
Siegrist J. GOLDHABER 1976D, PERUZZI 1976C, ABE 1994ZE, ABE 1995
Siegrist J.L. BANNER 1983B, BAGNAIA 1983B
Siegrist P. ABREU 1991B, DECAMP 1991S
Siemann R.H. ANDREWS 1980, ANDREWS 1980B, BEBEK 1981, CHADWICK 1981
Sigurdsson G. APEL 1975B
Sill A. ABE 1994ZE, ABE 1995
Silverman A. ANDREWS 1980, ANDREWS 1980B, BEBEK 1981, CHADWICK 1981, AVERY 1989C, ARTUSO 1989C, ALAM 1989B
Silvestrini V. EISLER 1958
Silvestris L. DECAMP 1991D, DECAMP 1991N, DECAMP 1991S, BUSKULIC 1993H
SiMohand D. BUSKULIC 1993H
Simon A. ALEXANDER 1991C, ACTON 1992B, DECAMP 1991S
Simonetti S. ABREU 1991B, DECAMP 1991S
Simonetto F. ABREU 1991B, DECAMP 1991S
Simopoulou E. DECAMP 1991D, DECAMP 1991N, DECAMP 1991S, BUSKULIC 1993H
Simpson K.M. BROBECK 1947
Sinclair C.K. PRESCOTT 1978C, PRESCOTT 1979C
Sinclair D. BIONTA 1987C
Sinervo P. ABE 1994ZE, ABE 1995
Singh J.B. ABACHI 1995G
Singh P. ALEXANDER 1991C, ACTON 1992B, DECAMP 1991S, ABE 1994ZE, ABE 1995
Sinram K. BARBER 1979F
Siotis I. ARNISON 1985D, ALBAJAR 1987C, BUSKULIC 1993H
Siroli G.P. ALEXANDER 1991C, ACTON 1992B, DECAMP 1991S
Sirotenko V.I. VOYVODIC 1986B, ABACHI 1995G
Sissakian A.N. ABREU 1991B, DECAMP 1991S
Sivertz M. BOHRINGER 1980B, FINOCCHIARO 1980
Skaali T.B. ABREU 1991B, DECAMP 1991S
Skard J.A. BERGER 1979C, BERGER 1980L
Skarha J. ABE 1994ZE, ABE 1995
Skillicorn I.O. BERTANZA 1962D, CONNOLLY 1963
Skjevling G. ABREU 1991B, DECAMP 1991S
Skobelzyn D.V. SKOBELZYN 1929
Skrinsky A.N. BUDKER 1976, SEREDNYAKOV 1976, DERBENEV 1978, DERBENEV 1980, ARTAMONOV 1984B
Skubic P. BUNCE 1976, ANDREWS 1980, ANDREWS 1980B, BEBEK 1981, CHADWICK 1981
Skuja A. BERGER 1979C, BERGER 1980L, ALEXANDER 1991C, ACTON 1992B, DECAMP 1991S
Skwarnicki T. AVERY 1989C, ARTUSO 1989C, ALAM 1989B
Slater W.E. PJERROU 1962, SCHLEIN 1963
Sleeman J.C. BAND 1988
Sliwa K. ABE 1994ZE, ABE 1995
Sloan T. AUBERT 1983, ASHMAN 1988B, DECAMP 1991D, DECAMP 1991N, DECAMP 1991S, BUSKULIC 1993H
Slobodyuk E.A. VOYVODIC 1986B
Smadja G. ABREU 1991B, DECAMP 1991S
Smart W.M. VOYVODIC 1986B, ABACHI 1995G
Smirnitsky V.A. VOYVODIC 1986B
Smirnov N.N. ADEVA 1990R, ADEVA 1991D, ABREU 1991B, DECAMP 1991S, DECAMP 1991S
Smith A. ABACHI 1995G
Smith A.M. BUESSER 1973, ALEXANDER 1991C, ACTON 1992B, DECAMP 1991S
Smith D. ARNISON 1983D, ARNISON 1984C, ARNISON 1984G, ARNISON 1985D, ALBAJAR 1987C
Smith D.A. ABE 1994ZE, ABE 1995
Smith F.M. BARKAS 1957
Smith G.A. KALBFLEISCH 1964
Smith G.R. ABREU 1991B, DECAMP 1991S
Smith J.G. BAND 1988, ABRAMS 1989B, WEIR 1989B
Smith J.R. BARNES 1964B, BARNES 1964E
Smith K. DECAMP 1991D, DECAMP 1991N, DECAMP 1991S, BUSKULIC 1993H
Smith L.T. SCHLEIN 1963
Smith M.G. BUSKULIC 1993H
Smith R.P. ABACHI 1995G
Smith S.D. KNAPP 1976B
Smith T.J. ALEXANDER 1991C, ACTON 1992B, DECAMP 1991S
Smith V.J. BIAGI 1983C, BIAGI 1985C
Smolik L. BUSKULIC 1993H
Smotritskii L.M. LOBASHOV 1966
Snell A.H. SNELL 1948, SNELL 1950
Snider F.D. ABE 1994ZE, ABE 1995
Snihur R. ABACHI 1995G
Snow G.A. ABRAMS 1964, ALEXANDER 1991C, ACTON 1992B, DECAMP 1991S, ABACHI 1995G
Snow S.W. DECAMP 1991D, DECAMP 1991N, DECAMP 1991S, BUSKULIC 1993H
Snyder A. ABRAMS 1989B, WEIR 1989B
Snyder H.D. HERB 1977, INNES 1977, UENO 1979
Snyder H.S. COURANT 1952
Snyder S. ABACHI 1995G
Sobel H.W. BIONTA 1987C
Sobie R. ARNISON 1985D, ALBAJAR 1987C, ALEXANDER 1991C, ACTON 1992B, DECAMP 1991S
Soderstrom E. ABRAMS 1989B, WEIR 1989B, ADEVA 1991D, DECAMP 1991S
Soding P. BRANDELIK 1979H, BRANDELIK 1980G

Soff G. JUNG 1992
Sogard M.R. BODEK 1983
Sokolov A.A. SOKOLOV 1963
Solmitz F.T. NORDIN 1958
Solomon J. ABACHI 1995G
Soloshenko V.A. ALBRECHT 1987P
Solovianov V.L. BEZNOGIKH 1969, FIDECARO 1978
Soloviev L.D. SOLOVIEV 1966, LOGUNOV 1967
Solovyev M.I. KANG-CHANG 1960
Solyanik V.I. BINON 1969, ANTIPOV 1970
Song L. ABE 1994ZE
Song T. ABE 1994ZE, ABE 1995
Sood P.M. ABACHI 1995G
Sopczak A. ADEVA 1990R, ADEVA 1991D, DECAMP 1991S
Sorrels J.D. SORRELS 1955
Sosebee M. ABACHI 1995G
Sosnowski R. ABREU 1991B, DECAMP 1991S
Souza M. ABACHI 1995G
Spaan B. ALBRECHT 1987P
Spadafora A.L. ABACHI 1995G
Spadtke P. JUNG 1992
Spagnolo P. BUSKULIC 1993H
Spalding J. ABE 1994ZE, ABE 1995
Spano M. BACCI 1974
Sparks K. ALAM 1989B
Spartiotis C. ADEVA 1990R, ADEVA 1991D, DECAMP 1991S
Spassoff T.S. ABREU 1991B, DECAMP 1991S
Spencer C. BACCI 1974
Spencer L.J. BOHRINGER 1980B, FINOCCHIARO 1980
Spengler J. ALBRECHT 1987P
Sphicas P. ALBAJAR 1987C, ALBAJAR 1991C, ALBAJAR 1991D, ABE 1994ZE, ABE 1995
Spickermann T. ADEVA 1990R, ADEVA 1991D, DECAMP 1991S
Spiegel L. ABE 1994ZE, ABE 1995
Spies A. ABE 1994ZE, ABE 1995
Spiess B. ADEVA 1990R, ADEVA 1991D, DECAMP 1991S
Spighel M. BUSHNIN 1969, BINON 1969, BINON 1970B
Spillantini P. ADEVA 1990R, ADEVA 1991D, DECAMP 1991S
Spinetti G.S.M. BACCI 1974
Spiriti E. ABREU 1991B, DECAMP 1991S
Spiro M. ARNISON 1983C, ARNISON 1983D, ARNISON 1984C, ARNISON 1984G
Spitzer H. BERGER 1978, BERGER 1979C, BERGER 1980L
Springer R.W. ALEXANDER 1991C, ACTON 1992B, DECAMP 1991S
Sproston M. ALEXANDER 1991C, ACTON 1992B, DECAMP 1991S
Squarcia S. ABREU 1991B, DECAMP 1991S
Sreekantan B.V. BRIDGE 1955
Staeck H. ABREU 1991B, DECAMP 1991S
Stahl A. DECAMP 1991D, DECAMP 1991N, DECAMP 1991S, BUSKULIC 1993H
Stahlbrandt C.A. BUSHNIN 1969
Staiano A. ASHMAN 1988B
Stanco L. ABE 1994ZE, ABE 1995
Stanescu C. ABREU 1991B, DECAMP 1991S
Stanghelini A. AMATI 1962
Stark J. STARK 1909
Starosta R. ADEVA 1990R, ADEVA 1991D, DECAMP 1991S
Stavinsky V.S. BALDIN 1973
Stavropoulos G. ABREU 1991B, DECAMP 1991S
StDenis R. DECAMP 1991D, DECAMP 1991N, DECAMP 1991S, BUSKULIC 1993H
Steck M. JUNG 1992
Steeg F. DECAMP 1991D, DECAMP 1991N, DECAMP 1991S, BUSKULIC 1993H
Steele J. ABE 1994ZE, ABE 1995
Stefanini A. ABE 1994ZE, ABE 1995
Stefanski R.J. BENVENUTI 1974F
Steffen P. BARTEL 1980D
Stein P.C. ANDREWS 1980, ANDREWS 1980B, BEBEK 1981, CHADWICK 1981
Stein S. PRESCOTT 1978C, PRESCOTT 1979C
Steinberger J. STEINBERGER 1950, DURBIN 1951C, CHINOWSKY 1955, PLANO 1957, IMPEDUGLIA 1958, EISLER 1958, PLANO 1959, DANBY 1962, BENNETT 1967, DECAMP 1991D, DECAMP 1991N, DECAMP 1991S, BUSKULIC 1993H
Steiner H. AGNEW 1958
Steiner H.M. BANNER 1983B, BAGNAIA 1983B
Steinmann E. DARDEN 1978, DARDEN 1978B
Stella B. BACCI 1974, BERGER 1979C, BERGER 1980L
Steller J. STEINBERGER 1950
Stenzler M. ARNISON 1985D
Stephens K. BARTEL 1980D, ALEXANDER 1991C, ACTON 1992B, DECAMP 1991S
Stephens R. ABACHI 1995G
Stern O. FRISCH 1933, ESTERMAN 1933, ESTERMAN 1934
Stern R.A. BRANDELIK 1980G
Steuer M. APEL 1975B, FIDECARO 1978, ADEVA 1990R, ADEVA 1991D, DECAMP 1991S
Stevenson E.G. STREET 1937, STREET 1937B
Stevenson M.L. ALVAREZ 1956, CRAWFORD 1958, BUTTON 1960, BUTTON 1961, MAGLIC 1961B, ABACHI 1995G
Stewart D. ABACHI 1995G
Stichelbaut F. ABREU 1991B, DECAMP 1991S
Stickland D.P. BIAGI 1983C, ADEVA 1990R, ADEVA 1991D, DECAMP 1991S
Sticozzi F. ADEVA 1991D, DECAMP 1991S
Stiegler U. DECAMP 1991D, DECAMP 1991N, DECAMP 1991S, BUSKULIC 1993H
Stier H.E. AUBERT 1983, ASHMAN 1988B, ALEXANDER 1991C, ACTON 1992B, DECAMP 1991S
Stierlin U. BLUM 1975, DECAMP 1991D, DECAMP 1991N, DECAMP 1991S, BUSKULIC 1993H
Stiewe J. ALBRECHT 1987P
Stiller B. GLASSER 1961
Stimpfl G. BANNER 1983B, BAGNAIA 1983B, DECAMP 1991D
Stimplf G. DECAMP 1991N, DECAMP 1991S
Stocchi A. ABREU 1991B, DECAMP 1991S
Stock J. ASHMAN 1988B
Stocker F. BANNER 1983B, BAGNAIA 1983B, ANSARI 1987E, ABACHI 1995G
Stockhausen W. AUBERT 1983
Stoeffl W. ADEVA 1990R, ADEVA 1991D, DECAMP 1991S
Stohlker T. JUNG 1992
Stohr B. ADEVA 1990R
Stoker D.P. ABRAMS 1989B, WEIR 1989B, ABACHI 1995G
Stone H. ADEVA 1990R, ADEVA 1991D, DECAMP 1991S
Stone J. BARNETT 1996
Stone J.L. BIONTA 1987C
Stone R. ANDREWS 1980, ANDREWS 1980B, BEBEK 1981, CHADWICK 1981
Stone S. ANDREWS 1980, ANDREWS 1980B, BEBEK 1981, CHADWICK 1981, AVERY 1989C, ARTUSO 1989C, ALAM 1989B
Stonehill D.L. STONEHILL 1961, BARNES 1964B, BARNES 1964E
Stork D.H. BIRGE 1955B, BARKAS 1957, PJERROU 1962, SCHLEIN 1963, ALBAJAR 1991C, ALBAJAR 1991D
Stoyanova D.A. BUSHNIN 1969, ANTIPOV 1970, DENISOV 1971F, ABACHI 1995G
Stradner H. FIDECARO 1978
Strahl K. ALBRECHT 1987P, ABE 1994ZE, ABE 1995

Strait J. ABE 1994ZE, ABE 1995
Strand R.C. PEVSNER 1961, BARNES 1964B, BARNES 1964E
Strauch K. CRETIEN 1962, ADEVA 1990R, ADEVA 1991D, DECAMP 1991S
Strauss J. ARNISON 1983C, ARNISON 1983D, ARNISON 1984C, ARNISON 1984G, ARNISON 1985D, ALBAJAR 1987C, ABREU 1991B, DECAMP 1991S
Street J.C. STREET 1937, STREET 1937B, CRETIEN 1962
Streets J. ARNISON 1983D, ARNISON 1984C, ARNISON 1984G, ARNISON 1985D, ALBAJAR 1987C
Streets K. ABACHI 1995G
Streit K.P. BIAGI 1983C, BIAGI 1985C
Stringfellow B.C. ADEVA 1990R, ADEVA 1991D, DECAMP 1991S
Strohbusch U. BIENLEIN 1978
Strohmer R. ACTON 1992B
Strom D. ALEXANDER 1991C, ACTON 1992B, DECAMP 1991S
Strong J. DECAMP 1991N, DECAMP 1991S
Strong J.A. DECAMP 1991D, BUSKULIC 1993H
Stroot J.P. BUSHNIN 1969, BINON 1969, BINON 1970B
Strovink M. ABACHI 1995G
Stroynowski R. ABRAMS 1989B, WEIR 1989B
Strub R. ABREU 1991B, DECAMP 1991S
Struminsky B.V. BOGOLYUBOV 1965
Strunov L.N. KIRILLOVA 1965
Stuart D. ABE 1994ZE, ABE 1995
Stubenrauch C. ARNISON 1985D, ALBAJAR 1987C, ABREU 1991B, ALBAJAR 1991C, ALBAJAR 1991D, DECAMP 1991S
Stückelberg E.C.G. STÜCKELBERG 1938, STÜCKELBERG 1953
Stump R. VOYVODIC 1986B
Sturm W. BRANDELIK 1977D
Suda T. BRANDELIK 1977D, BARTEL 1980D, HIRATA 1987C
Sudhakar K. ADEVA 1990R, ADEVA 1991D, DECAMP 1991S
Sukhina B.N. BUDKER 1976
Sulak L. BENVENUTI 1974F
Sulak L.R. BIONTA 1987C
Sullivan G. ABE 1994ZE, ABE 1995
Sultanov G.G. ADEVA 1990R, ADEVA 1991D, DECAMP 1991S
Sumarokov A. ABE 1995
Summerer K. JUNG 1992
Summers D. ARNISON 1985D, ALBAJAR 1987C
Sumner R.L. ADEVA 1990R, ADEVA 1991D, DECAMP 1991S
Sumorok K. ARNISON 1983C, ARNISON 1983D, ARNISON 1984C, ARNISON 1984G, ARNISON 1985D, ALBAJAR 1987C, ALBAJAR 1991C, ALBAJAR 1991D, ABE 1994ZE, ABE 1995
Sun C.R. AVERY 1989C, ARTUSO 1989C, ALAM 1989B
Sun L.Z. ADEVA 1990R, ADEVA 1991D, DECAMP 1991S
Sunderland J. BENNETT 1967
Sunyar A.W. GOLDHABER 1958C
Suter H. ADEVA 1990R, ADEVA 1991D, DECAMP 1991S
Sutton R.B. ADEVA 1990R, ADEVA 1991D, DECAMP 1991S
Suzuki A. HIRATA 1987C
Suzuki J. ABE 1995
Sviridov V. BEZNOGIKH 1969
Sviridov V.A. KIRILLOVA 1965
Svoboda R. BIONTA 1987C
Swain J.D. ALBRECHT 1987P, ADEVA 1990R, ADEVA 1991D, DECAMP 1991S
Swami M.S. FRY 1956
Swartz M. ANSARI 1987E, ABRAMS 1989B
Swartz R.L.Jr. ABE 1994ZE
Syed A.A. ADEVA 1990R, ADEVA 1991D, DECAMP 1991S
Symanzik K. LEHMANN 1955, SYMANZIK 1956
Symon K.R. KERST 1956
Szczekowski M. ABREU 1991B, DECAMP 1991S
Szeptycka M. ABREU 1991B, DECAMP 1991S
Szoncso F. ARNISON 1983C, ARNISON 1983D, ARNISON 1984C, ARNISON 1984G, ARNISON 1985D, ALBAJAR 1987C, ALBAJAR 1991C, ALBAJAR 1991D
Szymanski J.J. CRETIEN 1962
Szymanski P. ABREU 1991B, DECAMP 1991S
Ta-Chao T. KANG-CHANG 1960
Tabarelli T. ABREU 1991B, DECAMP 1991S
Taft H.D. NORDIN 1958, STONEHILL 1961, BROWN 1962
Takada T. ABE 1995
Takahashi K. HIRATA 1987C
Takahashi T. ABE 1994ZE, ABE 1995
Takano T. ABE 1995
Takashima M. DECAMP 1991D, DECAMP 1991N, DECAMP 1991S
Takeda H. BARTEL 1980D, ALEXANDER 1991C, ACTON 1992B, DECAMP 1991S
Takeshita T. ALEXANDER 1991C, ACTON 1992B, DECAMP 1991S
Taketani A. ABACHI 1995G
Takeuchi M. NISHINA 1937
Takikawa K. ABE 1994ZE, ABE 1995
Takita M. HIRATA 1987C
Talby M. DECAMP 1991D, DECAMP 1991N, DECAMP 1991S, BUSKULIC 1993H
Talman R. ANDREWS 1980, ANDREWS 1980B, BEBEK 1981, CHADWICK 1981
Tamburello P. ABACHI 1995G
Tamm I.E. TAMM 1930, ALTSHULER 1934, TAMM 1937
Tamura N. ABE 1995
Tan C.H. ALBAJAR 1991C, ALBAJAR 1991D
Tanaka M. BALTAY 1979E
Tanaka R. DECAMP 1991D, DECAMP 1991N, DECAMP 1991S, BUSKULIC 1993H
Tanenbaum W.M. AUGUSTIN 1974B, ABRAMS 1974, SCHWITTERS 1975, PERL 1975C, HANSON 1975, PERL 1976C, GOLDHABER 1976D, PERUZZI 1976C, PERL 1977F, ANDREWS 1980, ANDREWS 1980B, BEBEK 1981, CHADWICK 1981
Tang H.W. BARBER 1979F
Tang L.G. BARBER 1979F
Tang X.W. ADEVA 1990R, ADEVA 1991D, DECAMP 1991S
Tang Y.H. DECAMP 1991N, DECAMP 1991S
Tanikella V. AVERY 1989C
Tanimori T. HIRATA 1987C
Tao C. ARNISON 1983C, ARNISON 1983D, ARNISON 1984C, ARNISON 1984G, ARNISON 1985D, ALBAJAR 1987C
Taras P. ALEXANDER 1991C, ACTON 1992B, DECAMP 1991S
Tarazi J. ABACHI 1995G
Tarem S. ALEXANDER 1991C, ACTON 1992B, DECAMP 1991S
Tarkovskii E. ADEVA 1990R, ADEVA 1991D, DECAMP 1991S
Tartaglia M. ABACHI 1995G
Tartarelli F. ABE 1994ZE, ABE 1995
Tati T. KOBA 1947, KOBA 1947B, TATI 1948
Taurok A. ALBAJAR 1987C, ALBAJAR 1991C, ALBAJAR 1991D
Tavernier S. ABREU 1991B, DECAMP 1991S
Tavkhelidze A.N. LOGUNOV 1957, LOGUNOV 1963, BOGOLYUBOV 1965, LOGUNOV 1967, MATVEEV 1973
Taylor G. DECAMP 1991D, DECAMP 1991N, DECAMP 1991S, BUSKULIC 1993H
Taylor G.N. ASHMAN 1988B

Taylor L. ADEVA 1990R, ADEVA 1991D, ALBAJAR 1991C, ALBAJAR 1991D, DECAMP 1991S
Taylor R.E. BLOOM 1969B, BREIDENBACH 1969, MILLER 1972B, PRESCOTT 1978C, PRESCOTT 1979C
Taylor T.L. ABACHI 1995G
Taylor W. ABE 1994ZE, ABE 1995
Tayursky V.A. ARTAMONOV 1984B
Tchikilev O.G. ABREU 1991B, DECAMP 1991S
Tchistilin V.I. ALBRECHT 1987P
Teiger J. BANNER 1983B, BAGNAIA 1983B, ANSARI 1987E, ABACHI 1995G
Teixera-Dias P. ALEXANDER 1991C, ACTON 1992B, DECAMP 1991S
Tejessy W. DECAMP 1991D, DECAMP 1991N, DECAMP 1991S, BUSKULIC 1993H
Telegdi V.L. FRIEDMAN 1957
Teller E. GAMOW 1936
Telnov V.I. ARTAMONOV 1984B
Temnykh A.B. ARTAMONOV 1984B
Tempesta P. DECAMP 1991D, DECAMP 1991N, DECAMP 1991S, BUSKULIC 1993H
ten Have I. ALBAJAR 1987C, DECAMP 1991D, DECAMP 1991N, DECAMP 1991S, BUSKULIC 1993H
Tenchini R. DECAMP 1991D, DECAMP 1991N, DECAMP 1991S, BUSKULIC 1993H
Teng P.K. ABE 1995
Teramoto Y. ABE 1994ZE, ABE 1995
Ternov I.M. TERNOV 1961, SOKOLOV 1963
Terwilliger K.M. KERST 1956, LIN 1978
Tether S. ALBAJAR 1987C, ALBAJAR 1991C, ALBAJAR 1991D, ABE 1994ZE, ABE 1995
Teykal H. ALBAJAR 1991C, ALBAJAR 1991D
Thackray N.J. ALEXANDER 1991C, ACTON 1992B, DECAMP 1991S
Theodosiou G. ABREU 1991B, DECAMP 1991S
Theriot D. ABE 1994ZE, ABE 1995
Thirring W.E. GELL-MANN 1954
Thomas E. ADEVA 1990R
Thomas J. DECAMP 1991D, DECAMP 1991N, DECAMP 1991S, ABE 1994ZE
Thomas R.M. DECAMP 1991D, DECAMP 1991N, DECAMP 1991S
Thomas T.L. ABE 1995
Thompson A. DANYSZ 1963
Thompson A.S. DECAMP 1991D, DECAMP 1991N, DECAMP 1991S, BUSKULIC 1993H
Thompson G. ARNISON 1983C, ARNISON 1983D, ARNISON 1984C, ARNISON 1984G, ARNISON 1985D, ALBAJAR 1987C, ALBAJAR 1991C, ALBAJAR 1991D
Thompson J.C. AUBERT 1983, ASHMAN 1988B, DECAMP 1991D, DECAMP 1991N, DECAMP 1991S, BUSKULIC 1993H, ABACHI 1995G
Thompson L.F. DECAMP 1991D, DECAMP 1991N, DECAMP 1991S, BUSKULIC 1993H
Thompson R.L. BERGER 1978
Thompson R.W. THOMPSON 1953, THOMPSON 1953B
Thomson G.P. THOMSON 1928
Thomson J.J. THOMSON 1897
Thonemann H.G. ANDREWS 1980, ANDREWS 1980B, BEBEK 1981, CHADWICK 1981
Thorndahl L. CARRON 1978
Thorndike A.M. FOWLER 1953C, FOWLER 1953B, FOWLER 1954C, BROWN 1962, BARNES 1964B, BARNES 1964E
Thorndike E.H. ANDREWS 1980, ANDREWS 1980B, BEBEK 1981, CHADWICK 1981, ARTUSO 1989C, ALAM 1989B
Thornton G. BIONTA 1987C
Thornton R.L. LAWRENCE 1939, BROBECK 1947
Thresher J.J. BIAGI 1983C, BIAGI 1985C
Thun R. AMENDOLIA 1973B, ABRAMS 1989B, WEIR 1989B, ABE 1994ZE, ABE 1995
Thénard J.M. AUBERT 1983
Ticho H.K. CRAWFORD 1958, ALVAREZ 1959, ALSTON 1960, ALSTON 1961B, ALSTON 1961E, PJERROU 1962, SCHLEIN 1963
Tikhomirov I.N. ALBRECHT 1987P
Tikhonov Y.A. ARTAMONOV 1984B
Tilquin A. ABREU 1991B, DECAMP 1991S
Timko M. ABE 1994ZE, ABE 1995
Timm U. BERGER 1978, BERGER 1979C, BERGER 1980L
Timmer J. ARNISON 1983C, ARNISON 1983D, ARNISON 1984C, ARNISON 1984G
Timmermans C. ADEVA 1990R, ADEVA 1991D, DECAMP 1991S
Timmermans J. ABREU 1991B, DECAMP 1991S
Timofeev V.G. ABREU 1991B, DECAMP 1991S
Ting C.C. DORFAN 1965D
Ting S.C.C. AUBERT 1974D, BARBER 1979F, ADEVA 1990R, ADEVA 1991D, DECAMP 1991S
Ting S.M. ADEVA 1990R, ADEVA 1991D, DECAMP 1991S
Tipton P. ABE 1994ZE, ABE 1995
Titov A. ABE 1994ZE, ABE 1995
Tittel K. DECAMP 1991D, DECAMP 1991N, DECAMP 1991S, BUSKULIC 1993H
Tkaczyk S. ABE 1994ZE, ABE 1995
Tkatchev L.G. ABREU 1991B, DECAMP 1991S
Toback D. ABE 1995
Todorov T. KIRILLOVA 1965, ABREU 1991B, DECAMP 1991S
Toet D.Z. ABREU 1991B, DECAMP 1991S
Tokunaga S. MINAKAWA 1959
Tollefson K. ABE 1994ZE, ABE 1995
Tollestrup A. ABE 1994ZE, ABE 1995
Tollestrup A.V. FAZZINI 1958
Tomalin I.R. DECAMP 1991D, DECAMP 1991N, DECAMP 1991S, BUSKULIC 1993H
Tomasini G. BONETTI 1953, BONETTI 1953B, DAVIES 1955
Tomonaga S. TOMONAGA 1943, KOBA 1947, KOBA 1947B, TATI 1948
Tonelli G. DECAMP 1991D, DECAMP 1991N, DECAMP 1991S, BUSKULIC 1993H
Toner W.T. DANYSZ 1963
Tong Y.P. ADEVA 1990R, ADEVA 1991D, DECAMP 1991S
Tonisch F. ADEVA 1990R, ADEVA 1991D, DECAMP 1991S
Tonnison J. ABE 1994ZE, ABE 1995
Tonutti M. ADEVA 1990R, ADEVA 1991D, DECAMP 1991S
Tonwar S.C. ADEVA 1990R, ADEVA 1991D, DECAMP 1991S
Toohig T. PEVSNER 1961
Tornqvist N.A. BARNETT 1996
Torrente-Lujan E. ALBAJAR 1991C, ALBAJAR 1991D
Tortora L. BACCI 1974, ABREU 1991B, DECAMP 1991S
Toth J. ADEVA 1990R, ADEVA 1991D, DECAMP 1991S
Totsuka Y. BRANDELIK 1977D, BARTEL 1980D, HIRATA 1987C
Tovey S.N. BIAGI 1985C, ANSARI 1987E
Trainor M.T. ABREU 1991B, DECAMP 1991S
Transtromer G. ACTON 1992B
Trasatti L. BACCI 1974
Treille D. ABREU 1991B, DECAMP 1991S
Treiman S.B. HODSON 1954, PAIS 1957, GOLDBERGER 1958
Tretyak V.I. VOYVODIC 1986B
Trevisan U. ABREU 1991B, DECAMP 1991S
Triggiani G. DECAMP 1991D, DECAMP 1991N, DECAMP 1991S, BUSKULIC 1993H

Trilling G.H. FIRESTONE 1971C, AUGUSTIN 1974B, ABRAMS 1974, SCHWITTERS 1975, PERL 1975C, HANSON 1975, PERL 1976C, GOLDHABER 1976D, PERUZZI 1976C, ABRAMS 1989B, WEIR 1989B
Trines D. BRANDELIK 1979H, BRANDELIK 1980G
Tripp R.D. NORDIN 1958, TRIPP 1962, KALBFLEISCH 1964
Trippe T.G. ABACHI 1995G, BARNETT 1996
Trischuk W. ABREU 1991B, DECAMP 1991S
Tristram G. ABREU 1991B, DECAMP 1991S
Troncon C. ABREU 1991B, DECAMP 1991S
Trowitzsch G. ADEVA 1990R, ADEVA 1991D, DECAMP 1991S
Troya U. BACCI 1974
Truitt S. ABE 1995
Truong Bien BEZNOGIKH 1969
Tsang W.Y. ANSARI 1987E
Tscheslog E. ARNISON 1983C, ARNISON 1983D, ARNISON 1984C, ARNISON 1984G, ARNISON 1985D, ALBAJAR 1987C
Tschirhart R. WEIR 1989B
Tseng J. ABE 1994ZE, ABE 1995
Tsipolitis G. ALBRECHT 1987P
Tsirou A. ABREU 1991B, DECAMP 1991S
Tso T.T. CAZZOLI 1975E
Tsukamoto T. ALEXANDER 1991C, ACTON 1992B, DECAMP 1991S
Tsuzuki M. MINAKAWA 1959
Tsyganov E.N. ABREU 1991B, DECAMP 1991S
Tu N.D. KANG-CHANG 1960
Tubau E. DECAMP 1991D
Tuchscherer H. ALBAJAR 1991C, ALBAJAR 1991D
Tully C. ADEVA 1991D, DECAMP 1991S
Tumaikin G.M. SEREDNYAKOV 1976, DERBENEV 1978, DERBENEV 1980, ARTAMONOV 1984B
Tung K.L. BARBER 1979F, ADEVA 1990R, ADEVA 1991D, DECAMP 1991S
Tuominiemi J. ARNISON 1983D, ARNISON 1984C, ARNISON 1984G, ARNISON 1985D, ALBAJAR 1987C, ALBAJAR 1991C, ALBAJAR 1991D
Turala M. ABREU 1991B, DECAMP 1991S
Turchetta R. ABREU 1991B, DECAMP 1991S
Turcotte M. ABE 1994ZE
Turini N. ABE 1994ZE, ABE 1995
Turlay R. CHRISTENSON 1964
Turluer M.L. ABREU 1991B, DECAMP 1991S
Turnbull R.M. DECAMP 1991D, DECAMP 1991N, DECAMP 1991S, BUSKULIC 1993H
Turner M.F. ALEXANDER 1991C, ACTON 1992B, DECAMP 1991S
Tuts P.M. BOHRINGER 1980B, FINOCCHIARO 1980, ABACHI 1995G
Tuuva T. ABREU 1991B, DECAMP 1991S
Tuvdendorzh D. KIRILLOVA 1965
Tyapkin I.A. ABREU 1991B, DECAMP 1991S
Tyndel M. CALICCHIO 1980, ABREU 1991B, DECAMP 1991S
Tysarczyk-Niemeyer G. ALEXANDER 1991C, ACTON 1992B, DECAMP 1991S
Tyupkin Y.S. BELAVIN 1975
Tzamarias S. ABREU 1991B, DECAMP 1991S
't Hooft G. 'T HOOFT 1971, 'T HOOFT 1971B, 'T HOOFT 1972, 'T HOOFT 1972B
Uchida T. ABE 1995
Ueberschaer B. ABREU 1991B, DECAMP 1991S
Ueberschaer S. ABREU 1991B, DECAMP 1991S
Uemura N. ABE 1994ZE, ABE 1995
Ueno K. HERB 1977, INNES 1977, UENO 1979
Uhlenbeck G.E. UHLENBECK 1925
Ukegawa F. ABE 1994ZE, ABE 1995
Ulbricht J. ADEVA 1990R, ADEVA 1991D, DECAMP 1991S
Ullaland O. ABREU 1991B, DECAMP 1991S
Unal G. ABE 1994ZE, ABE 1995
Undrus A.E. ARTAMONOV 1984B
Urban L. AUBERT 1983, ADEVA 1990R, ADEVA 1991D, DECAMP 1991S
Urey H.C. UREY 1932, UREY 1932B
Uvarov V.A. ABREU 1991B, DECAMP 1991S
Uwer U. ADEVA 1990R, ADEVA 1991D, DECAMP 1991S
Vainstein A.I. SHIFMAN 1979
Vaisenberg A. VOYVODIC 1986B
Valdata M. AMENDOLIA 1973B
Valdata-Nappi M. ANSARI 1987E
Valente E. ADEVA 1990R, ADEVA 1991D, DECAMP 1991S
Valente V. BACCI 1974
Valenti G. BASILE 1981I, ABREU 1991B, DECAMP 1991S
Vallage B. DECAMP 1991N, DECAMP 1991S, BUSKULIC 1993H
Vallazza E. ABREU 1991B, DECAMP 1991S
Valls Ferrer J.A. ABREU 1991B, DECAMP 1991S
Van Apeldoorn G.W. ABREU 1991B, DECAMP 1991S
Van Atta L.C. VAN DE GRAAFF 1933
Van Berg R. HIRATA 1987C
Van Dalen G.J. ALEXANDER 1991C, ACTON 1992B, DECAMP 1991S
Van Dam P. ABREU 1991B, DECAMP 1991S
Van de Graaff R.J. VAN DE GRAAFF 1931, VAN DE GRAAFF 1933
Van de Guchte W. ALBAJAR 1991C, ALBAJAR 1991D
Van de Walle R.T. ADEVA 1990R, ADEVA 1991D, DECAMP 1991S
Van den Brink S.A. ABE 1994ZE, ABE 1995
Van den Plas D. ALEXANDER 1991C, ACTON 1992B, DECAMP 1991S
Van der Graaf H. ADEVA 1990R
Van der Meer S. CARRON 1978
Van der Velde C. ABREU 1991B, DECAMP 1991S
Van der Velde J.C. BIONTA 1987C
Van Dijk A. ALBAJAR 1991C, ALBAJAR 1991D
Van Doninck W.K. HASERT 1973, HASERT 1973B, BLIETSCHAU 1976, BLIETSCHAU 1977B, CALICCHIO 1980, ABREU 1991B, DECAMP 1991S
Van Dyck R.S. VAN DYCK 1986, VAN DYCK 1987
Van Eijk B. ARNISON 1983D, ARNISON 1984C, ARNISON 1984G, ARNISON 1985D, ALBAJAR 1987C
Van Eijndhoven N. ABREU 1991B, DECAMP 1991S
Van Kooten R. ABRAMS 1989B, WEIR 1989B, ALEXANDER 1991C, ACTON 1992B, DECAMP 1991S
Van Laak A. ADEVA 1990R
Van Lint V.A.J. ANDERSON 1953B, COWAN 1953
Van Norman Hilbery H. ROSSI 1939
Van Rossum L. BARKAS 1957
Van Staa R. BERGER 1978, BERGER 1979C, BERGER 1980L
Vander Velde-WilquetC CALICCHIO 1980
Vannini C. DECAMP 1991N, DECAMP 1991S, BUSKULIC 1993H
Vannini-Castaldi C. DECAMP 1991D
Vannucci F. AUGUSTIN 1974B, ABRAMS 1974, SCHWITTERS 1975, PERL 1975C, HANSON 1975, PERL 1976C, GOLDHABER 1976D, PERUZZI 1976C, BARBER 1979F
Vanoli F. BACCI 1974
Varela J. ABREU 1991B, DECAMP 1991S
Varelas N. ABACHI 1995G
Vargas M. ALBAJAR 1991C, ALBAJAR 1991D
Varnes E.W. ABACHI 1995G
Vascon M. FAISSNER 1978D
Vascotto A. FIDECARO 1978
Vasseur G. ALEXANDER 1991C, ACTON 1992B, DECAMP 1991S
Vavilov S.I. VAVILOV 1934

Vayaki A. DECAMP 1991D, DECAMP 1991N, DECAMP 1991S, BUSKULIC 1993H
Vaz P. ABREU 1991B, DECAMP 1991S
Veenhof R. BUSKULIC 1993H
Vegni G. ABREU 1991B, DECAMP 1991S
Veillet J.J. DECAMP 1991D, DECAMP 1991N, DECAMP 1991S, BUSKULIC 1993H
Veitch E. DECAMP 1991D, DECAMP 1991N, DECAMP 1991S, BUSKULIC 1993H
Vejcik S.III. ABE 1994ZE, ABE 1995
Veksler V.I. VEKSLER 1944, KANG-CHANG 1960
Velasco J. ABREU 1991B, DECAMP 1991S
Velev G. ABE 1995
Veltman H. WEIR 1989B
Veltman M. 'T HOOFT 1972, 'T HOOFT 1972B
Veneziano S. ALBAJAR 1991C, ALBAJAR 1991D
Ventura L. ABREU 1991B, DECAMP 1991S
Venturi A. DECAMP 1991D, DECAMP 1991N, DECAMP 1991S, BUSKULIC 1993H
Venus W. HASERT 1973, HASERT 1973B, CALICCHIO 1980, ABREU 1991B, DECAMP 1991S
Venuti J.P. BAND 1988
Verbeure F. ABREU 1991B, DECAMP 1991S
Vercesi V. BANNER 1983B, BAGNAIA 1983B, ANSARI 1987E
Verderi M. BUSKULIC 1993H
Verdier R. MILLER 1972B
Verdini P.G. BAND 1988, DECAMP 1991D, DECAMP 1991N, DECAMP 1991S, BUSKULIC 1993H
Verechia P. ARNISON 1985D
Vernon W. FITCH 1965B
Vertogradov L.S. ABREU 1991B, DECAMP 1991S
Vetlitsky I.A. ADEVA 1990R, ADEVA 1991D, DECAMP 1991S
Vialle J.P. HASERT 1973, HASERT 1973B, BLIETSCHAU 1976, BLIETSCHAU 1977B, ARNISON 1983C, ARNISON 1983D, ARNISON 1984C, ARNISON 1984G, ARNISON 1985D, ALBAJAR 1987C, ALBAJAR 1991C, ALBAJAR 1991D
Vibert L. ABREU 1991B, DECAMP 1991S
Vidal R. ABE 1994ZE, ABE 1995
Videau H. DECAMP 1991D, DECAMP 1991N, DECAMP 1991S, BUSKULIC 1993H
Videau I. DECAMP 1991D, DECAMP 1991N, DECAMP 1991S, BUSKULIC 1993H
Viertel G. ADEVA 1990R, ADEVA 1991D, DECAMP 1991S
Vikas P. ADEVA 1990R, ADEVA 1991D, DECAMP 1991S
Vikas U. ADEVA 1990R, ADEVA 1991D, DECAMP 1991S
Vilain P. HASERT 1973, HASERT 1973B
Vilanova D. ABREU 1991B, DECAMP 1991S
Vilchinska B. VOYVODIC 1986B
Vilchinski G. VOYVODIC 1986B
Villard P. VILLARD 1900, VILLARD 1900B
Villari A. FIDECARO 1978
Villasenor L. ALBAJAR 1987C
Vincelli M.L. APEL 1975B
Virador P.R.G. ABACHI 1995G
Virdee T.S. ARNISON 1985D, ALBAJAR 1987C, ALBAJAR 1991C, ALBAJAR 1991D
Virtue C.J. ALEXANDER 1991C, ACTON 1992B, DECAMP 1991S
Viryasov N.M. KANG-CHANG 1960
Vishnevsky N.K. BINON 1969, ANTIPOV 1970
Vismara G. ARNISON 1983C
Vitale S. BACCI 1974
Vititoe D. ABACHI 1995G
Vivargent M. ADEVA 1990R, ADEVA 1991D, DECAMP 1991S
Vlasov E.V. ABREU 1991B, DECAMP 1991S
Voci C. FIDECARO 1978
Vodopianov A.S. ABREU 1991B, DECAMP 1991S
Vogel A. BALTAY 1979E
Vogel H. BIENLEIN 1978, ADEVA 1990R, ADEVA 1991D, DECAMP 1991S
Vogl R. BUSKULIC 1993H
Vogt H. ADEVA 1990R, ADEVA 1991D, DECAMP 1991S
Vojsek B. VOYVODIC 1986B
Volkov A. ABACHI 1995G
Vollmar M. ADEVA 1990R
Vollmer M. ABREU 1991B, DECAMP 1991S
Volponi S. ABREU 1991B, DECAMP 1991S
Volter V. VOYVODIC 1986B
Von Dardel G. ADEVA 1990R, ADEVA 1991D, DECAMP 1991S
Von der Schmitt H. ARNISON 1985D, ALBAJAR 1987C, ALEXANDER 1991C, ACTON 1992B, DECAMP 1991S
Von Dratzig A.S. BRANDELIK 1979H, BRANDELIK 1980G
Von Gagern C. BRANDELIK 1977D
Von Goeler E. FRAUENFELDER 1957, BAND 1988, ABACHI 1995G
Von Harrach D. ASHMAN 1988B
Von Holtey G. AUBERT 1983
Von Krogh J. HASERT 1973B, BARTEL 1980D, ALEXANDER 1991C, ACTON 1992B, DECAMP 1991S
Von Ruden W. DECAMP 1991D
Von Schlippe W. ARNISON 1985D, ALBAJAR 1987C, ALBAJAR 1991C, ALBAJAR 1991D
Von Zanthier C. ABRAMS 1989B
Vondracek M. ABE 1994ZE, ABE 1995
Vonholtey G. ARNISON 1983C
Vorobev A.A. ADEVA 1990R, ADEVA 1991D, DECAMP 1991S
Vorobev A.P. ABACHI 1995G
Vorobev An.A. ADEVA 1990R, ADEVA 1991D, DECAMP 1991S
Vorobev I. ADEVA 1990R, ADEVA 1991D, DECAMP 1991S
Vorobiev A.I. ARTAMONOV 1984B
Voruganti P. ABRAMS 1989B, WEIR 1989B
Votruba M.F. CALICCHIO 1980
Voulgaris G. ABREU 1991B, DECAMP 1991S
Voutilainen M. ABREU 1991B, DECAMP 1991S
Voyvodic L. VOYVODIC 1986B
Vrana I. KANG-CHANG 1960
Vrana J. ARNISON 1983C, ARNISON 1983D, ARNISON 1984C, ARNISON 1984G, ARNISON 1985D, ALBAJAR 1987C, ALBAJAR 1991C, ALBAJAR 1991D
Vrba V. ABREU 1991B, DECAMP 1991S
Vucinic D. ABE 1995
Vuillemin V. ARNISON 1983C, ARNISON 1983D, ARNISON 1984C, ARNISON 1984G, ARNISON 1985D, ALBAJAR 1987C, ALBAJAR 1991C, ALBAJAR 1991D
Vuilleumier L. ADEVA 1990R, ADEVA 1991D, DECAMP 1991S
Wachsmuth H.W. HASERT 1973, HASERT 1973B, BLIETSCHAU 1976, BLIETSCHAU 1977B, DECAMP 1991D, DECAMP 1991N, DECAMP 1991S, BUSKULIC 1993H
Wacker K. BERGER 1978, BERGER 1979C, ALBAJAR 1987C, ALBAJAR 1991C, ALBAJAR 1991D
Wadhwa M. ADEVA 1990R, ADEVA 1991D, DECAMP 1991S
Wagman G.S. BARNETT 1996
Wagner A. BARTEL 1980D, ALEXANDER 1991C, ACTON 1992B, DECAMP 1991S
Wagner H. ALBAJAR 1991C, ALBAJAR 1991D
Wagner R.G. ABE 1994ZE, ABE 1995
Wagner R.L. ABE 1994ZE, ABE 1995
Wagner S.R. ABRAMS 1989B, WEIR 1989B
Wagner W. BERGER 1978, BERGER 1979C, BERGER 1980L

Wahl C. ALEXANDER 1991C, ACTON 1992B, DECAMP 1991S
Wahl H.D. ARNISON 1983C, ARNISON 1983D, ARNISON 1984C, ARNISON 1984G, ARNISON 1985D, ALBAJAR 1987C, ABACHI 1995G
Wahl J. ABE 1995
Wahlen H. AUBERT 1983, ABREU 1991B, DECAMP 1991S
Wainer N. ABE 1994ZE
Walcher T. ASHMAN 1988B
Walck C. ABREU 1991B, DECAMP 1991S
Waldi R. ALBRECHT 1987P
Waldner F. CRETIEN 1962, ABREU 1991B, DECAMP 1991S
Waldren D. BLIETSCHAU 1977B
Walker J.P. ALEXANDER 1991C, ACTON 1992B, DECAMP 1991S
Walker R.C. ABE 1994ZE, ABE 1995
Walker W.D. WALKER 1955D, ERWIN 1961C
Wallace R. BRABANT 1956C
Wallraff W. BRANDELIK 1977D, BRANDELIK 1979H, BRANDELIK 1980G, ADEVA 1990R, ADEVA 1991D, DECAMP 1991S
Waloschek P. BERGER 1978, BERGER 1979C, BERGER 1980L
Walsh A.M. BUSKULIC 1993H
Walsh J. DECAMP 1991D, DECAMP 1991N, DECAMP 1991S, BUSKULIC 1993H
Walsh M.A. DECAMP 1991D, DECAMP 1991N, DECAMP 1991S
Walther S.M. DECAMP 1991N, DECAMP 1991S, BUSKULIC 1993H
Walton E.T.S. COCKCROFT 1932
Wang C. ABE 1995
Wang C.H. ABE 1995
Wang C.R. ADEVA 1990R, ADEVA 1991D, DECAMP 1991S
Wang D.C. ABACHI 1995G
Wang G. ABE 1994ZE, ABE 1995
Wang G.H. ADEVA 1990R, ADEVA 1991D, DECAMP 1991S
Wang J. ABE 1994ZE, ABE 1995
Wang J.H. ADEVA 1990R, ADEVA 1991D, DECAMP 1991S
Wang L.Z. ABACHI 1995G
Wang M.J. ABE 1994ZE, ABE 1995
Wang Q.F. ADEVA 1990R, ADEVA 1991D, DECAMP 1991S, ABE 1994ZE, ABE 1995
Wang T. DECAMP 1991D, DECAMP 1991N, DECAMP 1991S, BUSKULIC 1993H
Wang X.L. ADEVA 1990R, ADEVA 1991D, DECAMP 1991S
Wang X.R. BARBER 1979F
Wang Y.F. ADEVA 1990R, ADEVA 1991D, DECAMP 1991S
Wang Z. ADEVA 1990R, ADEVA 1991D, DECAMP 1991S
Wang Z.M. ADEVA 1990R, ADEVA 1991D, DECAMP 1991S
Wanke R. BUSKULIC 1993H
Wanlass S.D. LEIGHTON 1951
Warburton A. ABE 1994ZE, ABE 1995
Warchol J. ABACHI 1995G
Ward C.P. ALEXANDER 1991C, ACTON 1992B, DECAMP 1991S
Ward D.R. ALEXANDER 1991C, ACTON 1992B, DECAMP 1991S
Ward J.C. WARD 1950, SALAM 1961, SALAM 1964
Warming P. BARTEL 1980D
Warren G. ANDREWS 1980, ANDREWS 1980B, BEBEK 1981
Wasserbaech S. DECAMP 1991D, DECAMP 1991N, DECAMP 1991S, BUSKULIC 1993H
Watanabe Y. BARTEL 1980D
Watkins P.M. ARNISON 1983D, ARNISON 1984C, ARNISON 1984G, ARNISON 1985D, ALBAJAR 1987C, ALEXANDER 1991C, ACTON 1992B, DECAMP 1991S
Watson A.T. ALEXANDER 1991C, ACTON 1992B, DECAMP 1991S
Watson K.M. BRUECKNER 1951
Watson M.B. TRIPP 1962
Watson N.K. ALEXANDER 1991C, ACTON 1992B, DECAMP 1991S
Watson S. ABRAMS 1989B
Wattenberg A. KNAPP 1976B
Watts G. ABE 1994ZE, ABE 1995
Watts T. ABE 1994ZE, ABE 1995
Wayne M. ABREU 1991B, DECAMP 1991S, ABACHI 1995G
Wear J.A. DECAMP 1991D, DECAMP 1991N, DECAMP 1991S, BUSKULIC 1993H
Webb R. ABE 1994ZE, ABE 1995
Weber A. ADEVA 1991D
Weber D. ANDREWS 1980, ANDREWS 1980B, BEBEK 1981, CHADWICK 1981
Weber F.V. DECAMP 1991D, DECAMP 1991N, DECAMP 1991S, BUSKULIC 1993H
Weber G. BARTEL 1980D
Weber J. ADEVA 1990R, ADEVA 1991D, DECAMP 1991S
Weber M. ALEXANDER 1991C, ACTON 1992B, DECAMP 1991S
Weber P. ABRAMS 1989B, WEIR 1989B, ACTON 1992B
Webster M.S. BROWN 1962, BARNES 1964B, BARNES 1964E
Wedemeyer R. BRANDELIK 1979H, BRANDELIK 1980G
Weerts H. HASERT 1973, HASERT 1973B, BLIETSCHAU 1976, BLIETSCHAU 1977B, ABACHI 1995G
Wegener D. DARDEN 1978, DARDEN 1978B, ALBRECHT 1987P
Wei C. ABE 1995
Wei P.S. BARBER 1979F
Weidberg A.R. BANNER 1983B, BAGNAIA 1983B, ANSARI 1987E
Weigend A. ABRAMS 1989B
Weilhammer P. BLUM 1975, ABREU 1991B, DECAMP 1991S
Weill R. BIAGI 1983C, BIAGI 1985C, ADEVA 1990R, ADEVA 1991D, DECAMP 1991S
Weinberg A. CRETIEN 1962
Weinberg S. GOLDSTONE 1961B, WEINBERG 1967, WEINBERG 1973
Weinrich M. GARWIN 1957
Weinstein A.J. ABRAMS 1989B, WEIR 1989B
Weinstein R. BAND 1988
Weir A.J. ABRAMS 1989B, WEIR 1989B
Weisz S. WEIR 1989B, ALEXANDER 1991C, ACTON 1992B, DECAMP 1991S
Welch G. BERGER 1979C, BERGER 1980L
Welch L. BLIETSCHAU 1976, BLIETSCHAU 1977B
Wells P.S. ALEXANDER 1991C, ACTON 1992B, DECAMP 1991S
Wenaus T.J. ADEVA 1990R, ADEVA 1991D, DECAMP 1991S
Wendt C. ABE 1994ZE, ABE 1995
Wenninger H. CALICCHIO 1980
Wenninger J. ADEVA 1990R, ADEVA 1991D, DECAMP 1991S
Wenzel H. ABE 1994ZE, ABE 1995
Wenzel W.A. BRABANT 1956C, CORK 1956, ABACHI 1995G
Wermes N. BRANDELIK 1979H, BRANDELIK 1980G, ALEXANDER 1991C, ACTON 1992B, DECAMP 1991S
Werner J. ABREU 1991B, DECAMP 1991S
Weseler S. ALBRECHT 1987P
West E. ERWIN 1961C
West L.R. DECAMP 1991D, DECAMP 1991N, DECAMP 1991S, BUSKULIC 1993H
Wester W.C.III. ABE 1994ZE, ABE 1995

Westgard J. BERTANZA 1962D
Westhusing T. ABE 1994ZE
Wetherell A.M. BUSHNIN 1969, AMALDI 1973E, ABREU 1991B, DECAMP 1991S
Wetjen G. BERGER 1978
Weymann M. ALEXANDER 1991C, ACTON 1992B, DECAMP 1991S
Whalley M. AUBERT 1983
Whalley M.A. ALEXANDER 1991C, ACTON 1992B, DECAMP 1991S
Whatley M.C. ABRAMS 1964
Wheeler J.A. WHEELER 1937
Wheeler S. ASHMAN 1988B, DECAMP 1991D, DECAMP 1991N, DECAMP 1991S
Whelan E.P. DECAMP 1991D, DECAMP 1991N, DECAMP 1991S, BUSKULIC 1993H
Whitaker J.S. AUGUSTIN 1974B, ABRAMS 1974, SCHWITTERS 1975, PERL 1975C, HANSON 1975, PERL 1976C, GOLDHABER 1976D, PERUZZI 1976C
White A. ABACHI 1995G
White H. AGNEW 1958
White H.S. GOOD 1961
White J.T. ABACHI 1995G
White M. BARBER 1979F, ADEVA 1990R, ADEVA 1991D, DECAMP 1991S
White S.L. WEIR 1989B
Whitehead M.N. BIRGE 1955B
Whitman R. ANDREWS 1980
Whitmore J. AVERY 1989C, ARTUSO 1989C, ALAM 1989B
Whitney M.H. DECAMP 1991D, DECAMP 1991N, DECAMP 1991S
Whittemore W.L. FOWLER 1953C, FOWLER 1953B, FOWLER 1954C
Wickens J.H. ABREU 1991B, DECAMP 1991S
Wicklund A.B. ABE 1994ZE, ABE 1995
Wicklund E. BRANDELIK 1980G, ABRAMS 1989B, WEIR 1989B, ABE 1995
Wiedenmann W. DECAMP 1991D, DECAMP 1991N, DECAMP 1991S, BUSKULIC 1993H
Wiegand C. CHAMBERLAIN 1955, CHAMBERLAIN 1956E, CHAMBERLAIN 1956F, AGNEW 1958
Wightman A.S. WIGHTMAN 1956
Wightman J.A. ABACHI 1995G
Wigner E.P. WIGNER 1927B, WIGNER 1932, BREIT 1936, WIGNER 1937
Wiik B.H. BRANDELIK 1977D, BRANDELIK 1979H, BRANDELIK 1980G
Wijangco A. KNAPP 1976B
Wikberg T. CARRON 1978
Wikne J. ABREU 1991B, DECAMP 1991S
Wilcox J. ABACHI 1995G
Wilczek F. GROSS 1973, GROSS 1973B
Wildish A. ALBAJAR 1987C
Wildish T. DECAMP 1991D, DECAMP 1991N, DECAMP 1991S, BUSKULIC 1993H
Wilhelm R. ADEVA 1990R
Wilkes R.J. VOYVODIC 1986B
Wilkinson G.R. ABREU 1991B, DECAMP 1991S
Wilkinson R. ABE 1994ZE, ABE 1995
Williams D. AUBERT 1983
Williams E.J. WILLIAMS 1940
Williams H.H. CAZZOLI 1975E, KIM 1981, ABE 1994ZE, ABE 1995
Williams W.S.C. AUBERT 1983, ASHMAN 1988B, ABREU 1991B, DECAMP 1991S
Willis S. ABACHI 1995G
Willis W.J. BARNES 1964B, BARNES 1964E
Willmott C. ADEVA 1990R, ADEVA 1991D, DECAMP 1991S
Wilquet G. CALICCHIO 1980
Wilson C.T.R. WILSON 1912, WILSON 1923
Wilson G.W. ALEXANDER 1991C, ACTON 1992B, DECAMP 1991S
Wilson J. ARNISON 1983C, ARNISON 1983D, ARNISON 1984C, ARNISON 1984G, ARNISON 1985D, ALBAJAR 1987C
Wilson J.A. ALEXANDER 1991C, ACTON 1992B, DECAMP 1991S
Wilson P. ABE 1994ZE, ABE 1995
Wilson R. CLARK 1951, ANDREWS 1980, ANDREWS 1980B, BEBEK 1981, CHADWICK 1981, AVERY 1989C, ARTUSO 1989C, ALAM 1989B
Wimpenny S.J. AUBERT 1983, ARNISON 1985D, ALBAJAR 1987C, ASHMAN 1988B, ABACHI 1995G
Windmolders R. ASHMAN 1988B
Winer B.L. ABE 1994ZE, ABE 1995
Wingerter I. ARNISON 1984G, ARNISON 1985D, ALBAJAR 1987C, ALEXANDER 1991C, ACTON 1992B, DECAMP 1991S
Winkelmann F.C. SCHWITTERS 1975, PERL 1975C, HANSON 1975
Winter G.G. BERGER 1978, BERGER 1979C, BERGER 1980L
Winter M. ABREU 1991B, DECAMP 1991S
Winterer V.H. ALEXANDER 1991C, ACTON 1992B, DECAMP 1991S
Wirjawan J.V.D. ABACHI 1995G
Wiser D.E. BAND 1988
Wiss J.E. AUGUSTIN 1974B, ABRAMS 1974, SCHWITTERS 1975, PERL 1975C, HANSON 1975, PERL 1976C, GOLDHABER 1976D, PERUZZI 1976C, PERL 1977F
Wittgenstein F. ADEVA 1990R, ADEVA 1991D, DECAMP 1991S
Witzeling W. DECAMP 1991D, DECAMP 1991N, DECAMP 1991S, BUSKULIC 1993H
Wohl C.G. BARNETT 1996
Wojcicki S.G. ALVAREZ 1959, ALSTON 1960, ALSTON 1961B, ALSTON 1961E, DORFAN 1967
Wolf B. DECAMP 1991N, DECAMP 1991S, BUSKULIC 1993H
Wolf G. BRANDELIK 1977D, BRANDELIK 1979H, BRANDELIK 1980G, DECAMP 1991D, DECAMP 1991N, DECAMP 1991S, BUSKULIC 1993H
Wolf Z. ABACHI 1995G
Wolfenstein L. WOLFENSTEIN 1964
Wolff S. BERGER 1978
Wolinski D. ABE 1995
Wolinski J. AVERY 1989C, ARTUSO 1989C, ALAM 1989B, ABE 1994ZE, ABE 1995
Wollstadt M. BRANDELIK 1979H, BRANDELIK 1980G
Wolsky G. ABRAMS 1964
Womersley J. ABACHI 1995G
Womersley W.J. ASHMAN 1988B
Won E. ABACHI 1995G
Wood D.R. WEIR 1989B, ABACHI 1995G
Wood N.C. ALEXANDER 1991C, ACTON 1992B, DECAMP 1991S
Woods M. ABRAMS 1989B
Woodworth P.L. BRANDELIK 1979H, BRANDELIK 1980G
Wooster W.A. ELLIS 1927, ELLIS 1927B
Worden H. AVERY 1989C, ARTUSO 1989C, ALAM 1989B
Workman R.L. BARNETT 1996
Wormald D. ABREU 1991B, DECAMP 1991S
Wormser G. ABREU 1991B, DECAMP 1991S
Worris M. AVERY 1989C, ARTUSO 1989C, ALAM 1989B
Woschnagg K. ABREU 1991B, DECAMP 1991S
Wotschack J. DECAMP 1991D, DECAMP 1991N, DECAMP 1991S, BUSKULIC 1993H
Wotton S. ALEXANDER 1991C, ACTON 1992B, DECAMP 1991S
Wriedt H. BARTEL 1980D
Wright A.G. DECAMP 1991D, DECAMP 1991N, DECAMP 1991S, BUSKULIC 1993H

Wright D. ADEVA 1990R, ADEVA 1991D, DECAMP 1991S
Wu C.S. WU 1957
Wu D.Y. ABRAMS 1989B, WEIR 1989B, ABE 1994ZE
Wu G.H. BARBER 1979F
Wu R.J. ADEVA 1990R, ADEVA 1991D, DECAMP 1991S
Wu S.L. AUBERT 1974D, BRANDELIK 1979H, BRANDELIK 1980G, ADEVA 1990R, DECAMP 1991D, ADEVA 1991D, DECAMP 1991N, DECAMP 1991S, DECAMP 1991S, BUSKULIC 1993H
Wu S.X. ADEVA 1990R, ADEVA 1991D, DECAMP 1991S
Wu T.W. BARBER 1979F
Wu W. DECAMP 1991N, DECAMP 1991S
Wu X. ALBAJAR 1987C, ALBAJAR 1991C, ALBAJAR 1991D, BUSKULIC 1993H, ABE 1994ZE, ABE 1995
Wu Y.G. ADEVA 1990R, ADEVA 1991D, DECAMP 1991S
Wuest C. BIONTA 1987C
Wulz C.E. ARNISON 1983D, ARNISON 1984C, ARNISON 1984G, ARNISON 1985D, ALBAJAR 1987C, ALBAJAR 1991C, ALBAJAR 1991D
Wunsch M. ANSARI 1987E, DECAMP 1991D, DECAMP 1991N, DECAMP 1991S, BUSKULIC 1993H
Wurth R. DARDEN 1978, DARDEN 1978B, ALBRECHT 1987P
Wyatt T.R. BRANDELIK 1980G, ARNISON 1985D, ALBAJAR 1987C, ALEXANDER 1991C, ACTON 1992B, DECAMP 1991S
Wyslouch B. ADEVA 1990R, ADEVA 1991D, DECAMP 1991S
Wyss J. ABE 1994ZE, ABE 1995
Xi J.P. BARBER 1979F
Xiao D. AVERY 1989C, ARTUSO 1989C, ALAM 1989B
Xie J. DECAMP 1991D, DECAMP 1991N, DECAMP 1991S
Xie Y. BUSKULIC 1993H
Xie Y.G. ARNISON 1983D, ARNISON 1984C
Xie Y.Y. ADEVA 1991D, DECAMP 1991S
Xu D. DECAMP 1991D, DECAMP 1991N, DECAMP 1991S, BUSKULIC 1993H
Xu R. DECAMP 1991D, DECAMP 1991N, DECAMP 1991S, BUSKULIC 1993H
Xu Y.D. ADEVA 1990R, ADEVA 1991D, DECAMP 1991S
Xu Z.Z. ADEVA 1990R, ADEVA 1991D, DECAMP 1991S
Xue Z.L. ADEVA 1990R, ADEVA 1991D, DECAMP 1991S
Xuong N.H. BUTTON 1961, XUONG 1961C
Yaari R. ALEXANDER 1991C, ACTON 1992B, DECAMP 1991S
Yagil A. ALBRECHT 1987P, ABE 1994ZE, ABE 1995
Yamada M. HIRATA 1987C
Yamada R. ABACHI 1995G
Yamada S. BRANDELIK 1977D, BARTEL 1980D
Yamamoto R.K. CRETIEN 1962
Yamamoto S.S. BERTANZA 1962D, CONNOLLY 1963, BARNES 1964B, BARNES 1964E, GOLDBERG 1964C
Yamanouchi H. MINAKAWA 1959
Yamanouchi T. HERB 1977, INNES 1977, UENO 1979
Yamdagni N. ABREU 1991B, DECAMP 1991S
Yamin P. BUNCE 1976, ABACHI 1995G
Yan D.S. ADEVA 1990R, ADEVA 1991D, DECAMP 1991S
Yan X.J. ADEVA 1991D, DECAMP 1991S
Yanagisawa C. BARTEL 1980D, ABACHI 1995G
Yang B.Z. ADEVA 1990R, ADEVA 1991D, DECAMP 1991S
Yang C.G. ADEVA 1990R, ADEVA 1991D, DECAMP 1991S
Yang C.N. LEE 1949, FERMI 1949, YANG 1954, LEE 1956, LEE 1956B, LEE 1957B, LEE 1957, LEE 1957C
Yang G. ADEVA 1990R, ADEVA 1991D, DECAMP 1991S
Yang J. ABACHI 1995G
Yang K.S. ADEVA 1990R, ADEVA 1991D, DECAMP 1991S
Yang P.C. BARBER 1979F
Yang Q.Y. ADEVA 1990R, ADEVA 1991D, DECAMP 1991S
Yang Y. ALEXANDER 1991C, ACTON 1992B, DECAMP 1991S
Yang Z.Q. ADEVA 1990R, ADEVA 1991D, DECAMP 1991S
Yao W. ABE 1994ZE, ABE 1995
Yao W.M. AVERY 1989C, ARTUSO 1989C, ALAM 1989B
Yarba V.A. VOYVODIC 1986B
Yarker S. BRANDELIK 1979H, BRANDELIK 1980G
Yasuda T. ABACHI 1995G
Yasuoka K. ABE 1994ZE, ABE 1995
Yatsuta M.I. BALDIN 1973
Ye C.H. ADEVA 1990R, ADEVA 1991D, DECAMP 1991S
Ye J.B. ADEVA 1990R, ADEVA 1991D, DECAMP 1991S
Ye Q. ADEVA 1990R, ADEVA 1991D, DECAMP 1991S
Ye Y. ABE 1994ZE, ABE 1995
Yeh G.P. ABE 1994ZE, ABE 1995
Yeh P. ABE 1994ZE, ABE 1995
Yeh S.C. ADEVA 1990R, ADEVA 1991D, DECAMP 1991S
Yekutieli G. ALEXANDER 1991C, ACTON 1992B, DECAMP 1991S
Yelton J.M. AVERY 1989C, ARTUSO 1989C, ALAM 1989B, WEIR 1989B
Yen W.L. BARTEL 1980D
Yennie D.R. YENNIE 1957
Yepes P. ABREU 1991B, DECAMP 1991S
Yin M. ABE 1994ZE, ABE 1995
Yin Z.W. ADEVA 1990R, ADEVA 1991D, DECAMP 1991S
Yodh G.B. ABRAMS 1964
Yoh J.K. BUESSER 1973, HERB 1977, INNES 1977, UENO 1979, BOHRINGER 1980B, FINOCCHIARO 1980, ABE 1994ZE, ABE 1995
Yoon T.S. ALBRECHT 1987P
Yordanov V. KIRILLOVA 1965
York C.M. ARMENTEROS 1952, YORK 1953
York H.F. BJORKLUND 1950
Yosef C. ABE 1995
Yoshida T. ABE 1994ZE, ABE 1995
Yoshikawa C. ABACHI 1995G
Yotch F.A. ANTIPOV 1970
You J.M. ADEVA 1990R, ADEVA 1991D, DECAMP 1991S
Young C.C. PRESCOTT 1979C
Youngman C. BRANDELIK 1979H, BRANDELIK 1980G
Yount D.E. KNAPP 1976B
Youssef S. ABACHI 1995G
Yovanovitch D.D. ABE 1994ZE, ABE 1995
Ypsilantis T. CHAMBERLAIN 1955, AGNEW 1958
Yu I. ABE 1994ZE, ABE 1995
Yu J. ABACHI 1995G
Yu X.H. BARBER 1979F
Yu Y. ABACHI 1995G
Yuan L.C.L. LINDENBAUM 1953
Yukawa H. YUKAWA 1935
Yun J.C. ALBRECHT 1987P, ABE 1994ZE, ABE 1995
Yurak A. VOYVODIC 1986B
Yurko M. ABRAMS 1989B, ACTON 1992B
Yurkov M.V. ARTAMONOV 1984B
Yvert M. ARNISON 1983C, ARNISON 1983D, ARNISON 1984C, ARNISON 1984G, ARNISON 1985D, ALBAJAR 1987C, ALBAJAR 1991C, ALBAJAR 1991D

Yzerman M. ADEVA 1990R, ADEVA 1991D, DECAMP 1991S
Zaccardelli C. ALBAJAR 1987C, ABRAMS 1989B, ADEVA 1990R, ADEVA 1991D, DECAMP 1991S
Zaccone H. BANNER 1983B, BAGNAIA 1983B, ANSARI 1987E
Zachariadou K. BUSKULIC 1993H
Zacharias J.R. KELLOGG 1939
Zachariazen F. FRAUTSCHI 1962
Zacharov I. ALBAJAR 1987C, ALEXANDER 1991C, ALBAJAR 1991C, ALBAJAR 1991D, ACTON 1992B, DECAMP 1991S
Zaganidis N. ARNISON 1985D, ALBAJAR 1987C
Zaitsev A.M. ANTIPOV 1970, ABREU 1991B, DECAMP 1991S
Zaitsev Y.M. ALBRECHT 1987P
Zakharov V.I. SHIFMAN 1979
Zakrzewski J.A. DANYSZ 1963, BAGNAIA 1983B, ANSARI 1987E
Zalewska A. ABREU 1991B, DECAMP 1991S
Zalewski P. ABREU 1991B, DECAMP 1991S
Zanello L. ARNISON 1985D, ALBAJAR 1987C, ALBAJAR 1991C, ALBAJAR 1991D
Zanetti A. ABE 1994ZE, ABE 1995
Zavattini E. CHRISTENSON 1970, BUESSER 1973
Zayachki V.I. BEZNOGIKH 1969
Zdarko R.W. BAND 1988
Zech G. BERGER 1978, BERGER 1979C, BERGER 1980L
Zehnder L. ADEVA 1990R, ADEVA 1991D, DECAMP 1991S
Zeldovich Y.B. GERSHTEIN 1955, ZELDOVICH 1959, GERSHTEIN 1966
Zeller W. BANNER 1983B, BAGNAIA 1983B
Zemp P. ADEVA 1991D, DECAMP 1991S
Zeng M. ADEVA 1990R, ADEVA 1991D, DECAMP 1991S
Zeng Y. ADEVA 1990R, ADEVA 1991D, DECAMP 1991S
Zetti F. ABE 1994ZE, ABE 1995
Zeuner W. ALEXANDER 1991C, ACTON 1992B, DECAMP 1991S
Zevgolatakos E. ABREU 1991B, DECAMP 1991S
Zhang D. ADEVA 1990R
Zhang D.H. ADEVA 1990R, ADEVA 1991D, DECAMP 1991S
Zhang G. ABREU 1991B, DECAMP 1991S
Zhang J. DECAMP 1991D, DECAMP 1991N, DECAMP 1991S, BUSKULIC 1993H
Zhang L. BUSKULIC 1993H, ABE 1995
Zhang N.L. BARBER 1979F
Zhang S. ABE 1994ZE
Zhang W. HIRATA 1987C, ABE 1994ZE, ABE 1995
Zhang Y. ABACHI 1995G
Zhang Z. DECAMP 1991D, DECAMP 1991N, DECAMP 1991S, BUSKULIC 1993H
Zhang Z.P. ADEVA 1990R, ADEVA 1991D, DECAMP 1991S
Zhao W. DECAMP 1991D, DECAMP 1991N, DECAMP 1991S, BUSKULIC 1993H
Zheng M. BUSKULIC 1993H
Zhidkov N.K. BEZNOGIKH 1969
Zhilchenkova G.D. ANTIPOV 1970
Zholentz A.A. ARTAMONOV 1984B
Zhou J.F. ADEVA 1990R, ADEVA 1991D, DECAMP 1991S
Zhou Y.H. ABACHI 1995G
Zhou Z.L. DECAMP 1991D
Zhu Q. ABACHI 1995G
Zhu R.Y. BARBER 1979F, ADEVA 1990R, ADEVA 1991D, DECAMP 1991S
Zhu Y.S. ABACHI 1995G
Zhu Z.H. ABACHI 1995G
Zhuang H.L. ADEVA 1990R, ADEVA 1991D, DECAMP 1991S
Zhuravleva L.I. ASTVACATUROV 1968
Zichichi A. BASILE 1981I, ADEVA 1990R, ADEVA 1991D, DECAMP 1991S
Zieminska D. ABACHI 1995G
Zieminski A. ABACHI 1995G
Ziemons K. ASHMAN 1988B
Zimin N.I. ABREU 1991B, DECAMP 1991S
Zimmermann W. LEHMANN 1955, BERGER 1978, BERGER 1979C, BERGER 1980L
Zinchenko A.I. ABACHI 1995G
Zipse J.E. AUGUSTIN 1974B, ABRAMS 1974
Zito G. DECAMP 1991D, DECAMP 1991N, DECAMP 1991S, BUSKULIC 1993H
Zito M. ABREU 1991B, DECAMP 1991S
Zitoun R. ABREU 1991B, DECAMP 1991S
Zlatanov Z. KIRILLOVA 1965
Zlateva A. KIRILLOVA 1965
Zobernig G. BRANDELIK 1979H, BRANDELIK 1980G, DECAMP 1991D, DECAMP 1991N, DECAMP 1991S, BUSKULIC 1993H
Zografou P. DECAMP 1991D, DECAMP 1991N, DECAMP 1991S
Zolin L.S. BEZNOGIKH 1969
Zolnierowski Y. ALBAJAR 1991C, ALBAJAR 1991D
Zolotorev M.S. BARKOV 1978, BARKOV 1979C
Zomer F. DECAMP 1991D, DECAMP 1991N, DECAMP 1991S
Zorn G.T. BERGER 1979C, BERGER 1980L, ALEXANDER 1991C, ACTON 1992B, DECAMP 1991S
Zotto P.L. ARNISON 1985D, ALBAJAR 1987C, ALBAJAR 1991C, ALBAJAR 1991D
Zubarev V.N. BALDIN 1973
Zucchelli S. ABE 1994ZE, ABE 1995
Zukanovich Funchal R. ABREU 1991B, DECAMP 1991S
Zumerle G. ABREU 1991B, DECAMP 1991S
Zuniga J. ABREU 1991B, DECAMP 1991S
Zupancic C. CHARPAK 1968
Zurfluh E. ARNISON 1983C, ARNISON 1983D, ARNISON 1984C
Zwanziger D. GREENBERG 1966
Zweig G. ZWEIG 1964, ZWEIG 1964B

V. Subject Index

accelerated protons, first evidence, 63
accelerator
  AGS synchrotron, 184
  Berkeley 184 inch synchrocyclotron, 92
  betatron, 84, 85
  betatron, maximal energy, 89
  Cockcroft–Walton, 63
  colliding beams, 144
  cyclotron, 60, 66
  cyclotron, medical application, 80
  electron cooling, 207, 230
  electrostatic, invention, 61, 67
  first pion production, 103
  linear, Stanford, 141
  neutrino beams, 170, 172
  phase stability, 88, 89
  phase stability principle, 92
  polarized particle, 237
  radiation self-polarization, 178, 192
  radiation self-polarization of beam, 230
  Sibirian snakes, 237
  stochastic cooling, 235
  storage rings, 207, 237
  strong focusing, 122, 184
  synchrotron, 90
  synchrotron, invention, 89
  Van de Graaff, 61, 67
aces (quarks), introduction, 194
adiabatic principle, quantum mechanics, 51
α decay, quantum theory, 55
α particles
  collision, 38, 39
  scattering, 32
α radioactivity
  evidence, 24
α rays
  doubly positive charged, 26
  scattering, 34
annihilation
  $e^+ e^-$, 239
  $e^+ e^-$ quark jets, 226
  $e^+ e^-$ three-jet, 240–242
  antiproton, first evidence, 143
  antiproton, first indication, 143
  antiproton, pion-pion correlations, 168
  antiproton-nucleon, 145, 152
anomalous magnetic moment
  electron, calculation, 99
  electron, first measurement, 95, 98
  neutron, prediction, 71
antideuterons, first evidence, 202
antideuteron production, confirmation, 213
antimatter production, 213
antineutrino, neutrino comparison, 133
antineutron
  confirmation, 164
  first evidence, 148
antiproton
  annihilation, first evidence, 143
  annihilation, first indication, 143
  annihilation, pion-pion correlations, 168
  collecting, 235
  confirmation, 142
  evidence, 138
  prediction, 60
antiproton-nucleon annihilation, 145, 152
anti-particle, discovery, 68
Argonne spin-effect, 234
associative production
  V particles, evidence, 125, 130
  V particles, explanation, 126
  V^0 particles, hypothesis, 121
  strange particles, 134
  strange particles, confirmation, 139
asymptotic freedom, 220, 221
atomic energy, 86
atomic model
  Bohr, 34, 35, 37
  Rutherford, 32, 36
atomic nucleus
  confirmation, 36
  evaluation of charge, 32, 36
  evidence, 32
  hypothesis, 32
  structure, neutron, 64
  structure, protons and neutrons, suggestion, 65
atomic number, nuclear charge, relation, 38
atomic spectrum, 48, 50
  quantum theory, 34, 35, 37
  quantum theory, invention, 35
atomic structure, 42
  Bohr's atom, 34, 35
  Rutherford's atom, 32, 36
Avogadro constant
  precise measurement, 33
axiomatic field theory
  S-matrix, beginnings, 140
  Wightman approach, 143

baryonic quantum number, conservation, 80
baryonic resonance
  $\Delta(1232\,P_{33})^{++}$ further evidence, 121
  $\Delta(1232\,P_{33})^{++}$, confirmation, 127
  $\Delta(1232\,P_{33})^{0}$ first indication, 121
beauty
  baryon Λ_b, confirmation, 262
  baryon Λ_b, first indication, 244
  first evidence, 233
  meson B, confirmation, 245
  meson B, first evidence, 244
  strange meson B_S, mass, precise measurement, 265
Becquerel, radioactivity, 22, 23, 25
Berkeley 184 inch synchrocyclotron, 92
betatron
  first, 85
  maximal energy, 89
  proposal, 84
β decay
  Λ, confirmation, 162
  Λ, first evidence, 162
  π^+, first observation, 184
  bound electron state, first observation, 263
  neutron, confirmation, 111
  neutron, first evidence, 100

neutron, first observation, 110
pion, confirmation, 161
pion, first evidence, 160
quantum field theory, 70
selection rules, 74
strong interactions corrections, 138
β particles, scattering, 32
β radioactivity, evidence, 24
β rays, ionization, 41
Bethe-Salpeter relativistic equation, bound state problem, 119
Big Bang model
chemical elements abundance, 89
Bjorken scaling
confirmation, 211
establishment, 217
evidence, 211
explanation, 213
invention, 211
black body radiation
explanation, 25, 26
Planck's formula, 25, 26
Planck's radiation law, 25
quantum hypothesis, 25, 26
Bohr
atomic model, 34, 35
atomic spectrum, 37
correspondence principle, 37
quantum theory, 34, 35, 37
Born approximation, 50
Bose, statistics, bosons, 41
Bose-Einstein
condensation, 42
correlations, charge pairs of pions, 173
quantum statistics, 41
statistics, discovery, 41, 42
bosons
light quanta, 41
second quantization, 52
bottonium, first evidence, 233
bound state problem, Bethe-Salpeter relativistic equation, 119
Breit-Wigner formula, resonance reactions amplitude, 73
bubble chamber
charged particle track, first observation, 125
invention, 125
B_S-$\overline{B}_S$ mixing
confirmation, 258, 259, 262, 264
first indication, 253
B^0-$\overline{B}^0$ mixing
confirmation, 256, 257
first evidence, 254

Cabbibo angle, 191
capture, atomic electron, first evidence, 79
cathode rays
electron discovery, 23
nature, material, 23
negatively charged particles, 21
phosphorescence, 21
properties, 21
causality condition, quantum field theory, 132, 134, 135
causality condition, quantum theory, 135
Chadwick
deuteron photodisintegration, 70
neutral particles, 36
neutron, 62, 63
neutron mass, 70
strong interaction, 39
charge
α rays, measurement, 26
conjugation, neutral particles behavior, 134
independence, nuclear forces, hypothesis, 74
charge-parity nonconservation, weak decays, 153
charmed
antibaryon $\overline{\Lambda}_c^-$, first evidence, 229
baryon Λ_c^+, confirmation, 238
baryon Λ_c^+, first indication, 225
baryon $\Sigma_c(2455)^+$, confirmation, 243
baryon $\Sigma_c(2455)^0$, first evidence, 252
baryon $\Sigma_c(2455)^{++}$, confirmation, 238
baryon $\Sigma_c(2455)^{++}$, first evidence, 225
doubly strange baryon Ω_c, first evidence, 250
hadrons, prediction, 194
meson D^+, first evidence, 229
meson D^-, first evidence, 229
meson D^0, first evidence, 228
quark existence, proposal, 194, 213
strange antibaryon $\overline{\Xi}_c(2460)^0$, first evidence, 257
strange baryons $\Xi_c(2460)^+$, 258
strange baryons $\Xi_c(2460)^0$, mass splitting, measurement, 258
strange baryon $\Xi_c(2460)^+$, first evidence, 246
strange baryon $\Xi_c(2460)^0$, first evidence, 257
strange mesons $D_S^\pm$, first evidence, 231
strange mesons $D_S^{*\pm}$, first evidence, 231
Vavilov-Čerenkov radiation
evidence, 71
explanation, 75
not a luminescence, evidences, 72
theory, confirmation, 77
Chew-Frautschi plot to classify hadrons, 183, 189
Chew-Low model, 144
classical electrodynamics, relativistic cut-off, 99
classification
Gell-Mann, 149
hadrons, 141, 183, 189, 193, 194
hadrons, conflict with Fermi statistics, 199, 201
hadrons, resolution conflict with Fermi statistics, 202
hadrons, $SU(6)$, 197, 199, 200
Nishijima, 141
cloud chamber, 41
development, 65
invention, 32
multiplate (MIT), 135
using with Geiger-Müller counters, 65
Cockcroft–Walton, accelerator, 63
coincidence method, invention, 56
colliding beams, first idea of using, 144
colored quarks and gluons, hypothesis, 201, 202
color quantum number, introduction, 201, 202
complex orbital momenta, theory of scattering, 171, 174
composite model of pions, first, 109
Compton effect, 40
confirmation, 41
condensation effect, monoatomic gases, 42
configuration space, 64
conservation
CP, 155, 157
CP, strong interaction, 232
baryonic quantum number, 80
law, 199
laws, quantum mechanics, 54, 67
leptons, 160
lepton number, hypothesis, 157
spatial parity, weak interaction, 148
strangeness, Gell-Mann classification, 149
strangeness, Nishijima classification, 141
conserved vector current hypothesis, evidence, 184
constituent quark model of hadrons, 193, 194
continuous β spectrum
confirmation, 52, 53
firm establishment, 58
neutrino hypothesis, 59, 70
observation, 36
corpuscular-wave dualism, 40
electron, 40, 41, 51, 54, 55
photon, 30
correspondence principle, invention, 37
cosmic rays

conclusive evidence, 32
confirmation, 33
firm establishment, 45
ionizing particles, 56
penetrating charged particles, confirmation, 65
shower, 56
shower, confirmation, 68
cosmological upper bound on ν_μ mass, 205
Cosmotron, first V^0 particles production, 124
Cowan, Ξ^- cascade decay, 130
CP conservation
strong interactions, explanation, 232
weak interaction, hypothesis, 155
weak interaction, proposal to test, 157
CP violation
K_L semilepton decays, first evidence, 207
baryonic asymmetry, universe, 206
evidence, 197
renormalizable theory, 218
superweak theory, 199
CPT invariance
local quantum field theory, 133
test, 255
cumulative effect, 218
CVC hypothesis validity, first experimental evidence, 184
cyclotron
further development, 66
medical applications, 80
phase stability principle, 92
proposal, 60
test, 60

Dalitz plot method, invention, 128
Dalitz, kaon zero spin, 131
decay
$K^+ \to \mu^+$ (neutrals), first evidence, 115
$K^+ \to \pi^+ \pi^0$, evidence, 132
$K^\pm \to \mu^\pm$ neutral, 129
$K^\pm \to \mu^\pm$ neutral, firmly established, 136
$K \to \pi\ \pi$ first evidence, 123
$\pi^- \to \mu^-$ neutral, first indication, 96
$\pi^0 \to \gamma\ \gamma$ confirmation, 114
$\pi^0 \to \gamma\ \gamma$ first evidence, 112
$\pi^\pm \to \mu^\pm$ neutral, 96
deep inelastic scattering
"proton spin crisis", 256
polarization phenomena, 256
polarized structure functions, 256
$\Delta(1232\, P_{33})^{++}$
confirmation, 127
further evidence, 121
$\Delta(1232\, P_{33})^0$ first indication, 121
detector
bubble chamber, 125
cloud chamber, 32, 41, 65
flash tube chambers, 140
Geiger-Müller counters, 65
liquid scintillation counters, 111
multiplate cloud chamber, 135
multiwire proportional chamber, 210
photographic emulsion, 92, 93
scintillation counters, 95
semi-conductor, 108
spark chamber, 109, 167
streamer chamber, 193
Wilson chamber with counters, 65
deuteron
discovery, 65
disintegration, 70
evidence, 64
magnetic moment, measurement, 72, 80
photodisintegration, evidence, 70
spin, first evidence, 72
de Broglie waves, 40, 41, 51, 54, 55
diffraction of electrons, 51, 54, 55
Dirac equation, 55, 56, 58–60
dispersion relations
asymptotic equality, 163
in two variables, 165
photoproduction amplitude, 157
pion-nucleon scattering, 135
quantum field theory, 132, 134, 135
quantum theory, 135

eightfold way, 181
Einstein
Bose-Einstein statistics, 42
photoelectric effect, 28, 30
special relativity theory, 28, 29
electromagnetic structure
nucleon, form factors, 159
nucleus, 98, 119
proton, first measurement, 145
electron
additional two-valued degree of freedom, 42
anomalous magnetic moment, calculation, 99
anomalous magnetic moment, first measurement, 95, 98
charge precise measurement, 31, 33
corpuscular-wave dualism, 40, 41, 51, 54, 55
de Broglie waves, 51
diffraction by crystals, 51, 54, 55
diffraction phenomena, 41
discovery, 23
g−2 factor, 95, 98, 252, 255
magnetic moment, 55, 56, 255
magnetic moment, measurement, 252
negative energy problem, 58–60
relativistic wave equation, 55, 56
sea holes, 59
sea holes, positive electrons, 58
spin, 43
wave nature, 40, 41
electron cooling
evidence, 230
proposal, 207
e^+e^-
pair production, first indication, 68
shower, 68
electron-nucleus scattering, 98, 119
electron-positron storage ring, resonance depolarization, 241
electron-positron symmetrical theory, 76
electron-proton elastic scattering, 145
electron-α elastic scattering, 145
electrostatic accelerator, invention, 61, 67
electroweak synthesis, Lagrangian, 201, 208, 209
elementary particle
electron, 23
photon, 40, 41
elementary particles
structure, 200
supermultiplets, 200
energy loss, ionization, Landau distribution, 88
energy quanta
discovery, 25, 26, 28
photoelectric effect, conclusive measurement, 36
energy spectrum, black body radiation, 25
energy-mass relation, 29
$E = mc^2$, 29
Esterman and Stern, magnetic moment, neutron and deuteron, 72
η meson
confirmation, 187
first evidence, 183
prediction, 163, 170
zero spin, evidence, 187
η' meson
confirmation, 197
evidence, 196
prediction, 163

exchange forces, nuclear
  further development of theory, 69
  theory, 65
exclusion principle, 42, 45, 46

Faddeev-Popov method, 209
Fermi
  first nuclear reactor, 86
  interaction, 74
  interaction, universality, 94, 106, 126
  neutrino mass, 70
fermions, 45, 46, 51
Fermi-Dirac statistics
  invention, 45, 46
  rediscovery, 51
Fermi–Yang, first composed model, 109
Feynman rules, 113
  for non-Abelian gauge theories, 209
flash tube chambers, invention, 140
Fock space, 64
form factors, nucleon electromagnetic structure, 159
Frank-Tamm theory, Vavilov-Čerenkov effect
  confirmation, 77
  explanation, 75
Froissart upper bound, hadronic collisions, 177
fundamental constants
  metrology, first step, 57
  metrology, further development, 81
  precise measurement, 33

γ radioactivity
  confirmation, 25
  discovery, 24
  evidence, 25
γ spin, experimental proof, 61
Gamow, Big Bang model, 89
Gamow-Teller selection rules, 74
gauge principle, invention, 175
gauge theories
  massive vector boson, renormalizability, 189
  regularization and renormalization method, 217
Geiger-Müller counters, 65
Gell-Mann
  isotopic multiplet, 126
  renormalization group, 131
Glaser, bubble chamber, 125
gluon jet
  confirmation, 240, 241
  first evidence, 239
gluon spin
  confirmation, 242
  first experimental determination, 241
Goldberger-Treiman relations, 164
Goldstone theorem, 175, 190, 195, 198, 199
Gribov-Lipatov-Altarelli-Parisi evolution equations, 216
g−2 factor
  electron, 252, 255
  positron, 255
G-parity, nonstrange mesons, invention, 149

hadronic collisions, Froissart upper bound, 177
hadronic shower
  limited transverse momenta, confirmation, 167
  limited transverse momenta, evidence, 158, 166
hadrons
  composite model, 151
  constituent quarks model, 193, 194
  model, weak interaction, 191, 213
hadrons classification
  $SU(6)$, 197, 199, 200
  aces, 194
  conflict with Fermi statistics, 201
  Gell-Mann, strangeness conservation, 149
  Nishijima, strangeness conservation, 141
  quarks, 193
  Regge poles, 189
  resolution conflict with Fermi statistics, 202
hadron nucleus collisions, 213
hadron production, quark-parton model, 226
hadron(antihadron)-hadron interaction
  asymptotic equality, 163
  dispersion relations, 163
heavy unstable particles, confirmation, 115, 116
Heisenberg
  S-matrix, 87
  isotopic spin, invention, 65
  matrix formalism, 44
  nuclear forces, theory, 65
  quantum mechanics, 43, 44
  rising total hadronic cross sections, 120
  uncertainty principle, 53
Hess, cosmic rays, 32, 33, 45
$^3\overline{\text{He}}$, evidence, 214
Higgs mechanism, gauge theories, 195, 198, 199, 204
Higgs-Kibble mechanism, non-Abelian gauge theories, 208
high transverse momentum hadrons, first observation, 220
hydrogen atom fine structure, first measurement, 94
hypernucleus
  double, first evidence, 192
  first evidence, 128
hyperon
  anti-lambda, $\overline{\Lambda}$, first evidence, 159
  neutral cascade, Ξ^0, first evidence, 168
hyperons, leptonic decay suppresion, explanation, 191

inclusive hadron spectra, scaling behavior
  confirmation, 214
  evidence, 212
  hypothesis, 212
infrared divergence, treatment, 77
integer spin, 41
intermediate charged gauge bosons
  confirmation, 249
  evidence, 246, 247
intermediate gauge bosons
  establishment of properties, 251, 253
  weak interaction, hypothesis, 158
intermediate neutral gauge boson, 184
  confirmation, 248, 250
  evidence, 248
ionization by photon, 41
ionization energy loss, Landau distribution, 88
ionizing particles, cosmic rays, 56
ionizing particle track visualization, 32
isotopic multiplet structure for V particles, 126
isotopic spin
  invention, 65
  local gauge invariance, 132
  nucleon, proposal, 74
Iwanenko, neutron, 64
Iwanenko–Pomeranchuk, betatron maximal energy, 89

jets in hadronic interactions, first evidence, 114
$J/\psi(1S)$
  confirmation, 224
  evidence, 223, 224

K_L
  confirmation, 147, 156
  first evidence, 147
  prediction, 134
K_L-K_S mass difference, first measurement, 159
K_L-K_S regeneration
  confirmation, 202
  experiment, proposal, 139
  first evidence, 177
kaons charged

decay branchings, measurement, 140
lifetimes comparison, 142, 143
masses comparison, 142
masses, measurement, 137
kaon decay, three-prong, first evidence, 103
$K^*(892)$ resonance
first evidence, 178
spin, first evidence, 188
K^+
confirmation, 118, 135
first evidence, 88
lifetime, first measurement, 137
lifetime, measurement from decay in flight, 137
mass, first precise measurement, 136
$K^\pm$
odd parity, first indication, 131
zero spin, first indication, 131
K^-, confirmation, 135
K^0
first evidence, 117, 118
mass, first measurement, 124
Kerst, betatron, 84, 85
Keuffel, spark chamber, 109
Kramers-Kronig generalized dispersion relations, 135
K-capture, first evidence, 79

Lagrangian, Minimal Standard Model, QCD, 221
Λ
beta decay, confirmation, 162
beta decay, first evidence, 162
confirmation, 125
evidence, 117, 118
$\overline{\Lambda}$
confirmation, 175
first evidence, 159
first indication, 130
$\Lambda(1405\ S_{01})$ resonance, first evidence, 180
Λ_b beauty baryon, 262
Lamb shift
calculation, 95
first measurement, 94
Landau distribution, ionization energy loss, 88
Landau singularities, perturbative amplitudes, 170
Lawrence, cyclotron, 60, 66, 80
Lebedev, light pressure, 26
lepton
family, 233
number conservation, hypothesis, 157
quantum number, invention, 126
leptonic decay, hyperons, explanation of suppresion, 191
lepton-quark symmetry, 213
lifetime
K^+, first measurement, 137
K^+, measurement from decay in flight, 137
π^0, 112
π^0, first conclusive measurements, 176
charged kaons, comparison, 142, 143
muon, first estimation, 81
muon, measurement, 84–86
neutron, 110, 111
light neutrinos number, 258, 260, 261, 263
light pressure, evidence, 26
linear accelerator, Stanford, 141
line-spectra theory, 37
Lippmann-Schwinger equation, 113
relativistic generalization, 192
liquid scintillation counters, invention, 111
local gauge invariance, isotopic spin, 132

magnetic moment
deuteron, measurement, 72, 80
electron, 55, 56, 255
electron, calculation, 99
electron, first measurement, 95, 98
electron, measurement, 252
muon, measurement, 153
neutron, first direct measurement, 83
neutron, measurement, 72
neutron, prediction, 71
positron, 255
proton, first measurement, 69
proton, further measurement, 70
proton, measurement, 80
magnetic monopole, possible existence, 60
Majorana neutrino, 76
Majorana, nuclear forces, 69
Mandelstam representation, 165
mass
B_S, precise measurement, 265
K^+, first precise measurement, 136
K^0, 124
K_S-K_L difference, 159
$W^\pm$, 208
$W^\pm$, estimation, 201
Z^0, 208
Σ^0, measurement, 152
ν_μ, cosmological upper bound, 205
π^0-π^- difference, 114
charged kaons, comparison, 142
charged kaons, measurement, 137
neutrino, first estimation, 70
neutrino, parity nonconservation, 152
neutron, 70
top quark, 265–267
massive muon pairs in hadron collisions, observation, 214
mass generation, gauge fields
Higgs mechanism, 198, 204
Higgs-Kibble mechanism, 208
mass-energy relation, 29
matrix formalism, Heisenberg quantum mechanics, invention, 44
McKay, semi-conductor detectors, 108
mesons, nonstrange, G-parity conception, 149
method
complex orbital momenta, 171, 174
extraction, pion-pion interaction, 161
Faddeev-Popov, 209
Regge poles, 171, 174
regularization, 217
renormalization, 217
metrology
fundamental constants, first step, 57
fundamental constants, further development, 81
particle physics, high precision measurements, 241, 250
particle properties data, 158, 159, 267
Millikan, precise measurement, 31, 33, 36
mixing
B^0-$\overline{B}^0$ confirmation, 256, 257
B^0-$\overline{B}^0$ first evidence, 254
B_S-$\overline{B}_S$ confirmation, 258, 259, 262, 264
B_S-$\overline{B}_S$ first indication, 253
model
Chew-Low, 144
constituent quarks, hadrons, 193, 194
Fermi–Yang, 109
hadrons, Nambu-Jona-Lasinio, nonlinear, 176, 177
paraquark, 199
pion-nucleon interaction, 144
quark-parton, 213, 226
Sakata, 151
Salam-Weinberg, 208, 209
three-triplet, 201
weak interaction, hadron, 191, 213
monoatomic gases
condensation effect, 42
quantization, 45, 46
multiplate cloud chamber (MIT), 135
multiple hadron production, nucleon-nucleon interactions, 114
multiwire proportional chamber, invention, 210

muon
- confirmation, 76, 78, 79
- decay, electron spectra continuity, confirmation, 106
- decay, electron spectra continuity, evidence, 105
- decay, first evidence, 81
- decay, first observation, 82
- decay, gamma-radiation search, negative result, 97, 99
- decay, search for gamma-radiation, 102
- exponential decay rate, 85, 86
- first evidence, 75
- lifetime, first estimation, 81
- lifetime, measurement, 84–86
- magnetic moment, measurement, 153
- not a strong interaction mediator, evidence, 91
- spin, 106

$\mu^- \to e^- \gamma$ decay suppression, 97

Nambu-Jona-Lasinio, hadrons nonlinear model, 176, 177
Neddermeyer and Anderson, muon, 75
negatively charged particle, 21
negative energy
- electron see, idea, 58
- problem, 58–60

neutral intermediate boson, 184
neutral penetrating particle, indirect evidence, 36
neutral weak current
- atomic, confirmation, 235, 239
- confirmation, 223, 231, 239
- first evidence, 219, 235
- leptonic, confirmation, 230, 236
- leptonic, first indication, 219

neutrino, 70
- astronomy, birth, 255
- beams feasibility with accelerators, 172
- beams with accelerators, proposal, 170
- burst, SN1987A, first observation, 255
- existence, 59
- hypothesis, 59
- light neutrinos number, first evidence, 258
- light neutrinos, number, confirmation, 260, 261, 263, 264
- Majorana, 76
- mass, first estimation, 70
- mass, relation to parity nonconservation, 152
- more than one kind, evidence, 186
- quantum numbers, 133
- radiochemical detection, 133
- radiochemical detection method, 91
- two-component theory, 154, 155

$\overline{\nu}_e$
- absorption by protons, measurement, 169
- confirmation, 146
- detection, confirmation, 169
- evidence, 125
- first detection, 146

ν_e, left handed, first evidence, 163
ν_e-ν_μ, distinguishability, proposal of experiment, 170
ν_μ mass, upper bound from cosmology, 205
ν_μ, first evidence, 186
$\nu - \overline{\nu}$ oscillations, lepton number nonconservation, 162
neutron
- β decay, confirmation, 111
- β decay, first evidence, 100
- β decay, first observation, 110
- anomalous magnetic moment, prediction, 71
- discovery, 63
- first evidence, 62
- hypothesis, 38
- lifetime, 110, 111
- magnetic moment, first direct measurement, 83
- magnetic moment, measurement, 72
- mass, 70
- nucleus constituent, suggestion, 64
- polarizability, estimation, 171
- spin, establishment, 79

Nobel prize
- 1901, W. C. Röntgen, 21
- 1903, H. Becquerel, 22, 23
- 1903, M. Curie, 22, 23
- 1903, P. Curie, 22, 23
- 1906, J. J. Thomson, 23
- 1908, E. Rutherford. (Chemistry), 24
- 1918, M. Planck, 25, 26
- 1921, A. Einstein, 28–30, 42
- 1922, N. Bohr, 34, 35, 37
- 1923, R. A. Millikan, 31, 33, 36
- 1926, J. B. Perrin, 21
- 1927, A. H. Compton, 40, 41
- 1927, C. T. R. Wilson, 32, 41
- 1929, L. De Broglie, 40, 41
- 1932, W. Heisenberg, 43
- 1933, E. Schrödinger, 46–48, 50, 55, 56
- 1933, P. A. M. Dirac, 46–48, 50, 55, 56
- 1935, J. Chadwick, 62, 63
- 1936, C. D. Anderson, 32, 33, 45, 62, 68
- 1936, V. F. Hess, 32, 33, 45, 62, 68
- 1937, C. J. Davisson, 51, 54, 55
- 1937, G. P. Thompson, 51, 54, 55
- 1939, E. O. Lawrence, 60, 66
- 1943, O. Stern, 69, 70
- 1944, I. I. Rabi, 80
- 1945, W. Pauli, 42
- 1948, P. M. S. Blackett, 65, 68
- 1949, H. Yukawa, 73
- 1950, C. F. Powell, 93, 96, 103
- 1951, E. T. S. Walton, 63
- 1951, J. D. Cockcroft, 63
- 1954, M. Born, 50, 51, 56
- 1954, W. Bothe, 50, 51, 56
- 1955, P. Kusch, 94, 95, 98
- 1955, W. E. Lamb, 94, 95, 98
- 1957, C. N. Yang, 148
- 1957, T. D. Lee, 148
- 1958, I. E. Tamm, 71, 75, 77
- 1958, I. M. Frank, 71, 75, 77
- 1958, P. A. Čerenkov, 71, 75, 77
- 1959, E. G. Segrè, 138, 143, 145
- 1959, O. Chamberlain, 138, 143, 145
- 1960, D. A. Glaser, 125
- 1961, R. Hofstadter, 145
- 1961, R. Mössbauer, 145
- 1963, E. P. Wigner, 54, 67, 75
- 1963, J. H. D. Jensen, 54, 67, 75
- 1963, M. Goeppert-Mayer, 54, 67, 75
- 1965, J. S. Schwinger, 86, 92, 93, 100–102, 104, 107, 108, 113
- 1965, R. P. Feynman, 86, 92, 93, 100–102, 104, 107, 108, 113
- 1965, S. Tomonaga, 86, 92, 93, 100–102, 104, 107, 108, 113
- 1968, L. W. Alvarez, 172, 178, 180, 181, 196
- 1969, M. Gell-Mann, 126, 181, 186, 193
- 1976, B. Richter, 223, 224
- 1976, S. C. C. Ting, 223, 224
- 1979, A. Salam, 184, 201, 208, 209
- 1979, S. L. Glashow, 184, 201, 208, 209
- 1979, S. Weinberg, 184, 201, 208, 209
- 1980, J. W. Cronin, 197
- 1980, V. L. Fitch, 197
- 1984, C. Rubbia, 235, 246, 248
- 1984, S. van der Meer, 235, 246, 248
- 1988, J. Steinberger, 172, 186
- 1988, L. M. Lederman, 172, 186
- 1988, M. Schwartz, 172, 186
- 1989, H. G. Dehmelt, 252, 255
- 1989, N. F. Ramsey, 252, 255
- 1989, W. Paul, 252, 255
- 1990, H. W. Kendall, 211, 217
- 1990, J. I. Friedman, 211, 217
- 1990, R. E. Taylor, 211, 217
- 1992, G. Charpak, 210
- 1995, F. Reines, 125, 146, 169, 226, 228, 232

1995, M. Perl, 125, 146, 169, 226, 228, 232
nonconservation
charge conjugation symmetry, weak interactions, 151, 156
charge-parity, 153
lepton number, $\nu - \overline{\nu}$ oscillations, 162
parity, 148, 152–156, 166, 235, 239
parity, weak nuclear interactions, 202, 204
spatial parity, 156
nonstrange mesons, G-parity conception, 149
non linear sigma model, 171
non-Abelian gauge theories
Feynman rules, 209
Higgs-Kibble mechanism, 208
renormalizability, 215, 216
ultraviolet behaviour, 220, 221
non-relativistic QED, 95
nuclear charge, atomic number, relation, 38
nuclear exchange forces
further development of theory, 69
theory, 65
nuclear forces, 73
charge independence, hypothesis, 74
nuclear property, radioactivity, 35
nuclear reactor, 86
nuclear spectroscopy, supermultiplet structure, 75
nuclear transmutation, 112
nucleon, 65
isotopic spin, proposal, 74
polarizability, estimation, 171
polarizability, measurement, 174
nucleus
cumulative effect, 218
electromagnetic structure, 98, 119
evidence, 32
hypothesis, 32
quantum theory of α decay, 55
nucleus structure
neutron constituent, 38
neutron constituent, suggestion, 64
protons and neutrons, suggestion, 65
proton constituent, 38

$\overline{\Omega}^+$, evidence, 215
ω meson resonance
first evidence, 181
Ω^- hyperon
confirmation, 200
first evidence, 195, 196
prediction, 186
one degree of freedom systems, 44

paraquark model, 199
parastatistics for quarks, 199
parity nonconservation, weak interaction
confirmation, 153, 154
first evidence, 153
neutrino mass, 152
neutrino two-component theory, 154, 155
proposal to test, 148, 166
weak decays, confirmation, 156
parity, invention, 54
particle energy loss, ionization, 88
particle physics data, first collection, 159
particle physics metrology, 267
particle properties data, first review, 158
particle properties, assessed, 267
particle-antiparticle conjugation, 133
partonic model
Gribov-Lipatov evolution equations, 216
path integral formalism, quantum mechanics, invention, 103
Pauli
exclusion principle, 42
neutrino, 70
principle, 42, 45, 46
spin and statistics, 83
penetrating charged particles, cosmic rays, confirmation, 65
Perrin, cathode rays, 21
perturbation theory, 48
phase stability principle, 92
invention, 88, 89
ϕ meson
$e^- e^+$ decay, 210
confirmation, 190
firm establishment, 191
first evidence, 188
photodisintegration
deuteron, evidence, 70
photoelectric effect
explanation, 30
law, 28
quantum hypothesis, 28, 30, 36
photographic emulsion method, 92, 93, 96, 103
photon
corpuscular-wave dualism, 30
elementary particle, 31, 41
elementary particle, direct confirmation, 40
statistics, 41
pion
β decay, confirmation, 161
β decay, first evidence, 160
first composite model, 109
mediator of strong interactions, 73
production by accelerator, first, 103
π^+
β decay, first observation, 184
spin $= 0$, 117
spin $= 0$, first evidence, 116
π^-
confirmation, 93, 96
first indication, 92
odd parity, confirmation, 115
parity, direct determination, 114
pion-nucleon
interaction, static model, 144
scattering, dispersion relations, 135
pion-pion
correlations, 168
interactions, method for extraction, 161, 171
system, $J = 1$ and T$= 1$ resonance, prediction, 175
π^0
confirmation, 112, 114
first evidence, 109
lifetime, first conclusive measurements, 176
lifetime, first estimation, 112
odd parity, first direct determination, 169
odd parity, first evidence, 139
production by photons, first evidence, 112
π^0-π^- mass difference, 114
Planck
constant, 35
constant, measurement, 36
energy quanta, 25
quantum hypothesis, 26
radiation law, 25, 41
Poincare relativity theory, 27
Pomeranchuk theorem, 163
positive particles as electron sea holes, 58, 60
positron
confirmation, 68
discovery, 68
first evidence, 62
g$-$2 factor, 255
magnetic moment, 255
prediction, 60
potential barrier, tunneling, 55
pressure of light, 26
principle of relativity, 27, 28
proton

as electron sea hole, difficulties, 59
discovery, 38
electromagnetic radius, first measurement, 145
electromagnetic structure, 145
first indication, 36
magnetic moment, first measurement, 69
magnetic moment, further measurement, 70
magnetic moment, measurement, 80
polarizability, estimation, 171
polarizability, measurement, 174

$J/\psi(1S)$
confirmation, 224
evidence, 223, 224

$\psi(2S)$, evidence, 225

QCD — quantum chromodynamics, 205, 206, 216, 221, 227, 239–242

QCD
beginnings, 206
Gribov-Lipatov-Altarelli-Parisi evolution equations, 216, 232
Lagrangian, final form, 221
Minimal Standard Model, Lagrangian, 221
origins, 205
parton densities evolution equations, 232
quarks strong interaction, vector gluon theory, 205
sum rules, 238
test, 239–242
theory, final form, 221
three triplet model, 206
vacuum topological structure, BPST-instanton, 227
Yang-Mills Lagrangian, invention, 221

QED — quantum electrodynamics, 52, 59, 86, 92, 93, 100–102, 104, 107, 108, 110, 113, 131, 149

QED
S-matrix of Heisenberg, 107
amplitudes, Feynman rules for calculations, 113
big distances, renormalization group, development, 149
Feynman method, 100, 107, 108, 113
foundation, 52
relativistically invariant formulation, 86, 92, 93, 100–102, 104, 107, 108
renormalization group, generalization, 149
Schwinger method, 101, 104
small distances, renormalization group exploration, 131
small distances, renormalization group, development, 149
Tomonaga method, 86, 92, 93, 102
Tomonaga-Schwinger — Feynman methods equivalence, 104
ultraviolet divergencies, first evidence, 59
Ward identity, 110

quantization of monoatomic perfect gases, 45, 46

quantum field theory
CPT invariance, 133
β decay, 70
causality condition, 132, 134, 135
dispersion relations, 135
dispersion relations, forward amplitude, 150
dispersion relations, general proofs, 150
first steps, 52
nuclear forces, Yukawa, 73
quasi-optical approach, 192
renormalization group, 149
ultraviolet divergencies, first evidence, 59

quantum hypothesis, 26
black body radiation, explanation, 25
photoelectric effect, explanation, 28, 30

quantum mechanics
applications, 50
conservation laws, 54, 67
foundation, 43
Heisenberg approach, 43
matrix formalism, invention, 44
path integral formalism, invention, 103
potential barrier, tunneling, 55
Schrödinger wave equation, 46, 48, 50
statistical interpretation, 50, 51
time inversion transformation, 67

quantum statistics, Fermi-Dirac, 46, 51

quantum systems, space of states, 64

quantum theory
atomic spectrum, 35, 37
atomic spectrum, invention, 34, 35
causality condition, 135
corpuscular-wave dualism, 40, 41
dispersion relations, 135
scattering, 50
scattering, variational principles, 113
scattering, S-matrix formalism, 78

quarks
introduction, 193, 194
with integral charge, 201

quark jets in $e^+ e^-$ annihilation, first evidence, 226

quark-parton model, confirmation, 226

quartet scheme, 213
failure to include CP violation, 218

radiation self-polarization, 230

radioactivity
α, 24
β, 24
γ, 24, 25
comparison with X rays, 23
components, 24–26
confirmation, 22, 23
electric properties, 22
evidence, 22
ionization, 24
nuclear property, evidence, 35
radium, 26
uranium, 23
uranium, establishment, 23

Regge poles
introduction, 174
method in the quantum theory of scattering, 171, 174
relativistic quantum scattering, 179

Regge trajectories, 189
hadrons classification, 183
linearity, confirmation, 227
Pomeranchuk vacuum trajectory – pomeron, 183

regularization method, gauge fields theories, 217

relations, Goldberger-Treiman, 164

relativistic
cut-off, classical electrodynamics, 99
field theories, 133
generalization, Lippmann-Schwinger equation, 192
quantum mechanics, 55, 56
wave equation for the electron, 55, 56

relativistically invariant formulation, QED, 86, 92, 93, 100–102, 104, 107, 108

relativity theory, special
invention, 28
mass-energy relation, 29

renormalizability
gauge theories, massive vector boson, 189
Yang-Mills field theory, 215, 216

renormalizable covariant QED, 101, 104, 108

renormalization group
development, 149
invention, 129, 131

renormalization method, gauge fields theories, 217

resonance
K-π system, 178
$\Lambda(1405\,S_{01})$, first evidence, 180
Σ-π system, 180
$\Xi(1530\,P_{13})$, first evidence, 187, 188
$\Xi\ \pi$ system, 187, 188
ω meson, confirmation, 182, 183
ω meson, first evidence, 181
π-π system, 179
ρ meson, 179

$f_4(2050)$ meson, evidence, 227
baryonic
$\Delta(1232\,P_{33})^{++}$ further evidence, 121
$\Delta(1232\,P_{33})^{++}$, confirmation, 127
$\Delta(1232\,P_{33})^0$ first indication, 121
reactions, amplitude, Breit-Wigner formula, 73
ρ meson resonance, 179
rising total hadronic cross sections
confirmation, 218, 219
first observation, 215
prediction, 120
Rutherford
α radioactivity, 26
atomic structure, 32, 36
neutron, 38
proton, 38
radioactivity, 24
Röntgen, X rays, 21

Sakata model
hadron structure, 151
Salam-Weinberg model, 208, 209
scaling behavior, inclusive hadron spectra
confirmation, 214
evidence, 212
hypothesis, 212
scattering, quantum theory, 50, 78, 113
Schrödinger wave equation
applications, 46
invention, 46
Schrödinger — Heisenberg quantum mechanics, equivalency, 47
Schwinger, neutron spin, 79
scintillation counters, invention, 95
second quantization, 64
invention, 52
selection rules
β decay, 74
semi-conductor detectors, invention, 108
shower, 56
Σ parity, determination, 185
$\overline{\Sigma}^-$, first evidence, 173
$\overline{\Sigma}^0$, first evidence, 172
$\Sigma(1385\,P_{13})$, evidence, 172
Σ^+
confirmation, 123
direct experimental evidence, 123
first evidence, 123
Σ^-
confirmation, 123, 130
first evidence, 123
$\Sigma_c(2455)^0$, first evidence, 252
Σ^0
confirmation, 152
first indication, 134
mass, measurement, 152
prediction, 141, 149
spark chamber
invention, 109
principle, 167
spatial parity
conservation in weak interaction, 148
invention, 54
nonconservation, 156
special relativity theory
invention, 28, 29
mass-energy relation, 29
spin
γ, 61
deuteron, 72
electron, hypothesis, 43
gluon, 241, 242
integer, 41
isotopic, 74
neutron, 79
spin and statistics connection theorem, 83
spin and unitary spin unification, 197, 200
spontaneously broken symmetry
unavoidable massless bosons, prediction, 175, 190
without massless bosons, 195, 198, 199
Stanford, GeV linear accelerator, 141
Stark, photon — elementary particle, 31
static model, pion-nucleon interaction, 144
statistical interpretation of quantum mechanics, 50, 51
statistics
Bose-Einstein, 41, 42
Fermi-Dirac, 45, 46, 51
statistics and spin connection theorem, 83
stochastic cooling, evidence, 235
Stokes's Law, correction, 31
storage rings
radiation self-polarization, 178, 192
Sibirian snakes, 237
strangeness conservation, Gell-Mann classification, 149
strangeness, Ω^-, 200
strange particles
associative production, 134
associative production, confirmation, 139
streamer chamber, invention, 193
strong focusing
principle, invention, 122
proton synchrotron, 184
strong interactions
CP conservation, 232
β decay corrections, 138
asymptotic energy dependences, 222
diffraction slope parameter, 212
dispersion sum rules, 209
dispersion sum rules, invention, 205
elastic amplitude, real part, 203
evidence, 39
high energy bounds, 209
inclusive approach, 190, 209, 212
mediator, 73, 91
multiperipheral model, 190
partonic model, 216
polarization experiment, Argonne spin effect, 234
polarization of inclusively measured hyperons, 229
polarization phenomena, 229, 234, 237
Regge asymptotics, 179
scaling behavior, 212
structure
elementary particles, 200
hadrons, 151
matter, discontinuous, 21, 23
structure function
bound nucleon, 247, 249
free nucleon, 247, 249
nucleon, 213, 247, 249
Stückelberg, renormalization group, 129
supermultiplet structure, nuclear spectroscopy, 75
superweak theory, CP violation, 199
$SU(3)$
decuplet structure, 186
octet structure, 186
octet structure, introduction, 181, 182
symmetry, hadrons, hypothesis, 170
$SU(6)$ classification, hadrons, 197, 199, 200
symmetry, nuclear Hamiltonian, 75
synchrocyclotron, the Berkeley 184 inch, 92
synchrotron
first strong focusing proton, 184
idea, further development, 90
invention, 89
strong focusing, invention, 122
S-matrix
axiomatic field theory, beginnings, 140
formalism, first proposal, 78

formalism, invention, 87

τ lepton
evidence, 228
indication, 226
properties, establishment, 232
third quark-lepton family, first indication, 233
Thomson, electron discovery, 23
three-jet, $e^+ e^-$ annihilations, 239–242
three-prong kaon decay, first evidence, 103
three-triplet model, 201
time inversion, 133
invention, 67
Tomonaga-Schwinger — Feynman methods equivalence, 107
$t\bar{t}$ production, 265–267
top quark
confirmation, 267
first evidence, 265
mass, 265–267
observation, 266
total hadronic cross sections
confirmation of rising, 218, 219
first observation of rising, 215
Froissart upper bound, 177
prediction of rising, 120
tunneling through a potential barrier, 55
two-component theory, neutrino, 154, 155

Uhlenbeck, spin, 43
ultraviolet behaviour of non-Abelian gauge theory, 220
ultraviolet divergencies in QED, first evidence, 59
uncertainty principle
discovery, 53
universe, baryonic asymmetry, 206
unstable particle, 97
$\Upsilon(1S)$ meson
confirmation, 234, 238
first evidence, 233
$\Upsilon(2S)$ meson
confirmation, 236–238
first evidence, 233
$\Upsilon(3S)$ meson
confirmation, 242
evidence, 238
$\Upsilon(4S)$ meson, 243, 244

Van de Graaff, accelerator, 61, 67
variational principles, quantum theory, scattering, 113
Vavilov-Čerenkov radiation, 72
vector coupling constant, 138
Villard, γ radioactivity, 24, 25
V events, 110
V events, first evidence, 97
V particles, 120
associative production, evidence, 125, 130
isotopic multiplet structure, 126
V−A theory of weak interaction, 160, 164
V^0 particles
associative production, hypothesis, 121
Cosmotron production, first, 124
two types, evidence, 117, 118

Ward identity, QED, 110
wave equation, 46
wave mechanics
applications, 46, 48, 50
creation, 46
generalization, 50
perturbation theory, 48
wave of matter, 40–42
weak and electromagnetic interactions, unification, 184
weak interaction
CP conservation, hypothesis, 155
CP conservation, proposal to test, 157
CP violation, 218
charge conjugation symmetry nonconservation, 151, 156
charge-parity nonconservation, 153
electron capture, 79
first steps, 70, 106, 126
hadron model, 191, 213
hyperons leptonic decays, suppresion, 165
intermediate gauge bosons, hypothesis, 158
neutrino, two-component theory, 154, 155
parity nonconservation, 153–155
parity nonconservation, confirmation, 154, 156
parity nonconservation, first evidence, 153
parity nonconservation, neutrino mass, 152
parity nonconservation, nuclear interactions, 202, 204
parity nonconservation, proposal to test, 148
selection rules, 74
V−A theory, 160, 164
weak neutral current
atomic, confirmation, 235, 239
confirmation, 223, 231, 239
first evidence, 219, 235
leptonic, confirmation, 230, 236
leptonic, first indication, 219
Weinberg angle, evaluation, 230, 231, 236, 239, 245
Wightman axiomatic field theory, 143
Wigner, supermultiplet structure, 75
Wilson chamber, using with Geiger-Müller counters, 65
Wilson, cloud chamber, 32, 41
W^+
confirmation, 249
establishment of properties, 251, 253
evidence, 246, 247
$W^\pm$ boson, 184
mass, estimation, 201, 208
W^-
confirmation, 249
establishment of properties, 251, 253
evidence, 246, 247

$\bar{\Xi}^+$
confirmation, 185
production, first observation, 185
$\Xi(1530\,P_{13})$ resonance, first evidence, 187, 188
Ξ^-
cascade decay, confirmation, 130
confirmation, 126, 127
first evidence, 122
Gell-Mann scheme, 139
$\bar{\Xi}_c(2460)^-$ state, 258
$\bar{\Xi}_c(2460)^0$ state, 258
$\Xi_c(2460)^+$ state, 258
$\Xi_c(2460)^+$ and $\Xi_c(2460)^0$, mass splitting measurement, 258
$\Xi_c(2460)^0$ state, 258
Ξ^0
first evidence, 168
prediction, 141, 149
X rays
discovery, 21
scattering, 40

Yang-Mills gauge theory, 132
BPST-instanton, 227
BPST-pseudoparticle, 227
Feynman rules, 209
renormalizability, 215, 216
Yukawa model, 73

Z^0
confirmation, 248, 250
establishment of properties, 251, 253
evidence, 248
mass, estimation, 208
Zeeman effect, 56

NFT
NFT
10

Berkeley
LIBRARY OF THE UNIVERSITY OF CALIFORNIA